W9-CCU-963

# STUDENT'S
# SOLUTIONS MANUAL

## GARRET J. ETGEN

*University of Houston*

# FINITE MATHEMATICS
## FOR BUSINESS, ECONOMICS,
## LIFE SCIENCES AND SOCIAL SCIENCES
### TWELFTH EDITION

## RAYMOND A. BARNETT

*Merritt College*

## MICHAEL R. ZIEGLER

*Marquette University*

## KARL E. BYLEEN

*Marquette University*

**Prentice Hall**
is an imprint of

The author and publisher of this book have used their best efforts in preparing this book. These efforts include the development, research, and testing of the theories and programs to determine their effectiveness. The author and publisher make no warranty of any kind, expressed or implied, with regard to these programs or the documentation contained in this book. The author and publisher shall not be liable in any event for incidental or consequential damages in connection with, or arising out of, the furnishing, performance, or use of these programs.

Reproduced by Prentice Hall from electronic files supplied by the author.

Copyright © 2011, 2008, 2005  Pearson Education, Inc.
Publishing as Pearson Prentice Hall, 75 Arlington Street, Boston, MA  02116.

All rights reserved. No part of this publication may be reproduced, stored in a retrieval system, or transmitted, in any form or by any means, electronic, mechanical, photocopying, recording, or otherwise, without the prior written permission of the publisher. Printed in the United States of America.

ISBN-13: 978-0-321-65511-0
ISBN-10: 0-321-65511-7

**Prentice Hall**
is an imprint of

www.pearsonhighered.com

# DIAGNOSTIC ALGEBRA TEST

## *DIAGNOSTIC ALGEBRA TEST*

1. (A) Commutative ($\cdot$): $x(y + z) = (y + z)x$
   (B) Associative (+): $2 + (x + y) = (2 + x) + y$
   (C) Distributive: $(2 + 3)x = 2x + 3x$                                      (A-1)

2. $(3x - 4) + (x + 2) + (2 - 3x^2) + (x^3 + 8)$
   $= 3x - 4 + x + 2 + 2 - 3x^2 + x^3 + 8 = x^3 - 3x^2 + 4x + 8$               (A-2)

3. $[(x + 2) + (x^3 + 8)] - [(3x - 4) + (2 - 3x^2)]$
   $= x^3 + x + 10 - [-3x^2 + 3x - 2] = x^3 + 3x^2 - 2x + 12$                  (A-2)

4. $(2 - 3x^2)(x^3 + 8) = 2x^3 - 3x^5 + 16 - 24x^2 = -3x^5 + 2x^3 - 24x^2 + 16$   (A-2)

5. (A) 1  (B) 1  (C) 2  (D) 3                                                 (A-2)

6. (A) 3  (B) 1  (C) –3  (D) 1                                                (A-2)

7. $5x^2 - 3x[4 - 3(x - 2)] = 5x^2 - 3x[4 - 3x + 6] = 5x^2 - 3x(-3x + 10) = 5x^2 + 9x^2 - 30x$
   $= 14x^2 - 30x$                                                            (A-2)

8. $(2x + y)(3x - 4y) = 6x^2 - 8xy + 3xy - 4y^2 = 6x^2 - 5xy - 4y^2$          (A-2)

9. $(2a - 3b)^2 = (2a)^2 - 2(2a)(3b) + (3b)^2 = 4a^2 - 12ab + 9b^2$           (A-2)

10. $(2x - y)(2x + y) - (2x - y)^2 = (2x)^2 - y^2 - (4x^2 - 4xy + y^2)$
    $= 4x^2 - y^2 - 4x^2 + 4xy - y^2 = 4xy - 2y^2$                            (A-2)

11. $(3x^3 - 2y)^2 = (3x^3)^2 - 2(3x^3)(2y) + (2y)^2 = 9x^6 - 12x^3y + 4y^2$  (A-2)

12. $(x - 2y)^3 = (x - 2y)(x - 2y)^2 = (x - 2y)(x^2 - 4xy + 4y^2) = x(x^2 - 4xy + 4y^2) - 2y(x^2 - 4xy + 4y^2)$
    $= x^3 - 4x^2y + 4xy^2 - 2x^2y + 8xy^2 - 8y^3 = x^3 - 6x^2y + 12xy^2 - 8y^3$   (A-2)

13. (A)  $4,065,000,000,000 = 4.065 \times 10^{12}$
    (B)  $0.0073 = 7.3 \times 10^{-3}$                                        (A-5)

14. (A)  $2.55 \times 10^8 = 255,000,000$   (B) $4.06 \times 10^{-4} = 0.000\ 406$   (A-5)

15. (A)  True; if $n$ is a natural number, then $n = \dfrac{n}{1}$
    (B)  False; a number with a repeating decimal expansion is a rational
         number.                                                             (A-1)

16. Integers that are not natural numbers: $0, -1, -2, \ldots$              (A-1)

17. $6(xy^3)^5 = 6x^5y^{15}$          (A-5)      18. $\dfrac{9u^8v^6}{3u^4v^8} = \dfrac{3^2u^8v^6}{3u^4v^8} = \dfrac{3^{2-1}u^{8-4}}{v^{8-6}} = \dfrac{3u^4}{v^2}$   (A-5)

19. $(2 \times 10^5)(3 \times 10^{-3}) = 6 \times 10^{5-3} = 6 \times 10^2 = 600$   (A-5)

20. $(x^{-3}y^2)^{-2} = x^6y^{-4} = \dfrac{x^6}{y^4}$   (A-5)      21. $u^{5/3}u^{2/3} = u^{5/3\ +\ 2/3} = u^{7/3}$   (A-6)

22. $(9a^4b^{-2})^{1/2} = (3^2a^4b^{-2})^{1/2} = (3^2)^{1/2}(a^4)^{1/2}(b^{-2})^{1/2} = 3a^2b^{-1} = \dfrac{3a^2}{b}$   (A-6)

Copyright © 2011 Pearson Education, Inc.  Publishing as Prentice Hall.

**23.** $\dfrac{5^0}{3^2} + \dfrac{3^{-2}}{2^{-2}} = \dfrac{1}{3^2} + \dfrac{\frac{1}{3^2}}{\frac{1}{2^2}} = \dfrac{1}{9} + \dfrac{1}{9} \cdot \dfrac{4}{1} = \dfrac{5}{9}$      (A-5)

**24.** $(x^{1/2} + y^{1/2})^2 = (x^{1/2})^2 + 2x^{1/2}y^{1/2} + (y^{1/2})^2 = x + 2x^{1/2}y^{1/2} + y$      (A-6)

**25.** $(3x^{1/2} - y^{1/2})(2x^{1/2} + 3y^{1/2}) = 6x + 9x^{1/2}y^{1/2} - 2x^{1/2}y^{1/2} - 3y = 6x + 7x^{1/2}y^{1/2} - 3y$      (A-6)

**26.** $12x^2 + 5x - 3$

$a = 12, b = 5, c = -3$

<u>Step 1.</u>    Use the $ac$-test

$ac = 12(-3) = -36$

                                               *pq*

                                        (1)(−36)

                                        (−1)(36)

                                        (2)(−18)

                                        (−2)(18)

                                        (3)(−12)

                                        (−3)(12)

                                        (4)(−9)

                                      (−4)(9)

                                      ⋮

$-4 + 9 = 5 = b$

$12x^2 + 5x - 3 = 12x^2 - 4x + 9x - 3 = (12x^2 - 4x) + (9x - 3)$

$= 4x(3x - 1) + 3(3x - 1) = (3x - 1)(4x + 3)$      (A-3)

**27.** $8x^2 - 18xy + 9y^2$

$a = 8, b = -18, c = 9$

$ac = (8)(9) = 72$

Note that $(-6)(-12) = 72$ and $-12 + (-6) = -18$. Thus

$8x^2 - 18xy + 9y^2 = 8x^2 - 12xy - 6xy + 9y^2 = (8x^2 - 12xy) - (6xy - 9y^2)$

$= 4x(2x - 3y) - 3y(2x - 3y) = (2x - 3y)(4x - 3y)$      (A-3)

**28.** $t^2 - 4t - 6$ This polynomial cannot be factored further.      (A-3)

**29.** $6n^3 - 9n^2 - 15n = 3n(2n^2 - 3n - 5) = 3n(2n - 5)(n + 1)$      (A-3)

**30.** $(4x - y)^2 - 9x^2 = (4x - y - 3x)(4x - y + 3x) = (x - y)(7x - y)$      (A-3)

**31.** $6x(2x + 1)^2 - 15x^2(2x + 1) = 3x(2x + 1)[2(2x + 1) - 5x] = 3x(2x + 1)(2 - x)$      (A-3)

**32.** $\dfrac{2}{5b} - \dfrac{4}{3a^3} - \dfrac{1}{6a^2b^2}$     LCD $= 30a^3b^2$

$= \dfrac{6a^3b}{6a^3b} \cdot \dfrac{2}{5b} - \dfrac{10b^2}{10b^2} \cdot \dfrac{4}{3a^3} - \dfrac{5a}{5a} \cdot \dfrac{1}{6a^2b^2} = \dfrac{12a^3b}{30a^3b^2} - \dfrac{40b^2}{30a^3b^2} - \dfrac{5a}{30a^3b^2} = \dfrac{12a^3b - 40b^2 - 5a}{30a^3b^2}$      (A-4)

**33.** $\dfrac{3x}{3x^2 - 12x} + \dfrac{1}{6x} = \dfrac{3x}{3x(x - 4)} + \dfrac{1}{6x} = \dfrac{1}{x - 4} + \dfrac{1}{6x} = \dfrac{6x}{6x(x - 4)} + \dfrac{x - 4}{6x(x - 4)} = \dfrac{6x + x - 4}{6x(x - 4)} = \dfrac{7x - 4}{6x(x - 4)}$      (A-4)

Copyright © 2011 Pearson Education, Inc. Publishing as Prentice Hall.

**34.**  $\dfrac{x}{x^2-16} - \dfrac{x+4}{x^2-4x} = \dfrac{x}{(x-4)(x+4)} - \dfrac{x+4}{x(x-4)}$    $[\text{LCD} = x(x-4)(x+4)]$

$$= \dfrac{x^2}{x(x-4)(x+4)} - \dfrac{(x+4)^2}{x(x-4)(x+4)} = \dfrac{x^2-(x+4)^2}{x(x-4)(x+4)}$$

$$= \dfrac{x^2-x^2-8x-16}{x(x-4)(x+4)} = \dfrac{-8(x+2)}{x(x-4)(x+4)} \hspace{2cm} \text{(A-4)}$$

**35.**  $\dfrac{(x+y)^2-x^2}{y} = \dfrac{x^2+2xy+y^2-x^2}{y} = \dfrac{2xy+y^2}{y} = 2x+y$ \hspace{1cm} (A-4)

**36.**  $\dfrac{\dfrac{1}{7+h}-\dfrac{1}{7}}{h} = \dfrac{\dfrac{7-(7+h)}{7(7+h)}}{h} = \dfrac{\dfrac{-h}{7(7+h)}}{\dfrac{h}{1}} = \dfrac{-\cancel{h}}{7(7+h)} \cdot \dfrac{1}{\cancel{h}} = \dfrac{-1}{7(7+h)}$ \hspace{0.5cm} (A-4)

**37.**  $\dfrac{x^{-1}+y^{-1}}{x^{-2}-y^{-2}} = \dfrac{\dfrac{1}{x}+\dfrac{1}{y}}{\dfrac{1}{x^2}-\dfrac{1}{y^2}} = \dfrac{\dfrac{x+y}{xy}}{\dfrac{y^2-x^2}{x^2y^2}} = \dfrac{x+y}{\cancel{xy}} \cdot \dfrac{\overset{xy}{\cancel{x^2y^2}}}{(y-x)\cancel{(y+x)}} = \dfrac{xy}{y-x}$ \hspace{0.3cm} (A-4)

**38.**  (A)   $(-7)-(-5) = -7 + [-(-5)]$   Subtraction

(B)   $5u + (3v+2) = (3v+2) + 5u$   Commutative (+)

(C)   $(5m-2)(2m+3) = (5m-2)2m + (5m-2)3$   Distributive

(D)   $9 \cdot (49) = (9 \cdot 4)y$   Associative ($\cdot$)

(E)   $\dfrac{u}{-(v-w)} = -\dfrac{u}{v-w}$   Negatives

(F)   $(x-y) + 0 = (x-y)$   Identity (+)  \hspace{2cm} (A-1)

**39.**  $6\sqrt[5]{x^2} - 7\sqrt[4]{(x-1)^3} = 6x^{2/5} - 7(x-1)^{3/4}$ \hspace{2cm} (A-6)

**40.**  $2x^{1/2} - 3x^{2/3} = 2\sqrt{x} - 3\sqrt[3]{x^2}$ \hspace{3cm} (A-6)

**41.**  $\dfrac{4\sqrt{x}-3}{2\sqrt{x}} = \dfrac{4x^{1/2}}{2x^{1/2}} - \dfrac{3}{2x^{1/2}} = 2 - \dfrac{3}{2}x^{-1/2}$ \hspace{2cm} (A-6)

**42.**  $\dfrac{3x}{\sqrt{3x}} = \dfrac{3x}{\sqrt{3x}} \cdot \dfrac{\sqrt{3x}}{\sqrt{3x}} = \dfrac{\cancel{3x}\sqrt{3x}}{\cancel{3x}} = \sqrt{3x}$ \hspace{2cm} (A-6)

**43.**  $\dfrac{x-5}{\sqrt{x}-\sqrt{5}} = \dfrac{x-5}{\sqrt{x}-\sqrt{5}} \cdot \dfrac{\sqrt{x}+\sqrt{5}}{\sqrt{x}+\sqrt{5}} = \dfrac{\cancel{(x-5)}(\sqrt{x}+\sqrt{5})}{\cancel{x-5}} = \sqrt{x}+\sqrt{5}$ \hspace{0.5cm} (A-6)

**44.**  $\dfrac{\sqrt{x-5}}{x-5} = \dfrac{\sqrt{x-5}}{x-5} \cdot \dfrac{\sqrt{x-5}}{\sqrt{x-5}} = \dfrac{\cancel{x-5}}{\cancel{(x-5)}\sqrt{x-5}} = \dfrac{1}{\sqrt{x-5}}$ \hspace{1cm} (A-6)

Copyright © 2011 Pearson Education, Inc.  Publishing as Prentice Hall.

**45.** $\dfrac{\sqrt{u+h}-\sqrt{u}}{h} = \dfrac{\sqrt{u+h}-\sqrt{u}}{h} \cdot \dfrac{\sqrt{u+h}+\sqrt{u}}{\sqrt{u+h}+\sqrt{u}} = \dfrac{u+h-u}{h(\sqrt{u+h}+\sqrt{u})}$

$$= \dfrac{\cancel{h}}{\cancel{h}(\sqrt{u+h}+\sqrt{u})} = \dfrac{1}{\sqrt{u+h}+\sqrt{u}} \qquad \text{(A-6)}$$

**46.** $\quad x^2 = 5x$

$x^2 - 5x = 0$   (solve by factoring)

$x(x-5) = 0$

$x = 0$   or   $x - 5 = 0$

$\qquad\qquad x = 5$      (A-7)

**47.** $\;3x^2 - 21 = 0$

$x^2 - 7 = 0$   (solve by the square root method)

$x^2 = 7$

$x = \pm\sqrt{7}$      (A-7)

**48.** $\quad x^2 - x - 20 = 0$   (solve by factoring)

$(x-5)(x+4) = 0$

$x - 5 = 0$   or   $x + 4 = 0$

$\quad x = 5 \qquad\qquad x = -4$

$\qquad\qquad\qquad\qquad\qquad$ (A-7)

**49.** $\;-6x^2 + 7x - 1 = (-6x+1)(x-1) = 0;\quad x = \dfrac{1}{6},\; 1$

$\qquad\qquad\qquad\qquad\qquad$ (A-7)

Copyright © 2011 Pearson Education, Inc. Publishing as Prentice Hall.

## 1    LINEAR EQUATIONS AND GRAPHS

---

### EXERCISE 1-1

*Things to remember:*

1.  FIRST DEGREE, OR LINEAR, EQUATIONS AND INEQUALITIES

    A FIRST DEGREE, or LINEAR, EQUATION in one variable $x$ is an equation that can be written in the form

    STANDARD FORM: $ax + b = 0$, $a \neq 0$

    If the equality symbol = is replaced by $<, >, \leq$, or $\geq$, then the resulting expression is called a FIRST DEGREE, or LINEAR, INEQUALITY.

2.  SOLUTIONS

    A SOLUTION OF AN EQUATION (or inequality) involving a single variable is a number that when substituted for the variable makes the equation (or inequality) true. The set of all solutions is called the SOLUTION SET. To SOLVE AN EQUATION (or inequality) we mean that we find the solution set. Two equations (or inequalities) are EQUIVALENT if they have the same solution set.

3.  EQUALITY PROPERTIES

    An equivalent equation will result if:

    a) The same quantity is added to or subtracted from each side of a given equation.

    b) Each side of a given equation is multiplied by or divided by the same nonzero quantity.

4.  INEQUALITY PROPERTIES

    An equivalent inequality will result and the SENSE OR DIRECTION WILL REMAIN THE SAME if each side of the original inequality:

    a) Has the same real number added to or subtracted from it.

    b) Is multiplied or divided by the same positive number.

    An equivalent inequality will result and the SENSE OR DIRECTION WILL REVERSE if each side of the original inequality:

    c) Is multiplied or divided by the same negative number.

    NOTE: Multiplication and division by 0 is not permitted.

5.  The double inequality $a < x < b$ means that $a < x$ and $x < b$. Other variations, as well as a useful interval notation, are indicated in the following table.

    | Interval Notation | Inequality Notation | Line Graph |
    |---|---|---|
    | | | 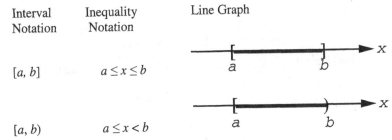 |
    | $[a, b]$ | $a \leq x \leq b$ | |
    | $[a, b)$ | $a \leq x < b$ | |

Copyright © 2011 Pearson Education, Inc. Publishing as Prentice Hall.

$(a, b]$        $a < x \le b$

$(a, b)$        $a < x < b$

$(-\infty, a]$        $x \le a$

$(-\infty, a)$        $x < a$

$[b, \infty)$        $x \ge b$

$(b, \infty)$        $x > b$

[Note: An endpoint on a line graph has a square bracket through it if it is included in the inequality and a parenthesis through it if it is not. An interval of the form $[a, b]$ is a CLOSED INTERVAL, an interval of the form $(a, b)$ is an OPEN INTERVAL. Intervals of the form $(a, b]$ and $[a, b)$ are HALF-OPEN and HALF-CLOSED.]

6.    PROCEDURE FOR SOLVING WORD PROBLEMS

   a.    Read the problem carefully and introduce a variable to represent an unknown quantity in the problem. Often the question asked in a problem will indicate the best way to introduce this variable.

   b.    Identify other quantities in the problem (known or unknown) and, whenever possible, express unknown quantities in terms of the variable you introduced in a.

   c.    Write a verbal statement using the conditions stated in the problem and then write an equivalent mathematical statement (equation or inequality).

   d.    Solve the equation or inequality and answer the questions posed in the problem.

   e.    Check the solution(s) in the original problem.

---

1.    $2m + 9 = 5m - 6$

   $2m + 9 - 9 = 5m - 6 - 9$        [using 3(a)]

   $2m = 5m - 15$

   $2m - 5m = 5m - 15 - 5m$        [using 3(a)]

   $-3m = -15$

   $\dfrac{-3m}{-3} = \dfrac{-15}{-3}$        [using 3(b)]

   $m = 5$

3.    $2x + 3 < -4$

   $2x + 3 - 3 < -4 - 3$        [using 4(a)]

   $2x < -7$

   $x < -\dfrac{7}{2}$        [using 4(b)]

5.    $-3x \ge -12$

   $\dfrac{-3x}{-3} \le \dfrac{-12}{-3}$        [using 4(c)]

   $x \le 4$

Copyright © 2011 Pearson Education, Inc. Publishing as Prentice Hall.

**7.**  $-4x - 7 > 5$

$-4x > 5 + 7$

$-4x > 12$

$x < -3$

Graph of $x < -3$ is:

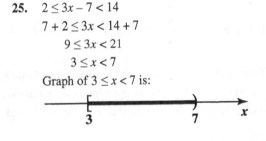

**9.**  $2 \leq x + 3 \leq 5$

$2 - 3 \leq x \leq 5 - 3$

$-1 \leq x \leq 2$

Graph of $-1 \leq x \leq 2$ is:

**11.**  $\dfrac{x}{4} + \dfrac{1}{2} = \dfrac{1}{8}$

Multiply both sides of the equation by 8. We obtain:

$2x + 4 = 1$    [using $\underline{3}$(b)]

$2x = -3$

**13.**  $\dfrac{y}{-5} > \dfrac{3}{2}$

Multiply both sides of the inequality by 10. We obtain:

$-2y > 15$

$y < -\dfrac{15}{2}$    [using $\underline{4}$(c)]

**15.**  $2u + 4 = 5u + 1 - 7u$

$2u + 4 = -2u + 1$

$4u = -3$

$u = -\dfrac{3}{4}$

**17.**  $10x + 25(x - 3) = 275$

$10x + 25x - 75 = 275$

$35x = 275 + 75$

$35x = 350$

$x = \dfrac{350}{35}$

$x = 10$

**19.**  $3 - y \leq 4(y - 3)$

$3 - y \leq 4y - 12$

$-5y \leq -15$

$y \geq 3$

[<u>Note</u>: Division by a negative number, $-3$.]

**21.**  $\dfrac{x}{5} - \dfrac{x}{6} = \dfrac{6}{5}$

Multiply both sides of the equation by 30. We obtain:

$6x - 5x = 36$

$x = 36$

**23.**  $\dfrac{m}{5} - 3 < \dfrac{3}{5} - \dfrac{m}{2}$

Multiply both sides of the inequality by 10. We obtain:

$2m - 30 < 6 - 5m$  [using $\underline{4}$(b)]

$7m < 36$

$m < \dfrac{36}{7}$

**25.**  $2 \leq 3x - 7 < 14$

$7 + 2 \leq 3x < 14 + 7$

$9 \leq 3x < 21$

$3 \leq x < 7$

Graph of $3 \leq x < 7$ is:

**27.**  $-4 \leq \dfrac{9}{5}C + 32 \leq 68$

$-36 \leq \dfrac{9}{5}C \leq 36$

$-36\left(\dfrac{5}{9}\right) \leq C \leq 36\left(\dfrac{5}{9}\right)$

$-20 \leq C \leq 20$

**29.**  $3x - 4y = 12$

$3x = 12 + 4y$

$3x - 12 = 4y$

$y = \dfrac{1}{4}(3x - 12)$

$y = \dfrac{3}{4}x - 3$

Copyright © 2011 Pearson Education, Inc.  Publishing as Prentice Hall.

Graph of $-20 \leq C \leq 20$ is:

**31.**  $Ax + By = C$
$$By = C - Ax$$
$$y = \frac{C}{B} - \frac{Ax}{B}, B \neq 0$$
or    $y = -\left(\frac{A}{B}\right)x + \frac{C}{B}$

**33.**  $F = \frac{9}{5}C + 32$
$$\frac{9}{5}C + 32 = F$$
$$\frac{9}{5}C = F - 32$$
$$C = \frac{5}{9}(F - 32)$$

**35.**  $-3 \leq 4 - 7x < 18$
$$-3 - 4 \leq -7x < 18 - 4$$
$$-7 \leq -7x < 14.$$

Dividing by –7, and recalling $\underline{4}$(c), we have
$1 \geq x > -2$ or $-2 < x \leq 1$
The graph is:

**37.**  (A)  $ab > 0$; $a > 0$ $\underline{\text{and}}$ $b > 0$, or $a < 0$ $\underline{\text{and}}$ $b < 0$

(B)  $ab < 0$; $a > 0$ $\underline{\text{and}}$ $b < 0$, or $a < 0$ $\underline{\text{and}}$ $b > 0$

(C)  $\dfrac{a}{b} > 0$; $a > 0$ $\underline{\text{and}}$ $b > 0$, or $a < 0$ $\underline{\text{and}}$ $b < 0$

(D)  $\dfrac{a}{b} < 0$; $a > 0$ $\underline{\text{and}}$ $b < 0$, or $a < 0$ $\underline{\text{and}}$ $b > 0$

**39.**  (A)   If $a - b = 2$, then $a > b$.

(B)   If $c - d = -1$, then $c < d$ $(d - c = 1$ so $d > c)$.

**41.**  Let $a, b > 0$. If $\dfrac{b}{a} > 1$, then $b > a$ so $a - b < 0$; $a - b$ is negative.

**43.**  True: Let $(a, b)$ and $(c, d)$ be open intervals such that $(a, b) \cap (c, d) \neq \varnothing$. If $(a, b) \subset (c, d)$, then $(a, b) \cap (c, d) = (a, b)$ an open interval. Similarly, if $(c, d) \subset (a, b)$. If neither interval is contained in the other, then  we can assume that $a < c < b < d$, and $(a, b) \cap (c, d) = (c, b)$, an open interval.

**45.**  False: $(0, 1) \cup (2, 3)$ is **not** an open interval.

**47.**  Let $x$ = number of $35 tickets.
Then the number of $55 tickets = $9500 - x$.
Now,
$$35x + 55(9500 - x) = 432,500$$
$$35x + 522,500 - 55x = 432,500$$

Copyright © 2011 Pearson Education, Inc.  Publishing as Prentice Hall.

$$-20x = -90,000$$
$$x = 4,500$$

Thus, $4,500$ $35$ tickets and $9,500 - 4,500 = 5,000$ $55$ tickets were sold.

**49.** Let $x =$ the amount invested in Fund $A$. Then $500,000 - x$ is the amount invested in Fund $B$. The annual interest income is

$$I = 0.052x + 0.077(500,000 - x)$$

Set $I = \$34,000$.

$$0.052x + 0.077(500,000 - x) = 34,000$$
$$52x + 77(500,000 - x) = 34,000,000$$
$$52x - 77x = 34,000,000 - 38,500,000$$
$$-25x = -450,000$$
$$x = 180,000$$

You should invest $180,000 in Fund $A$ and $320,000 in Fund $B$.

**51.** Let $x$ be the price of the car in 2005. Then

$$\frac{x}{10,000} = \frac{195.3}{130.7} \quad \text{(refer to Table 2, Example 10)}$$

$$x = 10,000 \cdot \frac{195.3}{130.7} = \approx \$14,943 \quad \text{(to the nearest dollar)}$$

**53.** (A) Wholesale price $300; mark-up $0.4(300) = 120$
   Retail price $300 + \$120 = \$420$
   (B) Let $x =$ wholesale price. Then

$$x + 0.4x = 77$$
$$1.4x = 77$$
$$x = 55$$

The wholesale price is $55.

**55.** The difference per round is $8. The number of rounds you must play to recover the cost of the clubs is:

$$\frac{270}{8} = 33.75$$

Since the number of rounds must be a whole number (positive integer), you will recover the cost of the clubs after 34 rounds.

**57.** The employee must earn $2000 in commission.
   If $x =$ sales over $7,000$, then

$$0.08x = 2000 \quad \text{and} \quad x = \frac{2000}{0.08} = 25,000$$

Therefore, the employee must sell $32,000 in the month.

**59.** Let $x =$ number of books produced. Then

Costs: $C = 1.60x + 55,000$

Revenue: $R = 11x$

To find the break-even point, set $R = C$:

Copyright © 2011 Pearson Education, Inc. Publishing as Prentice Hall.

$$11x = 1.60x + 55{,}000$$
$$9.40x = 55{,}000$$
$$x = 5851.06383$$

Thus, 5851 books will have to be sold for the publisher to break even.

**61.** Let $x$ = number of books produced.

Costs: $C = 55{,}000 + 2.10x$

Revenue: $R = 11x$

(A)    The obvious strategy is to raise the sales price of the book.

(B)    To find the break-even point, set $R(x) = C(x)$:
$$11x = 55{,}000 + 2.10x$$
$$8.90x = 55{,}000$$
$$x = 6179.78$$

The company must sell more than 6180 books.

(C)    From Problem 59, the production level at the break-even point is: 5,851 books. At this production
level, the costs are:
$$C = 55{,}000 + 2.10(5{,}851) = \$67{,}287.10.$$

If $p$ is the new price of the book, then we need
$$5851p = 67{,}287.10 \text{ and } p \approx 11.50.$$

The company should increase the price at least \$0.50 (50 cents).

**63.** Let $x$ = the number of rainbow trout in the lake. Then,

$$\frac{x}{200} = \frac{200}{8} \quad \text{(since proportions are the same)}$$

$$x = \frac{200}{8}(200)$$

$$x = 5{,}000$$

**65.** $\text{IQ} = \dfrac{\text{Mental age}}{\text{Chronological age}}(100)$

$$\frac{\text{Mental age}}{9}(100) = 140$$

$$\text{Mental age} = \frac{140}{100}(9)$$

$$= 12.6 \text{ years}$$

**67.** Cephalic index $= \dfrac{B}{L}$

Range: $\dfrac{12}{18} = \dfrac{2}{3} = 66.7\%$ to $\dfrac{18}{18} = 100\%$

---

## EXERCISE 1-2

*Things to remember:*

1.    CARTESIAN (RECTANGULAR) COORDINATE SYSTEM
The Cartesian coordinate system is formed by the intersection of a horizontal real number
line, called the **x-axis**, and a vertical real number line, called the **y-axis**. The two number
lines intersect at their origins. The axes, called the COORDINATE AXES, divide the plane
into four parts called QUADRANTS, which are numbered counterclockwise from I to IV.
(See the figure.)

Copyright © 2011 Pearson Education, Inc. Publishing as Prentice Hall.

COORDINATES are assigned to each point in the plane as follows. Given a point $P$ in the plane, pass a horizontal and a vertical line through $P$ (see the figure). The vertical line will intersect the $x$-axis at a point with coordinate $a$; the horizontal line will intersect the $y$-axis at a point with coordinate $b$. The ORDERED PAIR $(a, b)$ are the coordinates of $P$. The first coordinate, $a$, is called the ABSCISSA OF $P$; the second coordinate, $b$, is called the ORDINATE OF $P$. The point with coordinates $(0, 0)$ is called the ORIGIN. There is a one-to-one correspondence between the points in the plane and the set of all ordered pairs of real numbers.

2.  LINEAR EQUATIONS IN TWO VARIABLES

A LINEAR EQUATION IN TWO VARIABLES is an equation that can be written in the STANDARD FORM

$$Ax + By = C$$

where $A$, $B$, and $C$ are constants and $A$ and $B$ are not both zero.

3.  GRAPH OF A LINEAR EQUATION IN TWO VARIABLES

The graph of any equation of the form

$$Ax + By = C \quad (A \text{ and } B \text{ not both } 0)$$

is a line, and every line in a Cartesian coordinate system is the graph of a linear equation in two variables. The graph of the equation $x = a$ is a VERTICAL LINE; the graph of the equation $y = b$ is a HORIZONTAL LINE.

4.  INTERCEPTS

If a line crosses the $x$-axis at a point with $x$ coordinate $a$, then $a$ is called an $x$ intercept of the line; if it crosses the $y$-axis at a point with $y$ coordinate $b$, then $b$ is called the $y$ intercept.

5.  SLOPE OF A LINE

If a line passes through two distinct points $P_1(x_1, y_1)$ and $P_2(x_2, y_2)$, then its slope is given by the formula

$$m = \frac{y_2 - y_1}{x_2 - x_1} = \frac{\text{vertical change (rise)}}{\text{horizontal change (run)}}$$

Copyright © 2011 Pearson Education, Inc.  Publishing as Prentice Hall.

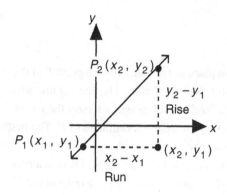

6. GEOMETRIC INTERPRETATION OF SLOPE

| Line | Slope | Example |
|------|-------|---------|
| Rising as $x$ moves from left to right | Positive | |
| Falling as $x$ moves from left to right | Negative | |
| Horizontal | 0 | |
| Vertical | Not defined | |

7. EQUATIONS OF A LINE; SPECIAL FORMS

| | | |
|------|------|------|
| Standard form | $Ax + By = C$ | $A$ and $B$ not both 0 |
| Slope-intercept form | $y = mx + b$ | Slope: $m$; $y$-intercept: $b$ |
| Point-slope form | $y - y_1 = m(x - x_1)$ | Slope: $m$; point: $(x_1, y_1)$ |
| Horizontal line | $y = b$ | Slope: 0 |
| Vertical line | $x = a$ | Slope: Undefined |

Copyright © 2011 Pearson Education, Inc. Publishing as Prentice Hall.

**1.** (D)

**3.** (C); The slope is 0.

**5.** $y = 2x - 3$

| x | y |
|---|---|
| 0 | -3 |
| 1 | -1 |
| 4 | 5 |

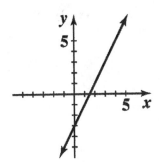

**7.** $2x + 3y = 12$

| x | y |
|---|---|
| 0 | 4 |
| 6 | 0 |
| 9 | -2 |

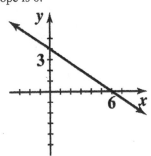

**9.** $y = 5x - 7$; slope: $m = 5$; y intercept: $b = -7$

**11.** $y = -\dfrac{5}{2}x - 9$; slope: $m = -\dfrac{5}{2}$; y intercept: $b = -9$

**13.** $y = \dfrac{x}{4} + \dfrac{2}{3} = \dfrac{1}{4}x + \dfrac{2}{3}$; slope: $m = \dfrac{1}{4}$; y intercept: $b = \dfrac{2}{3}$

**15.** $m = 2$, $b = 1$; using 7, equation: $y = 2x + 1$

**17.** $m = -\dfrac{1}{3}$, $b = 6$; using 7, equation: $y = -\dfrac{1}{3}x + 6$

**19.** x intercept: $-1$ [or $(-1, 0)$]

y intercept: $-2$ [or $(0, -2)$]

slope: $m = \dfrac{-2 - 0}{0 - (-1)} = \dfrac{-2}{1} = -2$

equation: $y = -2x - 2$

**21.** x intercept: $-3$ [or $(-3, 0)$]

y intercept: $1$ [or $(0, 1)$]

slope: $m = \dfrac{1 - 0}{0 - (-3)} = \dfrac{1}{3}$

equation: $y = \dfrac{1}{3}x + 1$

**23.** $y = -\dfrac{2}{3}x - 2$

$m = -\dfrac{2}{3}$, $b = -2$

| x | y |
|---|---|
| 0 | -2 |
| 3 | -4 |
| -3 | 0 |

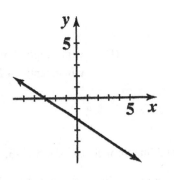

**25.** $3x - 2y = 10$

| x | y |
|---|---|
| 0 | -5 |
| 10 | 10 |
| -4 | 4 |

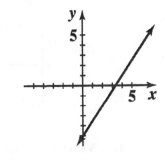

**27.**

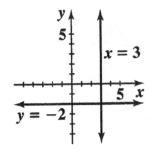

**29.** $4x + y = 3$

$y = -4x + 3$; slope: $m = -4$

**31.** $3x + 5y = 15$

$5y = -3x + 15$

$y = -\dfrac{3}{5}x + 3$; slope: $m = -\dfrac{3}{5}$

Copyright © 2011 Pearson Education, Inc. Publishing as Prentice Hall.

33.    $-4x + 2y = 9$
$$2y = 4x + 9$$
$$y = 2x + \frac{9}{2}; \quad \text{slope: } m = 2$$

35.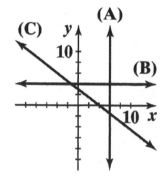

37.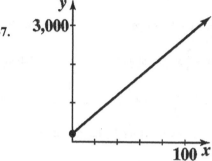

39.    (A)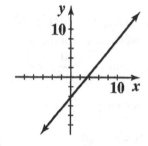

(B) $x$ intercept — set $y = 0$:
$$1.2x - 4.2 = 0$$
$$x = 3.5$$
$y$ intercept — set $x = 0$:
$$y = -4.2$$

(C)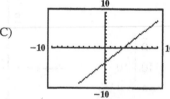

(D)    $x$-intercept: 3.5;
$y$-intercept: $-4.2$

41.    Vertical line through $(4, -3)$: $x = 4$; horizontal line through $(4, -3)$: $y = -3$.

43.    Vertical line through $(-1.5, -3.5)$: $x = -1.5$; horizontal line through $(-1.5, -3.5)$: $y = -3.5$.

45.    Slope: $m = -4$; point: $(2, -3)$. Using $\underline{5}$, equation: $y - (-3) = -4(x - 2)$
$$y + 3 = -4x + 8$$
$$y = -4x + 5$$

Copyright © 2011 Pearson Education, Inc. Publishing as Prentice Hall.

**47.**    Slope: $m = \dfrac{3}{2}$ ; point: $(-4, -5)$. Using $\underline{5}$, equation: $y - (-5) = \dfrac{3}{2}(x - [-4])$

$$y + 5 = \dfrac{3}{2}x + 6$$

$$y = \dfrac{3}{2}x + 1$$

**49.**    Slope: $m = 0$; point: $(-1.5, 4.6)$. Using $\underline{5}$, equation: $y - 4.6 = 0(x - [-1.5])$

$$y - 4.6 = 0$$

$$y = 4.6 \quad \text{(horizontal line)}$$

**51.**    Points $(2, 5)$, $(5, 7)$

(A) Using $\underline{5}$, slope: $m = \dfrac{7-5}{5-2} = \dfrac{2}{3}$

(B) Using the point-slope form:  $y - 5 = \dfrac{2}{3}(x - 2)$ ;

simplifying:  $3y - 15 = 2x - 4$  and  $-2x + 3y = 11$ .

(C) From (B),  $y = \dfrac{2}{3}x + \dfrac{11}{3}$

**53.**    Points $(-2, -1)$, $(2, -6)$

(A) Using $\underline{5}$, slope: $m = \dfrac{-6-(-1)}{2-(-2)} = -\dfrac{5}{4}$

(B)  Using the point-slope form:

$$y - (-1) = -\dfrac{5}{4}[x - (-2)]$$

$$y + 1 = -\dfrac{5}{4}(x + 2)$$

Simplifying:  $4y + 4 = -5x - 10$  and  $5x + 4y = -14$

(C)  From (B),  $y = -\dfrac{5}{4}x - \dfrac{7}{2}$

**55.**    Points $(5, 3)$, $(5, -3)$

(A) Using $\underline{5}$, slope: $m = m = \dfrac{-3-3}{5-5} = \dfrac{-6}{0}$,  not defined

(B) The line is vertical; equation $x = 5$.

(C) The slope is not defined; no slope-intercept form.

**57.**    Points $(-2, 5)$, $(3, 5)$

(A) Using $\underline{5}$, slope:  $m = \dfrac{5-5}{3-(-2)} = \dfrac{0}{5} = 0$

(B) The line is horizontal; equation $y = 5$.

(C)  $y = 0x + 5$   or   $y = 5$

**59.**    The graphs of $y = mx + 2$, $m$ any real number, all have the same $y$-intercept $(0, 2)$; for each real number $m$, $y = mx + 2$ is a non-vertical line that passes through the point $(0, 2)$.

Copyright © 2011 Pearson Education, Inc.  Publishing as Prentice Hall.

**61.** Fixed cost: $124; variable cost $0.12 per doughnut.

Total daily cost for producing $x$ doughnuts

$$C = 0.12x + 124$$

At a total daily cost of $250,

$$0.12x + 124 = 250$$
$$0.12x = 126$$
$$x = 1050$$

Thus, 1050 doughnuts can be produced at a total daily cost of $250.

**63.** (A) Since daily cost and production are linearly related,

$$C = mx + b$$

From the given information, the points $(80, 7647)$ and $(100, 9147)$ satisfy this equation. Therefore:

$$\text{slope } m = \frac{9147 - 7647}{100 - 80} = \frac{1500}{20} = 75$$

Using the point-slope form with $(x_1, C_1) = (80, 7647)$:

$$C - 7467 = 75(x - 80)$$
$$C = 75x - 6000 + 7647$$
$$C = 75x + 1647$$

(B)

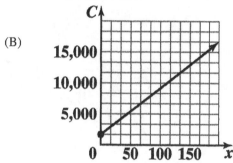

(C) The $y$-intercept, $1,647, is the fixed cost and the slope, $75, is the cost per club.

**65.** (A) $R = mC + b$

From the given information, the points $(85, 112)$ and $(175, 238)$ satisfy this equation. Therefore,

$$\text{slope } m = \frac{238 - 112}{175 - 85} = \frac{126}{90} = 1.4$$

Using the point-slope form with $(C_1, R_1) = (85, 112)$

$$R - 112 = 1.4(C - 85)$$
$$R = 1.4C - 119 + 112$$
$$R = 1.4C - 7$$

(B) Set $R = 185$. Then

$$185 = 1.4C - 7$$
$$1.4C = 192$$
$$C = 137.14$$

To the nearest dollar, the store pays $137.

**67.** (A) $V = mt + 157,000.$ At $t = 10,$

$$V = 10m + 157,000 = 82,000$$
$$10m = 82,000 - 157,000 = -75,000$$

Copyright © 2011 Pearson Education, Inc. Publishing as Prentice Hall.

$$m = -7,500$$
$$V = -7,500t + 157,000$$

(B) At $t = 6$, $V = -7,500(6) + 157,000 = 112,000$

The value of the tractor after 6 years is $112,000.

(C) Solve $-7,500t + 157,000 < 70,000$ for $t$:

$$-7,500t < 70,000 - 157,000 = -87,000$$
$$t > 11.6$$

The value of the tractor will fall below $70,000 in the 12th year.

(D)

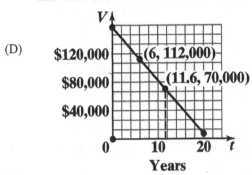

69. (A) $T = mx + b$

At $x = 0$, $T = 212$. Therefore, $b = 212$ and $T = mx + 212$.

At $x = 10$ (thousand)

$$193.6 = 10m + 212$$
$$10m = 193.6 - 212 = -18.4$$
$$m = -1.84x + 212$$

Therefore, $T = -1.84x + 212$

(B) At $t = 3.5$, $T = -1.84(3.5) + 212 = 205.56$

The boiling point at 3,500 feet is 205.56°F.

(C) Solve $200 = -1.84x + 212$ for $x$:

$$-1.84x = 200 - 212 = -12$$
$$x \approx 6.5217$$

The boiling point is 200°F at 6,522 feet.

(D)

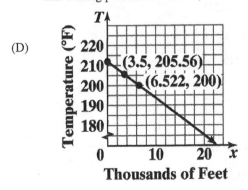

71. (A) $T = -3.6A + b$

At $A = 0$, $T = 70$. Therefore, $T = -3.6A + 70$.

(B) Solve $34 = -3.6A + 70$ for $A$:

$$-3.6A = 34 - 70 = -36$$

Copyright © 2011 Pearson Education, Inc. Publishing as Prentice Hall.

$A = 10$

The altitude of the aircraft is 10,000 feet.

**73.**  (A) $N = mt + b$

At $t = 0$, $N = 2.76$. Therefore, $N = mt + 2.76$.

At $t = 28$,

$2.55 = 28m + 2.76$

$28m = -0.21$

$m = -0.0075$

Therefore, $N = -0.0075t + 2.76$.

(B) At $t = 40$, $N = -0.0075(40) + 2.76 = 2.46$

The average number of persons per household in 2030 will be 2.46.

**75.**  (A) $f = mt + b$

At $t = 0$, $f = 21$. Therefore, $f = mt + 21$.

At $t = 6$,

$18 = 6m + 21$

$6m = -3$

$m = -0.5$

Therefore, $f = -0.5t + 21$.

(B) Solve $-0.5t + 21 < 10$ for $t$:

$-0.5t < -11$

$t > 22$

The percentage of female smokers will fall below 10% in 2022.

**77.**  (A) $p = mx + b$

At $x = 7,500$, $p = 2.28$; at $x = 7,900$, $p = 2.37$.

Therefore, slope $m = \dfrac{2.37 - 2.28}{7,900 - 7,500} = \dfrac{0.09}{400} = 0.000225$

Using the point-slope form with $(x_1, p_1) = (7,500, 2.28)$:

$p - 2.28 = 0.000225(x - 7,500) = 0.00025x + 0.5925$

$p = 0.000225x + 0.5925$  Price-supply equation

(B) $p = mx + b$

At $x = 7,900$, $p = 2.28$; at $x = 7,800$, $p = 2.37$.

Therefore, slope $m = \dfrac{2.37 - 2.28}{7,800 - 7,900} = -\dfrac{0.09}{100} = -0.0009$

Using the point-slope form with $(x_1, p_1) = (7,900, 2.37)$:

$p - 2.28 = -0.0009(x - 7,900) = -0.0009x + 9.39$

$p = -0.0009x + 9.39$  Price-demand equation

(C) To find the equilibrium point, solve

$0.000225x + 0.5925 = -0.0009x + 9.39$

Copyright © 2011 Pearson Education, Inc. Publishing as Prentice Hall.

$$0.001125x = 9.39 - 0.5925 = 8.7975$$
$$x = 7,820$$

At $x = 7,820$, $p = 0.000225(7,820) + 0.5925 = 2.352$

The equilibrium point is (7,820, 2.352).

(D)

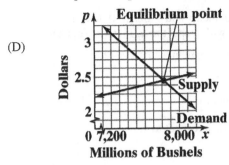

**Millions of Bushels**

**79.** (A)  $s = mw + b$

At $w = 0$, $s = 0$.  Therefore, $b = 0$ and $s = mw$.

At $w = 5$, $s = 2$.  Therefore,

$$2 = 5m \text{ and } m = \frac{2}{5}$$

Thus, $s = \frac{2}{5}w$.

(B)  At $w = 20$, $s = \frac{2}{5}(20) = 8$

The spring stretches 8 inches.

(C)  If $s = 3.6$,

$$3.6 = \frac{2}{5}w \text{ and } w = \frac{3.6(5)}{2} = 9$$

A 9 pound weight will stretch the spring 3.6 inches.

**81.** (A)  slope $m = \dfrac{29 - 9}{40 - 0} = \dfrac{20}{40} = 0.5$

The equation of the line is $y = 0.5x + 9$.

(B)  Slope $m = \dfrac{12 - 9}{10 - 0} = \dfrac{3}{10} = 0.3$

The equation of the line is $y = 0.3x + 9$.

(C)

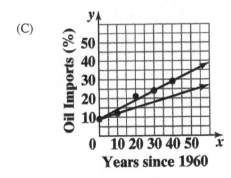

**Years since 1960**

(D) From part (A),

$$y = 0.5(60) + 9 = 39 \text{ or } 39\%$$

From part (B),

$$y = 0.3(60) + 9 = 27 \text{ or } 27\%$$

(E)  Clearly, the line in (A) gives a better representation of the data; the line in (B) predicts 27% in 2020 which is **below** the percentage in 2000.

Copyright © 2011 Pearson Education, Inc.  Publishing as Prentice Hall.

<u>EXERCISE 1-3</u>

*Things to remember:*

1.  LINEAR RELATION

    If the variables $x$ and $y$ are related by the equation

    $y = mx + b$ where $m$ and $b$ are constants with $m \neq 0$, then $x$ and $y$ are LINEARLY
    RELATED. The slope $m$ is the rate of change of $y$ with respect to $x$. If $P_1(x_1, y_1)$ and

    $P_2(x_2, y_2)$ are two distinct points on the line $y = mx + b$, then

    $$m = \frac{y_2 - y_1}{x_2 - x_1} = \frac{\text{change in } y}{\text{change in } x}$$

    This ratio is called the RATE OF CHANGE OF $y$ with respect to $x$.

2.  LINEAR REGRESSION

    REGRESSION ANALYSIS is a process for finding a function that models a given data set.
    Finding a linear model for a data set is called LINEAR REGRESSION and the line is called
    the REGRESSION LINE. A graph of the set of points in a data set is called a SCATTER
    PLOT. A regression model can be used to INTERPOLATE between points in a data set or
    to EXTRAPOLATE or predict points outside the data set.

---

1.  (A)  Let $h$ = height in inches over 5 ft.

    $w$ = weight in kilograms

    The linear model for Robinson's estimate is

    $w = 1.7h + 49$.

    (B)  The rate of change of weight with respect to height is 1.7 kilograms per inch.

    (C)  At $h = 4$,  $w = 1.7(4) + 49 = 55.8$ kg.

    (D)  Solve $60 = 1.7h + 49$ for $h$:

    $1.7h = 60 - 49 = 11$

    $h \approx 6.5$ inches

    The woman's height is approximately 5'6.5".

3.  (A)  $p = md + b$

    At $d = 0$,  $p = 14.7$. Therefore, $p = md + 14.7$.

    At $d = 33$,  $p = 29.4$.  Therefore,

    $$\text{slope } m = \frac{29.4 - 14.7}{33 - 0} = \frac{14.7}{33} = 0.44\overline{5}$$

    and $p = 0.44\overline{5} \, d + 14.7$.

    (B)  The rate of change of pressure with respect to depth is $0.44\overline{5}$ lbs/in$^2$ per foot.

    (C)  At $d = 50$,  $p = 0.44\overline{5} \, (50) + 14.7 \approx 37$ lbs/in$^2$.

    (D)  4 atmospheres $= 14.7(4) = 58.8$ lbs/in$^2$.

    Solve $58.8 = 0.44\overline{5} \, d + 14.7$ for $d$:

    $0.44\overline{5} \, d = 58.8 - 14.7 = 44.1$

    $d \approx 99$ feet

Copyright © 2011 Pearson Education, Inc. Publishing as Prentice Hall.

**5.**  (A)  $a = mt + b$

At time $t = 0$, $a = 2880$. Therefore, $b = 2880$.

At time $t = 120$, $a = 0$. The rate of change of altitude with respect to time is:

$$m = \frac{0 - 2880}{120 - 0} = -24$$

Therefore, $a = -24t + 2880$.

(B)  The rate of descent is $-24$ ft/sec.

(C)  The speed at landing is 24 ft/sec.

**7.**  $s = mt + b$

At temperature $t = 0$, $s = 331$. Therefore, $b = 331$.

At temperature $t = 20$, $s = 343$. Therefore,

$$\text{slope } m = \frac{343 - 331}{20 - 0} = \frac{12}{20} = 0.6$$

Therefore, $s = 0.6t + 331$; the rate of change of the speed of sound with respect to temperature is 0.6 m/sec per °C.

**9.**  $y = -0.3x + 84.6$

(A)

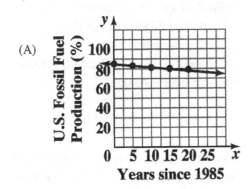

(B)  The rate of change of the percentage of fossil fuel with respect to time is $-0.3\%$ per year.

(C)  At $x = 35$, $y \approx -0.3(35) + 84.6 = 74.1$; fossil fuel production as a percentage of total production will be 74.1%.

(D) Solve  $-0.3x + 84.6 < 70$  for $x$:

$$-0.3x < 70 - 84.6 = -14.6$$

$$x > 48.\overline{6}$$

Fossil fuel production will be less than 70% of total energy production in 2034.

**11.**  $f = -0.47t + 22.19$

(A)

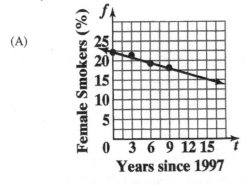

(B)  Solve  $-0.47t + 22.19 < 10$  for $t$:

$$-0.47t < 10 - 22.19 = -12.19$$

$$t > 25.94$$

The first year in which the percentage of female smokers will be less than 10% is 2023.

Copyright © 2011 Pearson Education, Inc. Publishing as Prentice Hall.

**13.**   $y = 0.17t + 10.1$

(A)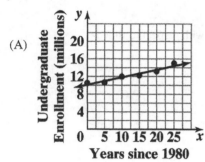

(B)   At $x = 36$, $y = 0.17(36) + 10.1 = 16.22$. The undergraduate enrollment on 2016 to the nearest hundred thousand will be 16,200,000.

(C)   Undergraduate enrollment is increasing at the rate of 170,000 students per year.

**15.**   $y = 0.72x + 0.03$

(A)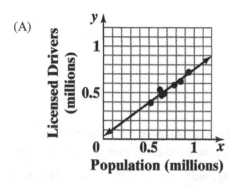

(B)   At $x = 1.5$, $y = 0.72(1.5) + 0.03 = 1.11$. There were approximately 1,110,000 licensed drivers in Idaho in 2006.

(C)   Solve $0.74 = 0.72x + 0.03$ for $x$:
$$0.72x = 0.74 - 0.03 = 0.71$$
$$x \approx 0.9861$$
The population of Rhode Island in 2006 was approximately 986,000.

**17.**   $S = 30.7t + 128$

(A)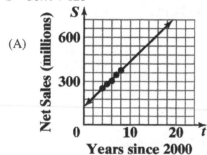

(B)   At $t = 17$, $S = 30.7(17) + 128 = 649.9 \approx 650$. Wal-Mart's net sales for 2017 will be $650 billion.

**19.**   $E = -0.55T + 31$

(A)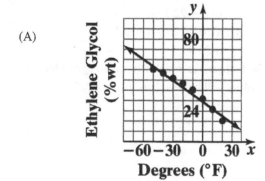

(B)   Solve $30 = -0.55T + 31$ for $T$:
$$-0.55T = 30 - 31 = -1$$
$$T \approx 1.\overline{81}$$
The freezing temperature of the solution is approximately 2°F.

(C)   at 15°F, $E = -0.55(15) + 31 = 22.75$.
A solution that freezes at 15°F is 22.75% ethylene glycol.

Copyright © 2011 Pearson Education, Inc. Publishing as Prentice Hall.

**21.** $y = 1.37x - 2.58$

   (A) The rate of change of height with respect to Dbh is 1.37 ft/in.

   (B) A 1 in. increase in Dbh produces a 1.37 foot increase in height.

   (C) At $x = 15$, $y = 1.37(15) - 2.58 = 17.97$. The spruce is approximately 18 feet tall.

   (D) Solve $25 = 1.37x - 2.58$ for $x$:

   $$1.37x = 25 + 2.58 = 27.8$$
   $$x \approx 20.3$$

   The Dbh is approximately 20 inches.

**23.** $y = 1.84x + 30.42$

   (A) The average monthly price is increasing at the rate of $1.84 per year.

   (B) At $x = 20$, $y = 1.84(20) + 30.42 = 67.22$. The average monthly price in 2020 will be $67.22.

**25.** Male enrollment:  $y = 0.05x + 5.3$;
   female enrollment:  $y = 0.17x + 4.0$

   (A) The male enrollment is increasing at the rate of $0.05(1,000,000) = 50,000$ students per year; the female enrollment is increasing at the rate of $0.17(1,000,000) = 170,000$ students per year.

   (B) Male enrollment in 2018:
   $$y = 0.05(48) + 5.3 = 7.7 \text{ or } 7.7 \text{ million};$$
   female enrollment in 2018:
   $$y = 0.17(48) + 4.0 \approx 12.2 \text{ or } 12.2 \text{ million}$$

   (C) Solve  $0.17x + 4.0 > 0.05x + 5.3 + 5$  for $x$.
   $$0.17x + 4.0 > 0.05x + 5.3 + 5$$
   $$0.12x > 6.3$$
   $$x > 52.5$$

   Female enrollment will exceed male enrollment by at least 5 million in 2023.

**27.** Linear regression $y = a + bx$

   Men: the regression formula is:   Women: the regression formula is:
   $y = 50.107 - 0.094x$
   $\qquad\qquad\qquad y = 55.512 - 0.076x$

   No; the men's times are decreasing at a faster rate than the women's times.

**29.** Linear regression $y = a + bx$

   Supply: 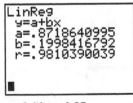    Demand: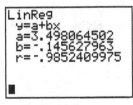

   $y = 0.20x + 0.87$ $\qquad\qquad\qquad$ $y = 0.15x + 3.5$

Copyright © 2011 Pearson Education, Inc.  Publishing as Prentice Hall.

To find the equilibrium price,
solve $0.87 + 0.20x = 3.5 - 0.15x$ for $x$:

$$0.87 + 0.20x = 3.5 - 0.15x$$
$$0.35x = 2.63$$
$$x \approx 7.51$$

At $x = 7.51$, $y = 0.87 + 0.20(7.51) = 2.37$. The equilibrium price is $2.37.

## CHAPTER 1 REVIEW

1.  $2x + 3 = 7x - 11$
    $-5x = -14$
    $x = \dfrac{14}{5} = 2.8$     (1-1)

2.  $\dfrac{x}{12} - \dfrac{x-3}{3} = \dfrac{1}{2}$

    Multiply each term by 12:
    $x - 4(x - 3) = 6$
    $x - 4x + 12 = 6$
    $-3x = 6 - 12$
    $-3x = -6$
    $x = 2$     (1-1)

3.  $2x + 5y = 9$
    $5y = 9 - 2x$
    $y = \dfrac{9}{5} - \dfrac{2}{5}x = 1.8 - 0.4x$     (1-1)

4.  $3x - 4y = 7$
    $3x = 7 + 4y$
    $x = \dfrac{7}{3} + \dfrac{4}{3}y$     (1-1)

5.  $4y - 3 < 10$
    $4y < 13$
    $y < \dfrac{13}{4}$  or  $\left(-\infty, \dfrac{13}{4}\right)$

    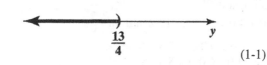
    (1-1)

6.  $-1 < -2x + 5 \le 3$
    $-6 < -2x \le -2$          (Divide the inequalities by $-2$ and reverse the direction.)
    $3 > x \ge 1$
    or $[1, 3)$

    (1-1)

7.  $1 - \dfrac{x-3}{3} \le \dfrac{1}{2}$

    Multiply both sides of the inequality by 6. We do not reverse the direction of the inequality, since $6 > 0$.

    $6 - 2(x - 3) \le 3$
    $6 - 2x + 6 \le 3$
    $-2x \le 3 - 12$
    $-2x \le -9$

    Divide both sides by $-2$ and reverse the direction of the inequality, since $-2 < 0$.

    $x \ge \dfrac{9}{2}$  or  $\left[\dfrac{9}{2}, \infty\right)$

    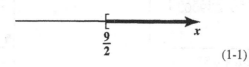
    (1-1)

Copyright © 2011 Pearson Education, Inc. Publishing as Prentice Hall.

**8.**   $3x + 2y = 9$

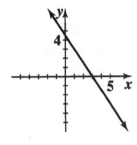

(1-2)

**9.**   The line passes through $(6, 0)$ and $(0, 4)$

slope  $m = \dfrac{4-0}{0-6} = -\dfrac{2}{3}$

From the slope-intercept form:  $y = -\dfrac{2}{3}x + 4$;  multiplying by  3  gives:  $3y = -2x + 12$,  so

$2x + 3y = 12$     (1-2)

**10.**   $x$-intercept:  $2x = 18$, $x = 9$;
$y$-intercept:  $-3y = 18$, $x = -6$;
slope-intercept form:

$y = \dfrac{2}{3}x - 6$;  slope $= \dfrac{2}{3}$

Graph:

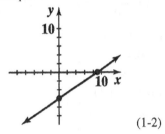

(1-2)

**11.**   $y = -\dfrac{2}{3}x + 6$     (1-2)

**12.**   Vertical line: $x = -6$;  horizontal line: $y = 5$     (1-2)

**13.**   Use the point-slope form:

(A) $y - 2 = -\dfrac{2}{3}[x - (-3)]$

$y - 2 = -\dfrac{2}{3}(x + 3)$

$y = -\dfrac{2}{3}x$

(B) $y - 3 = 0(x - 3)$

$y = 3$

(1-2)

**14.**   (A) Slope: $\dfrac{-1-5}{1-(-3)} = -\dfrac{3}{2}$

$y - 5 = -\dfrac{3}{2}(x + 3)$

$3x + 2y = 1$

(C) Slope: $\dfrac{-2-7}{-2-(-2)}$  not defined since  $2 - (-2) = 0$

$x = -2$

(B) Slope: $\dfrac{5-5}{4-(-1)} = 0$

$y - 5 = 0(x - 1)$

$y = 5$

(1-2)

Copyright © 2011 Pearson Education, Inc. Publishing as Prentice Hall.

**15.** $3x + 25 = 5x$

$-2x = -25$

$x = \dfrac{25}{2}$ (1-1)

**16.** $\dfrac{u}{5} = \dfrac{u}{6} + \dfrac{6}{5}$ (multiply by 30)

$6u = 5u + 36$

$u = 36$ (1-1)

**17.** $\dfrac{5x}{3} - \dfrac{4+x}{2} = \dfrac{x-2}{4} + 1$ (multiply by 12)

$20x - 6(4 + x) = 3(x - 2) + 12$

$20x - 24 - 6x = 3x - 6 + 12$

$11x = 30$

$x = \dfrac{30}{11}$ (1-1)

**18.** $0.05x + 0.25(30 - x) = 3.3$

$0.05x + 7.5 - 0.25x = 3.3$

$-0.20x = -4.2$

$x = \dfrac{-4.2}{-0.20} = 21$ (1-1)

**19.** $0.2(x - 3) + 0.05x = 0.4$

$0.2x - 0.6 + 0.05x = 0.4$

$0.25x = 1$

$x = 4$ (1-1)

**20.** $2(x + 4) > 5x - 4$

$2x + 8 > 5x - 4$

$2x - 5x > -4 - 8$

$-3x > -12$ (Divide both sides by –3 and reverse the inequality)

$x < 4$ or $(-\infty, 4)$

(1-1)

**21.** $3(2 - x) - 2 \le 2x - 1$

$6 - 3x \le 2x + 1$

$-5x \le -5$ (divide by –5 and reverse the inequality.)

$x \ge 1$ or $[1, \infty)$

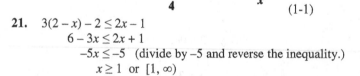

(1-1)

**22.** $\dfrac{x+3}{8} - \dfrac{4+x}{2} > 5 - \dfrac{2-x}{3}$ (multiply by 24)

$3(x + 3) - 12(4 + x) > 120 - 8(2 - x)$

$3x + 9 - 48 - 12x > 120 - 16 + 8x$

$-17x > 143$ (divide by –17 and reverse the inequality)

$x < -\dfrac{143}{17}$ or $\left(-\infty, -\dfrac{143}{17}\right)$

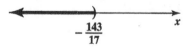

(1-1)

Copyright © 2011 Pearson Education, Inc. Publishing as Prentice Hall.

**23.**  $-5 \le 3 - 2x < 1$
$-8 \le -2x < -2$    (divide by $-2$ and reverse the directions of the inequalities.)
$4 \ge x > 1$    which is the same as $1 \le x \le 4$ or $(1, 4]$

(1-1)

**24.**  $-1.5 \le 2 - 4x \le 0.5$
$-3.5 \le -4x \le -1.5$    (divide by $-4$ and reverse the directions of the inequalities.)
$0.875 \ge x \ge 0.375$    which is the same as $0.375 \le x \le 0.875$

or  $[0.375, 0.875] = \left[\dfrac{3}{8}, \dfrac{7}{8}\right]$

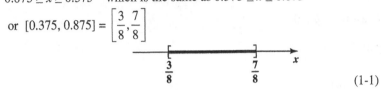

(1-1)

**25.**

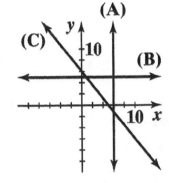

(1-2)

**26.**  The graph of $x = -3$ is a vertical line 3 units to the *left* of the $y$-axis; $y = 2$ is a horizontal line 2 units *above* the $x$-axis.

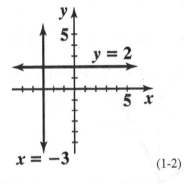

(1-2)

**27.**  (A) $3x + 4y = 0$;  $y = -\dfrac{3}{4}x$   an oblique line through the origin with slope $-\dfrac{3}{4}$

(B) $3x + 4 = 0$;  $x = -\dfrac{4}{3}$   a vertical line with $x$ intercept $-\dfrac{4}{3}$

(C) $4y = 0$;  $y = 0$   the $x$-axis

(D) $3x + 4y = 36$   an oblique line with $x$ intercept 12 and $y$-intercept 9.    (1-2)

Copyright © 2011 Pearson Education, Inc. Publishing as Prentice Hall.

**28.**  $A = \dfrac{1}{2}(a+b)h$; solve for $a$; assume $h \neq 0$

$A = \dfrac{1}{2}(ah + bh) = \dfrac{1}{2}ah + \dfrac{1}{2}bh$

$A - \dfrac{1}{2}bh = \dfrac{1}{2}ah$  (multiply by $\tfrac{2}{h}$)

$\dfrac{2A}{h} - b = a$  and  $a = \dfrac{2A - bh}{h}$    (1-1)

**29.**  $S = \dfrac{P}{1 - dt}$; solve for $d$; assume $dt \neq 0$

$S(1 - dt) = P$   (multiply by $1 - dt$)

$S - Sdt = P$

$-Sdt = P - S$  (divide by $-St$)

$d = \dfrac{P - S}{-St} = \dfrac{S - P}{St}$      (1-1)

**30.**  $a + b < b - a$

$2a < 0$

$a < 0$

The inequality is true for $a < 0$ and $b$ any number.    (1-1)

**31.**  $b < a < 0$   (divide by $b$; reverse the direction of the inequalities since $b < 0$).

$1 > \dfrac{a}{b} > 0$

Thus,  $0 < \dfrac{a}{b} < 1$;  $\dfrac{a}{b} < 1$;  $\dfrac{a}{b}$  is less than 1.    (1-1)

**32.**  The graphs of the pairs $\{y = 2x, \; y = -\dfrac{1}{2}x\}$ and

$\{y = \dfrac{2}{3}x + 2, \; y = -\dfrac{3}{2}x + 2\}$ are shown below:

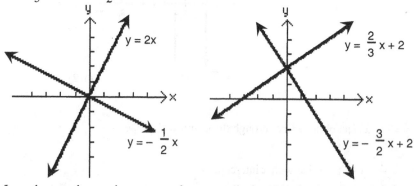

In each case, the graphs appear to be perpendicular to each other. It can be shown that two slant lines are perpendicular if and only if their slopes are negative reciprocals.    (1-2)

**33.**  Let $x =$ amount invested at 5%.

Then $300{,}000 - x =$ amount invested at 9%.

Copyright © 2011 Pearson Education, Inc. Publishing as Prentice Hall.

Yield = 300,000(0.08) = 24,000

Solve  $x(0.05) + (300,000 - x)(0.09) = 24,000$ for $x$:

$$0.05x + 27,000 - 0.09x = 24,000$$
$$-0.04x = -3,000$$
$$x = 75,000$$

Invest \$75,000 at 5%, \$225,000 at 9%.    (1-1)

**34.**  Let  $x$ = the number of CD's.

Cost:  $C(x) = 90,000 + 5.10x$

Revenue:  $R(x) = 14.70x$

Break-even point:  $C(x) = R(x)$
$$90,000 + 5.10x = 14.70x$$
$$90,000 = 9.60x$$
$$x = 9,375$$

9,375 CD's must be sold to break even.    (1-1)

**35.**  Let $x$ = person's age in years.

(A)  Minimum heart rate:  $m = (220 - x)(0.6) = 132 - 0.6x$

(B)  Maximum heart rate:  $M = (220 - x)(0.85) = 187 - 0.85x$

(C)  At $x = 20$,  $m = 132 - 0.6(20) = 120$
$$M = 187 - 0.85(20) = 170$$

range – between 120 and 170 beats per minute.

(D)  At $x = 50$,  $m = 132 - 0.6(50) = 102$
$$M = 187 - 0.85(50) = 144.5$$

range – between 102 and 144.5 beats per minute.    (1-3)

**36.**  $V = mt + b$

(A)  At  $t = 0$,  $V = 224,000$;  at  $t = 8$,   $V = 100,000$

$$\text{slope } m = \frac{100,000 - 224,000}{8 - 0} = -\frac{124,000}{8} = -15,500$$

$$V = -15,500t + 224,000$$

(B)  At  $t = 12$,  $V = -15,500(12) + 224,000 = 38,000$.

The bulldozer will be worth \$38,000 after 12 years.    (1-2)

**37.**  $R = mC + b$

(A)  From the given information, the points (50, 80) and (130, 208) satisfy this equation. Therefore:

$$\text{slope } m = \frac{208 - 80}{130 - 50} = \frac{128}{80} = 1.6$$

Using the point-slope form with $(C_1, R_1) = (50, 80)$:

$$R - 80 = 1.6(C - 50) = 1.6C - 80$$
$$R = 1.6C$$

(B)  At $C = 120$,  $R = 1.6(120) = 192$; \$192.

(C)  At $R = 176$, $176 = 1.6C$
$$C = \frac{176}{1.6} = 110; \$110$$

(D)  Slope  $m = 1.6$. The slope is the rate of change of retail price with respect to cost.    (1-2)

Copyright © 2011 Pearson Education, Inc. Publishing as Prentice Hall.

**38.** Let $x$ = weekly sales

$E = 400 + 0.10(x - 6,000)$ for $x \geq 6,000$

At $x = 4000$, $E = 400$; $400

At $x = 10,000$, $E = 400 + 0.10(10,000 - 6,000) = 400 + 0.10(4,000) = 800$; $800.   (1-1)

**39.** $p = mx + b$

From the given information, the points $(1,160, 3.79)$ and $(1,320, 3.59)$ satisfy this equation. Therefore,

slope $m = \dfrac{3.59 - 3.79}{1,320 - 1,160} = -\dfrac{0.20}{160} = -0.00125$

Using the point-slope form with $(x_1, p_1) = (1,160, 3.79)$

$p - 3.79 = -0.00125(x - 1,160) = -0.00125x + 1.45$

$p = -0.00125x + 5.24$

If $p = 3.29$, solve $3.29 = -0.00125x + 5.24$ for $x$:

$-0.00125x = 3.29 - 5.24 = -1.95$

$x = 1,560$

The stores would sell 1,560 bottles.   (1-2)

**40.** $T = 40 - 2M$

(B)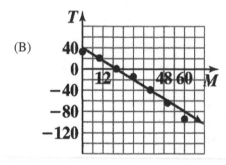

(A) At $M = 35$, $T = 40 - 2(35) = -30$; $-30°$F.

(C) At $T = -50$,

$-50 = 40 - 2M$

$-2M = -90$

$M = 45$; 45%   (1-3)

**41.** $r = -0.164t + 13.9$

(A)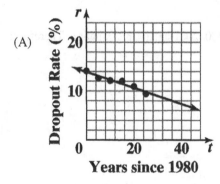

**Dropout Rate (%)**

**Years since 1980**

(C) Solve $8 = -0.164t + 13.9$ for $t$:

$-0.164t = -5.9$

$t = 35.97$

The dropout rate will be less than 8% in 2016.

(B) The dropout rate is decreasing at the rate of 0.164 percentage points per year.   (1-3)

Copyright © 2011 Pearson Education, Inc. Publishing as Prentice Hall.

**42.**   $y = 4.66x + 171.3$

(A)  The CPI is increasing at the rate of $4.66 per year.

(B)  At $x = 20$,   $y = 4.66(20) + 171.3 = 264.5$;   $264.50.      (1-3)

**43.**   $y = 0.74x + 2.83$

(A)  The rate of change of tree height with respect to Dbh is 0.74.

(B)  Tree height increases by 0.74 feet.

(C)  At $x = 25$, $y = 0.74(25) + 2.83 = 21.33$.  To the nearest foot, the tree is 21 feet high.

(D)  Solve  $15 = 0.74x + 2.83$  for $x$:

$$0.74x = 15 - 2.83 = 12.17$$
$$x = 16.446$$

To the nearest inch, the Dbh is 16 inches.     (1-3)

Copyright © 2011 Pearson Education, Inc.  Publishing as Prentice Hall.

## 2    FUNCTIONS AND GRAPHS

EXERCISE 2-1
_____

*Things to remember:*

1.  POINT-BY-POINT PLOTTING

    To sketch the graph of an equation in two variables, plot enough points from its solution set in a rectangular coordinate system so that the total graph is apparent and then connect these points with a smooth curve.

2.  A FUNCTION is a correspondence between one set of elements, called the DOMAIN, and a second set of elements, called the RANGE, such that to each element in the domain there corresponds one and only one element in the range.

3.  EQUATIONS AND FUNCTIONS

    Given an equation in two variables. If there corresponds exactly one value of the dependent variable (output) to each value of the independent variable (input), then the equation specifies a function. If there is more than one output for at least one input, then the equation does not specify a function.

4.  VERTICAL LINE TEST FOR A FUNCTION

    An equation specifies a function if each vertical line in the coordinate system passes through at most one point on the graph of the equation. If any vertical line passes through two or more points on the graph of an equation, then the equation does not specify a function.

5.  AGREEMENT ON DOMAINS AND RANGES

    If a function is specified by an equation and the domain is not given explicitly, then assume that the domain is the set of all real number replacements of the independent variable (inputs) that produce real values for the dependent variable (outputs). The range is the set of all outputs corresponding to input values.

    In many applied problems, the domain is determined by practical considerations within the problem.

6.  FUNCTION NOTATION   —   THE SYMBOL $f(x)$

    For any element $x$ in the domain of the function $f$, the symbol $f(x)$ represents the element in the range of $f$ corresponding to $x$ in the domain of $f$. If $x$ is an input value, then $f(x)$ is the corresponding output value. If $x$ is an element which is not in the domain of $f$, then $f$ is NOT DEFINED at $x$ and $f(x)$ DOES NOT EXIST.

_____

Copyright © 2011 Pearson Education, Inc. Publishing as Prentice Hall.

**1.** $y = x + 1$:

| $x$ | $-4$ | $-2$ | 0 | 2 | 4 |
|---|---|---|---|---|---|
| $y$ | $-3$ | $-1$ | 1 | 3 | 5 |

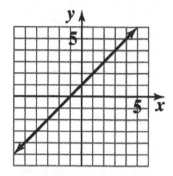

**3.** $x = y^2$:

| $x$ | 0 | 1 | 4 | 9 | 16 |
|---|---|---|---|---|---|
| $y$ | 0 | $\pm 1$ | $\pm 2$ | $\pm 3$ | $\pm 4$ |

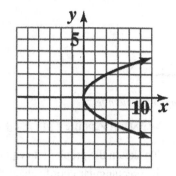

**5.** $y = x^3$:

| $x$ | $-2$ | $-1$ | 0 | 1 | 2 |
|---|---|---|---|---|---|
| $y$ | $-8$ | $-1$ | 0 | 1 | 8 |

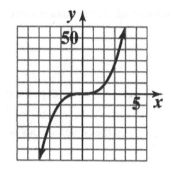

**7.** $xy = -6$:

| $x$ | $-6$ | $-3$ | $-1$ | 1 | 3 | 6 |
|---|---|---|---|---|---|---|
| $y$ | 1 | 2 | 6 | $-6$ | $-2$ | $-1$ |

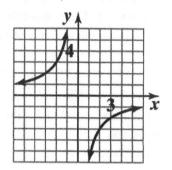

**9.** The table specifies a function, since for each domain value there corresponds one and only one range value.

**11.** The table does not specify a function, since more than one range value corresponds to a given domain value. (Range values 5, 6 correspond to domain value 3; range values 6, 7 correspond to domain value 4.)

**13.** This is a function.

**15.** The graph specifies a function; each vertical line in the plane intersects the graph in at most one point.

**17.** The graph does not specify a function. There are vertical lines which intersect the graph in more than one point. For example, the $y$-axis intersects the graph in three points.

**19.** The graph specifies a function.

**21.** $y - 2x = 7$ or $y = 2x + 7$; a linear function.

**23.** $xy - 4 = 0$ or $y = \dfrac{4}{x}$; neither a linear nor a constant function.

**25.** $y = 5x + \dfrac{1}{2}(7 - 10x) = 5x + \dfrac{7}{2} - 5x = \dfrac{7}{2}$; a constant function.

**27.** $3x + 4y = 5$ or $y = -\dfrac{3}{4}x + \dfrac{5}{4}$; a linear function

Copyright © 2011 Pearson Education, Inc. Publishing as Prentice Hall.

**29.** $f(x) = 1 - x$: Since $f$ is a linear function, we only need to plot two points.

| $x$ | $f(x)$ |
|-----|--------|
| $-2$ | $3$ |
| $2$ | $-1$ |

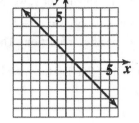

**31.** $f(x) = x^2 - 1$:

| $x$ | $-3$ | $-2$ | $-1$ | $0$ | $1$ | $2$ | $3$ |
|-----|------|------|------|-----|-----|-----|-----|
| $f(x)$ | $8$ | $3$ | $0$ | $-1$ | $0$ | $3$ | $8$ |

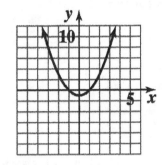

**33.** $f(x) = 4 - x^3$:

| $x$ | $-2$ | $-1$ | $0$ | $1$ | $2$ |
|-----|------|------|-----|-----|-----|
| $f(x)$ | $12$ | $5$ | $4$ | $3$ | $-4$ |

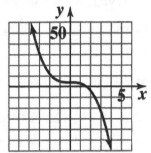

**35.** $f(x) = \dfrac{8}{x}$

| $x$ | $-8$ | $-4$ | $-2$ | $-1$ | $1$ | $2$ | $4$ | $8$ |
|-----|------|------|------|------|-----|-----|-----|-----|
| $f(x)$ | $-1$ | $-2$ | $-4$ | $-8$ | $8$ | $4$ | $2$ | $1$ |

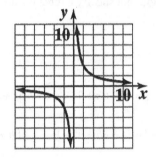

**37.** The graph of $f$ is:

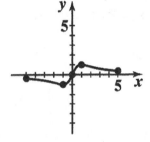

**39.** $y = f(-5) = 0$

**41.** $y = f(5) = 4$

**43.** $f(x) = 0$ at $x = -5, 0, 4$

**45.** $f(x) = -4$ at $x = -6$

**47.** $f(5) = 2 \cdot 5 - 3 = 10 - 3 = 7$

**49.** $g(-8) = (-8)^2 + 2(-8) = 64 - 16 = 48$

**51.** $f(10 + 4) = f(14) = 2(14) - 3 = 28 - 3 = 25$

**53.** $f(10) + f(4) = [2(10) - 3] + [2(4) - 3] = [20 - 3] + [8 - 3] = 17 + 5 = 22$

**55.** domain: all real numbers or $(-\infty, \infty)$

**57.** domain: all real numbers except $-4$

**59.** domain: $x \le 7$

Copyright © 2011 Pearson Education, Inc. Publishing as Prentice Hall.

**61.**  Given $x^2 - y = 1$. Solving for $y$, we have:

$$-y = -x^2 + 1 \quad \text{or} \quad y = x^2 - 1.$$

This equation specifies a function. The domain is $R$, the set of real numbers.

**63.**  Given $x + y^2 = 10$. Solving for $y$, we have:

$$y^2 = 10 - x$$
$$y = \pm\sqrt{10 - x}$$

This equation does not specify a function since each value of $x$,
$x < 10$, determines two values of $y$. For example, corresponding to
$x = 1$, we have $y = 3$ and $y = -3$; corresponding to $x = 6$, we have $y = 2$ and $y = -2$.

**65.**  Given $xy - 4y = 1$. Solving for $y$, we have:

$$(x - 4)y = 1 \quad \text{or} \quad y = \frac{1}{x - 4}$$

This equation specifies a function. The domain is all real numbers except $x = 4$.

**67.**  Given $x^2 + y^2 = 25$. Solving for $y$, we have:

$$y^2 = 25 - x^2 \quad \text{or} \quad y = \pm\sqrt{25 - x^2}$$

Thus, the equation does not specify a function since, for $x = 0$, we have $y = \pm 5$, when $x = 4$, $y = \pm 3$, and so on.

**69.**  $f(2x) = (2x)^2 - 1 = 4x^2 - 1$

**71.**  $f(x - 3) = (x - 3)^2 - 1 = x^2 - 6x + 9 - 1 = x^2 - 6x + 8$

**73.**  $f(4 + h) = (4 + h)^2 - 1 = 16 + 8h + h^2 - 1 = 15 + 8h + h^2$

**75.**  $f(4 + h) - f(4) = \left[(4 + h)^2 - 1\right] - \left[4^2 - 1\right] = 15 + 8h + h^2 - 15 = 8h + h^2$

**77.**  $f(x) = 4x - 3$

(A) $f(x + h) = 4(x + h) - 3 = 4x + 4h - 3$

(B) $f(x + h) - f(x) = 4x + 4h - 3 - (4x - 3) = 4h$

(C) $\dfrac{f(x + h) - f(x)}{h} = \dfrac{4h}{h} = 4$

**79.**  $f(x) = 4x^2 - 7x + 6$

(A) $f(x + h) = 4(x + h)^2 - 7(x + h) + 6$

$\qquad = 4(x^2 + 2xh + h^2) - 7x - 7h + 6$

$\qquad = 4x^2 + 8xh + 4h^2 - 7x - 7h + 6$

(B) $f(x + h) - f(x) = 4x^2 + 8xh + 4h^2 - 7x - 7h + 6 - (4x^2 - 7x + 6)$

$\qquad = 8xh + 4h^2 - 7h$

(C) $\dfrac{f(x + h) - f(x)}{h} = \dfrac{8xh + 4h^2 - 7h}{h} = \dfrac{h(8x + 4h - 7)}{h} = 8x + 4h - 7$

Copyright © 2011 Pearson Education, Inc. Publishing as Prentice Hall.

81.  $f(x) = x(20 - x) = 20x - x^2$

   (A) $f(x + h) = 20(x + h) - (x + h)^2 = 20x + 20h - x^2 - 2xh - h^2$

   (B) $f(x + h) - f(x) = 20x + 20h - x^2 - 2xh - h^2 - (20x - x^2)$

   $= 20h - 2xh - h^2$

   (C) $\dfrac{f(x+h) - f(x)}{h} = \dfrac{20h - 2xh - h^2}{h} = \dfrac{h(20 - 2x - h)}{h} = 20 - 2x - h$

83.  Given $A = \ell\, w = 25$.

   Thus, $\ell = \dfrac{25}{w}$. Now $P = 2\,\ell + 2w$

   $$= 2\left(\dfrac{25}{w}\right) + 2w = \dfrac{50}{w} + 2w.$$

   The domain is $w > 0$.

85.  Given $P = 2\,\ell + 2w = 100$ or $\ell + w = 50$ and $w = 50 - \ell$.

   Now $A = \ell\, w = \ell\,(50 - \ell)$ and $A = 50\,\ell - \ell^2$.

   The domain is $0 \le \ell \le 50$. [Note: $\ell \le 50$ since $\ell > 50$ implies $w < 0$.]

87.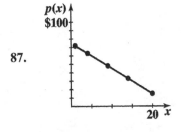

   (B)  $p(x) = 75 - 3x$
   $p(7) = 75 - 3(7) = 75 - 21 = 54$;
   $p(11) = 75 - 3(11) = 75 - 33 = 42$
   Estimated price per chip for a demand of 7 million chips: $54;
   for a demand of 11 million chips: $42

89.  (A)  $R(x) = x \cdot p(x) = x(75 - 3x) = 75x - 3x^2$, $1 \le x \le 20$

   (B)

   | $x$ | $R(x)$ |
   |-----|--------|
   | 1 | 72 |
   | 4 | 252 |
   | 8 | 408 |
   | 12 | 468 |
   | 16 | 432 |
   | 20 | 300 |

   (C)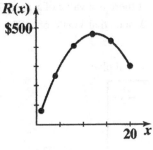

Copyright © 2011 Pearson Education, Inc.  Publishing as Prentice Hall.

**91.** (A) Profit: $P(x) = R(x) - C(x) = 75x - 3x^2 - [125 + 16x] = 59x - 3x^2 - 125, 1 \leq x \leq 20$

(B)

| $x$ | $P(x)$ |
|-----|--------|
| 1   | −69    |
| 4   | 63     |
| 8   | 155    |
| 12  | 151    |
| 16  | 51     |
| 20  | −145   |

(C)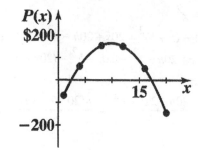

**93.**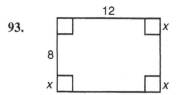

(A) $V = (\text{length})(\text{width})(\text{height})$

$V(x) = (12 - 2x)(8 - 2x)x$

$\quad\quad = x(8 - 2x)(12 - 2x)$

(B) Domain: $0 \leq x \leq 4$

(C) $V(1) = (12 - 2)(8 - 2)(1)$

$\quad\quad = (10)(6)(1) = 60$

$V(2) = (12 - 4)(8 - 4)(2)$

$\quad\quad = (8)(4)(2) = 64$

$V(3) = (12 - 6)(8 - 6)(3)$

$\quad\quad = (6)(2)(3) = 36$

Thus,

Volume

| $x$ | $V(x)$ |
|-----|--------|
| 1   | 60     |
| 2   | 64     |
| 3   | 36     |

(D)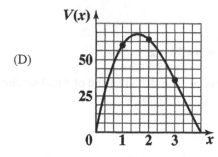

**95.** (A) The graph indicates that there is a value of $x$ near 2, and slightly less than 2, such that $V(x) = 65$. The table is shown at the right.

Thus, $x = 1.9$ to one decimal place.

(B)

| $x$ | $y_1$  |
|-----|--------|
| 1.7 | 67.252 |
| 1.8 | 66.528 |
| 1.9 | 65.436 |
| 2.0 | 64.000 |

(C)

| X    | Y1     |
|------|--------|
| 1.9  | 65.436 |
| 1.91 | 65.307 |
| 1.92 | 65.178 |
| 1.93 | 65.04  |
| 1.94 | 64.902 |
| 1.95 | 64.76  |
| 1.96 | 64.614 |

X=1.93

$x = 1.93$ to two decimal places.

Copyright © 2011 Pearson Education, Inc. Publishing as Prentice Hall.

**97.** Given $(w + a)(v + b) = c$. Let $a = 15, b = 1,$ and $c = 90$. Then: $(w + 15)(v + 1) = 90$
Solving for $v$, we have

$$v + 1 = \frac{90}{w+15} \quad \text{and} \quad v = \frac{90}{w+15} - 1 = \frac{90 - (w+15)}{w+15} = \frac{75 - w}{w+15}.$$

If $w = 16$, then $v = \frac{75-16}{16+15} = \frac{59}{31} \approx 1.9032$ cm/sec.

## EXERCISE 2-2

*Things to remember:*

<u>1.</u>    LIBRARY OF ELEMENTARY FUNCTIONS

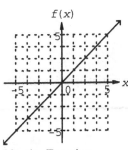

Identity Function

$f(x) = x$
Domain: All real numbers
Range: All real numbers
(a)

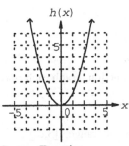

Square Function

$h(x) = x^2$
Domain: All real numbers
Range: $[0, \infty)$
(b)

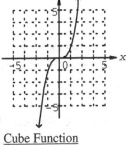

Cube Function

$m(x) = x^3$
Domain: All real numbers
Range: All real numbers
(c)

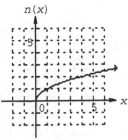

Square-Root Function

$n(x) = \sqrt{x}$
Domain: $[0, \infty)$
Range: $[0, \infty)$
(d)

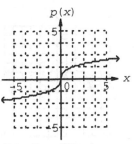

Cube-Root Function

$p(x) = \sqrt[3]{x}$
Domain: All real numbers
Range: All real numbers
(e)

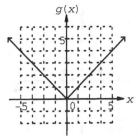

Absolute Value Function

$g(x) = |x|$
Domain: All real numbers
Range: $[0, \infty)$
(f)

*NOTE:* Letters used to designate the above functions may vary from context to context.

<u>2.</u>    GRAPH TRANSFORMATIONS SUMMARY

<u>Vertical Translation:</u>

$$y = f(x) + k \qquad \begin{cases} k > 0 & \text{Shift graph of } y = f(x) \text{ up } k \text{ units} \\ k < 0 & \text{Shift graph of } y = f(x) \text{ down } |k| \text{ units} \end{cases}$$

<u>Horizontal Translation:</u>

$$y = f(x + h) \qquad \begin{cases} h > 0 & \text{Shift graph of } y = f(x) \text{ left } h \text{ units} \\ h < 0 & \text{Shift graph of } y = f(x) \text{ right } |h| \text{ units} \end{cases}$$

<u>Reflection:</u>
$y = -f(x)$    Reflect the graph of $y = f(x)$ in the $x$ axis

Copyright © 2011 Pearson Education, Inc. Publishing as Prentice Hall.

Vertical Stretch and Shrink:

$$y = Af(x)\begin{cases} A > 1 & \text{Stretch graph of } y = f(x) \text{ vertically by multiplying each ordinate value by } A \\ \\ 0 < A < 1 & \text{Shrink graph of } y = f(x) \text{ vertically by multiplying each ordinate value by } A \end{cases}$$

3.   PIECEWISE-DEFINED FUNCTIONS

Functions whose definitions involve more than one rule are called PIECEWISE-DEFINED FUNCTIONS.

For example,

$$f(x) = |x| = \begin{cases} -x & \text{if} \quad x < 0 \\ x & \text{if} \quad x \geq 0 \end{cases}$$

is a piecewise-defined function.

To graph a piecewise-defined function, graph each rule over the appropriate portion of the domain.

---

1.   $f(x) = x^2 + 3$;   domain: all real numbers;   range: $[3, \infty)$

3.   $h(x) = -5|x|$;   domain: all real numbers;   range: $(-\infty, 0]$

5.   $m(x) = \sqrt{x} - 10$;   domain: $[0, \infty)$;   range: $[-10, \infty)$

7.   $r(x) = -\sqrt[3]{8x}$;   domain: all real numbers;   range: all real numbers

9.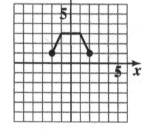

11.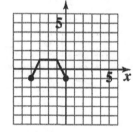

13.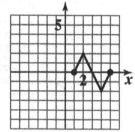

15.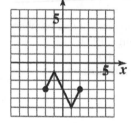

17.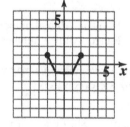

19.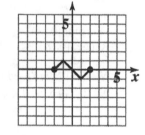

Copyright © 2011 Pearson Education, Inc. Publishing as Prentice Hall.

**21.** The graph of $g(x) = -|x + 3|$ is the graph of $y = |x|$ reflected in the $x$ axis and shifted 3 units to the left.

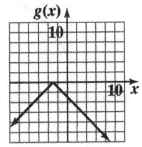

**23.** The graph of $f(x) = (x - 4)^2 - 3$ is the graph of $y = x^2$ shifted 4 units to the right and 3 units down.

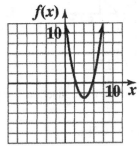

**25.** The graph of $f(x) = 7 - \sqrt{x}$ is the graph of $y = \sqrt{x}$ reflected in the $x$ axis and shifted 7 units up.

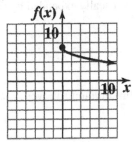

**27.** The graph of $h(x) = -3|x|$ is the graph of $y = |x|$ reflected in the $x$ axis and vertically expanded by a factor of 3.

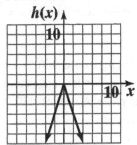

**29.** The graph of the basic function $y = x^2$ is shifted 2 units to the left and 3 units down.
Equation: $y = (x + 2)^2 - 3$.

**31.** The graph of the basic function $y = x^2$ is reflected in the $x$ axis, shifted 3 units to the right and 2 units up.
Equation: $y = 2 - (x - 3)^2$.

**33.** The graph of the basic function $y = \sqrt{x}$ is reflected in the $x$ axis and shifted 4 units up.
Equation: $y = 4 - \sqrt{x}$.

**35.** The graph of the basic function $y = x^3$ is shifted 2 units to the left and 1 unit down.
Equation: $y = (x + 2)^3 - 1$.

**37.** $g(x) = \sqrt{x - 2} - 3$

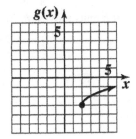

**39.** $g(x) = -|x + 3|$

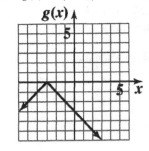

**41.** $g(x) = -(x - 2)^3 - 1$

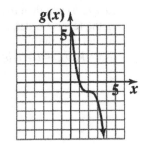

Copyright © 2011 Pearson Education, Inc. Publishing as Prentice Hall.

**43.**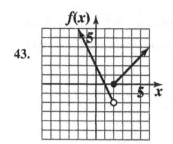

**45.**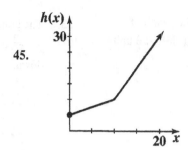

**47.**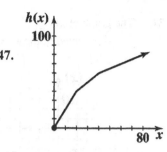

**49.** The graph of the basic function: $y = |x|$ is reflected in the $x$ axis and has a vertical contraction by the factor 0.5.

Equation: $y = -0.5|x|$.

**51.** The graph of the basic function $y = x^2$ is reflected in the $x$ axis and is vertically expanded by the factor 2.

Equation: $y = -2x^2$.

**53.** The graph of the basic function $y = \sqrt[3]{x}$ is reflected in the $x$ axis and is vertically expanded by the factor 3.

Equation: $y = -3\sqrt[3]{x}$.

**55.** Vertical shift, horizontal shift.

Reversing the order does not change the result. Consider a point $(a, b)$ in the plane. A vertical shift of $k$ units followed by a horizontal shift of $h$ units moves $(a, b)$ to $(a, b + k)$ and then to $(a + h, b + k)$.

In the reverse order, a horizontal shift of $h$ units followed by a vertical shift of $k$ units moves $(a, b)$ to $(a + h, b)$ and then to $(a + h, b + k)$. The results are the same.

**57.** Vertical shift, reflection in the $x$ axis.

Reversing the order can change the result. For example, let $(a, b)$ be a point in the plane with $b > 0$. A vertical shift of $k$ units, $k \neq 0$, followed by a reflection in the $x$ axis moves $(a, b)$ to $(a, b + k)$ and then to $(a, -[b + k]) = (a, -b - k)$.

In the reverse order, a reflection in the $x$ axis followed by the vertical shift of $k$ units moves $(a, b)$ to $(a, -b)$ and then to $(a, -b + k)$; $(a, -b - k) \neq (a, -b + k)$ when $k \neq 0$.

**59.** Horizontal shift, reflection in $y$ axis.

Reversing the order can change the result. For example, let $(a, b)$ be a point in the plane with $a > 0$. A horizontal shift of $h$ units followed by a reflection in the $y$ axis moves $(a, b)$ to the point $(a + h, b)$ and then to $(-[a + h], b) = (-a - h, b)$.

In the reverse order, a reflection in the $y$ axis followed by the horizontal sift of $h$ units moves $(a, b)$ to $(-a, b)$ and then the $(-a + h, b)$, $(-a - h, b) \neq (-a + h, b)$ when $h \neq 0$.

Copyright © 2011 Pearson Education, Inc. Publishing as Prentice Hall.

**61.** (A) The graph of the basic function $y = \sqrt{x}$ is reflected in the $x$ axis, vertically expanded by a factor of 4, and shifted up 115 units.

(B)

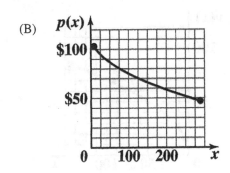

**63.** (A) The graph of the basic function $y = x^3$ is vertically contracted by a factor of 0.00048 and shifted right 500 units and up 60,000 units.

(B)

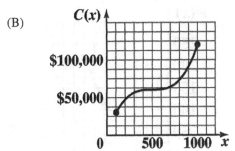

**65.** (A) $S(x) = \begin{cases} 8.50 + 0.0650x & \text{if } 0 \le x \le 700 \\ 8.50 + 0.0650(700) + 0.09(x - 700) & \text{if } x > 700 \end{cases}$

$\qquad = \begin{cases} 8.50 + 0.0650x & \text{if } 0 \le x \le 700 \\ -9 + 0.09x & \text{if } x > 700 \end{cases}$

(B)

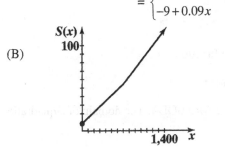

**67.** (A) If $0 \le x \le 30,000$, $T(x) = 0.035x$, and $T(30,000) = 1,050$.

If $30,000 < x \le 60,000$, $T(x) = 1,050 + 0.0625(x - 30,000) = 0.0625x - 825$, and $T(60,000) = 2,925$.

If $x > 60,000$, $T(x) = 2,925 + 0.0645(x - 60,000) = 0.0645x - 945$.

Thus, $T(x) = \begin{cases} 0.035x & \text{if} \quad 0 \le x \le 30,000 \\ 0.0625x - 825 & \text{if} \quad 30,000 < x \le 60,000 \\ 0.0645x - 945 & \text{if} \quad x > 60,000 \end{cases}$

(B)

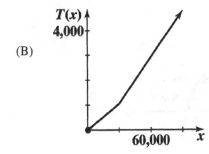

(C) $T(40,000) = 0.0625(40,000) - 825$
$\qquad = 1,675;$

$\$1,675$

$T(70,000) = 0.0645(70,000) - 945$
$\qquad = 3,570; \quad \$3,570$

Copyright © 2011 Pearson Education, Inc. Publishing as Prentice Hall.

**69.** (A) The graph of the basic function $y = x$ is vertically expanded by a factor of 5.5 and shifted down 220 units.

(B)

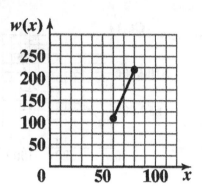

**71.** (A) The graph of the basic function $y = \sqrt{x}$ is vertically expanded by a factor of 7.08.

(B)

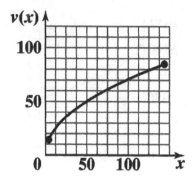

EXERCISE 2-3

*Things to remember:*

1. QUADRATIC FUNCTION

   If $a$, $b$, and $c$ are real numbers with $a \neq 0$, then the function

   $$f(x) = ax^2 + bx + c \qquad \text{STANDARD FORM}$$

   is a QUADRATIC FUNCTION and its graph is a PARABOLA. The domain of a quadratic function is the set of all real numbers.

2. PROPERTIES OF A QUADRATIC FUNCTION AND ITS GRAPH

   Given a quadratic function

   $$f(x) = ax^2 + bx + c, \quad a \neq 0,$$

   and the VERTEX FORM obtained by completing the square

   $$f(x) = a(x - h)^2 + k.$$

   The general properties of $f$ are as follows:

   a. The graph of $f$ is a parabola:

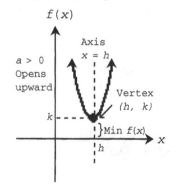

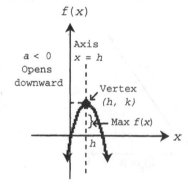

Copyright © 2011 Pearson Education, Inc. Publishing as Prentice Hall.

b.    Vertex: $(h, k)$ [parabola increases on one side of the vertex and decreases on the other]

c.    Axis (of symmetry): $x = h$ (parallel to $y$ axis)

d.    $f(h) = k$ is the minimum if $a > 0$ and the maximum if $a < 0$

e.    Domain: All real numbers

Range: $(-\infty, k]$ if $a < 0$ or $[k, \infty)$ if $a > 0$

f.    The graph of $f$ is the graph of $g(x) = ax^2$ translated horizontally $h$ units and vertically $k$ units.

---

1.   $f(x) = x^2 - 4x + 3$ ·          (standard form)

$\quad = (x^2 - 4x) + 3$

$\quad = (x^2 - 4x + 4) + 3 - 4$       (completing the square)

$\quad = (x - 2)^2 - 1$             (vertex form)

3.   $m(x) = -x^2 + 6x - 4$         (standard form)

$\quad = -(x^2 - 6x) - 4$

$\quad = -(x^2 - 6x + 9) - 4 + 9$     (completing the square)

$\quad = -(x - 3)^2 + 5$           (vertex form)

5.   From Problem 1, the graph of $f(x)$ is the graph of $y = x^2$ shifted right 2 units and down 1 unit.

7.   From Problem 3, the graph of $m(x)$ is the graph of $y = x^2$ reflected in the $x$ axis, then shifted right 3 units and up 5 units.

9.   (A) m    (B) g    (C) f    (D) n

11.   (A) $x$-intercepts: 1, 3;  $y$-intercept: –3    (B)  Vertex: $(2, 1)$

(C) Maximum: 1    (D) Range: $y \le 1$ or $(-\infty, 1]$

13.   (A) $x$-intercepts: –3, –1;  $y$intercept: 3    (B)  Vertex: $(-2, -1)$

(C) Minimum: –1    (D)  Range: $y \ge -1$ or $[-1, \infty)$

15.   $f(x) = -(x - 3)^2 + 2$

(A)    $y$-intercept:    $f(0) = -(0 - 3)^2 + 2 = -7$

$x$-intercepts:  $f(x) = 0$

$-(x - 3)^2 + 2 = 0$

$(x - 3)^2 = 2$

$x - 3 = \pm\sqrt{2}$

$x = 3 \pm \sqrt{2}$

(B) Vertex: $(3, 2)$    (C) Maximum: 2    (D) Range: $y \le 2$ or $(-\infty, 2]$

17.   $m(x) = (x + 1)^2 - 2$

(A)    $y$-intercept: $m(0) = (0 + 1)^2 - 2 = 1 - 2 = -1$

$x$-intercepts: $m(x) = 0$

Copyright © 2011 Pearson Education, Inc. Publishing as Prentice Hall.

$$(x + 1)^2 - 2 = 0$$
$$(x + 1)^2 = 2$$
$$x + 1 = \pm\sqrt{2}$$
$$x = -1 \pm \sqrt{2}$$

(B) Vertex: $(-1, -2)$  (C) Minimum: $-2$  (D) Range: $y \geq -2$ or $[-2, \infty)$

**19.** $y = -[x - (-2)]^2 + 5 = -(x + 2)^2 + 5$

**21.** $y = (x - 1)^2 - 3$

**23.** $f(x) = x^2 - 8x + 12 = (x^2 - 8x) + 12$
$$= (x^2 - 8x + 16) + 12 - 16$$
$$= (x - 4)^2 - 4 \quad \text{(vertex form)}$$

(A)  $y$-intercept: $f(0) = 0^2 - 8(0) + 12 = 12$
  $x$-intercepts: $f(x) = 0$
$$(x - 4)^2 - 4 = 0$$
$$(x - 4)^2 = 4$$
$$x - 4 = \pm 2$$
$$x = 2, 6$$

(B) Vertex: $(4, -4)$  (C) Minimum: $-4$  (D) Range: $y \geq -4$ or $[-4, \infty)$

**25.** $r(x) = -4x^2 + 16x - 15 = -4(x^2 - 4x) - 15$
$$= -4(x^2 - 4x + 4) - 15 + 16$$
$$= -4(x - 2)^2 + 1 \quad \text{(vertex form)}$$

(A)  $y$-intercept: $r(0) = -4(0)^2 + 16(0) - 15 = -15$
  $x$-intercepts: $r(x) = 0$
$$-4(x - 2)^2 + 1 = 0$$
$$(x - 2)^2 = \frac{1}{4}$$
$$x - 2 = \pm\frac{1}{2}$$
$$x = \frac{3}{2}, \frac{5}{2}$$

(B) Vertex: $(2, 1)$  (C) Maximum: $1$  (D) Range: $y \leq 1$ or $(-\infty, 1]$

**27.** $u(x) = 0.5x^2 - 2x + 5 = 0.5(x^2 - 4x) + 5$
$$= 0.5(x^2 - 4x + 4) + 3$$
$$= 0.5(x - 2)^2 + 3 \quad \text{(vertex form)}$$

(A)  $y$-intercept: $u(0) = 0.5(0)^2 - 2(0) + 5 = 5$
  $x$-intercepts: $u(x) = 0$
$$0.5(x - 2)^2 + 3 = 0$$

Copyright © 2011 Pearson Education, Inc. Publishing as Prentice Hall.

$(x-2)^2 = -6$; no solutions.

There are no $x$-intercepts.

(B) Vertex: $(2, 3)$    (C) Minimum: 3    (D) Range: $y \geq 3$ or $[3, \infty)$

29.  $f(x) = 0.3x^2 - x - 8$

(A)  $f(x) = 4$:  $0.3x^2 - x - 8 = 4$

$0.3x^2 - x - 12 = 0$

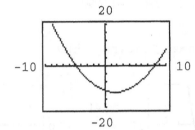

$x = -4.87, 8.21$

(B)  $f(x) = -1$:  $0.3x^2 - x - 8 = -1$

$0.3x^2 - x - 7 = 0$

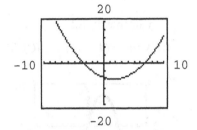

$x = -3.44, 6.78$

(C)  $f(x) = -9$:  $0.3x^2 - x - 8 = -9$

$0.3x^2 - x + 1 = 0$

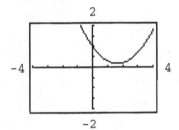

No solutions.

31.

maximum value $f(10.41667) = 651.0417$

33.  $g(x) = 0.25x^2 - 1.5x - 7 = 0.25(x^2 - 6x + 9) - 2.25 - 7 = 0.25(x-3)^2 - 9.25$

(A)  $x$-intercepts:  $0.25(x-3)^2 - 9.25 = 0$

$(x-3)^2 = 37$

$x - 3 = \pm\sqrt{37}$

$x = 3 + \sqrt{37} \approx 9.0828, \quad 3 - \sqrt{37} \approx -3.0828$

$y$-intercept: $-7$

(B) Vertex: $(3, -9.25)$    (C) Minimum: $-9.25$    (D) Range: $y \geq -9.25$ or $[-9.25, \infty)$

Copyright © 2011 Pearson Education, Inc. Publishing as Prentice Hall.

**35.**   $f(x) = -0.12x^2 + 0.96x + 1.2$

$= -0.12(x^2 - 8x + 16) + 1.92 + 1.2$

$= -0.12(x - 4)^2 + 3.12$

(A)  $x$-intercepts:  $-0.12(x - 4)^2 + 3.12 = 0$

$(x - 4)^2 = 26$

$x - 4 = \pm\sqrt{26}$

$x = 4 + \sqrt{26} \approx 9.0990, \ 4 - \sqrt{26} \approx -1.0990$

$y$-intercept:  1.2

(B)  Vertex: (4, 3.12)      (C)  Maximum: 3.12      (D)  Range: $y \le 3.12$ or $(-\infty, 3.12]$

**37.**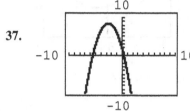

$x = -5.37, 0.37$

**39.**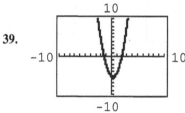

$-1.37 < x < 2.16$

**41.**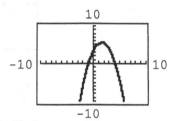

$x \le -0.74$ or $x \ge 4.19$

**43.**   $f$ is a quadratic function and min $f(x) = f(2) = 4$

Axis: $x = 2$

Vertex: (2, 4)

Range: $y \ge 4$ or $[4, \infty)$

$x$ intercepts: None

**45.**   (A)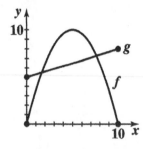

(B)  $f(x) = g(x)$

$-0.4x(x - 10) = 0.3x + 5$

$-0.4x^2 + 4x = 0.3x + 5$

$-0.4x^2 + 3.7x = 5$

$-0.4x^2 + 3.7x - 5 = 0$

$$x = \frac{-3.7 \pm \sqrt{3.7^2 - 4(-0.4)(-5)}}{2(-0.4)}$$

$$x = \frac{-3.7 \pm \sqrt{5.69}}{-0.8} \approx 1.64, 7.61$$

(C)  $f(x) > g(x)$ for $1.64 < x < 7.61$

(D)  $f(x) < g(x)$ for $0 \le x < 1.64$ or $7.61 < x \le 10$

Copyright © 2011 Pearson Education, Inc. Publishing as Prentice Hall.

**47.** (A)

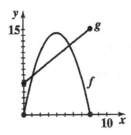

(B)  $f(x) = g(x)$

$$-0.9x^2 + 7.2x = 1.2x + 5.5$$

$$-0.9x^2 + 6x = 5.5$$

$$-0.9x^2 + 6x - 5.5 = 0$$

$$x = \frac{-6 \pm \sqrt{36 - 4(-0.9)(-5.5)}}{2(-0.9)}$$

$$x = \frac{-6 \pm \sqrt{16.2}}{-1.8} \approx 1.1,\ 5.57$$

(C)  $f(x) > g(x)$ for $1.10 < x < 5.57$

(D)  $f(x) < g(x)$ for $0 \le x < 1.10$ or $5.57 < x \le 8$

**49.**  A quadratic function has exactly one real zero if its graph is tangent to the $x$–axis.

**51.**  A quadratic function has two real zeros if $b^2 - 4ac > 0$.

**53.**  If $a$ and $k$ have the same sign, then the quadratic function has no real zeros.

**55.**

$$ax^2 + bx + c = a\left(x^2 + \frac{b}{a}x + \frac{c}{a}\right)$$

$$= a\left(x^2 + \frac{b}{a}x + \frac{b^2}{4a^2} - \frac{b^2}{4a^2} + \frac{c}{a}\right)$$

$$= a\left(\left[x^2 + \frac{b}{a}x + \frac{b^2}{4a^2}\right] + \frac{4ac - b^2}{4a^2}\right)$$

$$= a\left(x + \frac{b}{2a}\right)^2 + \frac{4ac - b^2}{4a}$$

Therefore,    $x = -\dfrac{b}{2a}$

**57.**  Mathematical model: $f(x) = -0.518x^2 + 33.3x - 481$

(A)

| $x$ | 28 | 30 | 32 | 34 | 36 |
|---|---|---|---|---|---|
| Mileage | 45 | 52 | 55 | 51 | 47 |
| $f(x)$ | 45.3 | 51.8 | 54.2 | 52.4 | 46.5 |

(B)

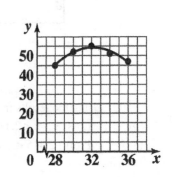

Copyright © 2011 Pearson Education, Inc.  Publishing as Prentice Hall.

(C) $x = 31: f(31) = -0.518(31)^2 + 33.3(31) - 481 = 53.502$

$f(31) \approx 53.50$ thousand miles

$x = 35: f(35) = -0.518(35)^2 + 33(35) - 481 \approx 49.95$ thousand miles

(D)     The maximum mileage is achieved at 32 lb/in$^2$ pressure. Increasing the pressure or decreasing the pressure reduces the mileage.

**59.** Quadratic regression using the data in Problem 53:

$f(x) = -0.518x^2 + 33.3x - 481$

**61.** $p(x) = 75 - 3x; \; R(x) = xp(x); \; 1 \le x \le 20$

$R(x) = x(75 - 3x) = 75x - 3x^2$

$\qquad = -3(x^2 - 25x)$

$\qquad = -3\left(x^2 - 25x + \dfrac{625}{4}\right) + \dfrac{1875}{4}$

$\qquad = -3\left(x - \dfrac{25}{2}\right)^2 + \dfrac{1875}{4}$

$\qquad = -3(x - 12.5)^2 + 468.75$

(A)

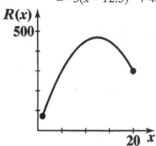

(B)    Output for maximum revenue: $x = 12.5$ (12,500,000 chips); maximum revenue: $468,750,000

(C)    Wholesale price per chip at maximum revenue:

$p(12.5) = 75 - 3(12.5) = 37.5$ or \$37.50

**63.** Revenue function: $R(x) = x(75 - 3x) = 75 - 3x^2$

Cost function: $C(x) = 125 + 16x$

(A)

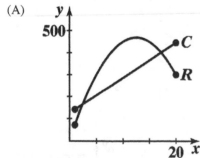

(B) Break-even points: $R(x) = C(x)$

$75x - 3x^2 = 125 + 16x$

$-3x^2 + 59x - 125 = 0$    or

$3x^2 - 59x + 125 = 0$

$x = \dfrac{59 \pm \sqrt{(59)^2 - 4(3)(125)}}{2(3)}$

$\quad = \dfrac{59 \pm \sqrt{1981}}{6} = \dfrac{59 \pm 44.508}{6}$

$x = 2.415$ or $17.251$

The company breaks even at $x = 2,451,000$ chips and $17,251,000$ chips.

(C)   Using the results from (A) and (B):

Loss: $1 \le x < 2.415$ or $17.251 < x \le 20$

Profit: $2.415 < x < 17.251$

Copyright © 2011 Pearson Education, Inc. Publishing as Prentice Hall.

**65.**   Revenue function: $R(x) = x(75 - 3x)$

Cost function: $C(x) = 125 + 16x$

(A)    Profit function: $P(x) = x(75 - 3x) - (125 + 16x)$

$$= 75x - 3x^2 - 125 - 16x;$$

$$P(x) = 59x - 3x^2 - 125$$

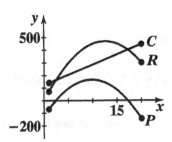

(B)  The $x$ coordinates of the intersection points of $R$ and $C$ are the same as the $x$-intercepts of $P$.

(C)  $x$-intercepts of $P$: $-3x^2 + 59x - 125$

$$x = \frac{-59 \pm \sqrt{(59)^2 - 4(-3)(-125)}}{-6}$$

$x = 2.415$ or $17.251$  (See Problem 61)

The break–even points are: 2,415,000 chips and 17,251,000 chips.

(D)    The maximum profit and the maximum revenue do not occur at the same output level and they are not

equal. The profit function involves both revenue and cost; the revenue function does not involve the production costs.

(E)    $P(x) = -3x^2 + 59x - 125$

$$= -3\left(x^2 - \frac{59}{3}x\right) - 125$$

$$= -3\left[x - \left(\frac{59}{6}\right)\right]^2 - 125 + 3\left(\frac{59}{6}\right)^2$$

$$= -3(x - 9.833)^2 + 165.083$$

The maximum profit is \$165,083,000; it occurs at an output level of 9,833,000 chips. From Problem 61(B), the maximum revenue is \$468,750,000. The maximum profit is much smaller than the maximum revenue.

**67.**   Solve:   $f(x) = 1,000(0.04 - x^2) = 20$

$$40 - 1000x^2 = 20$$

$$1000x^2 = 20$$

$$x^2 = 0.02$$

$$x = 0.14 \text{ or } -0.14$$

Since we are measuring distance, we take the positive solution:

$x = 0.14$ cm

**69.**   Quadratic regression model

```
QuadReg
 y=ax²+bx+c
 a=1.4E⁻6
 b=⁻.00266
 c=5.4

■
```

$y \approx 0.0000014x^2 - 0.00266 + 5.4 \approx 10.6$ mph.

Copyright © 2011 Pearson Education, Inc.  Publishing as Prentice Hall.

EXERCISE 2-4

*Things to remember:*

1.  POLYNOMIAL FUNCTION

    A POLYNOMIAL FUNCTION is a function that can be written in the form
    $$f(x) = a_n x^n + a_{n-1} x^{n-1} + \ldots + a_1 x + a_0$$
    for $n$ a nonnegative integer, called the DEGREE of the polynomial. The coefficients $a_0, a_1, \ldots, a_n$ are

    real numbers with $a_n \neq 0$, $a_n$ is called the LEADING COEFFICIENT of $f$. The DOMAIN of a poly-

    nomial function is the set of all real numbers. The graph of a polynomial function is continuous, with no

    holes or breaks.

2.  A RATIONAL FUNCTION is any function that can be written in the form
    $$f(x) = \frac{n(x)}{d(x)} \quad d(x) \neq 0$$
    where $n(x)$ and $d(x)$ are polynomials. The DOMAIN is the set of all real numbers such that $d(x) \neq 0$.

3.  PROCEDURE: VERTICAL AND HORIZONTAL ASYMPTOTES OF RATIONAL FUNCTIONS

    Consider the rational function
    $$f(x) = \frac{n(x)}{d(x)}$$
    where $n(x)$ and $d(x)$ are polynomials.

    VERTICAL ASYMPTOTES:

    Case 1. Suppose $n(x)$ and $d(x)$ have no real zeros in common. If $c$ is a number such that $d(c) = 0$,

    then the line $x = c$ is a vertical asymptote of the graph of $f$.

    Case 2. If $n(x)$ and $d(x)$ have one or more real zeros in common, cancel the common linear factors and

    apply Case 1 to the reduced function. (The reduced function has the same asymptotes as $f$.)

    HORIZONTAL ASYMPTOTES:

    Case 1. If degree $n(x) <$ degree $d(x)$, then $y = 0$ is the horizontal asymptote.

    Case 2. If degree $n(x) =$ degree $d(x)$, then $y = a/b$ is the horizontal asymptote, where $a$ is the leading

    coefficient of $n(x)$ and $b$ is the leading coefficient of $d(x)$.

    Case 3. If degree $n(x) >$ degree $d(x)$, then there is no horizontal asymptote.

Copyright © 2011 Pearson Education, Inc. Publishing as Prentice Hall.

1.  $f(x) = 5x + 3$
    (A) degree 1
    (B) $x$-intercept: $f(x) = 0$
    $$5x + 3 = 0$$
    $$x = -\frac{3}{5}$$
    (C) $y$-intercept: $f(0) = 3$

3.  $f(x) = x^2 - 9$
    (A) degree: 2
    (B) $x$-intercepts: $f(x) = 0$
    $$x^2 - 9 = 0$$
    $$(x - 3)(x + 3) = 0$$
    $$x = 3, -3$$
    (C) $y$-intercept: $f(0) = -9$

5.  $f(x) = (x - 2)(x + 3)(x - 5) = x^3 - 4x^2 - 11x + 30$
    (A) degree: 3
    (B) $x$-intercepts: $f(x) = 0$
    $$x = 2, -3, 5$$
    (C) $y$-intercept: $f(0) = 30$

7.  $f(x) = (2x - 9)(3x + 4) = 6x^2 - 19x - 36$
    (A) degree: 2
    (B) $x$-intercepts: $f(x) = 0$
    $$x = \frac{9}{2}, -\frac{4}{3}$$
    (C) $y$-intercept: $f(0) = -36$

9.  $f(x) = x^6 + 1$
    (A) degree: 6
    (B) $x$-intercepts: $f(x) = 0$
    $$x^6 + 1 = 0; \text{ no solutions; no } x\text{-intercepts}$$
    (C) $y$-intercept: $f(0) = 1$

11. (A) 4    (B) negative

13. (A) 5    (B) negative

15. (A) 1    (B) negative

17. (A) 6    (B) positive

19. 10

21. 1; polynomials of odd degreecross the $x$-axis at least once.

23. $f(x) = \dfrac{x + 2}{x - 2}$

    (A) *Intercepts:*

    $x$-intercepts:       $f(x) = 0$ only if $x + 2 = 0$ or $x = -2$.
    The $x$ intercept is $-2$.

    $y$-intercept:        $f(0) = \dfrac{0 + 2}{0 - 2} = -1$
    The $y$ intercept is $-1$.

    (B) *Domain:* The denominator is $0$ at $x = 2$. Thus, the domain is the set of all real numbers except 2.

    (C) *Asymptotes:*

    Vertical asymptotes:       $f(x) = \dfrac{x + 2}{x - 2}$    The denominator is 0 at $x = 2$. Therefore, the line $x = 2$ is a vertical asymptote.

    Horizontal asymptotes:  $f(x) = \dfrac{x + 2}{x - 2} = \dfrac{1 + \dfrac{2}{x}}{1 - \dfrac{2}{x}}$

    As $x$ increases or decreases without bound, the numerator tends to 1 and the denominator tends to 1. Therefore, the line $y = 1$ is a horizontal asymptote.

Copyright © 2011 Pearson Education, Inc. Publishing as Prentice Hall.

(D)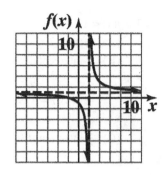

(E)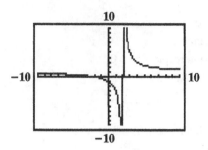

**25.** $f(x) = \dfrac{3x}{x+2}$

(A) *Intercepts:*

$x$-intercepts:    $f(x) = 0$ only if $3x = 0$ or $x = 0$.

The $x$-intercept is 0.

$y$-intercept:    $f(0) = \dfrac{3 \cdot 0}{0 + 2} = 0$

The $y$-intercept is 0.

(B) *Domain:*    The denominator is 0 at $x = -2$. Thus, the domain is the set of all real numbers except $-2$.

(C) *Asymptotes:*

Vertical asymptotes:    $f(x) = \dfrac{3x}{x + 2}$

The denominator is 0 at $x = -2$. Therefore, the line $x = -2$ is a vertical asymptote.

Horizontal asymptotes:    $f(x) = \dfrac{3x}{x + 2} = \dfrac{3}{1 + \dfrac{2}{x}}$

As $x$ increases or decreases without bound, the numerator is 3 and the denominator tends to 1. Therefore, the line $y = 3$ is a horizontal asymptote.

(D)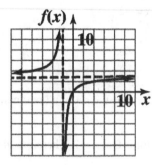

(E)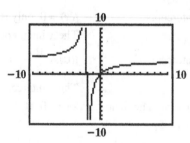

**27.** $f(x) = \dfrac{4 - 2x}{x - 4}$

(A) *Intercepts:*

$x$-intercepts:    $f(x) = 0$ only if $4 - 2x = 0$ or $x = 2$.

The $x$-intercept is 2.

Copyright © 2011 Pearson Education, Inc. Publishing as Prentice Hall.

y-intercept:  $$f(0) = \frac{4 - 2 \cdot 0}{0 - 4} = -1$$

The y-intercept is $-1$.

(B)   *Domain:* The denominator is 0 at $x = 4$. Thus, the domain is the set of all real numbers except 4.

(C)   *Asymptotes:*

Vertical asymptotes:  $$f(x) = \frac{4 - 2x}{x - 4}$$

The denominator is 0 at $x = 4$. Therefore, the line $x = 4$ is a vertical asymptote.

Horizontal asymptotes:  $$f(x) = \frac{4 - 2x}{x - 4} = \frac{\frac{4}{x} - 2}{1 - \frac{4}{x}}$$

As $x$ increases or decreases without bound, the numerator tends to $-2$ and the denominator tends to 1. Therefore, the line $y = -2$ is a horizontal asymptote.

(D)

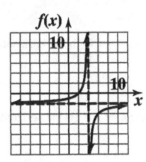

(E)

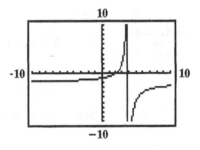

**29.**   (A)

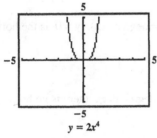

$y = 2x^4$

$y = 2x^4 - 5x^2 + x + 2$

(B)

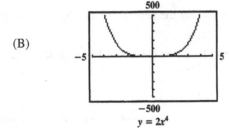

$y = 2x^4$

$y = 2x^4 - 5x^2 + x + 2$

Copyright © 2011 Pearson Education, Inc. Publishing as Prentice Hall.

**31.** (A)

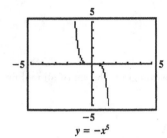

$y = -x^5$

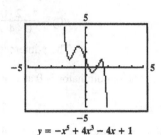

$y = -x^5 + 4x^3 - 4x + 1$

(B)

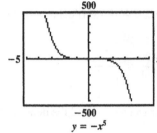

$y = -x^5$

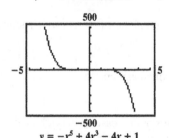

$y = -x^5 + 4x^3 - 4x + 1$

**33.** $f(x) = \dfrac{n(x)}{d(x)} = \dfrac{5x^3 + 2x - 3}{6x^3 - 7x + 1}$. Since degree $n(x) = 3 = $ degree $d(x)$, $y = \dfrac{5}{6}$ is the horizontal asymptote.

**35.** $f(x) = \dfrac{n(x)}{d(x)} = \dfrac{1 - 5x + x^2}{2 + 3x + 4x^2}$. Since degree $n(x) = 2 = $ degree $d(x)$, $y = \dfrac{1}{4}$ is the horizontal asymptote.

**37.** $f(x) = \dfrac{n(x)}{d(x)} = \dfrac{x^4 + 2x^2 + 1}{1 - x^5}$. Since degree $n(x) = 4 < 5 = $ degree $d(x)$, $y = 0$ is the horizontal asymptote.

**39.** $f(x) = \dfrac{n(x)}{d(x)} = \dfrac{x^2 + 6x + 1}{x - 5}$. Since degree $n(x) = 2 > 1 = $ degree $d(x)$, there is no horizontal asymptote.

**41.** $f(x) = \dfrac{n(x)}{d(x)} = \dfrac{x^2 + 1}{(x^2 - 1)(x^2 - 9)} = \dfrac{x^2 + 1}{(x-1)(x+1)(x-3)(x+3)}$. Since $n(x) = x^2 + 1$ has no real zeros and

$d(1) = d(-1) = d(3) = d(-3) = 0$, $x = 1$, $x = -1$, $x = 3$, $x = -3$ are the vertical asymptotes of the graph of $f$.

**43.** $f(x) = \dfrac{n(x)}{d(x)} = \dfrac{x^2 - x - 6}{x^2 - 3x - 10} = \dfrac{(x-3)(x+2)}{(x-5)(x+2)} = \dfrac{x-3}{x-5}$, $x \neq -2$. $x = 5$ is a vertical asymptote of the graph of

$f$.

**45.** $f(x) = \dfrac{n(x)}{d(x)} = \dfrac{x^2 + 3x}{x^3 - 36x} = \dfrac{x(x+3)}{x(x^2 - 36)} = \dfrac{x+3}{(x-6)(x+6)}$, $x \neq 0$. $x = 6$, $x = -6$ are the vertical asymptotes

of the graph of $f$.

Copyright © 2011 Pearson Education, Inc. Publishing as Prentice Hall.

**47.**  $f(x) = \dfrac{2x^2}{x^2 - x - 6}$

(A)  *Intercepts:*

$x$-intercepts:
$\qquad$ $f(x) = 0$ only if $2x^2 = 0$ or $x = 0$.
$\qquad$ The $x$-intercept is 0.

$y$-intercept:
$\qquad$ $f(0) = \dfrac{2 \cdot 0^2}{0^2 - 0 - 6} = 0$
$\qquad$ The $y$-intercept is 0.

(B)  *Asymptotes:*

Vertical asymptotes:
$\qquad$ $f(x) = \dfrac{2x^2}{x^2 - x - 6} = \dfrac{2x^2}{(x-3)(x+2)}$

The denominator is 0 at $x = -2$ and $x = 3$. Thus, the lines $x = -2$
$\qquad\qquad$ and $x = 3$ are vertical asymptotes.

Horizontal asymptotes:
$\qquad$ $f(x) = \dfrac{2x^2}{x^2 - x - 6} = \dfrac{2}{1 - \dfrac{1}{x} - \dfrac{6}{x^2}}$

As $x$ increases or decreases without bound, the numerator is 2 and the denominator tends to 1. Therefore, the line $y = 2$ is a horizontal asymptote.

(C) $\qquad\qquad\qquad\qquad\qquad\qquad$ (D)

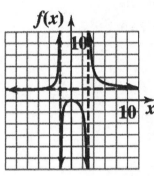
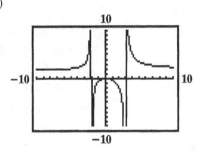

**49.**  $f(x) = \dfrac{6 - 2x^2}{x^2 - 9}$

(A)  *Intercepts:*

$x$-intercepts:
$\qquad$ $f(x) = 0$ only if $6 - 2x^2 = 0$
$\qquad\qquad\qquad\qquad\qquad 2x^2 = 6$
$\qquad\qquad\qquad\qquad\qquad x^2 = 3$
$\qquad\qquad\qquad\qquad\qquad x = \pm\sqrt{3}$
$\qquad$ The $x$-intercepts are $\pm\sqrt{3}$.

$y$-intercept:
$\qquad$ $f(0) = \dfrac{6 - 2 \cdot 0^2}{0^2 - 9} = -\dfrac{2}{3}$
$\qquad$ The $y$-intercept is $-\dfrac{2}{3}$.

(B)  *Asymptotes:*

Vertical asymptotes:
$\qquad$ $f(x) = \dfrac{6 - 2x^2}{x^2 - 9} = \dfrac{6 - 2x^2}{(x-3)(x+3)}$

Copyright © 2011 Pearson Education, Inc. Publishing as Prentice Hall.

The denominator is 0 at $x = -3$ and $x = 3$.  Thus, the lines $x = -3$ and $x = 3$ are vertical asymptotes.

Horizontal asymptotes: $$f(x) = \frac{6 - 2x^2}{x^2 - 9} = \frac{\dfrac{6}{x^2} - 2}{1 - \dfrac{9}{x^2}}$$

As $x$ increases or decreases without bound, the numerator tends to $-2$  and the denominator tends to 1. Therefore, the line $y = -2$ is a horizontal asymptote.

(C)
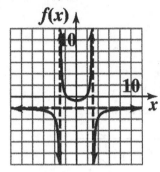

(D)
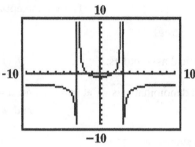

**51.**   $f(x) = \dfrac{-4x}{x^2 + x - 6}$

(A)   *Intercepts:*

$x$-intercepts:       $f(x) = 0$  only if $-4x = 0$  or  $x = 0$.
                                The $x$-intercept is 0.

$y$-intercept:        $f(0) = \dfrac{-4 \cdot 0}{0^2 + 0 - 6} = 0$

                                The $y$-intercept is 0.

(B)   *Asymptotes:*

Vertical asymptotes:     $f(x) = \dfrac{-4x}{x^2 + x - 6} = \dfrac{-4x}{(x+3)(x-2)}$

The denominator is 0 at  $x = -3$  and  $x = 2$. Thus, the lines $x = -3$  and  $x = 2$ are vertical asymptotes.

Horizontal asymptotes:      $$f(x) = \frac{-4x}{x^2 + x - 6} = \frac{-\dfrac{4}{x}}{1 + \dfrac{1}{x} + \dfrac{6}{x^2}}$$

As $x$ increases or decreases without bound, the numerator tends to 0 and the denominator tends to 1. Therefore, the line $y = 0$ (the $x$ axis) is a horizontal asymptote.

Copyright © 2011 Pearson Education, Inc.  Publishing as Prentice Hall.

(C)
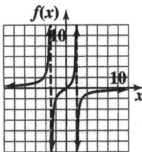

(D)
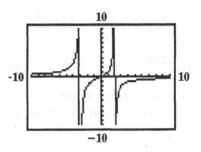

**53.** The graph has 1 turning point which implies degree $n = 2$. The $x$-intercepts are $x = -1$ and $x = 2$.
Thus, $f(x) = (x + 1)(x - 2) = x^2 - x - 2$.

**55.** The graph has 2 turning points which implies degree $n = 3$. The $x$-intercepts are $x = -2$, $x = 0$, and $x = 2$.
The direction of the graph indicates that leading coefficient is negative
$f(x) = -(x + 2)(x)(x - 2) = 4x - x^3$.

**57.** (A) Since $C(x)$ is a linear function of $x$, it can be written in the form
$C(x) = mx + b$
Since the fixed costs are $200, $b = 200$.
Also, $C(20) = 3800$, so
$3800 = m(20) + 200$

$20m = 3600$
$m = 180$
Therefore, $C(x) = 180x + 200$

(B) $\overline{C}(x) = \dfrac{C(x)}{x} = \dfrac{180x + 200}{x}$

(C)
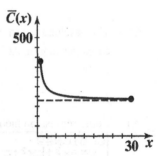

(D) $\overline{C}(x) = \dfrac{180x + 200}{x} = \dfrac{180 + \dfrac{200}{x}}{1}$

As $x$ increases, the numerator tends to 180 and the denominator is 1.
Therefore, $\overline{C}(x)$ tends to 180 or $180 per board.

**59.** (A) $\overline{C}(n) = \dfrac{2500 + 175n + 25n^2}{n}$

(B)
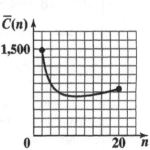

(C) Using the graph, we calculate

$\overline{C}(8) = \dfrac{2500 + 175(8) + 25(8)^2}{8} = 687.50$

$\overline{C}(9) = \dfrac{2500 + 175(9) + 25(9)^2}{9} = 677.78$

$\overline{C}(10) = \dfrac{2500 + 175(10) + 25(10)^2}{10} = 675.00$

Copyright © 2011 Pearson Education, Inc. Publishing as Prentice Hall.

$$\overline{C}(11) = \frac{2500 + 175(11) + 25(11)^2}{11} = 677.27$$

$$\overline{C}(12) = \frac{2500 + 175(12) + 25(12)^2}{12} = 683.33$$

Thus, it appears that the average cost per year is a minimum at $n = 10$ years; at 10 years, the average minimum cost is \$675.00 per year.

(D)    10 years; \$675.00 per year

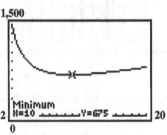

**61.**   (A)   $\overline{C}(x) = \dfrac{0.00048(x - 500)^3 + 60,000}{x}$        (B)

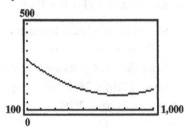

(C)   The caseload which yields the minimum average cost per case is 750 cases per month. At 750 cases per month, the average cost per case is \$90.

**63.**   (A)   Cubic regression model

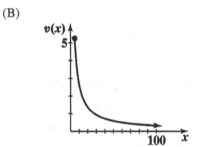

(B)   Per capita consumption of ice cream in 2020: $y(40) \approx 8.9$ lbs.

**65.**   (A)   $v(x) = \dfrac{26 + 0.06x}{x} = \dfrac{\dfrac{26}{x} + 0.06}{1}$        (B)

As $x$ increases, the numerator tends to 0.06 and the denominator is 1. Therefore, $v(x)$ approaches 0.06 centimeters per second as $x$ increases.

Copyright © 2011 Pearson Education, Inc.  Publishing as Prentice Hall.

**67.** (A)   Cubic regression model

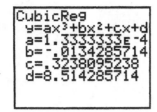

(B)   The marriage rate for 2020 will be 8.4.

## EXERCISE 2-5

*Things to remember:*

<u>1.</u>   **EXPONENTIAL FUNCTION**

The equation

$$f(x) = b^x, b > 0, b \neq 1$$

defines an EXPONENTIAL FUNCTION for each different constant $b$, called the BASE.  The DOMAIN of $f$ is all real numbers, and the RANGE of $f$ is the set of positive real numbers.

<u>2.</u>   **BASIC PROPERTIES OF THE GRAPH OF** $f(x) = b^x, b > 0, b \neq 1$

a.   All graphs pass through $(0,1)$; $b^0 = 1$ for any base $b$.

b.   All graphs are continuous curves; there are no holes or jumps.

c.   The $x$-axis is a horizontal asymptote.

d.   If $b > 1$, then $b^x$ increases as $x$ increases.

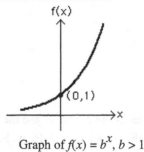

Graph of $f(x) = b^x, b > 1$

Copyright © 2011 Pearson Education, Inc.  Publishing as Prentice Hall.

e. If $0 < b < 1$, then $b^x$ decreases as $x$ increases.

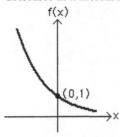

Graph of $f(x) = b^x$, $0 < b < 1$

3. PROPERTIES OF EXPONENTIAL FUNCTIONS

For $a, b > 0$, $a \neq 1$, $b \neq 1$, and $x, y$ real numbers:

a. EXPONENT LAWS

(i) $a^x a^y = a^{x+y}$

(ii) $\dfrac{a^x}{a^y} = a^{x-y}$

(iii) $(a^x)^y = a^{xy}$

(iv) $(ab)^x = a^x b^x$

(v) $\left(\dfrac{a}{b}\right)^x = \dfrac{a^x}{b^x}$

b. $a^x = a^y$ if and only if $x = y$.

c. For $x \neq 0$, $a^x = b^x$ if and only if $a = b$.

4. EXPONENTIAL FUNCTION WITH BASE $e = 2.71828...$

Exponential functions with base $e$ and base $1/e$ are respectively defined by $y = e^x$ and $y = e^{-x}$.

Domain: $(-\infty, \infty)$

Range: $(0, \infty)$

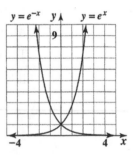

5. Functions of the form $y = ce^{kt}$, where $c$ and $k$ are constants and the independent variable $t$ represents time, are often used to model population growth and radioactive decay. Since $y(0) = c$, $c$ represents the initial population or initial amount. The constant $k$ represents the growth or decay rate; $k > 0$ in the case of population growth, $k < 0$ in the case of radioactive decay.

6. COMPOUND INTEREST

If a principal $P$ (present value) is invested at an annual rate $r$ (expressed as a decimal) compounded $m$ times per year, then the amount $A$ (future value) in the account at the end of $t$ years is given by:

$$A = P\left(1 + \frac{r}{m}\right)^{mt}.$$

Copyright © 2011 Pearson Education, Inc. Publishing as Prentice Hall.

If a principal $P$ is invested at an annual rate $r$ (expressed as a decimal) compounded continuously, then the amount in the account at the end of $t$ years is given by

$$A = Pe^{rt}$$

where $e \approx 2.71828$ is the base of the exponential function.

---

**1.** (A) $k$    (B) $g$    (C) $h$    (D) $f$

**3.** $y = 5^x$, $-2 \le x \le 2$

| $x$ | $y$ |
|-----|-----|
| $-2$ | $\frac{1}{25}$ |
| $-1$ | $\frac{1}{5}$ |
| $0$ | $1$ |
| $1$ | $5$ |
| $2$ | $25$ |

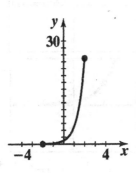

**5.** $y = \left(\dfrac{1}{5}\right)^x = 5^{-x}$, $-2 \le x \le 2$

| $x$ | $y$ |
|-----|-----|
| $-2$ | $25$ |
| $-1$ | $5$ |
| $0$ | $1$ |
| $1$ | $\frac{1}{5}$ |
| $2$ | $\frac{1}{25}$ |

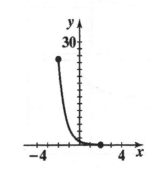

**7.** $f(x) = -5^x$, $-2 \le x \le 2$

| $x$ | $y$ |
|-----|-----|
| $-2$ | $-\frac{1}{25}$ |
| $-1$ | $-\frac{1}{5}$ |
| $0$ | $-1$ |
| $1$ | $-5$ |
| $2$ | $-25$ |

**9.** $y = -e^{-x}$, $-3 \le x \le 3$

| $x$ | $y$ |
|-----|-----|
| $-3$ | $\approx -20$ |
| $-2$ | $\approx -7.4$ |
| $-1$ | $\approx -2.7$ |
| $0$ | $-1$ |
| $1$ | $\approx -0.4$ |
| $2$ | $\approx -0.1$ |
| $3$ | $\approx 0.05$ |

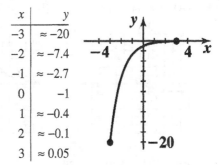

**11.** $g(x) = -f(x)$; the graph of $g$ is the graph of $f$ reflected in the $x$ axis.

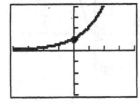

**13.** $g(x) = f(x + 1)$; the graph of $g$ is the graph of $f$ shifted one unit to the left.

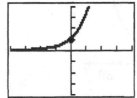

Copyright © 2011 Pearson Education, Inc.  Publishing as Prentice Hall.

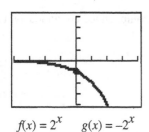

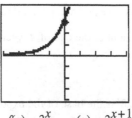

$$f(x) = 2^x \quad g(x) = -2^x$$

$$f(x) = 3^x \quad g(x) = 3^{x+1}$$

**15.** $g(x) = f(x) + 1$; the graph of $g$ is the graph of $f$ shifted one unit up.

**17.** $g(x) = 2f(x + 2)$; the graph of $g$ is the graph of $f$ vertically expanded by a factor of 2 and shifted to the left 2 units.

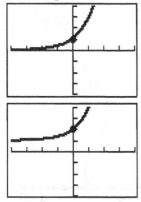

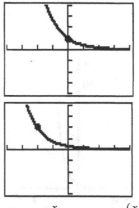

$$f(x) = e^x \quad g(x) = e^x + 1$$

$$f(x) = e^{-x} \quad g(x) = 2e^{-(x+2)}$$

**19.**   (A)   $y = f(x) - 1$     (B)   $y = f(x + 2)$     (C)   $y = 3f(x) - 2$     (D)   $y = 2 - f(x - 3)$

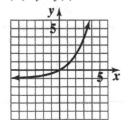  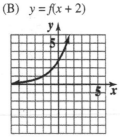  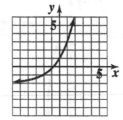  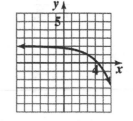

**21.** $f(t) = 2^{t/10}$, $-30 \le t \le 30$

**23.** $y = -3 + e^{1+x}$, $-4 \le x \le 2$

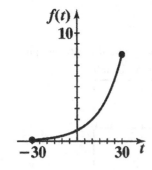

| $x$ | $y$ |
|----|-----|
| $-4$ | $\approx -3$ |
| $-2$ | $\approx -2.6$ |
| $-1$ | $-2$ |
| $0$ | $\approx 0.3$ |
| $1$ | $\approx 4.4$ |
| $2$ | $\approx 17.1$ |

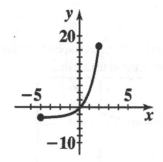

Copyright © 2011 Pearson Education, Inc.  Publishing as Prentice Hall.

**25.**  $y = e^{|x|}$,  $-3 \le x \le 3$

| $x$ | $y$ |
|-----|-----|
| $-3$ | $\approx 20.1$ |
| $-1$ | $\approx 2.7$ |
| $0$ | $1$ |
| $1$ | $\approx 2.7$ |
| $3$ | $\approx 20.1$ |

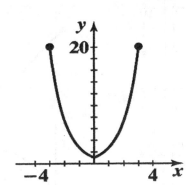

**27.**  Solve

$$a^2 = a^{-2}$$
$$a^2 = \frac{1}{a^2}$$
$$a^4 = 1$$
$$a^4 - 1 = 0$$
$$(a^2 - 1)(a^2 + 1) = 0$$

$a^2 - 1 = 0$ implies $a = 1, -1$

$a^2 + 1 = 0$ has no real solutions

The exponential function property: $a^x = a^y$ if and only if $x = y$ assumes $a > 0$ and $a \ne 1$. Our solutions are $a = 1, -1$; $1^x = 1^y$ for all real numbers $x, y$, $(-1)^x = (-1)^y$ for all even integers.

**29.**  $10^{2-3x} = 10^{5x-6}$ implies (see 3b)
$$2 - 3x = 5x - 6$$
$$-8x = -8$$
$$x = 1$$

**31.**  $4^{5x-x^2} = 4^{-6}$ implies
$$5x - x^2 = -6$$
or  $-x^2 + 5x + 6 = 0$
$$x^2 - 5x - 6 = 0$$
$$(x - 6)(x + 1) = 0$$
$$x = 6, -1$$

**33.**  $5^3 = (x + 2)^3$ implies (by property 3c)
$$5 = x + 2$$
Thus,  $x = 3$.

**35.**  $(x - 3)e^x = 0$
$$x - 3 = 0 \quad (\text{since } e^x \ne 0)$$
$$x = 3$$

**37.**  $3xe^{-x} + x^2 e^{-x} = 0$
$$e^{-x}(3x + x^2) = 0$$
$$3x + x^2 = 0 \quad (\text{since } e^{-x} \ne 0)$$
$$x(3 + x) = 0$$
$$x = 0, -3$$

Copyright © 2011 Pearson Education, Inc.  Publishing as Prentice Hall.

**39.** $h(x) = x2^x$, $-5 \le x \le 0$

| $x$ | $h(x)$ |
|---|---|
| $-5$ | $-\frac{5}{32}$ |
| $-4$ | $-\frac{1}{4}$ |
| $-3$ | $-\frac{3}{8}$ |
| $-2$ | $-\frac{1}{2}$ |
| $-1$ | $-\frac{1}{2}$ |
| $0$ | $0$ |

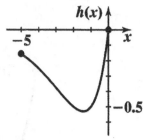

**41.** $N = \dfrac{100}{1 + e^{-t}}$, $0 \le t \le 5$

| $t$ | $N$ |
|---|---|
| $0$ | $50$ |
| $1$ | $\approx 73.1$ |
| $2$ | $\approx 88.1$ |
| $3$ | $\approx 95.3$ |
| $5$ | $\approx 99.3$ |

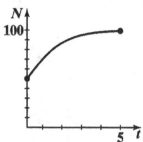

**43.** Use $A = Pe^{rt}$, $P = 10,000$, $r = 0.0395$, and $t = 12$.

$A = 10,000e^{0.0395(12)} \approx 10,000e^{0.474} \approx \$16,064.07$.

**45.** $A = P\left(1 + \dfrac{r}{m}\right)^{mt}$, we have:

(A) $P = 2,500$, $r = 0.07$, $m = 4$, $t = \dfrac{3}{4}$

$A = 2,500\left(1 + \dfrac{0.07}{4}\right)^{4 \cdot 3/4} = 2,500(1 + 0.0175)^3 = 2,633.56$

Thus, $A = \$2,633.56$.

(B) $A = 2,500\left(1 + \dfrac{0.07}{4}\right)^{4 \cdot 15} = 2,500(1 + 0.0175)^{60} = 7079.54$

Thus, $A = \$7,079.54$.

**47.** $A = P\left(1 + \dfrac{r}{m}\right)^{mt}$. With $A = 15,000$, $r = 0.0675$, and $m = 52$, we have:

$15,000 = P\left(1 + \dfrac{0.0675}{52}\right)^{52(5)} = P(1.0411)$.

Therefore, $P = \dfrac{15,000}{1.4011} \approx \$10,706$.

**49.** Use $A = P\left(1 + \dfrac{r}{m}\right)^{mt}$; $P = 10,000$, $t = 1$.

(A) Stonebridge Bank: $r = 0.0540$, $m = 12$

$A = 10,000\left(1 + \dfrac{0.0540}{12}\right)^{12(1)} = \$10,553.57$

Copyright © 2011 Pearson Education, Inc. Publishing as Prentice Hall.

(B)    Deep Green Bank: $r = 0.0495$, $m = 365$

$$A = 10,000\left(1 + \frac{0.0495}{365}\right)^{365(1)} = \$10,507.42$$

(C)    Provident Bank: $r = 0.0515$, $m = 4$

$$A = 10,000\left(1 + \frac{0.0515}{4}\right)^{4(1)} = \$10,525.03$$

**51.**    Given $N = 2(1 - e^{-0.037t})$, $0 \le t \le 50$

| $t$ | $N$ |
|-----|-----|
| 0 | 0 |
| 10 | $\approx 0.62$ |
| 30 | $\approx 1.34$ |
| 50 | $\approx 1.69$ |

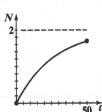

$N$ approaches 2 as $t$ increases without bound.

**53.**    (A)    Exponential regression model

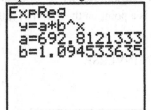

$y = ab^x$; 2020 is year 30;
$y(30) \approx \$10,411,000$

(B)    According to the model, $y(10) \approx \$1,710,000$. Inclusion of the year 2000 data would increase the estimated average salary in 2020 since the actual year 2000 average is higher than that estimated by the model. Inclusion of the data for 2000 gives an annual salary of \$10,721,000 in 2020.

**55.**    Given $I = I_0 e^{-0.23d}$

(A)    $I = I_0 e^{-0.23(10)} = I_0 e^{-2.3} \approx I_0(0.10)$

Thus, about 10% of the surface light will reach a depth of 10 feet.

(B)    $I = I_0 e^{-0.23(20)} = I_0 e^{-4.6} \approx I_0(0.010)$

Thus, about 1% of the surface light will reach a depth of 20 feet.

**57.**    (A)    Model: $P(t) = 6.8 e^{0.0114t}$

(B)    In the year 2020, $t = 11$;  $P(11) = 6.8 e^{0.0114(11)} = 6.8 e^{0.1254} \approx 7.7$ billion.
In the year 2030, $t = 21$;  $P(21) = 6.8 e^{0.0114(21)} = 6.8 e^{0.2394} \approx 8.6$ billion.

Copyright © 2011 Pearson Education, Inc. Publishing as Prentice Hall.

**59.** (A) Exponential regression model

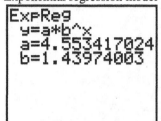

The model implies that the number of internet hosts will be almost 4 times larger than the estimated world population in 2018.

$y(24) \approx 28,652,000$

---

EXERCISE 2-6

*Things to remember:*

1. ONE-TO-ONE FUNCTIONS

   A function $f$ is said to be ONE-TO-ONE if each range value corresponds to exactly one domain value.

2. INVERSE OF A FUNCTION

   If $f$ is a one-to-one function, then the INVERSE of $f$ is the function formed by interchanging the independent and dependent variables for $f$. Thus, if $(a, b)$ is a point on the graph of $f$, then $(b, a)$ is a point on the graph of the inverse of $f$.

   *Note:* If $f$ is not one-to-one, then $f$ DOES NOT HAVE AN INVERSE.

3. LOGARITHMIC FUNCTIONS

   The inverse of an exponential function is called a LOGARITHMIC FUNCTION. For $b > 0$ and $b \neq 1$,

   | Logarithmic form | | Exponential form |
   |---|---|---|
   | $y = \log_b x$ | is equivalent to | $x = b^y$ |

   The LOG TO THE BASE $b$ OF $x$ is the exponent to which $b$ must be raised to obtain $x$. [Remember: A logarithm is an exponent.] The DOMAIN of the logarithmic function is the range of the corresponding exponential function, and the RANGE of the logarithmic function is the domain of the corresponding exponential function. Typical graphs of an exponential function and its inverse, a logarithmic function, for $b > 1$, are shown in the figure below:

   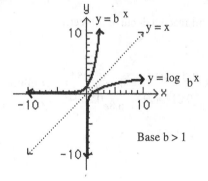

   Base $b > 1$

Copyright © 2011 Pearson Education, Inc. Publishing as Prentice Hall.

4.    PROPERTIES OF LOGARITHMIC FUNCTIONS

If $b$, $M$, and $N$ are positive real numbers, $b \neq 1$, and $p$ and $x$ are real numbers, then:

a. $\log_b 1 = 0$

e. $\log_b MN = \log_b M + \log_b N$

b. $\log_b b = 1$

f. $\log_b \dfrac{M}{N} = \log_b M - \log_b N$

c. $\log_b b^x = x$

g. $\log_b M^p = p \log_b M$

d. $b^{\log_b x} = x,\ x > 0$

h. $\log_b M = \log_b N$ if and only if $M = N$

5. LOGARITHMIC NOTATION; LOGARITHMIC-EXPONENTIAL RELATIONSHIPS

Common logarithm:    $\log x$ means $\log_{10} x$

Natural logarithm:    $\ln x$ means $\log_e x$

$\log x = y$    is equivalent to    $x = 10^y$

$\ln x = y$    is equivalent to    $x = e^y$

---

1.    $27 = 3^3$ (using 3)

3.    $1 = 10^0$

5.    $8 = 4^{3/2}$

7.    $\log_7 49 = 2$

9.    $\log_4 8 = \dfrac{3}{2}$

11.    $\log_b A = u$

13.    $\log_{10} 1 = y$ is equivalent to $10^y = 1$; $y = 0$.

15.    $\log_e e = y$ is equivalent to $e^y = e$; $y = 1$.

17.    $\log_2 2^{-3} = -3$

(using 2a)

19.    $\log_{10} 1{,}000 = \log_{10} 10^3 = 3$

21.    $\log_b \dfrac{P}{Q} = \log_b P - \log_b Q$

23.    $\log_b L^5 = 5 \log_b L$

25.    $3^{p \log_3 q} = 3^{\log_3 q^p} = q^p$ (using 4g and 4d)

27.    $\log_3 x = 2$

$x = 3^2$

$x = 9$

29.    $\log_7 49 = y$

$\log_7 7^2 = y$

$2 = y$

Thus, $y = 2$.

31.    $\log_b 10^{-4} = -4$

$10^{-4} = b^{-4}$

This equality implies $b = 10$ (since the exponents are the same).

33.    $\log_4 x = \dfrac{1}{2}$;    $x = 4^{1/2}$;    $x = 2$

35.    False; counterexample: $y = x^2$.

37.    True; if $g$ is the inverse of $f$, then $f$ is the inverse of $g$ so $g$ must be one-to-one.

39.    True; if $y = 2x$, then $x = 2y$ implies $y = \dfrac{x}{2}$.

Copyright © 2011 Pearson Education, Inc. Publishing as Prentice Hall.

**41.** False; $f(x) = \ln x$ is one-to-one; domain of $f = (0, \infty)$, range of $f = (-\infty, \infty)$.

**43.** $\log_b x = \dfrac{2}{3} \log_b 8 + \dfrac{1}{2} \log_b 9 - \log_b 6 = \log_b 8^{2/3} + \log_b 9^{1/2} - \log_b 6$

$\qquad = \log_b 4 + \log_b 3 - \log_b 6 = \log_b \dfrac{4 \cdot 3}{6}$

$\log_b x = \log_b 2$

$\qquad x = 2$

**45.** $\log_b x = \dfrac{3}{2} \log_b 4 - \dfrac{2}{3} \log_b 8 + 2 \log_b 2 = \log_b 4^{3/2} - \log_b 8^{2/3} + \log_b 2^2$

$\qquad = \log_b 8 - \log_b 4 + \log_b 4 = \log_b 8$

$\log_b x = \log_b 8$

$\qquad x = 8$

**47.** $\log_b x + \log_b(x - 4) = \log_b 21$

$\qquad \log_b x(x - 4) = \log_b 21$

Therefore, $\quad x(x - 4) = 21$

$\qquad x^2 - 4x - 21 = 0$

$\qquad (x - 7)(x + 3) = 0$

Thus, $\qquad\qquad x = 7.$

[Note: $x = -3$ is not a solution

since $\log_b(-3)$ is not defined.]

**49.** $\log_{10}(x - 1) - \log_{10}(x + 1) = 1$

$\qquad \log_{10}\left(\dfrac{x - 1}{x + 1}\right) = 1$

Therefore, $\dfrac{x - 1}{x + 1} = 10^1 = 10$

$\qquad x - 1 = 10(x + 1)$

$\qquad x - 1 = 10x + 10$

$\qquad -9x = 11$

$\qquad x = -\dfrac{11}{9}$

There is *no solution*, since

$\log_{10}\left(-\dfrac{11}{9} - 1\right) = \log_{10}\left(-\dfrac{20}{9}\right)$

is not defined.  Similarly,

$\log_{10}\left(-\dfrac{11}{9} + 1\right) = \log_{10}\left(-\dfrac{2}{9}\right)$

is not defined.

**51.** $y = \log_2(x - 2)$

$\quad x - 2 = 2^y$

$\quad x = 2^y + 2$

| $x$ | $y$ |
|---|---|
| $\frac{9}{4}$ | $-2$ |
| $\frac{5}{2}$ | $-1$ |
| 3 | 0 |
| 4 | 1 |
| 6 | 2 |
| 18 | 4 |

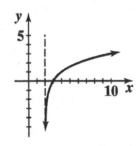

**53.** The graph of $y = \log_2(x - 2)$ is the graph of $y = \log_2 x$ shifted to the right 2 units.

**55.** Since logarithmic functions are defined only for positive "inputs", we must have $x + 1 > 0$ or $x > -1$; domain: $(-1, \infty)$. The range of $y = 1 + \ln(x + 1)$ is the set of all real numbers.

Copyright © 2011 Pearson Education, Inc.  Publishing as Prentice Hall.

**57.**  (A)   3.54743

(B)  −2.16032

(C)   5.62629

(D)  −3.19704

**59.**  (A)   $\log x = 1.1285$

$x = 13.4431$

(B)   $\log x = -2.0497$

$x = 0.0089$

(C)   $\ln x = 2.7763$

$x = 16.0595$

(D)   $\ln x = -1.8879$

$x = 0.1514$

**61.**  $10^x = 12$ (Take common logarithms of both sides)

$\log 10^x = \log 12 \approx 1.0792$

$x \approx 1.0792$ ($\log 10^x = x$   $\log 10 = x$; $\log 10 = 1$)

**63.**  $e^x = 4.304$ (Take natural logarithms of both sides)

$\ln e^x = \ln 4.304 \approx 1.4595$

$x \approx 1.4595$ ($\ln e^x = x$   $\ln e = x$; $\ln e = 1$)

**65.**  $1.005^{12t} = 3$    (Take either common or natural logarithms of both sides; here we'll use natural logarithms.)

$\ln 1.005^{12t} = \ln 3$

$12t = \dfrac{\ln 3}{\ln 1.005} \approx 220.2713$

$t = 18.3559$

**67.**  $y = \ln x, \; x > 0$

| $x$ | $y$ |
|-----|-----|
| 0.5 | $\approx -0.69$ |
| 1 | 0 |
| 2 | $\approx 0.69$ |
| 4 | $\approx 1.39$ |
| 5 | $\approx 1.61$ |

increasing $(0, \infty)$

**69.**  $y = |\ln x|, \; x > 0$

| $x$ | $y$ |
|-----|-----|
| 0.5 | $\approx 0.69$ |
| 1 | 0 |
| 2 | $\approx 0.69$ |
| 4 | $\approx 1.39$ |
| 5 | $\approx 1.6$ |

decreasing $(0, 1]$
increasing $[1, \infty)$

**71.**  $y = 2 \ln(x + 2), \; x > -2$

| $x$ | $y$ |
|-----|-----|
| −1.5 | $\approx -1.39$ |
| −1 | 0 |
| 0 | $\approx 1.39$ |
| 1 | $\approx 2.2$ |
| 5 | $\approx 3.89$ |
| 10 | $\approx 4.61$ |

increasing $(-2, \infty)$

**73.**  $y = 4 \ln x - 3, \; x > 0$

| $x$ | $y$ |
|-----|-----|
| 0.5 | $\approx -5.77$ |
| 1 | −3 |
| 5 | $\approx 3.44$ |
| 10 | $\approx 6.21$ |

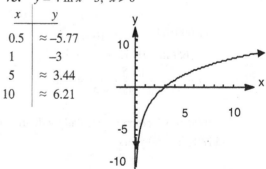

increasing $(0, \infty)$

Copyright © 2011 Pearson Education, Inc. Publishing as Prentice Hall.

**75.** For any number $b$, $b > 0$, $b \neq 1$, $\log_b 1 = y$ is equivalent to $b^y = 1$ which implies $y = 0$. Thus, $\log_b 1 = 0$ for any permissible base $b$.

**77.** 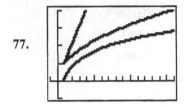 A function $f$ is "larger than" a function $g$ on an interval $[a, b]$ if $f(x) > g(x)$ for $a \leq x \leq b$.
$r(x) > q(x) > p(x)$ for $1 < x \leq 16$, that is $x > \sqrt{x} > \ln x$ for $1 < x \leq 16$

**79.** From the compound interest formula $A = P(1 + r)^t$, we have:

$2P = P(1 + 0.2136)^t$ or $(1.2136)^t = 2$

Take the natural log of both sides of this equation:

$\ln(1.2136)^t = \ln 2$          [Note: the common log could have been used instead of the natural log.]

$t \ln(1.2136) = \ln 2$

$t = \dfrac{\ln 2}{\ln 1.2136} \approx \dfrac{0.69135}{0.19359} = 3.58 \approx 4$ years

**81.** $A = A = P\left(1 + \dfrac{r}{m}\right)^{mt}$, $r = 0.06$, $P = 1000$, $A = 1800$.

Quarterly compounding: $m = 4$

$1800 = 1000\left(1 + \dfrac{0.06}{4}\right)^{4t} = 1000(1.015)^{4t}$

$(1.015)^{4t} = \dfrac{1800}{1000} = 1.8$

$4t \ln(1.015) = \ln(1.8)$

$t = \dfrac{\ln(1.8)}{4\ln(1.015)} \approx 9.87$

$1000 at 6% compounded quarterly will grow to $1800 in 9.87 years.

Daily compounding: $m = 365$

$1800 = 1000\left(1 + \dfrac{0.06}{365}\right)^{365t} = 1000(1.0001644)365t$

$(1.0001644)^{365t} = \dfrac{1800}{1000} = 1.8$

$365t \ln(1.0001644) = \ln 1.8$

$t = \dfrac{\ln 1.8}{365\ln(1.0001644)} \approx 9.80$

$1000 at 6% compounded daily will grow to $1800 in 9.80 years.

Copyright © 2011 Pearson Education, Inc. Publishing as Prentice Hall.

**83.**   $A = Pe^{rt}$;   $r = 0.0475$,   $P = 35,000$,   $A = 50,000$

$$50,000 = 35,000e^{0.0475t}$$

$$e^{0.0475t} = \frac{50,000}{35,000} = \frac{10}{7}$$

$$0.0475t = \ln\left(\frac{10}{7}\right)$$

$$t = \frac{\ln(10/7)}{0.0475} \approx 7.51$$

**85.**   (A) Logarithmic regression model,

Table 1:

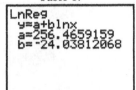

To estimate the demand at a price level of $50, we solve

$$a + b\ln x = 50$$

for $x$. The result is $x \approx 5,373$ screwdrivers per month.

(B) Logarithmic regression model,

Table 2:

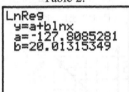

To estimate the supply at a price level of $50, we solve

$$a + b\ln x = 50$$

for $x$. The result is $x \approx 7,220$ screwdrivers per month.

(C) The condition is not stable, the price is likely to decrease since the demand at a price level of $50 is much lower than the supply at this level.

**87.**   $I = I_0 10^{N/10}$

Take the common log of both sides of this equation.  Then:

$$\log I = \log(I_0 10^{N/10}) = \log I_0 + \log 10^{N/10}$$

$$= \log I_0 + \frac{N}{10}\log\ 10 = \log I_0 + \frac{N}{10} \ \ (\text{since } \log 10 = 1)$$

So, $\dfrac{N}{10} = \log I - \log I_0 = \log\left(\dfrac{I}{I_0}\right)$ and $N = 10\log\left(\dfrac{I}{I_0}\right)$.

Copyright © 2011 Pearson Education, Inc. Publishing as Prentice Hall.

**89.** Logarithmic regression model

```
LnReg
 y=a+blnx
 a=-535.1958095
 b=145.237116

■
```

The yield in 2020: $y(120) \approx 160.1$ bushels/acre.

**91.** Assuming that the current population is 6.8 billion and that the growth rate is 1.14% compounded continuously, the population after $t$ years will be

$$P(t) = 6.8e^{0.0144t}$$

Given that there are $1.68 \times 10^{14} = 168{,}000$ billion square yards of land, solve

$$6.8e^{0.0114t} = 168{,}000$$

for $t$:

$$e^{0.0114t} = \frac{168{,}000}{6.8} \approx 24{,}706$$

$$0.0114t = \ln 24{,}706$$

$$t = \frac{\ln 24{,}706}{0.0114} \approx 887$$

It will take approximately 887 years.

# CHAPTER 2 REVIEW

**1.**

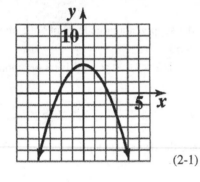

(2-1)

**2.** $x^2 = y^2$:

| $x$ | −3 | −2 | −1 | 0 | 1 | 2 | 3 |
|---|---|---|---|---|---|---|---|
| $y$ | ±3 | ±2 | ±1 | 0 | ±1 | ±2 | ±3 |

(2-1)

**3.** $y^2 = 4x^2$:

| $x$ | −3 | −2 | −1 | 0 | 1 | 2 | 3 |
|---|---|---|---|---|---|---|---|
| $y$ | ±6 | ±4 | ±2 | 0 | ±2 | ±4 | ±6 |

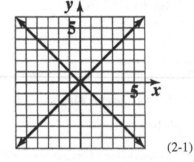

(2-1)

Copyright © 2011 Pearson Education, Inc. Publishing as Prentice Hall.

4.  (A) Not a function; fails vertical line test

    (B) A function

    (C) A function

    (D) Not a function; fails vertical line test    (2-1)

5.  $f(x) = 2x - 1$, $g(x) = x^2 - 2x$

    (A) $f(-2) + g(-1) = 2(-2) - 1 + (-1)^2 - 2(-1) = -2$

    (B) $f(0) \cdot g(4) = (2 \cdot 0 - 1)(4^2 - 2 \cdot 4) = -8$

    (C) $\dfrac{g(2)}{f(3)} = \dfrac{2^2 - 2 \cdot 2}{2 \cdot 3 - 1} = 0$

    (D) $\dfrac{f(3)}{g(2)}$ not defined because $g(2) = 0$    (2-1)

6.  $u = e^v$

    $v = \ln u$  (2-6)

7.  $x = 10^y$

    $y = \log x$  (2-6)

8.  $\ln M = N$

    $M = e^N$  (2-6)

9.  $\log u = v$

    $u = 10^v$    (2-6)

10.  $\log_3 x = 2$

    $x = 3^2 = 9$    (2-6)

11.  $\log_x 36 = 2$

    $x^2 = 36$

    $x = 6$

    (2-6)

12  $\log_2 16 = x$

    $2^x = 16$

    $x = 4$

    (2-6)

13.  $10^x = 143.7$

    $x = \log 143.7$

    $x \approx 2.157$

    (2-6)

14.  $e^x = 503{,}000$

    $x = \ln 503{,}000 \approx 13.128$    (2-6)

15.  $\log x = 3.105$

    $x = 10^{3.105} \approx 1273.503$  (2-6)

16.  $\ln x = -1.147$

    $x = e^{-1.147} \approx 0.318$    (2-6)

17.  (A) $y = 4$    (B) $x = 0$    (C) $y = 1$    (D) $x = -1$ or $1$

    (E) $y = -2$    (F) $x = -5$ or $5$    (2-1)

18.  (A)           (B)           (C)

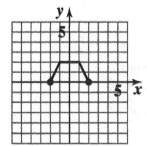
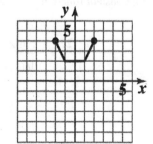
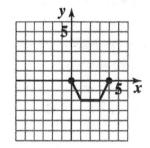

Copyright © 2011 Pearson Education, Inc.  Publishing as Prentice Hall.

(D)

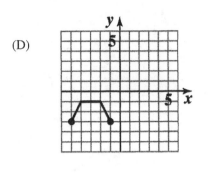

(2-2)

**19.** $f(x) = -x^2 + 4x = -(x^2 - 4x)$
$$= -(x^2 - 4x + 4) + 4$$
$$= -(x - 2)^2 + 4 \quad \text{(vertex form)}$$

The graph of $f(x)$ is the graph of $y = x^2$ reflected in the $x$ axis, then shifted right 2 units and up 4 units.
(2-3)

**20.** (A) $g$   (B) $m$   (C) $n$   (D) $f$     (2-2, 2-3)

**21.** $y = f(x) = (x + 2)^2 - 4$

(A) $x$ intercepts: $(x + 2)^2 - 4 = 0;$        $y$ intercept: 0
$$(x + 2)^2 = 4$$
$$x + 2 = -2 \text{ or } 2$$
$$x = -4, 0$$

(B) Vertex: $(-2, -4)$   (C) Minimum: $-4$   (D) Range: $y \geq -4$ or $[-4, \infty)$        (2-3)

**22.** $y = 4 - x + 3x^2 = 3x^2 - x + 4;$ quadratic function.     (2-3)

**23.** $y = \dfrac{1 + 5x}{6} = \dfrac{5}{6}x + \dfrac{1}{6};$   linear function.     (2-1, 2-3)

**24.** $y = \dfrac{7 - 4x}{2x} = \dfrac{7}{2x} - 2;$ none of these.     (2-1), (2-3)

**25.** $y = 8x + 2(10 - 4x) = 8x + 20 - 8x = 20;$ constant function       (2-1)

**26.** $\log(x + 5) = \log(2x - 3)$
$$x + 5 = 2x - 3$$
$$-x = -8$$
$$x = 8 \qquad \text{(2-6)}$$

**27.**  $2 \ln(x - 1) = \ln(x^2 - 5)$
$$\ln(x - 1)^2 = \ln(x^2 - 5)$$
$$(x - 1)^2 = x^2 - 5$$
$$x^2 - 2x + 1 = x^2 - 5$$
$$-2x = -6$$
$$x = 3 \qquad \text{(2-6)}$$

**28.**    $9^{x-1} = 3^{1+x}$
$$(3^2)^{x-1} = 3^{1+x}$$

**29.**      $e^{2x} = e^{x^2 - 3}$
$$2x = x^2 - 3$$

Copyright © 2011 Pearson Education, Inc. Publishing as Prentice Hall.

$$3^{2x-2} = 3^{1+x}$$
$$2x - 2 = 1 + x$$
$$x = 3 \qquad (2\text{-}5)$$

$$x^2 - 2x - 3 = 0$$
$$(x-3)(x-1) = 0$$
$$x = 3, \ -1 \qquad (2\text{-}5)$$

**30.** $2x^2 e^x = 3xe^x$

$$2x^2 = 3x$$

$$2x^2 - 3x = 0$$

$$x(2x - 3) = 0$$

$$x = 0, \ 3/2 \qquad (2\text{-}5)$$

**31.** $\log_{1/3} 9 = x$

$$\left(\frac{1}{3}\right)^x = 9$$

$$\frac{1}{3^x} = 9$$

$$3^x = \frac{1}{9}$$

$$x = -2 \qquad (2\text{-}6)$$

**32.** $\log_x 8 = -3$

$$x^{-3} = 8$$

$$\frac{1}{x^3} = 8$$

$$x^3 = \frac{1}{8}$$

$$x = \frac{1}{2} \qquad (2\text{-}6)$$

**33.** $\log_9 x = \frac{3}{2}$

$$9^{3/2} = x$$

$$x = 27 \qquad (2\text{-}6)$$

**34.** $x = 3(e^{1.49}) \approx 13.3113$

**35.** $x = 230(10^{-0.161}) \approx 158.7552 \qquad (2\text{-}5)$

**36.** $\log x = -2.0144$
$$x \approx 0.0097 \qquad (2\text{-}6)$$

**37.** $\ln x = 0.3618$
$$x \approx 1.4359 \qquad (2\text{-}6)$$

**38.** $35 = 7(3^x)$
$$3^x = 5$$
$$\ln 3^x = \ln 5$$
$$= \ln 5$$
$$x = \frac{\ln 5}{\ln 3} \approx 1.4650 \qquad (2\text{-}6)$$

**39.** $0.01 = e^{-0.05x}$
$$\ln(0.01) = \ln(e^{-0.05x}) = -0.05x$$
$$\text{Thus, } x = \frac{\ln(0.01)}{-0.05} \approx 92.1034 \qquad x \ln 3$$
$$(2\text{-}6)$$

**40.** $8{,}000 = 4{,}000(1.08)^x$
$$(1.08)^x = 2$$
$$\ln(1.08)^x = \ln 2$$
$$x \ln 1.08 = \ln 2$$
$$x = \frac{\ln 2}{\ln 1.08} \approx 9.0065 \qquad (2\text{-}6)$$

**41.** $5^{2x-3} = 7.08$
$$\ln(5^{2x-3}) = \ln 7.08$$
$$(2x - 3) \ln 5 = \ln 7.08$$
$$2x \ln 5 - 3 \ln 5 = \ln 7.08$$
$$x = \frac{\ln 7.08 + 3\ln 5}{2\ln 5}$$
$$x \approx 2.1081 \qquad (2\text{-}6)$$

Copyright © 2011 Pearson Education, Inc. Publishing as Prentice Hall.

**42.** (A) $x^2 - x - 6 = 0$ at $x = -2, 3$

Domain: all real numbers except $x = -2, 3$

(B) $5 - x > 0$ for $x < 5$

Domain: $x < 5$ or $(-\infty, 5)$    (2-1)

**43.** $f(x) = 4x^2 + 4x - 3 = 4(x^2 + x) - 3$

$$= 4\left(x^2 + x + \frac{1}{4}\right) - 3 - 1$$

$$= 4\left(x + \frac{1}{2}\right)^2 - 4 \ \text{(vertex form)}$$

Intercepts:

$y$ intercept: $f(0) = 4(0)^2 + 4(0) - 3 = -3$

$x$ intercepts: $f(x) = 0$

$$4\left(x + \frac{1}{2}\right)^2 - 4 = 0$$

$$\left(x + \frac{1}{2}\right)^2 = 1$$

$$x + \frac{1}{2} = \pm 1$$

$$x = -\frac{1}{2} \pm 1 = -\frac{3}{2}, \frac{1}{2}$$

Vertex: $\left(-\frac{1}{2}, -4\right)$;  minimum: $-4$;  range: $y \geq -4$ or $[-4, \infty)$     (2-3)

**44.** $f(x) = e^x - 1$, $g(x) = \ln(x + 2)$

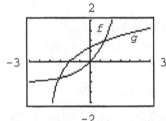

Points of intersection:

$(-1.54, -0.79)$, $(0.69, 0.99)$

(2-5, 2-6)

**45.** $f(x) = \dfrac{50}{x^2 + 1}$ :

| $x$ | -3 | -2 | -1 | 0 | 1 | 2 | 3 |
|---|---|---|---|---|---|---|---|
| $f(x)$ | 5 | 10 | 25 | 50 | 25 | 10 | 5 |

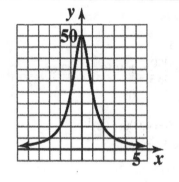

(2-1)

Copyright © 2011 Pearson Education, Inc.  Publishing as Prentice Hall.

**46.** $f(x) = \dfrac{-66}{2 + x^2}$ :

| $x$ | $-3$ | $-2$ | $-1$ | $0$ | $1$ | $2$ | $3$ |
|---|---|---|---|---|---|---|---|
| $f(x)$ | $-6$ | $-11$ | $-22$ | $-66$ | $-22$ | $-11$ | $-6$ |

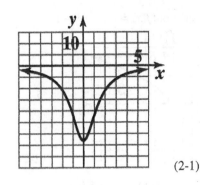

(2-1)

For Problems 47 – 50, $f(x) = 5x + 1$.

**47.** $f(f(0)) = f(5(0) + 1) = f(1) = 5(1) + 1 = 6$     (2-1)

**48.** $f(f(-1)) = f(5(-1) + 1) + f(-4) = 5(-4) + 1 = -19$     (2-1)

**49.** $f(2x - 1) = 5(2x - 1) + 1 = 10x - 4$     (2-1)

**50.** $f(4 - x) = 5(4 - x) + 1 = 20 - 5x + 1 = 21 - 5x$     (2-1)

**51.** $f(x) = 3 - 2x$
   (A) $f(2) = 3 - 2(2) = 3 - 4 = -1$
   (B) $f(2 + h) = 3 - 2(2 + h) = 3 - 4 - 2h = -1 - 2h$
   (C) $f(2 + h) - f(2) = -1 - 2h - (-1) = -2h$
   (D) $\dfrac{f(2+h) - f(2)}{h} = -\dfrac{2h}{h} = -2$     (2-1)

**52.** $f(x) = x^2 - 3x + 1$
   (A) $f(a) = a^2 - 3a + 1$
   (B) $f(a + h) = (a + h)^2 - 3(a + h) + 1 = a^2 + 2ah + h^2 - 3a - 3h + 1$
   (C) $f(a + h) - f(a) = a^2 + 2ah + h^2 - 3a - 3h + 1 - (a^2 - 3a + 1)$
          $= 2ah + h^2 - 3h$
   (D) $\dfrac{f(a+h) - f(a)}{h} = \dfrac{2ah + h^2 - 3h}{h} = \dfrac{h(2a + h - 3)}{h} = 2a + h - 3$     (2-1)

**53.** The graph of $m$ is the graph of $y = |x|$ reflected in the $x$ axis and shifted 4 units to the right.   (2–2)

**54.** The graph of $g$ is the graph of $y = x^3$ vertically contracted by a factor of 0.3 and shifted up 3 units.     (2-2)

**55.** The graph of $y = x^2$ is vertically expanded by a factor of 2, reflected in the $x$ axis and shifted to the left 3 units.
   Equation: $y = -2(x + 3)^2$     (2-2)

Copyright © 2011 Pearson Education, Inc. Publishing as Prentice Hall.

**56.** Equation: $f(x) = 2\sqrt{x+3} - 1$

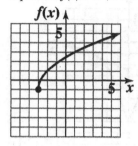

(2-2)

**57.** $f(x) = \dfrac{n(x)}{d(x)} = \dfrac{5x+4}{x^2-3x+1}$.  Since  degree $n(x) = 1 < 2 =$ degree $d(x)$,  $y = 0$  is the horizontal asymptote.

(2-4)

**58.** $f(x) = \dfrac{n(x)}{d(x)} = \dfrac{3x^2+2x-1}{4x^2-5x+3}$.  Since degree  $n(x) = 2 =$ degree $d(x)$,  $y = \dfrac{3}{4}$  is the horizontal asymptote.

(2-4)

**59.** $f(x) = \dfrac{n(x)}{d(x)} = \dfrac{x^2+4}{100x+1}$.  Since degree  $n(x) = 2 > 1 =$ degree $d(x)$,  there is no horizontal asymptote     (2-4)

**60.** $f(x) = \dfrac{n(x)}{d(x)} = \dfrac{x^2+100}{x^2-100} = \dfrac{x^2+100}{(x-10)(x+10)}$.  Since  $n(x) = x^2+100$  has no real zeros and

$d(10) = d(-10) = 0$,  $x = 10$  and  $x = -10$  are the vertical asymptotes of the graph of $f$.     (2-4)

**61.** $f(x) = \dfrac{n(x)}{d(x)} = \dfrac{x^2+3x}{x^2+2x} = \dfrac{x(x+3)}{x(x+2)} = \dfrac{x+3}{x+2}$,  $x \neq 0$.  $x = -2$  is a vertical asymptote of the graph of $f$.   (2-4)

**62.** True;  $p(x) = \dfrac{p(x)}{1}$  is a rational function for every polynomial $p$.       (2-4)

**63.** False;  $f(x) = \dfrac{1}{x} = x^{-1}$  is not a polynomial function.       (2-4)

**64.** False;  $f(x) = \dfrac{1}{x^2+1}$  has no vertical asymptotes.       (2-4)

**65.** True: let  $f(x) = b^x$, $(b > 0, b \neq 1)$, then the positive $x$-axis is a horizontal asymptote if $0 < b < 1$, and the negative $x$-axis is a horizontal asymptote if $b > 1$.       (2-5)

**66.** True: let  $f(x) = \log_b x$  $(b > 0, b \neq 1)$.  If $0 < b < 1$, then the positive $y$-axis is a vertical asymptote; if $b > 1$, then the negative $y$-axis is a vertical asymptote.       (2-6)

**67.** True;  $f(x) = \dfrac{x}{x-1}$  has vertical asymptote  $x = 1$  and horizontal asymptote  $y = 1$.       (2-4)

Copyright © 2011 Pearson Education, Inc.  Publishing as Prentice Hall.

**68.**

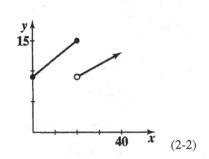

(2-2)

**69.**

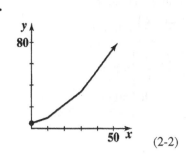

(2-2)

**70.**  $y = -(x-4)^2 + 3$    (2-2, 2-3)

**71.**  $f(x) = -0.4x^2 + 3.2x - 1.2 = -0.4(x^2 - 8x + 16) + 7.6$
$$= -0.4(x-4)^2 + 7.6$$

(A)  $y$ intercept: 1.2

   $x$ intercepts:  $-0.4(x-4)^2 + 7.6 = 0$
$$(x-4)^2 = 19$$
$$x = 4 + \sqrt{19} \approx 8.4, \, 4 - \sqrt{19} \approx -0.4$$

(B)  Vertex: (4.0, 7.6)    (C)  Maximum: 7.6    (D)  Range: $y \le 7.6$ or $(-\infty, 7.6]$    (2-3)

**72.**

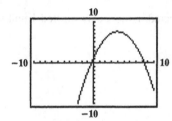

(A)  $y$ intercept: 1.2

   $x$ intercepts: $-0.4, 8.4$

(B)  Vertex: (4.0, 7.6)

(C)  Maximum: 7.6

(D)  Range: $y \le 7.6$ or $(-\infty, 7.6]$    (2-3)

**73.**  $\log 10^\pi = \pi \log 10 = \pi$ (see logarithm properties $\underline{4}$.b & g, Section 2-5)

$10^{\log \sqrt{2}} = y$  is equivalent to  $\log y = \log \sqrt{2}$

which implies $y = \sqrt{2}$

Similarly,  $\ln e^\pi = \pi \ln e = \pi$  (Section 2-5, $\underline{4}$.b & g)  and  $e^{\ln \sqrt{2}} = y$  implies  $\ln y = \ln \sqrt{2}$  and
$y = \sqrt{2}$ .    (2-6)

**74.**  $\log x - \log 3 = \log 4 - \log(x+4)$
$$\log \frac{x}{3} = \log \frac{4}{x+4}$$
$$\frac{x}{3} = \frac{4}{x+4}$$
$$x(x+4) = 12$$
$$x^2 + 4x - 12 = 0$$
$$(x+6)(x-2) = 0$$
$$x = -6, \, 2$$

Since $\log(-6)$ is not defined, $-6$ is not a solution.  Therefore, the solution is $x = 2$.    (2-6)

Copyright © 2011 Pearson Education, Inc.  Publishing as Prentice Hall.

**75.** $\ln(2x - 2) - \ln(x - 1) = \ln x$

$$\ln\left(\frac{2x-2}{x-1}\right) = \ln x$$

$$\ln\left[\frac{2(x-1)}{x-1}\right] = \ln x$$

$$\ln 2 = \ln x$$

$$x = 2 \qquad (2\text{-}6)$$

**76.** $\ln(x + 3) - \ln x = 2 \ln 2$

$$\ln\left(\frac{x+3}{x}\right) = \ln(2^2)$$

$$\frac{x+3}{x} = 4$$

$$x + 3 = 4x$$

$$3x = 3$$

$$x = 1 \qquad (2\text{-}6)$$

**77.**
$$\log 3x^2 = 2 + \log 9x$$

$$\log 3x^2 - \log 9x = 2$$

$$\log\left(\frac{3x^2}{9x}\right) = 2$$

$$\log\left(\frac{x}{3}\right) = 2$$

$$\frac{x}{3} = 10^2 = 100$$

$$x = 300 \qquad (2\text{-}6)$$

**78.**
$$\ln\ y = -5t + \ln\ c$$

$$\ln\ y - \ln\ c = -5t$$

$$\ln\frac{y}{c} = -5t$$

$$\frac{y}{c} = e^{-5t}$$

$$y = ce^{-5t} \qquad (2\text{-}6)$$

**79.** Let $x$ be *any* positive real number and suppose $\log_1 x = y$. Then $1^y = x$. But, $1^y = 1$, so $x = 1$, i.e., $x = 1$ for all positive real numbers $x$. This is clearly impossible. $\qquad (2\text{-}6)$

**80.** The graph of $y = \sqrt[3]{x}$ is vertically expanded by a factor of 2, reflected in the $x$ axis, shifted 1 unit to the left and 1 unit down.

Equation: $y = -2\sqrt[3]{x+1} - 1 \qquad (2\text{-}2)$

**81.** $G(x) = 0.3x^2 + 1.2x - 6.9 = 0.3(x^2 + 4x + 4) - 8.1$
$$= 0.3(x + 2)^2 - 8.1$$

(A) $y$ intercept: $-6.9$

$x$ intercepts: $0.3(x + 2)^2 - 8.1 = 0$

$$(x + 2)^2 = 27$$

$$x = -2 + \sqrt{27} \approx 3.2, \; -2 - \sqrt{27} \approx -7.2$$

(B) Vertex: $(-2, -8.1)$  (C) Minimum: $-8.1$  (D) Range: $y \geq -8.1$ or $[-8.1, \infty)$  $\qquad (2\text{-}3)$

**82.**

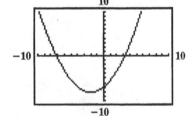

(A) $y$ intercept: $-6.9$

$x$ intercept: $-7.2, 3.2$

(B) Vertex: $(-2, -8.1)$

(C) Minimum: $-8.1$

(D) Range: $y \geq -8.1$ or $[-8.1, \infty)$  $\qquad (2\text{-}3)$

**83.** (A) $\quad S(x) = 3$ if $0 \leq x \leq 20$;

$\quad S(x) = 3 + 0.057(x - 20)$

$\qquad = 0.057x + 1.86$ if $20 < x \leq 200$;

$\quad S(200) = 13.36$

$\quad S(x) = 13.36 + 0.0346(x - 200)$

Copyright © 2011 Pearson Education, Inc. Publishing as Prentice Hall.

$$= 0.0346x + 6.34 \quad \text{if} \quad 200 < x \le 1000;$$

$$S(1000) = 40.94$$

$$S(x) = 40.94 + 0.0217(x - 1000)$$

$$= 0.0217x + 19.24 \quad \text{if} \quad x > 1000$$

$$\text{Therefore, } S(x) = \begin{cases} 3 & \text{if} \quad 0 \le x \le 20 \\ 0.057x + 1.86 & \text{if} \quad 20 < x \le 200 \\ 0.0346x + 6.34 & \text{if} \quad 200 < x \le 1000 \\ 0.0217x + 19.24 & \text{if} \quad x > 1000 \end{cases}$$

(B)

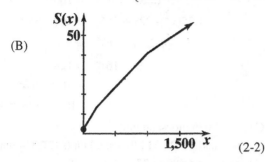

(2-2)

**84.**  $A = P\left(1 + \dfrac{r}{m}\right)^{mt}$;  $P = 5,000,\ r = 0.0535,\ m = 4,\ t = 5.$

$$A = 5000\left(1 + \frac{0.0535}{4}\right)^{4(5)} = 5000(1.013375)^{20} \approx 6,521.89$$

After 5 years, the CD will be worth \$6,521.89      (2-5)

**85.**  $A = \left(1 + \dfrac{r}{m}\right)^{mt}$;  $P = 5,000,\ r = 0.0482,\ m = 365,\ t = 5$

$$A = 5000\left(1 + \frac{0.0482}{365}\right)^{365(5)} \approx 5000(1.0001321) \approx 6,362.50$$

After 5 years, the CD will be worth \$6,362.50.      (2-5)

**86.**  $A = P\left(1 + \dfrac{r}{m}\right)^{mt},\ r = 0.0659,\ m = 12$

Solve $P\left(1 + \dfrac{0.0659}{12}\right)^{12t} = 3P$ or $(1.005492)^{12t} = 3$

for $t$:

$$12t \ln(1.005492) = \ln 3$$

$$t = \frac{\ln 3}{12 \ln(1.005492)} \approx 16.7 \text{ year.}\qquad (2\text{-}5)$$

**87.**  $A = Pe^{rt},\ r = 0.0739.$ Solve $2P = Pe^{0.0739t}$ for $t$.

$$2P = Pe^{0.0739t}$$

$$e^{0.0739t} = 2$$

$$0.0739t = \ln 2$$

$$t = \frac{\ln 2}{0.0739} \approx 9.38 \text{ years.}$$

(2-5)

Copyright © 2011 Pearson Education, Inc. Publishing as Prentice Hall.

**88.**   $p(x) = 50 - 1.25x$  Price-demand function

$C(x) = 160 + 10x$  Cost function

$R(x) = xp(x)$

$\quad = x(50 - 1.25x)$  Revenue function

(A)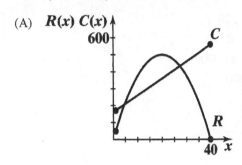

(B)   $R = C$

$$x(50 - 1.25x) = 160 + 10x$$
$$-1.25x^2 + 50x = 160 + 10x$$
$$-1.25x^2 + 40x = 160$$
$$-1.25(x^2 - 32x + 256) = 160 - 320$$
$$-1.25(x - 16)^2 = -160$$
$$(x - 16)^2 = 128$$
$$x = 16 + \sqrt{128} \approx 27.314,$$
$$16 - \sqrt{128} \approx 4.686$$

$R = C$ at $x = 4.686$ thousand units

(4,686 units) and $x = 27.314$ thousand units (27,314 units)

$R < C$ for $1 \le x < 4.686$ or $27.314 < x \le 40$

$R > C$ for $4.686 < x < 27.314$

(C)   Max Rev: $50x - 1.25x^2 = R$

$$-1.25(x^2 - 40x + 400) + 500 = R$$
$$-1.25(x - 20)^2 + 500 = R$$

Vertex at (20, 500)

Max. Rev. = 500 thousand ($500,000) occurs when <u>output</u> is 20 thousand (20,000 units)

<u>Wholesale price</u> at this output: $p(x) = 50 - 1.25x$

$$p(20) = 50 - 1.25(20) = \$25 \qquad (2\text{-}3)$$

**89.**   (A)   $P(x) = R(x) - C(x) = x(50 - 1.25x) - (160 + 10x)$

$$= -1.25x^2 + 40x - 160$$

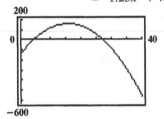

(B)   $P = 0$ for $x = 4.686$ thousand units (4,686 units) and $x = 27.314$ thousand units (27,314 units)

$P < 0$ for $1 \le x < 4.686$ or $27.314 < x \le 40$

$P > 0$ for $4.686 < x < 27.314$

(C)   Maximum profit is 160 thousand dollars ($160,000), and this occurs at $x = 16$ thousand units(16,000 units). The wholesale price at this output is $p(16) = 50 - 1.25(16) = \$30$, which is $5 greater than the $25 found in 88(C).   (2-3)

Copyright © 2011 Pearson Education, Inc.  Publishing as Prentice Hall.

**90.**   (A) The area enclosed by the pens is given by

$$A = (2y)x$$

Now, $3x + 4y = 840$

so    $y = 210 - \dfrac{3}{4}x$

Thus   $A(x) = 2\left(210 - \dfrac{3}{4}x\right)x$

$$= 420x - \dfrac{3}{2}x^2$$

(B)   Clearly $x$ and $y$ must be nonnegative; the fact that $y \geq 0$ implies

$$210 - \dfrac{3}{4}x \geq 0$$

and    $210 \geq \dfrac{3}{4}x$

$$840 \geq 3x$$
$$280 \geq x$$

Thus, domain $A$: $0 \leq x \leq 280$

(C)

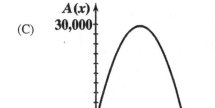

(D)   Graph $A(x) = 420x - \dfrac{3}{2}x^2$ and $y = 25{,}000$ together.

There are two values of $x$ that will produce storage areas with a combined area of 25,000 square feet, one near $x = 90$ and the other near $x = 190$.

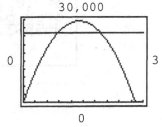

(E)   $x = 86,\ x = 194$

(F)   $A(x) = 420x - \dfrac{3}{2}x^2 = -\dfrac{3}{2}(x^2 - 280x)$

Completing the square, we have

$$A(x) = -\dfrac{3}{2}(x^2 - 280x + 19{,}600 - 19{,}600)$$

$$= -\dfrac{3}{2}[(x - 140)^2 - 19{,}600]$$

$$= -\dfrac{3}{2}(x - 140)^2 + 29{,}400$$

The dimensions that will produce the maximum combined area are: $x = 140$ ft, $y = 105$ ft. The maximum area is 29,400 sq. ft.        (2-3)

Copyright © 2011 Pearson Education, Inc. Publishing as Prentice Hall.

**91.**  (A)  Quadratic regression model,

To estimate the demand at price level of $180, we solve the equation

$$ax^2 + bx + c = 180$$

for $x$. The result is $x \approx 2{,}833$ sets.

(B)  Linear regression model,
Table 2:

To estimate the supply at a price level of $180, we solve the equation

$$ax + b = 180$$

for $x$. The result is $x \approx 4{,}836$ sets.

(C)  The condition is not stable; the price is likely to decrease since the supply at the price level of $180 exceeds the demand at this level.

(D)  Equilibrium price: $131.59
Equilibrium quantity: 3,587 cookware set.          (2-3)

**92.**  (A)  Cubic Regression

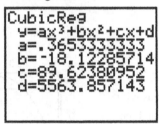

$$y = 0.36\overline{53}x^3 - 18.12285714x^2 + 89.62380952x + 5563.857143$$

(B)  $y(30) = 1806.$          (2-5)

**93.**  (A)  $N(0) = 1$

$$N\left(\frac{1}{2}\right) = 2$$

$$N(1) = 4 = 2^2$$

$$N\left(\frac{3}{2}\right) = 8 = 2^3$$

$$N(2) = 16 = 2^4$$

$$\vdots$$

Thus, we conclude that

$$N(t) = 2^{2t} \text{ or } N = 4^t.$$

(B)  We need to solve:

$$2^{2t} = 10^9$$

$$\log 2^{2t} = \log 10^9 = 9$$

$$2t \log 2 = 9$$

$$t = \frac{9}{2\log 2} \approx 14.95$$

Thus, the mouse will die in 15 days.

(2-6)

Copyright © 2011 Pearson Education, Inc.  Publishing as Prentice Hall.

**94.**  Given $I = I_0 e^{-kd}$. When $d = 73.6$, $I = \frac{1}{2} I_0$. Thus, we have:

$$\frac{1}{2} I_0 = I_0 e^{-k(73.6)}$$

$$e^{-k(73.6)} = \frac{1}{2}$$

$$-k(73.6) = \ln \frac{1}{2}$$

$$k = \frac{\ln(0.5)}{-73.6} \approx 0.00942$$

Thus,  $k \approx 0.00942$.

To find the depth at which 1% of the surface light remains, we set $I = 0.01 I_0$ and solve

$$0.01 I_0 = I_0 e^{-0.00942 d} \quad \text{for } d:$$

$$0.01 = e^{-0.00942 d}$$

$$-0.00942 d = \ln \ 0.01$$

$$d = \frac{\ln 0.01}{-0.00942} \approx 488.87$$

Thus, 1% of the surface light remains at approximately 489 feet.        (2-6)

**95.**  (A)  Logarithmic regression model:

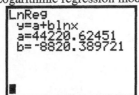

Year 2020 corresponds to  $x = 80$;  $y(80) \approx 5{,}569$ cows.

(B)    ln (0)  is not defined.                                              (2-6)

**96.**  Using the continuous compounding model, we have:

$$2P_0 = P_0 e^{0.03t}$$

$$2 = e^{0.03t}$$

$$0.03t = \ln \ 2$$

$$t = \frac{\ln 2}{0.03} \approx 23.1$$

Thus, the model predicts that the population will double in approximately 23.1 years.        (2-5)

Copyright © 2011 Pearson Education, Inc.  Publishing as Prentice Hall.

**97.**   (A)   Exponential regression model

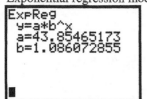

Year 2017 corresponds to $x = 37$;   $y(37) \approx \$931$ billion.

(B)   To find when the expenditures will reach one trillion, solve  $ab^x = 1{,}000$ for $x$. The result is
$x \approx 38$ years; that is, in 2018.        (2-5)

Copyright © 2011 Pearson Education, Inc.  Publishing as Prentice Hall.

**3    MATHEMATICS OF FINANCE**

---

**EXERCISE 3-1**

*Things to remember:*

1.   SIMPLE INTEREST              $I = Prt$

where $I$ = interest
   $P$ = principal
   $r$ = annual simple interest rate expressed as a decimal
   $t$ = time in years

2.   AMOUNT AT SIMPLE INTEREST

$A = P + Prt = P(1 + rt)$

where   $A$ = amount or *future value*
   $P$ = principal or *present value*
   $r$ = annual simple interest rate expressed as a decimal
   $t$ = time in years

---

**1.**   $3\% = 0.03$

**3.**   $0.105 = 10.5\%$

**5.**   $4.75\% = 0.0475$

**7.**   $0.21 = 21\%$

**9.**   4 months $= \dfrac{4}{12} = \dfrac{1}{3}$ year

**11.**   240 days $= \dfrac{240}{360} = \dfrac{2}{3}$ year

**13.**   12 weeks $= \dfrac{12}{52} = \dfrac{3}{13}$ year

**15.**   2 quarters $= \dfrac{2}{4} = \dfrac{1}{2}$ year

**17.**   $I = Prt$; $P = 300$, $r = 0.07$, $t = 2$
   $I = 300(0.07)2 = \$42$

**19.**   $I = Prt$; $I = 36$, $r = 0.04$, $t = 6$ months $= \dfrac{1}{2}$ year

$P = \dfrac{I}{rt} = \dfrac{36}{(0.04)(1/2)} = \$1{,}800$

**21.**   $I = Prt$; $I = 48$, $P = 600$, $t = 240$ days $= \dfrac{2}{3}$ year

$r = \dfrac{I}{Pt} = \dfrac{48}{600(2/3)} = 0.12;\ 12\%$

**23.**   $I = Prt$; $I = 60$, $P = 2{,}400$, $r = 0.05$
$t = \dfrac{I}{Pr} = \dfrac{60}{2400(0.05)} = \dfrac{1}{2}$ year

**25.**   $A = P(1 + rt)$; $P = 4{,}500$, $r = 0.10$, $t = \dfrac{1}{4}$ year

$A = 4{,}500(1 + (0.1)(0.25)) = 4500(1.025) = \$4612.50$

Copyright © 2011 Pearson Education, Inc. Publishing as Prentice Hall.

**27.** $A = P(1 + rt)$; $A = 910$, $r = 0.16$, $t = 13$ weeks $= \dfrac{1}{4}$ year

$910 = P(1 + 0.16[0.25]) = P(1.04)$; $P = \dfrac{910}{1.04} = \$875$

**29.** $A = P(1 + rt)$; $A = 14{,}560$, $P = 13{,}000$, $t = 4$ months $= \dfrac{1}{3}$ year

$14{,}560 = 13{,}000(1 + r/3) = 13{,}000 + \dfrac{13{,}000r}{3}$,

$\dfrac{13{,}000r}{3} = 1{,}560$; $r = \dfrac{3(1{,}560)}{13{,}000} = 0.36$ or $36\%$

**31.** $A = P(1 + rt)$; $A = 736$, $P = 640$, $r = 0.15$
$736 = 640(1 + 0.15t) = 640 + 96t$,
$96t = 96$, $t = 1$ year

**33.** $I = Prt$
Divide both sides by $Pt$.

$\dfrac{I}{Pt} = \dfrac{Prt}{Pt}$

$\dfrac{I}{Pt} = r$ or $r = \dfrac{I}{Pt}$

**35.** $A = P + Prt = P(1 + rt)$
Divide both sides by $(1 + rt)$.

$\dfrac{A}{1+rt} = \dfrac{P(1+rt)}{1+rt}$

$\dfrac{A}{1+rt} = P$ or $P = \dfrac{A}{1+rt}$

**37.** $A = P(1 + rt) = P + Prt$,
$Prt = A - P$

$t = \dfrac{A - P}{rP}$

**39.** Each of the graphs is a straight line; the $y$ intercept in each case is 1000 and their slopes are 40, 80, and 120.

**41.** $P = \$3000$, $r = 4.5\% = 0.045$, $t = 4$ months $= \dfrac{1}{3}$ year

$I = Prt = 3000(0.045)\left(\dfrac{1}{3}\right) = \$45$

**43.** $P = \$554$, $r = 20\% = 0.2$, $t = 1$ month $= \dfrac{1}{12}$ year

$I = Prt = 554(0.2)\left(\dfrac{1}{12}\right) = \$9.23$

**45.** $P = \$7260$, $r = 8\% = 0.08$, $t = 8$ months $= \dfrac{2}{3}$ year

$A = 7260\left[1 + 0.08\left(\dfrac{2}{3}\right)\right] = 7260[1.0533] = \$7647.20$

**47.** $P = \$4000$, $A = \$4270$, $t = 10$ months $= \dfrac{5}{6}$ year
The interest on the loan is $I = A - P = \$270$.

Copyright © 2011 Pearson Education, Inc. Publishing as Prentice Hall.

$r = \dfrac{I}{Pt} = \dfrac{270}{4000\left(\frac{5}{6}\right)} = 0.081.$ Thus, $r = 8.1\%$.

**49.** $P = \$1000$, $I = \$30$, $t = 60$ days $= \dfrac{1}{6}$ year

$r = \dfrac{I}{Pt} = \dfrac{30}{1000\left(\frac{1}{6}\right)} = 0.18.$ Thus, $r = 18\%$.

**51.** $P = \$1500$. The amount of interest paid is $I = (0.29)(3)(120) = \$104.40$. Thus, the total amount repaid is $\$1500 + \$104.40 = \$1604.40$. To find the annual interest rate, we let $t = 120$ days $= \dfrac{1}{3}$ year. Then

$r = \dfrac{I}{Pt} = \dfrac{104.40}{1500\left(\frac{1}{3}\right)} = 0.20880.$ Thus, $r = 20.880\%$.

**53.** Use Formula 2: $A = P(1+rt)$ with $A = 1,000$, $P = 989.37$ and $t = \dfrac{13}{52} = 0.25$.

$$1,000 = 989.37(1 + 0.25r)$$
$$= 989.37 + 247.3425r$$
$$247.3425r = 10.63$$
$$r = \dfrac{10.63}{247.3425} \approx 0.04298 \text{ or } 4.298\%$$

**55.** Use Formula 2: $A = P(1 + rt)$ with $A = 1,000$, $r = 5.53\% = 0.0553$ and $t = \dfrac{50}{360} = 0.1389$.

$$1,000 = P[1 + (0.0553)(0.1389)]$$
$$= P[1.00768]$$
$$P = \$992.38$$

**57.** Principal plus interest on the original note:

$$A = P(1 + rt) = \$5500\left[1 + 0.08\left(\dfrac{90}{360}\right)\right] = \$5610$$

The third party pays $\$5,560$ and will receive $\$5610$ in 60 days. We want to find $r$ given that $A = 5610$, $P = 5,560$ and $t = \dfrac{60}{360} = \dfrac{1}{6}$.

$$A = P + Prt$$
$$r = \dfrac{A - P}{Pt}$$
$$r = \dfrac{5610 - 5560}{5560\left(\frac{1}{6}\right)} \approx 0.05396 \text{ or } 5.396\%$$

**59.** The principal $P$ is the cost of the stock plus the firm's commission:
   Cost: $200(14.20) = 2840$
   $P$: $2840 + 25 + 0.018(2840) = 2916.12$
The investor sells the stock for $200(15.75) = 3150$, and the firm's commission is:
   $37 + 0.014(3150) = 81.10$
Thus, the investor has $3150 - 81.10 = \$3068.90$ after selling the stock; the investor has earned

Copyright © 2011 Pearson Education, Inc. Publishing as Prentice Hall.

$3068.90 - 2916.12 = \$152.78$.

Now, using formula $\underline{1}$, with $P = 2916.12$, $I = 152.78$, and $t = 39$ weeks $= \dfrac{3}{4}$ year, we have

$$r = \frac{152.78}{2916.12\left(\frac{3}{4}\right)} = 0.06986 \text{ or } 6.986\%$$

**61.**   The principal $P$ is the cost of the stock plus the firm's commission:

Cost: $215(45.75) = 9836.25$

$P$: $9836.25 + 56 + 0.01(9836.25) = 9990.61$

The investor sells the stock for $215(51.90) = 11,158.50$, and the firm's commission is:

$106 + 0.005(11,158.50) = 161.79$

Thus, the investor has $11,158.50 - 161.79 = 10,996.71$ after selling the stock; the investor has earned $10,996.71 - 9990.61 = \$1006.10$.

Now, using formula $\underline{1}$ with $P = 9990.61$, and $t = \dfrac{300}{360} = \dfrac{5}{6}$ year,

we have   $r = \dfrac{1006.10}{9990.61\left(\frac{5}{6}\right)} \approx 0.12085$ or $12.085\%$

**63.**   $P = \$475$, $I = \$29$, $t = \dfrac{20}{360} = \dfrac{1}{18}$ year

$$r = \frac{I}{Pt} = \frac{29}{475\left(\frac{1}{18}\right)} \approx 1.09895 \text{ or } 109.895\%$$

**65.**   $P = \$1,900$, $i = \$69$, $t = \dfrac{15}{360} = \dfrac{1}{24}$ year.

$$r = \frac{I}{Pt} = \frac{69}{1900\left(\frac{1}{24}\right)} \quad \bullet \quad 0.87158 \text{ or } 87.158\%$$

---

## EXERCISE 3-2

*Things to remember:*

$\underline{1}$.    AMOUNT—COMPOUND INTEREST

$A = P(1 + i)^n$, where $i = \dfrac{r}{m}$ and

$A$ = amount (future value) at the end of $n$ periods
$P$ = principal (present value)
$r$ = annual (quoted) rate
$m$ = number of compounding periods per year
$n$ = total number of compounding periods

$i = \dfrac{r}{m}$ = rate per compounding period

Copyright © 2011 Pearson Education, Inc.  Publishing as Prentice Hall.

2.   CONTINUOUS COMPOUND INTEREST FORMULA
     If a principal $P$ is invested at an annual rate $r$ (expressed as a decimal) compounded continuously, then the amount $A$ in the account at the end of $t$ years is given by

$$A = Pe^{rt}.$$

3.   ANNUAL PERCENTAGE YIELD
     If a principal is invested at the annual (nominal) rate $r$ compounded $m$ times per year, then the annual percentage yield is

$$APY = \left(1 + \frac{r}{m}\right)^m - 1$$

     If a prinicipal is invested at the annual (nominal) rate $r$ compounded continuously, then the annual percentage yield is

$$APY = e^r - 1$$

     The annual percentage yield is also referred to as the EFFECTIVE RATE or the TRUE INTEREST RATE.

---

1.   $P = \$5,000$; $I = 0.005$; $n = 36$. Find $A$.

     Using 1,   $A = (P(1+i)^n = 5,000(1+0.005)^{36} = 5,000(1.005)^{36} = \$5,983.40$.

3.   $A = 8,000$; $i = 0.02$; $n = 32$. Find $P$.

     Using 1,   $P = \dfrac{A}{(1+i)^n} = \dfrac{8,000}{(1+0.02)^{32}} = \dfrac{8,000}{(1.02)^{32}} = \$4,245.07$.

5.   $A = Pe^{rt}$; $P = 2,450$, $r = 0.0812$, $t = 3$
     $A = 2,450e^{0.0812(3)} = 2,450e^{0.2436} \approx \$3,125.79$

7.   $A = Pe^{rt}$; $A = 6,300$, $r = 0.0945$, $t = 8$
     $6,300 = Pe^{0.0945(8)} = Pe^{0.756} \approx P(2.12974)$
     $P = \dfrac{6,300}{2.12974} \approx \$2958.11$

9.   $A = Pe^{rt}$; $A = 88,000$, $P = 71,153$, $r = 0.085$
     $88,000 = 71,153e^{0.085t}$
     $e^{0.085t} = \dfrac{88,000}{71,153}$
     $0.085t = \ln\left(\dfrac{88,000}{71,153}\right) \approx 0.212504$
     $t = \dfrac{0.212504}{0.085} \approx 2.5$ years

11.  $A = Pe^{rt}$; $A = 15,875$, $P = 12,100$, $r = 48$ months $= 4$ years
     $15,875 = 12,100e^{r(4)}$
     $e^{4r} = \dfrac{15,875}{12,100} \approx 1.31198$

Copyright © 2011 Pearson Education, Inc. Publishing as Prentice Hall.

$$4r = \ln(1.31198)$$

$$r = \frac{\ln(1.31198)}{4} \approx 0.0679 \text{ or } 6.79\%$$

13.  $r = 9\% = 0.09$, $m = 12$; $i = \dfrac{r}{m} = \dfrac{0.09}{12} = 0.0075 = 0.75\%$ per month.

15.  $r = 14.6\% = 0.146$, $m = 360$; $i = \dfrac{r}{m} = \dfrac{0.146}{360} \approx 0.00040 = 0.04\%$ per day.

17.  $r = 4.8\% = 0.048$; $m = 4$; $i = \dfrac{r}{m} = \dfrac{0.048}{4} = 0.012 = 1.2\%$ per quarter.

19.  $i = 0.395\% = 0.00395$; $m = 12$; $r = im = (0.00395)12 = 0.0474 = 4.74\%$

21.  $i = 0.9\% = 0.009$; $m = 4$; $r = im = (0.009)4 = 0.036 = 3.6\%$

23.  $I = 2.1\% = 0.021$; $m = 2$; $r = im = (0.021)2 = 0.42 = 4.2\%$

25.  $P = \$100$, $r = 6\% = 0.06$

 (A)   $m = 1, i = 0.06, n = 4$

 $A = (1 + i)^n$

   $= 100(1 + 0.06)^4$

   $= 100(1.06)^4 = \$126.25$

 Interest $= 126.25 - 100 = \$26.25$

 (B)   $m = 4, i = \dfrac{0.06}{4} = 0.015$

 $n = 4(4) = 16$

   $A = 100(1 + 0.015)^{16}$

   $= 100(1.015)^{16} = \$126.90$

 Interest $= 126.90 - 100 = \$26.90$

 (C)   $m = 12$, $i = \dfrac{0.06}{12} = 0.005$, $n = 4(12) = 48$

 $A = 100(1 + 0.005)^{48} = 100(1.005)^{48} = \$127.05$

 Interest $= 127.05 - 100 = \$27.05$

27.  $P = \$5000$, $r = 5\%$, $m = 12$

 (A)   $n = 2(12) = 24$

 $i = \dfrac{0.05}{12} = 0.0042$

 $A = 5000\left(1 + \dfrac{0.05}{12}\right)^{24} = \$5,524.71$

 (B)   $n = 4(12) = 48$

 $i = \dfrac{0.05}{12} = 0.0042$

 $A = 5000\left(1 + \dfrac{0.05}{12}\right)^{48} = \$6,104.48$

29.  $A = Pe^{rt}$; $P = 8,000$, $r = 0.07$, $t = 6$
 $A = 8,000e^{0.07(6)} = 8,000e^{0.42} \approx \$12,175.69$

31.  Each of the graphs is increasing, curves upward and has $y$ intercept 1000. The greater the interest rate, the greater the increase. The amounts at the end of 8 years are:

 At 4%: $A = 1000\left(1 + \dfrac{0.04}{12}\right)^{96} = \$1376.40$

 At 8%: $A = 1000\left(1 + \dfrac{0.08}{12}\right)^{96} = \$1892.46$

Copyright © 2011 Pearson Education, Inc.  Publishing as Prentice Hall.

At 12%: $A = 1000\left(1 + \dfrac{0.12}{12}\right)^{96} = \$2599.27$

33. $P = 1,000$, $r = 9.75\% = 0.0975 = i$ since the interest is compounded annually.

1st year: $A = P(1 + i)^n = 1000(1 + 0.0975)^1 = \$1,097.50$
   Interest $97.50$

2nd year: $A = 1000(1 + 0.0975)^2 = \$1,204.51$
   Interest: $1,204.51 - 1,097.50 = \$107.01$

3rd year: $A = 1000(1 + 0.0975)^3 = \$1,321.95$
   Interest: $1,321.95 - 1,204.51 = \$117.44$

and so on. The results are:

| Period | Interest | Amount |
|--------|----------|--------|
| 0 |  | $1,000.00 |
| 1 | $97.50 | $1,097.50 |
| 2 | $107.01 | $1,204.51 |
| 3 | $117.44 | $1,321.95 |
| 4 | $128.89 | $1,450.84 |
| 5 | $141.46 | $1,592.29 |
| 6 | $155.25 | $1,747.54 |

35. $A = \$10,000$, $r = 6\% = 0.06$, $i = \dfrac{0.06}{2} = 0.03$

(A)  $n = 2(5) = 10$
$$A = P(1 + i)^n$$
$$10,000 = P(1 + 0.03)^{10}$$
$$= P(1.03)^{10}$$
$$P = \frac{10,000}{(1.03)^{10}} = \$7440.94$$

(B)  $n = 2(10) = 20$
$$P = \frac{A}{(1+i)^n} = \frac{10,000}{(1+0.03)^{20}}$$
$$= \frac{10,000}{(1.03)^{20}}$$
$$= \$5536.76$$

37. $A = Pe^{rt}$

(A) $A = 25,000$, $r = 9\% = 0.09$, $t = 3$
$$25,000 = Pe^{0.09(3)} = Pe^{0.27}$$
$$P = \frac{25,000}{e^{0.27}} \approx \$19,084.49$$

(B)  $25,000 = Pe^{0.09(9)} = Pe^{0.81}$
$$P = \frac{25,000}{e^{0.81}} \approx \$11,121.45$$

39. $APY = \left(1 + \dfrac{r}{m}\right)^m - 1$

(A)  $r = 3.9\% = 0.039$, $m = 12$
$$APY = \left(1 + \frac{0.039}{12}\right)^{12} - 1 = (1.00325)^{12} - 1 \approx 0.0397 = 3.97\%$$

Copyright © 2011 Pearson Education, Inc. Publishing as Prentice Hall.

(B)   $r = 2.3\% = 0.023,\ m = 4$

$$APY = \left(1 + \frac{0.023}{4}\right)^4 - 1 = (1.00575)^4 - 1 \approx 0.0232 = 2.32\%$$

**41.**   $APY = e^r - 1$

(A)   $r = 5.15\% = 0.0515$

$$APY = e^{0.0515} - 1 \approx 0.0528 = 5.28\%$$

(B)   $APY = \left(1 + \dfrac{r}{m}\right)^m - 1$;  $r = 5.20\% = 0.0520$;  $m = 2$

$$APY = \left(1 + \frac{0.0520}{2}\right)^2 - 1 = (1.026)^2 - 1 \approx 0.0527 = 5.27\%$$

**43.**   We have $P = \$4000,\ A = \$9000,\ r = 7\% = 0.07,\ m = 12,$ and

$i = \dfrac{0.07}{12} = 0.0058.$ Since $A = P(1 + i)^n$, we have:

$$9000 = 4000(1 + 0.0058)^n \text{ or } (1.0058)^n = 2.25$$

We solve for $n$ by taking logarithms of both sides:

$\ln(1.0058)^n = \ln(2.25)$
$n \ln(1.0058) = \ln(2.25)$

$$n = \frac{\ln(2.25)}{\ln(1.0058)} \approx \frac{0.8109}{0.0058} \approx 140.22$$

Thus, $n = 140$ months or 11 years and 8 months.

**45.**   $A = Pe^{rt},\ A = 8{,}600,\ P = 6{,}000,\ r = 9.6\% = 0.096$

$$8{,}600 = 6{,}000e^{0.096t}$$

$$e^{0.096t} = \frac{8{,}600}{6{,}000} = \frac{43}{30}\ ;\ 0.096t = \ln(43/30),\ t \approx 3.75 \text{ years}$$

**47.**   $A = 2P,\ i = 0.06$

$$A = P(1 + i)^n$$
$$2P = P(1 + 0.06)^n$$
$$(1.06)^n = 2$$
$$\ln(1.06)^n = \ln 2$$
$$n \ln(1.06) = \ln 2$$
$$n = \frac{\ln 2}{\ln 1.06} \approx \frac{0.6931}{0.0583} \approx 11.9 \approx 12$$

**49.**   We have $A = P(1 + i)^n$. To find the doubling time, set $A = 2P$. This yields:

$$2P = P(1 + i)^n \text{ or } (1 + i)^n = 2$$

Taking the natural logarithm of both sides, we obtain:

$\ln(1 + i)^n = \ln 2$
$n \ln(1 + i) = \ln 2$

Copyright © 2011 Pearson Education, Inc.  Publishing as Prentice Hall.

$$n = \frac{\ln 2}{\ln(1+i)}$$

(A)    $r = 10\% = 0.1$, $m = 4$. Thus,

$i = \dfrac{0.1}{4} = 0.025$ and $n = \dfrac{\ln 2}{\ln(1.025)} \approx 28.07$ quarters or $7\frac{1}{4}$ years.

(B)    $r = 12\% = 0.12$, $m = 4$. Thus,

$i = \dfrac{0.12}{4} = 0.03$ and $n = \dfrac{\ln 2}{\ln(1.03)} \approx 23.44$ quarters.

That is, 24 quarters or 6 years.

**51.**    Doubling time $T$: $2P = Pe^{rt}$, $e^{rt} = 2$, $rt = \ln 2$, $T = \dfrac{\ln 2}{r}$

(A) $T = \dfrac{\ln 2}{0.09} \approx 7.7$ years

(B) $T = \dfrac{\ln 2}{0.11} \approx 6.3$ years

**53.**    $P = 20{,}000$, $r = 7\% = 0.07$, $m = 4$, $i = \dfrac{0.07}{4} = 0.0175$,

$n = 17(4) = 68$

$A = P(1 + i)^n = 20{,}000(1.0175)^{68} \approx \$65{,}068.44$

**55.**    $P = 210{,}000$, $r = 3\% = 0.03$, $m = 1$, $i = \dfrac{0.03}{1} = 0.03$, $n = 10$

$A = P(1 + i)^n = 210{,}000(1.03)^{10} \approx \$282{,}222.44$

**57.**    $A = 25$, $r = 4.8\% = 0.048$, $m = 1$, $i = \dfrac{0.048}{1} = 0.048$, $n = 5$

$P = \dfrac{A}{(1+i)^n} = \dfrac{25}{(1.048)^5} \approx \$19.78$ per sq ft/mo.

**59.**    From Problem 51, the doubling time is

$T = \dfrac{\ln 2}{r} = \dfrac{\ln 2}{0.017} \approx 41$ years

**61.**    The effective rate, APY, of $r = 9\% = 0.09$ compounded monthly is:

$\text{APY} = \left(1 + \dfrac{0.09}{12}\right)^{12} - 1 = .0938$ or $9.38\%$

The effective rate of 9.3% compounded annually is 9.3%. Thus, 9% compounded monthly is better than 9.3% compounded annually.

**63.**    (A)    $2016 - 1776 = 240$ years.

We have $P = 100$, $r = 0.03$, $m = 4$, $i = \dfrac{0.03}{4} = 0.0075$, and in 2016,

$n = 240(4) = 960$.

$A = 100(1 + 0.0075)^{960} \approx \$130{,}392.50$

Copyright © 2011 Pearson Education, Inc. Publishing as Prentice Hall.

(B)   Monthly compounding:

$$n = (240)(12) = 2880; \quad i = \frac{0.03}{12} = 0.0025$$

$$A = 100(1.0025)^{2880} \approx \$132,744.98$$

Daily compounding:

$$n = 240(365) = 87,600; \quad i = \frac{0.03}{365} = 0.000082191$$

$$A = 100(1.000082191)^{87,600} \approx \$133,903.45$$

Continuous compounding

$$A = Pe^{rt}$$

$$= 100e^{0.03(240)} = 100e^{7.2} \approx \$133,943.08$$

(C)

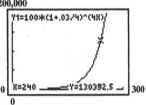

65.   $A = Pe^{rt}$; $A = 50,000$, $P = 28,000$, $t = 6$

$$50,000 = 28,000e^{6r}$$

$$e^{6r} = \frac{50,000}{28,000} = \frac{25}{14}$$

$$6r = \ln\left(\frac{25}{14}\right), \quad r = \frac{\ln(25/14)}{6} \approx 0.9664 \text{ or } 9.664\%$$

67.   $P = \$7000$, $A = \$9000$, $r = 9\% = 0.09$, $m = 12$, $i = \frac{0.09}{12} = 0.0075$

Since $A = P(1 + i)^n$, we have:

$$9000 = 7000(1 + 0.0075)^n \text{ or } (1.0075)^n = \frac{9}{7}$$

Therefore,   $\ln(1.0075)^n = \ln\left(\frac{9}{7}\right)$

$$n \ln(1.0075) = \ln\left(\frac{9}{7}\right)$$

$$n = \frac{\ln\left(\frac{9}{7}\right)}{\ln(1.0075)} \approx \frac{0.2513}{0.0075} \approx 33.6$$

Thus, it will take 34 months or 2 years and 10 months.

69.   $P = \$20,000$, $r = 6\% = 0.06$, $m = 365$, $i = \frac{0.06}{365}$,

$n = (365)35 = 12,775$

Since $A = P(1 + i)^n$, we have:

$$A = 20,000\left(1 + \frac{0.06}{365}\right)^{12,775} \approx \$163,295.21$$

Copyright © 2011 Pearson Education, Inc.  Publishing as Prentice Hall.

**71.** From Problem 49, the doubling time is:

$$n = \frac{\ln 2}{\ln(1+i)}$$

At 7% = 0.07 compounded daily, $i = \frac{0.07}{365} \approx 0.000191781$  and

$$n = \frac{\ln 2}{\ln(1.000191781)} \approx 3615 \text{ days or } 9.904 \text{ years}$$

From Problem 59, the doubling time is

$$T = \frac{\ln 2}{r}$$

At 8.2% = 0.082 compounded continuously

$$T = \frac{\ln 2}{0.082} \approx 8.453 \text{ years}$$

**73.** If an investment of $P$ dollars doubles in $n$ years at an interest rate $r$ compounded annually, then $r$ satisfies the equation:

$$2P = P(1 + r)^n$$

so    $1 + r = 2^{1/n}$

$$r = 2^{1/n} - 1$$

This is the exact value of $r$.

The approximate value of $r$ given by

$$r = \frac{72}{n}$$

is called the Rule of 72.

Setting $n = 6, 7, 8, 9, 10, 11, 12$ in these formulas gives:

| Years | Exact rate | Rule of 72 |
|-------|------------|------------|
| 6 | 12.2 | 12.0 |
| 7 | 10.4 | 10.3 |
| 8 | 9.1 | 9.0 |
| 9 | 8.0 | 8.0 |
| 10 | 7.2 | 7.2 |
| 11 | 6.5 | 6.5 |
| 12 | 5.9 | 6.0 |

**75.** $2,400 at 13% compounded quarterly:

Value after $n$ quarters: $A_1 = 2400\left(1 + \frac{0.13}{4}\right)^n = 2400(1.0325)^n$

$3,000 at 6% compounded quarterly:

Value after $n$ quarters: $A_2 = 3000\left(1 + \frac{0.06}{4}\right)^n = 3000(1.015)^n$

Graph $A_1$ and $A_2$:

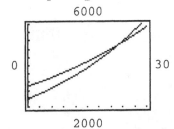

6000

0

30

2000

The graphs intersect at (13.05, 3643.56)
$A_2(n) > A_1(n)$ for $n < 13.05$, $A_1(n) > A_2(n)$ for $n > 13.05$.

Thus it will take 14 quarters for the $2,400 investment to be worth more than the $3,000 investment.

Copyright © 2011 Pearson Education, Inc.  Publishing as Prentice Hall.

**77.**   The value of $P$ dollars at 10% simple interest after $t$ years is given by
$$A_s = P(1 + 0.1t)$$
The value of $P$ dollars at 7% interest compounded annually after
$t$ years is given by
$$A_c = P(1.07)^t$$
Let $P = \$1$ and graph
$$A_s = 1 + 0.1t, \quad A_c = (1.07)^t.$$
The graphs intersect at the point where $t = 10.89 \approx 11$ years.

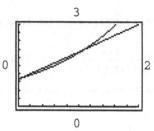

For investments of less than 11 years, simple interest at 10% is better; for investments of more than 11 years, interest compounded annually at 7% is better.

**79.**   The relationship between the annual percentage yield, APY, and the annual nominal rate $r$ is:
$$\text{APY} = \left(1 + \frac{r}{m}\right)^m - 1$$

In this case, APY = 0.068 and $m = 365$. Thus, we must solve
$$0.068 = \left(1 + \frac{r}{365}\right)^{365} - 1$$

for $r$:
$$\left(1 + \frac{r}{365}\right)^{365} = 1.068$$
$$1 + \frac{r}{365} = \sqrt[365]{1.068} = (1.068)^{1/365}$$
$$r = 365[(1.068)^{1/365} - 1] \approx 0.0658; \quad r = 6.58\%$$

**81.**   The APY corresponding to 7% = 0.07 compounded continuously is:
$$\text{APY} = e^r - 1 = e^{0.07} - 1 \approx 0.07251 \text{ or } 7.251\%$$
To find the annual nominal rate compounded monthly for the
APY = 7.251%, we solve
$$0.07251 = \left(1 + \frac{r}{12}\right)^{12} - 1$$

for $r$:
$$\left(1 + \frac{r}{12}\right)^{12} = 1.07251$$
$$1 + \frac{r}{12} = \sqrt[12]{1.07251} = (1.07251)^{1/12} \approx 1.005851$$
$$\frac{r}{12} = 0.005851$$
$$r = 0.0702 \text{ or } 7.02\%$$

Copyright © 2011 Pearson Education, Inc.  Publishing as Prentice Hall.

**83.**   $A = \$30,000$, $r = i = 4.348\% = 0.04348$, $n = 15$

From $A = P(1 + r)^n$, we have

$$P = \frac{A}{(1+r)^n} = \frac{30,000}{(1.04348)^{15}} \approx \$15,843.80$$

**85.**   $A = \$10,000$, $P = \$4,126$, $n = 20$

Using $A = P(1 + r)^n$, we have

$$10,000 = 4,126(1 + r)^{20}$$

$$(1 + r)^{20} = \frac{10,000}{4,126} \approx 2.4237$$

Therefore

$$1 + r = \sqrt[20]{2.4237} = (2.4237)^{1/20} \approx 1.04526$$

and    $r \approx 0.04526$ or $r = 4.526\%$

**87.**   (A)   Republic Bank: $r = 4.31\% = 0.0431$ compounded continuously

$$APY = e^{0.0431} - 1 \approx 0.04404 \text{ or } 4.04\%$$

   (B)   Chase Bank: $r = 0.0435$, $m = 365$

$$APY = \left(1 + \frac{0.0435}{365}\right)^{365} - 1 \approx 0.04446 = 4.446\%$$

   (C)   Bank First: $r = 0.0436$, $n = 12$

$$APY = \left(1 + \frac{0.0436}{12}\right)^{12} - 1 \approx 0.04448 = 4.448\%$$

**89.**   The principal $P$ is the cost of the stock plus the firm's commission:

   Cost: $100(65) = 6500$,

   $P = 6500 + 75 + 0.003(6500) = 6594.50$

The investor sells the stock for $100(125) = 12,500$ and the firm's commission is:

   $75 + 0.003(12,500) = 112.50$

Thus, the investor has $12,500 - 112.50 = 12,387.50$ after selling the stock. To find the annual compound rate of return, use

$A = P(1 + r)^n$ with $A = 12,387.50$, $P = 6594.50$, $n = 5$, and solve for $r$:

$$12,387.50 = 6594.50(1 + r)^5$$

$$(1 + r)^5 = \frac{12,387.50}{6594.50} \approx 1.8785$$

$$1 + r = \sqrt[5]{1.8785} = (1.8785)^{1/5} \approx 1.1344$$

$$r \approx 0.1344 = 13.44\%$$

**91.**   The principal $P$ is the cost of the stock plus the firm's commission:

   Cost: $200(28) = 5600$

   $P = 5600 + 57 + 0.006(5600) = 5690.60$

The investor sells the stock for $200(55) = 11,000$ and the firm's commission is:

   $75 + 0.003(11,000) = 108$.

Copyright © 2011 Pearson Education, Inc. Publishing as Prentice Hall.

Thus, the investor has $11,000 - 108 = 10,892$ after selling the stock. To find the annual compound rate of return, use

$A = P(1 + r)^n$ with $A = 10,892$, $P = 5690.60$, $n = 4$, and solve for $r$:

$$10,892 = 5690.60(1 + r)^4$$

$$(1 + r)^4 = \frac{10,892}{5690.60} \approx 1.9140$$

$$1 + r = \sqrt[4]{1.9140} = (1.9140)^{1/4} \approx 1.1762$$

$$r \approx 0.1762 = 17.62\%$$

## EXERCISE 3-3

*Things to remember:*

1.   FUTURE VALUE OF AN ORDINARY ANNUITY

$$FV = PMT\left(\frac{(1+i)^n - 1}{i}\right) = PMT\ s_{\overline{n}|i}$$

where

$FV$ = future value (amount)
$PMT$ = periodic payment
$i$ = rate per period
$n$ = number of payments (periods)

(Payments are made at the end of each period.)

2.   SINKING FUND PAYMENT

$$PMT = FV \frac{i}{(1+i)^n - 1}$$

3.   APPROXIMATING INTEREST RATES
Algebra can be used to solve the future value formula in 1 for $PMT$ or $n$, or the payment formula in 2 for $FV$ or $n$. Graphical techniques or equation solvers can be used to approximate $i$ to as many places as desired in each of these formulas.

1.   $r = 8\% = 0.08$, $m = 4$, $i = \dfrac{0.08}{4}$

$n = 20(4) = 80$

3.   $r = 7.5\% = 0.075$, $m = 2$, $i = \dfrac{0.075}{2} = 0.0375$

$n = 12(2) = 24$

5.   $r = 9\% = 0.09$, $m = 12$, $i = \dfrac{0.09}{12} = 0.0075$

$n = 4(12) = 48$

7.   $r = 5.95\% = 0.0595$, $m = 1$, $i = 0.0595$
$n = 12(1) = 12$

Copyright © 2011 Pearson Education, Inc. Publishing as Prentice Hall.

9.  $n = 20$, $i = 0.03$, $PMT = \$500$

$$FV = PMT\left(\frac{(1+i)^n - 1}{i}\right) = PMT\, s_{\overline{n}|i} = 500\frac{(1+0.03)^{20} - 1}{0.03} = 500\, s_{\overline{n}|i} = 500 s_{\overline{20}|0.03}$$

$$= 500(26.87037449) = \$13,435.19.$$

11. $FV = \$5,000$, $n = 15$, $i = 0.01$

$$PMT = FV\frac{i}{(1+i)^n - 1} = \frac{FV}{s_{\overline{n}|i}} = 5000\frac{0.01}{(1+0.01)^{15} - 1}$$

$$= \frac{5000}{16.09689554} \approx \$310.62$$

13. $FV = \$4000$, $i = 0.02$, $PMT = 200$, $n = ?$

$$FV = PMT\frac{(1+i)^n - 1}{i}$$

$$\frac{iFV}{PMT} = (1+i)^n - 1$$

$$(1+i)^n = \frac{iFV}{PMT} + 1$$

$$\ln(1+i)^n = \ln\left[\frac{iFV}{PMT} + 1\right]$$

$$n\ln(1+i) = \ln\left[\frac{iFV}{PMT} + 1\right]$$

$$n = \frac{\ln\left[\dfrac{iFV}{PMT} + 1\right]}{\ln(1+i)} = \frac{\ln\left[\dfrac{(0.02)4000}{200} + 1\right]}{\ln(1.02)} = \frac{\ln(1.4)}{\ln(1.02)} \approx \frac{0.3365}{0.01980} = 16.99 \text{ or } 17 \text{ periods.}$$

15. $FV = \$7,600$; $PMT = \$500$; $n = 10$, $i = ?$

$$FV = PMT\frac{(1+i)^n - 1}{i}$$

Substituting the given values into this formula gives

$$7600 = 500\frac{(1+i)^{10} - 1}{i}$$

and    $\dfrac{(1+i)^{10} - 1}{i} = 15.2$

Copyright © 2011 Pearson Education, Inc.  Publishing as Prentice Hall.

Graph $Y_1 = \dfrac{(1+x)^{10} - 1}{x}$,

$Y_2 = 15.2$.

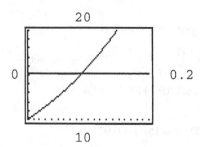

The graphs intersect at the point where $x \approx 0.09$. Thus, $i = 0.09$.

**17.** An ordinary annuity is an annuity in which the payments are made at the <u>end</u> of each time interval.

**19.** $FV = PMT\dfrac{(1+i)^n - 1}{i}$

$PMT\left[(1+i)^n - 1\right] = i\,FV$

$(1+i)^n - 1 = \dfrac{i\,FV}{PMT}$

$(1+i)^n = 1 + \dfrac{i\,FV}{PMT}$

$n\ln(1+i) = \ln\left[1 + \dfrac{i\,FV}{PMT}\right]$

$n = \dfrac{\ln\left[1 + \dfrac{i\,FV}{PMT}\right]}{\ln(1+i)}$

**21.** $PMT = \$500$, $n = 10(12) = 120$, $i = \dfrac{0.0665}{12}$

$FV = 500\,\dfrac{\left(1 + \frac{0.0665}{12}\right)^{120} - 1}{\frac{0.0665}{12}} = 500\,s_{\overline{120}|\,0.0665} = 500(169.7908065) = \$84{,}895.40$

Total deposits: $500(120) = \$60{,}000$

Interest: $FV - 60{,}000 = 84{,}895.40 - 60{,}000 = \$24{,}895.40$

**23.** $PMT = \$300$, $i = \dfrac{0.06}{12} = 0.005$, $n = 5(12) = 60$

$FV = 300\,\dfrac{(1 + 0.005)^{60} - 1}{0.005} = 300\,s_{\overline{60}|\,0.005}$   (using 1)

$= 300(69.77003051) = \$20{,}931.01$

After five years, \$20,931.01 will be in the account.

**25.** $FV = \$200{,}000$, $i = \dfrac{0.0635}{12}$, $n = 15(12) = 180$

$PMT = FV\,\dfrac{i}{(1+i)^{180} - 1} = 200{,}000(0.003337158) = \$667.43$ per month

Copyright © 2011 Pearson Education, Inc.  Publishing as Prentice Hall.

**27.**  $FV = \$100,000$, $i = \dfrac{0.075}{12} = 0.00625$, $n = 8(12) = 96$

$PMT = FV \dfrac{i}{(1+i)^{96} - 1} = 100,000(0.00763387) = \$763.39$ per month

**29.**  $PMT = \$1,000$, $i = \dfrac{0.0832}{1} = 0.0832$, $n = 5$

$FV = PMT \dfrac{(1+i)^n - 1}{i} = 1000 \dfrac{(1.0832)^n - 1}{0.0832}$

$n = 1$: $FV = \$1,000$

$n = 2$: $FV = 1000 \dfrac{(1.0832)^2 - 1}{0.0832} = \$2,083.20$

Interest: $2083.20 - 2000 = \$83.20$

$n = 3$: $FV = 1000 \dfrac{(1.0832)^3 - 1}{0.0832} = \$3,256.52$

Interest: $3256.52 - 2083.20 = \$173.32$

and so on.

Balance Sheet

| Period | Amount | Interest | Balance |
|--------|--------|----------|---------|
| 1 | $1,000.00 | $0.00 | $1,000.00 |
| 2 | $1,000.00 | $83.20 | $2,083.20 |
| 3 | $1,000.00 | $173.32 | $3,256.52 |
| 4 | $1,000.00 | $270.94 | $4,527.46 |
| 5 | $1,000.00 | $376.69 | $5,904.15 |

**31.**  $FV = PMT \dfrac{(1+i)^n - 1}{i} = 100 \dfrac{(1+0.0050)^{12} - 1}{0.0050}$ (after one year)

$= 100 \dfrac{(1.0050)^{12} - 1}{0.0050} \left[ \text{Note:} PMT = \$100, i = \dfrac{0.06}{12} = 0.0050, n = 12 \right]$

$= \$1233.56$

(1)

Total deposits in one year $= 12(100) = \$1200$.

Interest earned in first year $= FV - 1200 = 1233.56 - 1200 = \$33.56$.

At the end of the second year:

$FV = 100 \dfrac{(1+0.0050)^{24} - 1}{0.0050}$ [Note: $n = 24$]

$= 100 \dfrac{(1.0050)^{24} - 1}{0.0050} = \$2543.20$

(2)

Total deposits plus interest in the second year $= (2) - (1)$

$= 2543.20 - 1233.56 = \$1309.64$    (3)

Interest earned in the second year $= (3) - 1200$

$= 1309.64 - 1200 = \$109.64$

Copyright © 2011 Pearson Education, Inc. Publishing as Prentice Hall.

**(4)**

At the end of the third year,

$$FV = 100 \frac{(1+0.0050)^{36}-1}{0.0050} \quad [\text{Note: } n = 36]$$

$$= 100 \frac{(1.0050)^{36}-1}{0.0050} = \$3933.61$$

Total deposits plus interest in the third year  $= (4) - (2)$

$$= 3933.61 - 2543.20 = \$1390.41$$

**(5)**

Interest earned in the third year $= (5) - 1200$

$$= 1390.41 - 1200 = \$190.41$$

Thus

| Year | Interest earned |
|------|-----------------|
| 1 | $ 33.56 |
| 2 | $109.64 |
| 3 | $190.41 |

**33.**   $PMT = \$1000, \ i = 0.064, \ n = 12$

$$FV = 1000 \frac{(1+0.064)^{12}-1}{0.064} = 1000(17.26921764) = 17,269.21764$$

Thus, Bob will have \$17,269.22 in his IRA account on his 35th birthday. On his 65th birthday, he will have

$$A = 17,269.22(1 + 0.064)^{30} = 17,269.22(6.430561) = \$111,050.77$$

**35.**   $FV = \$111,050.77, \ i = 0.064, \ n = 30$

$$PMT = 111,050.77 \ \frac{0.064}{(1+0.064)^{30}-1}$$

$$= 111,050.77(0.011785155) = \$1308.75 \text{ per year}$$

**37.**   **(A)**   From Section 3.2, APY $= \left(1+\frac{r}{m}\right)^{m} - 1$

In this case, APY $= 4.86\% = 0.0486$ and $m = 12$. Thus, we must solve

$$0.0486 = \left(1+\frac{r}{12}\right)^{12} - 1$$

for $r$:

$$\left(1+\frac{r}{12}\right)^{12} = 1.0486$$

$$1 + \frac{r}{12} = \sqrt[12]{1.0486} = (1.0486)^{1/12} \approx 1.00396$$

$$r = 12[0.00396] \approx 0.0476; \quad r = 4.76\%$$

**(B)**   $FV = \$10,000, \ i = \dfrac{0.0475}{12}, n = 4(12) = 48$

$$PMT = FV \ \frac{i}{(1+i)^{48}-1} = 10,000(0.018957884) = \$189.58$$

Copyright © 2011 Pearson Education, Inc.  Publishing as Prentice Hall.

**39.** $PMT = \$200,\ FV = \$7,000,\ i = \dfrac{0.057}{12}$

From Problem 17:

$$n = \dfrac{\ln\left[\dfrac{FV(i)}{PMT}+1\right]}{\ln(1+i)} = \dfrac{\ln\left[\dfrac{7000\left(\frac{0.057}{12}\right)+1}{200}\right]}{\ln\left(1+\frac{0.057}{12}\right)} = \dfrac{0.153793473}{0.004738754} \approx 32.45$$

Thus, $n = 33$ months.

**41.** This problem was done with a graphics calculator.

Start with the equation $\dfrac{(1+i)^n - 1}{i} - \dfrac{FV}{PMT} = 0$

where $FV = 5840,\ PMT = 1000,\ n = 5$ and $i = \dfrac{r}{1} = r$,

where $r$ is the nominal annual rate. With these values, the equation is:

$$\dfrac{(1+r)^5 - 1}{r} - \dfrac{5840}{1000} = 0$$

or $(1+r)^5 - 1 - 5.840r = 0$

Set $y = (1+r)^5 - 1 - 5.840r$ and use your calculator to find the zero $r$ of the function $y$, where $0 < r < 1$. The result is $r = 0.077$ or $7.77\%$ to two decimal places.

**43.** Start with the equation $\dfrac{(1+i)^n - 1}{i} - \dfrac{FV}{PMT} = 0$

where $FV = 620,\ PMT = 50,\ n = 12$ and $i = \dfrac{r}{12}$,

where $r$ is the annual nominal rate. With these values, the equation becomes

$$\dfrac{(1+i)^{12} - 1}{i} - \dfrac{620}{50} = 0$$

or $(1+i)^{12} - 1 - 12.4i = 0$

Set $y = (1+i)^{12} - 1 - 12.4i$ and use your calculator to find the zero $i$ of the function $y$, where $0 < i < 1$. The result is $i = 0.005941$. Thus $r = 12(0.005941) = 0.0713$ or $r = 7.13\%$ to two decimal places.

**45.** Annuity: $PMT = 500,\ i = \dfrac{0.06}{4} = 0.015$

$$Y_1 = 500\dfrac{[(1+0.015)^{4x} - 1]}{0.015}$$

Simple interest: $P = 5000,\ r = 0.04$

$$Y_2 = 5000(1 + 0.04x)$$

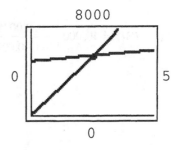

The graphs of $Y_1$ and $Y_2$ are shown in the figure at the right.

intersection: $x = 2.57;\ y = 5514$

The annuity will be worth more after 2.57 years, or 11 quarterly payments.

Copyright © 2011 Pearson Education, Inc. Publishing as Prentice Hall.

## EXERCISE 3-4

*Things to remember:*

1. PRESENT VALUE OF AN ORDINARY ANNUITY

$$PV = PMT\frac{1-(1+i)^{-n}}{i} = PMT\, a_{\overline{n}|i}$$

where

$PV$ = present value of all payments
$PMT$ = periodic payment
$i$ = rate per period
$n$ = number of periods

(Payments are made at the end of each period.)

2. AMORTIZATION FORMULA

$$PMT = PV\frac{i}{1-(1+i)^{-n}} = PV\frac{1}{a_{\overline{n}|i}}$$

---

1. $r = 7.2\% = 0.072$, $m = 12$, $i = \dfrac{0.072}{12} = 0.006$, $n = 4(12) = 48$

3. $r = 9.9\% = 0.099$, $m = 4$, $i = \dfrac{0.099}{4} = 0.02475$, $n = 10(4) = 40$

5. $r = 5.05\% = 0.0505$, $m = 2$, $i = \dfrac{0.0505}{2} = 0.02525$, $n = 16(2) = 32$

7. $r = 5.48\% = 0.0548$, $m = 1$, $i = 0.0548$, $n = 9(1) = 9$

9. $PV = 200\dfrac{1-(1+.04)^{-30}}{0.04} = PMT\, a_{\overline{30}|0.04} = 200(17.29203330)$ (using the table)

   $= \$3458.41$

11. $PMT = 40,000\dfrac{0.0075}{1-(1+0.0075)^{-96}} = \dfrac{40,000}{a_{\overline{96}|0.0075}} = \dfrac{40,000}{68.25843856} = \$586.01$

Copyright © 2011 Pearson Education, Inc. Publishing as Prentice Hall.

13.    $PV = \$5000$, $i = 0.01$, $PMT = 200$

We have, $PV = PMT \dfrac{1-(1+i)^{-n}}{i}$

$$5000 = 200\dfrac{1-(1+0.01)^{-n}}{0.01} = 20,000\left[1-(1.01)^{-n}\right]$$

$$\dfrac{1}{4} = 1-(1.01)^{-n}$$

$$(1.01)^{-n} = \dfrac{3}{4} = 0.75$$

$$\ln(1.01)^{-n} = \ln(0.75)$$

$$-n\ln(1.01) = \ln(0.75)$$

$$n = \dfrac{-\ln(0.75)}{\ln(1.01)} \approx 29$$

15.    $PV = \$9,000$, $PMT = \$600$, $n = 20$, $i = ?$

$PV = PMT \dfrac{1-(1+i)^{-n}}{i}$

Substituting the given values into this formula gives

$$9000 = 600 \ \dfrac{1-(1+i)^{-20}}{i}$$

$$15i = 1-(1+i)^{-20} = 1 - \dfrac{1}{(1+i)^{20}}$$

$$15i + \dfrac{1}{(1+i)^{20}} = 1$$

Graph $Y_1 = 15x + \dfrac{1}{(1+x)^{20}}$, $Y_2 = 1$.

The curves intersect at $x = 0$ and $x \approx 0.029$. Thus $i = 0.029$.

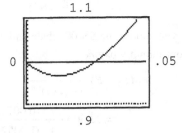

17.    The present value of an annuity is the current value of a series of equal payments made at equally spaced time intervals in the future.

19.    You are going to make monthly payments to a bank to pay off a loan of $P at an interest rate $r$. This is an annuity with present value $P.

21.    $PMT = \$5,000$, $i = 0.0665$, $n = 10$

$$PV = PMT \dfrac{1-(1+i)^{-n}}{i} = 5000 \ \dfrac{1-(1.0665)^{-10}}{0.0665}$$

$$= 5000(7.138636854) = \$35,693.18$$

Copyright © 2011 Pearson Education, Inc. Publishing as Prentice Hall.

**23.**  $PMT = 350, \ i = \dfrac{0.0756}{12} = 0.0063, \ n = 36$

$PV = PMT \ \dfrac{1-(1+i)^{-n}}{i} = 350 \dfrac{1-(1.0063)^{-36}}{0.0063} = 350(32.11945179) = 11{,}241.81$

You can borrow $11,241.81. The total amount paid on this loan is: 350(36) = $12,600. Thus, the total interest you will pay is

$12{,}600 - 11{,}241.81 = \$1{,}358.19$

**25.**  $PV = \$2{,}500, \ i = 0.0125, \ n = 48$

$PMT = PV \ \dfrac{i}{1-(1+i)^{-n}} = 2500 \dfrac{0.0125}{1-(1.0125)^{-48}} = 2500(0.027830748) = \$69.58$

Total amount paid: 69.58(48) = $3339.84

Total interest paid: 3335.04 − 2500 = $839.84

**27.**  0% financing for 72 months implies that the monthly payment should be $\dfrac{9330}{72} = \$129.58$ not $179. The interest rate actually being charged can be found by solving

$9330 = 179 \dfrac{1-(1+i)^{-72}}{i}$

for $i$. The result is:  $i \approx 0.009412$. The interest rate  $r = 12i \approx 0.1129$ or 11.29% compounded monthly.

**29.**  If you choose 0% financing, your monthly payment will be:

$PMT_1 = \dfrac{27{,}300}{60} = \$455$

If you choose the $5000 rebate and borrow $22,300 at 6.3% compounded monthly for 60 months, your monthly payment will be

$PMT_2 = PV \dfrac{i}{1-(1+i)^{-n}}$

$\qquad = 22{,}300 \cdot \dfrac{0.00525}{1-(1.00525)^{-60}} \quad (i = \dfrac{0.063}{12} = 0.00525)$

$\qquad = \$434.24$

You should choose the $5000 rebate. You will save $20.76 per month or 60(20.76) = $1245.60 over the life of the loan.

**31.**  Amortized amount: 35,000 − 0.20(35,000) = $28,000

We now have:  $PV = 28{,}000, \ i = \dfrac{0.0875}{12} \approx 0.007292, \ n = 12(12) = 144$

$PMT = PV \dfrac{i}{1-(1+i)^{-n}} = 28{,}000 \dfrac{0.007292}{1-(1.007292)^{-144}}$

$\qquad\qquad\qquad = 28{,}000(0.011240) = \$314.72$

Total amount paid: 144(314.72) = $45,319.68

Total interest paid: 45,319.68 − 28,000 = $17,319.68

Copyright © 2011 Pearson Education, Inc.  Publishing as Prentice Hall.

**33.** First, we compute the quarterly payment for $PV = \$5000$, $i = 0.028$ and $n = 8$:

$$PMT = PV \frac{i}{1-(1+i)^{-n}} = 5000 \frac{0.028}{1-(1.028)^{-8}} = 5000(0.14125701) = \$706.29$$

The amortization schedule is as follows:

| Payment number | Payment | Interest | Unpaid balance reduction | Unpaid balance |
|---|---|---|---|---|
| 0 | | | | $5,000.00 |
| 1 | $706.29 | $140.00 | $566.29 | 4,433.71 |
| 2 | 706.29 | 124.14 | 582.15 | 3,851.56 |
| 3 | 706.29 | 107.84 | 598.45 | 3,253.11 |
| 4 | 706.29 | 91.09 | 615.20 | 2,637.91 |
| 5 | 706.29 | 73.86 | 632.43 | 2,005.48 |
| 6 | 706.29 | 56.15 | 650.14 | 1,355.34 |
| 7 | 706.29 | 37.95 | 668.34 | 687.00 |
| 8 | 706.24 | 19.24 | 687.00 | 0.00 |
| Total | $5,650.27 | $650.27 | $5,000.00 | |

**35.** First, we compute the monthly payment for

$$PV = \$6000, \quad i = \frac{0.09}{12} = 0.0075, \quad n = 3(12) = 36:$$

$$PMT = PV \frac{i}{1-(1+i)^{-n}} = 6000 \frac{0.0075}{1-(1.0075)^{-36}}$$
$$= 6000(0.031799732) = \$190.80$$

Now, compute the unpaid balance after 12 payments by considering 24 unpaid payments:

$$PMT = 190.80, i = 0.0075, n = 24$$
$$PV = PMT \frac{1-(1+i)^{-n}}{i} = 190.80 \frac{1-(1.0075)^{-24}}{0.0075}$$
$$= 190.80(21.88914614) = \$4176.45$$

The amount of loan paid in 12 months is:
6000 − 4176.45 = $1823.55

The total amount paid during the first year is:
190.80(12) = $2289.60

Thus, the interest paid during the first year is:
2289.60 − 1823.55 = $466.05

Next, we compute the unpaid balance after 24 months by considering 12 unpaid payments:

$$PMT = \$190.80, \quad i = 0.0075, \quad n = 12$$
$$PV = PMT \frac{1-(1+i)^{-n}}{i} = 190.80 \frac{1-(1.0075)^{-12}}{0.0075}$$
$$= 190.80(11.43491267) = \$2181.78$$

The amount of the loan paid in 24 months is:
6000 − 2181.78 = $3818.22

Copyright © 2011 Pearson Education, Inc. Publishing as Prentice Hall.

and the amount of the loan paid during the second year is:

$3818.22 - 1823.55 = \$1994.67$

The total amount paid during the second year is $\$190.80(12) = \$2289.60$.

Therefore, the interest paid during the second year is:

$2289.60 - 1994.67 = \$294.93$

The total amount paid in 36 months is:

$190.80(36) = \$6868.80$

and the total interest paid is: $6868.80 - 6000 = \$868.80$

Therefore, the interest paid during the third year is:

$868.80 - [466.05 + 294.93] = \$107.82$

**37.** $PMT = \$525, \ i = \dfrac{0.078}{12} = 0.0065, \ n = 30(12) = 360$

The present value of these payments is:

$PV = PMT\ \dfrac{1 - (1+i)^{-n}}{i} = 525\ \dfrac{1 - (1.0065)^{-360}}{0.0065} = 525(138.9138739) = \$72,929.78$

Selling price = loan + down payment

$= 72,929.78 + 25,000 = \$97,929.78$

Total amount paid in 30 years: $525(360) = \$189,000$

Total interest paid: $189,000 - 72,929.78 = \$116,070.22$

**39.** $P = \$6000, \ n = 2(12) = 24, \ i = \dfrac{0.035}{12} \approx 0.0029167$

The total amount owed at the end of the two years is:

$A = P(1 + i)^n = 6000(1 + 0.0029167)^{24} = 6000(1.0029167)^{24} \approx 6434.39$

Now, the monthly payment is:

$PMT = PV\ \dfrac{i}{1 - (1+i)^{-n}}$

where $n = 4(12) = 48, \ PV = \$6434.39, \ i = \dfrac{0.035}{12} \approx 0.0029167$. Thus,

$PMT = 6434.39\ \dfrac{0.0029167}{1 - (1 + 0.0029167)^{-48}} = \$143.85$ per month

The total amount paid in 48 payments is $143.85(48) = \$6904.80$. Thus, the interest paid is $6904.80 - 6000 = \$904.80$.

**41.** Monthly payment: $PV = \$150,000, \ i = \dfrac{0.061}{12} = 0.005083\overline{3}, \ n = 30(12) = 360$

$PMT = PV\ \dfrac{i}{1 - (1+i)^{-n}} = 150,000\ \dfrac{0.005083\overline{3}}{1 - (1.005083\overline{3})^{-360}} = \$908.99$

(A)  To compute the balance after 10 years (with the balance of the loan to be paid in 20 years), use

$PMT = \$908.99, \ i = 0.005803, \ n = 20(12) = 240$

Balance after 10 years $= PMT\ \dfrac{1 - (1+i)^{-240}}{i} = PMT\ \dfrac{1 - (1.005083\overline{3})^{-240}}{0.005083\overline{3}} = 908.99\ \dfrac{1 - (1.005083\overline{3})^{-240}}{0.005083\overline{3}}$

Copyright © 2011 Pearson Education, Inc. Publishing as Prentice Hall.

$$= \$125,862$$

(B)  To compute the balance after 20 years, use

$PMT = \$908.99,\ i = 0.005803,\ n = 10(12) = 120$

$$\text{Balance after 20 years} = PMT\,\frac{1-(1+i)^{-120}}{i} = PMT\,\frac{1-(1.005083)^{-120}}{0.005083} = 908.99\,\frac{1-(1.005083)^{-120}}{0.005083}$$

$$= \$81,507$$

(C)  To compute the balance after 25 years, use

$PMT = \$908.99,\ i = 0.005803,\ n = 5(12) = 60$

$$\text{Balance after 25 years} = PMT\,\frac{1-(1+i)^{-60}}{i} = PMT\,\frac{1-(1.005083)^{-60}}{0.005083} = 908.99\,\frac{1-(1.005083)^{-60}}{0.005083}$$

$$= \$46,905$$

**43.**  (A)  $PV = \$129,000,\ i = \dfrac{0.072}{12} = 0.006,\ n = 20(12) = 240.$

Monthly payment:  $PMT = PV\,\dfrac{i}{1-(1+i)^{-n}} = 129,000\,\dfrac{0.006}{1-(1.006)^{-240}} = \$1,015.68$

Total amount paid in 240 payments:  $1,015.68(240) = \$243,763.34.$
Total interest paid:  $243,763.34 - 129,000 = \$114,763.34.$

(B)  New payment:  $PMT = 1,015.68 + 102.41 = \$1118.09$

$PV = \$129,000,\ i = \dfrac{0.072}{12} = 0.006$

We solve  $PMT = PV\,\dfrac{i}{1-(1+i)^{-n}}$  for $n$:

$$1,118.09 = 129,000\,\frac{0.006}{1-(1.006)^{-n}} = \frac{774}{1-(1.006)^{-n}}$$

$$1-(1.006)^{-n} = \frac{774}{1,118.09}$$

$$(1.006)^{-n} = 1 - \frac{774}{1,118.09}$$

$$-n\ln(1.006) = \ln\left(1 - \frac{774}{1,118.09}\right)$$

$$n = \frac{-\ln\left(1 - \dfrac{774}{1,118.09}\right)}{\ln(1.006)} \approx 197$$

The mortgage will be paid off in 197 months.

Total amount paid in 197 payments @ \$1,118.09:  $1,118.09\,(197) = \$220,263.73.$

Total interest paid:  $220,263.73 - 129,000 = \$91,263.73.$

Savings on interest:  $114,763 - 91,264 = \$23,499.$

Copyright © 2011 Pearson Education, Inc. Publishing as Prentice Hall.

**45.**   $PV = \$500,000, \ i = \dfrac{0.075}{12} = 0.00625$

$$500,000 = PMT \ \frac{1-(1.00625)^{-n}}{0.00625}$$

$$\frac{500,000(0.00625)}{PMT} = 1 - (1.00625)^{-n}$$

$$(1.00625)^{-n} = 1 - \frac{500,000(0.00625)}{PMT}$$

(A)   $PMT = \$5,000$

$$(1.00625)^{-n} = 1 - \frac{500,000(0.00625)}{5000} = 0.375$$

$$-n = \frac{\ln(0.375)}{\ln(1.00625)} = -157; \ \text{ and } \ n = 157$$

Thus, 157 withdrawals.

(B)   $PMT = \$4,000$

$$(1.00625)^{-n} = 1 - \frac{500,000(0.00625)}{4000} = 0.21875$$

$$-n = \frac{\ln(0.21875)}{\ln(1.00625)} = -243; \ \text{ and } \ n = 243$$

Thus, 243 withdrawals.

(C)   The interest per month on \$500,000 at 7.5% compounded monthly is greater than \$3,000. For example, the interest in the first month is

$$500,000\left(\frac{0.075}{12}\right) = \$3,125$$

Thus, the owner can withdraw \$3,000 per month forever.

**47.**   (A)   First, calculate the future value of the ordinary annuity:

$PMT = \$100, \ i = \dfrac{0.0744}{12} = 0.0062, \ n = 30(12) = 360$

$$FV = PMT \ \frac{(1+i)^n - 1}{i} = 100 \ \frac{(1.0062)^{360} - 1}{0.0062} = \$133,137$$

The interest earned during the 30 year period is:

$\quad 133,137 - 360(100) = 133,137 - 36,000 = \$97,137$

Next, using \$133,137 as the present value, determine the monthly payment:

$PV = \$133,137, \ i = 0.0062, \ n = 15(12) = 180$

$$PMT = PV \ \frac{i}{1-(1+i)^{-n}} = 133,137 \ \frac{0.0062}{1-(1.0062)^{-180}} = 1229.66$$

The monthly withdrawals are \$1229.66. Interest earned during the 15 year period is: $(1229.66)(180) - 133,137 = \$88,201.80$

Total interest: $97,137 + 88,201.80 = \$185,338.80$

(B)   First find the present value of the ordinary annuity:

$PMT = \$2,000, \ i = 0.0062, \ n = 180$

Copyright © 2011 Pearson Education, Inc.  Publishing as Prentice Hall.

$$PV = 2000 \, \frac{1-(1.0062)^{-180}}{0.0062} = \$216{,}542.54$$

Next, using \$216,542.54, as the future value of an ordinary annuity, calculate the monthly payment.

$$FV = \$216{,}542.54, \ i = 0.0062, \ n = 360$$

$$PMT = 216{,}542.54 \, \frac{0.0062}{(1.0062)^{360}-1} = 162.65$$

The monthly deposit is \$162.65.

**49.** $PV = (\$179{,}000)(0.80) = \$143{,}200, \ i = \dfrac{0.084}{12} = 0.007, \ n = 12(30) = 360.$

Monthly payment: $PMT = PV \dfrac{i}{1-(1+i)^{-n}} = 143{,}200\dfrac{0.007}{1-(1+0.007)^{-360}} = \dfrac{632}{1-(1.007)^{-360}} = \$1{,}090.95$

Next, we find the present value of a \$1,090.95 per month, 18-year annuity.

$PMT = \$1{,}090.95, \ i = 0.007, \ $ and $ \ n = 12(18) = 216.$

$$PV = PMT\frac{1-(1+i)^{-n}}{i} = 1{,}090.95\frac{1-(1.007)^{-216}}{0.007} = \$121{,}308.50$$

Finally,  equity = (current market value) − (unpaid loan balance)

$$= 215{,}000 - 121{,}308.50 = \$93{,}691.50$$

The couple can borrow (\$93,691.50)(0.70) = \$65,584.

**51.** Original mortgage: $145{,}000 - 0.20(145{,}000) = \$116{,}000.$

With $PV = \$116{,}000, \ i = \dfrac{0.079}{12} = 0.00658\overline{3}, \ $ and $ \ n = 30(12) = 360,$

the monthly payment is:

$$PMT = PV \frac{i}{1-(1+i)^{-n}} 116{,}000 \frac{0.00658\overline{3}}{1-(1.00685\overline{3})^{-360}} = \$843.10 \, .$$

To compute the unpaid balance after 10 years, use

$PMT = \$843.10, \ i = 0.00658\overline{3}, \ n = 20(12) = 240.$

Unpaid balance: $PMT\dfrac{1-(1+i)^{-240}}{i} = 843.10\dfrac{1-(1.006583)^{-240}}{0.00658\overline{3}} = \$101{,}550.51 \, .$

Now, refinancing this amount at the new interest rate for 20 years, we find the new monthly payment:

$$PV = \$101{,}550.51, \ i = \frac{0.055}{12} = 0.00458\overline{3}, \ n = 20(12) = 240$$

$$PMT = PV \frac{i}{1-(1+i)^{-n}} = 101{,}550.51\frac{0.00458\overline{3}}{1-(1.00458\overline{3})^{-240}} = \$698.55$$

If the original mortgage had been completed, the interest paid would have been

$843.10(360) - 116{,}000 = 303{,}516 - 116{,}000 = \$187{,}516.$

Copyright © 2011 Pearson Education, Inc.  Publishing as Prentice Hall.

The amount of the mortgage paid during the first 10 years is:
116,000 − 101,550.51=$14,449.49.

Thus, the interest paid during the first 10 years is:
843.10(120)−14,449.49 = 101,172 − 14,449.49 = $86,722.51.

The interest paid on the new 20-year mortgage will be
698.55(240) − 101,550.51 = 167,652 − 101,550.51 = $66,101.49.

Therefore, the total interest on the new mortgage will be
86,722.51 + 66,101.49 = $152,824.

The savings on interest is
187,516 − 152,824 = $34,692.

*Problems 53 thru 57 start from the equation*

$$(*)\quad \frac{1-(1+i)^{-n}}{i} - \frac{PV}{PMT} = 0$$

*A graphics calculator was used to solve these problems.*

53.  The graphs are decreasing, curve downward, and have $x$ intercept 30. The unpaid balances are always in the ratio 2:3:4. The monthly payments and total interest in each case are:

(use $PMT = PV\, \dfrac{i}{1-(1+i)^{-n}}$  where  $i = \dfrac{0.09}{12} = 0.0075$, $n = 360$)

$50,000 mortgage

$$PMT = 50,000\, \frac{0.0075}{1-(1.0075)^{-360}}$$

$$= 50,000(0.0080462)$$

$$= \$402.31 \text{ per month}$$

Total interest paid = 360(402.31) − 50,000 = $94,831.60

$75,000 mortgage

$$PMT = 75,000\, \frac{0.0075}{1-(1.0075)^{-360}}$$

$$= 75,000(0.0080462)$$

$$= \$603.47 \text{ per month}$$

Total interest paid = 360(603.47) − 75,000 = $142,249.20

$100,000 mortgage

$$PMT = 100,000\, \frac{0.0075}{1-(1.0075)^{-360}}$$

$$= 100,000(0.0080462)$$

$$= \$804.62 \text{ per month}$$

Total interest paid = 360(804.62) − 100,000 = $189,663.20

Copyright © 2011 Pearson Education, Inc. Publishing as Prentice Hall.

**55.**  $PV = 1000$, $PMT = 90$, $n = 12$, $i = \dfrac{r}{12}$ where $r$ is the annual nominal rate.

With these values, the equation (*) becomes

$$\frac{1-(1+i)^{-12}}{i} - \frac{1000}{90} = 0$$

or      $1 - (1 + i)^{-12} - 11.11i = 0$

Put  $y = 1 - (1 + i)^{-12} - 11.11i$  and use your calculator to find the zero $i$ of $y$, where $0 < i < 1$.  The result is $i \approx 0.01204$  and  $r = 12(0.01204) = 0.14448$.  Thus, $r = 14.45\%$ (two decimal places).

**57.**  $PV = 90{,}000$, $PMT = 1200$, $n = 12(10) = 120$, $i = \dfrac{r}{12}$ where $r$ is the annual nominal rate.  With these

values, the equation (*) becomes

$$\frac{1-(1+i)^{-120}}{i} - \frac{90{,}000}{1200} = 0$$

or      $1 - (1 + i)^{-120} - 75i = 0$

This equation can be written

$$(1 + i)^{120} - (1 - 75i)^{-1} = 0$$

Put  $y = (1 + x)^{120} - (1 - 75x)^{-1}$  and use your calculator to find the zero $x$ of $y$ where $0 < x < 1$.  The result is  $x \approx 0.00851$  and  $r = 12(0.00851) = 0.10212$.  Thus $r = 10.21\%$ (two decimal places).

## CHAPTER 3 REVIEW

**1.**  $A = 100\left(1 + 0.09 \cdot \dfrac{1}{2}\right)$

   $= 100(1.045) = \$104.50$    (3-1)

**2.**  $808 = P\left(1 + 0.12 \cdot \dfrac{1}{12}\right)$

   $P = \dfrac{808}{1.01} = \$800$    (3-1)

**3.**  $212 = 200(1 + 0.08 \cdot t)$

   $1 + 0.08t = \dfrac{212}{200}$

   $0.08t = \dfrac{212}{200} - 1 = \dfrac{12}{200}$

   $t = \dfrac{0.06}{0.08} = 0.75$ yr. or 9 mos.

   (3-1)

**4.**  $4120 = 4000\left(1 + r \cdot \dfrac{1}{2}\right)$

   $1 + \dfrac{r}{2} = \dfrac{4120}{4000}$

   $\dfrac{r}{2} = \dfrac{4120}{4000} - 1 = \dfrac{120}{4000} = 0.03$

   $r = 0.06$  or  6%    (3-1)

**5.**  $A = 1200(1 + 0.005)^{30}$

   $= 1200(1.005)^{30} = \$1393.68$

   (3-2)

**6.**  $P = \dfrac{A}{(1+i)^n} = \dfrac{5000}{(1+0.0075)^{60}} = \dfrac{5000}{(1.0075)^{60}}$

   $= \$3193.50$

   (3-2)

**7.**  $A = Pe^{rt}$; $P = 4{,}750$, $r = 6.8\% = 0.068$, $t = 3$

   $A = 4{,}750e^{0.068(3)} \approx 4{,}750e^{0.204} \approx \$5{,}824.92$    (3-2)

Copyright © 2011 Pearson Education, Inc. Publishing as Prentice Hall.

8.   $A = Pe^{rt}$;  $A = 36{,}000$,  $r = 9.3\% = 0.093$,  $t = 60$ months $= 5$ years

   $36{,}000 = Pe^{0.093(5)}$,   $P = \dfrac{36{,}000}{e^{0.465}} \approx \$22{,}612.86$   (3-2)

9.   $FV = 1000\ s_{\overline{60}|0.005}$

   $= 1000 \cdot 69.77003051$

   $= \$69{,}770.03$   (3-3)

10.   $PMT = \dfrac{FV}{s_{\overline{n}|i}} = \dfrac{8{,}000}{s_{\overline{48}|0.015}}$

   $= \dfrac{8{,}000}{69.56321929} = \$115.00$   (3-3)

11.   $PV = PMT\ a_{\overline{n}|i} = 2500\ a_{\overline{16}|0.02}$

   $= 2500 \cdot 13.57770931$

   $= \$33{,}944.27$   (3-4)

12.   $PMT = \dfrac{FV}{s_{\overline{n}|i}} = \dfrac{8{,}000}{s_{\overline{60}|0.0075}}$

   $= \dfrac{8{,}000}{48.17337352} = \$166.07$   (3-4)

13.   (A)   $2500 = 1000(1.06)^n$

   $(1.06)^n = 2.5$

   $\ln(1.06)^n = \ln 2.5$

   $n \ln 1.06 = \ln 2.5$

   $n = \dfrac{\ln 2.5}{\ln 1.06} = 15.73 \approx 16$

   (B)   We find the intersection of
   $Y_1 = 1000(1.06)\text{\textasciicircum}X$  and  $Y_2 = 2500$

   The graphs are shown in the figure at the right;
   intersection:  $x = 15.73$;  $y = 2500$

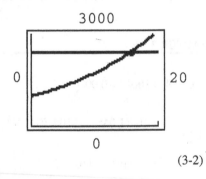

3000

0                    20

0

(3-2)

14.   (A)   $5000 = 100\,\dfrac{(1.01)^n - 1}{0.01}$

   $5000 = 10{,}000[(1.01)^n - 1]$

   $0.5 = (1.01)^n - 1$

   $(1.01)^n = 1.5$

   $n \ln(1.01) = \ln(1.5)$

   $n = \dfrac{ln(1.5)}{ln(1.01)} = 40.75 \approx 41$

   (B)   We find the intersection of
   $Y_1 = 100\,\dfrac{(1.01)^x - 1}{0.01} = 10{,}000[(1.01)\text{\textasciicircum}X - 1]$  and  $Y_2 = 5000$

Copyright © 2011 Pearson Education, Inc.  Publishing as Prentice Hall.

The graphs are shown in the figure at the right.
intersection: $x = 40.75$; $y = 5000$

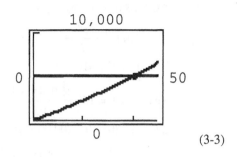

(3-3)

15. $P = \$3000$, $r = 0.14$, $t = \dfrac{10}{12}$

$A = 3000\left(1 + 0.14 \cdot \dfrac{10}{12}\right)$  [using $A = P(1 + rt)$]

$= \$3350$

Interest $= 3350 - 3000 = \$350$    (3-1)

16. $P = \$6,000$, $r = 7\% = 0.07$, $i = \dfrac{0.07}{12} = 0.00583$, $n = 17(12) = 204$

$A = P(1 + i)^n = 6000(1 + 0.00583)^{204} = 6000(1.00583)^{204} = \$19,654.42$  or  $\$19,654$  rounded to the nearest dollar.    (3-2)

17. $A = \$25,000$, $r = 6.6\% = 0.066$, $i = \dfrac{0.066}{12} = 0.0055$, $n = 10(12) = 120$

$P = \dfrac{A}{(1+i)^n} = \dfrac{25,000}{(1.0055)^{120}} = \$12,944.67$    (3-2)

18. (A)  $P = \$400$, $i = 0.054$

$A = P(1 + i)^n$

Year 1: $A = 400(1.054)^1 = \$421.60$
Interest: $\$21.60$

Year 2: $A = 400(1.054)^2 = \$444.37$
Interest: $444.37 - 421.60 = \$22.77$

Year 3: $A = 400(1.054)^3 = \$468.36$
Interest: $468.36 - 444.37 = \$23.99$

Year 4: $A = 400(1.054)^4 = \$493.65$
Interest: $493.95 - 468.36 = \$25.29$

| Period | Interest | Amount |
|--------|----------|----------|
| 0 |  | $400.00 |
| 1 | $21.60 | $421.60 |
| 2 | $22.77 | $444.37 |
| 3 | $24.00 | $468.36 |
| 4 | $25.29 | $493.65 |

Copyright © 2011 Pearson Education, Inc. Publishing as Prentice Hall.

(B)    $PMT = \$100$, $i = 0.054$, $n = 4$

$$FV = PMT\,\frac{(1+i)^n - 1}{i}$$

Year 1:  $FV = \$100$

Year 2:  $FV = 100\,\dfrac{(1.054)^2 - 1}{0.054} = \$205.40$

Interest:  $205.40 - 200.00 = \$5.40$

Year 3:  $FV = 100\,\dfrac{(1.054)^3 - 1}{0.054} = \$316.49$

Interest:  $316.49 - 205.40 = \$11.09$

Year 4:  $FV = 100\,\dfrac{(1.054)^4 - 1}{0.054} = \$433.58$

Interest:  $433.58 - 316.49 = \$17.09$

| Period | Interest | Payment | Balance |
|---|---|---|---|
| 1 |  | $100.00 | $100.00 |
| 2 | $5.40 | $100.00 | $205.40 |
| 3 | $11.09 | $100.00 | $316.49 |
| 4 | $17.09 | $100.00 | $433.58 |

(3-2, 3-3)

19.  The value of $1 at 13% simple interest after $t$ years is:
$$A_s = 1(1 + 0.13t) = 1 + 0.13t$$

The value of $1 at 9% interest compounded annually for $t$ years is:

$$A_c = 1(1 + 0.09)^t = (1.09)^t$$

Graph $Y_1 = 1 + 0.13x$, $Y_2 = (1.09)^x$

The graphs intersect at the point where $x \approx 9$. For investments lasting less than 9 years, simple interest at 13% is better; for investments lasting more than 9 years, interest

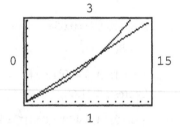

compounded annually at 9% is better.          (3-2)

20.  $P = \$10,000$, $r = 7\% = 0.07$, $m = 365$, $i = \dfrac{0.07}{365} = 0.0001918$, and $n = 40(365) = 14,600$

$A = P(1 + i)^n = 10,000(1 + 0.0001918)^{14,600}$

$\quad = 10,000(1.0001918)^{14,600} = \$164,402$     (3-2)

21.  $A = Pe^{rt}$;  $A = 40,000$, $P = 25,000$, $t = 6$

$40,000 = 25,000e^{6r}$

$\quad e^{6r} = \dfrac{40,000}{25,000} = \dfrac{8}{5}$

$\quad 6r = \ln(8/5)$,  $r = \dfrac{\ln(8/5)}{6} \approx 0.7833$ or $7.833\%$     (3-2)

Copyright © 2011 Pearson Education, Inc.  Publishing as Prentice Hall.

**22.** The effective rate for 9% compounded quarterly is:

$$\text{APY} = \left(1 + \frac{r}{m}\right)^m - 1, \ r = 0.09, \ m = 4$$

$$= \left(1 + \frac{0.09}{4}\right)^4 - 1 = (1.0225)^4 - 1 \approx 0.0931 \ \text{ or } \ 9.31\%$$

The effective rate for 9.25% compounded annually is 9.25%.

Thus, 9% compounded quarterly is the better investment.    (3-2)

**23.** $PMT = \$200, \ r = 7.2\% = 0.072, \ i = \dfrac{0.072}{12} = 0.006, \ n = 8(12) = 96$

$$FV = PMT \frac{(1+i)^n - 1}{i} = 200 \frac{(1.006)^{96} - 1}{0.006} = 200(129.308244) = \$25,861.65$$

The total amount invested is: $200(96) = \$19,200$.

Thus, the interest earned with this annuity is:

$I = 25,861.65 - 19,200 = \$6,661.65.$    (3-3)

**24.** $P = \$635, \ r = 22\% = 0.22, \ t = \dfrac{1}{12}, \ i = Prt = 635(0.22)\dfrac{1}{12} = \$11.64$    (3-1)

**25.** $A = P(1+i)^n; \ P = \$23,000, \ r = 5\% = 0.05, \ m = 1, \ i = \dfrac{r}{m} = 0.05, \ n = 5$

$$A = 23,000(1.05)^5 \approx \$29,354$$    (3-2)

**26.** $A = P(1+i)^n; \ A = \$23,000, \ r = 5\% = 0.05, \ m = 1, \ i = \dfrac{r}{m} = 0.05, \ n = 5$

$$23,000 = P(1.05)^5; \quad P = \frac{23,000}{(1.05)^5} = \$18,021$$    (3-2)

**27.** The interest paid was $\$2812.50 - \$2500 = \$312.50. \ P = \$2500,$

$$t = \frac{10}{12} = \frac{5}{6}$$

Solving $I = Prt$ for $r$, we have:

$$r = \frac{I}{Pt} = \frac{312.50}{2500\left(\frac{5}{6}\right)} = 0.15 \ \text{ or } \ 15\%$$    (3-1)

**28.** If you choose 0% financing, your monthly payment will be

$$PMT_1 = \frac{21,600}{48} = \$450$$

If you choose the $3000 rebate and borrow $18,600 at 4.8% compounded monthly for 48 months, your monthly payment will be:

$$PMT_2 = PV \frac{i}{1-(1+i)^{-n}}$$

$$= 18,600 \frac{0.004}{1-(1.004)^{-48}} \quad \left(i = \frac{0.048}{12} = 0.004\right)$$

$$= \$426.66$$

Copyright © 2011 Pearson Education, Inc. Publishing as Prentice Hall.

You should choose the $3000 rebate. You will save $23.34 per month or 48(23.34) = $1120.32 over the life of the loan.   (3-4)

**29.**   (A)   $r = 6.25\% = 0.0625$, $m = 12$,

$$\text{APY} = \left(1 + \frac{0.0625}{12}\right)^{12} - 1 = 0.06432 \text{ or } 6.432\%$$

(B)   $r = 0.0625$ compounded continuously

$$\text{APY} = e^{0.0625} - 1 \approx 0.06449 \text{ or } 6.449\%   \quad (3-2)$$

**30.**   $P = \$5,000$, $r = 9\% = 0.09$, $m = 4$, $i = \dfrac{0.09}{4} = 0.0225$, $A = \$6000$

$$A = P(1+i)^n$$
$$6000 = 5000(1.0225)^n$$
$$(1.0225)^n = \frac{6000}{5000} = \frac{6}{5}$$
$$n\ln(1.0225) = \ln(6/5)$$
$$n = \frac{\ln(6/5)}{\ln(1.0225)} = 8.19$$

Thus, it will take 9 quarters, or 2 years and 3 months.   (3-2)

**31.**   (A)   $r = 6\% = 0.06$, $m = 12$, $i = \dfrac{0.06}{12} = 0.005$

If we invest $P$ dollars, then we want to know how long it will take to have $2P$ dollars:

$$A = P(1+i)^n$$
$$2P = P(1.005)^n$$
$$(1.005)^n = 2$$
$$\ln(1.005)^n = \ln 2$$
$$n \ln(1.005) = \ln 2$$
$$n = \frac{\ln 2}{\ln(1.005)} \approx 138.98 \text{ or } 139 \text{ months}$$

Thus, it will take 139 months, or 11 years and 6 months, for an investment to double at 6% interest compounded monthly.

(B)   $r = 9\% = 0.09$, $m = 12$, $i = \dfrac{0.09}{12} = 0.0075$

$$2P = P(1.0075)^n$$
$$(1.0075)^n = 2$$
$$\ln(1.0075)^n = \ln 2$$
$$n = \frac{\ln 2}{\ln(1.0075)} \approx 92.77 \text{ or } 93 \text{ months}$$

Thus, it will take 93 months, or 7 years and 9 months, for an investment to double at 9% compounded monthly.   (3-2)

Copyright © 2011 Pearson Education, Inc. Publishing as Prentice Hall.

**32.** (A)  $PMT = \$2000,\ m = 1,\ r = i = 7\% = 0.07,\ n = 45$

$$FV = PMT\ \frac{(1+i)^n - 1}{i}$$

$$= 2000\ \frac{(1+0.07)^{45} - 1}{0.07} = 2000\ \frac{(1.07)^{45} - 1}{0.07} \approx \$571,499$$

(B)  $PMT = \$2000,\ m = 1,\ r = i = 11\% = 0.11,\ n = 45$

$$FV = PMT\ \frac{(1+i)^n - 1}{i}$$

$$= 2000\ \frac{(1+0.11)^{45} - 1}{0.11} = 2000\ \frac{(1.11)^{45} - 1}{0.11} \approx \$1,973,277 \quad (3\text{-}3)$$

**33.**  $A = \$17,388.17,\ P = \$12,903.28,\ m = 1,\ r = i,\ n = 3$

$$A = P(1 + i)^n$$

$$17,388.17 = 12,903.28(1 + i)^3$$

$$(1 + i)^3 = \frac{17,388.17}{12,903.28} \approx 1.3475775$$

$$3\ln(1 + i) = \ln(1.3475775)$$

$$\ln(1 + i) \approx \frac{0.2983085}{3} \approx 0.0994362$$

$$1 + i = e^{0.0994362} \approx 1.1045$$

$$i = 0.1045\ \text{or}\ 10.45\% \quad (3\text{-}2)$$

**34.** (A)  $P = \$400,\ t = \dfrac{15}{360},\ I = \$29.00$

$$I = Prt;\ r = \frac{I}{Pt} = \frac{29}{400\left(\frac{15}{360}\right)} = 1.74$$

The annual rate of interest is 174%.

(B)  $P = \$1,800,\ t = \dfrac{21}{360},\ I = \$69.00$

$$r = \frac{I}{Pt} = \frac{69}{1,800\left(\frac{21}{360}\right)} \approx 0.6571$$

The annual rate of interest is 65.71%.    (3-1)

**35.**  $FV = \$50,000,\ i = \dfrac{0.055}{12} = 0.004583\overline{3},\ n = 5(12) = 60$

$$PMT = FV\ \frac{i}{(1+i)^n - 1} = 50,000\ \frac{0.004583\overline{3}}{(1.004583\overline{3})^{60} - 1}$$

$$= 50,000(0.014517828) = \$725.89$$

The monthly deposit should be \$725.89.    (3-3)

Copyright © 2011 Pearson Education, Inc. Publishing as Prentice Hall.

**36.** **(A)**    The present value of an annuity which provides for quarterly withdrawals of $5,000 for 10 years at 7.32% interest compounded quarterly is given by:

$$PV = PMT \frac{1-(1+i)^{-n}}{i} \quad \text{with } PMT = \$5,000, \ i = \frac{0.0732}{4} = 0.0183, \text{ and } n = 10(4) = 40$$

$$= 5,000 \frac{1-(1.0183)^{-40}}{0.0183} = 5,000(28.18911469) = \$140,945.57$$

This amount will have to be in the account when he retires.

**(B)**    To determine the quarterly deposit to accumulate the amount in part (A), we use the formula:

$$PMT = FV \frac{i}{(1+i)^n - 1} \quad \text{where } FV = \$140,945.57, \ i = 0.0183 \text{ and } n = 4(20) = 80$$

$$= \$140,945.57 \frac{0.0183}{(1.0183)^{80} - 1} = 140,945.57(0.005602533)$$

$$= 789.65 \text{ (quarterly payment)}$$

**(C)**    The amount collected during the 10-year period is
($5000)40 = $200,000.
The amount deposited during the 20-year period is:
(789.65)80 ≈ $63,172.
Thus, the interest earned during the 30-year period is:
$200,000 – 63,172 = $136,828.    (3-3, 3-4)

**37.**    $PV = \$4,000, \ i = 0.009, \ n = 48$

$$PMT = PV \frac{i}{1-(1+i)^{-n}} = 4,000 \frac{0.009}{1-(1.009)^{-48}} = 4,000(0.025748504) = \$102.99$$

The monthly payment is $102.99.
The total amount paid is: $102.99(48) = $4,943.52.
Therefore, the interest paid is 4,943.52 – 4,000 = $943.52.    (3-4)

**38.**    $FV = \$50,000, \ r = 6.12\% = 0.0612, \ m = 12, \ i = \frac{0.0612}{12} = 0.0051, \ n = 12(6) = 72$

$$PMT = FV \frac{i}{(1+i)^n - 1} = 50,000 \frac{0.0051}{(1.0051)^{72} - 1} = \$576.48 \text{ per month.}    (3-3)$$

**39.**    To determine how long it will take money to double, we need to solve the equation $2P = P(1 + i)^n$ for $n$.
From this equation, we obtain:

$$(1 + i)^n = 2$$
$$\ln(1 + i)^n = \ln 2$$
$$n \ln(1 + i) = \ln 2$$
$$n = \frac{\ln 2}{\ln(1+i)}$$

Copyright © 2011 Pearson Education, Inc.  Publishing as Prentice Hall.

(A)  $i = \dfrac{0.075}{365} = 0.000205479$

Thus, $n = \dfrac{\ln 2}{\ln(1.000205479)} \approx 3373.67$ days or 9.24 years

(B)  $i = 0.075$

Thus, $n = \dfrac{\ln 2}{\ln(1.075)} \approx 9.58$ years or 10 years to the nearest year.    (3-2)

**40.**  First, we must calculate the future value of $8000 at 5.5% interest compounded monthly for 2.5 years.

$A = P(1 + i)^n$ where $P = \$8000$, $i = \dfrac{0.055}{12}$ and $n = 30$

$= 8000\left(1 + \dfrac{0.055}{12}\right)^{30} = \$9176.33$

Now, we calculate the monthly payment to amortize this debt at 5.5% interest compounded monthly over 5 years.

$PMT = PV\dfrac{i}{1 + (1 + i)^{-n}}$ where $PV = \$9176.33$, $i = \dfrac{0.055}{12} \approx 0.0045833$, and $n = 12(5) = 60$

$= 9176.33\dfrac{0.0045833}{1 - (1 + 0.0045833)^{-60}} = \dfrac{42.058179}{1 - (1.0045833)^{-60}} \approx \$175.28$

The total amount paid on the loan is:
$175.28(60) = \$10,516.80.$
Thus, the interest paid is:
$I = \$10,516.80 - \$8000 = \$2516.80.$    (3-4)

**41.**  $A = Pe^{rt}$; $P = 5,650$, $r = 8.65\% = 0.0865$, $t = 10$
$A = 5,650e^{0.0865(10)} = 5,650e^{0.865} \approx \$13,418.78$    (3-2)

**42.**  Use $FV = PMT\dfrac{(1 + i)^n - 1}{i}$ where $PMT = 1200$ and $i = \dfrac{0.06}{12} = 0.005$.
The graphs of

$Y_1 = 1200\dfrac{(1.005)^x - 1}{0.005} = 240,000[(1.005)^x - 1]$

$Y_2 = 100,000$

are shown in the figure at the right.

Intersection: $x \approx 70$; $y = 100,000$

The fund will be worth $100,000 after 70 payments, that is,
after 5 years, 10 months.    (3-3)

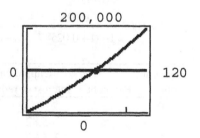

Copyright © 2011 Pearson Education, Inc.  Publishing as Prentice Hall.

**43.**    We first find the monthly payment:  $PV = \$50,000,\ i = \dfrac{0.09}{12} = 0.0075,\ n = 12(20) = 240$

$$PMT = PV\,\frac{i}{1-(1+i)^{-n}}$$

$$= 50,000\,\frac{0.0075}{1-(1.0075)^{-240}} = \$449.86 \text{ per month}$$

The present value of the \$449.86 per month, 20 year annuity at 9%, after $x$ years, is given by

$$y = 449.86\,\frac{1-(1.0075)^{-12(20-x)}}{0.0075}$$

$$= 59,981.33\left[1-(1.0075)^{-12(20-x)}\right]$$

The graphs of
$$Y_1 = 59,981.33\left[1-(1.0075)^{-12(20-x)}\right]$$
$$Y_2 = 10,000$$

are shown in the figure at the right.

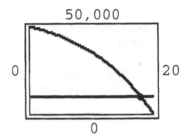

Intersection: $x \approx 18;\ y = 10,000$

The unpaid balance will be below \$10,000 after 18 years.    (3-4)

**44.**    $P = \$100,\ I = \$0.08,\ t = \dfrac{1}{360}$

$$r = \frac{I}{Pt} = \frac{0.08}{100\left(\frac{1}{360}\right)} = 0.288 \text{ or } 28.8\%.    (3\text{-}1)$$

**45.**    $PV = \$1000,\ i = 0.025,\ n = 4$
The quarterly payment is:

$$PMT = PV\,\frac{i}{1-(1+i)^{-n}}$$

$$= 1000\,\frac{0.025}{1-(1+0.025)^{-4}} = \frac{25}{1-(1.025)^{-4}} \approx \$265.82$$

| Payment number | Payment | Interest | Unpaid balance reduction | Unpaid balance |
|---|---|---|---|---|
| 0 | | | | $1000.00 |
| 1 | $265.82 | $25.00 | $240.82 | 759.18 |
| 2 | 265.82 | 18.98 | 246.84 | 512.34 |
| 3 | 265.82 | 12.81 | 253.01 | 259.33 |
| 4 | 265.81 | 6.48 | 259.33 | 0.00 |
| Totals | $1063.27 | $63.27 | $1000.00 | |

(3-4)

Copyright © 2011 Pearson Education, Inc.  Publishing as Prentice Hall.

**46.** $PMT = \$300$, $FV = \$9,000$, $i = \dfrac{0.0798}{12} = 0.00665$; $\quad FV = PMT\,\dfrac{(1+i)^n - 1}{i}$

$$9000 = 300\,\frac{(1.00665)^n - 1}{0.00665}$$

$$(1.00665)^n - 1 = \frac{9000(0.00665)}{300} = 0.1995$$

$$(1.00665)^n = 1.1995$$

$$n\ln(1.00665) = \ln(1.1995)$$

$$n = \frac{\ln(1.1995)}{\ln(1.00665)} \approx 27.44$$

It will take 28 months.      (3-3)

**47.** $FV = \$850,000$, $r = 8.76\% = 0.0876$, $m = 2$, $i = \dfrac{0.0876}{2} = 0.0438$, $n = 2(6) = 12$

$$PMT = FV\,\frac{i}{(1+i)^n - 1}$$

$$= 850,000\,\frac{0.0438}{(1+0.0438)^{12} - 1} = \frac{37,230}{(1.0438)^{12} - 1} \approx \$55,347.48$$

The total amount invested is:

$12(55,347.48) = \$664,169.76$.

Thus, the interest earned with this annuity is:

$I = \$850,000 - \$664,169.76 = \$185,830.24$.      (3-3)

**48.** $APY = 6.48\% = 0.0648$, $m = 12$;

$$APY = \left(1 + \frac{r}{m}\right)^m - 1$$

We solve $0.0648 = \left(1 + \dfrac{r}{12}\right)^{12} - 1$ for $r$:

$$\left(1 + \frac{r}{12}\right)^{12} = 1.0648$$

$$1 + \frac{r}{12} = \sqrt[12]{1.0648} = (1.0648)^{1/12} \approx 1.00525$$

$$r = 12(0.00525) \approx 0.0630 \text{ or } 6.30\%.      (3-2)$$

**49.** The interest earned is $I = \$5,000 - \$4,922.15 = \$77.85$.

So $P = \$4,922.15$, $t = \dfrac{13}{52} = 0.25$ and $I = Prt$. Solving for $r$, we have

$$r = \frac{I}{Pt} = \frac{77.85}{4,922.15(0.25)} \approx 0.0633 \text{ or } 6.33\%.      (3-1)$$

**50.** Using the sinking fund formula

$$PMT = FV\,\frac{i}{(1+i)^n - 1}$$

Copyright © 2011 Pearson Education, Inc.  Publishing as Prentice Hall.

with $PMT = \$200.00$, $FV = \$10,000$, and $i = \dfrac{0.0702}{12} = 0.00585$, we have

$$200 = 10,000 \frac{0.00585}{(1.00585)^n - 1} = \frac{58.5}{(1.00585)^n - 1}$$

Therefore,   $(1.00585)^n - 1 = \dfrac{58.5}{200} = 0.2925$

$$(1.00585)^n = 1.2925$$

$$\ln(1.00585)^n = \ln(1.2925)$$

$$n = \frac{\ln(1.2925)}{\ln(1.00585)} \approx 43.99$$

The couple will have to make 44 deposits.     (3-3)


**51.**   $PV = \$80,000$, $i = \dfrac{0.0942}{12} = 0.00785$, $n = 8(12) = 96$

(A)   $PMT = PV \dfrac{i}{1 - (1+i)^{-n}}$

$$= 80,000 \frac{0.00785}{1 - (1.00785)^{-96}} = 80,000(0.014869003) = \$1,189.52 \text{ monthly payment}$$

(B)   Now use $PMT = \$1,189.52$, $i = 0.00785$, and $n = 96 - 12 = 84$ to calculate the unpaid balance.

$$PV = PMT \frac{1 - (1+i)^{-n}}{i}$$

$$= 1,189.52 \frac{1 - (1.00785)^{-84}}{0.00785} = 1,189.52(61.33824469)$$

$$= \$72,963.07 \text{ unpaid balance after the first year}$$

(C)   Amount of loan paid during the first year:
$\$80,000 - 72,963.07 = \$7,036.93$.
Amount of payments during the first year:
$12(\$1,189.52) = \$14,274.24$.
Thus, the interest paid during the first year is:
$\$14,274.24 - 7,036.93 = \$7,237.31$.     (3-4)


**52.**   <u>Certificate of Deposit</u>: $\$10,000$ at $7\% = 0.07$ compounded monthly for $360 - 72 = 288$ months:
$P = \$10,000$, $i = \dfrac{0.07}{12} = 0.00583\overline{3}$, $n = 288$

$A = P(1 + i)^n$
$$= 10,000(1.00583\overline{3})^{288} = 10,000(5.339429847) = \$53,394.30$$

Copyright © 2011 Pearson Education, Inc.  Publishing as Prentice Hall.

Reduce the Principal:

Step 1: Find the monthly payment on the mortgage:

$PV = \$60,000$, $i = \dfrac{0.082}{12} = 0.006\overline{83}$, $n = 30(12) = 360$

$PMT = PV \dfrac{i}{1 - (1+i)^{-n}}$

$\phantom{PMT} = 60,000 \dfrac{0.006\overline{83}}{1 - (1.006\overline{83})^{-360}} = 60,000(0.007477544) = \$448.65$ per month

Step 2: Find the unpaid balance after 72 payments, that is, find the present value of a $448.65 per month annuity at 8.2% for 288 payments:

$PMT = \$448.65$, $i = \dfrac{0.082}{12} = 0.006\overline{83}$, $n = 288$

$PV = 448.65 \dfrac{1 - (1.006\overline{83})^{-288}}{0.006\overline{83}} = 448.65(125.7549535) = \$56,419.96$

Step 3: Reduce the principal by $10,000 and determine the time to pay off the loan, that is find out how long it will take to pay off $56,419.96 - 10,000 = \$46,419.96$ with monthly payments of $448.65:

$46,419.96 = 448.65 \dfrac{1 - (1.006\overline{83})^{-n}}{0.006\overline{83}}$

$1 - \left(1.006\overline{83}\right)^{-n} = \dfrac{46,419.96(0.006\overline{83})}{448.65}$

$1 - \left(1.006\overline{83}\right)^{-n} = 0.292983295$

$-n \ln(1.006\overline{83}) = \ln(0.292983295)$

$n = \dfrac{-\ln(0.292983295)}{\ln(1.006\overline{83})} \approx 180$

The loan will be paid off after 180 payments. Thus, by reducing the principal after 72 payments, the entire mortgage will be paid after $72 + 180 = 252$ payments.

Step 4: Calculate the future value of a $448.65 per month annuity at 7% for $360 - 252 = 108$ months.

$PMT = \$448.65$, $i = \dfrac{0.07}{12} = 0.005\overline{83}$, $n = 108$

$FV = PMT \dfrac{(1+i)^n - 1}{i} = 448.65 \dfrac{(1.005\overline{83})^{108} - 1}{0.005\overline{83}}$

$\phantom{FV} = 448.65(149.8589179) = \$67,234.20$

Conclusion: Use the $10,000 to reduce the principal and invest the monthly payment at 7% for 108 months. (3-2, 3-3, 3-4)

**53.**   We find the monthly payment and the total interest for each of the options. The monthly payment is given by:

$PMT = PV \dfrac{i}{1 - (1+i)^{-n}}$,   $PV = \$75.000$, $n = 12(30) = 360$

7.54% mortgage: $i = \dfrac{0.0754}{12} = 0.00628\overline{3}$

Copyright © 2011 Pearson Education, Inc.  Publishing as Prentice Hall.

$$PMT = 75{,}000 \ \frac{0.006283\overline{3}}{1-(1.006283\overline{3})^{-360}}$$

$$= 75{,}000(0.007019555) = \$526.47 \text{ per month}$$
Total interest paid $= 360(526.47) - 75{,}000 = \$114{,}529.20$

6.87% mortgage: $i = \dfrac{0.0687}{12} = 0.005725$

$$PMT = 75{,}000 \ \frac{0.005725}{1-(1.005725)^{-360}}$$

$$= 75{,}000(0.006565947) = \$492.45 \text{ per month}$$
Total interest paid $= 360(492.45) - 75{,}000 = \$102{,}282.00$
The lower rate would save over \$12,247.20 in interest.      (3-4)

**54.**   $A = \$5000, \ r = i = 5.6\% = 0.056, \ n = 5$

$$P = \frac{A}{(1+i)^n} = \frac{5000}{(1.056)^5} = \$3{,}807.59. \qquad (3\text{-}2)$$

**55.**   $P = \$5{,}695, \ A = \$10{,}000, \ n = 10, \ m = 1, \ r = i$

$$A = P(1+i)^n$$

$$10{,}000 = 5{,}695(1+i)^{10}$$

$$(1+i)^{10} = \frac{10{,}000}{5{,}695} = 1.755926251$$

$$10\ln(1+i) = \ln(1.755926251) = 0.0562996496$$

$$\ln(1+i) = 0.0562996496$$
$$1 + i = e^{0.0562996496} \approx 1.057914639$$
$$I \approx 0.0579 \text{ or } 5.79\% \qquad (3\text{-}2)$$

**56.**   $A = \$5000, \ r = 6.4\% = 0.064, \ t = \dfrac{26}{52} = 0.5$

$$P = \frac{A}{1+rt} = \frac{5000}{1+(0.064)(0.5)} = \frac{5{,}000}{1{,}032} = \$4{,}844.96 \qquad (3\text{-}1)$$

**57.**   We first compute the monthly payment using $PV = \$10{,}000, \ i = \dfrac{0.12}{12} = 0.01, \ \text{and} \ n = 5(12) = 60.$

$$PMT = PV \ \frac{i}{1-(1+i)^{-n}}$$

$$= 10{,}000 \ \frac{0.01}{1-(1+0.01)^{-60}} = \frac{100}{1-(1.01)^{-60}} = \$222.44 \text{ per month}$$

Now, we calculate the unpaid balance after 24 payments by using
$PMT = \$222.44, \ i = 0.01, \ \text{and} \ n = 60 - 24 = 36.$

$$PV \ = PMT \ \frac{1-(1+i)^{-n}}{i}$$

Copyright © 2011 Pearson Education, Inc.  Publishing as Prentice Hall.

$$= 222.44 \frac{1-(1+0.01)^{-36}}{0.01} = 22{,}244[1-(1.01)^{-36}] = \$6697.11$$

Thus, the unpaid balance after 2 years is $6697.11.        (3-4)

58.  First find the annual percentage yield for 7.28% compounded quarterly:
$r = 7.28\% = 0.0728$, $m = 4$

$$APY = \left(1+\frac{r}{m}\right)^m - 1 = \left(1+\frac{0.0728}{4}\right)^4 - 1 = (1.0182)^4 - 1$$
$$= 0.074812 \text{ or } 7.4812\%$$

Now find the rate $r$ compounded monthly that has the APY of 7.4812%:
$APY = 0.074812$, $m = 12$

$$0.074812 = \left(1+\frac{r}{12}\right)^{12} - 1$$

$$\left(1+\frac{r}{12}\right)^{12} = 1.074812$$

$$1 + \frac{r}{12} = \sqrt[12]{1.074812} = (1.074812)^{1/12} \approx 1.00603$$

$$r = 12(0.00603) \approx 0.0724; \quad r = 7.24\% \qquad (3-2)$$

59.  (A)  We first calculate the future value of an annuity of $2000 at 8% compounded annually for 9 years.

$$FV = PMT\frac{(1+i)^n - 1}{i} \text{ where } PMT = \$2000, \ i = 0.08, \text{ and } n = 9$$

$$= 2000\frac{(1+0.08)^9 - 1}{0.08} = 25{,}000[(1.08)^9 - 1] \approx \$24{,}975.12$$

Now, we calculate the future value of this amount at 8% compounded annually for 36 years.
$A = P(1+i)^n$, where $P = \$24{,}975.12$, $i = 0.08$, and $n = 36$
$$= 24{,}975.12(1+0.08)^{36} = 24{,}975.12(1.08)^{36} \approx \$398{,}807$$

(B)  This is the future value of a $2000 annuity at 8% compounded annually for 35 years.

$$FV = PMT\frac{(1+i)^n - 1}{i} \text{ where } PMT = \$2000, \ i = 0.08, \text{ and } n = 36$$

$$= 2000\frac{(1+0.08)^{36} - 1}{0.08} = 25{,}000[(1.08)^{36} - 1] \approx \$374{,}204 \qquad (3-3)$$

60.  $A = Pe^{rt}$; $A = 27{,}000$, $r = 5.5\% = 0.055$, $t = 10$
$$27{,}000 = Pe^{0.055(10)} = Pe^{0.550}$$

$$P = \frac{27{,}000}{e^{0.550}} \approx \$15{,}577.64 \qquad (3-2)$$

Copyright © 2011 Pearson Education, Inc.  Publishing as Prentice Hall.

**61.** The amount of the loan is 0.8(100,00) = $80,000, and

$$PMT = PV \frac{i}{1-(1+i)^{-n}} .$$

(A)  First let $i = \dfrac{0.0768}{12} = 0.0064$, $n = 30(12) = 360$. Then

$$PMT = 80,000 \frac{0.0064}{1-(1.0064)^{-360}} = 80,000(0.00711581) = 569.26$$

The monthly payment on the 30 year mortgage is: $569.26.

Next, let $i = \dfrac{0.0768}{12} = 0.0064$, $n = 15(12) = 180$. Then

$$PMT = 80,000 \frac{0.0064}{1-(1.0064)^{-180}} = 80,000(0.00937271) = 749.82$$

The monthly payment on the 15 year mortgage is: $749.82.

(B)  To find the unpaid balance after 10 years, we use

$$PV = PMT \frac{1-(1+i)^{-n}}{i} .$$

For the 30 year mortgage, there are 20 years remaining at $569.26 per month.

$PMT = \$569.26$, $i = 0.0064$, $n = 20(12) = 240$

$$PV = 569.26 \frac{1-(1.0064)^{-240}}{0.0064} = 569.26(122.4536903) = 69,707.99$$

The unpaid balance for the 30 year mortgage is $69,707.99.

For the 15 year mortgage, there are 5 years remaining at $749.82 per month:

$PMT = \$749.82$, $i = 0.0064$, $n = 5(12) = 60$

$$PV = 749.82 \frac{1-(1.0064)^{-60}}{0.0064} = 749.82(49.69291289) = 37,260.74$$

The unpaid balance for the 15 year mortgage is: $37,260.74.    (3-4)

**62.** The amount of the mortgage is: 0.8(83,000) = $66,400.
The monthly payment is given by:

$$PMT = PV \frac{i}{1-(1+i)^{-n}} \quad \text{where } PV = \$66,400, \ i = \frac{0.084}{12} = 0.007, \ n = 30(12) = 360$$

$$PMT = 66,400 \frac{0.007}{1-(1.007)^{-360}} = 66,400(0.007618376) = 505.86$$

The monthly payment is: $505.86.

Next, we find the present value of a $505.86 per month,
22 year annuity at 8.4%.

$$PMT = \$505.86, \ i = \frac{0.084}{12} = 0.007, \ \text{and} \ n = 22(12) = 264$$

$$PV = PMT \frac{1-(1+i)^{-n}}{i} = 505.86 \frac{1-(1.007)^{-264}}{0.007}$$

$$= 505.86(120.204351) = 60,806.57$$

Copyright © 2011 Pearson Education, Inc.  Publishing as Prentice Hall.

The unpaid loan balance is $60,806.57.

Finally, equity = (current market value) – (unpaid loan balance)
$$= 95,000 - 60,806.57 = \$34,193.43.$$

The family can borrow up to 0.6(34,193.43) = $20,516.    (3-4)

**63.**  $PV = \$600$, $PMT = 110$, $n = 6$

Solve $600 = 110\dfrac{1-(1+i)^{-6}}{i}$ for $i$ and multiply the result

by 12 to find $r$.

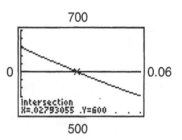

$i \approx 0.02793055$
$r \approx 0.33517$ or $r = 33.52\%$
(3-4)

**64.  (A)**    $FV = \$220,000$, $PMT = \$2,000$, $n = 25$.

Solve $220,000 = 2,000\dfrac{(1+i)^{25}-1}{i}$ for $i$:

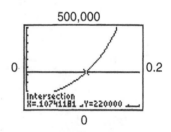

$i \approx 0.10741$ or $i = 10.74\%$

**(B)**    Withdrawals at $30,000 per year:

Solve $220,000 = 30,000\dfrac{1-(1.10741)^{-n}}{0.10741}$ for $n$

$n \approx 15$ years

Withdrawals $24,000 per year:

Solve $220,000 = 24,000\dfrac{1-(1.10741)^{-n}}{0.10741}$

$n \approx 41$ years    (3-3, 3-4)

Copyright © 2011 Pearson Education, Inc. Publishing as Prentice Hall.

**4    SYSTEMS OF LINEAR EQUATIONS; MATRICES**

*Things to remember:*

<u>1.</u>    SYSTEMS OF TWO EQUATIONS IN TWO VARIABLES

Given the LINEAR SYSTEM

$$ax + by = h$$
$$cx + dy = k$$

where $a$, $b$, $c$, $d$, $h$, and $k$ are real constants, a pair of numbers $x = x_0$ and $y = y_0$ [also written as an ordered pair $(x_0, y_0)$] is a SOLUTION to this system if each equation is satisfied by the pair. The set of all such ordered pairs is called the SOLUTION SET for the system. To SOLVE a system is to find its solution set.

<u>2.</u>    SYSTEMS OF LINEAR EQUATIONS: BASIC TERMS

A system of linear equations is CONSISTENT if it has one or more solutions and INCONSISTENT if no solutions exist. Furthermore, a consistent system is said to be INDEPENDENT if it has exactly one solution (often referred to as the UNIQUE SOLUTION) and DEPENDENT if it has more than one solution. Two systems of equations are EQUIVALENT if they have the same solution set.

<u>3.</u>    The system of two linear equations in two variables

$$ax + by = h$$
$$cx + dy = k$$

can be solved by:
(a)    graphing;
(b)    substitution;
(c)    elimination by addition.

<u>4.</u>    POSSIBLE SOLUTIONS TO A LINEAR SYSTEM

The linear system

$$ax + by = h$$
$$cx + dy = k$$

must have:

(a)    exactly one solution (consistent and independent); or
(b)    no solution (inconsistent); or
(c)    infinitely many solutions (consistent and dependent).

Copyright © 2011 Pearson Education, Inc. Publishing as Prentice Hall.

5.   OPERATIONS THAT PRODUCE EQUIVALENT SYSTEMS
A system of linear equations is transformed into an equivalent system if:

(a)   two equations are interchanged;

(b)   an equation is multiplied by a nonzero constant;

(c)   a constant multiple of one equation is added to another equation.

---

1.   (B); no solution

3.   (A); $x = -3$, $y = 1$

5.   $3x - y = 2$
   $x + 2y = 10$
   Point of intersection: $(2, 4)$
   Solution: $x = 2$; $y = 4$

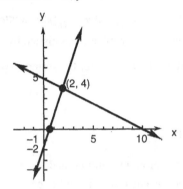

7.   $m + 2n = 4$
   $2m + 4n = -8$
   Since the graphs of the given equations are parallel lines, there is no solution.

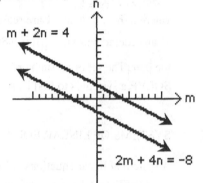

9.       $y = 2x - 3$   (1)
   $x + 2y = 14$       (2)

By substituting $y$ from (1) into (2), we get:

$x + 2(2x - 3) = 14$
   $x + 4x - 6 = 14$
       $5x = 20$
        $x = 4$

Now, substituting $x = 4$ into (1), we have:

$y = 2(4) - 3$
$y = 5$

Solution: $x = 4$, $y = 5$

11.   $2x + y = 6$       (1)
    $x - y = -3$       (2)

Solve (2) for $y$ to obtain the system:

$2x + y = 6$       (3)
    $y = x + 3$   (4)

Substitute $y$ from (4) into (3):

$2x + x + 3 = 6$
       $3x = 3$
        $x = 1$

Now, substituting $x = 1$ into (4), we get:

$y = 1 + 3$
$y = 4$

Solution: $x = 1$, $y = 4$

Copyright © 2011 Pearson Education, Inc. Publishing as Prentice Hall.

**13.**   $3u - 2v = 12$   (1)
$7u + 2v = 8$   (2)

Add (1) and (2):

$10u = 20$
$u = 2$

Substituting $u = 2$ into (2),
we get:

$7(2) + 2v = 8$
$2v = -6$
$v = -3$

Solution: $u = 2,\ v = -3$

**15.**   $2m - n = 10$   (1)
$m - 2n = -4$   (2)

Multiply (1) by $-2$ and add to (2) to obtain:

$-3m = -24$
$m = 8$

Substituting $m = 8$ into (2),
we get:

$8 - 2n = -4$
$-2n = -12$
$n = 6$

Solution: $m = 8,\ n = 6$

**17.**   $9x - 3y = 24$   (1)
$11x + 2y = 1$   (2)

Solve (1) for $y$ to obtain:

$y = 3x - 8$   (3)

and substitute into (2):

$11x + 2(3x - 8) = 1$
$11x + 6x - 16 = 1$
$17x = 17$
$x = 1$

Now, substitute $x = 1$ into (3):

$y = 3(1) - 8$
$y = -5$

Solution: $x = 1,\ y = -5$

**19.**   $2x - 3y = -2$   (1)
$-4x + 6y = 7$   (2)

Multiply (1) by 2 and add to (2) to get:

$0 = 3$

This implies that the system is inconsistent,
and thus there is no solution.

**21.**   $3x + 8y = 4$   (1)
$15x + 10y = -10$   (2)

Multiply (1) by $-5$ and add to (2) to get:

$-30y = -30$
$y = 1$

Substituting $y = 1$ into (1),
we get:

$3x + 8(1) = 4$
$3x = -4$
$x = -\dfrac{4}{3}$

Solution: $x = -\dfrac{4}{3},\ y = 1$

**23.**   $-6x + 10y = -30$   (1)
$3x - 5y = 15$   (2)

Multiply (2) by 2 and add to (1).  This yields:

$0 = 0$

which implies that (1) and (2) are equivalent
equations and there are infinitely many
solutions.  Geometrically, the two lines are
coincident.  The system is dependent.

Copyright © 2011 Pearson Education, Inc.  Publishing as Prentice Hall.

**25.**        $x + y = 1$    (1)

$0.3x - 0.4y = 0$    (2)

Multiply equation (2) by 10 to remove the decimals

$x + y = 1$    (1)

$3x - 4y = 0$    (3)

Multiply (1) by 4 and add to (2) to get

$7x = 4$

$x = \dfrac{4}{7}$

Now substitute $x = \dfrac{4}{7}$ in (1):

$\dfrac{4}{7} + y = 1$

$y = 1 - \dfrac{4}{7} = \dfrac{3}{7}$

Solution: $x = \dfrac{4}{7}$, $y = \dfrac{3}{7}$

**27.**  $x + 0y = 7$    implies  $x = 7$

$0x + y = 3$    implies  $y = 3$

**29.**  $5x + 0y = 4$    implies  $x = 4/5$

$0x + 3y = -2$  implies  $y = -2/3$

**31.**  $x + y = 0$    (1)

$x - y = 0$    (2)

Adding (1) and (2) gives  $2x = 0$,  so  $x = 0$,  which implies  $y = 0$.

**33.**  $x - 2y = 4$    (1)

$0x + y = 5$    implies  $y = 5$

Substituting  $y = 5$  into (1) gives  $x - 2(5) = 4$,  which implies  $x = 14$.

**35.**  The price will tend to go down.

**37.**  If $m \neq n$, then the system has a unique solution independent of the values of $b$ and $c$.

**39.**  If the system has infinitely many solutions, then there is a non-zero number $k$ such that $n = km$ and $c = kb$.

**41.**  Solution: $x = -14$, $y = -37$; the lines intersect at the point $(-14, -37)$.

**43.**  No solution; the lines are parallel.

**45.**  First solve each equation for $y$:

$y = \dfrac{3}{2}x - \dfrac{15}{2}$

$y = -\dfrac{4}{3}x + \dfrac{13}{3}$

The graphs of the equations are shown at the right.

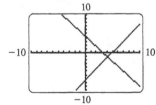

intersection: $x = 4.176$,  $y = -1.235$

$(4.176, -1.235)$

Copyright © 2011 Pearson Education, Inc.  Publishing as Prentice Hall.

**47.** Multiply each equation by 10 and then solve for $y$:

$$y = \frac{24}{35}x + \frac{1}{35}$$

$$y = \frac{17}{26}x - \frac{1}{13}$$

The graphs of these equations are almost indistinguishable.

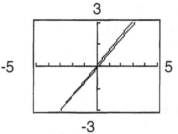

intersection: $x = -3.310$,  $y = -2.241$

$(-3.310, -2.241)$

**49.** $x - 2y = -6$ $\qquad (L_1)$

$2x + y = 8$ $\qquad (L_2)$

$x + 2y = -2$ $\qquad (L_3)$

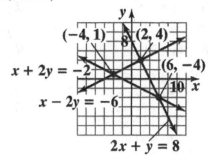

(A)   $L_1$ and $L_2$ intersect:

$x - 2y = -6$ $\qquad$ (1)

$2x + y = 8$ $\qquad$ (2)

Multiply (2) by 2 and add to (1):

$5x = 10$

$x = 2$

Substitute $x = 2$ in (1) to get

$2 - 2y = -6$

$\quad -2y = -8$

$\qquad y = 4$

Solution: $x = 2$,  $y = 4$

(B)   $L_1$ and $L_3$ intersect:

$x - 2y = -6$ $\qquad$ (3)

$x + 2y = -2$ $\qquad$ (4)

Add (3) and (4):

$2x = -8$

$x = -4$

Substitute $x = -4$ in (3)     to get

$-4 - 2y = -6$

$\quad\ -2y = -2$

$\qquad\ \ y = 1$

Solution: $x = -4$,  $y = 1$

(C)   $L_2$ and $L_3$ intersect:

$2x + y = 8$ $\qquad$ (5)

$x + 2y = -2$ $\qquad$ (6)

Multiply (6) by $-2$ and add   to (5)

$-3y = 12$

$\ y = -4$

Substitute $y = -4$ in (5) to get

$2x - 4 = 8$

$\ 2x = 12$

$\quad x = 6$

Solution: $x = 6$,  $y = -4$

**51.**  $x + y = 1$ $\qquad\qquad (L_1)$

$x - 2y = -8$ $\qquad\quad (L_2)$

$3x + y = -3$ $\qquad\quad (L_3)$

(A)   $L_1$ and $L_2$ intersect

$x + y = 1$ $\qquad$ (1)

$x - 2y = -8$ $\qquad$ (2)

Subtract (2) from (1):

$3y = 9$

$\ y = 3$

Substitute $y = 3$ in (1) to get  $x + 3 = 1$,   $x = -2$

Solution: $x = -2$,  $y = 3$

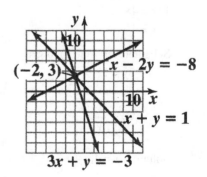

Copyright © 2011 Pearson Education, Inc. Publishing as Prentice Hall.

(B)    $L_1$ and $L_3$ intersect:

$x + y = 1$          (3)
$3x + y = -3$          (4)

Subtract (4) from (3):

$-2x = 4$
$x = -2$

Substitute $x = -2$ in (3) to get

$-2 + y = 1$
$y = 3$

Solution: $x = -2$, $y = 3$

(C)    It follows from (A) and (B), that $L_2$ and $L_3$ intersect at

$x = -2$, $y = 3$.

**53.**    $4x - 3y = -24$          $(L_1)$

$2x + 3y = 12$          $(L_2)$

$8x - 6y = 24$          $(L_3)$

(A)    $L_1$ and $L_2$ intersect:

$4x - 3y = -24$          (1)
$2x + 3y = 12$          (2)

Add (1) and (2):

$6x = -12$
$x = -2$

Substitute $x = -2$ in (2) to get

$2(-2) + 3y = 12$
$3y = 16$
$y = \dfrac{16}{3}$

Solution: $x = -2$, $y = \dfrac{16}{3}$

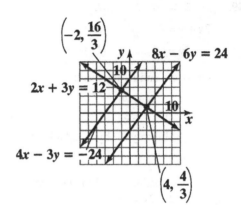

(B)    In slope-intercept form, $(L_1)$ and $(L_3)$ have equations

$y = \dfrac{4}{3}x + 8$          $(L_1)$

$y = \dfrac{4}{3}x - 4$          $(L_2)$

Thus, $(L_1)$ and $(L_3)$ have the same slope and different

$y$-intercepts; $(L_1)$ and $(L_3)$ are parallel; they do not intersect.

(C)    $L_2$ and $L_3$ intersect:

$2x + 3y = 12$          (3)
$8x - 6y = 24$          (4)

Multiply (3) by 2 and add to (4):

$12x = 48$
$x = 4$

Substitute $x = 4$ in (3) to get

$2(4) + 3y = 12$
$3y = 4$
$y = \dfrac{4}{3}$

Solution: $x = 4$, $y = \dfrac{4}{3}$

Copyright © 2011 Pearson Education, Inc. Publishing as Prentice Hall.

**55.** (A)    $5x + 4y = 4$
$11x + 9y = 4$

Multiply the top equation by 9 and the bottom equation by −4.

$45x + 36y = 36$
$\underline{-44x - 36y = -16}$
$x = 20$

Add the equations.

$5(20) + 4y = 4$
$4y = -96$
$y = -24$

Substitute $x = 20$ in the first equation.

Solution: (20, −24)

(B)    $5x + 4y = 4$
$11x + 8y = 4$

Multiply the top equation by 8 and the bottom equation by −4.

$40x + 32y = 32$
$\underline{-44x - 32y = -16}$
$-4x = 16$
$x = -4$

Add the equations.

$5(-4) + 4y = 4$
$4y = 24$
$y = 6$

Substitute $x = 4$ in the first equation.

Solution: (−4, 6)

(C)    $5x + 4y = 4$
$10x + 8y = 4$

Multiply the top equation by 8 and the bottom equation by −4.

$40x + 32y = 32$
$\underline{-40x - 32y = -16}$
$0 = -16$

Add the equations.

This system has no solutions.

**57.** $p = 0.7q + 3$ Supply equation
$p = -1.7q + 15$ Demand equation

(A)   $p = \$4$

Supply: $4 = 0.7q + 3$
$0.7q = 1$
$q = \dfrac{1}{0.7} \approx 1.429$

Thus, the supply will be 143 T-shirts.

Demand: $4 = -1.7q + 15$
$1.7q = 11$
$q = \dfrac{11}{1.7} \approx 6.471$

Copyright © 2011 Pearson Education, Inc. Publishing as Prentice Hall.

Thus, the demand will be 647 T-shirts. At this price level, the demand exceeds the supply; the price will rise.

(B)   $p = \$9$

Supply: $9 = 0.7q + 3$

$\quad\quad 0.7q = 6$

$\quad\quad\quad q = \dfrac{6}{0.7} \approx 8.571$

Thus, the supply will be 857 T-shirts.

Demand:   $9 = -1.7q + 15$

$\quad\quad\quad 1.7q = 6$

$\quad\quad\quad\quad q = \dfrac{6}{1.7} \approx 3.529$

Thus, the demand will be 353 T-shirts. At this price level, the supply exceeds the demand; the price will fall.

(C)   Solve the pair of equations to find the equilibrium price and the equilibrium quantity.

$\quad\quad 0.7q + 3 = -1.7q + 15$

$\quad\quad\quad\quad 2.4q = 12$

$\quad\quad\quad\quad\quad q = \dfrac{12}{2.4} = 5$

The equilibrium quantity is 500 T-shirts. Substitute $q = 5$ in either of the two equations to find $p$.

$\quad\quad p = (0.7)5 + 3 = 3.5 + 3$

$\quad\quad p = 6.50$

The equilibrium price is $6.50.

(D)

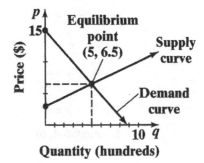

59.   (A) Supply equation: $p = aq + b$

At $p = 4.80$, $q = 1.9 : 4.80 = 1.9a + b$

At $p = 5.10$, $q = 2.1 : 5.10 = 2.1a + b$

Solve

$\quad 1.9a + b = 4.80$   (1)

$\quad 2.1a + b = 5.10$   (2)

Subtract equation (1) from equation (2):

$\quad 0.2a = 0.3$

$\quad\quad a = \dfrac{3}{2} = 1.5$

Substitute $a = 1.5$ into equation (1)  (or equation (2)) to find $b$:

$1.9(1.5) + b = 4.80$

$\quad\quad\quad b = 4.80 - 2.85 = 1.95$

The supply equation is:

$\quad p = 1.5q + 1.95$

(B)   Demand equation: $p = aq + b$

At 4.80,  $q = 2.0$:  $4.80 = 2.0a + b$

Copyright © 2011 Pearson Education, Inc. Publishing as Prentice Hall.

At 5.10, $q = 1.8$: $5.10 = 1.8a + b$

Solve    $2.0a + b = 4.80$  (1)

$1.8a + b = 5.10$  (2)

Subtract equation (1) from equation (2):

$-0.2a = 0.3$

$$a = -\frac{3}{2} = -1.5$$

Substitute $a = -1.5$ into equation (1) (or into equation (2)) to find $b$:

$2.0(-1.5) + b = 4.80$

$b = 4.80 + 2.0(1.5) = 7.80$

The demand equation is:

$p = -1.5q + 7.80$

(C)  Equilibrium price and quantity

Solve  $1.5q + 1.95 = -1.5q + 7.80$  for $q$:

$3q = 7.80 - 1.95 = 5.85$

$q = 1.95$

Substitute $q = 1.95$ into either the supply equation or the demand equation:

$p = 1.5(1.95) + 1.95 = 4.875$

The equilibrium price is \$4.875 or \$4.88; the equilibrium quantity is 1.95 billion bushels.

(D)

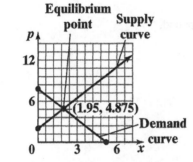

61.  (A)  The company breaks even when:

Cost = Revenue

$48,000 + 1400x = 1800x$

$48,000 = 1800x - 1400x$

or        $400x = 48,000$

$$x = \frac{48,000}{400}$$

$x = 120$

Thus, 120 units must be manufactured and sold to break even.

Cost $= 48,000 + 1400(120)$

$= \$216,000 =$ Revenue

(B)

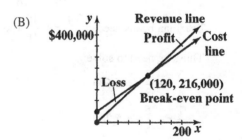

63.  Let $x =$ number of CD's marketed per month

(A)  Revenue: $R = 19.95x$

Cost:  $C = 7.45x + 24,000$

(B)  At the break-even point Revenue = Cost, that is

$19.95x = 7.45x + 24,000$

$12.50x = 24,000$

$x = 1920$

Copyright © 2011 Pearson Education, Inc. Publishing as Prentice Hall.

Thus, 1920 CD's must be sold per month to break even.

Cost = Revenue = \$38,304 at the break-even point.

(C)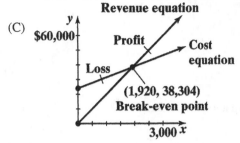

**65.** Let $x$ = base price

$y$ = surcharge

5 pound package: $x + 4y = 27.75$

20 pound package: $x + 19y = 64.50$

Solve the two equations:

$$x + 4y = 27.75 \qquad (1)$$
$$x + 19y = 64.50 \qquad (2)$$

Multiply (1) by $(-1)$ and add to (2):

$$15y = 36.75$$
$$y = 2.45$$

Substitute $y = 2.45$ into (1):

$$x + 4(2.45) = 27.75$$
$$x = 27.75 - 9.80 = 17.95$$

Thus, the base price is \$17.95; the surcharge is \$2.45 per pound.

**67.** Let $x$ = number of pounds of robust blend

$y$ = number of pounds of mild blend

Total amount of Columbian beans: $132 \times 50 = 6,600$ lbs.

Total amount of Brazilian beans: $132 \times 40 = 5,280$ lbs.

Columbian beans needed: $\dfrac{12}{16}x + \dfrac{6}{16}y$ or $\dfrac{3}{4}x + \dfrac{3}{8}y$

Brazilian beans needed: $\dfrac{4}{16}x + \dfrac{10}{16}y$ or $\dfrac{1}{4}x + \dfrac{5}{8}y$

Thus, we need to solve:

$$\frac{3}{4}x + \frac{3}{8}y = 6,600 \qquad (1)$$

$$\frac{1}{4}x + \frac{5}{8}y = 5,280 \qquad (2)$$

Multiply (2) by $-3$ and add to (1):

$$-\frac{12}{8}y = -9,240$$

$$y = 6,160$$

Substitute $y = 6,160$ into (1):

$$\frac{3}{4}x + \frac{3}{8}(6,160) = 6,600$$

$$\frac{3}{4}x = 6,600 - 2,310 = 4,290$$

$$x = 5,720$$

Copyright © 2011 Pearson Education, Inc. Publishing as Prentice Hall.

Therefore, the manufacturer should produce 5,720 pounds of the robust blend and 6,160 pounds of the mild blend

**69.**     Let $x$ = amount of mix A, and

$y$ = amount of mix B.

We want to solve the following system of equations:
$$0.1x + 0.2y = 20 \qquad (1)$$
$$0.06x + 0.02y = 6 \qquad (2)$$

Clear the decimals from (1) and (2) by multiplying both sides of (1) by 10 and both sides of (2) by 100.
$$x + 2y = 200 \qquad (3)$$
$$6x + 2y = 600 \qquad (4)$$
Multiply (3) by $-1$ and add to (4):
$$5x = 400$$
$$x = 80$$
Now substitute $x = 80$ into (3):
$$80 + 2y = 200$$
$$2y = 120$$
$$y = 60$$
Solution: $x$ = mix A = 80 grams; $y$ = mix B = 60 grams

**71.**     Let $x$ = number of hours of Mexico plant
$y$ = number of hours of Taiwan plant
Then we need to solve
$$40x + 20y = 4000 \qquad (1)$$
$$32x + 32y = 4000 \qquad (2)$$
Multiply (1) by 32, and (2) by ($-20$) and add the resulting equations:
$$640x = 48,000$$
$$x = 75$$
Substitute $x = 75$ into (1) (or (2))
$$40(75) + 20y = 4000$$
$$20y = 4000 - 3000 = 1000$$
$$y = 50$$
The plant in Mexico should be operated 75 hours, the plant in Taiwan should be operated 50 hours.

**73.**     $s = a + bt^2$
(A) At $t = 1$, $s = 180$ :    $a + b = 180$
At $t = 2$, $s = 132$ :    $a + 4b = 132$
Solve the equations
$$a + b = 180 \qquad (1)$$
$$a + 4b = 132 \qquad (2)$$
Subtract (1) from (2):
$$3b = -48$$
$$b = -16$$
Substitute $b = -16$ in (1):
$$a - 16 = 180$$
$$a = 196$$
Thus, $s = 196 - 16t^2$

(B) At $t = 0$, $s = 196$; the building is 196 feet high

Copyright © 2011 Pearson Education, Inc. Publishing as Prentice Hall.

(C) At $s = 0$, $196 - 16t^2 = 0$

$$16t^2 = 196$$
$$t^2 = 12.25$$
$$t = 3.5$$

The object falls for 3.5 seconds.

**75.** Let $d$ = the distance to the earthquake. Then it took $d/5$ seconds for the primary wave to reach the station and $d/3$ seconds for the secondary wave to reach the station. We are given that

$$\frac{d}{3} - \frac{d}{5} = 16$$
$$5d - 3d = 240$$
$$2d = 240$$
$$d = 120$$

The earthquake was 120 miles from the station. The primary wave traveled $\dfrac{120}{5} = 24$ seconds; the secondary

wave traveled $\dfrac{120}{3} = 40$ seconds.

**77.** $p = -\dfrac{1}{5}d + 70$          [Approach equation]

$p = -\dfrac{4}{3}d + 230$          [Avoidance equation]

(A)    The figure shows the graphs of the two equations.

(B)    Setting the two equations equal to each other, we have

$$-\frac{1}{5}d + 70 = -\frac{4}{3}d + 230$$

$$-\frac{1}{5}d + \frac{4}{3}d = 230 - 70$$

$$\frac{17}{15}d = 160$$

$$d = 141 \text{ cm (approx.)}$$

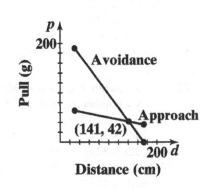

(C)    The rat would be very confused (!); it would vacillate.

---

**EXERCISE 4-2**

*Things to remember:*

1.    MATRICES

A MATRIX is a rectangular array of numbers written within brackets. Each number in a matrix is called an ELEMENT. If a matrix has $m$ rows and $n$ columns, it is called an $m \times n$ MATRIX; $m \times n$ is the SIZE; $m$ and $n$ are the DIMENSIONS. A matrix with $n$ rows and $n$ columns is a SQUARE MATRIX OF ORDER $n$. A matrix with only one column is a COLUMN MATRIX; a matrix with only one row is a ROW MATRIX. The element in the $i$th row and $j$th column of a matrix $A$ is denoted $a_{ij}$. The PRINCIPAL DIAGONAL of a

matrix $A$ consists of the elements
$a_{11}, a_{22}, a_{33}, \ldots$

2.    Associated with the linear system

Copyright © 2011 Pearson Education, Inc. Publishing as Prentice Hall.

$$a_1x_1 + b_1x_2 = k_1$$
$$a_2x_1 + b_2x_2 = k_2 \qquad \text{(I)}$$

is the AUGMENTED MATRIX of the system

$$\begin{bmatrix} a_1 & b_1 & \bigm| & k_1 \\ a_2 & b_2 & \bigm| & k_2 \end{bmatrix}. \qquad \text{(II)}$$

<u>3.</u>    OPERATIONS THAT PRODUCE ROW-EQUIVALENT MATRICES

An augmented matrix is transformed into a row-equivalent matrix if:

(a)    two rows are interchanged ($R_i \leftrightarrow R_j$);

(b)    a row is multiplied by a nonzero constant ($kR_i \rightarrow R_i$);

(c)    a constant multiple of one row is added to another row
($kR_j + R_i \rightarrow R_i$).

(Note: The arrow $\rightarrow$ means "replaces.")

<u>4.</u>    Given the system of linear equations (I) and its associated augmented matrix (II). If (II) is row equivalent to a matrix of the form:

(1) $\begin{bmatrix} 1 & 0 & \bigm| & m \\ 0 & 1 & \bigm| & n \end{bmatrix}$,    then (I) has a unique solution (consistent and independent);

(2) $\begin{bmatrix} 1 & m & \bigm| & n \\ 0 & 0 & \bigm| & 0 \end{bmatrix}$,    then (I) has infinitely many solutions (consistent and dependent);

(3) $\begin{bmatrix} 1 & m & \bigm| & n \\ 0 & 0 & \bigm| & p \end{bmatrix}$,    $p \neq 0$, then (I) has no solution (inconsistent).

**1.**    $A$ is $2 \times 3$; $C$ is $1 \times 3$          **3.** $C$          **5.** $B$

**7.**    $a_{12} = -4$, $a_{23} = -5$          **9.**    $-1, 8, 0$

**11.**    (A)    $E = \begin{bmatrix} 1 & -2 & 3 & 9 \\ -5 & 0 & 7 & -8 \end{bmatrix}$; size $2 \times 4$

(B)    $F = \begin{bmatrix} 4 & -6 \\ 2 & 3 \\ -5 & 7 \end{bmatrix}$; size $3 \times 2$. $F$ would have to have one more column to be square.

(C)    $e_{23} = 7$; $f_{12} = -6$

Copyright © 2011 Pearson Education, Inc. Publishing as Prentice Hall.

**13.** Interchange row 1 and row 2.

$$\begin{bmatrix} 4 & -6 & | & -8 \\ 1 & -3 & | & 2 \end{bmatrix}$$

**15.** Multiply row 1 by –4.

$$\begin{bmatrix} -4 & 12 & | & -8 \\ 4 & -6 & | & -8 \end{bmatrix}$$

**17.** Multiply row 2 by 2.

$$\begin{bmatrix} 1 & -3 & | & 2 \\ 8 & -12 & | & -16 \end{bmatrix}$$

**19.** Replace row 2 by the sum of row 2 and –4 times row 1.

$$\begin{bmatrix} 1 & -3 & | & 2 \\ 0 & 6 & | & -16 \end{bmatrix}$$

**21.** Replace row 2 by the sum of row 2 and –2 times row 1.

$$\begin{bmatrix} 1 & -3 & | & 2 \\ 2 & 0 & | & -12 \end{bmatrix}$$

**23.** Replace row 2 by the sum of row 2 and –1 times row 1.

$$\begin{bmatrix} 1 & -3 & | & 2 \\ 3 & -3 & | & -10 \end{bmatrix}$$

**25.** $\begin{bmatrix} -1 & 2 & | & -3 \\ 6 & -3 & | & 12 \end{bmatrix} \dfrac{1}{3}R_2 \rightarrow R_2 \sim \begin{bmatrix} -1 & 2 & | & -3 \\ 2 & -1 & | & 4 \end{bmatrix}$

**27.** $\begin{bmatrix} -1 & 2 & | & -3 \\ 6 & -3 & | & 12 \end{bmatrix} 6R_1 + R_2 \rightarrow R_2 \sim \begin{bmatrix} -1 & 2 & | & 3 \\ 0 & 9 & | & -6 \end{bmatrix}$

**29.** $\begin{bmatrix} -1 & 2 & | & -3 \\ 6 & -3 & | & 12 \end{bmatrix} \dfrac{1}{3}R_2 + R_1 \rightarrow R_1 \sim \begin{bmatrix} 1 & 1 & | & 1 \\ 6 & -3 & | & 12 \end{bmatrix}$

**31.** $\begin{bmatrix} -1 & 2 & | & -3 \\ 6 & -3 & | & 12 \end{bmatrix} R_1 \leftrightarrow R_2 \sim \begin{bmatrix} 6 & -3 & | & 12 \\ -1 & 2 & | & -3 \end{bmatrix}$

**33.** $\begin{bmatrix} 3 & -2 & | & 6 \\ 4 & -3 & | & 6 \end{bmatrix} \sim \begin{bmatrix} 1 & -1 & | & 0 \\ 4 & -3 & | & 6 \end{bmatrix} \sim \begin{bmatrix} 1 & -1 & | & 0 \\ 0 & 1 & | & 6 \end{bmatrix} \sim \begin{bmatrix} 1 & 0 & | & 6 \\ 0 & 1 & | & 6 \end{bmatrix}$

$R_2 - R_1 \rightarrow R_1 \quad -4R_1 + R_2 \rightarrow R_2 \quad R_2 + R_1 \rightarrow R_1$

Thus, $x_1 = 6$ and $x_2 = 6$.

The graph of each equation in the system passes through the point (6, 6).

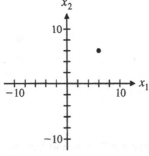

**35.** $\begin{bmatrix} 3 & -2 & | & -3 \\ -6 & 4 & | & 6 \end{bmatrix} \sim \begin{bmatrix} 3 & -2 & | & -3 \\ 0 & 0 & | & 0 \end{bmatrix} \sim \begin{bmatrix} 1 & -\frac{2}{3} & | & -1 \\ 0 & 0 & | & 0 \end{bmatrix}$

$2R_1 + R_2 \rightarrow R_2 \quad \dfrac{1}{3}R_1 \rightarrow R_1$

Copyright © 2011 Pearson Education, Inc. Publishing as Prentice Hall.

From <u>4</u>, Form (2), the system has infinitely many solutions (consistent and dependent).

If $x_2 = t$, then $x_1 = \frac{2}{3}t - 1$;

solution set: $\{\frac{2}{3}t - 1, t \mid t$ any real number$\}$

The graph of the solution set is the same as the graph of each equation in the system.

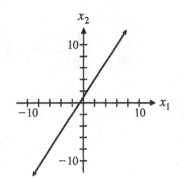

**37.** 

System    Augmented matrix    Graphs:

$x_1 + x_2 = 5$    $\begin{bmatrix} 1 & 1 & | & 5 \\ 1 & -1 & | & 1 \end{bmatrix}$

$x_1 - x_2 = 1$

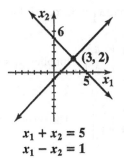

$x_1 + x_2 = 5$
$x_1 - x_2 = 1$

$\begin{bmatrix} 1 & 1 & | & 5 \\ 1 & -1 & | & 1 \end{bmatrix} (-1)R_1 + R_2 \rightarrow R_2 \begin{bmatrix} 1 & 1 & | & 5 \\ 0 & -2 & | & -4 \end{bmatrix}$

$x_1 + x_2 = 5$
$-2x_2 = -4$

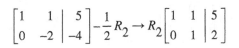

$x_1 + x_2 = 5$
$-2x_2 = -4$

$\begin{bmatrix} 1 & 1 & | & 5 \\ 0 & -2 & | & -4 \end{bmatrix} -\frac{1}{2}R_2 \rightarrow R_2 \begin{bmatrix} 1 & 1 & | & 5 \\ 0 & 1 & | & 2 \end{bmatrix}$

$x_1 + x_2 = 5$
$\quad x_2 = 2$

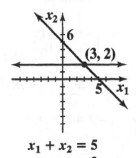

$x_1 + x_2 = 5$
$\quad x_2 = 2$

Copyright © 2011 Pearson Education, Inc. Publishing as Prentice Hall.

$$\begin{bmatrix} 1 & 1 & | & 5 \\ 0 & 1 & | & 2 \end{bmatrix} \quad (-1)R_2 + R_1 \to R_1 \begin{bmatrix} 1 & 0 & | & 3 \\ 0 & 1 & | & 2 \end{bmatrix}$$

$x_1 = 3$
$x_2 = 2$

$x_1 = 3$
$x_2 = 2$

Solution: $x_1 = 3$, $x_2 = 2$. Each pair of lines has the same intersection point.

**39.** $\begin{bmatrix} 1 & 0 & | & -4 \\ 0 & 1 & | & 6 \end{bmatrix}$   implies $x_1 = -4$, $x_2 = 6$.

**41.** $\begin{bmatrix} 1 & 3 & | & 2 \\ 0 & 0 & | & 4 \end{bmatrix}$   The second equation is $0x_1 + 0x_2 = 0$  which has no solution.

**43.** $\begin{bmatrix} 1 & -2 & | & 15 \\ 0 & 0 & | & 0 \end{bmatrix}$   The second equation is $0x_1 + 0x_2 = 0$  which has infinitely many solutions.  Set $x_2 = t$.
Then  $x_1 = 2t + 15$  for any real number  $t$.

**45.** $\begin{bmatrix} 1 & -2 & | & 1 \\ 2 & -1 & | & 5 \end{bmatrix} \sim \begin{bmatrix} 1 & -2 & | & 1 \\ 0 & 3 & | & 3 \end{bmatrix} \sim \begin{bmatrix} 1 & -2 & | & 1 \\ 0 & 1 & | & 1 \end{bmatrix} \sim \begin{bmatrix} 1 & 0 & | & 3 \\ 0 & 1 & | & 1 \end{bmatrix}$

$(-2)R_1 + R_2 \to R_2 \quad \frac{1}{3}R_2 \to R_2 \quad 2R_2 + R_1 \to R_1$

Thus,  $x_1 = 3$ and $x_2 = 1$.

**47.** $\begin{bmatrix} 1 & -4 & | & -2 \\ -2 & 1 & | & -3 \end{bmatrix} \sim \begin{bmatrix} 1 & -4 & | & -2 \\ 0 & -7 & | & -7 \end{bmatrix} \sim \begin{bmatrix} 1 & -4 & | & -2 \\ 0 & 1 & | & 1 \end{bmatrix} \sim \begin{bmatrix} 1 & 0 & | & 2 \\ 0 & 1 & | & 1 \end{bmatrix}$

$2R_1 + R_2 \to R_2 \quad \left(-\frac{1}{7}\right)R_2 \to R_2 \quad 4R_2 + R_1 \to R_1$

Thus, $x_1 = 2$ and $x_2 = 1$.

**49.** $\begin{bmatrix} 3 & -1 & | & 2 \\ 1 & 2 & | & 10 \end{bmatrix} \sim \begin{bmatrix} 1 & 2 & | & 10 \\ 3 & -1 & | & 2 \end{bmatrix} \sim \begin{bmatrix} 1 & 2 & | & 10 \\ 0 & -7 & | & -28 \end{bmatrix} \sim \begin{bmatrix} 1 & 2 & | & 10 \\ 0 & 1 & | & 4 \end{bmatrix} \sim \begin{bmatrix} 1 & 0 & | & 2 \\ 0 & 1 & | & 4 \end{bmatrix}$

$R_1 \leftrightarrow R_2 \quad (-3)R_1 + R_2 \to R_2 \quad \left(-\frac{1}{7}\right)R_2 \to R_2 \quad (-2)R_2 + R_1 \to R_1$

Thus, $x_1 = 2$ and $x_2 = 4$.

Copyright © 2011 Pearson Education, Inc.  Publishing as Prentice Hall.

**51.** $\begin{bmatrix} 1 & 2 & | & 4 \\ 2 & 4 & | & -8 \end{bmatrix} \sim \begin{bmatrix} 1 & 2 & | & 4 \\ 0 & 0 & | & -16 \end{bmatrix}$     From $\underline{4}$, Form (3), the system is inconsistent; there is no solution.

$(-2)R_1 + R_2 \to R_2$

**53.** $\begin{bmatrix} 2 & 1 & | & 6 \\ 1 & -1 & | & -3 \end{bmatrix} \sim \begin{bmatrix} 1 & -1 & | & -3 \\ 2 & 1 & | & 6 \end{bmatrix} \sim \begin{bmatrix} 1 & -1 & | & -3 \\ 0 & 3 & | & 12 \end{bmatrix} \sim \begin{bmatrix} 1 & -1 & | & -3 \\ 0 & 1 & | & 4 \end{bmatrix} \sim \begin{bmatrix} 1 & 0 & | & 1 \\ 0 & 1 & | & 4 \end{bmatrix}$

   $R_1 \leftrightarrow R_2$     $(-2)R_1 + R_2 \to R_2$     $\frac{1}{3}R_2 \to R_2$     $R_2 + R_1 \to R_1$

Thus, $x_1 = 1$ and $x_2 = 4$.

**55.** $\begin{bmatrix} 3 & -6 & | & -9 \\ -2 & 4 & | & 6 \end{bmatrix} \sim \begin{bmatrix} 1 & -2 & | & -3 \\ -2 & 4 & | & 6 \end{bmatrix} \sim \begin{bmatrix} 1 & -2 & | & -3 \\ 0 & 0 & | & 0 \end{bmatrix}$

   $\frac{1}{3}R_1 \to R_1$     $2R_1 + R_2 \to R_2$

From $\underline{4}$, Form (2), the system has infinitely many solutions (consistent and dependent). If $x_2 = s$, then

$x_1 - 2s = -3$   or   $x_1 = 2s - 3$.

Thus, $x_2 = s$, $x_1 = 2s - 3$, for any real number $s$, are the solutions.

**57.** $\begin{bmatrix} 4 & -2 & | & 2 \\ -6 & 3 & | & -3 \end{bmatrix} \sim \begin{bmatrix} 1 & -\frac{1}{2} & | & \frac{1}{2} \\ -6 & 3 & | & -3 \end{bmatrix} \sim \begin{bmatrix} 1 & -\frac{1}{2} & | & \frac{1}{2} \\ 0 & 0 & | & 0 \end{bmatrix}$

   $\frac{1}{4}R_1 \to R_1$     $6R_1 + R_2 \to R_2$

Thus, the system has infinitely many solutions (consistent and dependent). Let $x_2 = s$. Then

$x_1 - \frac{1}{2}s = \frac{1}{2}$   or   $x_1 = \frac{1}{2}s + \frac{1}{2}$.

The set of solutions is $x_2 = s$, $x_1 = \frac{1}{2}s + \frac{1}{2}$ for any real number $s$.

**59.** $\begin{bmatrix} 2 & 1 & | & 1 \\ 4 & -1 & | & -7 \end{bmatrix} \sim \begin{bmatrix} 1 & \frac{1}{2} & | & \frac{1}{2} \\ 4 & -1 & | & -7 \end{bmatrix} \sim \begin{bmatrix} 1 & \frac{1}{2} & | & \frac{1}{2} \\ 0 & -3 & | & -9 \end{bmatrix} \sim \begin{bmatrix} 1 & \frac{1}{2} & | & \frac{1}{2} \\ 0 & 1 & | & 3 \end{bmatrix} \sim \begin{bmatrix} 1 & 0 & | & -1 \\ 0 & 1 & | & 3 \end{bmatrix}$

   $\frac{1}{2}R_1 \to R_1$     $(-4)R_1 + R_2 \to R_2$     $\left(-\frac{1}{3}\right)R_2 \to R_2$     $\left(-\frac{1}{2}\right)R_2 + R_1 \to R_1$

Thus, $x_1 = -1$ and $x_2 = 3$.

**61.** $\begin{bmatrix} 4 & -6 & | & 8 \\ -6 & 9 & | & -10 \end{bmatrix} \sim \begin{bmatrix} 1 & -\frac{3}{2} & | & 2 \\ -6 & 9 & | & -10 \end{bmatrix} \sim \begin{bmatrix} 1 & -\frac{3}{2} & | & 2 \\ 0 & 0 & | & 2 \end{bmatrix}$

   $\frac{1}{4}R_1 \to R_1$     $6R_1 + R_2 \to R_2$

The second row of the final augmented matrix corresponds to the equation

$0x_1 + 0x_2 = 2$ which has no solution. Thus, the system has no solution; it is inconsistent.

Copyright © 2011 Pearson Education, Inc. Publishing as Prentice Hall.

**63.** $\begin{bmatrix} -4 & 6 & | & -8 \\ 6 & -9 & | & 12 \end{bmatrix} \sim \begin{bmatrix} 1 & -\frac{3}{2} & | & 2 \\ 6 & -9 & | & 12 \end{bmatrix} \sim \begin{bmatrix} 1 & -\frac{3}{2} & | & 2 \\ 0 & 0 & | & 0 \end{bmatrix}$

$\left(-\frac{1}{4}\right)R_1 \to R_1 \quad (-6)R_1 + R_2 \to R_2$

The system has infinitely many solutions (consistent and dependent). If $x_2 = t$, then

$x_1 - \frac{3}{2}t = 2$ or $x_1 = \frac{3}{2}t + 2$

Thus, the set of solutions is $x_2 = t$, $x_1 = \frac{3}{2}t + 2$ for any real number $t$.

**65.** $\begin{bmatrix} 3 & -1 & | & 7 \\ 2 & 3 & | & 1 \end{bmatrix} \sim \begin{bmatrix} 1 & -\frac{1}{3} & | & \frac{7}{3} \\ 2 & 3 & | & 1 \end{bmatrix} \sim \begin{bmatrix} 1 & -\frac{1}{3} & | & \frac{7}{3} \\ 0 & \frac{11}{3} & | & -\frac{11}{3} \end{bmatrix} \sim \begin{bmatrix} 1 & -\frac{1}{3} & | & \frac{7}{3} \\ 0 & 1 & | & -1 \end{bmatrix} \sim \begin{bmatrix} 1 & 0 & | & 2 \\ 0 & 1 & | & -1 \end{bmatrix}$

$\frac{1}{3}R_1 \to R_1 \ (-2)R_1 + R_2 \to R_2 \ \frac{3}{11}R_2 \to R_2 \ \frac{1}{3}R_2 + R_1 \to R_1$

Thus $x_1 = 2$, $x_2 = -1$.

**67.** $\begin{bmatrix} 3 & 2 & | & 4 \\ 2 & -1 & | & 5 \end{bmatrix} \sim \begin{bmatrix} 1 & \frac{2}{3} & | & \frac{4}{3} \\ 2 & -1 & | & 5 \end{bmatrix} \sim \begin{bmatrix} 1 & \frac{2}{3} & | & \frac{4}{3} \\ 0 & -\frac{7}{3} & | & \frac{7}{3} \end{bmatrix} \sim \begin{bmatrix} 1 & \frac{2}{3} & | & \frac{4}{3} \\ 0 & 1 & | & -1 \end{bmatrix} \sim \begin{bmatrix} 1 & 0 & | & 2 \\ 0 & 1 & | & -1 \end{bmatrix}$

$\frac{1}{3}R_1 \to R_1 \ (-2)R_1 + R_2 \to R_2 \ \left(-\frac{3}{7}\right)R_2 \to R_2 \ \left(-\frac{2}{3}\right)R_2 + R_1 \to R_1$

Thus $x_1 = 2$, $x_2 = -1$.

**69.** $\begin{bmatrix} 0.2 & -0.5 & | & 0.07 \\ 0.8 & -0.3 & | & 0.79 \end{bmatrix} \sim \begin{bmatrix} 1 & -2.5 & | & 0.35 \\ 0.8 & -0.3 & | & 0.79 \end{bmatrix} \sim \begin{bmatrix} 1 & -2.5 & | & 0.35 \\ 0 & 1.7 & | & 0.51 \end{bmatrix}$

$\frac{1}{0.2}R_1 \to R_1 \quad (-0.8)R_1 + R_2 \to R_2 \quad \frac{1}{1.7}R_2 \to R_2$

$\sim \begin{bmatrix} 1 & -2.5 & | & 0.35 \\ 0 & 1 & | & 0.3 \end{bmatrix} \sim \begin{bmatrix} 1 & 0 & | & 1.1 \\ 0 & 1 & | & 0.3 \end{bmatrix}$ Thus $x_1 = 1.1$, $x_2 = 0.3$.

$2.5R_2 + R_1 \to R_1$

**71.**  $0.8x_1 + 2.88x_2 = 4$
$1.25x_1 + 4.34x_2 = 5$

$\begin{bmatrix} 0.8 & 2.88 & | & 4 \\ 1.25 & 4.34 & | & 5 \end{bmatrix} \sim \begin{bmatrix} 1 & 3.6 & | & 5 \\ 1.25 & 4.34 & | & 5 \end{bmatrix} \sim \begin{bmatrix} 1 & 3.6 & | & 5 \\ 0 & -0.16 & | & -1.25 \end{bmatrix}$
$(1.25)R_1 \to R_1 \qquad (-1.25)R_1 + R_2 \to R_2 \qquad (-6.25)R_2 \to R_2$

$\sim \begin{bmatrix} 1 & 3.6 & | & 5 \\ 0 & 1 & | & 7.8125 \end{bmatrix} \sim \begin{bmatrix} 1 & 0 & | & -23.125 \\ 0 & 1 & | & 7.8125 \end{bmatrix}$
$(-3.6)R_2 + R_1 \to R_1$
Solution: $x_1 = -23.125$, $x_2 = 7.8125$

Copyright © 2011 Pearson Education, Inc.  Publishing as Prentice Hall.

**73.**  $4.8x_1 - 40.32x_2 = 295.2$
$-3.75x_1 + 28.7x_2 = -211.2$

$$\begin{bmatrix} 4.8 & -40.32 & | & 295.2 \\ -3.75 & 28.7 & | & -211.2 \end{bmatrix} \sim \begin{bmatrix} 1 & -8.4 & | & 61.5 \\ -3.75 & 28.7 & | & -211.2 \end{bmatrix} \sim \begin{bmatrix} 1 & -8.4 & | & 61.5 \\ 0 & -2.8 & | & 19.425 \end{bmatrix}$$
$(0.20833)R_1 \rightarrow R_1 \qquad\qquad (3.75)R_1 + R_2 \rightarrow R_2 \qquad (-0.35714)R_2 \rightarrow R_2$

$$\sim \begin{bmatrix} 1 & -8.4 & | & 61.5 \\ 0 & 1 & | & -6.9375 \end{bmatrix} \sim \begin{bmatrix} 1 & 0 & | & 3.225 \\ 0 & 1 & | & -6.9375 \end{bmatrix}$$
$\qquad (8.4)R_2 + R_1 \rightarrow R_1$

Solution:  $x_1 = 3.225, \quad x_2 = -6.9375$

---

## EXERCISE 4-3

*Things to remember:*

1.  A matrix is said to be in REDUCED ROW ECHELON FORM or, more simply, in REDUCED FORM if

    (a)  each row consisting entirely of zeros is below any row having at least one nonzero element;

    (b)  the left-most nonzero element in each row is 1;

    (c)  all other elements in the column containing the left-most 1 of a given row are zeros;

    (d)  the left-most 1 in any row is to the right of the left-most 1 in any row above.

2.  GAUSS-JORDAN ELIMINATION

    *Step 1.* Choose the leftmost nonzero column and use appropriate row operations to get a 1 at the top.

    *Step 2.* Use multiples of the row containing the 1 from step 1 to get zeros in all remaining places in the column containing this 1.

    *Step 3.* Repeat step 1 with the SUBMATRIX formed by (mentally) deleting the row used in step 2 and all rows above this row.

    *Step 4.* Repeat step 2 with the ENTIRE MATRIX, including the mentally deleted rows. Continue this process until the entire matrix is in reduced form.

    [*Note:* If at any point in this process we obtain a row with all zeros to the left of the vertical line and a nonzero number to the right, we can stop before we find the reduced form, since we will have a contradiction: $0 = n$, $n \neq 0$. We can then conclude that the system has no solution.]

---

**1.**  $\begin{bmatrix} 1 & 0 & | & 2 \\ 0 & 1 & | & -1 \end{bmatrix}$
is in reduced form

**3.**  $\begin{bmatrix} 1 & 0 & 2 & | & 3 \\ 0 & 0 & 0 & | & 0 \\ 0 & 1 & -1 & | & 4 \end{bmatrix}$
is not in reduced form: condition (a) is violated; the second row should be at the bottom.

Copyright © 2011 Pearson Education, Inc. Publishing as Prentice Hall.

$$R_2 \leftrightarrow R_3$$

**5.** $\begin{bmatrix} 0 & 1 & 0 & | & 2 \\ 0 & 0 & 3 & | & -1 \\ 0 & 0 & 0 & | & 0 \end{bmatrix}$

is not in reduced form: The first non-zero element in the second row is not 1; condition (b) is violated

$$\frac{1}{3}R_2 \rightarrow R_2$$

**7.** $\begin{bmatrix} 1 & 1 & 0 & | & 1 \\ 0 & 0 & 1 & | & 1 \\ 0 & 0 & 0 & | & 0 \end{bmatrix}$

is in reduced form.

**9.** $\begin{bmatrix} 1 & 0 & -2 & 0 & | & 1 \\ 0 & 0 & 1 & 1 & | & 0 \end{bmatrix}$

$$2R_2 + R_1 \rightarrow R_1$$

not in reduced form: The column containing the left-most 1 in row 2 has a nonzero element; condition (c) is violated

**11.** $x_1 = -2$

$\quad x_2 = 3$

$\quad\quad x_3 = 0$

**13.** $x_1 \quad - 2x_3 = 3 \quad (1)$

$\quad\quad x_2 + x_3 = -5 \quad (2)$

Let $x_3 = t$. From (2), $x_2 = -5 - t$.

From (1), $x_1 = 3 + 2t$. Thus, the

solution is

$x_1 = 2t + 3$

$x_2 = -t - 5$

$x_3 = t$

$t$ any real number.

**15.** $x_1 = 0$

$\quad x_2 = 0$

$\quad\quad 0 = 1$

Inconsistent; no solution.

**17.** $x_1 - 3x_3 = 5$

$\quad x_2 + 2x_3 = -7$

Let $x_3 = t$. Then $x_1 = 3t + 5$,

$x_2 = -2t - 7$, $t$ any real number.

**19.** $x_1 - 2x_2 \quad - 3x_4 = -5$

$\quad\quad x_3 + 3x_4 = 2$

Let $x_2 = s$ and $x_4 = t$. Then

$x_1 = 2s + 3t - 5$

$x_2 = s$

$x_3 = -3t + 2$

$x_4 = t$

$s$ and $t$ any real numbers.

**21.** Problem 11.

**23.** Problems 13, 17, 19.

**25.** False. Counterexample: Problem 15

**27.** True. For example, the reduced form of the augmented matrix in the 3x3 case will be

Copyright © 2011 Pearson Education, Inc. Publishing as Prentice Hall.

$$\begin{bmatrix} 1 & 0 & 0 & | & r_1 \\ 0 & 1 & 0 & | & r_2 \\ 0 & 0 & 1 & | & r_3 \end{bmatrix} \quad \text{which has exactly one solution.}$$

**29.** False.  Counterexample:

$$x_1 + x_2 = 1$$
$$x_1 - x_2 = 2$$
$$2x_1 + 2x_2 = 3$$

**31.** $\begin{bmatrix} 1 & 2 & | & -1 \\ 0 & 1 & | & 3 \end{bmatrix} \sim \begin{bmatrix} 1 & 0 & | & -7 \\ 0 & 1 & | & 3 \end{bmatrix}$

$(-2)R_2 + R_1 \rightarrow R_1$

**33.** $\begin{bmatrix} 1 & 1 & 1 & | & 16 \\ 2 & 3 & 4 & | & 25 \end{bmatrix} \sim \begin{bmatrix} 1 & 1 & 1 & | & 16 \\ 0 & 1 & 2 & | & -7 \end{bmatrix} \sim \begin{bmatrix} 1 & 0 & -1 & | & 23 \\ 0 & 1 & 2 & | & -7 \end{bmatrix}$

$(-2)R_1 + R_2 \rightarrow R_2 \quad (-1)R_2 + R_1 \rightarrow R_1$

**35.** $\begin{bmatrix} 1 & 0 & -3 & | & 1 \\ 0 & 1 & 2 & | & 0 \\ 0 & 0 & 3 & | & -6 \end{bmatrix} \sim \begin{bmatrix} 1 & 0 & -3 & | & 1 \\ 0 & 1 & 2 & | & 0 \\ 0 & 0 & 1 & | & -2 \end{bmatrix} \sim \begin{bmatrix} 1 & 0 & 0 & | & -5 \\ 0 & 1 & 0 & | & 4 \\ 0 & 0 & 1 & | & -2 \end{bmatrix}$

$\frac{1}{3}R_3 \rightarrow R_3 \qquad 3R_3 + R_1 \rightarrow R_1$
$\qquad\qquad (-2)R_3 + R_2 \rightarrow R_2$

**37.** $\begin{bmatrix} 1 & 2 & -2 & | & -1 \\ 0 & 3 & -6 & | & 1 \\ 0 & -1 & 2 & | & -\frac{1}{3} \end{bmatrix} \sim \begin{bmatrix} 1 & 2 & -2 & | & -1 \\ 0 & 1 & -2 & | & \frac{1}{3} \\ 0 & -1 & 2 & | & -\frac{1}{3} \end{bmatrix} \sim \begin{bmatrix} 1 & 2 & -2 & | & -1 \\ 0 & 1 & -2 & | & \frac{1}{3} \\ 0 & 0 & 0 & | & 0 \end{bmatrix} \sim \begin{bmatrix} 1 & 0 & 2 & | & -\frac{5}{3} \\ 0 & 1 & -2 & | & \frac{1}{3} \\ 0 & 0 & 0 & | & 0 \end{bmatrix}$

$\frac{1}{3}R_2 \rightarrow R_2 \qquad R_2 + R_3 \rightarrow R_3 \qquad (-2)R_2 + R_1 \rightarrow R_1$

**39.** The corresponding augmented matrix is:

$\begin{bmatrix} 2 & 4 & -10 & | & -2 \\ 3 & 9 & -21 & | & 0 \\ 1 & 5 & -12 & | & 1 \end{bmatrix} \sim \begin{bmatrix} 1 & 2 & -5 & | & -1 \\ 3 & 9 & -21 & | & 0 \\ 1 & 5 & -12 & | & 1 \end{bmatrix} \sim \begin{bmatrix} 1 & 2 & -5 & | & -1 \\ 0 & 3 & -6 & | & 3 \\ 0 & 3 & -7 & | & 2 \end{bmatrix}$

$\frac{1}{2}R_1 \rightarrow R_1 \qquad (-3)R_1 + R_2 \rightarrow R_2 \qquad \frac{1}{3}R_2 \rightarrow R_2$
$\qquad\qquad (-1)R_1 + R_3 + \rightarrow R_3$

$\sim \begin{bmatrix} 1 & 2 & -5 & | & -1 \\ 0 & 1 & -2 & | & 1 \\ 0 & 3 & -7 & | & 2 \end{bmatrix} \sim \begin{bmatrix} 1 & 0 & -1 & | & -3 \\ 0 & 1 & -2 & | & 1 \\ 0 & 0 & -1 & | & -1 \end{bmatrix} \sim \begin{bmatrix} 1 & 0 & -1 & | & -3 \\ 0 & 1 & -2 & | & 1 \\ 0 & 0 & 1 & | & 1 \end{bmatrix} \sim \begin{bmatrix} 1 & 0 & 0 & | & -2 \\ 0 & 1 & 0 & | & 3 \\ 0 & 0 & 1 & | & 1 \end{bmatrix}$

$(-3)R_2 + R_3 \rightarrow R_3 \quad (-1)R_3 \rightarrow R_3 \quad 2R_3 + R_2 + \rightarrow R_2$

Copyright © 2011 Pearson Education, Inc. Publishing as Prentice Hall.

$$(-2)R_2 + R_1 \rightarrow R_1 \qquad\qquad\qquad R_3 + R_1 \rightarrow R_1$$

Thus, $x_1 = -2$; $x_2 = 3$; $x_3 = 1$.

**41.** The corresponding augmented matrix is:

$$\begin{bmatrix} 3 & 8 & -1 & | & -18 \\ 2 & 1 & 5 & | & 8 \\ 2 & 4 & 2 & | & -4 \end{bmatrix} \sim \begin{bmatrix} 2 & 4 & 2 & | & -4 \\ 2 & 1 & 5 & | & 8 \\ 3 & 8 & -1 & | & -18 \end{bmatrix} \sim \begin{bmatrix} 1 & 2 & 1 & | & -2 \\ 2 & 1 & 5 & | & 8 \\ 3 & 8 & -1 & | & -18 \end{bmatrix}$$

$$\qquad R_1 \leftrightarrow R_3 \qquad\qquad \frac{1}{2}R_1 \rightarrow R_1 \qquad (-2)R_1 + R_2 \rightarrow R_2$$
$$\qquad\qquad\qquad\qquad\qquad\qquad\qquad (-3)R_1 + R_3 \rightarrow R_3$$

$$\sim \begin{bmatrix} 1 & 2 & 1 & | & -2 \\ 0 & -3 & 3 & | & 12 \\ 0 & 2 & -4 & | & -12 \end{bmatrix} \sim \begin{bmatrix} 1 & 2 & 1 & | & -2 \\ 0 & 1 & -1 & | & -4 \\ 0 & 2 & -4 & | & -12 \end{bmatrix} \sim \begin{bmatrix} 1 & 0 & 3 & | & 6 \\ 0 & 1 & -1 & | & -4 \\ 0 & 0 & -2 & | & -4 \end{bmatrix}$$

$$\left(-\frac{1}{3}\right)R_2 \rightarrow R_2 \qquad (-2)R_2 + R_3 \rightarrow R_3 \qquad \left(-\frac{1}{2}\right)R_3 \rightarrow R_3$$
$$\qquad\qquad\qquad (-2)R_2 + R_1 \rightarrow R_1$$

$$\sim \begin{bmatrix} 1 & 0 & 3 & | & 6 \\ 0 & 1 & -1 & | & -4 \\ 0 & 0 & 1 & | & 2 \end{bmatrix} \sim \begin{bmatrix} 1 & 0 & 0 & | & 0 \\ 0 & 1 & 0 & | & -2 \\ 0 & 0 & 1 & | & 2 \end{bmatrix} \qquad \text{Thus} \quad x_1 = 0, \ x_2 = -2, \ x_3 = 2.$$

$$(-3)R_3 + R_1 \rightarrow R_1$$
$$\ \ R_3 + R_2 \rightarrow R_2$$

**43.** $\begin{bmatrix} 2 & -1 & -3 & | & 8 \\ 1 & -2 & 0 & | & 7 \end{bmatrix} \sim \begin{bmatrix} 1 & -2 & 0 & | & 7 \\ 2 & -1 & -3 & | & 8 \end{bmatrix} \sim \begin{bmatrix} 1 & -2 & 0 & | & 7 \\ 0 & 3 & -3 & | & -6 \end{bmatrix}$

$$\qquad R_1 \leftrightarrow R_2 \qquad (-2)R_1 + R_2 \rightarrow R_2 \qquad \frac{1}{3}R_2 \rightarrow R_2$$

$$\sim \begin{bmatrix} 1 & -2 & 0 & | & 7 \\ 0 & 1 & -1 & | & -2 \end{bmatrix} \sim \begin{bmatrix} 1 & 0 & -2 & | & 3 \\ 0 & 1 & -1 & | & -2 \end{bmatrix} \qquad\qquad \text{Thus, } \begin{aligned} x_1 \quad - 2x_3 &= 3 \quad (1) \\ x_2 - \ x_3 &= -2 \quad (2) \end{aligned}$$

$$\qquad 2R_2 + R_1 \rightarrow R_1$$

Let $x_3 = t$, where $t$ is any real number. Then: $x_1 = 2t + 3$, $x_2 = t - 2$, $x_3 = t$

**45.** $\begin{bmatrix} 2 & -1 & | & 0 \\ 3 & 2 & | & 7 \\ 1 & -1 & | & -1 \end{bmatrix} \sim \begin{bmatrix} 1 & -1 & | & -1 \\ 3 & 2 & | & 7 \\ 2 & -1 & | & 0 \end{bmatrix} \sim \begin{bmatrix} 1 & -1 & | & -1 \\ 0 & 5 & | & 10 \\ 0 & 1 & | & 2 \end{bmatrix} \sim \begin{bmatrix} 1 & -1 & | & -1 \\ 0 & 1 & | & 2 \\ 0 & 5 & | & 10 \end{bmatrix}$

$$\quad R_1 \leftrightarrow R_3 \qquad (-3)R_1 + R_2 \rightarrow R_2 \quad R_2 \leftrightarrow R_3 \qquad R_2 + R_1 \rightarrow R_1$$
$$\qquad\qquad\qquad (-2)R_1 + R_3 \rightarrow R_3 \qquad\qquad\qquad (-5)R_2 + R_3 \rightarrow R_3$$

$$\sim \begin{bmatrix} 1 & 0 & | & 1 \\ 0 & 1 & | & 2 \\ 0 & 0 & | & 0 \end{bmatrix} \qquad \text{Thus} \quad x_1 = 1, \ x_2 = 2.$$

Copyright © 2011 Pearson Education, Inc.  Publishing as Prentice Hall.

**47.** $\begin{bmatrix} 3 & -4 & -1 & | & 1 \\ 2 & -3 & 1 & | & 1 \\ 1 & -2 & 3 & | & 2 \end{bmatrix} \sim \begin{bmatrix} 1 & -2 & 3 & | & 2 \\ 2 & -3 & 1 & | & 1 \\ 3 & -4 & -1 & | & 1 \end{bmatrix} \sim \begin{bmatrix} 1 & -2 & 3 & | & 2 \\ 0 & 1 & -5 & | & -3 \\ 0 & 2 & -10 & | & -5 \end{bmatrix} \sim \begin{bmatrix} 1 & -2 & 3 & | & 2 \\ 0 & 1 & -5 & | & -3 \\ 0 & 0 & 0 & | & 1 \end{bmatrix}$

$R_1 \leftrightarrow R_3$        $(-2)R_1 + R_2 \to R_2$    $(-2)R_2 + R_3 \to R_3$
                $(-3)R_1 + R_3 \to R_3$

From the last row, we conclude that there is no solution; the system is inconsistent.

**49.** $\begin{bmatrix} 3 & -2 & 1 & | & -7 \\ 2 & 1 & -4 & | & 0 \\ 1 & 1 & -3 & | & 1 \end{bmatrix} \sim \begin{bmatrix} 1 & 1 & -3 & | & 1 \\ 2 & 1 & -4 & | & 0 \\ 3 & -2 & 1 & | & -7 \end{bmatrix} \sim \begin{bmatrix} 1 & 1 & -3 & | & 1 \\ 0 & -1 & 2 & | & -2 \\ 0 & -5 & 10 & | & -10 \end{bmatrix} \sim \begin{bmatrix} 1 & 1 & -3 & | & 1 \\ 0 & 1 & -2 & | & 2 \\ 0 & -5 & 10 & | & -10 \end{bmatrix}$

$R_1 \leftrightarrow R_3$        $(-2)R_1 + R_2 \to R_2$        $(-1)R_2 \to R_2$        $(-1)R_2 + R_1 \to R_1$
                $(-3)R_1 + R_3 \to R_3$                        $5R_2 + R_3 \to R_3$

$\begin{bmatrix} 1 & 1 & -3 & | & 1 \\ 0 & 1 & -2 & | & 2 \\ 0 & -5 & 10 & | & -10 \end{bmatrix} \sim \begin{bmatrix} 1 & 0 & -1 & | & -1 \\ 0 & 1 & -2 & | & 2 \\ 0 & 0 & 0 & | & 0 \end{bmatrix}$

$(-1)R_2 + R_1 \to R_1$
$5R_2 + R_3 \to R_3$

From this matrix, $x_1 - x_3 = -1$ and $x_2 - 2x_3 = 2$. Let $x_3 = t$ be any real number, then $x_1 = t - 1$, $x_2 = 2t + 2$, and $x_3 = t$.

**51.** $\begin{bmatrix} 2 & 4 & -2 & | & 2 \\ -3 & -6 & 3 & | & -3 \end{bmatrix} \sim \begin{bmatrix} 1 & 2 & -1 & | & 1 \\ -3 & -6 & 3 & | & -3 \end{bmatrix} \sim \begin{bmatrix} 1 & 2 & -1 & | & 1 \\ 0 & 0 & 0 & | & 0 \end{bmatrix}$

$\frac{1}{2}R_1 \to R_1$        $3R_1 + R_2 \to R_2$

From this matrix, $x_1 + 2x_2 - x_3 = 1$. Let $x_2 = s$ and $x_3 = t$. Then
$x_1 = -2s + t + 1$, $x_2 = s$, and $x_3 = t$,   $s$ and $t$ any real numbers.

**53.** $\begin{bmatrix} 4 & -1 & 2 & | & 3 \\ -4 & 1 & -3 & | & -10 \\ 8 & -2 & 9 & | & -1 \end{bmatrix} \begin{array}{c} \frac{1}{4}R_1 \to R_1 \end{array} \sim \begin{bmatrix} 1 & -\frac{1}{4} & \frac{1}{2} & | & \frac{3}{4} \\ -4 & 1 & -3 & | & -10 \\ 8 & -2 & 9 & | & -1 \end{bmatrix} \begin{array}{c} 4R_1 + R_2 \to R_2 \\ (-8)R_1 + R_3 \to R_3 \end{array}$

$\sim \begin{bmatrix} 1 & -\frac{1}{4} & \frac{1}{2} & | & \frac{3}{4} \\ 0 & 0 & -1 & | & -7 \\ 0 & 0 & 5 & | & -7 \end{bmatrix} (-1)R_2 \to R_2 \sim \begin{bmatrix} 1 & -\frac{1}{4} & \frac{1}{2} & | & \frac{3}{4} \\ 0 & 0 & 1 & | & 7 \\ 0 & 0 & 5 & | & -7 \end{bmatrix} (-5)R_2 + R_3 \to R_3$

$\sim \begin{bmatrix} 1 & -\frac{1}{4} & \frac{1}{2} & | & \frac{3}{4} \\ 0 & 0 & 1 & | & 7 \\ 0 & 0 & 0 & | & -42 \end{bmatrix}$ No solution.

**55.** (A) The system is dependent with two parameters and an infinite number of solutions.

(B) The system is dependent with one parameter and an infinite number of solutions.

(C) The system is independent with a unique solution.

Copyright © 2011 Pearson Education, Inc. Publishing as Prentice Hall.

(D) Impossible

**57.**
$$\begin{bmatrix} 1 & 2 & -4 & -1 & | & 7 \\ 2 & 5 & -9 & -4 & | & 16 \\ 1 & 5 & -7 & -7 & | & 13 \end{bmatrix} \sim \begin{bmatrix} 1 & 2 & -4 & -1 & | & 7 \\ 0 & 1 & -1 & -2 & | & 2 \\ 0 & 3 & -3 & -6 & | & 6 \end{bmatrix} \sim \begin{bmatrix} 1 & 0 & -2 & 3 & | & 3 \\ 0 & 1 & -1 & -2 & | & 2 \\ 0 & 0 & 0 & 0 & | & 0 \end{bmatrix}$$

$(-2)R_1 + R_2 \rightarrow R_2 \qquad (-2)R_2 + R_1 \rightarrow R_1$

$(-1)R_1 + R_3 \rightarrow R_3 \qquad (-3)R_2 + R_3 \rightarrow R_3$

Thus, $x_1 - 2x_3 + 3x_4 = 3$ and $x_2 - x_3 - 2x_4 = 2$. Let $x_3 = s$ and $x_4 = t$. Then $x_1 = 2s - 3t + 3$, $x_2 = s + 2t + 2$, $x_3 = s$, $x_4 = t$, where $s$, $t$ are any real numbers.

**59.**
$$\begin{bmatrix} 1 & -1 & 3 & -2 & | & 1 \\ -2 & 4 & -3 & 1 & | & 0.5 \\ 3 & -1 & 10 & -4 & | & 2.9 \\ 4 & -3 & 8 & -2 & | & 0.6 \end{bmatrix} \quad \begin{matrix} 2R_1 + R_2 \rightarrow R_2 \\ (-3)R_1 + R_3 \rightarrow R_3 \\ (-4)R_1 + R_4 \rightarrow R_4 \end{matrix} \quad \sim \begin{bmatrix} 1 & -1 & 3 & -2 & | & 1 \\ 0 & 2 & 3 & -3 & | & 2.5 \\ 0 & 2 & 1 & 2 & | & -0.1 \\ 0 & 1 & -4 & 6 & | & -3.4 \end{bmatrix} R_2 \leftrightarrow R_4$$

$$\sim \begin{bmatrix} 1 & -1 & 3 & -2 & | & 1 \\ 0 & 1 & -4 & 6 & | & -3.4 \\ 0 & 2 & 1 & 2 & | & -0.1 \\ 0 & 2 & 3 & -3 & | & 2.5 \end{bmatrix} \quad \begin{matrix} R_2 + R_1 \rightarrow R_1 \\ (-2)R_2 + R_3 \rightarrow R_3 \\ (-2)R_2 + R_4 \rightarrow R_4 \end{matrix}$$

$$\sim \begin{bmatrix} 1 & 0 & -1 & 4 & | & -2.4 \\ 0 & 1 & -4 & 6 & | & -3.4 \\ 0 & 0 & 9 & -10 & | & 6.7 \\ 0 & 0 & 11 & -15 & | & 9.3 \end{bmatrix} \quad (-1)R_4 + R_3 \rightarrow R_3 \text{ (to simplify arithmetic)}$$

$$\sim \begin{bmatrix} 1 & 0 & -1 & 4 & | & -2.4 \\ 0 & 1 & -4 & 6 & | & -3.4 \\ 0 & 0 & -2 & 5 & | & -2.6 \\ 0 & 0 & 11 & -15 & | & 9.3 \end{bmatrix} \left(-\frac{1}{2}\right) R_3 \rightarrow R_3$$

$$\sim \begin{bmatrix} 1 & 0 & -1 & 4 & | & -2.4 \\ 0 & 1 & -4 & 6 & | & -3.4 \\ 0 & 0 & 1 & -2.5 & | & 1.3 \\ 0 & 0 & 11 & -15 & | & 9.3 \end{bmatrix} \quad \begin{matrix} R_3 + R_1 \rightarrow R_1 \\ 4R_3 + R_2 \rightarrow R_2 \\ (-11)R_3 + R_4 \rightarrow R_4 \end{matrix}$$

$$\sim \begin{bmatrix} 1 & 0 & 0 & 1.5 & | & -1.1 \\ 0 & 1 & 0 & -4 & | & 1.8 \\ 0 & 0 & 1 & -2.5 & | & 1.3 \\ 0 & 0 & 0 & 12.5 & | & -5 \end{bmatrix} \frac{1}{12.5} R_4 \rightarrow R_4$$

$$\sim \begin{bmatrix} 1 & 0 & 0 & 1.5 & | & -1.1 \\ 0 & 1 & 0 & -4 & | & 1.8 \\ 0 & 0 & 1 & -2.5 & | & 1.3 \\ 0 & 0 & 0 & 1 & | & -0.4 \end{bmatrix} \quad \begin{matrix} (-1.5)R_4 + R_1 \rightarrow R_1 \\ 4R_4 + R_2 \rightarrow R_2 \\ 2.5R_4 + R_3 \rightarrow R_3 \end{matrix}$$

$$\sim \begin{bmatrix} 1 & 0 & 0 & 0 & -0.5 \\ 0 & 1 & 0 & 0 & 0.2 \\ 0 & 0 & 1 & 0 & 0.3 \\ 0 & 0 & 0 & 1 & -0.4 \end{bmatrix}$$

Solution: $x_1 = -0.5$, $x_2 = 0.2$, $x_3 = 0.3$, $x_2 = -0.4$

**61.** $\begin{bmatrix} 1 & -2 & 1 & 1 & 2 & 2 \\ -2 & 4 & 2 & 2 & -2 & 0 \\ 3 & -6 & 1 & 1 & 5 & 4 \\ -1 & 2 & 3 & 1 & 1 & 3 \end{bmatrix}$  $\begin{matrix} 2R_1 + R_2 \to R_2 \\ (-3)R_1 + R_3 \to R_3 \\ R_1 + R_4 \to R_4 \end{matrix}$

$\sim \begin{bmatrix} 1 & -2 & 1 & 1 & 2 & 2 \\ 0 & 0 & 4 & 4 & 2 & 4 \\ 0 & 0 & 2 & -2 & -1 & -2 \\ 0 & 0 & 4 & 2 & 3 & 5 \end{bmatrix}$  $\dfrac{1}{4}R_2 \to R_2$

$\sim \begin{bmatrix} 1 & -2 & 1 & 1 & 2 & 2 \\ 0 & 0 & 1 & 1 & \frac{1}{2} & 1 \\ 0 & 0 & -2 & -2 & -1 & -2 \\ 0 & 0 & 4 & 2 & 3 & 5 \end{bmatrix}$  $\begin{matrix} (-1)R_2 + R_1 \to R_1 \\ 2R_2 + R_3 \to R_3 \\ (-4)R_2 + R_4 \to R_4 \end{matrix}$

$\sim \begin{bmatrix} 1 & -2 & 1 & 1 & \frac{3}{2} & 1 \\ 0 & 0 & 1 & 1 & \frac{1}{2} & 1 \\ 0 & 0 & 0 & 0 & 0 & 0 \\ 0 & 0 & 0 & -2 & 1 & 1 \end{bmatrix}$  $R_3 \leftrightarrow R_4 \sim \begin{bmatrix} 1 & -2 & 0 & 0 & \frac{3}{2} & 1 \\ 0 & 0 & 1 & 1 & \frac{1}{2} & 1 \\ 0 & 0 & 0 & -2 & 1 & 1 \\ 0 & 0 & 0 & 0 & 0 & 0 \end{bmatrix}$  $\left(-\dfrac{1}{2}\right)R_3 \to R_3$

$\sim \begin{bmatrix} 1 & -2 & 0 & 0 & \frac{3}{2} & 1 \\ 0 & 0 & 1 & 1 & \frac{1}{2} & 1 \\ 0 & 0 & 0 & 1 & -\frac{1}{2} & -\frac{1}{2} \\ 0 & 0 & 0 & 0 & 0 & 0 \end{bmatrix}$  $(-1)R_3 + R_2 \to R_2 \sim \begin{bmatrix} 1 & -2 & 0 & 0 & \frac{3}{2} & 1 \\ 0 & 0 & 1 & 0 & 1 & \frac{3}{2} \\ 0 & 0 & 0 & 1 & -\frac{1}{2} & -\frac{1}{2} \\ 0 & 0 & 0 & 0 & 0 & 0 \end{bmatrix}$

The system of equations is:

$$x_1 - 2x_2 \qquad\qquad + \frac{3}{2}x_5 = 1$$
$$x_3 \qquad + x_5 = \frac{3}{2}$$
$$x_4 - \frac{1}{2}x_5 = -\frac{1}{2}$$

or $x_1 = 2x_2 - \dfrac{3}{2}x_5 + 1$, $x_3 = -x_5 + \dfrac{3}{2}$, $x_4 = \dfrac{1}{2}x_5 - \dfrac{1}{2}$

Let $x_2 = s$ and $x_5 = t$. Then $x_1 = 2s - \dfrac{3}{2}t + 1$, $x_2 = s$, $x_3 = -t + \dfrac{3}{2}$, $x_4 = \dfrac{1}{2}t - \dfrac{1}{2}$, $x_5 = t$ for any real numbers $s$ and $t$.

**63.** The coordinates of the points must satisfy the quadratic equation: $y = ax^2 + bx + c$.

Copyright © 2011 Pearson Education, Inc. Publishing as Prentice Hall.

$(-2, 9)$:   $9 = a(-2)^2 + b(-2) + c$   or     $4a - 2b + c = 9$

$(1, -9)$:   $-9 = a(1)^2 + b(1) + c$   or       $a + b + c = -9$

$(4, 9)$:   $9 = a(4)^2 + b(4) + c$   or     $16a + 4b + c = 9$

The system of equations is:

$$4a - 2b + c = 9$$
$$a + b + c = -9$$
$$16a + 4b + c = 9$$

The augmented matrix is:

$$\begin{pmatrix} 4 & -2b & 1 & | & 9 \\ 1 & 1 & 1 & | & -9 \\ 16 & 4 & 1 & | & 9 \end{pmatrix} \sim \begin{pmatrix} 1 & 1 & 1 & | & -9 \\ 4 & -2 & 1 & | & 9 \\ 16 & 4 & 1 & | & 9 \end{pmatrix} \sim \begin{pmatrix} 1 & 1 & 1 & | & -9 \\ 0 & -6 & -3 & | & 45 \\ 0 & -12 & -15 & | & 153 \end{pmatrix}$$

$\quad R_1 \to R_2 \qquad\qquad -4R_1 + R_2 \to R_2 \quad \left(-\frac{1}{6}\right)R_2 \to R_2$
$\qquad\qquad\qquad\qquad -16R_1 + R_3 \to R_3$

$$\sim \begin{pmatrix} 1 & 1 & 1 & | & -9 \\ 0 & 1 & \frac{1}{2} & | & -\frac{15}{2} \\ 0 & -12 & -15 & | & 153 \end{pmatrix} \sim \begin{pmatrix} 1 & 1 & 1 & | & -9 \\ 0 & 1 & \frac{1}{2} & | & -\frac{15}{2} \\ 0 & 0 & -9 & | & 63 \end{pmatrix} \sim \begin{pmatrix} 1 & 1 & 1 & | & -9 \\ 0 & 1 & \frac{1}{2} & | & -\frac{15}{2} \\ 0 & 0 & 1 & | & -7 \end{pmatrix}$$

$12R_2 + R_3 \to R_3 \qquad \left(-\frac{1}{9}\right)R_3 \to R_3 \qquad \left(-\frac{1}{2}\right)R_3 + R_2 \to R_2$
$\qquad\qquad\qquad\qquad\qquad\qquad\qquad (-1)R_3 + R_1 \to R_1$

$$\sim \begin{pmatrix} 1 & 1 & 0 & | & -2 \\ 0 & 1 & 0 & | & -4 \\ 0 & 0 & 1 & | & -7 \end{pmatrix} \sim \begin{pmatrix} 1 & 0 & 0 & | & 2 \\ 0 & 1 & 0 & | & -4 \\ 0 & 0 & 1 & | & -7 \end{pmatrix}$$   Thus  $a = 2,\ b = -4,\ c = -7$.

$(-1)R_2 + R_1 \to R_1$

**65.**  Let $x_1$ = Number of one-person boats

$\qquad x_2$ = Number of two-person boats

$\qquad x_3$ = Number of four-person boats.

(A)   The mathematical model is:

$$0.5x_1 + x_2 + 1.5x_3 = 380$$
$$0.6x_1 + 0.9x_2 + 1.2x_3 = 330$$
$$0.2x_1 + 0.3x_2 + 0.5x_3 = 120$$

$$\begin{bmatrix} 0.5 & 1 & 1.5 & | & 380 \\ 0.6 & 0.9 & 1.2 & | & 330 \\ 0.2 & 0.3 & 0.5 & | & 120 \end{bmatrix} \sim \begin{bmatrix} 1 & 2 & 3 & | & 760 \\ 0.6 & 0.9 & 1.2 & | & 330 \\ 0.2 & 0.3 & 0.5 & | & 120 \end{bmatrix} \sim \begin{bmatrix} 1 & 2 & 3 & | & 760 \\ 0 & -0.3 & -0.6 & | & -126 \\ 0 & -0.1 & -0.1 & | & -32 \end{bmatrix}$$

$\quad 2R_1 \to R_1 \qquad (-0.6)R_1 + R_2 \to R_2 \quad \left(-\dfrac{1}{0.3}\right)R_2 \to R_2$

$\qquad\qquad\qquad (-0.2)R_1 + R_3 \to R_3$

$$\sim \begin{bmatrix} 1 & 2 & 3 & | & 760 \\ 0 & 1 & 2 & | & 420 \\ 0 & -0.1 & -0.1 & | & -32 \end{bmatrix} \sim \begin{bmatrix} 1 & 0 & -1 & | & 760 \\ 0 & 1 & 2 & | & 420 \\ 0 & 0 & 0.1 & | & 10 \end{bmatrix} \sim \begin{bmatrix} 1 & 0 & -1 & | & -80 \\ 0 & 1 & 2 & | & 420 \\ 0 & 0 & 1 & | & 10 \end{bmatrix}$$

$$(0.1)R_2 + R_3 \rightarrow R_3 \quad 10R_3 \rightarrow R_3 \quad R_3 + R_1 \rightarrow R_1$$
$$(-2)R_2 + R_1 \rightarrow R_1 \qquad\qquad (-2)R_3 + R_2 \rightarrow R_2$$

$$\sim \begin{bmatrix} 1 & 0 & 0 & 20 \\ 0 & 1 & 0 & 220 \\ 0 & 0 & 1 & 100 \end{bmatrix}$$

Thus, $x_1 = 20$, $x_2 = 220$, and $x_3 = 100$, or 20 one-person boats, 220 two-person boats, and 100 four-person boats.

(B)    The mathematical model is:

$$0.5x_1 + x_2 + 1.5x_3 = 380$$
$$0.6x_1 + 0.9x_2 + 1.2x_3 = 330$$

$$\begin{bmatrix} 0.5 & 1 & 1.5 & 380 \\ 0.6 & 0.9 & 1.2 & 330 \end{bmatrix} \sim \begin{bmatrix} 1 & 2 & 3 & 760 \\ 0.6 & 0.9 & 1.2 & 330 \end{bmatrix} \sim \begin{bmatrix} 1 & 2 & 3 & 760 \\ 0 & -0.3 & -0.6 & -126 \end{bmatrix}$$

$$2R_1 \rightarrow R_1 \qquad (-0.6)R_1 + R_2 \rightarrow R_2 \qquad \left(-\dfrac{1}{0.3}\right)R_2 \rightarrow R_2$$

$$\sim \begin{bmatrix} 1 & 2 & 3 & 760 \\ 0 & 1 & 2 & 420 \end{bmatrix} \sim \begin{bmatrix} 1 & 0 & -1 & -80 \\ 0 & 1 & 2 & 420 \end{bmatrix}$$

$$(-2)R_2 + R_1 \rightarrow R_1$$

Thus,
$$x_1 \quad - x_3 = -80 \qquad (1)$$
$$x_2 + 2x_3 = 420 \qquad (2)$$

Let $x_3 = t$  ($t$ any real number). Then, $x_2 = 420 - 2t$ [from (2)] and $x_1 = t - 80$  [from (1)].

In order to keep $x_1$ and $x_2$ positive, $t \le 210$ and $t \ge 80$.

Thus,
$$x_1 = t - 80 \quad \text{(one-person boats)}$$
$$x_2 = 420 - 2t \quad \text{(two-person boats)}$$
$$x_3 = t \qquad \text{(four-person boats)}$$

where $80 \le t \le 210$ and $t$ is an integer.

(C)    The mathematical model is:
$$0.5x_1 + x_2 = 380$$
$$0.6x_1 + 0.9x_2 = 330$$
$$0.2x_1 + 0.3x_2 = 120$$

$$\begin{bmatrix} 0.5 & 1 & 380 \\ 0.6 & 0.9 & 330 \\ 0.2 & 0.3 & 120 \end{bmatrix} \sim \begin{bmatrix} 1 & 2 & 760 \\ 0.6 & 0.9 & 330 \\ 0.2 & 0.3 & 120 \end{bmatrix} \sim \begin{bmatrix} 1 & 1 & 760 \\ 0 & -0.3 & -126 \\ 0 & -0.1 & -32 \end{bmatrix} \sim \begin{bmatrix} 1 & 2 & 760 \\ 0 & 1 & 420 \\ 0 & -0.1 & -32 \end{bmatrix}$$

Copyright © 2011 Pearson Education, Inc. Publishing as Prentice Hall.

$$2R_1 \to R_1 \qquad (-0.6R_1) + R_2 \to R_2 \qquad \left(-\frac{1}{0.3}\right)R_2 \to R_2 \qquad 0.1R_2 + R_3 \to R_3$$
$$(-0.2R_1) + R_3 \to R_3$$

$$\sim \begin{bmatrix} 1 & 2 & | & 760 \\ 0 & 1 & | & 420 \\ 0 & 0 & | & 10 \end{bmatrix}$$   From this matrix, we conclude that there is no solution; there is no production schedule that will use all the labor-hours in all departments.

**67.**   Let $x_1$ = number of 8,000 gallon tank cars

$x_2$ = number of 16,000 gallon tank cars

$x_3$ = number of 24,000 gallon tank cars

Then, the mathematical model is:
$$x_1 + \quad x_2 + \quad x_3 = 24$$
$$8,000x_1 + 16,000x_2 + 24,000x_3 = 520,000$$

Dividing the second equation by 8,000, we get the equivalent system:
$$x_1 + x_2 + x_3 = 24$$
$$x_1 + 2x_2 + 3x_3 = 65$$

The augmented matrix corresponding to this system is
$$\begin{pmatrix} 1 & 1 & 1 & | & 24 \\ 1 & 2 & 3 & | & 65 \end{pmatrix} \sim \begin{pmatrix} 1 & 1 & 1 & | & 24 \\ 0 & 1 & 2 & | & 41 \end{pmatrix} \sim \begin{pmatrix} 1 & 0 & -1 & | & -17 \\ 0 & 1 & 2 & | & 41 \end{pmatrix}$$
$$(-1)R_1 + R_2 \to R_2 \qquad (-1)R_2 + R_1 \to R_1$$
Thus, $x_1 \quad - x_3 = -17$
$$x_2 + 2x_3 = 41$$

Let $x_3 = t$. Then $x_1 = t - 17$ and $x_2 = 41 - 2t$. Thus, $(t - 17)$ 8,000-gallon tank cars, $(41 - 2t)$ 16,000-gallon tank cars and $(t)$ 24,000-gallon tank cars should be purchased. Also, since $t$, $41 - 2t$ and $t - 17$ must each be non-negative integers, it follows that $t = 17, 18, 19,$ or $20$.

**69.**   The cost $C(t)$ of leasing $(t)$ 24,000 gallon tank cars, $(41 - 2t)$ 16,000-gallon tank cars, and $(t - 17)$ 8,000-gallon tank cars is:

$C(t) = 450(t - 17) + 650(41 - 2t) + 1,150t$ dollars per month.

$C(17) = 650(41 - 34) + 1,150(17) = 650(7) + 1,150(17) = \$24,100$
$C(18) = 450(1) + 650(5) + 1,150(18) = \$24,400$
$C(19) = 450(2) + 650(3) + 1,150(19) = \$24,700$
$C(20) = 450(3) + 650(1) + 1,150(20) = \$25,000$

The minimum monthly cost is $24,100 when 17 24,000-gallon tank cars and 7 16,000-gallon tank cars are leased.

**71.**   Let $x_1$ = federal income tax

$x_2$ = state income tax

$x_3$ = local income tax

Then, the mathematical model is:
$$x_1 = 0.50[7,650,000 - (x_2 + x_3)]$$
$$x_2 = 0.20[7,650,000 - (x_1 + x_3)]$$

Copyright © 2011 Pearson Education, Inc.  Publishing as Prentice Hall.

$$x_3 = 0.10[7,650,000 - (x_1 + x_2)]$$

and

$$x_1 + 0.5x_2 + 0.5x_3 = 3,825,000$$
$$0.2x_1 + x_2 + 0.2x_3 = 1,530,000$$
$$0.1x_1 + 0.1x_2 + x_3 = 765,000$$

The corresponding augmented matrix is:

$$\begin{bmatrix} 1 & 0.5 & 0.5 & | & 3,825,000 \\ 0.2 & 1 & 0.2 & | & 1,530,000 \\ 0.1 & 0.1 & 1 & | & 765,000 \end{bmatrix} \begin{array}{l} (-0.2)R_1 + R_2 \to R_2 \\ (-0.1)R_1 + R_3 \to R_3 \end{array}$$

$$\sim \begin{bmatrix} 1 & 0.5 & 0.5 & | & 3,825,000 \\ 0 & 0.9 & 0.1 & | & 765,000 \\ 0 & 0.05 & 0.95 & | & 382,500 \end{bmatrix} 20R_3 \to R_3 \text{ (simplify arithmetic)}$$

$$\sim \begin{bmatrix} 1 & 0.5 & 0.5 & | & 3,825,000 \\ 0 & 0.9 & 0.1 & | & 765,000 \\ 0 & 1 & 19 & | & 7,650,000 \end{bmatrix} R_2 \leftrightarrow R_3$$

$$\sim \begin{bmatrix} 1 & 0.5 & 0.5 & | & 3,825,000 \\ 0 & 1 & 19 & | & 7,650,000 \\ 0 & 0.9 & 0.1 & | & 765,000 \end{bmatrix} \begin{array}{l} (-0.5)R_2 + R_1 \to R_1 \\ (-0.9)R_2 + R_3 \to R_3 \end{array}$$

$$\sim \begin{bmatrix} 1 & 0 & -9 & | & 0 \\ 0 & 1 & 19 & | & 7,650,000 \\ 0 & 0 & -17 & | & -6,120,000 \end{bmatrix} \left(-\frac{1}{17}\right)R_3 \to R_3$$

$$\sim \begin{bmatrix} 1 & 0 & -9 & | & 0 \\ 0 & 1 & 19 & | & 7,650,000 \\ 0 & 0 & 1 & | & 360,000 \end{bmatrix} \begin{array}{l} 9R_3 + R_1 \to R_1 \\ (-19)R_3 + R_2 \to R_2 \end{array}$$

$$\sim \begin{bmatrix} 1 & 0 & 0 & | & 3,240,000 \\ 0 & 1 & 0 & | & 810,000 \\ 0 & 0 & 1 & | & 360,000 \end{bmatrix}$$

Thus, $x_1 = \$3,240,000$, $x_2 = \$810,000$, $x_3 = \$360,000$. The total tax liability is

$x_1 + x_2 + x_3 = \$4,410,000$ which is 57.65% of the taxable income $\left(\dfrac{4,410,000}{7,650,000} = 0.5765 \text{ or } 57.65\%\right)$.

**73.** Let $x_1$ = Taxable income of company $A$

$x_2$ = Taxable income of company $B$

$x_3$ = Taxable income of company $C$

$x_4$ = Taxable income of company $D$

The taxable income of each company is given by the system of equations:

$$x_1 = 0.71(3.2) + 0.08x_2 + 0.03x_3 + 0.07x_4$$
$$x_2 = 0.12x_1 + 0.81(2.6) + 0.11x_3 + 0.13x_4$$
$$x_3 = 0.11x_1 + 0.09x_2 + 0.72(3.8) + 0.08x_4$$

Copyright © 2011 Pearson Education, Inc. Publishing as Prentice Hall.

$$x_4 = 0.06x_1 + 0.02x_2 + 0.14x_3 + 0.72(4.4)$$

which is the same as:

$$x_1 - 0.08x_2 - 0.03x_3 - 0.07x_4 = 2.272$$
$$-0.12x_1 + x_2 - 0.11x_3 - 0.13x_4 = 2.106$$
$$-0.11x_1 - 0.09x_2 + x_3 - 0.08x_4 = 2.736$$
$$-0.06x_1 - 0.02x_2 - 0.14x_3 + x_4 = 3.168$$

The corresponding augmented coefficient matrix is

$$\begin{bmatrix} 1 & -0.08 & -0.03 & -0.07 & | & 2.272 \\ -0.12 & 1 & -0.11 & -0.13 & | & 2.106 \\ -0.11 & -0.09 & 1 & -0.08 & | & 2.736 \\ -0.06 & -0.02 & -0.14 & 1 & | & 3.168 \end{bmatrix}$$

and the reduced form (from a graphing utility) is

$$\begin{bmatrix} 1 & 0 & 0 & 0 & | & 2.927 \\ 0 & 1 & 0 & 0 & | & 3.372 \\ 0 & 0 & 1 & 0 & | & 3.675 \\ 0 & 0 & 0 & 1 & | & 3.926 \end{bmatrix}$$

The taxable incomes are: Company $A$ - \$2,927,000, Company $B$ - \$3,372,000, Company $C$ - \$3,675,000, Company $D$ - \$3,926,000

**75.** Let $x_1$ = number of ounces of food A,

$x_2$ = number of ounces of food B,

$x_3$ = number of ounces of food C.

(A)   The mathematical model is:

$$30x_1 + 10x_2 + 20x_3 = 340$$
$$10x_1 + 10x_2 + 20x_3 = 180$$
$$10x_1 + 30x_2 + 20x_3 = 220$$

$$\begin{bmatrix} 30 & 10 & 20 & | & 340 \\ 10 & 10 & 20 & | & 180 \\ 10 & 30 & 20 & | & 220 \end{bmatrix} \quad \sim \begin{bmatrix} 10 & 10 & 20 & | & 180 \\ 30 & 10 & 20 & | & 340 \\ 10 & 30 & 20 & | & 220 \end{bmatrix} \quad \sim \begin{bmatrix} 1 & 1 & 2 & | & 18 \\ 3 & 1 & 2 & | & 34 \\ 1 & 3 & 2 & | & 22 \end{bmatrix}$$

$$R_1 \leftrightarrow R_2 \qquad\qquad \frac{1}{10}R_1 \to R_1 \qquad\qquad (-3)R_1 + R_2 \to R_2$$
$$\qquad\qquad\qquad\qquad \frac{1}{10}R_2 \to R_2 \qquad\qquad (-1)R_1 + R_3 \to R_3$$
$$\qquad\qquad\qquad\qquad \frac{1}{10}R_3 \to R_3$$

$$\sim \begin{bmatrix} 1 & 1 & 2 & | & 18 \\ 0 & -2 & -4 & | & -20 \\ 0 & 2 & 0 & | & 4 \end{bmatrix} \quad \sim \begin{bmatrix} 1 & 1 & 2 & | & 18 \\ 0 & 1 & 2 & | & 10 \\ 0 & 2 & 0 & | & 4 \end{bmatrix} \quad \sim \begin{bmatrix} 1 & 0 & 0 & | & 8 \\ 0 & 1 & 2 & | & 10 \\ 0 & 0 & -4 & | & -16 \end{bmatrix}$$

$$-\frac{1}{2}R_2 \to R_2 \qquad (-1)R_2 + R_1 \to R_1 \qquad -\frac{1}{4}R_3 \to R_3$$
$$\qquad\qquad\qquad (-2)R_2 + R_3 \to R_3$$

Copyright © 2011 Pearson Education, Inc. Publishing as Prentice Hall.

$$\sim \begin{bmatrix} 1 & 0 & 0 & | & 8 \\ 0 & 1 & 2 & | & 10 \\ 0 & 0 & 1 & | & 4 \end{bmatrix} \sim \begin{bmatrix} 1 & 0 & 0 & | & 8 \\ 0 & 1 & 0 & | & 2 \\ 0 & 0 & 1 & | & 4 \end{bmatrix}$$

$(-2)R_3 + R_2 \rightarrow R_2$

Thus, $x_1 = 8$, $x_2 = 2$, $x_3 = 4$, or 8 ounces of food A, 2 ounces of food B, and 4 ounces of food C.

(B)    The mathematical model is:

$30x_1 + 10x_2 = 340$

$10x_1 + 10x_2 = 180$

$10x_1 + 30x_2 = 220$

$$\begin{bmatrix} 30 & 10 & | & 340 \\ 10 & 10 & | & 180 \\ 10 & 30 & | & 220 \end{bmatrix} \sim \begin{bmatrix} 3 & 1 & | & 34 \\ 1 & 1 & | & 18 \\ 1 & 3 & | & 22 \end{bmatrix} \sim \begin{bmatrix} 1 & 1 & | & 18 \\ 3 & 1 & | & 34 \\ 1 & 3 & | & 22 \end{bmatrix} \sim \begin{bmatrix} 1 & 1 & | & 18 \\ 0 & -2 & | & -20 \\ 0 & 2 & | & 4 \end{bmatrix}$$

$\frac{1}{10}R_1 \rightarrow R_1$        $R_1 \leftrightarrow R_2$        $(-3)R_1 + R_2 \rightarrow R_2$        $\left(-\frac{1}{2}\right)R_2 \rightarrow R_2$

$\frac{1}{10}R_2 \rightarrow R_2$                                $(-1)R_1 + R_3 \rightarrow R_3$

$\frac{1}{10}R_3 \rightarrow R_3$

$$\sim \begin{bmatrix} 1 & 1 & | & 18 \\ 0 & 1 & | & 10 \\ 0 & 2 & | & 4 \end{bmatrix} \sim \begin{bmatrix} 1 & 1 & | & 18 \\ 0 & 1 & | & 10 \\ 0 & 0 & | & -16 \end{bmatrix}$$        From this matrix, we conclude that there is no solution.

$(-2)R_2 + R_3 \rightarrow R_3$

(C)    The mathematical model is:

$30x_1 + 10x_2 + 20x_3 = 340$

$10x_1 + 10x_2 + 20x_3 = 180$

$$\begin{bmatrix} 30 & 10 & 20 & | & 340 \\ 10 & 10 & 20 & | & 180 \end{bmatrix} \sim \begin{bmatrix} 10 & 10 & 20 & | & 180 \\ 30 & 10 & 20 & | & 340 \end{bmatrix} \sim \begin{bmatrix} 1 & 1 & 2 & | & 18 \\ 3 & 1 & 2 & | & 34 \end{bmatrix} \sim \begin{bmatrix} 1 & 1 & 2 & | & 18 \\ 0 & -2 & -4 & | & -20 \end{bmatrix}$$

$R_1 \leftrightarrow R_2$                $\frac{1}{10}R_1 \rightarrow R_1$        $(-3)R_1 + R_2 \rightarrow R_2$        $\left(-\frac{1}{2}\right)R_2 \rightarrow R_2$

$\frac{1}{10}R_2 \rightarrow R_2$

$$\sim \begin{bmatrix} 1 & 1 & 2 & | & 18 \\ 0 & 1 & 2 & | & 10 \end{bmatrix} \sim \begin{bmatrix} 1 & 0 & 0 & | & 8 \\ 0 & 1 & 2 & | & 10 \end{bmatrix}$$        Thus,    $x_1 \qquad = 8$

$(-1)R_2 + R_1 \rightarrow R_1$                                            $x_2 + 2x_3 = 10$

Let $x_3 = t$ ($t$ any real number). Then, $x_2 = 10 - 2t$, $0 \le t \le 5$, for $x_2$ to be positive.

The solution is: $x_1 = 8$ ounces of food A; $x_2 = 10 - 2t$ ounces of food B; $x_3 = t$ ounces of food C, $0 \le t \le 5$.

Copyright © 2011 Pearson Education, Inc. Publishing as Prentice Hall.

**77.** Let $x_1$ = number of barrels of mix A,

$x_2$ = number of barrels of mix B,

$x_3$ = number of barrels of mix C,

$x_4$ = number of barrels of mix D,

The mathematical model is:

$30x_1 + 30x_2 + 30x_3 + 60x_4 = 900$ (1)

$50x_1 + 75x_2 + 25x_3 + 25x_4 = 750$ (2)

$30x_1 + 20x_2 + 20x_3 + 50x_4 = 700$ (3)

Divide each side of equation (1) by 30, each side of equation (2) by 25, and each side of equation (3) by 10. This yields the system of linear equations:

$x_1 + x_2 + x_3 + 2x_4 = 30$

$2x_1 + 3x_2 + x_3 + x_4 = 30$

$3x_1 + 2x_2 + 2x_3 + 5x_4 = 70$

$$\begin{bmatrix} 1 & 1 & 1 & 2 & | & 30 \\ 2 & 3 & 1 & 1 & | & 30 \\ 3 & 2 & 2 & 5 & | & 70 \end{bmatrix} \sim \begin{bmatrix} 1 & 1 & 1 & 2 & | & 30 \\ 0 & 1 & -1 & -3 & | & -30 \\ 0 & -1 & -1 & -1 & | & -20 \end{bmatrix} \sim \begin{bmatrix} 1 & 0 & 2 & 5 & | & 60 \\ 0 & 1 & -1 & -3 & | & -30 \\ 0 & 0 & -2 & -4 & | & -50 \end{bmatrix}$$

$(-2)R_1 + R_2 \rightarrow R_2$      $R_3 + R_2 \rightarrow R_3$      $\left(-\dfrac{1}{2}\right)R_3 \rightarrow R_3$

$(-3)R_1 + R_3 \rightarrow R_3$      $(-1)R_2 + R_1 \rightarrow R_1$

$$\sim \begin{bmatrix} 1 & 0 & 2 & 5 & | & 60 \\ 0 & 1 & -1 & -3 & | & -30 \\ 0 & 0 & 1 & 2 & | & 25 \end{bmatrix} \sim \begin{bmatrix} 1 & 0 & 0 & 1 & | & 10 \\ 0 & 1 & 0 & -1 & | & -5 \\ 0 & 0 & 1 & 2 & | & 25 \end{bmatrix}$$

$R_3 + R_2 \rightarrow R_2$

$(-2)R_3 + R_1 \rightarrow R_1$

Thus, $x_1 \quad\quad + x_4 = 10$

$x_2 \quad - x_4 = -5$

$x_3 + 2x_4 = 25$

Let $x_4 = t$ = number of barrels of mix D. Then $x_1 = 10 - t$ = number of barrels of mix A, $x_2 = t - 5$ = number of barrels of mix B, and $x_3 = 25 - 2t$ = number of barrels of mix C. Since the number of barrels of each mix must be nonnegative, $5 \leq t \leq 10$. Also, $t$ is an integer.

**79.** From Problem 77, the cost of the four mixes is

$C(t) = 46(10 - t) + 72(t - 5) + 57(25 - 2t) + 63t$

$= -25t + 1525, 5 \leq t \leq 10.$

The graph of $C$ is a straight line with negative slope. Therefore, the minimum occurs at $t = 10$. The minimizing solution is:  0 barrels of mix $A$,   5 barrels of mix $B$,   5 barrels of mix $C$, and  10 barrels of mix $D$.

Copyright © 2011 Pearson Education, Inc.  Publishing as Prentice Hall.

**81.**   $y = ax^2 + bx + c$

Let 1900 correspond to $x = 0$. Then   $P(0) = 75 = a(0)^2 + b(0) + c$.
Thus, $c = 75$.

1950:   $a(50)^2 + b(50) + 75 = 150$     or     $50a + b = \dfrac{3}{2}$

2000:   $a(100)^2 + b(100) + 75 = 275$     or     $100a + b = 2$

Substitute $b = 2 - 100a$   into   $50a + b = \dfrac{3}{2}$.   This gives

$$50a + (2 - 100a) = \frac{3}{2}$$

$$-50a = -\frac{1}{2}$$

$$a = \frac{1}{100} = 0.01$$

Therefore,   $b = 2 - 100\left(\dfrac{1}{100}\right) = 2 - 1 = 1$ and   $P = 0.01x^2 + x + 75$.

The year 2050 corresponds to $t = 150$:   $P(150) = 0.01(150)^2 + 150 + 75 = 450$.  The model estimates a population of 450 million in 2050.

**83.**   $L = ax^2 + bx + c$

Let 1980 – 1985 correspond to $t = 0$. Then
   $L(0) = c = 77.6$.

1985–1990:   $L(5) = a(5)^2 + b(5) + 77.6 = 78$        or     $25a + 5b = 0.4$
1990–1995:   $L(10) = a(10)^2 + b(10) + 77.6 = 78.6$     or   $100a + 10b = 1$

The augmented matrix for the equations is:

$$\begin{pmatrix} 25 & 5 & | & 0.4 \\ 100 & 10 & | & 1 \end{pmatrix} \rightarrow \begin{pmatrix} 1 & \frac{1}{5} & | & 0.016 \\ 100 & 10 & | & 1 \end{pmatrix} \rightarrow \begin{pmatrix} 1 & \frac{1}{5} & | & 0.016 \\ 0 & -10 & | & -0.6 \end{pmatrix}$$

$$\left(\frac{1}{25}\right)R_1 \rightarrow R_1 \qquad (-100)R_1 + R_2 \rightarrow R_2 \qquad \left(-\frac{1}{10}\right)R_2 \rightarrow R_2$$

$$\begin{pmatrix} 1 & \frac{1}{5} & | & 0.016 \\ 0 & 1 & | & 0.06 \end{pmatrix} \rightarrow \begin{pmatrix} 1 & 0 & | & 0.004 \\ 0 & 1 & | & 0.06 \end{pmatrix}$$

$$\left(-\frac{1}{5}\right)R_2 + R_1 \rightarrow R_1$$

Therefore, $a = 0.004$, $b = 0.06$, and $L = 0.004x^2 + 0.06x + 77.6$.
Life expectancy 1995–2000

$$L(15) = 0.004(15)^2 + 0.06(15) + 77.6 \approx 79.4 \text{ years}$$

Life expectancy 2000 – 2005

$$L(20) = 0.004(20)^2 + 0.06(20) + 77.6 \approx 80.4 \text{ years}$$

Copyright © 2011 Pearson Education, Inc. Publishing as Prentice Hall.

**85.** Quadratic Regression

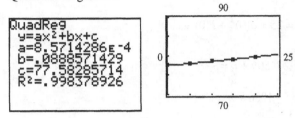

**87.** Let $x_1$ = number of hours for Company A,

and $x_2$ = number of hours for Company B.

The mathematical model is: $30x_1 + 20x_2 = 600$

$$10x_1 + 20x_2 = 400$$

Divide each side of each equation by 10. This yields the system of linear equations:

$3x_1 + 2x_2 = 60$

$x_1 + 2x_2 = 40$

$$\begin{bmatrix} 3 & 2 & | & 60 \\ 1 & 2 & | & 40 \end{bmatrix} \sim \begin{bmatrix} 1 & 2 & | & 40 \\ 3 & 2 & | & 60 \end{bmatrix} \sim \begin{bmatrix} 1 & 2 & | & 40 \\ 0 & -4 & | & -60 \end{bmatrix} \sim \begin{bmatrix} 1 & 2 & | & 40 \\ 0 & 1 & | & 15 \end{bmatrix} \sim \begin{bmatrix} 1 & 0 & | & 10 \\ 0 & 1 & | & 15 \end{bmatrix}$$

$R_1 \leftrightarrow R_2 \qquad (-3)R_1 + R_2 \to R_2 \qquad \left(-\dfrac{1}{4}\right)R_2 \to R_2 \qquad (-2)R_2 + R_1 \to R_1$

Thus, $x_1 = 10$ and $x_2 = 15$, or 10 hours for Company A and 15 hours for Company B.

**89.** **(A)** 6th and Washington Ave.: $x_1 + x_2 = 1200$

6th and Lincoln Ave.: $x_2 + x_3 = 1000$

5th and Lincoln Ave.: $x_3 + x_4 = 1300$

**(B)** The system of equations is:

$x_1 \qquad\quad + x_4 = 1500$

$x_1 + x_2 \qquad\quad = 1200$

$\quad x_2 + x_3 \quad = 1000$

$\qquad x_3 + x_4 = 1300$

$$\begin{bmatrix} 1 & 0 & 0 & 1 & | & 1500 \\ 1 & 1 & 0 & 0 & | & 1200 \\ 0 & 1 & 1 & 0 & | & 1000 \\ 0 & 0 & 1 & 1 & | & 1300 \end{bmatrix} \sim \begin{bmatrix} 1 & 0 & 0 & 1 & | & 1500 \\ 0 & 1 & 0 & -1 & | & -300 \\ 0 & 1 & 1 & 0 & | & 1000 \\ 0 & 0 & 1 & 1 & | & 1300 \end{bmatrix} \sim \begin{bmatrix} 1 & 0 & 0 & 1 & | & 1500 \\ 0 & 1 & 0 & -1 & | & -300 \\ 0 & 0 & 1 & 1 & | & 1300 \\ 0 & 0 & 1 & 1 & | & 1300 \end{bmatrix}$$

$(-1)R_1 + R_2 \to R_2 \qquad\qquad (-1)R_2 + R_3 \to R_3 \qquad\qquad (-1)R_3 + R_4 \to R_4$

$$\sim \begin{bmatrix} 1 & 0 & 0 & 1 & | & 1500 \\ 0 & 1 & 0 & -1 & | & -300 \\ 0 & 0 & 1 & 1 & | & 1300 \\ 0 & 0 & 0 & 0 & | & 0 \end{bmatrix}$$

Thus $x_1 \qquad\quad + x_4 = 1500$

$\qquad x_2 \qquad - x_4 = -300$

$\qquad\qquad x_3 + x_4 = 1300$

Let $x_4 = t$. Then $x_1 = 1500 - t$, $x_2 = t - 300$ and $x_3 = 1300 - t$. Since $x_1$, $x_2$, $x_3$, and $x_4$ must be nonnegative integers, we have $300 \le t \le 1300$.

Copyright © 2011 Pearson Education, Inc. Publishing as Prentice Hall.

(C)    The flow from Washington Ave. to Lincoln Ave. on 5th Street is given by $x_4 = t$. As shown in part (B), $300 \le t \le 1300$, that is, the maximum number of vehicles is 1300 and the minimum number is 300.

(D)    If $x_4 = t = 1000$, then   Washington Ave.:   $x_1 = 1500 - 1000 = 500$,   6th St.:   $x_2 = 1000 - 300 = 700$,   Lincoln Ave.: $x_3 = 1300 - 1000 = 300$.

---

## EXERCISE 4-4

*Things to remember:*

1.    A matrix with $m$ rows and $n$ columns is said to have SIZE $m \times n$. If a matrix has the same number of rows and columns, then it is called a SQUARE MATRIX. A matrix with only one column is a COLUMN MATRIX, and a matrix with only one row is a ROW MATRIX.

2.    Two matrices are EQUAL if they have the same size and their corresponding elements are equal.

3.    The SUM of two matrices of the same size, $m \times n$, is an $m \times n$ matrix whose elements are the sum of the corresponding elements of the two given matrices. Addition is not defined for matrices with different sizes. Matrix addition is commutative: $A + B = B + A$, and associative: $(A + B) + C = A + (B + C)$.

4.    A matrix with all elements equal to zero is called a ZERO MATRIX.

5.    The NEGATIVE OF A MATRIX $M$, denoted by $-M$, is the matrix whose elements are the negatives of the elements of $M$.

6.    If $A$ and $B$ are matrices of the same size, then subtraction is defined by $A - B = A + (-B)$. Thus, to subtract $B$ from $A$, simply subtract corresponding elements.

7.    If $M$ is a matrix and $k$ is a number, then $kM$ is the matrix formed by multiplying each element of $M$ by $k$.

8.    PRODUCT OF A ROW MATRIX AND A COLUMN MATRIX

The product of a $1 \times n$ row matrix and an $n \times 1$ column matrix is the $1 \times 1$ matrix given by

$$\underset{1 \times n}{\begin{bmatrix} a_1 & a_2 & \cdots a_n \end{bmatrix}} \underset{n \times 1}{\begin{bmatrix} b_1 \\ b_2 \\ \vdots \\ b_n \end{bmatrix}} = \begin{bmatrix} a_1 b_1 + a_2 b_2 + \cdots a_n b_n \end{bmatrix}$$

Note that the number of elements in the row matrix and the number of elements in the column matrix must be the same for the product to be defined.

Copyright © 2011 Pearson Education, Inc. Publishing as Prentice Hall.

9.    Let $A$ be an $m \times p$ matrix and $B$ be a $p \times n$ matrix. The MATRIX PRODUCT of $A$ and $B$, denoted $AB$, is the $m \times n$ matrix whose element in the $i$th row and the $j$th column is the real number obtained from the product of the $i$th row of $A$ and the $j$th column of $B$. If the number of columns in $A$ does not equal the number of rows in $B$, then the matrix product $AB$ is not defined.

NOTE: Matrix multiplication is *not* commutative. That is $AB$ does not always equal $BA$, even when both multiplications are defined.

---

1.  $\begin{bmatrix} 2 & -1 \\ 3 & 0 \end{bmatrix} + \begin{bmatrix} -3 & 1 \\ 2 & -3 \end{bmatrix} = \begin{bmatrix} 2+(-3) & -1+1 \\ 3+2 & 0+(-3) \end{bmatrix} = \begin{bmatrix} -1 & 0 \\ 5 & -3 \end{bmatrix}$

3.  Addition not defined; the matrices have different sizes.

5.  $\begin{bmatrix} 4 & -5 \\ 1 & 0 \\ 1 & -3 \end{bmatrix} - \begin{bmatrix} -1 & 2 \\ 6 & -2 \\ 1 & -7 \end{bmatrix} = \begin{bmatrix} 4-(-1) & -5-2 \\ 1-6 & 0-(-2) \\ 1-1 & -3-(-7) \end{bmatrix} = \begin{bmatrix} 5 & -7 \\ -5 & 2 \\ 0 & 4 \end{bmatrix}$

7.  $5\begin{bmatrix} 1 & -2 & 0 & 4 \\ -3 & 2 & -1 & 6 \end{bmatrix} = \begin{bmatrix} 5(1) & 5(-2) & 5(0) & 5(4) \\ 5(-3) & 5(2) & 5(-1) & 5(6) \end{bmatrix} = \begin{bmatrix} 5 & -10 & 0 & 20 \\ -15 & 10 & -5 & 30 \end{bmatrix}$

9.  $\begin{bmatrix} 3 & 4 \\ -1 & -2 \end{bmatrix}\begin{bmatrix} -1 \\ 2 \end{bmatrix} = \begin{bmatrix} [3 \ 4]\begin{bmatrix} -1 \\ 2 \end{bmatrix} \\ [-1 \ -2]\begin{bmatrix} -1 \\ 2 \end{bmatrix} \end{bmatrix} = \begin{bmatrix} -3+8 \\ 1-4 \end{bmatrix} = \begin{bmatrix} 5 \\ -3 \end{bmatrix}$

11. $\begin{bmatrix} 2 & -3 \\ 1 & 2 \end{bmatrix}\begin{bmatrix} 1 & -1 \\ 0 & -2 \end{bmatrix} = \begin{bmatrix} [2 \ -3]\begin{bmatrix} 1 \\ 0 \end{bmatrix} & [2 \ -3]\begin{bmatrix} -1 \\ -2 \end{bmatrix} \\ [1 \ 2]\begin{bmatrix} 1 \\ 0 \end{bmatrix} & [1 \ 2]\begin{bmatrix} -1 \\ -2 \end{bmatrix} \end{bmatrix} = \begin{bmatrix} 2+0 & -2+6 \\ 1+0 & -1-4 \end{bmatrix} = \begin{bmatrix} 2 & 4 \\ 1 & -5 \end{bmatrix}$

13. $\begin{bmatrix} 1 & -1 \\ 0 & -2 \end{bmatrix}\begin{bmatrix} 2 & -3 \\ 1 & 2 \end{bmatrix}$

$= \begin{bmatrix} [1 \ -1]\begin{bmatrix} 2 \\ 1 \end{bmatrix} & [1 \ -1]\begin{bmatrix} -3 \\ 2 \end{bmatrix} \\ [0 \ -2]\begin{bmatrix} 2 \\ 1 \end{bmatrix} & [0 \ -2]\begin{bmatrix} -3 \\ 2 \end{bmatrix} \end{bmatrix}$

$= \begin{bmatrix} 2-1 & -3-2 \\ 0-2 & 0-4 \end{bmatrix} = \begin{bmatrix} 1 & -5 \\ -2 & -4 \end{bmatrix}$

15. $\begin{bmatrix} 5 & 0 \\ 0 & 5 \end{bmatrix}\begin{bmatrix} 1 & 3 \\ 2 & 4 \end{bmatrix} = \begin{bmatrix} 5 & 10 \\ 15 & 20 \end{bmatrix}$       17. $\begin{bmatrix} 1 & 3 \\ 2 & 4 \end{bmatrix}\begin{bmatrix} 5 & 0 \\ 0 & 5 \end{bmatrix} = \begin{bmatrix} 5 & 10 \\ 15 & 20 \end{bmatrix}$

19. $\begin{bmatrix} 0 & 1 \\ 0 & 0 \end{bmatrix}\begin{bmatrix} 3 & 7 \\ 5 & 9 \end{bmatrix} = \begin{bmatrix} 7 & 9 \\ 0 & 0 \end{bmatrix}$       21. $\begin{bmatrix} 3 & 7 \\ 5 & 9 \end{bmatrix}\begin{bmatrix} 0 & 1 \\ 0 & 0 \end{bmatrix} = \begin{bmatrix} 0 & 3 \\ 0 & 7 \end{bmatrix}$

23. $\begin{bmatrix} 5 & -2 \end{bmatrix}\begin{bmatrix} -3 \\ -4 \end{bmatrix} = [-15+8] = [-7]$

Copyright © 2011 Pearson Education, Inc.  Publishing as Prentice Hall.

25. $\begin{bmatrix} -3 \\ -4 \end{bmatrix}[5 \ -2] = \begin{bmatrix} (-3)(5) & (-3)(-2) \\ (-4)(5) & (-4)(-2) \end{bmatrix} = \begin{bmatrix} -15 & 6 \\ -20 & 8 \end{bmatrix}$    27. $[3 \ -2 \ -4]\begin{bmatrix} 1 \\ 2 \\ -3 \end{bmatrix} = [(3 - 4 + 12)] = [11]$

29. $\begin{bmatrix} 1 \\ 2 \\ -3 \end{bmatrix}[3 \ -2 \ -4] = \begin{bmatrix} (1)(3) & (1)(-2) & (1)(-4) \\ (2)(3) & (2)(-2) & (2)(-4) \\ (-3)(3) & (-3)(-2) & (-3)(-4) \end{bmatrix} = \begin{bmatrix} 3 & -2 & -4 \\ 6 & -4 & -8 \\ -9 & 6 & 12 \end{bmatrix}$

31. $AC = \begin{bmatrix} 2 & -1 & 3 \\ 0 & 4 & -2 \end{bmatrix}\begin{bmatrix} -1 & 0 & 2 \\ 4 & -3 & 1 \\ -2 & 3 & 5 \end{bmatrix} = \begin{bmatrix} -12 & 12 & 18 \\ 20 & -18 & -6 \end{bmatrix}$

33. $AB$ is not defined; the number of columns of $A$ (3) does not equal the number of rows of $B$ (2).

35. $B^2 = BB = \begin{bmatrix} -3 & 1 \\ 2 & 5 \end{bmatrix}\begin{bmatrix} -3 & 1 \\ 2 & 5 \end{bmatrix} = \begin{bmatrix} 11 & 2 \\ 4 & 27 \end{bmatrix}$

37. $B + AD = \begin{bmatrix} -3 & 1 \\ 2 & 5 \end{bmatrix} + \begin{bmatrix} 2 & -1 & 3 \\ 0 & 4 & -2 \end{bmatrix}\begin{bmatrix} 3 & -2 \\ 0 & -1 \\ 1 & 2 \end{bmatrix}$

   $= \begin{bmatrix} -3 & 1 \\ 2 & 5 \end{bmatrix} + \begin{bmatrix} 9 & 3 \\ -2 & -8 \end{bmatrix} = \begin{bmatrix} 6 & 4 \\ 0 & -3 \end{bmatrix}$

39. $0.1DB = 0.1\begin{bmatrix} 3 & -2 \\ 0 & -1 \\ 1 & 2 \end{bmatrix}\begin{bmatrix} -3 & 1 \\ 2 & 5 \end{bmatrix} = 0.1\begin{bmatrix} -13 & -7 \\ -2 & -5 \\ 1 & 11 \end{bmatrix} = \begin{bmatrix} -1.3 & -0.7 \\ -0.2 & -0.5 \\ 0.1 & 1.1 \end{bmatrix}$

41. $3BA + 4AC = 3\begin{bmatrix} -3 & 1 \\ 2 & 5 \end{bmatrix}\begin{bmatrix} 2 & -1 & 3 \\ 0 & 4 & -2 \end{bmatrix} + 4\begin{bmatrix} 2 & -1 & 3 \\ 0 & 4 & -2 \end{bmatrix}\begin{bmatrix} -1 & 0 & 2 \\ 4 & -3 & 1 \\ -2 & 3 & 5 \end{bmatrix}$

   $= 3\begin{bmatrix} -6 & 7 & -11 \\ 4 & 18 & -4 \end{bmatrix} + 4\begin{bmatrix} -12 & 12 & 18 \\ 20 & -18 & -6 \end{bmatrix}$

   $= \begin{bmatrix} -18 & 21 & -33 \\ 12 & 54 & -12 \end{bmatrix} + \begin{bmatrix} -48 & 48 & 72 \\ 80 & -72 & -24 \end{bmatrix} = \begin{bmatrix} -66 & 69 & 39 \\ 92 & -18 & -36 \end{bmatrix}$

43. $(-2)BA + 6CD$ is not defined; $-2BA$ is $2 \times 3$, $6CD$ is $3 \times 2$.

Copyright © 2011 Pearson Education, Inc. Publishing as Prentice Hall.

**45.**  $ACD = A(CD) = \begin{bmatrix} 2 & -1 & 3 \\ 0 & 4 & -2 \end{bmatrix} \left( \begin{bmatrix} -1 & 0 & 2 \\ 4 & -3 & 1 \\ -2 & 3 & 5 \end{bmatrix} \begin{bmatrix} 3 & -2 \\ 0 & -1 \\ 1 & 2 \end{bmatrix} \right)$

$$= \begin{bmatrix} 2 & -1 & 3 \\ 0 & 4 & -2 \end{bmatrix} \begin{bmatrix} -1 & 6 \\ 13 & -3 \\ -1 & 11 \end{bmatrix} = \begin{bmatrix} -18 & 48 \\ 54 & -34 \end{bmatrix}$$

**47.**  $DBA = D(BA) = \begin{bmatrix} 3 & -2 \\ 0 & -1 \\ 1 & 2 \end{bmatrix} \left( \begin{bmatrix} -3 & 1 \\ 2 & 5 \end{bmatrix} \begin{bmatrix} 2 & -1 & 3 \\ 0 & 4 & -2 \end{bmatrix} \right)$

$$= \begin{bmatrix} 3 & -2 \\ 0 & -1 \\ 1 & 2 \end{bmatrix} \begin{bmatrix} -6 & 7 & -11 \\ 4 & 18 & -4 \end{bmatrix} = \begin{bmatrix} -26 & -15 & -25 \\ -4 & -18 & 4 \\ 2 & 43 & -19 \end{bmatrix}$$

**49.**  $AB = \begin{bmatrix} a & a \\ b & b \end{bmatrix} \begin{bmatrix} a & a \\ -a & -a \end{bmatrix} = \begin{bmatrix} a^2 - a^2 & a^2 - a^2 \\ ab - ab & ab - ab \end{bmatrix} = \begin{bmatrix} 0 & 0 \\ 0 & 0 \end{bmatrix}$

$BA = \begin{bmatrix} a & a \\ -a & -a \end{bmatrix} \begin{bmatrix} a & a \\ b & b \end{bmatrix} = \begin{bmatrix} a^2 + ab & a^2 + ab \\ -a^2 - ab & -a^2 - ab \end{bmatrix}$

**51.**  $A^2 = \begin{bmatrix} ab & b^2 \\ -a^2 & -ab \end{bmatrix} \begin{bmatrix} ab & b^2 \\ -a^2 & -ab \end{bmatrix} = \begin{bmatrix} a^2b^2 - a^2b^2 & ab^3 - ab^3 \\ -a^3b + a^3b & -a^2b^2 + a^2b^2 \end{bmatrix} = \begin{bmatrix} 0 & 0 \\ 0 & 0 \end{bmatrix}$

**53.**  $A = [0.3 \quad 0.7], B = \begin{bmatrix} 0.4 & 0.6 \\ 0.2 & 0.8 \end{bmatrix}$

$B^2 = \begin{bmatrix} 0.28 & 0.72 \\ 0.24 & 0.76 \end{bmatrix}, \quad B^3 = \begin{bmatrix} 0.256 & 0.744 \\ 0.248 & 0.752 \end{bmatrix} \cdots,$

$B^8 = \begin{bmatrix} 0.250 & 0.74999 \\ 0.24999 & 0.75000 \end{bmatrix}, \quad B^n \rightarrow \begin{bmatrix} 0.25 & 0.75 \\ 0.25 & 0.75 \end{bmatrix}$

$AB = [0.26 \quad 0.74], AB^2 = [0.252 \quad 0.748],$

$AB^3 = [0.2504 \quad 0.7496] \cdots,$

$AB^{10} = [0.2500\ldots \quad 0.7499\ldots], \quad AB^n \rightarrow [0.25 \quad 0.75]$

**55.**  $\begin{bmatrix} a & b \\ c & d \end{bmatrix} + \begin{bmatrix} 2 & -3 \\ 0 & 1 \end{bmatrix} = \begin{bmatrix} a+2 & b-3 \\ c+0 & c+1 \end{bmatrix} = \begin{bmatrix} 1 & -2 \\ 3 & -4 \end{bmatrix}$

Thus,  $a + 2 = 1, \qquad a = -1$

$\qquad b - 3 = -2, \qquad b = 1$

$\qquad c + 0 = 3, \qquad c = 3$

$\qquad d + 1 = -4, \qquad d = -5$

Copyright © 2011 Pearson Education, Inc.  Publishing as Prentice Hall.

57. $\begin{bmatrix} 1 & -2 \\ 2 & -3 \end{bmatrix} \begin{bmatrix} a & b \\ c & d \end{bmatrix} = \begin{bmatrix} 1 & 0 \\ 3 & 2 \end{bmatrix}$

$\begin{bmatrix} a-2c & b-2d \\ 2a-3c & 2b-3d \end{bmatrix} = \begin{bmatrix} 1 & 0 \\ 3 & 2 \end{bmatrix}$

implies     $a - 2c = 1$     $b - 2d = 0$

$2a - 3c = 3$     $2b - 3d = 2$

The augmented matrix for the first system is:

$\begin{bmatrix} 1 & -2 & | & 1 \\ 2 & -3 & | & 3 \end{bmatrix} (-2)R_1 + R_2 \rightarrow R_2 \sim \begin{bmatrix} 1 & -2 & | & 1 \\ 0 & 1 & | & 1 \end{bmatrix} 2R_2 + R_1 \rightarrow R_1$

$\sim \begin{bmatrix} 1 & 0 & | & 3 \\ 0 & 1 & | & 1 \end{bmatrix}$   Thus,   $a = 3$,   $c = 1$.

For the second system, substitute $b = 2d$ from the first equation into the second equation:

$2(2d) - 3d = 2$

$d = 2$

$b = 4$

Solution:   $a = 3$,   $b = 4$,   $c = 1$,   $d = 2$

59. False.  A 1x1 matrix is equivalent to a real number; multiplication of real numbers is commutative.

61. True.  For example,  $A = \begin{bmatrix} 0 & 0 \\ 1 & 2 \end{bmatrix}$,  $B = \begin{bmatrix} 2 & 0 \\ -1 & 0 \end{bmatrix}$;  $AB = \begin{bmatrix} 0 & 0 \\ 0 & 0 \end{bmatrix}$

63. Let $A = \begin{bmatrix} a_1 & 0 \\ 0 & a_2 \end{bmatrix}$ and $B = \begin{bmatrix} b_1 & 0 \\ 0 & b_2 \end{bmatrix}$

(A)   Always true:

$A + B = \begin{bmatrix} a_1 & 0 \\ 0 & a_2 \end{bmatrix} + \begin{bmatrix} b_1 & 0 \\ 0 & b_2 \end{bmatrix} = \begin{bmatrix} a_1 + b_1 & 0 \\ 0 & a_2 + b_2 \end{bmatrix}$

(B)   Always true: matrix addition is commutative, $A + B = B + A$ for *any* pair of matrices of the same size.

(C)   Always true:

$AB = \begin{bmatrix} a_1 & 0 \\ 0 & a_2 \end{bmatrix} \begin{bmatrix} b_1 & 0 \\ 0 & b_2 \end{bmatrix} = \begin{bmatrix} a_1 b_1 & 0 \\ 0 & a_2 b_2 \end{bmatrix}$

(D)   Always true:

$BA = \begin{bmatrix} b_1 & 0 \\ 0 & b_2 \end{bmatrix} \begin{bmatrix} a_1 & 0 \\ 0 & a_2 \end{bmatrix} = \begin{bmatrix} b_1 a_1 & 0 \\ 0 & b_2 a_2 \end{bmatrix} = \begin{bmatrix} a_1 b_1 & 0 \\ 0 & a_2 b_2 \end{bmatrix} = AB$

Copyright © 2011 Pearson Education, Inc.  Publishing as Prentice Hall.

**65.** $A + B = \begin{bmatrix} \$47 & \$39 \\ \$90 & \$125 \end{bmatrix} + \begin{bmatrix} \$56 & \$42 \\ \$84 & \$115 \end{bmatrix} = \begin{bmatrix} \$103 & \$81 \\ \$174 & \$240 \end{bmatrix}$

$\frac{1}{2}(A + B) = \frac{1}{2}\begin{matrix} \text{Guitar} & \text{Banjo} \\ \begin{bmatrix} \$103 & \$81 \\ \$174 & \$240 \end{bmatrix} \end{matrix} = \begin{bmatrix} \$51.50 & \$40.50 \\ \$87.00 & \$120.00 \end{bmatrix}\begin{matrix} \text{Materials} \\ \text{Labor} \end{matrix}$

**67.** The dealer is increasing the retail prices by 10%. Thus, the new retail price matrix $M'$ is:

$M' = M + 0.1M = (1.1)M = (1.1)\begin{bmatrix} 35{,}075 & 2560 & 1070 & 640 \\ 39{,}045 & 1840 & 770 & 460 \\ 45{,}535 & 3400 & 1415 & 850 \end{bmatrix}$

$= \begin{bmatrix} 38{,}582.50 & 2816 & 1177 & 704 \\ 42{,}949.50 & 2024 & 847 & 506 \\ 50{,}088.50 & 3740 & 1556.50 & 935 \end{bmatrix}$

$N' = N + 0.15N = (1.15)N = (1.15)\begin{bmatrix} 30{,}996 & 2050 & 850 & 510 \\ 34{,}857 & 1585 & 660 & 395 \\ 41{,}667 & 2890 & 1200 & 725 \end{bmatrix}$

$= \begin{bmatrix} 35{,}645.40 & 2357.50 & 977.50 & 586.50 \\ 40{,}085.55 & 1822.75 & 759 & 454.25 \\ 47{,}917.05 & 3323.50 & 1380 & 833.75 \end{bmatrix}$

The new markup (to the nearest dollar) is:

$M' - N' = \begin{matrix} & \text{Basic Car} & \text{Air Cond} & \text{AM/FM} & \text{Cruise} \\ \text{Model } A & \\ \text{Model } B & \\ \text{Model } C & \end{matrix}\begin{bmatrix} \$2937 & \$459 & \$200 & \$118 \\ \$2864 & \$201 & \$88 & \$52 \\ \$2171 & \$417 & \$177 & \$101 \end{bmatrix}$

**69.** (A) $[0.6 \ \ 0.6 \ \ 0.2]\begin{bmatrix} 17.30 \\ 12.22 \\ 10.63 \end{bmatrix} = \$19.84$

The labor cost per boat for one-person boats at the Massachusetts plant is: $19.84.

(B) $[1.5 \ \ 1.2 \ \ 0.4]\begin{bmatrix} 14.65 \\ 10.29 \\ 9.66 \end{bmatrix} = \$38.19$

The labor cost per boat for four-person boats at the Virginia plant is: $38.19.

(C) $MN$ gives the labor cost per boat at each plant; $NM$ is not defined.

(D) $MN = \begin{matrix} \text{MA} & \text{VA} \\ \begin{bmatrix} 19.84 & 16.90 \\ 31.49 & 26.81 \\ 44.87 & 38.19 \end{bmatrix} \end{matrix}\begin{matrix} \text{One-person boat} \\ \text{Two-person boat} \\ \text{Four-person boat} \end{matrix}$

Copyright © 2011 Pearson Education, Inc. Publishing as Prentice Hall.

71. (A)    $[4 \; 2] \begin{bmatrix} 15 \\ 5 \end{bmatrix} = 70$    There are 70 g of protein in Mix X.

   (B)    $[3 \; 1] \begin{bmatrix} 5 \\ 15 \end{bmatrix} = 30$    There are 30 g of fat in Mix Z.

   (C)    $MN$ gives the amount (in grams) of protein, carbohydrates and fat in 20 ounces of each mix. The product $NM$ is not defined.

   (D)    $MN = \begin{bmatrix} 4 & 2 \\ 20 & 16 \\ 3 & 1 \end{bmatrix} \begin{bmatrix} 15 & 10 & 5 \\ 5 & 10 & 15 \end{bmatrix} = \begin{array}{c} \\ \\ \\ \end{array} \begin{matrix} \text{Mix X} & \text{Mix Y} & \text{Mix Z} \\ \begin{bmatrix} 70 & 60 & 50 \\ 380 & 360 & 340 \\ 50 & 40 & 30 \end{bmatrix} & & \end{matrix} \begin{array}{l} \text{Protein} \\ \text{Carbohydrate} \\ \text{Fat} \end{array}$

73. (A)    $[1000 \; 500 \; 5000] \begin{bmatrix} 1.20 \\ 3.00 \\ 1.45 \end{bmatrix} = 9{,}950;$  total amount spent in Berkeley = \$9,950.

   (B)    $[2000 \; 800 \; 8000] \begin{bmatrix} 1.20 \\ 3.00 \\ 1.45 \end{bmatrix} = 16{,}400;$  total amount spent in Oakland = \$16,400.

   (C)    $MN$ is not defined; $NM$ gives the total cost for each city.

   (D)    $NM = \begin{bmatrix} 1000 & 500 & 5000 \\ 2000 & 800 & 8000 \end{bmatrix} \begin{bmatrix} 1.20 \\ 3.00 \\ 1.45 \end{bmatrix} = \begin{matrix} \text{Cost/City} \\ \begin{bmatrix} \$9{,}950 \\ \$16{,}400 \end{bmatrix} \end{matrix} \begin{array}{l} \text{Berkeley} \\ \text{Oakland} \end{array}$

   (E)    $[1 \; 1] \cdot N = [1 \; 1] \begin{bmatrix} 1000 & 500 & 5000 \\ 2000 & 800 & 8000 \end{bmatrix}$

   $\begin{matrix} \text{Telephone} & \text{House} & \text{letters} \\ = [3{,}000 & 1{,}300 & 13{,}000] \end{matrix}$

   (F)    $N \cdot \begin{bmatrix} 1 \\ 1 \\ 1 \end{bmatrix} = \begin{bmatrix} 1{,}000 & 500 & 5{,}000 \\ 2{,}000 & 800 & 8{,}000 \end{bmatrix} \begin{bmatrix} 1 \\ 1 \\ 1 \end{bmatrix} = \begin{matrix} \text{Total contacts} \\ \begin{bmatrix} 6{,}500 \\ 10{,}800 \end{bmatrix} \end{matrix} \begin{array}{l} \text{Berkeley} \\ \text{Oakland} \end{array}$

## EXERCISE 4-5

*Things to remember:*

1. The IDENTITY element for multiplication for the set of square matrices of order $n$ (dimension $n \times n$) is the square matrix $I$ of order $n$ which has 1's on the principal diagonal (upper left corner to lower right corner) and 0's elsewhere. The identity matrices of order 2

Copyright © 2011 Pearson Education, Inc. Publishing as Prentice Hall.

and 3, respectively, are

$$I = \begin{bmatrix} 1 & 0 \\ 0 & 1 \end{bmatrix} \text{ and } I = \begin{bmatrix} 1 & 0 & 0 \\ 0 & 1 & 0 \\ 0 & 0 & 1 \end{bmatrix}.$$

2.  If $M$ is any square matrix of order $n$ and $I$ is the identity matrix of order $n$, then

$$IM = MI = M.$$

3.  INVERSE OF A SQUARE MATRIX

Let $M$ be a square matrix of order $n$ and $I$ be the identity matrix of order $n$. If there exists a matrix $M^{-1}$ such that

$$MM^{-1} = M^{-1}M = I$$

then $M^{-1}$ is called the MULTIPLICATIVE INVERSE OF $M$ or, more simply, the INVERSE OF $M$. $M^{-1}$ is read "$M$ inverse."

4.  If the augmented matrix $[M \,|\, I]$ is transformed by row operations into $[I \,|\, B]$, then the resulting matrix $B$ is $M^{-1}$. However, if all zeros are obtained in one or more rows to the left of the vertical line during the row transformation procedure, then $M^{-1}$ does not exist. Matrices that do not have inverses are called SINGULAR MATRICES.

---

1.  (a) $\begin{bmatrix} 1 & 0 \\ 0 & 0 \end{bmatrix}\begin{bmatrix} 2 & -3 \\ 4 & 5 \end{bmatrix} = \begin{bmatrix} 2 & -3 \\ 0 & 0 \end{bmatrix}$
    (b) $\begin{bmatrix} 2 & -3 \\ 4 & 5 \end{bmatrix}\begin{bmatrix} 1 & 0 \\ 0 & 0 \end{bmatrix} = \begin{bmatrix} 2 & 0 \\ 4 & 0 \end{bmatrix}$

3.  (a) $\begin{bmatrix} 0 & 0 \\ 0 & 1 \end{bmatrix}\begin{bmatrix} 2 & -3 \\ 4 & 5 \end{bmatrix} = \begin{bmatrix} 0 & 0 \\ 4 & 5 \end{bmatrix}$
    (b) $\begin{bmatrix} 2 & -3 \\ 4 & 5 \end{bmatrix}\begin{bmatrix} 0 & 0 \\ 0 & 1 \end{bmatrix} = \begin{bmatrix} 0 & -3 \\ 0 & 5 \end{bmatrix}$

5.  (a) $\begin{bmatrix} 1 & 0 \\ 0 & 1 \end{bmatrix}\begin{bmatrix} 2 & -3 \\ 4 & 5 \end{bmatrix} = \begin{bmatrix} 2 & -3 \\ 4 & 5 \end{bmatrix}$
    (b) $\begin{bmatrix} 2 & -3 \\ 4 & 5 \end{bmatrix}\begin{bmatrix} 1 & 0 \\ 0 & 1 \end{bmatrix} = \begin{bmatrix} 2 & -3 \\ 4 & 5 \end{bmatrix}$

7.  $\begin{bmatrix} 1 & 0 & 0 \\ 0 & 1 & 0 \\ 0 & 0 & 1 \end{bmatrix}\begin{bmatrix} -2 & 1 & 3 \\ 2 & 4 & -2 \\ 5 & 1 & 0 \end{bmatrix}$

$$= \begin{bmatrix} 1(-2)+0\cdot 2+0\cdot 5 & 1\cdot 1+0\cdot 4+0\cdot 1 & 1\cdot 3+0(-2)+0\cdot 0 \\ 0(-2)+1\cdot 2+0\cdot 5 & 0\cdot 1+1\cdot 4+0\cdot 1 & 0\cdot 3+1(-2)+0\cdot 0 \\ 0(-2)+0\cdot 2+1\cdot 5 & 0\cdot 1+0\cdot 4+1\cdot 1 & 0\cdot 3+0(-2)+1\cdot 0 \end{bmatrix} = \begin{bmatrix} -2 & 1 & 3 \\ 2 & 4 & -2 \\ 5 & 1 & 0 \end{bmatrix}$$

9.  $\begin{bmatrix} -2 & 1 & 3 \\ 2 & 4 & -2 \\ 5 & 1 & 0 \end{bmatrix}\begin{bmatrix} 1 & 0 & 0 \\ 0 & 1 & 0 \\ 0 & 0 & 1 \end{bmatrix}$

$$= \begin{bmatrix} (-2)\cdot 1+1\cdot 0+3\cdot 0 & (-2)0+1\cdot 1+3\cdot 0 & (-2)0+1\cdot 0+3\cdot 1 \\ 2\cdot 1+4\cdot 0+(-2)0 & 2\cdot 0+4\cdot 1+(-2)0 & 2\cdot 0+4\cdot 0+(-2)1 \\ 5\cdot 1+1\cdot 0+0\cdot 0 & 5\cdot 0+1\cdot 1+0\cdot 0 & 5\cdot 0+1\cdot 0+0\cdot 1 \end{bmatrix} = \begin{bmatrix} -2 & 1 & 3 \\ 2 & 4 & -2 \\ 5 & 1 & 0 \end{bmatrix}$$

Copyright © 2011 Pearson Education, Inc. Publishing as Prentice Hall.

**11.** $\begin{bmatrix} 3 & -4 \\ -2 & 3 \end{bmatrix} \begin{bmatrix} 3 & 4 \\ 2 & 3 \end{bmatrix} = \begin{bmatrix} 1 & 0 \\ 0 & 1 \end{bmatrix}$ Yes

**13.** $\begin{bmatrix} 2 & 2 \\ -1 & -1 \end{bmatrix} \begin{bmatrix} 1 & 1 \\ -1 & -1 \end{bmatrix} = \begin{bmatrix} 0 & 0 \\ 0 & 0 \end{bmatrix}$ No

**17.** No. The second matrix has a column of zeros; it does not have an inverse.

**15.** $\begin{bmatrix} -5 & 2 \\ -8 & 3 \end{bmatrix} \begin{bmatrix} 3 & -2 \\ 8 & -5 \end{bmatrix} = \begin{bmatrix} 1 & 0 \\ 0 & 1 \end{bmatrix}$ Yes

**19.** $\begin{bmatrix} 1 & -1 & 1 \\ 0 & 2 & -1 \\ 2 & 3 & 0 \end{bmatrix} \begin{bmatrix} 3 & 3 & -1 \\ -2 & -2 & 1 \\ -4 & -5 & 2 \end{bmatrix} = \begin{bmatrix} 1 & 0 & 0 \\ 0 & 1 & 0 \\ 0 & 0 & 1 \end{bmatrix}$ Yes

**21.** The matrix is not square.

**23.** The matrix is not square.

**25.** The matrix has a row of zeros.

**27.** The matrix has a column of zeros.

**29.** $D = 1 \cdot 6 - 2 \cdot 3 = 0$

**31.** $\begin{bmatrix} -1 & 0 & | & 1 & 0 \\ -3 & 1 & | & 0 & 1 \end{bmatrix} \sim \begin{bmatrix} 1 & 0 & | & -1 & 0 \\ -3 & 1 & | & 0 & 1 \end{bmatrix} \sim \begin{bmatrix} 1 & 0 & | & -1 & 0 \\ 0 & 1 & | & -3 & 1 \end{bmatrix}$

$(-1)R_1 \to R_1 \qquad 3R_1 + R_2 \to R_2$

Thus, $M^{-1} = \begin{bmatrix} -1 & 0 \\ -3 & 1 \end{bmatrix}$

Check:

$M \cdot M^{-1} = \begin{bmatrix} -1 & 0 \\ -3 & 1 \end{bmatrix} \begin{bmatrix} -1 & 0 \\ -3 & 1 \end{bmatrix} = \begin{bmatrix} (-1)(-1)+0(-3) & (-1)0+0\cdot1 \\ (-3)(-1)+1(-3) & (-3)0+1\cdot1 \end{bmatrix} = \begin{bmatrix} 1 & 0 \\ 0 & 1 \end{bmatrix}$

**33.** $\begin{bmatrix} 1 & 2 & | & 1 & 0 \\ 1 & 3 & | & 0 & 1 \end{bmatrix} \sim \begin{bmatrix} 1 & 2 & | & 1 & 0 \\ 0 & 1 & | & -1 & 1 \end{bmatrix} \sim \begin{bmatrix} 1 & 0 & | & 3 & -2 \\ 0 & 1 & | & -1 & 1 \end{bmatrix}$

$(-1)R_1 + R_2 \to R_2 \qquad (-2)R_2 + R_1 \to R_1$

Thus, $M^{-1} = \begin{bmatrix} 3 & -2 \\ -1 & 1 \end{bmatrix}$.

Check:

$M \cdot M^{-1} = \begin{bmatrix} 1 & 2 \\ 1 & 3 \end{bmatrix} \begin{bmatrix} 3 & -2 \\ -1 & 1 \end{bmatrix} = \begin{bmatrix} 7 & -3 \\ -2 & 1 \end{bmatrix} = \begin{bmatrix} 1 & 0 \\ 0 & 1 \end{bmatrix}$

**35.** $\begin{bmatrix} 1 & 3 & | & 1 & 0 \\ 2 & 7 & | & 0 & 1 \end{bmatrix} \sim \begin{bmatrix} 1 & 3 & | & 1 & 0 \\ 0 & 1 & | & -2 & 1 \end{bmatrix} \sim \begin{bmatrix} 1 & 0 & | & 7 & -3 \\ 0 & 1 & | & -2 & 1 \end{bmatrix}$

$(-2)R_1 + R_2 \to R_2 \qquad (-3)R_2 + R_1 \to R_1$

Thus, $M^{-1} = \begin{bmatrix} 7 & -3 \\ -2 & 1 \end{bmatrix}$.

Check:

$\begin{bmatrix} 1 & 3 \\ 2 & 7 \end{bmatrix} \begin{bmatrix} 7 & -3 \\ -2 & 1 \end{bmatrix} = \begin{bmatrix} 1\cdot7+3(-2) & 1(-3)+3\cdot1 \\ 2\cdot7+7(-2) & 2(-3)+7\cdot1 \end{bmatrix} = \begin{bmatrix} 1 & 0 \\ 0 & 1 \end{bmatrix}$

Copyright © 2011 Pearson Education, Inc. Publishing as Prentice Hall.

**37.** $\begin{bmatrix} 1 & -3 & 0 & | & 1 & 0 & 0 \\ 0 & 1 & 1 & | & 0 & 1 & 0 \\ 2 & -1 & 4 & | & 0 & 0 & 1 \end{bmatrix} (-2)R_1 + R_3 \to R_3 \sim \begin{bmatrix} 1 & -3 & 0 & | & 1 & 0 & 0 \\ 0 & 1 & 1 & | & 0 & 1 & 0 \\ 0 & 5 & 4 & | & -2 & 0 & 1 \end{bmatrix} \begin{array}{l} 3R_2 + R_1 \to R_1 \\ (-5)R_2 + R_3 \to R_3 \end{array}$

$\sim \begin{bmatrix} 1 & 0 & 3 & | & 1 & 3 & 0 \\ 0 & 1 & 1 & | & 0 & 1 & 0 \\ 0 & 0 & -1 & | & -2 & -5 & 1 \end{bmatrix} (-1)R_3 \to R_3 \sim \begin{bmatrix} 1 & 0 & 3 & | & 1 & 3 & 0 \\ 0 & 1 & 1 & | & 0 & 1 & 0 \\ 0 & 0 & -1 & | & 2 & 5 & -1 \end{bmatrix} \begin{array}{l} (-3)R_3 + R_1 \to R_1 \\ (-1)R_3 + R_2 \to R_2 \end{array}$

$\sim \begin{bmatrix} 1 & 0 & 0 & | & -5 & -12 & 3 \\ 0 & 1 & 0 & | & -2 & -4 & 1 \\ 0 & 0 & 1 & | & 2 & 5 & -1 \end{bmatrix}; \quad M^{-1} = \begin{bmatrix} -5 & -12 & 3 \\ -2 & -4 & 1 \\ 2 & 5 & -1 \end{bmatrix}$

$M \cdot M^{-1} = \begin{bmatrix} 1 & -3 & 0 \\ 0 & 1 & 1 \\ 2 & -1 & 4 \end{bmatrix} \begin{bmatrix} -5 & -12 & 3 \\ -2 & -4 & 1 \\ 2 & 5 & -1 \end{bmatrix} = \begin{bmatrix} 1 & 0 & 0 \\ 0 & 1 & 0 \\ 0 & 0 & 1 \end{bmatrix} = \begin{bmatrix} 1 & 0 & 0 \\ 0 & 1 & 0 \\ 0 & 0 & 1 \end{bmatrix}$

**39.** $\begin{bmatrix} 1 & 1 & 0 & | & 1 & 0 & 0 \\ 2 & 3 & -1 & | & 0 & 1 & 0 \\ 1 & 0 & 2 & | & 0 & 0 & 1 \end{bmatrix} \begin{array}{l} (-2)R_1 + R_2 \to R_2 \\ (-1)R_1 + R_3 \to R_3 \end{array} \sim \begin{bmatrix} 1 & 1 & 0 & | & 1 & 0 & 0 \\ 0 & 1 & -1 & | & -2 & 1 & 0 \\ 0 & -1 & 2 & | & -1 & 0 & 1 \end{bmatrix} \begin{array}{l} (-1)R_2 + R_1 \to R_1 \\ R_2 + R_3 \to R_3 \end{array}$

$\sim \begin{bmatrix} 1 & 0 & 1 & | & 3 & -1 & 0 \\ 0 & 1 & -1 & | & -2 & 1 & 0 \\ 0 & 0 & 1 & | & -3 & 1 & 1 \end{bmatrix} \begin{array}{l} (-1)R_3 + R_1 \to R_1 \\ R_3 + R_2 \to R_2 \end{array} \sim \begin{bmatrix} 1 & 0 & 0 & | & 6 & -2 & -1 \\ 0 & 1 & 0 & | & -5 & 2 & 1 \\ 0 & 0 & 1 & | & -3 & 1 & 1 \end{bmatrix}; \quad M^{-1} = \begin{bmatrix} 6 & -2 & -1 \\ -5 & 2 & 1 \\ -3 & 1 & 1 \end{bmatrix}$

$M \cdot M^{-1} = \begin{bmatrix} 1 & 1 & 0 \\ 2 & 3 & -1 \\ 1 & 0 & 2 \end{bmatrix} \begin{bmatrix} 6 & -2 & -1 \\ -5 & 2 & 1 \\ -3 & 1 & 1 \end{bmatrix} = \begin{bmatrix} 1 & 0 & 0 \\ 0 & 1 & 0 \\ 0 & 0 & 1 \end{bmatrix}$

**41.** $\begin{bmatrix} 4 & 3 & | & 1 & 0 \\ -3 & -2 & | & 0 & 1 \end{bmatrix} R_2 + R_1 \to R_1 \sim \begin{bmatrix} 1 & 1 & | & 1 & 1 \\ -3 & -2 & | & 0 & 1 \end{bmatrix} 3R_1 + R_2 \to R_2$

$\sim \begin{bmatrix} 1 & 1 & | & 1 & 1 \\ 0 & 1 & | & 3 & 4 \end{bmatrix} (-1)R_2 + R_1 \to R_1 \sim \begin{bmatrix} 1 & 0 & | & -2 & -3 \\ 0 & 1 & | & 3 & 4 \end{bmatrix} (-1)R_2 + R_1 \to R_1$

$M^{-1} = \begin{bmatrix} -2 & -3 \\ 3 & 4 \end{bmatrix}$

**43.** $\begin{bmatrix} 2 & 6 & | & 1 & 0 \\ 3 & 9 & | & 0 & 1 \end{bmatrix} \frac{1}{2}R_1 \to R_1 \sim \begin{bmatrix} 1 & 3 & | & \frac{1}{2} & 0 \\ 3 & 9 & | & 0 & 1 \end{bmatrix} (-3)R_1 + R_2 \to R_2 \sim \begin{bmatrix} 1 & 3 & | & \frac{1}{2} & 0 \\ 0 & 0 & | & -\frac{3}{2} & 1 \end{bmatrix}$

The inverse does not exist.

Copyright © 2011 Pearson Education, Inc.  Publishing as Prentice Hall.

**45.** $\begin{bmatrix} 2 & 1 & | & 1 & 0 \\ 4 & 3 & | & 0 & 1 \end{bmatrix} \dfrac{1}{2}R_1 \to R_1 \sim \begin{bmatrix} 1 & \frac{1}{2} & | & \frac{1}{2} & 0 \\ 4 & 3 & | & 0 & 1 \end{bmatrix} (-4)R_1 + R_2 \to R_2$

$\sim \begin{bmatrix} 1 & \frac{1}{2} & | & \frac{1}{2} & 0 \\ 0 & 1 & | & -2 & 1 \end{bmatrix} \left(-\dfrac{1}{2}\right)R_2 + R_1 \to R_1 \sim \begin{bmatrix} 1 & 0 & | & \frac{3}{2} & -\frac{1}{2} \\ 0 & 1 & | & -2 & 1 \end{bmatrix}$

$M^{-1} = \begin{bmatrix} \frac{3}{2} & -\frac{1}{2} \\ -2 & 1 \end{bmatrix} = \begin{bmatrix} 1.5 & -0.5 \\ -2 & 1 \end{bmatrix}$

**47.** $\begin{bmatrix} 2 & 0 \\ 0 & 2 \end{bmatrix}^{-1} = \begin{bmatrix} \frac{1}{2} & 0 \\ 0 & \frac{1}{2} \end{bmatrix}$

**49.** $\begin{bmatrix} 3 & 0 \\ 0 & -5 \end{bmatrix}^{-1} = \begin{bmatrix} \frac{1}{3} & 0 \\ 0 & -\frac{1}{5} \end{bmatrix}$

**51.** $\begin{bmatrix} 4 & 0 & 0 \\ 0 & 2 & 0 \\ 0 & 0 & -8 \end{bmatrix}^{-1} = \begin{bmatrix} \frac{1}{4} & 0 & 0 \\ 0 & \frac{1}{2} & 0 \\ 0 & 0 & -\frac{1}{8} \end{bmatrix}$

**53.** $\begin{bmatrix} -5 & -2 & -2 & | & 1 & 0 & 0 \\ 2 & 1 & 0 & | & 0 & 1 & 0 \\ 1 & 0 & 1 & | & 0 & 0 & 1 \end{bmatrix} \sim \begin{bmatrix} 1 & 0 & 1 & | & 0 & 0 & 1 \\ 2 & 1 & 0 & | & 0 & 1 & 0 \\ -5 & -2 & -2 & | & 1 & 0 & 0 \end{bmatrix} \sim \begin{bmatrix} 1 & 0 & 1 & | & 0 & 0 & 1 \\ 0 & 1 & -2 & | & 0 & 1 & -1 \\ 0 & -2 & 3 & | & 1 & 0 & 5 \end{bmatrix}$

$\qquad R_1 \leftrightarrow R_3 \qquad\qquad (-2)R_1 + R_2 \to R_2 \qquad\qquad 2R_2 + R_3 \to R_3$
$\qquad\qquad\qquad\qquad\qquad\quad 5R_1 + R_3 \to R_3$

$\sim \begin{bmatrix} 1 & 0 & 1 & | & 0 & 0 & 1 \\ 0 & 1 & -2 & | & 0 & 1 & -2 \\ 0 & 0 & -1 & | & 1 & 2 & 1 \end{bmatrix} \sim \begin{bmatrix} 1 & 0 & 1 & | & 0 & 0 & 1 \\ 0 & 1 & -2 & | & 0 & 1 & -2 \\ 0 & 0 & 1 & | & -1 & -2 & -1 \end{bmatrix} \sim \begin{bmatrix} 1 & 0 & 0 & | & 1 & 2 & 2 \\ 0 & 1 & 0 & | & -2 & -3 & -4 \\ 0 & 0 & 1 & | & -1 & -2 & -1 \end{bmatrix}$

$\qquad (-1)R_3 \to R_3 \qquad\qquad 2R_3 + R_2 \to R_2$
$\qquad\qquad\qquad\qquad\quad (-1)R_3 + R_1 \to R_1$

Thus, the inverse is $\begin{bmatrix} 1 & 2 & 2 \\ -2 & -3 & -4 \\ -1 & -2 & -1 \end{bmatrix}$.

**55.** $\begin{bmatrix} 2 & 1 & 1 & | & 1 & 0 & 0 \\ 1 & 1 & 0 & | & 0 & 1 & 0 \\ -1 & -1 & 0 & | & 0 & 0 & 1 \end{bmatrix} \to \begin{bmatrix} 1 & 1 & 0 & | & 0 & 1 & 0 \\ 2 & 1 & 1 & | & 1 & 0 & 0 \\ -1 & -1 & 0 & | & 0 & 0 & 1 \end{bmatrix} \to \begin{bmatrix} 1 & 1 & 0 & | & 0 & 1 & 0 \\ 0 & -1 & 1 & | & 1 & -2 & 0 \\ 0 & 0 & 0 & | & 0 & 1 & 1 \end{bmatrix}$

$\qquad R_1 \leftrightarrow R_2 \qquad\qquad (-2)R_1 + R_2 \to R_2$
$\qquad\qquad\qquad\qquad\quad R_1 + R_3 \to R_3$

From this matrix, we conclude that the inverse does not exist.

**57.** $\begin{bmatrix} -1 & -2 & 2 & | & 1 & 0 & 0 \\ 4 & 3 & 0 & | & 0 & 1 & 0 \\ 4 & 0 & 4 & | & 0 & 0 & 1 \end{bmatrix} (-1)R_1 \to R_1 \sim \begin{bmatrix} 1 & 2 & -2 & | & -1 & 0 & 0 \\ 4 & 3 & 0 & | & 0 & 1 & 0 \\ 4 & 0 & 4 & | & 0 & 0 & 1 \end{bmatrix} \begin{matrix} (-4)R_1 + R_2 \to R_2 \\ (-4)R_1 + R_3 \to R_3 \end{matrix}$

Copyright © 2011 Pearson Education, Inc.  Publishing as Prentice Hall.

$$\sim \begin{bmatrix} 1 & 2 & -2 & | & -1 & 0 & 0 \\ 0 & -5 & 8 & | & 4 & 1 & 0 \\ 0 & -8 & 12 & | & 4 & 0 & 1 \end{bmatrix} \left(-\frac{1}{5}\right)R_2 \to R_2 \sim \begin{bmatrix} 1 & 2 & -2 & | & -1 & 0 & 0 \\ 0 & 1 & -\frac{8}{5} & | & -\frac{4}{5} & -\frac{1}{5} & 0 \\ 0 & -8 & 12 & | & 4 & 0 & 1 \end{bmatrix} \begin{array}{l} (-2)R_2 + R_1 \to R_1 \\ 8R_2 + R_3 \to R_3 \end{array}$$

$$\sim \begin{bmatrix} 1 & 0 & \frac{6}{5} & | & \frac{3}{5} & \frac{2}{5} & 0 \\ 0 & 1 & -\frac{8}{5} & | & -\frac{4}{5} & -\frac{1}{5} & 0 \\ 0 & 0 & -\frac{4}{5} & | & -\frac{12}{5} & -\frac{8}{5} & 1 \end{bmatrix} \left(-\frac{5}{4}\right)R_3 \to R_3$$

$$\sim \begin{bmatrix} 1 & 0 & \frac{6}{5} & | & \frac{3}{5} & \frac{2}{5} & 0 \\ 0 & 1 & -\frac{8}{5} & | & -\frac{4}{5} & -\frac{1}{5} & 0 \\ 0 & 0 & 1 & | & 3 & 2 & -\frac{5}{4} \end{bmatrix} \begin{array}{l} \left(-\frac{6}{5}\right)R_3 + R_1 \to R_1 \\ \left(\frac{8}{5}\right)R_3 + R_2 \to R_2 \end{array}$$

$$\sim \begin{bmatrix} 1 & 0 & 0 & | & -3 & -2 & \frac{3}{2} \\ 0 & 1 & 0 & | & 4 & 3 & -2 \\ 0 & 0 & 1 & | & 3 & 2 & -\frac{5}{4} \end{bmatrix}$$

$$M^{-1} = \begin{bmatrix} -3 & -2 & \frac{3}{2} \\ 4 & 3 & -2 \\ 3 & 2 & -\frac{5}{4} \end{bmatrix} = \begin{bmatrix} -3 & -2 & 1.5 \\ 4 & 3 & -2 \\ 3 & 2 & -1.25 \end{bmatrix}$$

**59.** $\begin{bmatrix} 2 & -1 & -2 & | & 1 & 0 & 0 \\ -4 & 2 & 8 & | & 0 & 1 & 0 \\ 6 & -2 & -1 & | & 0 & 0 & 1 \end{bmatrix} \dfrac{1}{2}R_1 \to R_1$

$$\sim \begin{bmatrix} 1 & -\frac{1}{2} & -1 & | & \frac{1}{2} & 0 & 0 \\ -4 & 2 & 8 & | & 0 & 1 & 0 \\ 6 & -2 & -1 & | & 0 & 0 & 1 \end{bmatrix} \begin{array}{l} 4R_1 + R_2 \to R_2 \\ (-6)R_1 + R_3 \to R_3 \end{array}$$

$$\sim \begin{bmatrix} 1 & -\frac{1}{2} & -1 & | & \frac{1}{2} & 0 & 0 \\ 0 & 0 & 4 & | & 2 & 1 & 0 \\ 0 & 1 & 5 & | & -3 & 0 & 1 \end{bmatrix} R_2 \leftrightarrow R_3$$

$$\sim \begin{bmatrix} 1 & -\frac{1}{2} & -1 & | & \frac{1}{2} & 0 & 0 \\ 0 & 1 & 5 & | & -3 & 0 & 1 \\ 0 & 0 & 4 & | & 2 & 1 & 0 \end{bmatrix} \begin{array}{l} \frac{1}{2}R_2 + R_1 \to R_1 \\ \frac{1}{4}R_3 \to R_3 \end{array}$$

$$\sim \begin{bmatrix} 1 & 0 & \frac{3}{2} & | & -1 & 0 & \frac{1}{2} \\ 0 & 1 & 5 & | & -3 & 0 & 1 \\ 0 & 0 & 1 & | & \frac{1}{2} & \frac{1}{4} & 0 \end{bmatrix} \begin{array}{l} \left(-\frac{3}{2}\right)R_3 + R_1 \to R_1 \\ (-5)R_3 + R_2 \to R_2 \end{array}$$

$$\sim \begin{bmatrix} 1 & 0 & 0 & | & -\frac{7}{4} & -\frac{3}{8} & \frac{1}{2} \\ 0 & 1 & 0 & | & -\frac{11}{2} & -\frac{5}{4} & 1 \\ 0 & 0 & 1 & | & \frac{1}{2} & \frac{1}{4} & 0 \end{bmatrix}$$

$$M^{-1} = \begin{bmatrix} -\frac{7}{4} & -\frac{3}{8} & \frac{1}{2} \\ -\frac{11}{2} & -\frac{5}{4} & 1 \\ \frac{1}{2} & \frac{1}{4} & 0 \end{bmatrix} = \begin{bmatrix} -1.75 & -0.375 & 0.5 \\ -5.5 & -1.25 & 1 \\ 0.5 & 0.25 & 0 \end{bmatrix}$$

Copyright © 2011 Pearson Education, Inc. Publishing as Prentice Hall.

**61.** $A = \begin{bmatrix} 4 & 3 \\ 3 & 2 \end{bmatrix}$;    $\begin{bmatrix} 4 & 3 & | & 1 & 0 \\ 3 & 2 & | & 0 & 1 \end{bmatrix} (-1)R_2 + R_1 \rightarrow R_1$

$\sim \begin{bmatrix} 1 & 1 & | & 1 & -1 \\ 3 & 2 & | & 0 & 1 \end{bmatrix} (-3)R_1 + R_2 \rightarrow R_2 \sim \begin{bmatrix} 1 & 1 & | & 1 & -1 \\ 0 & -1 & | & -3 & 4 \end{bmatrix} (-1)R_2 \rightarrow R_2$

$\sim \begin{bmatrix} 1 & 1 & | & 1 & -1 \\ 0 & 1 & | & 3 & -4 \end{bmatrix} (-1)R_2 + R_1 \rightarrow R_1 \sim \begin{bmatrix} 1 & 0 & | & -2 & 3 \\ 0 & 1 & | & 3 & -4 \end{bmatrix}$

$A^{-1} = \begin{bmatrix} -2 & 3 \\ 3 & -4 \end{bmatrix}$;    $\begin{bmatrix} -2 & 3 & | & 1 & 0 \\ 3 & -4 & | & 0 & 1 \end{bmatrix} R_2 + R_1 \rightarrow R_1$

$\sim \begin{bmatrix} 1 & -1 & | & 1 & 1 \\ 3 & -4 & | & 0 & 1 \end{bmatrix} (-3)R_1 + R_2 \rightarrow R_2 \sim \begin{bmatrix} 1 & -1 & | & 1 & 1 \\ 0 & -1 & | & -3 & -2 \end{bmatrix} (-1)R_2 \rightarrow R_2$

$\sim \begin{bmatrix} 1 & -1 & | & 1 & 1 \\ 0 & 1 & | & 3 & 2 \end{bmatrix} R_2 + R_1 \rightarrow R_1 \sim \begin{bmatrix} 1 & 0 & | & 4 & 3 \\ 0 & 1 & | & 3 & 2 \end{bmatrix}$

Thus, $(A^{-1})^{-1} = \begin{bmatrix} 4 & 3 \\ 3 & 2 \end{bmatrix} = A$

**63.** $\begin{bmatrix} a & 0 & | & 1 & 0 \\ 0 & d & | & 0 & 1 \end{bmatrix} \frac{1}{a}R_1 \rightarrow R_1$, provided $a \neq 0$
    $\frac{1}{d}R_2 \rightarrow R_2$, provided $d \neq 0$

$\begin{bmatrix} 1 & 0 & | & \frac{1}{a} & 0 \\ 0 & 1 & | & 0 & \frac{1}{d} \end{bmatrix}$. $M^{-1}$ exists and equals $\begin{bmatrix} \frac{1}{a} & 0 \\ 0 & \frac{1}{d} \end{bmatrix}$ if and only if $a \neq 0$, $d \neq 0$. In general, the inverse of a

diagonal matrix exists if and only if each of the diagonal elements is non-zero.

**65.** $A = \begin{bmatrix} 3 & 2 \\ -4 & -3 \end{bmatrix}$

$A^{-1}$: $\begin{bmatrix} 3 & 2 & | & 1 & 0 \\ -4 & -3 & | & 0 & 1 \end{bmatrix} \sim \begin{bmatrix} -1 & -1 & | & 1 & 1 \\ -4 & -3 & | & 0 & 1 \end{bmatrix} \sim \begin{bmatrix} 1 & 1 & | & -1 & -1 \\ -4 & -3 & | & 0 & 1 \end{bmatrix}$
$\qquad R_1 + R_2 \rightarrow R_1 \qquad\quad (-1)R_1 \rightarrow R_1 \qquad\quad 4R_1 + R_2 \rightarrow R_2$

$\sim \begin{bmatrix} 1 & 1 & | & -1 & -1 \\ 0 & 1 & | & -4 & -3 \end{bmatrix} \sim \begin{bmatrix} 1 & 0 & | & 3 & 2 \\ 0 & 1 & | & -4 & -3 \end{bmatrix}$
$\quad (-1)R_2 + R_1 \rightarrow R_1$

$A^{-1} = \begin{bmatrix} 3 & 2 \\ -4 & -3 \end{bmatrix} = A$; $A^2 = AA = AA^{-1} = \begin{bmatrix} 1 & 0 \\ 0 & 1 \end{bmatrix} = I$

**67.** $A = \begin{bmatrix} 4 & 3 \\ -5 & -4 \end{bmatrix}$

$A^{-1}$: $\begin{bmatrix} 4 & 3 & | & 1 & 0 \\ -5 & -4 & | & 0 & 1 \end{bmatrix} \sim \begin{bmatrix} -1 & -1 & | & 1 & 0 \\ -5 & -4 & | & 0 & 1 \end{bmatrix} \sim \begin{bmatrix} 1 & 1 & | & -1 & -1 \\ -5 & -4 & | & 0 & 1 \end{bmatrix}$
$\qquad R_2 + R_1 \rightarrow R_1 \qquad\quad (-1)R_1 \rightarrow R_1 \qquad\quad 5R_1 + R_2 \rightarrow R_2$

$$\sim \begin{bmatrix} 1 & 1 & -1 & -1 \\ 0 & 1 & -5 & -4 \end{bmatrix} \sim \begin{bmatrix} 1 & 0 & 4 & 3 \\ 0 & 1 & -5 & -4 \end{bmatrix}$$

$$(-1)R_2 + R_1 \to R_1$$

$$A^{-1} = \begin{bmatrix} 4 & 3 \\ -5 & -4 \end{bmatrix} = A; \quad A^2 = AA = AA^{-1} = I$$

**69.** $A = \begin{bmatrix} 1 & 2 \\ 1 & 3 \end{bmatrix}$

Assign the numbers 1—26 to the letters of the alphabet, in order, and let 0 correspond to a blank space.
Then the message "THE SUN ALSO RISES" corresponds to the sequence

20 8 5 0 19 21 14 0 1 12 19 15 0 18 9 19 5 19

To encode this message, divide the numbers into groups of two and use the groups as columns of a matrix $B$
with two rows

$$B = \begin{bmatrix} 20 & 5 & 19 & 14 & 1 & 19 & 0 & 9 & 5 \\ 8 & 0 & 21 & 0 & 12 & 15 & 18 & 19 & 19 \end{bmatrix}$$

Now

$$AB = \begin{bmatrix} 1 & 2 \\ 1 & 3 \end{bmatrix} \begin{bmatrix} 20 & 5 & 19 & 14 & 1 & 19 & 0 & 9 & 5 \\ 8 & 0 & 21 & 0 & 12 & 15 & 18 & 19 & 19 \end{bmatrix}$$

$$= \begin{bmatrix} 36 & 5 & 61 & 14 & 25 & 49 & 36 & 47 & 43 \\ 44 & 5 & 82 & 14 & 37 & 64 & 54 & 66 & 62 \end{bmatrix}$$

The coded message is:

36 44 5 5 61 82 14 14 25 37 49 64 36 54 47 66 43 62.

**71.** First we must find the inverse of $A = \begin{bmatrix} 1 & 2 \\ 1 & 3 \end{bmatrix}$

$$\begin{bmatrix} 1 & 2 & 1 & 0 \\ 1 & 3 & 0 & 1 \end{bmatrix} \sim \begin{bmatrix} 1 & 2 & 1 & 0 \\ 0 & 1 & -1 & 1 \end{bmatrix} \sim \begin{bmatrix} 1 & 0 & 3 & -2 \\ 0 & 1 & -1 & 1 \end{bmatrix}$$

$$(-1)R_1 + R_2 \to R_2 \qquad (-2)R_2 + R_1 \to R_1$$

Thus, $A^{-1} = \begin{bmatrix} 3 & -2 \\ -1 & 1 \end{bmatrix}$

Now $\begin{bmatrix} 3 & -2 \\ -1 & 1 \end{bmatrix} \begin{bmatrix} 37 & 24 & 46 & 49 & 8 & 36 & 5 & 41 & 22 \\ 52 & 29 & 69 & 69 & 8 & 44 & 5 & 50 & 26 \end{bmatrix}$

$$= \begin{bmatrix} 7 & 14 & 0 & 9 & 8 & 20 & 5 & 23 & 14 \\ 15 & 5 & 23 & 20 & 0 & 8 & 0 & 9 & 4 \end{bmatrix}$$

Thus, the decoded message is

7 15 14 5 0 23 9 20 8 0 20 8 5 0 23 9 14 4

which corresponds to  GONE WITH THE WIND

Copyright © 2011 Pearson Education, Inc.  Publishing as Prentice Hall.

**73.** "THE BEST YEARS OF OUR LIVES" corresponds to the sequence

20 8  5 0  2 5 19 20 0  25 5 1  18 19 0  15 6 0  15 21  18 0  12 9 22 5 19

We divide the numbers in the sequence into groups of 5 and use these groups as the columns of a matrix with 5 rows, adding 3 blanks at the end to make the columns come out even. Then we multiply this matrix on the left by the given matrix $B$.

$$\begin{bmatrix} 1 & 0 & 1 & 0 & 1 \\ 0 & 1 & 1 & 0 & 3 \\ 2 & 1 & 1 & 1 & 1 \\ 0 & 0 & 1 & 0 & 2 \\ 1 & 1 & 1 & 2 & 1 \end{bmatrix} \begin{bmatrix} 20 & 5 & 5 & 15 & 18 & 5 \\ 8 & 19 & 1 & 6 & 0 & 19 \\ 5 & 20 & 18 & 0 & 12 & 0 \\ 0 & 0 & 19 & 15 & 9 & 0 \\ 2 & 25 & 0 & 21 & 22 & 0 \end{bmatrix} = \begin{bmatrix} 27 & 50 & 23 & 36 & 52 & 5 \\ 19 & 114 & 19 & 69 & 78 & 19 \\ 55 & 74 & 48 & 72 & 79 & 29 \\ 9 & 70 & 18 & 42 & 56 & 0 \\ 35 & 69 & 62 & 72 & 70 & 24 \end{bmatrix}$$

The encoded message is:

27 19 55  9 35 50 114 74 70 69 23 19 48 18 62 36 69

72 42 72 52 78 79  56 70 5 19 29 0 24

**75.** First, we must find the inverse of $B$:

$$B^{-1} = \begin{bmatrix} -2 & -1 & 2 & 2 & -1 \\ 3 & 2 & -2 & -4 & 1 \\ 6 & 2 & -4 & -5 & 2 \\ -2 & -1 & 1 & 2 & 0 \\ -3 & -1 & 2 & 3 & -1 \end{bmatrix}$$

Now

$$B^{-1} \begin{bmatrix} 32 & 24 & 27 & 29 & 21 & 8 \\ 34 & 21 & 44 & 65 & 66 & 0 \\ 60 & 67 & 85 & 82 & 44 & 16 \\ 19 & 11 & 16 & 28 & 41 & 0 \\ 40 & 69 & 85 & 82 & 62 & 8 \end{bmatrix} = \begin{bmatrix} 20 & 18 & 19 & 15 & 0 & 8 \\ 8 & 5 & 20 & 23 & 5 & 0 \\ 5 & 1 & 0 & 0 & 1 & 0 \\ 0 & 20 & 19 & 15 & 18 & 0 \\ 7 & 5 & 8 & 14 & 20 & 0 \end{bmatrix}$$

The decoded message:

20 8 5 0  7 18 5 1 20 5 19 20 0 19 8 15 23 0 15 14 0 5 1 18 20 8

which corresponds to:   THE GREATEST SHOW ON EARTH

---

## EXERCISE 4-6

*Things to remember:*

1.  **BASIC PROPERTIES OF MATRICES**

    Assuming all products and sums are defined for the indicated matrices $A$, $B$, $C$, $I$, and $O$, then

    **ADDITION PROPERTIES**

    | | | |
    |---|---|---|
    | Associative: | : | $(A + B) + C = A + (B + C)$ |
    | Commutative: | | $A + B = B + A$ |
    | Additive Identity: | | $A + 0 = 0 + A = A$ |
    | Additive Inverse: | | $A + (-A) = (-A) + A = 0$ |

Copyright © 2011 Pearson Education, Inc. Publishing as Prentice Hall.

MULTIPLICATION PROPERTIES

Associative Property:      $A(BC) = (AB)C$

Multiplicative Identity:      $AI = IA = A$

Multiplicative Inverse:      If $A$ is a square matrix and $A^{-1}$ exists, then $AA^{-1} = A^{-1}A = I$.

COMBINED PROPERTIES

Left Distributive:      $A(B + C) = AB + AC$

Right Distributive:      $(B + C)A = BA + CA$

EQUALITY

Addition:      If $A = B$, then $A + C = B + C$.

Left Multiplication:      If $A = B$, then $CA = CB$.

Right Multplication:      If $A = B$, then $AC = BC$.

2.    USING INVERSE METHODS TO SOLVE SYSTEMS OF EQUATIONS

If the number of equations in a system equals the number of variables and the coefficient matrix has an inverse, then the system will always have a unique solution that can be found by using the inverse of the coefficient matrix to solve the corresponding matrix equation.

Matrix Equation          Solution

$$AX = B \qquad\qquad X = A^{-1}B$$

**1.** $\begin{bmatrix} 3 & 1 \\ 2 & -1 \end{bmatrix} \begin{bmatrix} x_1 \\ x_2 \end{bmatrix} = \begin{bmatrix} 5 \\ -4 \end{bmatrix}$

$\begin{bmatrix} 3x_1 + x_2 \\ 2x_1 - x_2 \end{bmatrix} = \begin{bmatrix} 5 \\ -4 \end{bmatrix}$

Thus, $3x_1 + x_2 = 5$
$\phantom{Thus, } 2x_1 - x_2 = -4$

**3.** $\begin{bmatrix} -3 & 1 & 0 \\ 2 & 0 & 1 \\ -1 & 3 & -2 \end{bmatrix} \begin{bmatrix} x_1 \\ x_2 \\ x_3 \end{bmatrix} = \begin{bmatrix} 3 \\ -4 \\ 2 \end{bmatrix}$

$\begin{bmatrix} -3x_1 + x_2 \\ 2x_1 + x_3 \\ -x_1 + 3x_2 - 2x_3 \end{bmatrix} = \begin{bmatrix} 3 \\ -4 \\ 2 \end{bmatrix}$

Thus, $\quad -3x_1 + x_2 \phantom{+x_3} = 3$
$\phantom{Thus, } 2x_1 \phantom{+x_2} + x_3 = -4$
$\phantom{Thus, } -x_1 + 3x_2 - 2x_3 = 2$

**5.** $3x_1 - 4x_2 = 1$
$\phantom{5.} 2x_1 + \phantom{4}x_2 = 5$

$\begin{bmatrix} 3x_1 - 4x_2 \\ 2x_1 + x_2 \end{bmatrix} = \begin{bmatrix} 1 \\ 5 \end{bmatrix}$ and $\begin{bmatrix} 3 & -4 \\ 2 & 1 \end{bmatrix} \begin{bmatrix} x_1 \\ x_2 \end{bmatrix} = \begin{bmatrix} 1 \\ 5 \end{bmatrix}$

**7.** $\phantom{-}x_1 - 3x_2 + 2x_3 = -3$
$-2x_1 + 3x_2 \phantom{+ 2x_3} = 1$
$\phantom{-}x_1 + \phantom{3}x_2 + 4x_3 = -2$

Copyright © 2011 Pearson Education, Inc. Publishing as Prentice Hall.

$$\begin{bmatrix} x_1 & -3x_2 & +2x_3 \\ -2x_1 & +3x_2 & \\ x_1 & +x_2 & +4x_3 \end{bmatrix} = \begin{bmatrix} -3 \\ 1 \\ -2 \end{bmatrix} \text{ and } \begin{bmatrix} 1 & -3 & 2 \\ -2 & 3 & 0 \\ 1 & 1 & 4 \end{bmatrix}\begin{bmatrix} x_1 \\ x_2 \\ x_3 \end{bmatrix} = \begin{bmatrix} -3 \\ 1 \\ -2 \end{bmatrix}$$

**9.** $\begin{bmatrix} x_1 \\ x_2 \end{bmatrix} = \begin{bmatrix} 3 & -2 \\ 1 & 4 \end{bmatrix}\begin{bmatrix} -2 \\ 1 \end{bmatrix} = \begin{bmatrix} 3(-2)+(-2)1 \\ 1(-2)+4\cdot1 \end{bmatrix} = \begin{bmatrix} -8 \\ 2 \end{bmatrix}$   Thus, $x_1 = -8$ and $x_2 = 2$

**11.** $\begin{bmatrix} x_1 \\ x_2 \end{bmatrix} = \begin{bmatrix} -2 & 3 \\ 2 & -1 \end{bmatrix}\begin{bmatrix} 3 \\ 2 \end{bmatrix} = \begin{bmatrix} (-2)3+3\cdot2 \\ 2\cdot3+(-1)2 \end{bmatrix} = \begin{bmatrix} 0 \\ 4 \end{bmatrix}$   Thus, $x_1 = 0$ and $x_2 = 4$

**13.** $\begin{bmatrix} 1 & -1 \\ 1 & -2 \end{bmatrix}\begin{bmatrix} x_1 \\ x_2 \end{bmatrix} = \begin{bmatrix} 5 \\ 7 \end{bmatrix}$

If $A = \begin{bmatrix} 1 & -1 \\ 1 & -2 \end{bmatrix}$ has an inverse, then $\begin{bmatrix} x_1 \\ x_2 \end{bmatrix} = A^{-1}\begin{bmatrix} 5 \\ 7 \end{bmatrix}$.

$$\left[\begin{array}{cc|cc} 1 & -1 & 1 & 0 \\ 1 & -2 & 0 & 1 \end{array}\right] \sim \left[\begin{array}{cc|cc} 1 & -1 & 1 & 0 \\ 0 & -1 & -1 & 1 \end{array}\right] \sim \left[\begin{array}{cc|cc} 1 & -1 & 1 & 0 \\ 0 & 1 & 1 & -1 \end{array}\right]$$
$$(-1)R_1 \to R_2 \qquad (-1)R_2 \to R_2 \qquad R_2 + R_1 \to R_1$$
$$\sim \left[\begin{array}{cc|cc} 1 & 0 & 2 & -1 \\ 0 & 1 & 1 & -1 \end{array}\right]; \quad A^{-1} = \begin{bmatrix} 2 & -1 \\ 1 & -1 \end{bmatrix}, \quad \begin{bmatrix} x_1 \\ x_2 \end{bmatrix} = \begin{bmatrix} 2 & -1 \\ 1 & -1 \end{bmatrix}\begin{bmatrix} 5 \\ 7 \end{bmatrix} = \begin{bmatrix} 3 \\ -2 \end{bmatrix}$$
Therefore, $x_1 = 3$, $x_2 = -2$.

**15.** $\begin{bmatrix} 1 & 1 \\ 2 & -3 \end{bmatrix}\begin{bmatrix} x_1 \\ x_2 \end{bmatrix} = \begin{bmatrix} 15 \\ 10 \end{bmatrix}$

If $A = \begin{bmatrix} 1 & 1 \\ 2 & -3 \end{bmatrix}$ has an inverse, then $\begin{bmatrix} x_1 \\ x_2 \end{bmatrix} = A^{-1}\begin{bmatrix} 15 \\ 10 \end{bmatrix}$

$$\left[\begin{array}{cc|cc} 1 & 1 & 1 & 0 \\ 2 & -3 & 0 & 1 \end{array}\right] \sim \left[\begin{array}{cc|cc} 1 & 1 & 1 & 0 \\ 0 & -5 & -2 & 1 \end{array}\right] \sim \left[\begin{array}{cc|cc} 1 & 1 & 1 & 0 \\ 0 & 1 & \frac{2}{5} & -\frac{1}{5} \end{array}\right]$$
$$(-2)R_1 + R_2 \to R_2 \qquad \left(-\frac{1}{5}\right)R_2 \to R_2 \qquad (-1)R_2 + R_1 \to R_1$$
$$\sim \left[\begin{array}{cc|cc} 1 & 0 & \frac{3}{5} & \frac{1}{5} \\ 0 & 1 & \frac{2}{5} & -\frac{1}{5} \end{array}\right]; \quad A^{-1} = \begin{bmatrix} \frac{3}{5} & \frac{1}{5} \\ \frac{2}{5} & -\frac{1}{5} \end{bmatrix}, \quad \begin{bmatrix} x_1 \\ x_2 \end{bmatrix} = \begin{bmatrix} \frac{3}{5} & \frac{1}{5} \\ \frac{2}{5} & -\frac{1}{5} \end{bmatrix}\begin{bmatrix} 15 \\ 10 \end{bmatrix} = \begin{bmatrix} 11 \\ 4 \end{bmatrix}$$
Therefore, $x_1 = 11$, $x_2 = 4$.

**17.** $\begin{bmatrix} 1 & 2 \\ 1 & 1 \end{bmatrix}\begin{bmatrix} x_1 \\ x_2 \end{bmatrix} + \begin{bmatrix} 3 \\ 4 \end{bmatrix} = \begin{bmatrix} 9 \\ 9 \end{bmatrix}$

$\begin{bmatrix} 1 & 2 \\ 1 & 1 \end{bmatrix}\begin{bmatrix} x_1 \\ x_2 \end{bmatrix} = \begin{bmatrix} 9 \\ 9 \end{bmatrix} - \begin{bmatrix} 3 \\ 4 \end{bmatrix} = \begin{bmatrix} 6 \\ 5 \end{bmatrix}$

If $A = \begin{bmatrix} 1 & 2 \\ 1 & 1 \end{bmatrix}$ has an inverse, then $\begin{bmatrix} x_1 \\ x_2 \end{bmatrix} = A^{-1}\begin{bmatrix} 6 \\ 5 \end{bmatrix}$

Copyright © 2011 Pearson Education, Inc. Publishing as Prentice Hall.

$$\begin{bmatrix} 1 & 2 & | & 1 & 0 \\ 1 & 1 & | & 0 & 1 \end{bmatrix} \sim \begin{bmatrix} 1 & 2 & | & 1 & 0 \\ 0 & -1 & | & -1 & 1 \end{bmatrix} \sim \begin{bmatrix} 1 & 2 & | & 1 & 0 \\ 0 & 1 & | & 1 & -1 \end{bmatrix} \sim \begin{bmatrix} 1 & 0 & | & -1 & 2 \\ 0 & 1 & | & 1 & -1 \end{bmatrix}$$

$(-1)R_1 + R_2 \to R_2 \quad (-1)R_2 \to R_2 \quad (-2)R_2 + R_1 \to R_1$

$$A^{-1} = \begin{bmatrix} -1 & 2 \\ 1 & -1 \end{bmatrix}; \quad \begin{bmatrix} x_1 \\ x_2 \end{bmatrix} = \begin{bmatrix} -1 & 2 \\ 1 & -1 \end{bmatrix}\begin{bmatrix} 6 \\ 5 \end{bmatrix} = \begin{bmatrix} 4 \\ 1 \end{bmatrix}.$$

Therefore, $x_1 = 4$, $x_2 = 1$.

**19.**
$$\begin{bmatrix} 2 & 2 \\ 2 & 3 \end{bmatrix}\begin{bmatrix} x_1 \\ x_2 \end{bmatrix} + \begin{bmatrix} -5 \\ 2 \end{bmatrix} = \begin{bmatrix} 2 \\ 3 \end{bmatrix}$$

$$\begin{bmatrix} 2 & 2 \\ 2 & 3 \end{bmatrix}\begin{bmatrix} x_1 \\ x_2 \end{bmatrix} = \begin{bmatrix} 2 \\ 3 \end{bmatrix} - \begin{bmatrix} -5 \\ 2 \end{bmatrix} = \begin{bmatrix} 7 \\ 1 \end{bmatrix}$$

If $A = \begin{bmatrix} 2 & 2 \\ 2 & 3 \end{bmatrix}$ has an inverse, then $\begin{bmatrix} x_1 \\ x_2 \end{bmatrix} = A^{-1}\begin{bmatrix} 7 \\ 1 \end{bmatrix}$

$$\begin{bmatrix} 2 & 2 & | & 1 & 0 \\ 2 & 3 & | & 0 & 1 \end{bmatrix} \sim \begin{bmatrix} 2 & 2 & | & 0 & 1 \\ 0 & 1 & | & -1 & 1 \end{bmatrix} \sim \begin{bmatrix} 2 & 0 & | & 3 & -2 \\ 0 & 1 & | & -1 & 1 \end{bmatrix} \sim \begin{bmatrix} 1 & 0 & | & \frac{3}{2} & -1 \\ 0 & 1 & | & -1 & 1 \end{bmatrix}$$

$(-1)R_1 + R_2 \to R_2 \quad (-2)R_2 + R_1 \to R_1 \quad \left(\frac{1}{2}\right)R_1 \to R_1$

$$A^{-1} = \begin{bmatrix} \frac{3}{2} & -1 \\ -1 & 1 \end{bmatrix}; \quad \begin{bmatrix} x_1 \\ x_2 \end{bmatrix} = \begin{bmatrix} \frac{3}{2} & -1 \\ -1 & 1 \end{bmatrix}\begin{bmatrix} 7 \\ 1 \end{bmatrix} = \begin{bmatrix} \frac{19}{2} \\ -6 \end{bmatrix}.$$

Therefore, $x_1 = \dfrac{19}{2}$, $x_2 = -6$.

**21.**
$$\begin{bmatrix} 3 & -2 \\ 6 & -4 \end{bmatrix}\begin{bmatrix} x_1 \\ x_2 \end{bmatrix} + \begin{bmatrix} 0 \\ -3 \end{bmatrix} = \begin{bmatrix} 1 \\ -2 \end{bmatrix}$$

$$\begin{bmatrix} 3 & -2 \\ 6 & -4 \end{bmatrix}\begin{bmatrix} x_1 \\ x_2 \end{bmatrix} = \begin{bmatrix} 1 \\ -2 \end{bmatrix} - \begin{bmatrix} 0 \\ -3 \end{bmatrix} = \begin{bmatrix} 1 \\ 1 \end{bmatrix}$$

If $A = \begin{bmatrix} 3 & -2 \\ 6 & -4 \end{bmatrix}$ has an inverse, then $\begin{bmatrix} x_1 \\ x_2 \end{bmatrix} = A^{-1}\begin{bmatrix} 1 \\ 1 \end{bmatrix}.$

$$\begin{bmatrix} 3 & -2 & | & 1 & 0 \\ 6 & -4 & | & 0 & 1 \end{bmatrix} \sim \begin{bmatrix} 3 & -2 & | & 1 & 0 \\ 0 & 0 & | & -2 & 1 \end{bmatrix}; \; A \text{ does not have an inverse.}$$

$(-2)R_1 + R_2 \to R_2$

We conclude that the equation has either no solution or infinitely many solutions. We row reduce the augmented matrix:

$$\begin{bmatrix} 3 & -2 & | & 1 \\ 6 & -4 & | & 1 \end{bmatrix} \sim \begin{bmatrix} 3 & -2 & | & 1 \\ 0 & 0 & | & -1 \end{bmatrix}; \quad \text{there is no solution.}$$

$(-2)R_1 + R_2 \to R_2$

**23.**   The matrix equation for the given system is:

$$\begin{bmatrix} 1 & 2 \\ 1 & 3 \end{bmatrix}\begin{bmatrix} x_1 \\ x_2 \end{bmatrix} = \begin{bmatrix} k_1 \\ k_2 \end{bmatrix}$$

Copyright © 2011 Pearson Education, Inc. Publishing as Prentice Hall.

From Exercise 4-5, Problem 31, $\begin{bmatrix} 1 & 2 \\ 1 & 3 \end{bmatrix}^{-1} = \begin{bmatrix} 3 & -2 \\ -1 & 1 \end{bmatrix}$

Thus, $\begin{bmatrix} x_1 \\ x_2 \end{bmatrix} = \begin{bmatrix} 3 & -2 \\ -1 & 1 \end{bmatrix} \begin{bmatrix} k_1 \\ k_2 \end{bmatrix}$

(A) $\begin{bmatrix} x_1 \\ x_2 \end{bmatrix} = \begin{bmatrix} 3 & -2 \\ -1 & 1 \end{bmatrix} \begin{bmatrix} 1 \\ 3 \end{bmatrix} = \begin{bmatrix} 1 \\ 3 \end{bmatrix}$    Thus, $x_1 = -3$
and $x_2 = 2$

(B) $\begin{bmatrix} x_1 \\ x_2 \end{bmatrix} = \begin{bmatrix} 3 & -2 \\ -1 & 1 \end{bmatrix} \begin{bmatrix} 3 \\ 5 \end{bmatrix} = \begin{bmatrix} -1 \\ 2 \end{bmatrix}$    Thus, $x_1 = -1$
and $x_2 = 2$

(C) $\begin{bmatrix} x_1 \\ x_2 \end{bmatrix} = \begin{bmatrix} 3 & -2 \\ -1 & 1 \end{bmatrix} \begin{bmatrix} -2 \\ 1 \end{bmatrix} = \begin{bmatrix} -8 \\ 3 \end{bmatrix}$    Thus, $x_1 = -8$
and $x_2 = 3$

**25.** The matrix equation for the given system is:

$\begin{bmatrix} 1 & 3 \\ 2 & 7 \end{bmatrix} \begin{bmatrix} x_1 \\ x_2 \end{bmatrix} = \begin{bmatrix} k_1 \\ k_2 \end{bmatrix}$

From Exercise 4-5, Problem 33, $\begin{bmatrix} 1 & 3 \\ 2 & 7 \end{bmatrix}^{-1} = \begin{bmatrix} 7 & -3 \\ -2 & 1 \end{bmatrix}$

Thus, $\begin{bmatrix} x_1 \\ x_2 \end{bmatrix} = \begin{bmatrix} 7 & -3 \\ -2 & 1 \end{bmatrix} \begin{bmatrix} k_1 \\ k_2 \end{bmatrix}$

(A) $\begin{bmatrix} x_1 \\ x_2 \end{bmatrix} = \begin{bmatrix} 7 & -3 \\ -2 & 1 \end{bmatrix} \begin{bmatrix} 2 \\ -1 \end{bmatrix} = \begin{bmatrix} 17 \\ -5 \end{bmatrix}$    Thus, $x_1 = 17$
and $x_2 = -5$

(B) $\begin{bmatrix} x_1 \\ x_2 \end{bmatrix} = \begin{bmatrix} 7 & -3 \\ -2 & 1 \end{bmatrix} \begin{bmatrix} 1 \\ 0 \end{bmatrix} = \begin{bmatrix} 7 \\ -2 \end{bmatrix}$    Thus, $x_1 = 7$
and $x_2 = -2$

(C) $\begin{bmatrix} x_1 \\ x_2 \end{bmatrix} = \begin{bmatrix} 7 & -3 \\ -2 & 1 \end{bmatrix} \begin{bmatrix} 3 \\ -1 \end{bmatrix} = \begin{bmatrix} 24 \\ -7 \end{bmatrix}$    Thus, $x_1 = 24$
and $x_2 = -7$

**27.** The matrix equation for the given system is:

$\begin{bmatrix} 1 & -3 & 0 \\ 0 & 1 & 1 \\ 2 & -1 & 4 \end{bmatrix} \begin{bmatrix} x_1 \\ x_2 \\ x_3 \end{bmatrix} = \begin{bmatrix} k_1 \\ k_2 \\ k_3 \end{bmatrix}$

From Exercise 4-5, Problem 35, $\begin{bmatrix} 1 & -3 & 0 \\ 0 & 1 & 1 \\ 2 & -1 & 4 \end{bmatrix}^{-1} = \begin{bmatrix} -5 & -12 & 3 \\ -2 & -4 & 1 \\ 2 & 5 & -1 \end{bmatrix}$

Thus,

$\begin{bmatrix} x_1 \\ x_2 \\ x_3 \end{bmatrix} = \begin{bmatrix} -5 & -12 & 3 \\ -2 & -4 & 1 \\ 2 & 5 & -1 \end{bmatrix} \begin{bmatrix} k_1 \\ k_2 \\ k_3 \end{bmatrix}$

Copyright © 2011 Pearson Education, Inc. Publishing as Prentice Hall.

(A) $\begin{bmatrix} x_1 \\ x_2 \\ x_3 \end{bmatrix} = \begin{bmatrix} -5 & -12 & 3 \\ -2 & -4 & 1 \\ 2 & 5 & -1 \end{bmatrix} \begin{bmatrix} 1 \\ 0 \\ 2 \end{bmatrix} = \begin{bmatrix} 1 \\ 0 \\ 0 \end{bmatrix}$;   $x_1 = 1$, $x_2 = 0$, $x_3 = 0$

(B) $\begin{bmatrix} x_1 \\ x_2 \\ x_3 \end{bmatrix} = \begin{bmatrix} -5 & -12 & 3 \\ -2 & -4 & 1 \\ 2 & 5 & -1 \end{bmatrix} \begin{bmatrix} -1 \\ 1 \\ 0 \end{bmatrix} = \begin{bmatrix} -7 \\ -2 \\ 3 \end{bmatrix}$;   $x_1 = -7$, $x_2 = -2$, $x_3 = 3$

(C) $\begin{bmatrix} x_1 \\ x_2 \\ x_3 \end{bmatrix} = \begin{bmatrix} -5 & -12 & 3 \\ -2 & -4 & 1 \\ 2 & 5 & -1 \end{bmatrix} \begin{bmatrix} 2 \\ -2 \\ 1 \end{bmatrix} = \begin{bmatrix} 17 \\ 5 \\ -7 \end{bmatrix}$;   $x_1 = 17$, $x_2 = 5$, $x_3 = -7$

**29.**   The matrix equation for the given system is:

$$\begin{bmatrix} 1 & 1 & 0 \\ 2 & 3 & -1 \\ 1 & 0 & 2 \end{bmatrix} \begin{bmatrix} x_1 \\ x_2 \\ x_3 \end{bmatrix} = \begin{bmatrix} k_1 \\ k_2 \\ k_3 \end{bmatrix}$$

From Exercise 4-5, Problem 37, $\begin{bmatrix} 1 & 1 & 0 \\ 2 & 3 & -1 \\ 1 & 0 & 2 \end{bmatrix}^{-1} = \begin{bmatrix} 6 & -2 & -1 \\ -5 & 2 & 1 \\ -3 & 1 & 1 \end{bmatrix}$

Thus,

$$\begin{bmatrix} x_1 \\ x_2 \\ x_3 \end{bmatrix} = \begin{bmatrix} 6 & -2 & -1 \\ -5 & 2 & 1 \\ -3 & 1 & 1 \end{bmatrix} \begin{bmatrix} k_1 \\ k_2 \\ k_3 \end{bmatrix}$$

(A) $\begin{bmatrix} x_1 \\ x_2 \\ x_3 \end{bmatrix} = \begin{bmatrix} 6 & -2 & -1 \\ -5 & 2 & 1 \\ -3 & 1 & 1 \end{bmatrix} \begin{bmatrix} 2 \\ 0 \\ 4 \end{bmatrix} = \begin{bmatrix} 8 \\ -6 \\ -2 \end{bmatrix}$;   $x_1 = 8$, $x_2 = -6$, $x_3 = -2$

(B) $\begin{bmatrix} x_1 \\ x_2 \\ x_3 \end{bmatrix} = \begin{bmatrix} 6 & -2 & -1 \\ -5 & 2 & 1 \\ -3 & 1 & 1 \end{bmatrix} \begin{bmatrix} 0 \\ 4 \\ -2 \end{bmatrix} = \begin{bmatrix} -6 \\ 6 \\ 2 \end{bmatrix}$;   $x_1 = -6$, $x_2 = 6$, $x_3 = 2$

(C) $\begin{bmatrix} x_1 \\ x_2 \\ x_3 \end{bmatrix} = \begin{bmatrix} 6 & -2 & -1 \\ -5 & 2 & 1 \\ -3 & 1 & 1 \end{bmatrix} \begin{bmatrix} 4 \\ 2 \\ 0 \end{bmatrix} = \begin{bmatrix} 20 \\ -16 \\ -10 \end{bmatrix}$;   $x_1 = 20$, $x_2 = -16$, $x_3 = -10$

**31.**   $AX = B$, $X = \dfrac{B}{A}$, the expression $\dfrac{B}{A}$, a quotient of matrices is not defined.

**33.**   $XA = B$, $X = A^{-1}B$

Solution:

Copyright © 2011 Pearson Education, Inc.  Publishing as Prentice Hall.

$(XA)A^{-1} = BA^{-1}$    multiply on the right by $A^{-1}$

$X(AA^{-1}) = BA^{-1}$    associative property

$XI = BA^{-1}$    $AA^{-1} = I$, the identity matrix

$X = BA^{-1}$

$X \neq A^{-1}B$ because matrix multiplication is not commutative; $A^{-1}B \neq BA^{-1}$ in general.

**35.**  $AX = B, \quad X = A^{-1}BA, \quad X = B$

Solution:

$A^{-1}(AX) = A^{-1}(BA)$    multiply on the left by $A^{-1}$

$(A^{-1}A)X = A^{-1}BA$    associative property

$IX = A^{-1}BA$    $AA^{-1} = I$, the identity matrix

$X = A^{-1}BA$

$A^{-1}(BA) \neq (BA)A^{-1} = B(AA^{-1}) = B$ because matrix multiplication is not commutative.

**37.**  $-2x_1 + 4x_2 = 5$

$6x_1 - 12x_2 = 15$

The second equation is a multiple (−3) of the first. Therefore, the system has infinitely many solutions. The solutions are:

$$x_1 = 2t + \frac{5}{2}, \quad x_2 = t, \quad t \text{ any real number.}$$

**39.**  $x_1 - 3x_2 - 2x_3 = -1$

$-2x_1 + 6x_2 + 4x_3 = 3$

The system is not "square" − 2 equations in 3 unknowns. The matrix of coefficients is $2 \times 3$; it does not have an inverse.

Solve the system by Gauss-Jordan elimination:

$$\begin{pmatrix} 1 & -3 & -2 & | & -1 \\ -2 & 6 & 4 & | & 3 \end{pmatrix} \sim \begin{pmatrix} 1 & -3 & -2 & | & -1 \\ 0 & 0 & 0 & | & 1 \end{pmatrix}$$

$2R_1 + R_2 \rightarrow R_2$

The system has no solution.

**41.**  $e_1: \quad x_1 - 2x_2 + 3x_3 = 1$

$e_2: \quad 2x_1 - 3x_2 - 2x_3 = 3$

$e_3: \quad x_1 - x_2 - 5x_3 = 2$

Note that $e_3 = (-1)e_1 + e_2$. This implies that the system has infinitely many solutions; the coefficient matrix does not have an inverse. Solve the system by Gauss-Jordan elimination:

$$\begin{bmatrix} 1 & -2 & 3 & | & 1 \\ 2 & -3 & -2 & | & 3 \\ 1 & -1 & -5 & | & 2 \end{bmatrix} \sim \begin{bmatrix} 1 & -2 & 3 & | & 1 \\ 0 & 1 & -8 & | & 1 \\ 0 & 1 & -8 & | & 1 \end{bmatrix} \sim \begin{bmatrix} 1 & -2 & 3 & | & 1 \\ 0 & 1 & -8 & | & 1 \\ 0 & 0 & 0 & | & 0 \end{bmatrix} \sim \begin{bmatrix} 1 & 0 & -13 & | & 3 \\ 0 & 1 & -8 & | & 1 \\ 0 & 0 & 0 & | & 0 \end{bmatrix}$$

$(-2)R_1 + R_2 \rightarrow R_2$    $(-1)R_2 + R_3 \rightarrow R_3$    $2R_2 + R_1 \rightarrow R_1$

$(-1)R_1 + R_3 \rightarrow R_3$

Solutions: $x_1 = 3 + 13t$, $x_2 = 1 + 8t$, $x_3 = t$, $t$ any real number.

Copyright © 2011 Pearson Education, Inc. Publishing as Prentice Hall.

**43.** $AX - BX = C$

$(A - B)X = C$

$\quad X = (A - B)^{-1}C$

**45.** $AX + X = C$

$(A + I)X = C$, where $I$ is the identity matrix of order $n$

$\quad X = (A + I)^{-1}C$

**47.** $AX - C = D - BX$

$AX + BX = C + D$

$(A + B)X = C + D$

$\quad X = (A + B)^{-1}(C + D)$

**49.** The matrix equation for the given system is:

$$\begin{bmatrix} 1 & 2.001 \\ 1 & 2 \end{bmatrix} \begin{bmatrix} x_1 \\ x_2 \end{bmatrix} = \begin{bmatrix} k_1 \\ k_2 \end{bmatrix}$$

First we compute the inverse of $\begin{bmatrix} 1 & 2.001 \\ 1 & 2 \end{bmatrix}$

$$\begin{bmatrix} 1 & 2.001 & | & 1 & 0 \\ 1 & 2 & | & 0 & 1 \end{bmatrix} \sim \begin{bmatrix} 1 & 2.001 & | & 1 & 0 \\ 0 & -0.001 & | & -1 & 1 \end{bmatrix} \sim \begin{bmatrix} 1 & 2.001 & | & 1 & 0 \\ 0 & 1 & | & 1000 & -1000 \end{bmatrix}$$

$(-1)R_1 + R_2 \rightarrow R_2 \qquad (-1000)R_2 \rightarrow R_2 \qquad (-2.001)R_2 + R_1 \rightarrow R_1$

$$\sim \begin{bmatrix} 1 & 0 & | & -2000 & 2001 \\ 0 & 1 & | & 1000 & -1000 \end{bmatrix}$$

Thus, $\begin{bmatrix} 1 & 2.001 \\ 1 & 2 \end{bmatrix}^{-1} = \begin{bmatrix} -2000 & 2001 \\ 1000 & -1000 \end{bmatrix}$ and $\begin{bmatrix} x_1 \\ x_2 \end{bmatrix} = \begin{bmatrix} -2000 & 2001 \\ 1000 & -1000 \end{bmatrix} \begin{bmatrix} k_1 \\ k_2 \end{bmatrix}$

(A) $\begin{bmatrix} x_1 \\ x_2 \end{bmatrix} = \begin{bmatrix} -2000 & 2001 \\ 1000 & -1000 \end{bmatrix} \begin{bmatrix} 1 \\ 1 \end{bmatrix} = \begin{bmatrix} 1 \\ 0 \end{bmatrix}$; $\quad x_1 = 1, \ x_2 = 0$

(B) $\begin{bmatrix} x_1 \\ x_2 \end{bmatrix} = \begin{bmatrix} -2000 & 2001 \\ 1000 & -1000 \end{bmatrix} \begin{bmatrix} 1 \\ 0 \end{bmatrix} = \begin{bmatrix} -2000 \\ 1000 \end{bmatrix}$; $\quad x_1 = -2,000, \ x_2 = 1,000$

(C) $\begin{bmatrix} x_1 \\ x_2 \end{bmatrix} = \begin{bmatrix} -2000 & 2001 \\ 1000 & -1000 \end{bmatrix} \begin{bmatrix} 0 \\ 1 \end{bmatrix} = \begin{bmatrix} 2001 \\ -1000 \end{bmatrix}$; $\quad x_1 = 2,001, \ x_2 = -1,000$

**51.** The matrix equation for the given system is:

$$\begin{bmatrix} 1 & 8 & 7 \\ 6 & 6 & 8 \\ 3 & 4 & 6 \end{bmatrix} \begin{bmatrix} x_1 \\ x_2 \\ x_3 \end{bmatrix} = \begin{bmatrix} 135 \\ 155 \\ 75 \end{bmatrix}$$

Thus, $\begin{bmatrix} x_1 \\ x_2 \\ x_3 \end{bmatrix} = \begin{bmatrix} 1 & 8 & 7 \\ 6 & 6 & 8 \\ 3 & 4 & 6 \end{bmatrix}^{-1} \begin{bmatrix} 135 \\ 155 \\ 75 \end{bmatrix} = \begin{bmatrix} -0.08 & 0.4 & -0.44 \\ 0.24 & 0.3 & -0.68 \\ -0.12 & -0.4 & 0.84 \end{bmatrix} \begin{bmatrix} 135 \\ 155 \\ 75 \end{bmatrix} = \begin{bmatrix} 18.2 \\ 27.9 \\ -15.2 \end{bmatrix}$ and

$x_1 = 18.2, \ x_2 = 27.9, \ x_3 = -15.2$

Copyright © 2011 Pearson Education, Inc. Publishing as Prentice Hall.

**53.** The matrix equation for the given system is:

$$\begin{bmatrix} 6 & 9 & 7 & 5 \\ 6 & 4 & 7 & 3 \\ 4 & 5 & 3 & 2 \\ 4 & 3 & 8 & 2 \end{bmatrix} \begin{bmatrix} x_1 \\ x_2 \\ x_3 \\ x_4 \end{bmatrix} = \begin{bmatrix} 250 \\ 195 \\ 145 \\ 125 \end{bmatrix}$$

Thus

$$\begin{bmatrix} x_1 \\ x_2 \\ x_3 \\ x_4 \end{bmatrix} = \begin{bmatrix} 6 & 9 & 7 & 5 \\ 6 & 4 & 7 & 3 \\ 4 & 5 & 3 & 2 \\ 4 & 3 & 8 & 2 \end{bmatrix}^{-1} \begin{bmatrix} 250 \\ 195 \\ 145 \\ 125 \end{bmatrix} = \begin{bmatrix} -0.25 & 0.37 & 0.28 & -0.21 \\ 0 & -0.4 & 0.4 & 0.2 \\ 0 & -0.16 & -0.04 & 0.28 \\ 0.5 & 0.5 & -1 & -0.5 \end{bmatrix} \begin{bmatrix} 250 \\ 195 \\ 145 \\ 125 \end{bmatrix} = \begin{bmatrix} 24 \\ 5 \\ -2 \\ 15 \end{bmatrix}$$

and $x_1 = 24$, $x_2 = 5$, $x_3 = -2$, $x_4 = 15$.

**55.** (A) Let $x_1$ = Number of \$25 tickets sold

$\qquad x_2$ = Number of \$35 tickets sold

The mathematical model is:

$$x_1 + \quad x_2 = 10{,}000$$
$$25x_1 + 35x_2 = k \quad , \; k = 275{,}000, \; 300{,}000, \; 325{,}000$$

The corresponding matrix equation is:

$$\begin{bmatrix} 1 & 1 \\ 25 & 35 \end{bmatrix} = \begin{bmatrix} 10{,}000 \\ k \end{bmatrix}$$

Compute the inverse of the coefficient matrix $A$.

$$\left[\begin{array}{cc|cc} 1 & 1 & 1 & 0 \\ 25 & 35 & 0 & 1 \end{array}\right] \sim \left[\begin{array}{cc|cc} 1 & 1 & 1 & 0 \\ 0 & 10 & -25 & 1 \end{array}\right] \sim \left[\begin{array}{cc|cc} 1 & 1 & 1 & 0 \\ 0 & 1 & -\frac{5}{2} & \frac{1}{10} \end{array}\right]$$

$(-25)R_1 + R_2 \to R_2 \qquad \frac{1}{10}R_2 \to R_2 \qquad (-1)R_2 + R_1 \to R_1$

$$\sim \left[\begin{array}{cc|cc} 1 & 0 & \frac{7}{2} & -\frac{1}{10} \\ 0 & 1 & -\frac{5}{2} & \frac{1}{10} \end{array}\right] \quad \text{Thus,} \quad A^{-1} = \begin{bmatrix} \frac{7}{2} & -\frac{1}{10} \\ -\frac{5}{2} & \frac{1}{10} \end{bmatrix}$$

<u>Concert 1</u>: Return \$275,000

$$\begin{bmatrix} x_1 \\ x_2 \end{bmatrix} = \begin{bmatrix} \frac{7}{2} & -\frac{1}{10} \\ -\frac{5}{2} & \frac{1}{10} \end{bmatrix} = \begin{bmatrix} 10{,}000 \\ 275{,}000 \end{bmatrix} = \begin{bmatrix} 7{,}500 \\ 2{,}500 \end{bmatrix}$$

Thus, 7,500 \$25 tickets and 2,500 \$35 tickets must be sold.

<u>Concert 2</u>: Return \$300,000

$$\begin{bmatrix} x_1 \\ x_2 \end{bmatrix} = \begin{bmatrix} \frac{7}{2} & -\frac{1}{10} \\ -\frac{5}{2} & \frac{1}{10} \end{bmatrix} \begin{bmatrix} 10{,}000 \\ 300{,}000 \end{bmatrix} = \begin{bmatrix} 5{,}000 \\ 5{,}000 \end{bmatrix}$$

Thus, 5,000 \$25 tickets and 5,000 \$35 tickets must be sold.

Copyright © 2011 Pearson Education, Inc. Publishing as Prentice Hall.

<u>Concert 3</u>: Return $325,000

$$\begin{bmatrix} x_1 \\ x_2 \end{bmatrix} = \begin{bmatrix} \frac{7}{2} & -\frac{1}{10} \\ -\frac{5}{2} & \frac{1}{10} \end{bmatrix} \begin{bmatrix} 10,000 \\ 325,000 \end{bmatrix} = \begin{bmatrix} 2,500 \\ 7,500 \end{bmatrix}$$

Thus, 2,500 $25 tickets and 7,500 $35 tickets must be sold.

(B)    $200,000 Return?

$$\begin{bmatrix} x_1 \\ x_2 \end{bmatrix} \stackrel{?}{=} \begin{bmatrix} \frac{7}{2} & -\frac{1}{10} \\ -\frac{5}{2} & \frac{1}{10} \end{bmatrix} \begin{bmatrix} 10,000 \\ 200,000 \end{bmatrix} = \begin{bmatrix} 15,000 \\ -5,000 \end{bmatrix}$$

This is not possible; $x_2$ cannot be negative.

$400,000 Return?

$$\begin{bmatrix} x_1 \\ x_2 \end{bmatrix} \stackrel{?}{=} \begin{bmatrix} \frac{7}{2} & -\frac{1}{10} \\ -\frac{5}{2} & \frac{1}{10} \end{bmatrix} \begin{bmatrix} 10,000 \\ 400,000 \end{bmatrix} = \begin{bmatrix} -5,000 \\ 15,000 \end{bmatrix}$$

This is not possible; $x_1$ cannot be negative.

(C)    Fix a return $k$. Then

$$\begin{bmatrix} x_1 \\ x_2 \end{bmatrix} = \begin{bmatrix} \frac{7}{2} & -\frac{1}{10} \\ -\frac{5}{2} & \frac{1}{10} \end{bmatrix} \begin{bmatrix} 10,000 \\ k \end{bmatrix} = \begin{bmatrix} 35,000 - \frac{k}{10} \\ -25,000 + \frac{k}{10} \end{bmatrix}$$

Thus, $x_1 = 35,000 - \dfrac{k}{10}$ and $x_2 = -25,000 + \dfrac{k}{10}$.

Since $x_1 \geq 0$, $35,000 - \dfrac{k}{10} \geq 0$, $-\dfrac{k}{10} \geq -35,000$, $k \leq 350,000$

Since $x_2 \geq 0$, $-25,000 + \dfrac{k}{10} \geq 0$, $\dfrac{k}{10} \geq 25,000$, $k \geq 250,000$.

Thus, $250,000 \leq k \leq 350,000$; any number between (and including) $250,000 and $350,000 is a possible return.

57.    Let $x_1$ = number of hours at Plant A

and $x_2$ = number of hours at Plant B

Then, the mathematical model is:

$10x_1 + 8x_2 = k_1$  (number of car frames)

$5x_1 + 8x_2 = k_2$  (number of truck frames)

The corresponding matrix equation is:

$$\begin{bmatrix} 10 & 8 \\ 5 & 8 \end{bmatrix} \begin{bmatrix} x_1 \\ x_2 \end{bmatrix} = \begin{bmatrix} k_1 \\ k_2 \end{bmatrix}$$

First we compute the inverse of $\begin{bmatrix} 10 & 8 \\ 5 & 8 \end{bmatrix}$

$$\left[\begin{array}{cc|cc} 10 & 8 & 1 & 0 \\ 5 & 8 & 0 & 1 \end{array}\right] \sim \left[\begin{array}{cc|cc} 1 & \frac{4}{5} & \frac{1}{10} & 0 \\ 5 & 8 & 0 & 1 \end{array}\right] \sim \left[\begin{array}{cc|cc} 1 & \frac{4}{5} & \frac{1}{10} & 0 \\ 0 & 4 & -\frac{1}{2} & 1 \end{array}\right] \sim \left[\begin{array}{cc|cc} 1 & \frac{4}{5} & \frac{1}{10} & 0 \\ 0 & 1 & -\frac{1}{8} & \frac{1}{4} \end{array}\right]$$

$\dfrac{1}{10}R_1 \rightarrow R_1$    $(-5)R_1 + R_2 \rightarrow R_2$    $\dfrac{1}{4}R_2 \rightarrow R_2$    $\left(-\dfrac{4}{5}\right)R_2 + R_1 \rightarrow R_1$

Copyright © 2011 Pearson Education, Inc.  Publishing as Prentice Hall.

$$\sim \begin{bmatrix} 1 & 0 & | & \frac{1}{5} & -\frac{1}{5} \\ 0 & 1 & | & -\frac{1}{8} & \frac{1}{4} \end{bmatrix}$$

Thus $\begin{bmatrix} 10 & 8 \\ 5 & 8 \end{bmatrix}^{-1} = \begin{bmatrix} \frac{1}{5} & -\frac{1}{5} \\ -\frac{1}{8} & \frac{1}{4} \end{bmatrix}$ and $\begin{bmatrix} x_1 \\ x_2 \end{bmatrix} = \begin{bmatrix} \frac{1}{5} & -\frac{1}{5} \\ -\frac{1}{8} & \frac{1}{4} \end{bmatrix} \begin{bmatrix} k_1 \\ k_2 \end{bmatrix}$

Now, for order 1:

$\begin{bmatrix} x_1 \\ x_2 \end{bmatrix} = \begin{bmatrix} \frac{1}{5} & -\frac{1}{5} \\ -\frac{1}{8} & \frac{1}{4} \end{bmatrix} \begin{bmatrix} 3000 \\ 1600 \end{bmatrix} = \begin{bmatrix} 280 \\ 25 \end{bmatrix}$ and $x_1 = 280$ hours at Plant A
$x_2 = 25$ hours at Plant B

For order 2:

$\begin{bmatrix} x_1 \\ x_2 \end{bmatrix} = \begin{bmatrix} \frac{1}{5} & -\frac{1}{5} \\ -\frac{1}{8} & \frac{1}{4} \end{bmatrix} \begin{bmatrix} 2800 \\ 2000 \end{bmatrix} = \begin{bmatrix} 160 \\ 150 \end{bmatrix}$ and $x_1 = 160$ hours at Plant A
$x_2 = 150$ hours at Plant B

For order 3:

$\begin{bmatrix} x_1 \\ x_2 \end{bmatrix} = \begin{bmatrix} \frac{1}{5} & -\frac{1}{5} \\ -\frac{1}{8} & \frac{1}{4} \end{bmatrix} \begin{bmatrix} 2600 \\ 2200 \end{bmatrix} = \begin{bmatrix} 80 \\ 225 \end{bmatrix}$ and $x_1 = 80$ hours at Plant A
$x_2 = 225$ hours at Plant B

**59.** Let $x_1 =$ President's bonus

$x_2 =$ Executive Vice President's bonus

$x_3 =$ Associate Vice President's bonus

$x_4 =$ Assistant Vice President's bonus

Then, the mathematical model is:

$x_1 = 0.03(2,000,000 - x_2 - x_3 - x_4)$
$x_2 = 0.025(2,000,000 - x_1 - x_3 - x_4)$
$x_3 = 0.02(2,000,000 - x_1 - x_2 - x_4)$
$x_4 = 0.015(2,000,000 - x_1 - x_2 - x_3)$

or

$$\begin{aligned}
x_1 + 0.03x_2 + 0.03x_3 + 0.03x_4 &= 60,000 \\
0.025x_1 + x_2 + 0.025x_3 + 0.025x_4 &= 50,000 \\
0.02x_1 + 0.02x_2 + x_3 + 0.02x_4 &= 40,000 \\
0.015x_1 + 0.015x_2 + 0.015x_3 + x_4 &= 30,000
\end{aligned}$$

and $\begin{bmatrix} 1 & 0.03 & 0.03 & 0.03 \\ 0.025 & 1 & 0.025 & 0.025 \\ 0.02 & 0.02 & 1 & 0.02 \\ 0.015 & 0.015 & 0.015 & 1 \end{bmatrix} \begin{bmatrix} x_1 \\ x_2 \\ x_3 \\ x_4 \end{bmatrix} = \begin{bmatrix} 60,000 \\ 50,000 \\ 40,000 \\ 30,000 \end{bmatrix}$

Thus $\begin{bmatrix} x_1 \\ x_2 \\ x_3 \\ x_4 \end{bmatrix} = \begin{bmatrix} 1 & 0.03 & 0.03 & 0.03 \\ 0.025 & 1 & 0.025 & 0.025 \\ 0.02 & 0.02 & 1 & 0.02 \\ 0.015 & 0.015 & 0.015 & 1 \end{bmatrix}^{-1} \begin{bmatrix} 60,000 \\ 50,000 \\ 40,000 \\ 30,000 \end{bmatrix} \approx \begin{bmatrix} 56,600 \\ 47,000 \\ 37,400 \\ 27,900 \end{bmatrix}$

or $x_1 = \$56,600$, $x_2 = \$47,000$, $x_3 = \$37,400$, $x_4 = \$27,900$ to the nearest hundred dollars.

Copyright © 2011 Pearson Education, Inc. Publishing as Prentice Hall.

**61.** Let $x_1$ = number of ounces of mix $A$.

$x_2$ = number of ounces of mix $B$.

The mathematical model is:

$0.2x_1 + 0.14x_2 = k_1$ (protein)

$0.04x_1 + 0.03x_2 = k_2$ (fat)

The corresponding matrix equation is:

$$\begin{bmatrix} 0.2 & 0.14 \\ 0.04 & 0.03 \end{bmatrix} \begin{bmatrix} x_1 \\ x_2 \end{bmatrix} = \begin{bmatrix} k_1 \\ k_2 \end{bmatrix}$$

Next, compute the inverse of the coefficient matrix $A$:

$$\begin{bmatrix} 0.2 & 0.14 & | & 1 & 0 \\ 0.04 & 0.03 & | & 0 & 1 \end{bmatrix} \sim \begin{bmatrix} 1 & 0.7 & | & 5 & 0 \\ 0.04 & 0.03 & | & 0 & 1 \end{bmatrix} \sim \begin{bmatrix} 1 & 0.7 & | & 5 & 0 \\ 0 & 0.002 & | & -0.2 & 1 \end{bmatrix}$$

$5R_1 \rightarrow R_1 \qquad (-0.04)R_1 + R_2 \rightarrow R_2 \qquad 500R_2 \rightarrow R_2$

$$\sim \begin{bmatrix} 1 & 0.7 & | & 5 & 0 \\ 0 & 1 & | & -100 & 500 \end{bmatrix} \sim \begin{bmatrix} 1 & 0 & | & 75 & -350 \\ 0 & 1 & | & -100 & 500 \end{bmatrix}$$

$(-0.7)R_2 + R_1 \rightarrow R_1$

Thus, $A^{-1} = \begin{bmatrix} 75 & -350 \\ -100 & 500 \end{bmatrix}$

(A)  Diet 1: Protein - 80 oz., Fat - 17 oz.

$$\begin{bmatrix} x_1 \\ x_2 \end{bmatrix} = \begin{bmatrix} 75 & -350 \\ -100 & 500 \end{bmatrix} \begin{bmatrix} 80 \\ 17 \end{bmatrix} = \begin{bmatrix} 50 \\ 500 \end{bmatrix}$$

Thus, 50 ounces of mix $A$, 500 ounces of mix $B$.

Diet 2: Protein - 90 oz., Fat - 18 oz.

$$\begin{bmatrix} x_1 \\ x_2 \end{bmatrix} = \begin{bmatrix} 75 & -350 \\ -100 & 500 \end{bmatrix} \begin{bmatrix} 90 \\ 18 \end{bmatrix} = \begin{bmatrix} 450 \\ 0 \end{bmatrix}$$

Thus, 450 ounces of mix $A$, 0 ounces of mix $B$.

Diet 3: Protein - 100 oz., Fat - 21 oz.

$$\begin{bmatrix} x_1 \\ x_2 \end{bmatrix} = \begin{bmatrix} 75 & -350 \\ -100 & 500 \end{bmatrix} \begin{bmatrix} 100 \\ 21 \end{bmatrix} = \begin{bmatrix} 150 \\ 500 \end{bmatrix}$$

Thus, 150 ounces of mix $A$, 500 ounces of mix $B$.

(B)  Protein - 100 oz., Fat - 22 oz.

$$\begin{bmatrix} x_1 \\ x_2 \end{bmatrix} = \begin{bmatrix} 75 & -350 \\ -100 & 500 \end{bmatrix} \begin{bmatrix} 100 \\ 22 \end{bmatrix} = \begin{bmatrix} -200 \\ -1000 \end{bmatrix}$$

This is not possible; $x_1$ and $x_2$ must both be non-negative.

Protein - 80 oz., Fat - 15 oz.

$$\begin{bmatrix} x_1 \\ x_2 \end{bmatrix} = \begin{bmatrix} 75 & -350 \\ -100 & 500 \end{bmatrix} \begin{bmatrix} 80 \\ 15 \end{bmatrix} = \begin{bmatrix} -750 \\ -500 \end{bmatrix}$$

This is not possible; $x_1$ and $x_2$ must both be non-negative.

Copyright © 2011 Pearson Education, Inc. Publishing as Prentice Hall.

EXERCISE 4-7

*Things to remember:*

<u>1</u>.   Given two industries $C_1$ and $C_2$, with

$$M = \begin{matrix} & C_1 & C_2 \\ \begin{matrix} C_1 \\ C_2 \end{matrix} & \begin{bmatrix} a_{11} & a_{12} \\ a_{21} & a_{22} \end{bmatrix} \end{matrix}, \quad X = \begin{bmatrix} x_1 \\ x_2 \end{bmatrix}, \quad D = \begin{bmatrix} d_1 \\ d_2 \end{bmatrix},$$

Technology    Output    Final Demand
Matrix    Matrix    Matrix

where $a_{ij}$ is the input required from $C_i$ to produce a dollar's worth of output for $C_j$. The solution to the input-output matrix equation

$$X = MX + D \quad \text{is} \quad X = (I - M)^{-1}D,$$

where $I$ is the identity matrix, assuming $I - M$ has an inverse.

---

1.   40¢ from $A$ and 20¢ from $E$ are required to produce a dollar's worth of output for $A$.

3.   $I - M = \begin{bmatrix} 1 & 0 \\ 0 & 1 \end{bmatrix} - \begin{bmatrix} 0.4 & 0.2 \\ 0.2 & 0.1 \end{bmatrix} = \begin{bmatrix} 0.6 & -0.2 \\ -0.2 & 0.9 \end{bmatrix}$

Converting the decimals to fractions to calculate the inverse, we have:

$$\begin{bmatrix} \frac{3}{5} & -\frac{1}{5} & | & 1 & 0 \\ \frac{1}{5} & \frac{9}{10} & | & 0 & 1 \end{bmatrix} \sim \begin{bmatrix} 1 & -\frac{1}{3} & | & \frac{5}{3} & 0 \\ -\frac{1}{5} & \frac{9}{10} & | & 0 & 1 \end{bmatrix} \sim \begin{bmatrix} 1 & -\frac{1}{3} & | & \frac{5}{3} & 0 \\ 0 & \frac{5}{6} & | & \frac{1}{3} & 1 \end{bmatrix} \sim \begin{bmatrix} 1 & -\frac{1}{3} & | & \frac{5}{3} & 0 \\ 0 & 1 & | & \frac{2}{5} & \frac{6}{5} \end{bmatrix}$$

$$\frac{5}{3}R_1 \to R_1 \qquad \frac{1}{5}R_1 + R_2 \to R_2 \qquad \frac{6}{5}R_2 \to R_2 \qquad \frac{1}{3}R_2 + R_1 \to R_1$$

$$\sim \begin{bmatrix} 1 & 0 & | & \frac{9}{5} & \frac{2}{5} \\ 0 & 1 & | & \frac{2}{5} & \frac{6}{5} \end{bmatrix} \quad \text{Thus,} \quad I - M = \begin{bmatrix} 0.6 & -0.2 \\ -0.2 & 0.9 \end{bmatrix} \quad \text{and} \quad (I - M)^{-1} = \begin{bmatrix} 1.8 & 0.4 \\ 0.4 & 1.2 \end{bmatrix}.$$

5.   $X = (I - M)^{-1}D_2 = \begin{bmatrix} 1.8 & 0.4 \\ 0.4 & 1.2 \end{bmatrix} \begin{bmatrix} 8 \\ 5 \end{bmatrix}$   Thus,   $\begin{bmatrix} x_1 \\ x_2 \end{bmatrix} = \begin{bmatrix} 16.4 \\ 9.2 \end{bmatrix}$   and $x_1 = 16.4$, $x_2 = 9.2$.

7.   20¢ from $A$, 10¢ from $B$, and 10¢ from $E$ are required to produce a dollar's worth of output for $B$.

9.   $\begin{bmatrix} 1 & 0 & 0 \\ 0 & 1 & 0 \\ 0 & 0 & 1 \end{bmatrix} - \begin{bmatrix} 0.3 & 0.2 & 0.2 \\ 0.1 & 0.1 & 0.1 \\ 0.2 & 0.1 & 0.1 \end{bmatrix} = \begin{bmatrix} 0.7 & -0.2 & -0.2 \\ -0.1 & 0.9 & -0.1 \\ -0.2 & -0.1 & -0.9 \end{bmatrix}$

Copyright © 2011 Pearson Education, Inc. Publishing as Prentice Hall.

**11.** $X = (I - M)^{-1}D_1$

Therefore, $\begin{bmatrix} x_1 \\ x_2 \\ x_3 \end{bmatrix} = \begin{bmatrix} 1.6 & 0.4 & 0.4 \\ 0.22 & 1.18 & 0.18 \\ 0.38 & 0.22 & 1.22 \end{bmatrix} \begin{bmatrix} 5 \\ 10 \\ 15 \end{bmatrix} = \begin{bmatrix} 18 \\ 15.6 \\ 22.4 \end{bmatrix}$

Thus, agriculture, \$18 billion; building, \$15.6 billion; and energy, \$22.4 billion.

**13.** $I - M = \begin{bmatrix} 1 & 0 \\ 0 & 1 \end{bmatrix} - \begin{bmatrix} 0.2 & 0.2 \\ 0.3 & 0.3 \end{bmatrix} = \begin{bmatrix} 0.8 & -0.2 \\ -0.3 & 0.7 \end{bmatrix} = \begin{bmatrix} \frac{4}{5} & -\frac{1}{5} \\ -\frac{3}{10} & \frac{7}{10} \end{bmatrix},$

converting the decimals to fractions.

$\begin{bmatrix} \frac{4}{5} & -\frac{1}{5} & | & 1 & 0 \\ -\frac{3}{10} & \frac{7}{10} & | & 0 & 1 \end{bmatrix} \sim \begin{bmatrix} 1 & -\frac{1}{4} & | & \frac{5}{4} & 0 \\ -\frac{3}{10} & \frac{7}{10} & | & 0 & 1 \end{bmatrix} \sim \begin{bmatrix} 1 & -\frac{1}{4} & | & \frac{5}{4} & 0 \\ 0 & \frac{5}{8} & | & \frac{3}{8} & 1 \end{bmatrix}$

$\quad\quad \dfrac{5}{4}R_1 \rightarrow R_1 \quad\quad\quad \dfrac{3}{10}R_1 + R_2 \rightarrow R_2 \quad\quad \dfrac{8}{5}R_2 \rightarrow R_2$

$\sim \begin{bmatrix} 1 & -\frac{1}{4} & | & \frac{5}{4} & 0 \\ 0 & 1 & | & \frac{3}{5} & \frac{8}{5} \end{bmatrix} \sim \begin{bmatrix} 1 & 0 & | & \frac{7}{5} & \frac{2}{5} \\ 0 & 1 & | & \frac{3}{5} & \frac{8}{5} \end{bmatrix}$ Thus, $(I - M)^{-1} = \begin{bmatrix} 1.4 & 0.4 \\ 0.6 & 1.6 \end{bmatrix}.$

$\quad\quad \dfrac{1}{4}R_2 + R_1 \rightarrow R_1$

Now, $X = (I - M)^{-1}D = \begin{bmatrix} 1.4 & 0.4 \\ 0.6 & 1.6 \end{bmatrix} \begin{bmatrix} 10 \\ 25 \end{bmatrix} = \begin{bmatrix} 24 \\ 46 \end{bmatrix}.$

**15.** $I - M = \begin{bmatrix} 1 & 0 \\ 0 & 1 \end{bmatrix} - \begin{bmatrix} 0.7 & 0.8 \\ 0.3 & 0.2 \end{bmatrix} = \begin{bmatrix} 0.3 & -0.8 \\ -0.3 & 0.8 \end{bmatrix}$

$I - M$ is singular ($R_2 = (-1)R_1$); $X$ does not exist.

**17.** $I - M = \begin{bmatrix} 1 & 0 & 0 \\ 0 & 1 & 0 \\ 0 & 0 & 1 \end{bmatrix} - \begin{bmatrix} 0.3 & 0.1 & 0.3 \\ 0.2 & 0.1 & 0.2 \\ 0.1 & 0.1 & 0.1 \end{bmatrix} = \begin{bmatrix} 0.7 & -0.1 & -0.3 \\ -0.2 & 0.9 & -0.2 \\ -0.1 & -0.1 & 0.9 \end{bmatrix}$

$\begin{bmatrix} 0.7 & -0.1 & -0.3 & | & 1 & 0 & 0 \\ -0.2 & 0.9 & -0.2 & | & 0 & 1 & 0 \\ -0.1 & -0.1 & 0.9 & | & 0 & 0 & 1 \end{bmatrix} \sim \begin{bmatrix} 7 & -1 & -3 & | & 10 & 0 & 0 \\ -2 & 9 & -2 & | & 0 & 10 & 0 \\ 1 & 1 & -9 & | & 0 & 0 & -10 \end{bmatrix}$

$\quad\quad 10R_1 \rightarrow R_1 \quad\quad\quad\quad\quad R_1 \leftrightarrow R_3$
$\quad\quad 10R_2 \rightarrow R_2$
$\quad\quad -10R_3 \rightarrow R_3$

$\sim \begin{bmatrix} 1 & 1 & -9 & | & 0 & 0 & -10 \\ -2 & 9 & -2 & | & 0 & 10 & 0 \\ 7 & -1 & -3 & | & 10 & 0 & 0 \end{bmatrix} \sim \begin{bmatrix} 1 & 1 & -9 & | & 0 & 0 & -10 \\ 0 & 11 & -20 & | & 0 & 10 & -20 \\ 0 & -8 & 60 & | & 10 & 0 & 70 \end{bmatrix}$

$\quad\quad 2R_1 + R_2 \rightarrow R_2 \quad\quad\quad \dfrac{1}{11}R_2 \rightarrow R_2$
$\quad\quad (-7)R_1 + R_3 \rightarrow R_3$

Copyright © 2011 Pearson Education, Inc. Publishing as Prentice Hall.

$$\sim \begin{bmatrix} 1 & 1 & -9 & 0 & 0 & -10 \\ 0 & 1 & -1.82 & 0 & 0.91 & -1.82 \\ 0 & -8 & 60 & 10 & 0 & 70 \end{bmatrix} \sim \begin{bmatrix} 1 & 0 & -7.18 & 0 & -0.91 & -8.18 \\ 0 & 1 & -1.82 & 0 & 0.91 & -1.82 \\ 0 & 0 & 45.44 & 10 & 7.28 & 55.44 \end{bmatrix}$$

$$8R_2 + R_3 \rightarrow R_3 \qquad\qquad \frac{1}{45.44} R_3 \rightarrow R_3$$

$$\sim \begin{bmatrix} 1 & 0 & -7.18 & 0 & -0.91 & -8.18 \\ 0 & 1 & -1.82 & 0 & 0.91 & -1.82 \\ 0 & 0 & 1 & 0.22 & 0.16 & 1.22 \end{bmatrix} \sim \begin{bmatrix} 1 & 0 & 0 & 1.58 & 0.24 & 0.58 \\ 0 & 1 & 0 & 0.4 & 1.2 & 0.4 \\ 0 & 0 & 1 & 0.22 & 0.16 & 1.22 \end{bmatrix}$$

$$1.82R_3 + R_2 \rightarrow R_2$$
$$7.18R_3 + R_1 \rightarrow R_1$$

Thus, $(I-M)^{-1} = \begin{bmatrix} 1.58 & 0.24 & 0.58 \\ 0.4 & 1.2 & 0.4 \\ 0.22 & 0.16 & 1.22 \end{bmatrix}$,

and $X = (I-M)^{-1}D = \begin{bmatrix} 1.58 & 0.24 & 0.58 \\ 0.4 & 1.2 & 0.4 \\ 0.22 & 0.16 & 1.22 \end{bmatrix} \begin{bmatrix} 20 \\ 5 \\ 10 \end{bmatrix} = \begin{bmatrix} 38.6 \\ 18 \\ 17.4 \end{bmatrix}$.

**19.** **(A)**   The technology matrix $M = \begin{bmatrix} 0.3 & 0.25 \\ 0.1 & 0.25 \end{bmatrix}$ and the final demand matrix

$D = \begin{bmatrix} 40 \\ 40 \end{bmatrix}$. The input-output matrix equation is $X = MX + D$ or

$X = \begin{bmatrix} 0.3 & 0.25 \\ 0.1 & 0.25 \end{bmatrix} X + \begin{bmatrix} 40 \\ 40 \end{bmatrix}$, where $X = \begin{bmatrix} x_1 \\ x_2 \end{bmatrix}$.

The solution is $X = (I-M)^{-1}D$, provided $I-M$ has an inverse. Now,

$I - M = \begin{bmatrix} 1 & 0 \\ 0 & 1 \end{bmatrix} - \begin{bmatrix} 0.3 & 0.25 \\ 0.1 & 0.25 \end{bmatrix} = \begin{bmatrix} 0.7 & -0.25 \\ -0.1 & 0.75 \end{bmatrix}$

$(I-M)^{-1}:$   $\begin{bmatrix} 0.7 & -0.25 & 1 & 0 \\ -0.1 & 0.75 & 0 & 1 \end{bmatrix} -10R_2 \rightarrow R_2 \sim \begin{bmatrix} 0.7 & -0.25 & 1 & 0 \\ 1 & -7.5 & 0 & -10 \end{bmatrix} R_1 \leftrightarrow R_2$

$\sim \begin{bmatrix} 1 & -7.5 & 0 & -10 \\ 0.7 & -0.25 & 1 & 0 \end{bmatrix} (-0.7)R_1 + R_2 \rightarrow R_2 \sim \begin{bmatrix} 1 & -7.5 & 0 & -10 \\ 0 & 5 & 1 & 7 \end{bmatrix} (0.2)R_2 \rightarrow R_2$

$\sim \begin{bmatrix} 1 & -7.5 & 0 & -10 \\ 0 & 1 & 0.2 & 1.4 \end{bmatrix} (7.5)R_2 + R_1 \rightarrow R_1 \sim \begin{bmatrix} 1 & 0 & 1.5 & 0.5 \\ 0 & 1 & 0.2 & 1.4 \end{bmatrix}$

Thus, $(I-M)^{-1} = \begin{bmatrix} 1.5 & 0.5 \\ 0.2 & 1.4 \end{bmatrix}$ and $X = \begin{bmatrix} 1.5 & 0.5 \\ 0.2 & 1.4 \end{bmatrix} \begin{bmatrix} 40 \\ 40 \end{bmatrix} = \begin{bmatrix} 80 \\ 64 \end{bmatrix}$

Thus, the output for each sector is:

Agriculture: $80 million; Manufacturing: $64 million

Copyright © 2011 Pearson Education, Inc. Publishing as Prentice Hall.

(B)    If the agricultural output is increased by \$20 million and the manufacturing output remains at \$64 million, then the final demand $D$ is given by

$$D = (I - M)X = \begin{bmatrix} 0.7 & -0.25 \\ -0.1 & 0.75 \end{bmatrix} \begin{bmatrix} 100 \\ 64 \end{bmatrix} = \begin{bmatrix} 54 \\ 38 \end{bmatrix}$$

The final demand for agriculture increases to \$54 million and the final demand for manufacturing decreases to \$38 million.

21.   From Problem 19, the technology matrix $T$ in this case is:

$$T = \begin{bmatrix} 0.25 & 0.1 \\ 0.25 & 0.3 \end{bmatrix}.$$

The final demand matrix $D = \begin{bmatrix} 40 \\ 40 \end{bmatrix}$.

The input-output matrix is $X = TX + D$ and the solution is $X = (I - T)^{-1}D$ provided

$$I - T = \begin{bmatrix} 1 & 0 \\ 0 & 1 \end{bmatrix} - \begin{bmatrix} 0.25 & 0.1 \\ 0.25 & 0.3 \end{bmatrix} = \begin{bmatrix} 0.75 & -0.1 \\ -0.25 & 0.7 \end{bmatrix}$$

has an inverse

$$\begin{bmatrix} 0.75 & -0.1 & | & 1 & 0 \\ -0.25 & 0.7 & | & 0 & 1 \end{bmatrix} \sim \begin{bmatrix} 1 & -0.8 & | & 1 & -1 \\ -0.25 & 0.7 & | & 0 & 1 \end{bmatrix} \sim \begin{bmatrix} 1 & -0.8 & | & 1 & -1 \\ 0 & 0.5 & | & 0.25 & 0.75 \end{bmatrix}$$

$$(-1)R_2 + R_1 \rightarrow R_1 \qquad\qquad (0.25)R_1 + R_2 \rightarrow R_2 \qquad\qquad 2R_2 \rightarrow R_2$$

$$\sim \begin{bmatrix} 1 & -0.8 & | & 1 & -1 \\ 0 & 1 & | & 0.5 & 1.5 \end{bmatrix} \sim \begin{bmatrix} 1 & 0 & | & 1.4 & 0.2 \\ 0 & 1 & | & 0.5 & 1.5 \end{bmatrix}$$

$$(0.8)R_2 + R_1 \rightarrow R_1$$

Thus $(I - T)^{-1} = \begin{bmatrix} 1.4 & 0.2 \\ 0.5 & 1.5 \end{bmatrix}$ and $X = \begin{bmatrix} 1.4 & 0.2 \\ 0.5 & 1.5 \end{bmatrix} \begin{bmatrix} 40 \\ 40 \end{bmatrix} = \begin{bmatrix} 64 \\ 80 \end{bmatrix}$

The output for each sector is: manufacturing: \$64 million; agriculture: \$80 million.

23.   Let $x_1$ = total output of energy

$x_2$ = total output of mining

Then the final demand matrix $D = \begin{pmatrix} 0.4x_1 \\ 0.4x_2 \end{pmatrix}$ and the input-output matrix equation is:

$$\begin{bmatrix} x_1 \\ x_2 \end{bmatrix} = \begin{bmatrix} 0.2 & 0.3 \\ 0.4 & 0.3 \end{bmatrix} \begin{bmatrix} x_1 \\ x_2 \end{bmatrix} + \begin{bmatrix} 0.4x_1 \\ 0.4x_2 \end{bmatrix}$$

This yields the dependent system of equations

$$0.6x_1 = 0.2x_1 + 0.3x_2$$
$$0.6x_2 = 0.4x_1 + 0.3x_2$$

which is equivalent to

$$0.4x_1 - 0.3x_2 = 0$$

or $\qquad\qquad x_1 = \dfrac{3}{4}x_2$

Thus, the total output of the energy sector should be 75% of the total output of the mining sector.

Copyright © 2011 Pearson Education, Inc. Publishing as Prentice Hall.

**25.** Each element of a technology matrix represents the input needed from $C_i$ to produce \$1 dollar's worth of output for $C_j$. Hence, each element must be a number between 0 and 1, inclusive.

**27.** The technology matrix $M = \begin{bmatrix} 0.1 & 0.2 \\ 0.2 & 0.4 \end{bmatrix}$ and the final demand matrix $D = \begin{bmatrix} 20 \\ 10 \end{bmatrix}$.

The input-output matrix equation is $X = MX + D$ or

$X = \begin{bmatrix} 0.1 & 0.2 \\ 0.2 & 0.4 \end{bmatrix} X + \begin{bmatrix} 20 \\ 10 \end{bmatrix}$ where $X = \begin{bmatrix} x_1 \\ x_2 \end{bmatrix}$.

The solution is $X = (I - M)^{-1} D$, provided $(I - M)$ has an inverse. Now,

$I - M = \begin{bmatrix} 1 & 0 \\ 0 & 1 \end{bmatrix} - \begin{bmatrix} 0.1 & 0.2 \\ 0.2 & 0.4 \end{bmatrix} = \begin{bmatrix} 0.9 & -0.2 \\ -0.2 & 0.6 \end{bmatrix} = \begin{bmatrix} \frac{9}{10} & -\frac{1}{5} \\ -\frac{1}{5} & \frac{3}{5} \end{bmatrix}$

$\begin{bmatrix} \frac{9}{10} & -\frac{1}{5} & | & 1 & 0 \\ -\frac{1}{5} & \frac{3}{5} & | & 0 & 1 \end{bmatrix} \sim \begin{bmatrix} 1 & -\frac{2}{9} & | & \frac{10}{9} & 0 \\ -\frac{1}{5} & \frac{3}{5} & | & 0 & 1 \end{bmatrix} \sim \begin{bmatrix} 1 & -\frac{2}{9} & | & \frac{10}{9} & 0 \\ 0 & \frac{5}{9} & | & \frac{2}{9} & 1 \end{bmatrix} \sim \begin{bmatrix} 1 & -\frac{2}{9} & | & \frac{10}{9} & 0 \\ 0 & 1 & | & \frac{2}{5} & \frac{9}{5} \end{bmatrix}$

$\qquad \frac{10}{9} R_1 \rightarrow R_1 \qquad\qquad \frac{1}{5} R_1 + R_2 \rightarrow R_2 \qquad \frac{9}{5} R_2 \rightarrow R_2 \qquad \frac{2}{9} R_2 + R_1 \rightarrow R_1$

$\sim \begin{bmatrix} 1 & 0 & | & \frac{6}{5} & \frac{2}{5} \\ 0 & 1 & | & \frac{2}{5} & \frac{9}{5} \end{bmatrix}$ Thus, $(I - M)^{-1} = \begin{bmatrix} \frac{6}{5} & \frac{2}{5} \\ \frac{2}{5} & \frac{9}{5} \end{bmatrix} = \begin{bmatrix} 1.2 & 0.4 \\ 0.4 & 1.8 \end{bmatrix}$, and

$X = \begin{bmatrix} 1.2 & 0.4 \\ 0.4 & 1.8 \end{bmatrix} \begin{bmatrix} 20 \\ 10 \end{bmatrix} = \begin{bmatrix} 28 \\ 26 \end{bmatrix}$.

Therefore, the output for each sector is: coal, \$28 billion; steel, \$26 billion.

**29.** The technology matrix $M = \begin{bmatrix} 0.20 & 0.40 \\ 0.15 & 0.30 \end{bmatrix} = \begin{bmatrix} \frac{1}{5} & \frac{2}{5} \\ \frac{3}{20} & \frac{3}{10} \end{bmatrix}$ and the final demand matrix $D = \begin{bmatrix} 60 \\ 80 \end{bmatrix}$. The

input-output matrix equation is $X = MX + D$ or

$X = \begin{bmatrix} \frac{1}{5} & \frac{2}{5} \\ \frac{3}{20} & \frac{3}{10} \end{bmatrix} X + \begin{bmatrix} 60 \\ 80 \end{bmatrix}$ where $X = \begin{bmatrix} x_1 \\ x_2 \end{bmatrix}$. The solution is $X = (I - M)^{-1} D$, provided $(I - M)$ has an

inverse. Now

$I - M = \begin{bmatrix} 1 & 0 \\ 0 & 1 \end{bmatrix} - \begin{bmatrix} \frac{1}{5} & \frac{2}{5} \\ \frac{3}{20} & \frac{3}{10} \end{bmatrix} = \begin{bmatrix} \frac{4}{5} & -\frac{2}{5} \\ -\frac{3}{20} & \frac{7}{10} \end{bmatrix}$.

$\begin{bmatrix} \frac{4}{5} & -\frac{2}{5} & | & 1 & 0 \\ -\frac{3}{20} & \frac{7}{10} & | & 0 & 1 \end{bmatrix} \sim \begin{bmatrix} 1 & -\frac{1}{2} & | & \frac{5}{4} & 0 \\ -\frac{3}{20} & \frac{7}{10} & | & 0 & 1 \end{bmatrix} \sim \begin{bmatrix} 1 & -\frac{1}{2} & | & \frac{5}{4} & 0 \\ 0 & \frac{5}{8} & | & \frac{3}{16} & 1 \end{bmatrix}$

$\qquad \frac{5}{4} R_1 \rightarrow R_1 \qquad\qquad \frac{3}{20} R_1 + R_2 \rightarrow R_2 \qquad \frac{8}{5} R_2 \rightarrow R_2$

$\sim \begin{bmatrix} 1 & -\frac{1}{2} & | & \frac{5}{4} & 0 \\ 0 & 1 & | & \frac{3}{10} & \frac{8}{5} \end{bmatrix} \sim \begin{bmatrix} 1 & 0 & | & \frac{7}{5} & \frac{4}{5} \\ 0 & 1 & | & \frac{3}{10} & \frac{8}{5} \end{bmatrix}$

$\qquad \frac{1}{2} R_2 + R_1 \rightarrow R_1$

Copyright © 2011 Pearson Education, Inc. Publishing as Prentice Hall.

Thus $(I-M)^{-1} = \begin{bmatrix} \frac{7}{5} & \frac{4}{5} \\ \frac{3}{10} & \frac{8}{5} \end{bmatrix}$ and $X = \begin{bmatrix} \frac{7}{5} & \frac{4}{5} \\ \frac{3}{10} & \frac{8}{5} \end{bmatrix}\begin{bmatrix} 60 \\ 80 \end{bmatrix} = \begin{bmatrix} 148 \\ 146 \end{bmatrix}$

Therefore, the output for each sector is: Agriculture—\$148 million; Tourism—\$146 million

**31.** The technology matrix $M = \begin{bmatrix} 0.2 & 0.4 & 0.3 \\ 0.2 & 0.1 & 0.1 \\ 0.2 & 0.1 & 0.1 \end{bmatrix}$ and the final demand matrix $D = \begin{bmatrix} 10 \\ 15 \\ 20 \end{bmatrix}$.

The input-output matrix equation is $X = MX + D$ or

$$X = \begin{bmatrix} 0.2 & 0.4 & 0.3 \\ 0.2 & 0.1 & 0.1 \\ 0.2 & 0.1 & 0.1 \end{bmatrix} X + \begin{bmatrix} 10 \\ 15 \\ 20 \end{bmatrix}.$$

The solution is $X = (I-M)^{-1}D$, provided $I-M$ has an inverse. Now,

$$I-M = \begin{bmatrix} 1 & 0 & 0 \\ 0 & 1 & 0 \\ 0 & 0 & 1 \end{bmatrix} - \begin{bmatrix} 0.2 & 0.4 & 0.3 \\ 0.2 & 0.1 & 0.1 \\ 0.2 & 0.1 & 0.1 \end{bmatrix} = \begin{bmatrix} 0.8 & -0.4 & -0.3 \\ -0.2 & 0.9 & -0.1 \\ -0.2 & -0.1 & 0.9 \end{bmatrix}.$$

$$\begin{bmatrix} 0.8 & -0.4 & -0.3 & | & 1 & 0 & 0 \\ -0.2 & 0.9 & -0.1 & | & 0 & 1 & 0 \\ -0.2 & -0.1 & 0.9 & | & 0 & 0 & 1 \end{bmatrix} \sim \begin{bmatrix} 8 & -4 & -3 & | & 10 & 0 & 0 \\ -2 & 9 & -1 & | & 0 & 10 & 0 \\ -2 & -1 & 9 & | & 0 & 0 & 10 \end{bmatrix}$$

$$\begin{matrix} 10R_1 \to R_1 \\ 10R_2 \to R_2 \\ 10R_3 \to R_3 \end{matrix} \qquad \left(-\frac{1}{2}\right)R_2 \to R_2$$

$$\sim \begin{bmatrix} 8 & -4 & -3 & | & 10 & 0 & 0 \\ 1 & -\frac{9}{2} & \frac{1}{2} & | & 0 & -5 & 0 \\ -2 & 9 & 9 & | & 0 & 0 & 10 \end{bmatrix} \sim \begin{bmatrix} 1 & -\frac{9}{2} & \frac{1}{2} & | & 0 & -5 & 0 \\ 8 & -4 & -3 & | & 10 & 0 & 0 \\ -2 & 1 & 9 & | & 0 & 0 & 10 \end{bmatrix}$$

$$R_1 \leftrightarrow R_2 \qquad \begin{matrix} (-8)R_1 + R_2 \to R_2 \\ 2R_1 + R_3 \to R_3 \end{matrix}$$

$$\sim \begin{bmatrix} 1 & -\frac{9}{2} & \frac{1}{2} & | & 0 & -5 & 0 \\ 0 & 32 & -7 & | & 10 & 40 & 0 \\ 0 & -10 & 10 & | & 0 & -10 & 10 \end{bmatrix} \sim \begin{bmatrix} 1 & -\frac{9}{2} & \frac{1}{2} & | & 0 & -5 & 0 \\ 0 & 32 & -7 & | & 10 & 40 & 0 \\ 0 & 1 & -1 & | & 0 & 1 & -1 \end{bmatrix}$$

$$\left(-\frac{1}{10}\right)R_3 \qquad\qquad R_2 \leftrightarrow R_3$$

Copyright © 2011 Pearson Education, Inc. Publishing as Prentice Hall.

$$\sim \begin{bmatrix} 1 & -\frac{9}{2} & \frac{1}{2} & 0 & -5 & 0 \\ 0 & 1 & -1 & 0 & 1 & -1 \\ 0 & 32 & -7 & 10 & 40 & 0 \end{bmatrix} \sim \begin{bmatrix} 1 & 0 & -4 & 0 & -\frac{1}{2} & -\frac{9}{2} \\ 0 & 1 & -1 & 0 & 1 & -1 \\ 0 & 0 & 25 & 10 & 8 & 32 \end{bmatrix}$$

$$(-32)R_2 + R_3 \to R_3 \qquad\qquad \frac{1}{25} R_3 \to R_3$$

$$\frac{9}{2} R_2 + R_1 \to R_1$$

$$\begin{bmatrix} 1 & 0 & -4 & 0 & -\frac{1}{2} & -\frac{9}{2} \\ 0 & 1 & -1 & 0 & 1 & -1 \\ 0 & 0 & 1 & 0.4 & 0.32 & 1.28 \end{bmatrix} = \begin{bmatrix} 1 & 0 & 0 & 1.6 & 0.78 & 0.62 \\ 0 & 1 & 0 & 0.4 & 1.32 & 0.28 \\ 0 & 0 & 1 & 0.4 & 0.32 & 1.28 \end{bmatrix}$$

$$R_3 + R_2 \to R_2$$
$$4R_3 + R_1 \to R_1$$

Thus, $(I-M)^{-1} = \begin{bmatrix} 1.6 & 0.78 & 0.62 \\ 0.4 & 1.32 & 0.28 \\ 0.4 & 0.32 & 1.28 \end{bmatrix}$, and $X = (I-M)^{-1}D = \begin{bmatrix} 1.6 & 0.78 & 0.62 \\ 0.4 & 1.32 & 0.28 \\ 0.4 & 0.32 & 1.28 \end{bmatrix}\begin{bmatrix} 10 \\ 15 \\ 20 \end{bmatrix} = \begin{bmatrix} 40.1 \\ 29.4 \\ 34.4 \end{bmatrix}$

Therefore, agriculture, \$40.1 billion; manufacturing, \$29.4 billion; and energy, \$34.4 billion.

**33.** The technology matrix is $M = \begin{bmatrix} 0.05 & 0.17 & 0.23 & 0.09 \\ 0.07 & 0.12 & 0.15 & 0.19 \\ 0.25 & 0.08 & 0.03 & 0.32 \\ 0.11 & 0.19 & 0.28 & 0.16 \end{bmatrix}$.

The input-output matrix equation is $X = MX + D$

where $X = \begin{bmatrix} A \\ E \\ L \\ M \end{bmatrix}$ and $D$ is the final demand matrix. Thus $X = (I-M)^{-1}D$, where

$$I - M = \begin{bmatrix} 0.95 & -0.17 & -0.23 & -0.09 \\ -0.07 & 0.88 & -0.15 & -0.19 \\ -0.25 & -0.08 & 0.97 & -0.32 \\ -0.11 & -0.19 & -0.28 & 0.84 \end{bmatrix}$$

Now, $X = \begin{bmatrix} 1.25 & 0.37 & 0.47 & 0.40 \\ 0.26 & 1.33 & 0.41 & 0.48 \\ 0.47 & 0.36 & 1.39 & 0.66 \\ 0.38 & 0.47 & 0.62 & 1.57 \end{bmatrix} D$

Year 1: $D = \begin{bmatrix} 23 \\ 41 \\ 18 \\ 31 \end{bmatrix}$ and $(I-M)^{-1}D \approx \begin{bmatrix} 65 \\ 83 \\ 71 \\ 88 \end{bmatrix}$

Copyright © 2011 Pearson Education, Inc. Publishing as Prentice Hall.

Agriculture: $65 billion; Energy: $83 billion; Labor: $71 billion; Manufacturing: $88 billion

Year 2: $D = \begin{bmatrix} 32 \\ 48 \\ 21 \\ 33 \end{bmatrix}$ and $(I - M)^{-1}D \approx \begin{bmatrix} 81 \\ 97 \\ 83 \\ 99 \end{bmatrix}$

Agriculture: $81 billion; Energy: $97 billion; Labor: $83 billion; Manufacturing: $99 billion

Year 3: $D = \begin{bmatrix} 55 \\ 62 \\ 25 \\ 35 \end{bmatrix}$ and $(I - M)^{-1}D \approx \begin{bmatrix} 117 \\ 124 \\ 106 \\ 120 \end{bmatrix}$

Agriculture: $117 billion; Energy: $124 billion; Labor: $106 billion; Manufacturing: $120 billion

## CHAPTER 4 REVIEW

**1.** $y = 2x - 4$

(1)

$y = \dfrac{1}{2}x + 2$

(2)

The point of intersection is the solution. This is
$x = 4$, $y = 4$.     (4-1)

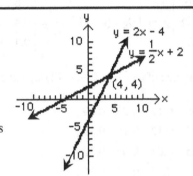

**2.** Substitute equation (1) into (2):

$$2x - 4 = \frac{1}{2}x + 2$$

$$\frac{3}{2}x = 6$$

$$x = 4$$

Substitute $x = 4$ into (1):

$$y = 2 \cdot 4 - 4 = 4$$

Solution: $x = 4$, $y = 4$

(4-1)

**3.** (A) $\begin{bmatrix} 0 & 1 & | & 2 \\ 1 & 0 & | & 3 \end{bmatrix}$ is not in reduced form; the left-most 1 in the second row is not to the right of the left-most 1 in the first row. [condition (d)]   $R_1 \leftrightarrow R_2$

(B) $\begin{bmatrix} 1 & 0 & | & 2 \\ 0 & 3 & | & 3 \end{bmatrix}$ is not in reduced form; the left-most nonzero element in row 2 is not 1. [condition (b)]

$\dfrac{1}{3}R_2 \to R_2$

(C) $\begin{bmatrix} 1 & 0 & 1 & 2 \\ 0 & 1 & 1 & 3 \end{bmatrix}$ is in reduced form.

(D) $\begin{bmatrix} 1 & 1 & 0 & 2 \\ 0 & 1 & 1 & 3 \end{bmatrix}$ is not in reduced form; the left-most 1 in the second row is not the only non-zero element in its column. [condition (c)]

$(-1)R_2 + R_1 \to R_1$     (4-3)

Copyright © 2011 Pearson Education, Inc. Publishing as Prentice Hall.

4. $A = \begin{bmatrix} 5 & 3 & -1 & 0 & 2 \\ -4 & 8 & 1 & 3 & 0 \end{bmatrix}$, $B = \begin{bmatrix} -3 & 2 \\ 0 & 4 \\ -1 & 7 \end{bmatrix}$

(A)   $A$ is $2 \times 5$, $B$ is $3 \times 2$

(B)   $a_{24} = 3$, $a_{15} = 2$, $b_{31} = -1$, $b_{22} = 4$

(C)   $AB$ is not defined; the number of columns of $A \neq$ the number of rows of $B$. $BA$ is defined.
(4-2, 4-4)

5.   (A)   $\begin{bmatrix} 1 & -2 \\ 1 & -3 \end{bmatrix} \begin{bmatrix} x_1 \\ x_2 \end{bmatrix} = \begin{bmatrix} 4 \\ 2 \end{bmatrix}$

$\begin{bmatrix} 1 & -2 & | & 4 \\ 1 & -3 & | & 2 \end{bmatrix} \sim \begin{bmatrix} 1 & -2 & | & 4 \\ 0 & -1 & | & -2 \end{bmatrix} \sim \begin{bmatrix} 1 & -2 & | & 4 \\ 0 & 1 & | & 2 \end{bmatrix}$
$(-1)R_1 + R_2 \rightarrow R_2 \quad (-1)R_2 \rightarrow R_2 \qquad (2)R_2 + R_1 \rightarrow R_1$

$\sim \begin{bmatrix} 1 & 0 & | & 8 \\ 0 & 1 & | & 2 \end{bmatrix}$   Therefore, $x_1 = 8$, $x_2 = 2$.

or calculate the inverse of the coefficient matrix $A$:

$$A^{-1} = \begin{bmatrix} 3 & -2 \\ 1 & -1 \end{bmatrix} \text{ and } \begin{bmatrix} x_1 \\ x_2 \end{bmatrix} = \begin{bmatrix} 3 & -2 \\ 1 & -1 \end{bmatrix} \begin{bmatrix} 4 \\ 2 \end{bmatrix} = \begin{bmatrix} 8 \\ 2 \end{bmatrix}; \ x_1 = 8, \ x_2 = 2.$$

(B)   $\begin{bmatrix} 5 & 3 \\ 1 & 1 \end{bmatrix} \begin{bmatrix} x_1 \\ x_2 \end{bmatrix} + \begin{bmatrix} 25 \\ 14 \end{bmatrix} = \begin{bmatrix} 18 \\ 22 \end{bmatrix}$

$\begin{bmatrix} 5 & 3 \\ 1 & 1 \end{bmatrix} \begin{bmatrix} x_1 \\ x_2 \end{bmatrix} = \begin{bmatrix} 18 \\ 22 \end{bmatrix} - \begin{bmatrix} 25 \\ 14 \end{bmatrix} = \begin{bmatrix} -7 \\ 8 \end{bmatrix}$

Augmented matrix

$\begin{bmatrix} 5 & 3 & | & -7 \\ 1 & 1 & | & 8 \end{bmatrix} \sim \begin{bmatrix} 1 & 1 & | & 8 \\ 5 & 3 & | & -7 \end{bmatrix} \sim \begin{bmatrix} 1 & 1 & | & 8 \\ 0 & -2 & | & -47 \end{bmatrix} \sim \begin{bmatrix} 1 & 1 & | & 8 \\ 0 & 1 & | & \frac{47}{2} \end{bmatrix}$

$R_1 \leftrightarrow R_2 \quad (-5)R_1 + R_2 \rightarrow R_2 \quad \left(-\frac{1}{2}\right)R_2 \rightarrow R_2 \quad (-1)R_2 + R_1 \rightarrow R_1$

$\sim \begin{bmatrix} 1 & 0 & | & -\frac{31}{2} \\ 0 & 1 & | & \frac{47}{2} \end{bmatrix}$.  Therefore, $x_1 = -\frac{31}{2}$, $x_2 = \frac{47}{2}$.

Inverse matrix

$A^{-1} = \begin{bmatrix} 0.5 & -1.5 \\ -0.5 & 2.5 \end{bmatrix}$ and $\begin{bmatrix} x_1 \\ x_2 \end{bmatrix} = \begin{bmatrix} 0.5 & -1.5 \\ -0.5 & 2.5 \end{bmatrix} \begin{bmatrix} -7 \\ 8 \end{bmatrix} = \begin{bmatrix} -15.5 \\ 23.5 \end{bmatrix}$;

$x_1 = -15.5 = -\frac{31}{2}$, $x_2 = 23.5 = \frac{47}{2}$.    (4-2, 4-6)

6.   $A + B = \begin{bmatrix} 1+2 & 2+1 \\ 3+1 & 1+1 \end{bmatrix} = \begin{bmatrix} 3 & 3 \\ 4 & 2 \end{bmatrix}$

(4-4)

7.   $B + D = \begin{bmatrix} 2 & 1 \\ 1 & 1 \end{bmatrix} + \begin{bmatrix} 1 \\ 2 \end{bmatrix}$

The matrices $B$ and $D$ cannot be added because their dimensions are different.   (4-4)

Copyright © 2011 Pearson Education, Inc. Publishing as Prentice Hall.

8. $A - 2B = \begin{bmatrix} 1 & 2 \\ 3 & 1 \end{bmatrix} - 2\begin{bmatrix} 2 & 1 \\ 1 & 1 \end{bmatrix} = \begin{bmatrix} 1 & 2 \\ 3 & 1 \end{bmatrix} + \begin{bmatrix} -4 & -2 \\ -2 & -2 \end{bmatrix} = \begin{bmatrix} -3 & 0 \\ 1 & -1 \end{bmatrix}$  (4-4)

9. $AB = \begin{bmatrix} 1 & 2 \\ 3 & 1 \end{bmatrix}\begin{bmatrix} 2 & 1 \\ 1 & 1 \end{bmatrix} = \begin{bmatrix} \begin{bmatrix} 1 & 2 \end{bmatrix}\begin{bmatrix} 2 \\ 1 \end{bmatrix} & \begin{bmatrix} 1 & 2 \end{bmatrix}\begin{bmatrix} 1 \\ 1 \end{bmatrix} \\ \begin{bmatrix} 3 & 1 \end{bmatrix}\begin{bmatrix} 2 \\ 1 \end{bmatrix} & \begin{bmatrix} 3 & 1 \end{bmatrix}\begin{bmatrix} 1 \\ 1 \end{bmatrix} \end{bmatrix} = \begin{bmatrix} 4 & 3 \\ 7 & 4 \end{bmatrix}$  (4-4)

10. $AC$ is *not defined* because the dimension of $A$ is $2 \times 2$ and the dimension of $C$ is $1 \times 2$. So, the number of columns in $A$ is not equal to the number of rows in $C$.  (4-4)

11. $AD = \begin{bmatrix} 1 & 2 \\ 3 & 1 \end{bmatrix}\begin{bmatrix} 1 \\ 2 \end{bmatrix} = \begin{bmatrix} \begin{bmatrix} 1 & 2 \end{bmatrix}\begin{bmatrix} 1 \\ 2 \end{bmatrix} \\ \begin{bmatrix} 3 & 1 \end{bmatrix}\begin{bmatrix} 1 \\ 2 \end{bmatrix} \end{bmatrix} = \begin{bmatrix} 5 \\ 5 \end{bmatrix}$  (4-4)

12. $DC = \begin{bmatrix} 1 \\ 2 \end{bmatrix}\begin{bmatrix} 2 & 3 \end{bmatrix} = \begin{bmatrix} (1)\cdot(2) & (1)\cdot(3) \\ (2)\cdot(2) & (2)\cdot(3) \end{bmatrix} = \begin{bmatrix} 2 & 3 \\ 4 & 6 \end{bmatrix}$  (4-4)

13. $CD = \begin{bmatrix} 2 & 3 \end{bmatrix}\begin{bmatrix} 1 \\ 2 \end{bmatrix} = [2 + 6] = [8]$  (4-4)

14. $C + D = \begin{bmatrix} 2 & 3 \end{bmatrix} + \begin{bmatrix} 1 \\ 2 \end{bmatrix}$

    Not defined because the dimensions of $C$ and $D$ are different.  (4-4)

15. $\begin{bmatrix} 4 & 3 & | & 1 & 0 \\ 3 & 2 & | & 0 & 1 \end{bmatrix} (-1)R_2 + R_1 \rightarrow R_1 \sim \begin{bmatrix} 1 & 1 & | & 1 & -1 \\ 3 & 2 & | & 0 & 1 \end{bmatrix} (-3)R_1 + R_2 \rightarrow R_2$

    $\sim \begin{bmatrix} 1 & 1 & | & 1 & -1 \\ 0 & -1 & | & -3 & 4 \end{bmatrix} (-1)R_2 \rightarrow R_2 \sim \begin{bmatrix} 1 & 1 & | & 1 & -1 \\ 0 & 1 & | & 3 & -4 \end{bmatrix} (-1)R_2 + R_1 \rightarrow R_1$

    $\sim \begin{bmatrix} 1 & 0 & | & -2 & 3 \\ 0 & 1 & | & 3 & -4 \end{bmatrix}$

    Thus, $A^{-1} = \begin{bmatrix} -2 & 3 \\ 3 & -4 \end{bmatrix}$ and $A^{-1}A = \begin{bmatrix} -2 & 3 \\ 3 & -4 \end{bmatrix}\begin{bmatrix} 4 & 3 \\ 3 & 2 \end{bmatrix} = \begin{bmatrix} 1 & 0 \\ 0 & 1 \end{bmatrix}$  (4-5)

16. $\begin{aligned} 4x_1 + 3x_2 &= 3 \quad (1) \\ 3x_1 + 2x_2 &= 5 \quad (2) \end{aligned}$  Multiply (1) by 2 and (2) by –3.

    $\begin{aligned} 8x_1 + 6x_2 &= 6 \\ -9x_1 - 6x_2 &= -15 \end{aligned}$  Add the two equations.

Copyright © 2011 Pearson Education, Inc. Publishing as Prentice Hall.

$$-x_1 = -9$$
$$x_1 = 9$$

Substitute $x_1 = 9$ into either (1) or (2); we choose (2).

$$3(9) + 2x_2 = 5$$
$$27 + 2x_2 = 5$$
$$2x_2 = -22$$
$$x_2 = -11$$

Solution: $x_1 = 9$, $x_2 = -11$.    (4-1)

**17.** The augmented matrix of the system is:

$$\begin{bmatrix} 4 & 3 & | & 3 \\ 3 & 2 & | & 5 \end{bmatrix} (-1)R_2 + R_1 \to R_1 \sim \begin{bmatrix} 1 & 1 & | & -2 \\ 3 & 2 & | & 5 \end{bmatrix} (-3)R_1 + R_2 \to R_2 \sim \begin{bmatrix} 1 & 1 & | & -2 \\ 0 & -1 & | & 11 \end{bmatrix} (-1)R_2 \to R_2$$

$$\sim \begin{bmatrix} 1 & 1 & | & -2 \\ 0 & 1 & | & -11 \end{bmatrix} (-1)R_2 + R_1 \to R_1$$

$$\sim \begin{bmatrix} 1 & 0 & | & 9 \\ 0 & 1 & | & -11 \end{bmatrix}$$

The system of equations is:
$$x_1 = 9$$
$$x_2 = -11$$

Solution: $x_1 = 9$, $x_2 = -11$.    (4-2)

**18.** The system of equations in matrix form is:

$$\begin{bmatrix} 4 & 3 \\ 3 & 2 \end{bmatrix} \begin{bmatrix} x_1 \\ x_2 \end{bmatrix} = \begin{bmatrix} 3 \\ 5 \end{bmatrix}$$

Thus, $\begin{bmatrix} x_1 \\ x_2 \end{bmatrix} = \begin{bmatrix} 4 & 3 \\ 3 & 2 \end{bmatrix}^{-1} \begin{bmatrix} 3 \\ 5 \end{bmatrix} = \begin{bmatrix} -2 & 3 \\ 3 & -4 \end{bmatrix} \begin{bmatrix} 3 \\ 5 \end{bmatrix}$ (by Problem 15) $= \begin{bmatrix} 9 \\ -11 \end{bmatrix}$

Solution: $x_1 = 9$, $x_2 = -11$.

Replacing the constants 3, 5 by 7, 10, respectively:

$$\begin{bmatrix} x_1 \\ x_2 \end{bmatrix} = \begin{bmatrix} -2 & 3 \\ 3 & -4 \end{bmatrix} \begin{bmatrix} 7 \\ 10 \end{bmatrix} = \begin{bmatrix} 16 \\ -19 \end{bmatrix}$$

Solution: $x_1 = 16$, $x_2 = -19$

Replacing the constants 3, 5 by 4, 2, respectively:

$$\begin{bmatrix} x_1 \\ x_2 \end{bmatrix} = \begin{bmatrix} -2 & 3 \\ 3 & -4 \end{bmatrix} \begin{bmatrix} 4 \\ 2 \end{bmatrix} = \begin{bmatrix} -2 \\ 4 \end{bmatrix}$$

Solution: $x_1 = -2$, $x_2 = 4$.    (4-6)

**19.** $A + D = \begin{bmatrix} 2 & -2 \\ 1 & 0 \\ 3 & 2 \end{bmatrix} + \begin{bmatrix} 3 & -2 & 1 \\ -1 & 1 & 2 \end{bmatrix}$    Not defined because the dimensions of $A$ and $D$ are different.

(4-4)

**20.** $E + DA = \begin{bmatrix} 3 & -4 \\ -1 & 0 \end{bmatrix} + \begin{bmatrix} 3 & -2 & 1 \\ -1 & 1 & 2 \end{bmatrix} \begin{bmatrix} 2 & -2 \\ 1 & 0 \\ 3 & 2 \end{bmatrix} = \begin{bmatrix} 3 & -4 \\ -1 & 0 \end{bmatrix} + \begin{bmatrix} 7 & -4 \\ 5 & 6 \end{bmatrix} = \begin{bmatrix} 10 & -8 \\ 4 & 6 \end{bmatrix}$    (4-4)

Copyright © 2011 Pearson Education, Inc. Publishing as Prentice Hall.

**21.** From Problem 20, $DA = \begin{bmatrix} 7 & -4 \\ 5 & 6 \end{bmatrix}$. Thus,

$$DA - 3E = \begin{bmatrix} 7 & -4 \\ 5 & 6 \end{bmatrix} - 3\begin{bmatrix} 3 & -4 \\ -1 & 0 \end{bmatrix} = \begin{bmatrix} 7 & -4 \\ 5 & 6 \end{bmatrix} + \begin{bmatrix} -9 & 12 \\ 3 & 0 \end{bmatrix} = \begin{bmatrix} -2 & 8 \\ 8 & 6 \end{bmatrix} \qquad (4\text{-}4)$$

**22.** $BC = \begin{bmatrix} -1 \\ 2 \\ 3 \end{bmatrix}[2 \ 1 \ 3] = \begin{bmatrix} -2 & -1 & -3 \\ 4 & 2 & 6 \\ 6 & 3 & 9 \end{bmatrix} \qquad (4\text{-}4)$

**23.** $CB = [2 \ 1 \ 3]\begin{bmatrix} -1 \\ 2 \\ 3 \end{bmatrix} = [-2 + 2 + 9] = [9]$ (a $1 \times 1$ matrix) $\qquad (4\text{-}4)$

**24.** $AD - BC$

$$AD = \begin{bmatrix} 2 & -2 \\ 1 & 0 \\ 3 & 2 \end{bmatrix}\begin{bmatrix} 3 & -2 & 1 \\ -1 & 1 & 2 \end{bmatrix} = \begin{bmatrix} [2 \ -2]\begin{bmatrix} 3 \\ -1 \end{bmatrix} & [2 \ -2]\begin{bmatrix} -2 \\ 1 \end{bmatrix} & [2 \ -2]\begin{bmatrix} 1 \\ 2 \end{bmatrix} \\ [1 \ 0]\begin{bmatrix} 3 \\ -1 \end{bmatrix} & [1 \ 0]\begin{bmatrix} -2 \\ 1 \end{bmatrix} & [1 \ 0]\begin{bmatrix} 1 \\ 2 \end{bmatrix} \\ [3 \ 2]\begin{bmatrix} 3 \\ -1 \end{bmatrix} & [3 \ 2]\begin{bmatrix} -2 \\ 1 \end{bmatrix} & [3 \ 2]\begin{bmatrix} 1 \\ 2 \end{bmatrix} \end{bmatrix}$$

$$= \begin{bmatrix} 8 & -6 & -2 \\ 3 & -2 & 1 \\ 7 & -4 & 7 \end{bmatrix}$$

$$BC = \begin{bmatrix} -1 \\ 2 \\ 3 \end{bmatrix}[2 \ 1 \ 3] = \begin{bmatrix} -2 & -1 & -3 \\ 4 & 2 & 6 \\ 6 & 3 & 9 \end{bmatrix}$$

$$AD - BC = \begin{bmatrix} 8 & -6 & -2 \\ 3 & -2 & 1 \\ 7 & -4 & 7 \end{bmatrix} - \begin{bmatrix} -2 & -1 & -3 \\ 4 & 2 & 6 \\ 6 & 3 & 9 \end{bmatrix} = \begin{bmatrix} 8-(-2) & -6-(-1) & -2-(-3) \\ 3-4 & -2-2 & 1-6 \\ 7-6 & -4-3 & 7-9 \end{bmatrix}$$

$$= \begin{bmatrix} 10 & -5 & 1 \\ -1 & -4 & -5 \\ 1 & -7 & -2 \end{bmatrix} \qquad (4\text{-}4)$$

**25.** $\begin{bmatrix} 1 & 2 & 3 & | & 1 & 0 & 0 \\ 2 & 3 & 4 & | & 0 & 1 & 0 \\ 1 & 2 & 1 & | & 0 & 0 & 1 \end{bmatrix} \sim \begin{bmatrix} 1 & 2 & 3 & | & 1 & 0 & 0 \\ 0 & -1 & -2 & | & -2 & 1 & 0 \\ 0 & 0 & -2 & | & -2 & 0 & 1 \end{bmatrix} \sim \begin{bmatrix} 1 & 2 & 3 & | & 1 & 0 & 0 \\ 0 & 1 & 2 & | & -2 & -1 & 0 \\ 0 & 0 & -2 & | & -1 & 0 & 1 \end{bmatrix}$

$(-2)R_1 + R_2 \to R_2 \qquad\qquad (-1)R_1 \to R_2 \qquad\qquad (-2)R_2 + R_1 \to R_1$

$(-1)R_1 + R_3 \to R_3 \qquad\qquad\qquad\qquad\qquad \left(-\dfrac{1}{2}\right)R_3 \to R_3$

Copyright © 2011 Pearson Education, Inc. Publishing as Prentice Hall.

$$\sim \begin{bmatrix} 1 & 0 & -1 & | & -3 & 2 & 0 \\ 0 & 1 & 2 & | & 2 & -1 & 0 \\ 0 & 0 & 1 & | & \frac{1}{2} & 0 & -\frac{1}{2} \end{bmatrix} \sim \begin{bmatrix} 1 & 0 & 0 & | & -\frac{5}{2} & 2 & -\frac{1}{2} \\ 0 & 1 & 0 & | & 1 & -1 & 1 \\ 0 & 0 & 1 & | & \frac{1}{2} & 0 & -\frac{1}{2} \end{bmatrix}; \quad A^{-1} = \begin{bmatrix} -\frac{5}{2} & 2 & -\frac{1}{2} \\ 1 & -1 & 1 \\ \frac{1}{2} & 0 & -\frac{1}{2} \end{bmatrix}$$

$$R_3 + R_1 \to R_1$$
$$(-2)R_3 + R_2 \to R_2$$

Check:

$$A^{-1}A = \begin{bmatrix} -\frac{5}{2} & 2 & -\frac{1}{2} \\ 1 & -1 & 1 \\ \frac{1}{2} & 0 & -\frac{1}{2} \end{bmatrix} \begin{bmatrix} 1 & 2 & 3 \\ 2 & 3 & 4 \\ 1 & 2 & 1 \end{bmatrix} = \begin{bmatrix} -\frac{5}{2}+4-\frac{1}{2} & -5+6-1 & -\frac{15}{2}+8-\frac{1}{2} \\ 1-2+1 & 2-3+2 & 3-4+1 \\ \frac{1}{2}+0-\frac{1}{2} & 1+0-1 & \frac{3}{2}+0-\frac{1}{2} \end{bmatrix} = \begin{bmatrix} 1 & 0 & 0 \\ 0 & 1 & 0 \\ 0 & 0 & 1 \end{bmatrix} \qquad (4\text{-}5)$$

**26.** (A)   The augmented matrix corresponding to the given system is:

$$\begin{bmatrix} 1 & 2 & 3 & | & 1 \\ 2 & 3 & 4 & | & 3 \\ 1 & 2 & 1 & | & 3 \end{bmatrix} \sim \begin{bmatrix} 1 & 2 & 3 & | & 1 \\ 0 & -1 & -2 & | & 1 \\ 0 & 0 & -2 & | & 2 \end{bmatrix} \sim \begin{bmatrix} 1 & 2 & 3 & | & 1 \\ 0 & 1 & 2 & | & -1 \\ 0 & 0 & -2 & | & 2 \end{bmatrix}$$

$$(-2)R_1+R_2 \to R_2 \qquad (-1)R_2 \to R_2 \qquad (-2)R_2+R_1 \to R_1$$
$$(-1)R_1+R_3 \to R_3$$

$$\sim \begin{bmatrix} 1 & 0 & -1 & | & 3 \\ 0 & 1 & 2 & | & -1 \\ 0 & 0 & -2 & | & 2 \end{bmatrix} \sim \begin{bmatrix} 1 & 0 & -1 & | & 3 \\ 0 & 1 & 2 & | & -1 \\ 0 & 0 & 1 & | & -1 \end{bmatrix} \sim \begin{bmatrix} 1 & 0 & 0 & | & 2 \\ 0 & 1 & 0 & | & 1 \\ 0 & 0 & 1 & | & -1 \end{bmatrix}$$

$$\left(-\frac{1}{2}\right)R_3 \to R_3 \qquad \begin{array}{c}(-2)R_3+R_2 \to R_2 \\ R_3+R_1 \to R_1\end{array}$$

Thus, the solution is: $x_1 = 2$, $x_2 = 1$, $x_3 = -1$.

(B)   The augmented matrix corresponding to the given system is:

$$\begin{bmatrix} 1 & 2 & -1 & | & 2 \\ 2 & 3 & 1 & | & -3 \\ 3 & 5 & 0 & | & -1 \end{bmatrix} \sim \begin{bmatrix} 1 & 2 & -1 & | & 2 \\ 0 & -1 & 3 & | & -7 \\ 0 & -1 & 3 & | & -7 \end{bmatrix} \sim \begin{bmatrix} 1 & 2 & -1 & | & 2 \\ 0 & 1 & -3 & | & 7 \\ 0 & -1 & 3 & | & -7 \end{bmatrix}$$

$$(-2)R_1+R_2 \to R_2 \qquad (-1)R_2 \to R_2 \qquad R_2+R_3 \to R_3$$
$$(-3)R_1+R_3 \to R_3 \qquad\qquad\qquad (-2)R_2+R_1 \to R_1$$

$$\sim \begin{bmatrix} 1 & 0 & 5 & | & -12 \\ 0 & 1 & -3 & | & 7 \\ 0 & 0 & 0 & | & 0 \end{bmatrix}$$

Thus, $x_1 \qquad + 5x_3 = -12 \quad (1)$
$$x_2 - 3x_3 = 7 \quad (2)$$

Let $x_3 = t$ ($t$ any real number). Then, from (1), $x_1 = -5t - 12$ and, from (2), $x_2 = 3t + 7$.

Thus, the solution is $x_1 = -5t - 12$, $x_2 = 3t + 7$, $x_3 = t$.

(C)   The augmented matrix corresponding to the given system is:

$$\begin{bmatrix} 1 & 1 & 1 & | & 8 \\ 3 & 2 & 4 & | & 21 \end{bmatrix} \sim \begin{bmatrix} 1 & 1 & 1 & | & 8 \\ 0 & -1 & 1 & | & -3 \end{bmatrix} \sim \begin{bmatrix} 1 & 1 & 1 & | & 8 \\ 0 & 1 & -1 & | & 3 \end{bmatrix} \sim \begin{bmatrix} 1 & 0 & 2 & | & 5 \\ 0 & 1 & -1 & | & 3 \end{bmatrix}$$

$$(-3)R_1+R_2 \to R_2 \qquad (-1)R_2 \to R_2 \qquad (-1)R_2+R_1 \to R_1$$

Copyright © 2011 Pearson Education, Inc. Publishing as Prentice Hall.

Thus, $x_1 \quad + 2x_3 = 5$

$\qquad x_2 - x_3 = 3$

Let $x_3 = t$. Then, $x_1 = 5 - 2t$, $x_2 = 3 + t$ and the solution is: $x_1 = 5 - 2t$, $x_2 = 3 + t$,

$x_3 = t$, $t$ any real number $\qquad$ (4-3)

27. (A) The matrix equation for the given system is:

$$\begin{bmatrix} 1 & 2 & 3 \\ 2 & 3 & 4 \\ 1 & 2 & 1 \end{bmatrix} \begin{bmatrix} x_1 \\ x_2 \\ x_3 \end{bmatrix} = \begin{bmatrix} 1 \\ 3 \\ 3 \end{bmatrix}$$

The inverse matrix of the coefficient matrix of the system, from Problem 25, is:

$$\begin{bmatrix} -\frac{5}{2} & 2 & -\frac{1}{2} \\ 1 & -1 & 1 \\ \frac{1}{2} & 0 & -\frac{1}{2} \end{bmatrix} \qquad \text{Thus,} \begin{bmatrix} x_1 \\ x_2 \\ x_3 \end{bmatrix} = \begin{bmatrix} -\frac{5}{2} & 2 & -\frac{1}{2} \\ 1 & -1 & 1 \\ \frac{1}{2} & 0 & -\frac{1}{2} \end{bmatrix} \begin{bmatrix} 1 \\ 3 \\ 3 \end{bmatrix} = \begin{bmatrix} 2 \\ 1 \\ -1 \end{bmatrix}$$

Solution: $x_1 = 2$, $x_2 = 1$, $x_3 = -1$.

(B) $\begin{bmatrix} x_1 \\ x_2 \\ x_3 \end{bmatrix} = \begin{bmatrix} -\frac{5}{2} & 2 & -\frac{1}{2} \\ 1 & -1 & 1 \\ \frac{1}{2} & 0 & -\frac{1}{2} \end{bmatrix} \begin{bmatrix} 0 \\ 0 \\ -2 \end{bmatrix} = \begin{bmatrix} 1 \\ -2 \\ 1 \end{bmatrix}$ 
Solution: $x_1 = 1$,
$x_2 = -2$
$x_3 = 1$.

(C) $\begin{bmatrix} x_1 \\ x_2 \\ x_3 \end{bmatrix} = \begin{bmatrix} -\frac{5}{2} & 2 & -\frac{1}{2} \\ 1 & -1 & 1 \\ \frac{1}{2} & 0 & -\frac{1}{2} \end{bmatrix} \begin{bmatrix} -3 \\ -4 \\ 1 \end{bmatrix} = \begin{bmatrix} -1 \\ 2 \\ -2 \end{bmatrix}$ 
Solution: $x_1 = -1$,
$x_2 = 2$,
$x_3 = -2$. $\qquad$ (4-6)

28. $2x_1 - 6x_2 = 4$

$-x_1 + kx_2 = -2$

If $k = 3$, then the first equation is a multiple $(-2)$ of the second equation which implies there are infinitely many solutions. If $k \neq 3$, then the system has a unique solution. $\qquad$ (4-3)

29. $M = \begin{bmatrix} 0.2 & 0.15 \\ 0.4 & 0.3 \end{bmatrix}$; $\quad I - M = \begin{bmatrix} 0.8 & -0.15 \\ -0.4 & 0.7 \end{bmatrix} = \begin{bmatrix} \frac{4}{5} & -\frac{3}{20} \\ -\frac{2}{5} & \frac{7}{10} \end{bmatrix}$

$(I - M)^{-1}$: $\quad \begin{bmatrix} \frac{4}{5} & -\frac{3}{20} & | & 1 & 0 \\ -\frac{2}{5} & \frac{7}{10} & | & 0 & 1 \end{bmatrix} \sim \begin{bmatrix} \frac{4}{5} & -\frac{3}{20} & | & 1 & 0 \\ 0 & \frac{5}{8} & | & \frac{1}{2} & 1 \end{bmatrix}$

$\qquad \left(\frac{1}{2}\right) R_1 + R_2 \rightarrow R_2 \qquad \left(\frac{5}{4}\right) R_1 \rightarrow R_1, \left(\frac{8}{5}\right) R_2 \rightarrow R_2$

$\qquad \sim \begin{bmatrix} 1 & -\frac{3}{16} & | & \frac{5}{4} & 0 \\ 0 & 1 & | & \frac{4}{5} & \frac{8}{5} \end{bmatrix} \sim \begin{bmatrix} 1 & 0 & | & \frac{7}{5} & \frac{3}{10} \\ 0 & 1 & | & \frac{4}{5} & \frac{8}{5} \end{bmatrix}$

$\qquad \left(\frac{3}{16}\right) R_2 + R_1 \rightarrow R_1$

Thus, $(I - M)^{-1} = \begin{bmatrix} \frac{7}{5} & \frac{3}{10} \\ \frac{4}{5} & \frac{8}{5} \end{bmatrix} = \begin{bmatrix} 1.4 & 0.3 \\ 0.8 & 1.6 \end{bmatrix}$

Copyright © 2011 Pearson Education, Inc. Publishing as Prentice Hall.

The output matrix $X$ is given by

$$X = (I - M)^{-1}D = \begin{bmatrix} 1.4 & 0.3 \\ 0.8 & 1.6 \end{bmatrix} \begin{bmatrix} 30 \\ 20 \end{bmatrix} = \begin{matrix} A \\ E \end{matrix}\begin{bmatrix} 48 \\ 56 \end{bmatrix}$$

Agriculture: \$48 billion; Energy: \$56 billion.        (4-7)

**30.**  $T = \begin{bmatrix} 0.3 & 0.4 \\ 0.15 & 0.2 \end{bmatrix}, \quad D = \begin{bmatrix} 20 \\ 30 \end{bmatrix}$

$I - T = \begin{bmatrix} 0.7 & -0.4 \\ -0.15 & 0.8 \end{bmatrix} = \begin{bmatrix} \frac{7}{10} & -\frac{2}{5} \\ -\frac{3}{20} & \frac{4}{5} \end{bmatrix}$

$(I - T)^{-1}$:     $\begin{bmatrix} \frac{7}{10} & -\frac{2}{5} & | & 1 & 0 \\ -\frac{3}{20} & \frac{4}{5} & | & 0 & 1 \end{bmatrix} \sim \begin{bmatrix} 1 & -\frac{4}{7} & | & \frac{10}{7} & 0 \\ -\frac{3}{20} & \frac{4}{5} & | & 0 & 1 \end{bmatrix} \sim \begin{bmatrix} 1 & -\frac{4}{7} & | & \frac{10}{7} & 0 \\ 0 & \frac{5}{7} & | & \frac{3}{14} & 1 \end{bmatrix}$

$\left(\frac{10}{7}\right)R_1 \to R_1 \qquad \left(\frac{3}{20}\right)R_1 + R_2 \to R_2 \qquad \left(\frac{7}{5}\right)R_2 \to R_2$

$\sim \begin{bmatrix} 1 & -\frac{4}{7} & | & \frac{10}{7} & 0 \\ 0 & 1 & | & \frac{3}{10} & \frac{7}{5} \end{bmatrix} \sim \begin{bmatrix} 1 & 0 & | & \frac{8}{5} & \frac{4}{5} \\ 0 & 1 & | & \frac{3}{10} & \frac{7}{5} \end{bmatrix}$

$\left(\frac{4}{7}\right)R_2 + R_1 \to R_1$

Thus, $(I - T)^{-1} = \begin{bmatrix} \frac{8}{5} & \frac{4}{5} \\ \frac{3}{10} & \frac{7}{5} \end{bmatrix} = \begin{bmatrix} 1.6 & 0.8 \\ 0.3 & 1.4 \end{bmatrix}.$

The output matrix $X$ is given by

$$X = (I - T)^{-1}D = \begin{bmatrix} 1.6 & 0.8 \\ 0.3 & 1.4 \end{bmatrix} \begin{bmatrix} 20 \\ 30 \end{bmatrix} = \begin{matrix} E \\ A \end{matrix}\begin{bmatrix} 56 \\ 48 \end{bmatrix}$$

Energy: \$56 billion; Agriculture: \$48 billion.        (4-7)

**31.**  $M = \begin{bmatrix} 0.45 & 0.65 \\ 0.55 & 0.35 \end{bmatrix}; \quad I - M = \begin{bmatrix} 0.55 & -0.65 \\ -0.55 & 0.65 \end{bmatrix}$

Since $R_2 = (-1)R_1$, $(I - M)$ is singular and $X$ does not exist.        (4-5)

**32.**  The graphs of the two equations are:

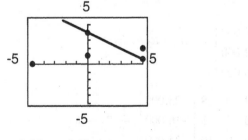

$x \approx 3.46, y \approx 1.69$        (4-1)

Copyright © 2011 Pearson Education, Inc. Publishing as Prentice Hall.

**33.**
$$\begin{bmatrix} 4 & 5 & 6 & | & 1 & 0 & 0 \\ 4 & 5 & -4 & | & 0 & 1 & 0 \\ 1 & 1 & 1 & | & 0 & 0 & 1 \end{bmatrix} R_1 \leftrightarrow R_3 \sim \begin{bmatrix} 1 & 1 & 1 & | & 0 & 0 & 1 \\ 4 & 5 & -4 & | & 0 & 1 & 0 \\ 4 & 5 & 6 & | & 1 & 0 & 0 \end{bmatrix} \begin{matrix} (-4)R_1 + R_2 \to R_2 \\ (-4)R_1 + R_3 \to R_3 \end{matrix}$$

$$\sim \begin{bmatrix} 1 & 1 & 1 & | & 0 & 0 & 1 \\ 0 & 1 & -8 & | & 0 & 1 & -4 \\ 0 & 1 & 2 & | & 1 & 0 & -4 \end{bmatrix} \begin{matrix} (-1)R_2 + R_1 \to R_1 \\ (-1)R_2 + R_3 \to R_3 \end{matrix} \sim \begin{bmatrix} 1 & 0 & 9 & | & 0 & -1 & 5 \\ 0 & 1 & -8 & | & 0 & 1 & -4 \\ 0 & 0 & 10 & | & 1 & -1 & 0 \end{bmatrix} \frac{1}{10}R_3 \to R_3$$

$$\sim \begin{bmatrix} 1 & 0 & 9 & | & 0 & -1 & 5 \\ 0 & 1 & -8 & | & 0 & 1 & -4 \\ 0 & 0 & 1 & | & \frac{1}{10} & -\frac{1}{10} & 0 \end{bmatrix} \begin{matrix} (-9)R_3 + R_1 \to R_1 \\ 8R_3 + R_2 \to R_2 \end{matrix} \sim \begin{bmatrix} 1 & 0 & 0 & | & -\frac{9}{10} & -\frac{1}{10} & 5 \\ 0 & 1 & 0 & | & \frac{8}{10} & \frac{2}{10} & -4 \\ 0 & 0 & 1 & | & \frac{1}{10} & -\frac{1}{10} & 0 \end{bmatrix}$$

Thus, $A^{-1} = \begin{bmatrix} -0.9 & -0.1 & 5 \\ 0.8 & 0.2 & -4 \\ 0.1 & -0.1 & 0 \end{bmatrix}$;

$$A^{-1}A = \begin{bmatrix} -0.9 & -0.1 & 5 \\ 0.8 & 0.2 & -4 \\ 0.1 & -0.1 & 0 \end{bmatrix}\begin{bmatrix} 4 & 5 & 6 \\ 4 & 5 & -4 \\ 1 & 1 & 1 \end{bmatrix} = \begin{bmatrix} 1 & 0 & 0 \\ 0 & 1 & 0 \\ 0 & 0 & 1 \end{bmatrix} \qquad (4\text{-}5)$$

**34.**   The given system is equivalent to:
$$4x_1 + 5x_2 + 6x_3 = 36,000$$
$$4x_1 + 5x_2 - 4x_3 = 12,000$$
$$x_1 + x_2 + x_3 = 7,000$$

In matrix form, this system is:
$$\begin{bmatrix} 4 & 5 & 6 \\ 4 & 5 & -4 \\ 1 & 1 & 1 \end{bmatrix}\begin{bmatrix} x_1 \\ x_2 \\ x_3 \end{bmatrix} = \begin{bmatrix} 36,000 \\ 12,000 \\ 7,000 \end{bmatrix}$$

Thus,
$$\begin{bmatrix} x_1 \\ x_2 \\ x_3 \end{bmatrix} = \begin{bmatrix} 4 & 5 & 6 \\ 4 & 5 & -4 \\ 1 & 1 & 1 \end{bmatrix}^{-1}\begin{bmatrix} 36,000 \\ 12,000 \\ 7,000 \end{bmatrix} = \begin{bmatrix} -0.9 & -0.1 & 5 \\ 0.8 & 0.2 & -4 \\ 0.1 & -0.1 & 0 \end{bmatrix}\begin{bmatrix} 36,000 \\ 12,000 \\ 7,000 \end{bmatrix} = \begin{bmatrix} 1,400 \\ 3,200 \\ 2,400 \end{bmatrix}$$

Solution: $x_1 = 1,400,\ \ x_2 = 3,200,\ \ x_3 = 2,400.$   (4-6)

**35.**   First, multiply the first two equations of the system by 100. Then the augmented matrix of the resulting system is:

$$\begin{bmatrix} 4 & 5 & 6 & | & 36,000 \\ 4 & 5 & -4 & | & 12,000 \\ 1 & 1 & 1 & | & 7,000 \end{bmatrix} R_1 \leftrightarrow R_3 \sim \begin{bmatrix} 1 & 1 & 1 & | & 7,000 \\ 4 & 5 & -4 & | & 12,000 \\ 4 & 5 & 6 & | & 36,000 \end{bmatrix} \begin{matrix} (-4)R_1 + R_2 \to R_2 \\ (-4)R_1 + R_3 \to R_3 \end{matrix}$$

$$\sim \begin{bmatrix} 1 & 1 & 1 & | & 7,000 \\ 0 & 1 & -8 & | & -16,000 \\ 0 & 1 & 2 & | & 8,000 \end{bmatrix} \begin{matrix} (-1)R_2 + R_1 \to R_1 \\ (-1)R_2 + R_3 \to R_3 \end{matrix} \sim \begin{bmatrix} 1 & 0 & 9 & | & 23,000 \\ 0 & 1 & -8 & | & -16,000 \\ 0 & 0 & 10 & | & 24,000 \end{bmatrix} \frac{1}{10}R_3 \to R_3$$

Copyright © 2011 Pearson Education, Inc.  Publishing as Prentice Hall.

$$\sim \begin{bmatrix} 1 & 0 & 9 & | & 23{,}000 \\ 0 & 1 & -8 & | & -16{,}000 \\ 0 & 0 & 1 & | & 2{,}400 \end{bmatrix} \begin{array}{l} (-9)R_3 + R_1 \to R_1 \\ \\ 8R_3 + R_2 \to R_2 \end{array} \sim \begin{bmatrix} 1 & 0 & 0 & | & 1{,}400 \\ 0 & 1 & 0 & | & 3{,}200 \\ 0 & 0 & 1 & | & 2{,}400 \end{bmatrix} \begin{array}{l} x_1 = 1{,}400 \\ x_2 = 3{,}200 \\ x_3 = 2{,}400 \end{array}$$

Solution: $x_1 = 1{,}400$, $x_2 = 3{,}200$, $x_3 = 2{,}400$.    (4-3)

**36.** $M = \begin{bmatrix} 0.2 & 0 & 0.4 \\ 0.1 & 0.3 & 0.1 \\ 0 & 0.4 & 0.2 \end{bmatrix} = \begin{bmatrix} \frac{1}{5} & 0 & \frac{2}{5} \\ \frac{1}{10} & \frac{3}{10} & \frac{1}{10} \\ 0 & \frac{2}{5} & \frac{1}{5} \end{bmatrix}$ and $D = \begin{bmatrix} 40 \\ 20 \\ 30 \end{bmatrix}$

$$I - M = \begin{bmatrix} 1 & 0 & 0 \\ 0 & 1 & 0 \\ 0 & 0 & 1 \end{bmatrix} - \begin{bmatrix} \frac{1}{5} & 0 & \frac{2}{5} \\ \frac{1}{10} & \frac{3}{10} & \frac{1}{10} \\ 0 & \frac{2}{5} & \frac{1}{5} \end{bmatrix} = \begin{bmatrix} \frac{4}{5} & 0 & -\frac{2}{5} \\ -\frac{1}{10} & \frac{7}{10} & -\frac{1}{10} \\ 0 & -\frac{2}{5} & \frac{4}{5} \end{bmatrix}.$$

$$\begin{bmatrix} \frac{4}{5} & 0 & -\frac{2}{5} & | & 1 & 0 & 0 \\ -\frac{1}{10} & \frac{7}{10} & -\frac{1}{10} & | & 0 & 1 & 0 \\ 0 & -\frac{2}{5} & \frac{4}{5} & | & 0 & 0 & 1 \end{bmatrix} \sim \begin{bmatrix} 1 & 0 & -\frac{1}{2} & | & \frac{5}{4} & 0 & 0 \\ -\frac{1}{10} & \frac{7}{10} & -\frac{1}{10} & | & 0 & 1 & 0 \\ 0 & -\frac{2}{5} & \frac{4}{5} & | & 0 & 0 & 1 \end{bmatrix}$$

$$\frac{5}{4}R_1 \to R_1 \qquad\qquad \frac{1}{10}R_1 + R_2 \to R_2$$

$$\sim \begin{bmatrix} 1 & 0 & -\frac{1}{2} & | & \frac{5}{4} & 0 & 0 \\ 0 & \frac{7}{10} & -\frac{3}{20} & | & \frac{1}{8} & 1 & 0 \\ 0 & -\frac{2}{5} & \frac{4}{5} & | & 0 & 0 & 1 \end{bmatrix} \sim \begin{bmatrix} 1 & 0 & -\frac{1}{2} & | & \frac{5}{4} & 0 & 0 \\ 0 & -\frac{2}{5} & \frac{4}{5} & | & 0 & 0 & 1 \\ 0 & \frac{7}{10} & -\frac{3}{20} & | & \frac{1}{8} & 1 & 0 \end{bmatrix}$$

$$R_2 \leftrightarrow R_3 \qquad\qquad \left(-\frac{5}{2}\right)R_2 \to R_2$$

$$\sim \begin{bmatrix} 1 & 0 & -\frac{1}{2} & | & \frac{5}{4} & 0 & 0 \\ 0 & -\frac{2}{5} & \frac{4}{5} & | & 0 & 0 & 1 \\ 0 & \frac{7}{10} & -\frac{3}{20} & | & \frac{1}{8} & 1 & 0 \end{bmatrix} \sim \begin{bmatrix} 1 & 0 & -\frac{1}{2} & | & \frac{5}{4} & 0 & 0 \\ 0 & 1 & -2 & | & 0 & 0 & -\frac{5}{2} \\ 0 & 0 & \frac{5}{4} & | & \frac{1}{8} & 1 & \frac{7}{4} \end{bmatrix}$$

$$\left(-\frac{7}{10}\right)R_2 + R_3 \to R_3 \qquad\qquad \frac{4}{5}R_3 \to R_3$$

$$\sim \begin{bmatrix} 1 & 0 & -\frac{1}{2} & | & \frac{5}{4} & 0 & 0 \\ 0 & 1 & -2 & | & 0 & 0 & -\frac{5}{2} \\ 0 & 0 & 1 & | & \frac{1}{10} & \frac{4}{5} & \frac{7}{5} \end{bmatrix} \sim \begin{bmatrix} 1 & 0 & 0 & | & \frac{13}{10} & \frac{2}{5} & \frac{7}{10} \\ 0 & 1 & 0 & | & \frac{1}{5} & \frac{8}{5} & \frac{3}{10} \\ 0 & 0 & 1 & | & \frac{1}{10} & \frac{4}{5} & \frac{7}{5} \end{bmatrix}$$

$$2R_3 + R_2 \to R_2, \ \frac{1}{2}R_3 + R_1 \to R_1$$

Thus $(I - M)^{-1} = \begin{bmatrix} \frac{13}{10} & \frac{2}{5} & \frac{7}{10} \\ \frac{1}{5} & \frac{8}{5} & \frac{3}{10} \\ \frac{1}{10} & \frac{4}{5} & \frac{7}{5} \end{bmatrix} = \begin{bmatrix} 1.3 & 0.4 & 0.7 \\ 0.2 & 1.6 & 0.3 \\ 0.1 & 0.8 & 1.4 \end{bmatrix}$ and

$$X = (I - M)^{-1}D = \begin{bmatrix} 1.3 & 0.4 & 0.7 \\ 0.2 & 1.6 & 0.3 \\ 0.1 & 0.8 & 1.4 \end{bmatrix} \begin{bmatrix} 40 \\ 20 \\ 30 \end{bmatrix} = \begin{bmatrix} 81 \\ 49 \\ 62 \end{bmatrix}$$    (4-7)

**37.** (A)    The system has a unique solution.

Copyright © 2011 Pearson Education, Inc. Publishing as Prentice Hall.

(B) The system either has no solutions or infinitely many solutions. (4-6)

**38.** (A) The system has a unique solution.

(B) The system has **no** solutions.

(C) The system has infinitely many solutions. (4-3)

**39.** The third step in (A) is incorrect:
$$X - MX = (I - M)X \quad \textbf{not} \quad X(I - M)$$
Each step in (B) is correct. (4-6)

**40.** Let $x$ = the number of machines produced.

(A) $C(x) = 243{,}000 + 22.45x$
$R(x) = 59.95x$

(B) Set $C(x) = R(x)$:
$$59.95x = 243{,}000 + 22.45x$$
$$37.5x = 243{,}000$$
$$x = 6{,}480$$

If 6,480 machines are produced, $C = R = \$388{,}476$; break-even point (6,480, 388,476).

(C) A profit occurs if $x > 6{,}480$; a loss occurs if $x < 6{,}480$

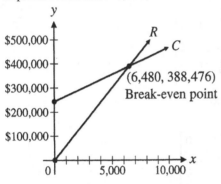

(4-1)

**41.** Let $x_1$ = Number of tons of Voisey's Bay ore

$x_2$ = number of tons of Hawk Ridge ore

Then, we have the following system of equations
$$0.02x_1 + 0.03x_2 = 6$$
$$0.04x_1 + 0.02x_2 = 8$$

Multiply each equation by 100. This yields
$$2x_1 + 3x_2 = 600$$
$$4x_1 + 2x_2 = 800$$

The augmented matrix corresponding to this system is:
$$\begin{bmatrix} 2 & 3 & | & 600 \\ 4 & 2 & | & 800 \end{bmatrix} \sim \begin{bmatrix} 2 & 3 & | & 600 \\ 0 & -4 & | & -400 \end{bmatrix} \sim \begin{bmatrix} 2 & 3 & | & 600 \\ 0 & 1 & | & 100 \end{bmatrix} \sim \begin{bmatrix} 2 & 0 & | & 300 \\ 0 & 1 & | & 100 \end{bmatrix} \sim \begin{bmatrix} 1 & 0 & | & 150 \\ 0 & 1 & | & 100 \end{bmatrix}$$

$(-2)R_1 + R_2 \rightarrow R_2 \quad \left(-\dfrac{1}{4}\right)R_2 \rightarrow R_2 \quad (-3)R_2 + R_1 \rightarrow R_1 \quad \left(\dfrac{1}{2}\right)R_1 \rightarrow R_1$

Thus, the solution is: $x_1 = 150$ tons of Voisey's Bay ore, $x_2 = 100$ tons of Hawk Ridge ore. (4-3)

Copyright © 2011 Pearson Education, Inc. Publishing as Prentice Hall.

**42.** (A)  The matrix equation for Problem 41 is:

$$\begin{bmatrix} 0.02 & 0.03 \\ 0.04 & 0.02 \end{bmatrix} \begin{bmatrix} x_1 \\ x_2 \end{bmatrix} = \begin{bmatrix} 6 \\ 8 \end{bmatrix}$$

First, we compute the inverse of $\begin{bmatrix} 0.02 & 0.03 \\ 0.04 & 0.02 \end{bmatrix}$;

$$\begin{bmatrix} 0.02 & 0.03 & | & 1 & 0 \\ 0.04 & 0.02 & | & 0 & 1 \end{bmatrix} \sim \begin{bmatrix} 1 & 1.5 & | & 50 & 0 \\ 0.04 & 0.02 & | & 0 & 1 \end{bmatrix} \sim \begin{bmatrix} 1 & 1.5 & | & 50 & 0 \\ 0 & -0.04 & | & -2 & 1 \end{bmatrix}$$

$$50R_1 \to R_1 \qquad (0.04)R_1 + R_2 \to R_2 \qquad \left(-\frac{1}{0.04}\right)R_2 \to R_2$$

$$\sim \begin{bmatrix} 1 & 1.5 & | & 50 & 0 \\ 0 & 1 & | & 50 & -25 \end{bmatrix} \sim \begin{bmatrix} 1 & 0 & | & -25 & 37.5 \\ 0 & 1 & | & 50 & -25 \end{bmatrix}$$

$$(-1.5)R_2 + R_1 \to R_1$$

Thus, the inverse of the coefficient matrix is:

$$\begin{bmatrix} -25 & 37.5 \\ 50 & -25 \end{bmatrix}$$

Now, $\begin{bmatrix} x_1 \\ x_2 \end{bmatrix} = \begin{bmatrix} -25 & 37.5 \\ 50 & -25 \end{bmatrix} \begin{bmatrix} 6 \\ 8 \end{bmatrix} = \begin{bmatrix} 150 \\ 100 \end{bmatrix}$

Again, the solution is: $x_1 = 150$ tons of Voisey's Bay ore, $x_2 = 100$ tons of Hawk Ridge ore.

(B)  $\begin{bmatrix} x_1 \\ x_2 \end{bmatrix} = \begin{bmatrix} -25 & 37.5 \\ 50 & -25 \end{bmatrix} \begin{bmatrix} 7.5 \\ 7 \end{bmatrix} = \begin{bmatrix} 75 \\ 200 \end{bmatrix}$

Now the solution is: $x_1 = 75$ tons of Voisey's Bay ore, $x_2 = 200$ tons of Hawk Ridge ore.

(4-6)

**43.** (A)  Let $x_1$ = number of 3,000 cubic foot hoppers

  $x_2$ = number of 4,500 cubic foot hoppers

  $x_3$ = number of 6,000 cubic foot hoppers

Then

$$x_1 + x_2 + x_3 = 20$$
$$3,000x_1 + 4,500x_2 + 6,000x_3 = 108,000$$

or

$$x_1 + x_2 + x_3 = 20$$
$$3x_1 + 4.5x_2 + 6x_3 = 108$$

The augmented matrix for this system is:

$$\begin{bmatrix} 1 & 1 & 1 & | & 20 \\ 3 & 4.5 & 6 & | & 108 \end{bmatrix}$$

Now

$$\begin{bmatrix} 1 & 1 & 1 & | & 20 \\ 3 & 4.5 & 6 & | & 108 \end{bmatrix} \sim \begin{bmatrix} 1 & 1 & 1 & | & 20 \\ 0 & 1.5 & 3 & | & 48 \end{bmatrix} \sim \begin{bmatrix} 1 & 1 & 1 & | & 20 \\ 0 & 1 & 2 & | & 32 \end{bmatrix}$$

Copyright © 2011 Pearson Education, Inc. Publishing as Prentice Hall.

$$(-3)R_1 + R_2 \rightarrow R_2 \qquad \frac{2}{3}R_2 \rightarrow R_2 \qquad (-1)R_2 + R_1 \rightarrow R_1$$

$$\sim \begin{bmatrix} 1 & 0 & -1 & -12 \\ 0 & 1 & 2 & 32 \end{bmatrix}$$

The corresponding system of equations is:

$$x_1 \qquad - x_3 = -12$$
$$x_2 + 2x_3 = 32$$

and the solutions are: $x_1 = t - 12$   3,000 cubic foot hoppers, $x_2 = 32 - 2t$   4,500 cubic foot hoppers, $x_3 = t$   6,000 cubic foot hoppers.

Since $x_1, x_2$, and $x_3$ are non-negative integers, it follows that $t = 12, 13, 14, 15$, or 16.

(B)   Cost: $C(t) = 180(t - 12) + 225(32 - 2t) + 325(t)$
$C(12) = 225(8) + 325(12) = \$5,700$
$C(13) = 180(1) + 225(6) + 325(13) = \$5,755$
$C(14) = 180(2) + 225(4) + 325(14) = \$5,810$
$C(15) = 180(3) + 225(2) + 325(15) = \$5,865$
$C(16) = 180(4) + 325(16) = \$5,920$

The minimum monthly cost is \$5,700 when 8 4,500 cubic foot hoppers and 12 6,000 cubic foot hoppers are leased.   (4-3)

**44.**   (A)   The elements of *MN* give the cost of materials for each alloy from each supplier. The product *NM* is also defined, but does not have an interpretation in the context of this problem.

(B)   $MN = \begin{bmatrix} 4,800 & 600 & 300 \\ 6,000 & 1,400 & 700 \end{bmatrix} \begin{bmatrix} 0.75 & 0.70 \\ 6.50 & 6.70 \\ 0.40 & 0.50 \end{bmatrix}$

$$= \begin{matrix} & \text{Supplier A} \quad \text{Supplier B} \\ \begin{bmatrix} \$7,620 & \$7,530 \\ \$13,880 & \$13,930 \end{bmatrix} & \begin{matrix} \text{Alloy 1} \\ \text{Alloy 2} \end{matrix} \end{matrix}$$

(C)   The total costs of materials from Supplier *A* is:
\$7,620 + \$13,880 = \$21,500

The total costs of materials from Supplier *B* is:
\$7,530 + \$13,930 = \$21,460

These values can be obtained from the matrix product
$$\begin{bmatrix} 1 & 1 \end{bmatrix} \begin{bmatrix} 7,620 & 7,530 \\ 13,880 & 13,930 \end{bmatrix}$$

Supplier *B* will provide the materials at lower cost.   (4-4)

Copyright © 2011 Pearson Education, Inc. Publishing as Prentice Hall.

**45.** (A)  $[0.25 \ 0.20 \ 0.05]\begin{bmatrix} 12 \\ 15 \\ 7 \end{bmatrix} = 6.35$

The labor cost for one model $B$ calculator at the California plant is $6.35.

(B)  The elements of $MN$ give the total labor costs for each calculator at each plant. The product $NM$ is also defined, but does not have an interpretation in the context of this problem.

(C)  $MN = \begin{bmatrix} 0.15 & 0.10 & 0.05 \\ 0.25 & 0.20 & 0.05 \end{bmatrix}\begin{bmatrix} 12 & 10 \\ 15 & 12 \\ 7 & 6 \end{bmatrix} = \begin{matrix} \quad CA \quad\ TX \\ \begin{bmatrix} \$3.65 & \$3.00 \\ \$6.35 & \$5.20 \end{bmatrix} \end{matrix}\begin{matrix} \text{Model } A \\ \text{Model } B \end{matrix}$

**46.** Let $x_1$ = amount invested at 5%
and $x_2$ = amount invested at 10%.

Then,  $x_1 + x_2 = 5000$
$0.05x_1 + 0.1x_2 = 400$

The augmented matrix for the system given above is:

$\begin{bmatrix} 1 & 1 & | & 5000 \\ 0.05 & 0.1 & | & 400 \end{bmatrix} \sim \begin{bmatrix} 1 & 1 & | & 5000 \\ 0 & 0.05 & | & 150 \end{bmatrix} \sim \begin{bmatrix} 1 & 1 & | & 5000 \\ 0 & 1 & | & 3000 \end{bmatrix} \sim \begin{bmatrix} 1 & 0 & | & 2000 \\ 0 & 1 & | & 3000 \end{bmatrix}$

$(-0.05)R_1 + R_2 \rightarrow R_1 \qquad \frac{1}{0.05}R_2 \rightarrow R_2 \qquad (-1)R_2 + R_1 \rightarrow R_1$

Hence,  $x_1 = \$2000$ at 5%,  $x_2 = \$3000$ at 10%.    (4-3)

**47.** The matrix equation corresponding to the system in Problem 46 is:

$\begin{bmatrix} 1 & 1 \\ 0.05 & 0.1 \end{bmatrix}\begin{bmatrix} x_1 \\ x_2 \end{bmatrix} = \begin{bmatrix} 5000 \\ 400 \end{bmatrix}$

Now we compute the inverse matrix of $\begin{bmatrix} 1 & 1 \\ 0.05 & 0.1 \end{bmatrix}$.

$\begin{bmatrix} 1 & 1 & | & 1 & 0 \\ 0.05 & 0.1 & | & 0 & 1 \end{bmatrix} \sim \begin{bmatrix} 1 & 1 & | & 1 & 0 \\ 0 & 0.05 & | & -0.05 & 1 \end{bmatrix} \sim \begin{bmatrix} 1 & 1 & | & 1 & 0 \\ 0 & 1 & | & -1 & 20 \end{bmatrix}$

$(-0.05)R_1 + R_2 \rightarrow R_1 \qquad \frac{1}{0.05}R_2 \rightarrow R_2 \qquad\qquad (-1)R_2 + R_1 \rightarrow R_1$

$\sim \begin{bmatrix} 1 & 0 & | & 2 & -20 \\ 0 & 1 & | & -1 & 20 \end{bmatrix}$

Thus, the inverse of the coefficient matrix is $\begin{bmatrix} 2 & -20 \\ -1 & 20 \end{bmatrix}$, and

$\begin{bmatrix} x_1 \\ x_2 \end{bmatrix} = \begin{bmatrix} 2 & -20 \\ -1 & 20 \end{bmatrix}\begin{bmatrix} 5000 \\ 400 \end{bmatrix} = \begin{bmatrix} 10,000 - 8,000 \\ -5,000 + 8,000 \end{bmatrix} = \begin{bmatrix} 2000 \\ 3000 \end{bmatrix}$. So, $x_1 = \$2000$ at 5%,    (4-6)
$x_2 = \$3000$ at 10%.

Copyright © 2011 Pearson Education, Inc. Publishing as Prentice Hall.

**48.** From Problem 46, the system of equations is:

$$x_1 + x_2 = 5000$$
$$0.05x_1 + 0.1x_2 = k \text{ (annual return)}$$

Calculate the inverse of the coefficient matrix $A$.

$$\begin{bmatrix} 1 & 1 & | & 1 & 0 \\ 0.05 & 0.1 & | & 0 & 1 \end{bmatrix} \sim \begin{bmatrix} 1 & 1 & | & 1 & 0 \\ 0 & 0.05 & | & -0.05 & 1 \end{bmatrix} \sim \begin{bmatrix} 1 & 1 & | & 1 & 0 \\ 0 & 1 & | & -1 & 20 \end{bmatrix} \sim \begin{bmatrix} 1 & 0 & | & 2 & -20 \\ 0 & 1 & | & -1 & 20 \end{bmatrix}$$

$(-0.05)R_1 + R_2 \rightarrow R_2 \qquad \dfrac{1}{0.05}R_2 \rightarrow R_2 \qquad (-1)R_2 + R_1 \rightarrow R_1$

Thus, $A^{-1} = \begin{bmatrix} 2 & -20 \\ -1 & 20 \end{bmatrix}$

$k = \$200?$

$$\begin{bmatrix} x_1 \\ x_2 \end{bmatrix} = \begin{bmatrix} 2 & -20 \\ -1 & 20 \end{bmatrix}\begin{bmatrix} 5000 \\ 200 \end{bmatrix} = \begin{bmatrix} 6000 \\ -1000 \end{bmatrix}; \quad x_1 = \$6{,}000, \quad x_2 = -\$1{,}000$$

This is not possible, $x_2$ cannot be negative.

$k = \$600?$

$$\begin{bmatrix} x_1 \\ x_2 \end{bmatrix} = \begin{bmatrix} 2 & -20 \\ -1 & 20 \end{bmatrix}\begin{bmatrix} 5000 \\ 600 \end{bmatrix} = \begin{bmatrix} -2000 \\ 7000 \end{bmatrix}; \quad x_1 = -\$2{,}000, \quad x_2 = \$7{,}000$$

This is not possible, $x_1$ cannot be negative.

Fix a return $k$. Then

$$\begin{bmatrix} x_1 \\ x_2 \end{bmatrix} = \begin{bmatrix} 2 & -20 \\ -1 & 20 \end{bmatrix}\begin{bmatrix} 5000 \\ k \end{bmatrix} = \begin{bmatrix} 10{,}000 - 20k \\ -5{,}000 + 20k \end{bmatrix}$$

so $x_1 = 10{,}000 - 20k$, $x_2 = -5{,}000 + 20k$

Since $x_1 \geq 0$, we have $\quad 10{,}000 - 20k \geq 0$

$$20k \leq 10{,}000$$
$$k \leq 500$$

Since $x_2 \geq 0$, we have $\quad -5{,}000 + 20k \geq 0$

$$20k \geq 5{,}000$$
$$k \geq 250$$

The possible annual yields must satisfy $250 \leq k \leq 500$.   (4-6)

**49.** Let $x_1$ = number of \$8 tickets

$\qquad x_2$ = number of \$12 tickets

$\qquad x_3$ = number of \$20 tickets

Since the number of \$8 tickets must equal the number of \$20 tickets, we have

$$x_1 = x_3 \text{ or } x_1 - x_3 = 0$$

Also, since all seats are sold

$$x_1 + x_2 + x_3 = 25{,}000$$

Finally, the return is

$$8x_1 + 12x_2 + 20x_3 = R \text{ (where } R \text{ is the return required).}$$

Thus, the system of equations is:

Copyright © 2011 Pearson Education, Inc. Publishing as Prentice Hall.

$$x_1 \qquad - \; x_3 = 0$$
$$x_1 + \; x_2 + \; x_3 = 25{,}000 \qquad \text{or}$$
$$8x_1 + 12x_2 + 20x_3 = R$$

$$\begin{bmatrix} 1 & 0 & -1 \\ 1 & 1 & 1 \\ 8 & 12 & 20 \end{bmatrix} \begin{bmatrix} x_1 \\ x_2 \\ x_3 \end{bmatrix} = \begin{bmatrix} 0 \\ 25{,}000 \\ R \end{bmatrix}$$

First, we compute the inverse of the coefficient matrix

$$\left[ \begin{array}{ccc|ccc} 1 & 0 & -1 & 1 & 0 & 0 \\ 1 & 1 & 1 & 0 & 1 & 0 \\ 8 & 12 & 20 & 0 & 0 & 1 \end{array} \right] \sim \left[ \begin{array}{ccc|ccc} 1 & 0 & -1 & 1 & 0 & 0 \\ 0 & 1 & 2 & -1 & 1 & 0 \\ 0 & 12 & 28 & -8 & 0 & 1 \end{array} \right]$$

$$(-1)R_1 + R_2 \rightarrow R_2 \qquad (-12)R_2 + R_3 \rightarrow R_3$$
$$(-8)R_1 + R_3 \rightarrow R_3$$

$$\sim \left[ \begin{array}{ccc|ccc} 1 & 0 & -1 & 1 & 0 & 0 \\ 0 & 1 & 2 & -1 & 1 & 0 \\ 0 & 0 & 4 & 4 & -12 & 1 \end{array} \right] \sim \left[ \begin{array}{ccc|ccc} 1 & 0 & -1 & 1 & 0 & 0 \\ 0 & 1 & 2 & -1 & 1 & 0 \\ 0 & 0 & 1 & 1 & -3 & \frac{1}{4} \end{array} \right]$$

$$\tfrac{1}{4}R_3 \rightarrow R_3 \qquad (-2)R_3 + R_2 \rightarrow R_2$$
$$R_3 + R_1 \rightarrow R_1$$

$$\sim \left[ \begin{array}{ccc|ccc} 1 & 0 & 0 & 2 & -3 & \frac{1}{4} \\ 0 & 1 & 0 & -3 & 7 & -\frac{1}{2} \\ 0 & 0 & 1 & 1 & -3 & \frac{1}{4} \end{array} \right]. \quad \text{Thus, the inverse is} \quad \begin{bmatrix} 2 & -3 & \frac{1}{4} \\ -3 & 7 & -\frac{1}{2} \\ 1 & -3 & \frac{1}{4} \end{bmatrix}.$$

Concert 1:

$$x_1 \qquad - \; x_3 = 0$$
$$x_1 + \; x_2 + \; x_3 = 25{,}000 \qquad \text{or}$$
$$8x_1 + 12x_2 + 20x_3 = 320{,}000$$

$$\begin{bmatrix} 1 & 0 & -1 \\ 1 & 1 & 1 \\ 8 & 12 & 20 \end{bmatrix} \begin{bmatrix} x_1 \\ x_2 \\ x_3 \end{bmatrix} = \begin{bmatrix} 0 \\ 25{,}000 \\ 320{,}000 \end{bmatrix}$$

$$\text{Thus } \begin{bmatrix} x_1 \\ x_2 \\ x_3 \end{bmatrix} = \begin{bmatrix} 2 & -3 & \frac{1}{4} \\ -3 & 7 & -\frac{1}{2} \\ 1 & -3 & \frac{1}{4} \end{bmatrix} \begin{bmatrix} 0 \\ 25{,}000 \\ 320{,}000 \end{bmatrix} = \begin{bmatrix} 5{,}000 \\ 15{,}000 \\ 5{,}000 \end{bmatrix}$$

and $x_1 = 5{,}000$  $8 tickets

$\quad x_2 = 15{,}000$ $12 tickets

$\quad x_3 = 5{,}000$ $20 tickets

Concert 2:

$$x_1 \qquad - \; x_3 = 0$$
$$x_1 + \; x_2 + \; x_3 = 25{,}000 \qquad \text{or}$$
$$8x_1 + 12x_2 + 20x_3 = 330{,}000$$

$$\begin{bmatrix} 1 & 0 & -1 \\ 1 & 1 & 1 \\ 8 & 12 & 20 \end{bmatrix} \begin{bmatrix} x_1 \\ x_2 \\ x_3 \end{bmatrix} = \begin{bmatrix} 0 \\ 25{,}000 \\ 330{,}000 \end{bmatrix}$$

$$\text{Thus } \begin{bmatrix} x_1 \\ x_2 \\ x_3 \end{bmatrix} = \begin{bmatrix} 2 & -3 & \frac{1}{4} \\ -3 & 7 & -\frac{1}{2} \\ 1 & -3 & \frac{1}{4} \end{bmatrix} \begin{bmatrix} 0 \\ 25{,}000 \\ 330{,}000 \end{bmatrix} = \begin{bmatrix} 7{,}500 \\ 10{,}000 \\ 7{,}500 \end{bmatrix}$$

Copyright © 2011 Pearson Education, Inc. Publishing as Prentice Hall.

and $x_1 = 7{,}500$  $8 tickets

$x_2 = 10{,}000$  $12 tickets

$x_3 = 7{,}500$  $20 tickets

Concert 3:

$$\begin{aligned} x_1 \quad - \quad x_3 &= 0 \\ x_1 + x_2 + x_3 &= 25{,}000 \\ 8x_1 + 12x_2 + 20x_3 &= 340{,}000 \end{aligned} \qquad \text{or} \qquad \begin{bmatrix} 1 & 0 & -1 \\ 1 & 1 & 1 \\ 8 & 12 & 20 \end{bmatrix} \begin{bmatrix} x_1 \\ x_2 \\ x_3 \end{bmatrix} = \begin{bmatrix} 0 \\ 25{,}000 \\ 340{,}000 \end{bmatrix}$$

Thus $\begin{bmatrix} x_1 \\ x_2 \\ x_3 \end{bmatrix} = \begin{bmatrix} 2 & -3 & \frac{1}{4} \\ -3 & 7 & -\frac{1}{2} \\ 1 & -3 & \frac{1}{4} \end{bmatrix} \begin{bmatrix} 0 \\ 25{,}000 \\ 340{,}000 \end{bmatrix} = \begin{bmatrix} 10{,}000 \\ 5{,}000 \\ 10{,}000 \end{bmatrix}$

and $x_1 = 10{,}000$  $8 tickets

$x_2 = 5{,}000$  $12 tickets

$x_3 = 10{,}000$  $20 tickets     (4-6)

**50.** From Problem 49, if it is not required to have an equal number of $8 tickets and $12 tickets, then the new mathematical model is:

$$\begin{aligned} x_1 + x_2 + x_3 &= 25{,}000 \\ 8x_1 + 12x_2 + 20x_3 &= k \quad \text{(return requested)} \end{aligned}$$

The augmented matrix is:

$$\begin{bmatrix} 1 & 1 & 1 & | & 25{,}000 \\ 8 & 12 & 20 & | & k \end{bmatrix} \sim \begin{bmatrix} 1 & 1 & 1 & | & 25{,}000 \\ 0 & 4 & 12 & | & k - 200{,}000 \end{bmatrix}$$

$(-8)R_1 + R_2 \rightarrow R_2 \qquad \frac{1}{4}R_2 \rightarrow R_2$

$$\sim \begin{bmatrix} 1 & 1 & 1 & | & 25{,}000 \\ 0 & 1 & 3 & | & \frac{k}{4} - 50{,}000 \end{bmatrix} \sim \begin{bmatrix} 1 & 0 & -2 & | & -\frac{k}{4} + 75{,}000 \\ 0 & 1 & 3 & | & \frac{k}{4} - 50{,}000 \end{bmatrix}$$

$(-1)R_2 + R_1 \rightarrow R_1$

Concert 1: $k = \$320{,}000$;  $\frac{1}{4}k = 80{,}000$

$\begin{bmatrix} 1 & 0 & -2 & | & -5{,}000 \\ 0 & 1 & 3 & | & 30{,}000 \end{bmatrix};$

$x_1 = 2t - 5{,}000$   $8 tickets

$x_2 = 30{,}000 - 3t$  $12 tickets

$x_3 = t$  $20 tickets, $t$ an integer

Since $x_1, x_2 \geq 0$, $t$ must satisfy $2{,}500 \leq t \leq 10{,}000$.

Concert 2: $k = \$330{,}000$;  $\frac{1}{4}k = 82{,}500$

$\begin{bmatrix} 1 & 0 & -2 & | & -7{,}500 \\ 0 & 1 & 3 & | & 32{,}500 \end{bmatrix};$

$x_1 = 2t - 7{,}500$   $8 tickets

$x_2 = 32{,}500 - 3t$  $12 tickets

$x_3 = t$  $20 tickets, $t$ an integer

Since $x_1, x_2 \geq 0$, $t$ must satisfy $3{,}750 \leq t \leq 10{,}833$.

Concert 3: $k = \$340{,}000$;  $\frac{1}{4}k = 85{,}000$

Copyright © 2011 Pearson Education, Inc. Publishing as Prentice Hall.

$$\begin{bmatrix} 1 & 0 & -2 & | & -10{,}000 \\ 0 & 1 & 3 & | & 35{,}000 \end{bmatrix};$$

$x_1 = 2t - 10{,}000$  \$8 tickets

$x_2 = 35{,}000 - 3t$  \$12 tickets

$x_3 = t$  \$20 tickets, $t$ an integer

Since $x_1, x_2 \geq 0$, $t$ must satisfy $5{,}000 \leq t \leq 11{,}666$.    (4-3)

**51.** The technology matrix is

$$M = \begin{array}{c} \\ \text{Agriculture} \\ \text{Fabrication} \end{array} \begin{array}{c} \text{Agriculture Fabrication} \\ \begin{bmatrix} 0.30 & 0.10 \\ 0.20 & 0.40 \end{bmatrix} \end{array}$$

Now $I - M = \begin{bmatrix} 1 & 0 \\ 0 & 1 \end{bmatrix} - \begin{bmatrix} 0.30 & 0.10 \\ 0.20 & 0.40 \end{bmatrix} = \begin{bmatrix} 0.70 & -0.10 \\ -0.20 & 0.60 \end{bmatrix} = \begin{bmatrix} \frac{7}{10} & -\frac{1}{10} \\ -\frac{1}{5} & \frac{3}{5} \end{bmatrix}$

Next, we calculate the inverse of $I - M$

$$\begin{bmatrix} \frac{7}{10} & -\frac{1}{10} & | & 1 & 0 \\ -\frac{1}{5} & \frac{3}{5} & | & 0 & 1 \end{bmatrix} \sim \begin{bmatrix} 1 & -3 & | & 0 & -5 \\ \frac{7}{10} & -\frac{1}{10} & | & 1 & 0 \end{bmatrix} \sim \begin{bmatrix} 1 & -3 & | & 0 & -5 \\ 0 & 2 & | & 1 & \frac{7}{2} \end{bmatrix} \sim \begin{bmatrix} 1 & -3 & | & 0 & -5 \\ 0 & 1 & | & \frac{1}{2} & \frac{7}{4} \end{bmatrix}$$

$\begin{array}{c} -5R_2 \to R_2 \\ R_1 \leftrightarrow R_2 \end{array}$     $\left(-\dfrac{7}{10}\right)R_1 + R_2 \to R_2$     $\dfrac{1}{2}R_2 \to R_2$     $3R_2 + R_1 \to R_1$

$$\sim \begin{bmatrix} 1 & 0 & | & \frac{3}{2} & \frac{1}{4} \\ 0 & 1 & | & \frac{1}{2} & \frac{7}{4} \end{bmatrix} \quad \text{Thus,} \quad (I - M)^{-1} = \begin{bmatrix} \frac{3}{2} & \frac{1}{4} \\ \frac{1}{2} & \frac{7}{4} \end{bmatrix}.$$

Let $x_1$ = output for agriculture and $x_2$ = output for fabrication.

Then the output $X = \begin{bmatrix} x_1 \\ x_2 \end{bmatrix}$ needed to satisfy a final demand $D = \begin{bmatrix} d_1 \\ d_2 \end{bmatrix}$ for agriculture and fabrication is given by $X = (I - M)^{-1}D$.

(A)   Let $D = \begin{bmatrix} 50 \\ 20 \end{bmatrix}$. Then $X = \begin{bmatrix} \frac{3}{2} & \frac{1}{4} \\ \frac{1}{2} & \frac{7}{4} \end{bmatrix}\begin{bmatrix} 50 \\ 20 \end{bmatrix} = \begin{bmatrix} 75 + 5 \\ 25 + 35 \end{bmatrix} = \begin{bmatrix} 80 \\ 60 \end{bmatrix}$

Thus, the total output for agriculture is \$80 billion; the total output for fabrication is \$60 billion.

(B)   Let $D = \begin{bmatrix} 80 \\ 60 \end{bmatrix}$. Then $X = \begin{bmatrix} \frac{3}{2} & \frac{1}{4} \\ \frac{1}{2} & \frac{7}{4} \end{bmatrix}\begin{bmatrix} 80 \\ 60 \end{bmatrix} = \begin{bmatrix} 120 + 15 \\ 40 + 105 \end{bmatrix} = \begin{bmatrix} 135 \\ 145 \end{bmatrix}$

Thus, the total output for agriculture is \$135 billion; the total output for fabrication is \$145 billion.
(4-7)

**52.** First we find the inverse of $B = \begin{bmatrix} 1 & 1 & 0 \\ 1 & 0 & 1 \\ 1 & 1 & 1 \end{bmatrix}$.

$$\begin{bmatrix} 1 & 1 & 0 & | & 1 & 0 & 0 \\ 1 & 0 & 1 & | & 0 & 1 & 0 \\ 1 & 1 & 1 & | & 0 & 0 & 1 \end{bmatrix} \sim \begin{bmatrix} 1 & 1 & 0 & | & 1 & 0 & 0 \\ 0 & -1 & 1 & | & -1 & 1 & 0 \\ 0 & 0 & 1 & | & -1 & 0 & 1 \end{bmatrix} \sim \begin{bmatrix} 1 & 1 & 0 & | & 1 & 0 & 0 \\ 0 & 1 & -1 & | & 1 & -1 & 0 \\ 0 & 0 & 1 & | & -1 & 0 & 1 \end{bmatrix}$$

$\begin{array}{c} (-1)R_1 + R_2 \to R_2 \\ (-1)R_1 + R_3 \to R_3 \end{array}$     $(-1)R_2 \to R_2$     $(-1)R_2 + R_1 \to R_1$

Copyright © 2011 Pearson Education, Inc. Publishing as Prentice Hall.

$$\sim \begin{bmatrix} 1 & 0 & 1 & | & 0 & 1 & 0 \\ 0 & 1 & -1 & | & 1 & -1 & 0 \\ 0 & 0 & 1 & | & -1 & 0 & 1 \end{bmatrix} \sim \begin{bmatrix} 1 & 0 & 1 & | & 1 & 1 & -1 \\ 0 & 1 & 0 & | & 0 & -1 & 1 \\ 0 & 0 & 1 & | & -1 & 0 & 1 \end{bmatrix} \quad \text{Thus, } B^{-1} = \begin{bmatrix} 1 & 1 & -1 \\ 0 & -1 & 1 \\ -1 & 0 & 1 \end{bmatrix}.$$

$$(-1)R_3 + R_1 \rightarrow R_1$$
$$R_3 + R_2 \rightarrow R_2$$

Now, $\begin{bmatrix} 1 & 1 & -1 \\ 0 & -1 & 1 \\ -1 & 0 & 1 \end{bmatrix} \begin{bmatrix} 25 & 24 & 21 & 41 & 21 & 25 \\ 8 & 25 & 14 & 30 & 32 & 25 \\ 26 & 33 & 28 & 50 & 41 & 25 \end{bmatrix} = \begin{bmatrix} 7 & 16 & 14 & 21 & 12 & 25 \\ 18 & 8 & 7 & 20 & 9 & 0 \\ 1 & 9 & 0 & 9 & 20 & 0 \end{bmatrix}$

Thus, the decoded message is
  7 18 1 16 8 9 14 7 0 21 20 9 12 9 20 25 0 0
which corresponds to GRAPHING UTILITY.     (4-5)

**53.** (A)  1st & Elm:  $x_1 + x_4 = 1300$

   2nd & Elm:  $x_1 - x_2 = 400$

   2nd & Oak:  $x_2 + x_3 = 700$

   1st & Oak:  $x_3 - x_4 = -200$

(B)  The augmented matrix for the system in part (A) is:

$$\begin{bmatrix} 1 & 0 & 0 & 1 & | & 1300 \\ 1 & -1 & 0 & 0 & | & 400 \\ 0 & 1 & 1 & 0 & | & 700 \\ 0 & 0 & 1 & -1 & | & -200 \end{bmatrix} \sim \begin{bmatrix} 1 & 0 & 0 & 1 & | & 1300 \\ 0 & -1 & 0 & -1 & | & -900 \\ 0 & 1 & 1 & 0 & | & 700 \\ 0 & 0 & 1 & -1 & | & -200 \end{bmatrix}$$

$$(-1)R_1 + R_2 \rightarrow R_2 \qquad (-1)R_2 \rightarrow R_2$$

$$\sim \begin{bmatrix} 1 & 0 & 0 & 1 & | & 1300 \\ 0 & 1 & 0 & 1 & | & 900 \\ 0 & 1 & 1 & 0 & | & 700 \\ 0 & 0 & 1 & -1 & | & -200 \end{bmatrix} \sim \begin{bmatrix} 1 & 0 & 0 & 1 & | & 1300 \\ 0 & 1 & 0 & 1 & | & 900 \\ 0 & 0 & 1 & -1 & | & -200 \\ 0 & 0 & 1 & -1 & | & -200 \end{bmatrix} \sim \begin{bmatrix} 1 & 0 & 0 & 1 & | & 1300 \\ 0 & 1 & 0 & 1 & | & 900 \\ 0 & 0 & 1 & -1 & | & -200 \\ 0 & 0 & 0 & 0 & | & 0 \end{bmatrix}$$

$$(-1)R_2 + R_3 \rightarrow R_3 \qquad (-1)R_3 + R_4 \rightarrow R_4$$

The corresponding system of equations is
$$x_1 \qquad\quad + x_4 = 1300$$
$$\quad x_2 \quad\ + x_4 = 900$$
$$\qquad\ x_3 - x_4 = -200$$

Let $x_4 = t$. Then, $x_1 = 1300 - t$, $x_2 = 900 - t$, $x_3 = t - 200$, $x_4 = t$ where $200 \leq t \leq 900$.

*($t \geq 200$  so that $x_3$ is non-negative; $t \leq 900$ so that $x_2$ is non-negative)

(C)  maximum: 900,  minimum: 200

(D)  Elm St.: $x_1 = 800$;  2nd St.: $x_2 = 400$;  Oak St.: $x_3 = 300$.     (4-3)

Copyright © 2011 Pearson Education, Inc. Publishing as Prentice Hall.

## 5   LINEAR INEQUALITIES AND LINEAR PROGRAMMING

### EXERCISE 5-1

*Things to remember:*

1.  A line divides the plane into two sets called HALF-PLANES. A vertical line divides the plane into LEFT and RIGHT HALF-PLANES; a nonvertical line divides the plane into UPPER and LOWER HALF-PLANES. In either case, the dividing line is called the BOUNDARY LINE of each half-plane.

2.  The graph of the linear inequality
    $$Ax + By < C \text{ or } Ax + By > C$$
    with $B \neq 0$ is either the upper half-plane or the lower half-plane (but not both) determined by the line $Ax + By = C$.
    If $B = 0$, the graph of
    $$Ax < C \text{ or } Ax > C$$
    is either the right half-plane or the left half-plane (but not both) determined by the vertical line $Ax = C$.

3.  For strict inequalities ("<" or ">"), the line is not included in the graph. For weak inequalities ("≤" or "≥"), the line is included in the graph.

4.  PROCEDURE FOR GRAPHING LINEAR INEQUALITIES
    *Step 1.* First graph $Ax + By = C$ as a dashed line if equality is not included in the original statement or as a solid line if equality is included.
    *Step 2.* Choose a test point anywhere in the plane not on the line [the origin (0, 0) often requires the least computation] and substitute the coordinates into the inequality.
    *Step 3.* Does the test point satisfy the original inequality? If so, shade the half-plane that contains the test point. If not, shade the opposite half-plane.

1.  $y \leq x - 1$
    Graph $y = x - 1$ as a solid line.
    Test point (0, 0):
    $0 \leq 0 - 1$
    $0 \leq -1$
    The inequality is false. Thus, the graph is the half-plane below the line $y = x - 1$, including the line.

    | x | y  |
    |---|----|
    | 0 | -1 |
    | 1 | 0  |

    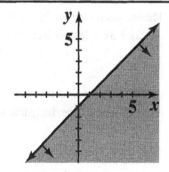

3.  $3x - 2y > 6$
    Graph $3x - 2y = 6$ as a dashed line.
    Test point (0, 0):
    $3 \cdot 0 - 2 \cdot 0 > 6$
    $0 > 6$

5.  $x \geq -4$
    Graph $x = -4$ [the vertical line through (-4, 0)] as a solid line.
    Test point (0, 0):
    $0 \geq -4$

Copyright © 2011 Pearson Education, Inc. Publishing as Prentice Hall.

The inequality is false. Thus, the graph is the half-plane below the line $3x - 2y = 6$, not including the line.

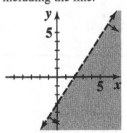

The inequality is true. Thus, the graph is the half-plane to the right of the line $x = -4$, including the line.

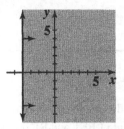

7.    $6x + 4y \geq 24$

Graph the line $6x + 4y = 24$ as a solid line.

Test point $(0, 0)$:

$6 \cdot 0 + 4 \cdot 0 \geq 24$

$0 \geq 24$

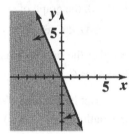

The inequality is false. Thus, the graph is the half-plane above the line, including the line.

9.    $5x \leq -2y$  or  $5x + 2y \leq 0$

Graph the line $5x + 2y = 0$ as a solid line.  Since the line passes through the origin
$(0, 0)$, we use $(1, 0)$ as a test point:

$5 \cdot 1 + 2 \cdot 0 \leq 0$

$5 \leq 0$

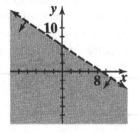

This inequality is false. Thus, the graph

is the half-plane below the line $5x + 2y = 0$, including the line.

11.   (A) Graph $2x + 3y = 18$ as a dashed line

Test point $(0, 0)$:

$2 \cdot 0 + 3 \cdot 0 < 18$

$0 < 18$

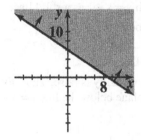

The inequality is true. Thus, the graph is the half-plane below the line $2x + 3y = 18$, not including the line.

(B) The set of points that do not satisfy the inequality is the half-plane above the line, including the line.

Copyright © 2011 Pearson Education, Inc.  Publishing as Prentice Hall.

**13.**  (A) Graph $5x - 2y = 20$ as a solid line

Test point $(0, 0)$:

$$5 \cdot 0 - 2 \cdot 0 \geq 20$$

$$0 \geq 20$$

This inequality is false. Thus, the graph is the half-plane below the line $5x - 2y = 20$, including the line.

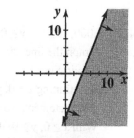

(B) The set of points that do not satisfy the inequality is the half-plane above the line, not including the line.

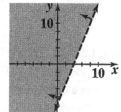

**15.**  Let $h$ = number of overtime hours; $h < 20$.

**17.**  Let $s$ = annual salary; $s \geq \$65,000$.

**19.**  Let $a$ = number of freshmen admitted; $a \leq 1,700$.

**21.**  The boundary line passes through the points $(-3, 0)$ and $(0, -2)$.

slope: $m = \dfrac{-2 - 0}{0 - (-3)} = -\dfrac{2}{3}$ ;  $y$ intercept: $b = -2$

Boundary line equation: $y = -\dfrac{2}{3}x - 2$   or   $2x + 3y = -6$

Since $(0, 0)$ is in the shaded region and the boundary line is solid, the graph is the graph of $2x + 3y \geq -6$.

**23.**  The equation of the boundary line is $y = 3$. Since $(0, 0)$ is in the shaded region and the boundary line is dashed, the graph is the graph of $y < 3$.

**25.**  The boundary line passes through the origin $(0, 0)$ and the point $(5, 4)$.

Slope $m = \dfrac{4}{5}$

Boundary line equation: $y = \dfrac{4}{5}x$   or   $4x - 5y = 0$.

Since $(1, 0)$ is in the shaded region and the boundary line is solid, the graph is the graph of $4x - 5y \geq 0$.

**27.**  Let $x$ = enrollment in finite mathematics;  let  $y$ = enrollment in calculus;   $x + y < 300$.

**29.**  Let $x$ = revenue;  let  $y$ = cost;   $x + 20,000 \leq y$  or  $x \leq y - 20,000$.

**31.**  Let $x$ = number of grams of saturated fat;  let  $y$ = number of grams of unsaturated fat;

$x > 3y$.

Copyright © 2011 Pearson Education, Inc. Publishing as Prentice Hall.

**33.**  $25x + 40y \leq 3{,}000, \ x \geq 0, \ y \geq 0$

    Step 1.    Graph the line $25x + 40y = 3000$ as
                   a solid line.

    Step 2.    Substituting $x = 0$, $y = 0$ in the inequality produces a
                   true statement,  so $(0, 0)$ is in the solution set.

    Step 3.    With $x \geq 0$, $y \geq 0$, the graph is shown
                   at the right.

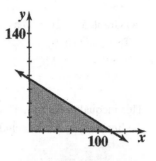

**35.**  $15x - 50y < 1{,}500, \ x \geq 0, \ y \geq 0$

    Step 1. Graph the line $15x - 50y = 1{,}500$
                  as a dashed line.

    Step 2.    Substituting $x = 0$, $y = 0$ in the
                   inequality produces a true statement,
                   so $(0, 0)$ is in the solution set.

    Step 3.    With $x \geq 0$, $y \geq 0$, the graph is shown
                   at the right.

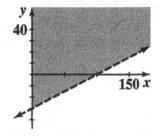

**37.**  $-18x + 30y \geq 2{,}700, \ x \geq 0, \ y \geq 0$

    Step 1.    Graph the line $-18x + 30y = 2{,}700$
                   as a solid line.

    Step 2.    Substituting $x = 0$, $y = 0$ in the
                   inequality produces a false statement,
                   so $(0, 0)$ is not in the solution set.

    Step 3.    With $x \geq 0$, $y \geq 0$, the graph is shown
                   at the right.

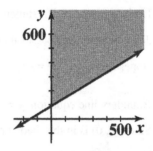

**39.**  $40x - 55y > 0, \ x \geq 0, \ y \geq 0$

    Step 1.    Graph the line $40x - 55y = 0$
                   as a dashed line.

    Step 2.    Since the line passes through the
                   origin choose $(1, 0)$ as a test point;
                   substituting $x = 1$, $y = 0$ in the
                   inequality produces a true statement,
                   so $(1, 0)$ is in the solution set.

    Step 3.    With $x \geq 0$, $y \geq 0$, the graph is shown
                   at the right.

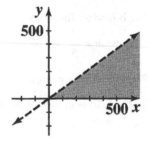

**41.**  $25x + 75y < -600, \ x \geq 0, \ y \geq 0$

    Step 1.    Graph the line $25x + 75y = -600$ as a dashed line.

    Step 2.    Substituting $x = 0$, $y = 0$ into the inequality produces a false statement, so $(0, 0)$ is not in
                   the solution set.

    Step 3.    With $x \geq 0$, $y \geq 0$, the solution set is empty and has no graph.

Copyright © 2011 Pearson Education, Inc.  Publishing as Prentice Hall.

**43.**  Let $x$ = number of acres for corn

$y$ = number of acres for soybeans

Then $40x + 32y \le 5,000$.

Dividing by 8, we get the inequality

$5x + 4y \le 625$.

We must also have $x \ge 0$, $y \ge 0$.

Graph the line $5x + 4y = 625$ as a solid line. Since $x = 0$, $y = 0$

satisfies the inequality, $(0, 0)$, is in the solution set. The graph is

shown at the right.

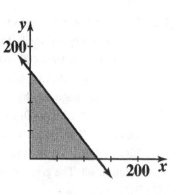

**45.**  Let $x$ = number of pounds of Brand A;

$y$ = number of pounds of Brand B.

(A)  Number of pounds of nitrogen: $0.26x + 0.16y$.

She wants

$0.26x + 0.16y \ge 120$, $x \ge 0$, $y \ge 0$

Multiplying the inequality by 100, we get

$26x + 16y \ge 12,000$, $x \ge 0$, $y \ge 0$

or   $13x + 8y \ge 6,000$, $x \ge 0$, $y \ge 0$

Graph the line $13x + 8y = 6,000$ as a solid line.

Since $x = 0$, $y = 0$ does not satisfy the inequality $(0, 0)$ is not in the

solution set.

The graph is shown at the right.

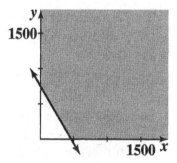

(B)  Number of pounds of phosphate: $0.03x + 0.08y$.

She wants

$0.03x + 0.08y \le 28$, $x \ge 0$, $y \ge 0$

Multiplying by 100, we get

$3x + 8y \le 2800$, $x \ge 0$, $y \ge 0$

Graph the line $3x + 8y = 2800$ as a solid line.

Since $x = 0$, $y = 0$ satisfies the inequality,

$(0, 0)$ is in the solution set. The graph is shown at the right.

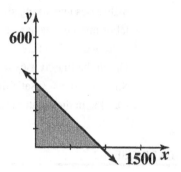

**47.**  Let $x$ = number of pounds of standard blend

$y$ = number of pounds of deluxe blend

The mill wants to produce a fabric that is at least 20% acrylic.

Therefore,

$0.30x + 0.09y \ge 0.20(x + y)$,   $x \ge 0$,   $y \ge 0$.

Multiplying by 100,  we get

$30x + 9y \ge 20(x + y)$    which implies    $y \le \dfrac{10}{11}x$

Graph the line  $y = \dfrac{10}{11}x$  as a solid line. Since $x = 1$, $y = 0$ satisfies the

inequality, $(1, 0)$ is in the solution set. The graph is shown at the right.

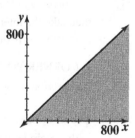

Copyright © 2011 Pearson Education, Inc. Publishing as Prentice Hall.

**49.**  Let $x$ = number of weeks to operate plant A,
        $y$ = number of weeks to operate plant B.
To produce at least 400 sedans, we must have
$10x + 8y \geq 400, x \geq 0, y \geq 0$
Divide the inequality by 2:
$5x + 4y \geq 200, x \geq 0, y \geq 0$
Graph the line $5x + 4y = 200$ as a solid line.
Since $x = 0, y = 0$ does not satisfy the inequality, $(0, 0)$ is not in the
solution set. The graph is shown at the right.

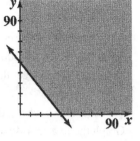

**51.**  Let $x$ = number of radio ads,
        $y$ = number of television ads.
Then
$200x + 800y \leq 10,000$ or $x + 4y \leq 50,$
$x \geq 0, y \geq 0$

Graph the line $x + 4y = 50$ as a solid line.
Since $x = 0, y = 0$ satisfies the inequality
$(0, 0)$ is in the solution set.

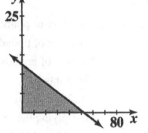

**53.**  Let $x$ = number of regular mattresses,
        $y$ = number of king size mattresses
Then $5x + 6y$ is the number of minutes required to cut $x$ regular
mattresses and $y$ king size mattresses. 50 labor hours = 50(60) = 3000
labor minutes are available. Therefore,
$5x + 6y \leq 3,000, x \geq 0, y \geq 0$
Graph the line $5x + 6y = 3,000$ as a solid line.
Since $x = 0, y = 0$ satisifies the inequality
$(0, 0)$ is in the solution set.

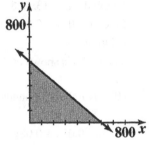

---

## EXERCISE 5-2

*Things to remember:*

1.  To solve a system of linear inequalities graphically, graph each inequality in the system and
    then take the intersection of all the graphs. The resulting graph is called the SOLUTION
    REGION, or FEASIBLE REGION.

2.  A CORNER POINT of a solution region is a point in the solution region that is the
    intersection of two boundary lines.

3.  The solution region of a system of linear inequalities is BOUNDED if it can be enclosed
    within a circle; if it cannot be enclosed within a circle, then it is UNBOUNDED.

Copyright © 2011 Pearson Education, Inc.  Publishing as Prentice Hall.

1.  The graph of $x + 2y \leq 8$ is the half-plane below the line $x + 2y = 8$
    [e.g., (0, 0) satisfies the inequality].  The graph of $3x - 2y \geq 0$ is the half-plane below the line $3x - 2y = 0$
    [e.g., (1, 0) satisfies the inequality].  The intersection of these two regions is region IV.

3.  The graph of $x + 2y \geq 8$ is the half-plane above the line $x + 2y = 8$ [e.g., (0, 0) does not satisfy the
    inequality].  The graph of $3x - 2y \geq 0$ is the half-plane below the line $3x - 2y = 0$ [e.g., (1, 0) satisfies the
    inequality].  The intersection of these two regions is region I.

5.  The graphs of the inequalities $3x + y \geq 6$ and $x \leq 4$ are:

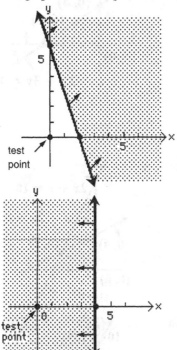

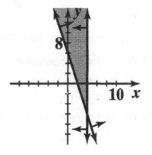

The intersection of these regions (drawn on the same coordinate
plane) is shown in the graph at the right.

7.  The graphs of the inequalities $x - 2y \leq 12$ and $2x + y \geq 4$ are:

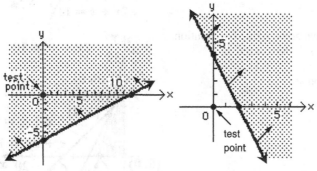

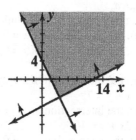

The intersection of these regions (drawn on the same coordinate
plane) is shown in the graph at the right.

Copyright © 2011 Pearson Education, Inc.  Publishing as Prentice Hall.

9. The graph of $x + 3y \leq 18$ is the region below the line $x + 3y = 18$ and the graph of $2x + y \geq 16$ is the region above the line $2x + y = 16$. The graph of $x \geq 0$, $y \geq 0$ is the first quadrant. The intersection of these regions is region IV. The corner points are $(8, 0)$, $(18, 0)$, and $(6, 4)$.

11. The graph of $x + 3y \geq 18$ is the region above the line $x + 3y = 18$ and the graph of $2x + y \geq 16$ is the region above the line $2x + y = 16$. The graph of $x \geq 0$, $y \geq 0$ is the first quadrant. The intersection of these regions is region I. The corner points are $(0, 16)$, $(6, 4)$, and $(18, 0)$.

13. The graphs of the inequalities are shown at the right. The solution region is indicated by the shaded region. The solution region is *bounded*.

    The corner points of the solution region are:

    $(0, 0)$, the intersection of $x = 0$, $y = 0$;
    $(0, 4)$, the intersection of $x = 0$,
        $2x + 3y = 12$;
    $(6, 0)$, the intersection of $y = 0$,
        $2x + 3y = 12$.

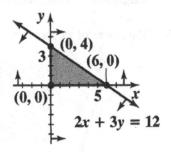

15. The graphs of the inequalities are shown at the right. The solution region is shaded. The solution region is *bounded*.

    The corner points of the solution region are:

    $(0, 0)$, the intersection of $x = 0$, $y = 0$;
    $(0, 4)$, the intersection of $x = 0$, $x + 2y = 8$;
    $(4, 2)$, the intersection of $x + 2y = 8$,
        $2x + y = 10$;
    $(5, 0)$, the intersection of $y = 0$,
        $2x + y = 10$.

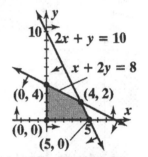

17. The graphs of the inequalities are shown at the right. The solution region is shaded. The solution region is *unbounded*.

    The corner points of the solution region are:

    $(0, 10)$, the intersection of $x = 0$,
        $2x + y = 10$;
    $(4, 2)$, the intersection of $x + 2y = 8$,
        $2x + y = 10$;
    $(8, 0)$, the intersection of $y = 0$, $x + 2y = 8$.

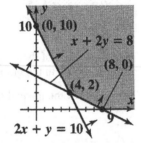

19. The graphs of the inequalities are shown at the right. The solution is indicated by the shaded region. The solution region is *bounded*.
    The corner points of the solution region are:
    $(0, 0)$, the intersection of $x = 0$, $y = 0$,
    $(0, 6)$, the intersection of $x = 0$,
        $x + 2y = 12$;
    $(2, 5)$, the intersection of $x + 2y = 12$,
        $x + y = 7$;
    $(3, 4)$, the intersection of $x + y = 7$,
        $2x + y = 10$;
    $(5, 0)$, the intersection of $y = 0$,
        $2x + y = 10$.
    Note that the point of intersection of the lines $2x + y = 10$, $x + 2y = 12$ is not a corner point because it is not in the solution region.

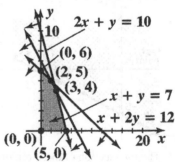

Copyright © 2011 Pearson Education, Inc. Publishing as Prentice Hall.

**21.** The graphs of the inequalities are shown at the right. The solution is indicated by the shaded region, which is *unbounded*.

The corner points are:

(0, 16), the intersection of $x = 0$,
$\qquad 2x + y = 16$;
(4, 8), the intersection of $2x + y = 16$,
$\qquad x + y = 12$;
(10, 2), the intersection of $x + y = 12$,
$\qquad x + 2y = 14$;
(14, 0), the intersection of $y = 0$,
$\qquad x + 2y = 14$.

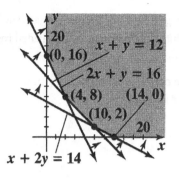

The intersection of $x + 2y = 14$, $2x + y = 16$ is not a corner point because it is not in the solution region.

**23.** The graphs of the inequalities are shown at the right. The solution is indicated by the shaded region, which is *bounded*.

The corner points are (8, 6), (4, 7), and (9, 3).

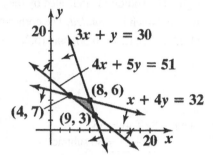

**25.** The graphs of the inequalities are shown at the right. The system of inequalities does not have a solution because the intersection of the graphs is empty.

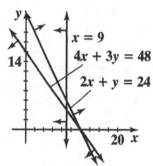

**27.** The graphs of the inequalities are shown at the right. The solution is indicated by the shaded region, which is *unbounded*.

The corner points are (0, 0), (4, 4), and (8, 12).

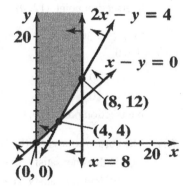

Copyright © 2011 Pearson Education, Inc. Publishing as Prentice Hall.

**29.** The graphs of the inequalities are shown at the right. The solution is indicated by the shaded region, which is *bounded*.

The corner points are $(2, 1)$, $(3, 6)$, $(5, 4)$, and $(5, 2)$.

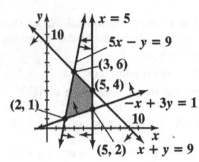

**31.** The graphs of the inequalities are shown at the right. The solution is indicated by the shaded region, which is *bounded*. The corner points are $(1.24, 5.32)$, $(2.17, 6.56)$, and $(6.2, 1.6)$.

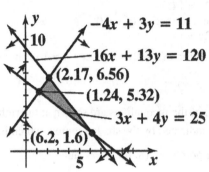

**33.** (A)

$$3x + 4y = 36$$
$$\underline{3x + 2y = 30} \text{ subtract}$$
$$2y = 6$$
$$y = 3$$
$$x = 8$$
intersection point: $(8, 3)$

$$3x + 4y = 36$$
$$\underline{y = 0}$$
$$3x = 36$$
$$x = 12$$
intersection point: $(12, 0)$

$$3x + 2y = 30$$
$$\underline{y = 0}$$
$$3x = 30$$
$$x = 10$$
intersection point: $(10, 0)$

$$3x + 4y = 36$$
$$\underline{x = 0}$$
$$4y = 36$$
$$y = 9$$
intersection point: $(0, 9)$

$$3x + 2y = 30$$
$$\underline{x = 0}$$
$$2y = 30$$
$$y = 15$$
intersection point: $(0, 15)$

$$x = 0$$
$$y = 0$$

intersection point: $(0, 0)$

(B) The corner points are: $(8, 3)$, $(0, 9)$, $(10, 0)$, $(0, 0)$;
$(0, 15)$ does not satisfy $3x + 4y \leq 36$,
$(12, 0)$ does not satisfy $3x + 2y \leq 30$.

Copyright © 2011 Pearson Education, Inc. Publishing as Prentice Hall.

**35.** Let $x$ = the number of trick skis and $y$ = the number of slalom skis produced per day. The information is summarized in the following table.

| | Hours per ski | | Maximum labor-hours per day available |
|---|---|---|---|
| | Trick ski | Slalom ski | |
| Fabrication | 6 hrs | 4 hrs | 108 hrs |
| Finishing | 1 hr | 1 hr | 24 hrs |

We have the following inequalities:

$6x + 4y \leq 108$  for fabrication

$x + y \leq 24$  for finishing

Also, $x \geq 0$ and $y \geq 0$.

The graphs of these inequalities are shown at the right. The shaded region indicates the set of feasible solutions.

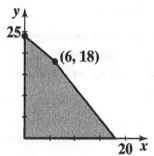

**37.** (A) If $x$ is the number of trick skis and $y$ is the number of slalom skis per day, then the profit per day is given by

$$P(x, y) = 50x + 60y$$

All the production schedules in the feasible region that lie on the graph of the line

$$50x + 60y = 1,100$$

will provide a profit of $1,100.

(B) There are many possible choices. For example, producing 5 trick skis and 15 slalom skis per day will produce a profit of

$$P(5, 15) = 50(5) + 60(15) = 1,150$$

All the production schedules in the feasible region that lie on the graph of $50x + 60y = 1,150$ will provide a profit of $1,150.

**39.** Let $x$ = the number of cubic yards of mix A and $y$ = the number of cubic yards of mix B. The information is summarized in the following table:

| | Amount of substance per cubic yard | | Minimum Monthly requirement |
|---|---|---|---|
| | Mix A | Mix B | |
| Phosphoric acid | 20 lbs | 10 lbs | 460 lbs |
| Nitrogen | 30 lbs | 30 lbs | 960 lbs |
| Potash | 5 lbs | 10 lbs | 220 lbs |

Copyright © 2011 Pearson Education, Inc. Publishing as Prentice Hall.

We have the following inequalities:

$20x + 10y \geq 460$

$30x + 30y \geq 960$

$5x + 10y \geq 220$

Also, $x \geq 0$ and $y \geq 0$.

The graphs of these inequalities are shown at the right. The shaded region indicates the set of feasible solutions.

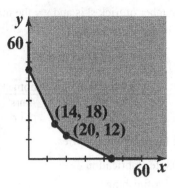

**41.** Let $x$ = the number of mice used and $y$ = the number of rats used. The information is summarized in the following table.

|       | Mice   | Rats   | Maximum time available per day |
|-------|--------|--------|--------------------------------|
| Box A | 10 min | 20 min | 800 min                        |
| Box B | 20 min | 10 min | 640 min                        |

We have the following inequalities:

$10x + 20y \leq 800$  for box A

$20x + 10y \leq 640$  for box B

Also, $x \geq 0$ and $y \geq 0$.

The graphs of these inequalities are shown at the right. The shaded region indicates the set of feasible solutions.

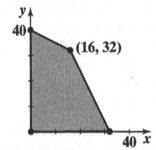

## EXERCISE 5-3

*Things to remember:*

1.  A LINEAR PROGRAMMING PROBLEM is a problem that is concerned with finding the OPTIMAL VALUE (maximum or minimum value) of a linear OBJECTIVE FUNCTION of the form

$$z = c_1 x_1 + c_2 x_2 + \cdots + c_n x_n,$$

where the DECISION VARIABLES $x_1, x_2, \ldots, x_n$ are subject to PROBLEM CONSTRAINTS in the form of linear inequalities and equations. In addition, the decision variables must satisfy the NONNEGATIVE CONSTRAINTS $x_i \geq 0$, for $i = 1, 2, \ldots, n$. The set of points satisfying both the problem constraints and the nonnegative constraints is called the FEASIBLE REGION for the problem. Any point in the feasible region that produces the optimal value of the objective function over the feasible region is called an OPTIMAL SOLUTION.

2.  CONSTRUCTING A MATHEMATICAL MODEL FOR AN APPLIED LINEAR PROGRAMMING PROBLEM

    *Step 1.* Introduce decision variables

Copyright © 2011 Pearson Education, Inc. Publishing as Prentice Hall.

*Step 2.* Summarize relevant material in table form, relating columns to the decision variables, if possible.

*Step 3.* Determine the objective and write a linear objective function.

*Step 4.* Write problem constraints using linear equations and/or inequalities.

*Step 5.* Write non-negative constraints.

3.    FUNDAMENTAL THEOREM OF LINEAR PROGRAMMING

If the optimal value of the objective function in a linear programming problem exists, then that value must occur at one (or more) of the corner points of the feasible region.

4.    EXISTENCE OF SOLUTIONS

(A)    If the feasible region for a linear programming problem is bounded, then both the maximum value and the minimum value of the objective function always exist.

(B)    If the feasible region is unbounded, and the coefficients of the objective function are positive, then the minimum value of the objective function exists, but the maximum value does not.

(C)    If the feasible region is empty (that is, there are no points that satisfy all the constraints), then both the maximum value and the minimum value of the objective function do not exist.

5.    GEOMETRIC METHOD FOR SOLVING A LINEAR PROGRAMMING PROBLEM WITH TWO DECISION VARIABLES.

*Step 1.*    Graph the feasible region. Then, if according to 4 an optimal solution exists, find the coordinates of each corner point.

*Step 2.*    Construct a CORNER POINT TABLE listing the value of the objective function at each corner point.

*Step 3.*    Determine the optimal solution(s) from the table in Step (2).

*Step 4.*    For an applied problem, interpret the optimal solution(s) in terms of the original problem.

---

1.

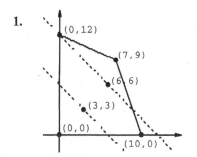

From the figure:

maximum profit $P = 16$ at $x = 7$, $y = 9$.

Copyright © 2011 Pearson Education, Inc. Publishing as Prentice Hall.

Step (2):          Evaluate the objective function at each corner point.

| Corner Point | $P = x + y$ |
|---|---|
| (0, 0) | 0 |
| (0, 12) | 12 |
| (7, 9) | 16 |
| (10, 0) | 10 |

Step (3):          Determine the optimal solution from Step (2).

The maximum value of $P$ is 16 at (7, 9).

**3.**

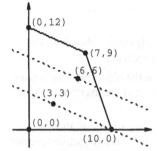

From the figure:

maximum profit $P = 84$ at $x = 0$, $y = 12$; at $x = 7$, $y = 9$;
and at every point on the line segment joining (0, 12) and
(7, 9).

Step (2):          Evaluate the objective function at each corner point.

| Corner Point | $P = 3x + 7y$ |
|---|---|
| (0, 0) | 0 |
| (0, 12) | 84 |
| (7, 9) | 84 |
| (10, 0) | 30 |

Step (3):          Determine the optimal solution from Step (2).

The maximum value of $P$ is 84 at (0, 12) *and* (7, 9).

This is a multiple optimal solution.

Copyright © 2011 Pearson Education, Inc. Publishing as Prentice Hall.

**5.**    $C = 7x + 4y$

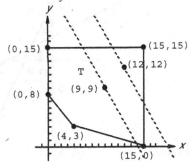

From the figure:   minimum cost $C = 32$ at $x = 0$, $y = 8$.

Step (2):    Evaluate the objective function at each corner point.

| Corner Point | $C = 7x + 4y$ |
| --- | --- |
| (15, 15) | 165 |
| (0, 15) | 60 |
| (0, 8) | 32 |
| (4, 3) | 40 |
| (15, 0) | 105 |

Step (3):    Determine the optimal solution from Step (2).

The minimum value of $C$ is 32 at $x = 0$, $y = 8$.

**7.**    $C = 3x + 8y$

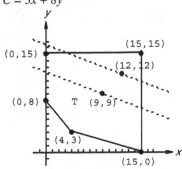

From the figure:   minimum cost $C = 36$ at $x = 4$, $y = 3$.

Step (2):    Evaluate the objective function at each corner point.

| Corner Point | $C = 3x + 8y$ |
| --- | --- |
| (15, 15) | 165 |
| (0, 15) | 120 |
| (0, 8) | 64 |
| (4, 3) | 36 |
| (15, 0) | 45 |

Step (3):    Determine the optimal solution from Step (2).

The minimum value of $C$ is 36 at $x = 4$, $y = 3$.

Copyright © 2011 Pearson Education, Inc. Publishing as Prentice Hall.

9.  <u>Step (1)</u>:        Graph the feasible region and find the corner points.

The feasible region $S$ is the solution set of the given inequalities.  This region is indicated by the shading in the graph at the right.

The corner points are $(0, 0)$, $(0, 4)$, $(4, 2)$, and $(5, 0)$.

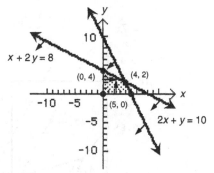

Since $S$ is bounded, it follows from <u>4</u>(a) that $P$ has a maximum value

<u>Step (2)</u>:        Evaluate the objective function at each corner point.

The value of $P$ at each corner point is given in the following table.

| Corner Point | $P = 5x + 5y$ |
|---|---|
| $(0,0)$ | $P = 5(0) + 5(0) = 0$ |
| $(0,4)$ | $P = 5(0) + 5(4) = 20$ |
| $(4,2)$ | $P = 5(4) + 5(2) = 30$ |
| $(5,0)$ | $P = 5(5) + 5(0) = 25$ |

<u>Step (3)</u>:        Determine the optimal solution.

The maximum value of $P$ is 30 at $x = 4$, $y = 2$.

11.  <u>Step (1)</u>:  Graph the feasible region and find the corner points.

The feasible region $S$ is the solution set of the given inequalities.  This region is indicated by the shading in the graph at the right.

The corner points are $(0, 10)$,  $(4, 2)$, and $(8, 0)$.

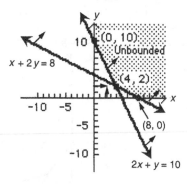

Since $S$ is unbounded and $a = 2 > 0$, $b = 3 > 0$, it follows from <u>4</u>(b) that $z$ has a minimum value but not a maximum value.

<u>Step (2)</u>:        Evaluate the objective function at each corner point.

The value of $z$ at each corner point is given in the following table:

| Corner Point | $z = 2x + 3y$ |
|---|---|
| $(0,10)$ | $z = 2(0) + 3(10) = 30$ |
| $(4,2)$ | $z = 2(4) + 3(2) = 14$ |
| $(8,0)$ | $z = 2(8) + 3(0) = 16$ |

<u>Step (3)</u>:        Determine the optimal solutions.

The minimum occurs at $x = 4$, $y = 2$, and the minimum value is $z = 14$; $z$ does not have a maximum value.

Copyright © 2011 Pearson Education, Inc.  Publishing as Prentice Hall.

**13.**  Step (1):  Graph the feasible region and find the corner points.

The feasible region $S$ is the solution set of the given inequalities. This region is indicated by the shading in the graph at the right.

The corner points are $(0, 0)$, $(0, 6)$, $(2, 5)$, $(3, 4)$, and $(5, 0)$.

Since $S$ is bounded, it follows from $\underline{4}$(a) that $P$ has a maximum value.

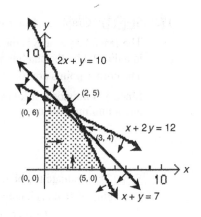

Step (2):   Evaluate the objective function at each corner point.

The value of $P$ at each corner point is:

| Corner Point | $P = 30x + 40y$ |
|---|---|
| $(0,0)$ | $P = 30(0) + 40(0) = 0$ |
| $(0,6)$ | $P = 30(0) + 40(6) = 240$ |
| $(2,5)$ | $P = 30(2) + 40(5) = 260$ |
| $(3,4)$ | $P = 30(3) + 40(4) = 250$ |
| $(5,0)$ | $P = 30(5) + 40(0) = 150$ |

Step (3):   Determine the optimal solution.

The maximum occurs at $x = 2$, $y = 5$, and the maximum value is $P = 260$.

**15.**  Step (1):   Graph the feasible region and find the corner points.

The feasible region $S$ is the solution set of the given inequalities. This region is indicated by the shading in the graph at the right.

The corner points are $(0, 16)$, $(4, 8)$, $(10, 2)$, and $(14, 0)$.

Since $S$ is unbounded and $a = 10 > 0$, $b = 30 > 0$, it follows from $\underline{4}$(b) that $z$ has a minimum value but not a maximum value.

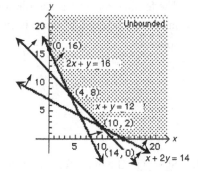

Step (2):   Evaluate the objective function at each corner point.
The value of $z$ at each corner point is:

| Corner Point | $z = 10x + 30y$ |
|---|---|
| $(0,16)$ | $z = 10(0) + 30(16) = 480$ |
| $(4,8)$ | $z = 10(4) + 30(8) = 280$ |
| $(10,2)$ | $z = 10(10) + 30(2) = 160$ |
| $(14,0)$ | $z = 10(14) + 30(0) = 140$ |

Step (3):   Determine the optimal solution.

The minimum occurs at $x = 14$, $y = 0$, and the minimum value is $z = 140$; $z$ does not have a maximum value.

Copyright © 2011 Pearson Education, Inc. Publishing as Prentice Hall.

17. <u>Step (1)</u>: Graph the feasible region and find the corner points.

    The feasible region $S$ is the solution set of the given inequalities, and is indicated by the shading in the graph at the right.

    The corner points are $(0, 2)$, $(0, 9)$, $(2, 6)$, $(5, 0)$, and $(2, 0)$.

    Since $S$ is bounded, it follows from <u>4</u>(a) that $P$ has a maximum value and a minimum value.

    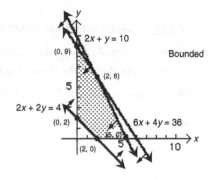

    <u>Step (2)</u>: Evaluate the objective function at each corner point.
    The value of $P$ at each corner point is given in the following table:

    | Corner Point | $P = 30x + 10y$ |
    |:---:|:---|
    | $(0, 2)$ | $P = 30(0) + 10(2) = 20$ |
    | $(0, 9)$ | $P = 30(0) + 10(9) = 90$ |
    | $(2, 6)$ | $P = 30(2) + 10(6) = 120$ |
    | $(5, 0)$ | $P = 30(5) + 10(0) = 150$ |
    | $(2, 0)$ | $P = 30(2) + 10(0) = 60$ |

    <u>Step (3)</u>: Determine the optimal solutions.
    The maximum occurs at $x = 5$, $y = 0$, and the maximum value is $P = 150$; the minimum occurs at $x = 0$, $y = 2$, and the minimum value is $P = 20$.

19. <u>Step (1)</u>: Graph the feasible region and find the corner points.

    The feasible region $S$ is the solution set of the given inequalities. As indicated, the feasible region is empty. Thus, by <u>4</u>(c), there are no optimal solutions.

    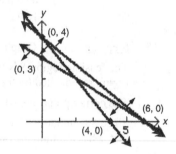

21. <u>Step (1)</u>: Graph the feasible region and find the corner points.

    The feasible region $S$ is the solution set of the given inequalities, and is indicated by the shading in the graph at the right.

    The corner points are $(3, 8)$, $(8, 10)$, and $(12, 2)$.

    Since $S$ is bounded, it follows from <u>4</u>(a) that $P$ has a maximum value and a minimum value.

    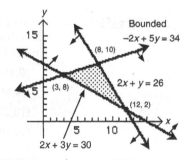

    <u>Step (2)</u>: Evaluate the objective function at each corner point.
    The value of $P$ at each corner point is:

    | Corner Point | $P = 20x + 10y$ |
    |:---:|:---|
    | $(3, 8)$ | $P = 20(3) + 10(8) = 140$ |
    | $(8, 10)$ | $P = 20(8) + 10(10) = 260$ |
    | $(12, 2)$ | $P = 20(12) + 10(2) = 260$ |

Copyright © 2011 Pearson Education, Inc. Publishing as Prentice Hall.

Step (3): Determine the optimal solutions.

The minimum occurs at $x = 3$, $y = 8$, and the minimum value is $P = 140$; the maximum occurs at $x = 8$, $y = 10$, at $x = 12$, $y = 2$, and at any point along the line segment joining (8, 10) and (12, 2). The maximum value is $P = 260$.

**23.** Step (1): Graph the feasible region and find the corner points. The feasible region $S$ is the set of solutions of the given inequalities, and is indicated by the shading in the graph at the right.

The corner points are (0, 0), (0, 800), (400, 600), (600, 450), and (900, 0). Since $S$ is bounded, it follows from 4(a) that $P$ has a maximum value.

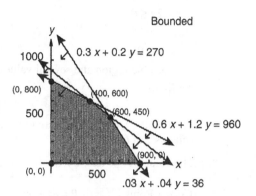

Step (2): Evaluate the objective function at each corner point. The value of $P$ at each corner point is:

| Corner Point | $P = 20x + 30y$ |
|---|---|
| (0, 0) | $P = 20(0) + 30(0) = 0$ |
| (0, 800) | $P = 20(0) + 30(800) = 24,000$ |
| (400, 600) | $P = 20(400) + 30(600) = 26,000$ |
| (600, 450) | $P = 20(600) + 30(450) = 25,500$ |
| (900, 0) | $P = 20(900) + 30(0) = 18,000$ |

Step (3): Determine the optimal solution.

The maximum occurs at $x = 400$, $y = 600$, and the maximum value is $P = 26,000$.

**25.** $\ell_1 : 275x + 322y = 3,381$

$\ell_2 : 350x + 340y = 3,762$

$\ell_3 : 425x + 306y = 4,114$.

Step (1): Graph the feasible region and find the corner points. The feasible region $S$ is the solution set of the given inequalities, and is indicated by the shading in the graph at the right. The corner points are (0, 0), (0, 10.5), (3.22, 7.75), (6.62, 4.25), (9.68, 0).

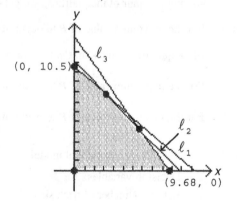

Step (2): Evaluate the objective function at each corner point. The value of $P$ at each corner point is

| Corner Point | $P = 525x + 478y$ |
|---|---|
| (0, 0) | $P = 525(0) + 478(0) = 0$ |
| (0, 10.5) | $P = 525(0) + 478(10.5) = 5,019$ |
| (3.22, 7.75) | $P = 525(3.22) + 478(7.75) = 5,395$ |
| (6.62, 4.25) | $P = 525(6.62) + 478(4.25) = 5,507$ |
| (9.68, 0) | $P = 525(9.68) + 478(0) = 5,082$ |

Copyright © 2011 Pearson Education, Inc. Publishing as Prentice Hall.

Step (3): Determine the optimal solution.

The maximum occurs at $x = 6.62$, $y = 4.25$, and the maximum value is $P = 5{,}507$.

27. Minimize and maximize $z = x - y$

    Subject to
    $$x - 2y \le 0$$
    $$2x - y \le 6$$
    $$x, y \ge 0$$

    The feasible region and several values of the objective function are shown in the figure.

    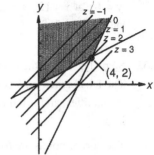

    The points $(0, 0)$ and $(4, 2)$ are the corner points; $z = x - y$ does not have a minimum value. Its maximum value is 2 at $(4, 2)$.

29. The value of $P = ax + by$, $a > 0$, $b > 0$, at each corner point is:

    | Corner Point | $P = ax + by$ |
    |---|---|
    | $O: (0,0)$ | $P = a(0) + b(0) = 0$ |
    | $A: (0,5)$ | $P = a(0) + b(5) = 5b$ |
    | $B: (4,3)$ | $P = a(4) + b(3) = 4a + 3b$ |
    | $C: (5,0)$ | $P = a(5) + b(0) = 5a$ |

    (A) For the maximum value of $P$ to occur at $A$ only, we must have $5b > 4a + 3b$ and $5b > 5a$. Solving the first inequality, we get $2b > 4a$ or $b > 2a$; from the second inequality, we get $b > a$. Therefore, we must have $b > 2a$ or $2a < b$ in order for $P$ to have its maximum value at $A$ only.

    (B) For the maximum value of $P$ to occur at $B$ only, we must have $4a + 3b > 5b$ and $4a + 3b > 5a$.

    Solving this pair of inequalities, we get $4a > 2b$ and $3b > a$, which is the same as $\dfrac{a}{3} < b < 2a$.

    (C) For the maximum value of $P$ to occur at $C$ only, we must have $5a > 4a + 3b$ and $5a > 5b$. This pair of inequalities implies that $a > 3b$ or $b < \dfrac{a}{3}$.

    (D) For the maximum value of $P$ to occur at both $A$ and $B$, we must have $5b = 4a + 3b$ or $b = 2a$.

    (E) For the maximum value of $P$ to occur at both $B$ and $C$, we must have $4a + 3b = 5a$ or $b = \dfrac{a}{3}$.

31. (A) Construct the mathematical model

    a. Decision variables:

       Let $x$ = Number of trick skis

       $y$ = Number of slalom skis

    b. Relevant material in table form:

    |  | Trick ski | Slalom ski | Labor-hours available |
    |---|---|---|---|
    | Fabricating | 6 | 4 | 108 |
    | Finishing | 1 | 1 | 24 |
    | Profit | $40/ski | $30/ski | |

    c. Objective function:

       Maximize profit $P = 40x + 30y$

Copyright © 2011 Pearson Education, Inc. Publishing as Prentice Hall.

d.  Problem constraints:

$6x + 4y \leq 108$    [Fabricating constraint]

$x + y \leq 24$    [Finishing constraint]

e.  Non-negativity constraints

$x \geq 0, y \geq 0$

The mathematical model for this problem is:

$$\text{Maximize } P = 40x + 30y$$
$$\text{Subject to: } 6x + 4y \leq 108$$
$$x + y \leq 24$$
$$x \geq 0, y \geq 0$$

Step (1):    Graph the feasible region and find the corner points.

The feasible region $S$ is the solution set of the given system of inequalities, and is indicated by the shading in the graph below.

The corner points are (0, 0), (0, 24), (6, 18), and (18, 0).

Since $S$ is bounded, $P$ has a maximum value by 4(a).

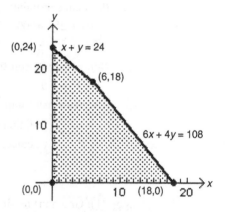

Step (2): Evaluate the objective function at each corner point.

The value of $P$ at each corner point is:

| Corner Point | $P = 40x + 30y$ |
|---|---|
| (0,0) | $P = 40(0) + 30(0) = 0$ |
| (0,24) | $P = 40(0) + 30(24) = 720$ |
| (6,18) | $P = 40(6) + 30(18) = 780$ |
| (18,0) | $P = 40(18) + 30(0) = 720$ |

Step (3): Determine the optimal solution.

The maximum occurs when $x = 6$ (trick skis) and

$y = 18$ (slalom skis) are produced. The maximum profit is $P = \$780$.

| Corner Point | $P = 40x + 25y$ |
|---|---|
| (0,0) | $P = 40(0) + 25(0) = 0$ |
| (B)    (0,24) | $P = 40(0) + 25(24) = 600$ |
| (6,18) | $P = 40(6) + 25(18) = 690$ |
| (18,0) | $P = 40(18) + 25(0) = 720$ |

The maximum profit decreases to $720 when 18 trick skis and no slalom skis are produced.

Copyright © 2011 Pearson Education, Inc.  Publishing as Prentice Hall.

| Corner Point | $P = 40x + 45y$ |
|---|---|
| $(0,0)$ | $P = 40(0) + 45(0) = 0$ |
| $(0,24)$ | $P = 40(0) + 45(24) = 1080$ |
| $(6,18)$ | $P = 40(6) + 45(18) = 1050$ |
| $(18,0)$ | $P = 40(18) + 45(0) = 720$ |

(C)

The maximum profit increases to $1,080 when no trick skis and 24 slalom skis are produced.

33. (A)   Construct the mathematical model

a.   Decision variables:

Let $x$ = Number of days to operate Plant $A$

$y$ = Number of days to operate Plant $B$

b.   Relevant material in table form:

|  | Plant $A$ | Plant $B$ | Amount required |
|---|---|---|---|
| Tables | 20 | 25 | 200 |
| Chairs | 60 | 50 | 500 |
| Cost/day | $1000 | $900 |  |

c.   Objective function:

Minimize the cost $C = 1000x + 900y$

d.   Problem constraints:

$20x + 25y \geq 200$    [Table constraint]

$60x + 50y \geq 500$    [Chair constraint]

e.   Non-negativity constraints

$x \geq 0, y \geq 0$

The mathematical model for this problem is:

Minimize $C = 1000x + 900y$

Subject to: $20x + 25y \geq 200$

$60x + 50y \geq 500$

$x \geq 0, y \geq 0$

Step (1): Graph the feasible region and find the corner points.

The feasible region $S$ is the solution set of the system of inequalities, and is indicated by the shading in the graph shown below.

The corner points are (0, 10), (5, 4), and (10, 0).

Since $S$ is unbounded and $a = 1000 > 0, b = 900 > 0, C$ has a minimum value by 4(b).

Copyright © 2011 Pearson Education, Inc. Publishing as Prentice Hall.

Step (2): Evaluate the objective function at each corner point.

The value of $C$ at each corner point is:

| Corner Point | $C = 1000x + 900y$ |
|---|---|
| $(0,10)$ | $C = 1000(0) + 900(10) = 9,000$ |
| $(5,4)$ | $C = 1000(5) + 900(4) = 8,600$ |
| $(10,0)$ | $C = 1000(10) + 900(0) = 10,000$ |

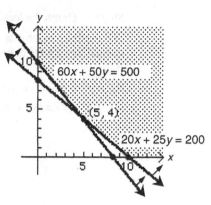

Step (3): Determine the optimal solution.
The minimum occurs when $x = 5$ and $y = 4$.

That is, Plant A should be operated five days and Plant B should be operated four days. The minimum cost is $C = \$8600$.

| Corner Point | $C = 600x + 900y$ |
|---|---|
| $(0,10)$ | $C = 600(0) + 900(10) = 9,000$ |
| $(5,4)$ | $C = 600(5) + 900(4) = 6,600$ |
| $(10,0)$ | $C = 600(10) + 900(0) = 6,000$ |

(B)

The minimum cost decreases to $6,000 per day when Plant $A$ is operated 10 days and Plant $B$ is operated 0 days.

| Corner Point | $C = 1000x + 800y$ |
|---|---|
| $(0,10)$ | $C = 1000(0) + 800(10) = 8,000$ |
| $(5,4)$ | $C = 1000(5) + 800(4) = 8,200$ |
| $(10,0)$ | $C = 1000(10) + 800(0) = 10,000$ |

(C)

The minimum cost decreases to $8,000 per day when Plant $A$ is operated 0 days and Plant $B$ is operated 10 days.

**35.** Construct the mathematical model:
   a.  Decision variables:
       Let $x$ = Number of buses
       $y$ = Number of vans
   b.  Relevant material in table form:

|  | Buses | Vans | Number to accommodate |
|---|---|---|---|
| Students | 40 | 8 | 400 |
| Chaperones | 3 | 1 | 36 |
| Rental cost | $1200/bus | $100/van | |

   c.  Objective function:
       Minimize the cost $C = 1200x + 100y$
   d.  Problem constraints:
       $40x + 8y \geq 400$  [Student constraint]
       $3x + y \leq 36$  [Chaperone constraint]
   e.  Non-negative constraints
       $x \geq 0, y \geq 0$
       The mathematical model for this problem is:
       Minimize $C = 1200x + 100y$
       Subject to: $40x + 8y \geq 400$
       $3x + y \leq 36$
       $x \geq 0, y \geq 0$

Copyright © 2011 Pearson Education, Inc. Publishing as Prentice Hall.

Step (1): Graph the feasible region and find the corner points.

The feasible region $S$ is the solution set of the system of inequalities, and is indicated by the shading in the graph at the right.

The corner points are $(10, 0)$, $(7, 15)$, and $(12, 0)$.

Since $S$ is bounded, $C$ has a minimum value by 4(a).

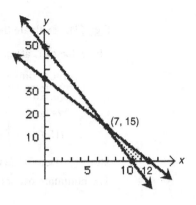

Step (2): Evaluate the objective function at each corner point.
The value of $C$ at each corner point is:

| Corner Point | $C = 1200x + 100y$ |
|---|---|
| $(10, 0)$ | $C = 1200(10) + 100(0) = 12,000$ |
| $(7, 15)$ | $C = 1200(7) + 100(15) = 9,900$ |
| $(12, 0)$ | $C = 1200(12) + 100(0) = 14,400$ |

Step (3): Determine the optimal solution.
The minimum occurs when $x = 7$ and $y = 15$. That is, the officers should rent 7 buses and 15 vans at the minimum cost of $9900.

**37.** (A)  Construct the mathematical model
   a.  Decision variables:
       Let $x$ = Amount invested in the CD
           $y$ = Amount invested in the mutual fund
   c.  Objective function:
       Maximize the return $P = 0.05x + 0.09y$
   d.  Problem constraints:
       $x + y \leq 60,000$   [Amount available constraint]
       $y \geq 10,000$   [Mutual fund constraint]
       $x \geq 2y$      [Investor constraint]
   e.  Non-negative constraints
       $x \geq 0,\ y \geq 0$

The mathematical model for this problem is
Maximize $P = 0.05x + 0.09y$
Subject to:  $x + y \leq 60,000$
                 $y \geq 10,000$
                 $x \geq 2y$
             $x, y \geq 0$

The feasible region $S$ is the solution set of the system of inequalities and is indicated by the shading in the graph.

The corner points are $(20,000, 10,000)$, $(40,000, 20,000)$ and $(50,000, 10,000)$.

Since $S$ is bounded, $P$ has a maximum value by 4a.

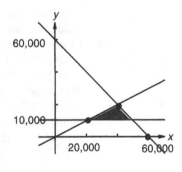

The value of $P$ at each corner point is given in the table below.

Copyright © 2011 Pearson Education, Inc. Publishing as Prentice Hall.

| Corner Point | $P = 0.05x + 0.09y$ |
|---|---|
| $(20{,}000, 10{,}000)$ | $P = 0.05(20{,}000) + 0.09(10{,}000) = 1900$ |
| $(40{,}000, 20{,}000)$ | $P = 0.05(40{,}000) + 0.09(20{,}000) = 3800$ |
| $(50{,}000, 10{,}000)$ | $P = 0.05(50{,}000) + 0.09(10{,}000) = 3400$ |

Thus, the maximum return is $3,800 when $40,000 is invested in the CD and $20,000 is invested in the mutual fund.

**39.** Construct the mathematical model

a.  Decision variables:

Let $x$ = Number of gallons produced by old process

$y$ = Number of gallons produced by new process

b.  Relevant material in table form:

| | Grams/gallon old process | Grams/gallon new process | Maximum allowed |
|---|---|---|---|
| Sulfur dioxide | 20 | 5 | 16,000 |
| Particulate | 40 | 20 | 30,000 |
| Profit | 60¢/gal | 20¢/gal | |

c.  Objective function:

Maximize the profit function $P = 60x + 20y$

d.  Problem constraints:

$20x + 5y \le 16{,}000$   [Sulfur dioxide constraint]

$40x + 20y \le 30{,}000$   [Particulate constraint]

e.  Non-negative constraints

$x \ge 0, y \ge 0$

The mathematical model for this problem is:

$$\text{Maximize } P = 60x + 20y$$
$$\text{Subject to: } 20x + 5y \le 16{,}000$$
$$40x + 20y \le 30{,}000$$
$$x \ge 0, y \ge 0$$

Step (1): Graph the feasible region and find the corner points.

The feasible region $S$ is the solution set of the given inequalities, and is indicated by the shading in the graph at the right.

The corner points are $(0, 0)$, $(0, 1500)$, and $(750, 0)$.

Since $S$ is bounded, $P$ has a maximum value by 4(a).

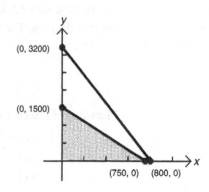

Copyright © 2011 Pearson Education, Inc. Publishing as Prentice Hall.

<u>Step (2)</u>: Evaluate the objective function at each corner point.

The value of $P$ at each corner point is:

| Corner Point | $P = 60x + 20y$ |
|---|---|
| $(0,0)$ | $P = 60(0) + 20(0) = 0 \text{ (cents)}$ |
| $(0,1500)$ | $P = 60(0) + 20(1500) = 30,000 \text{ (cents)}$ |
| $(750,0)$ | $P = 60(750) + 20(0) = 45,000 \text{ (cents)}$ |

<u>Step (3)</u>: The maximum profit is \$450 when 750 gallons are produced using the old process exclusively.

(B) The mathematical model for this problem is:

Maximize $P = 60x + 20y$

Subject to: $20x + 5y \le 11,500$

$40x + 20y \le 30,000$

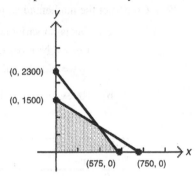

The feasible region $S$ for this problem is indicated by the shading in the graph at the right.

The corner points are $(0, 0)$, $(0, 1500)$, $(400, 700)$, and $(575, 0)$.

The value $P$ at each corner point is:

| Corner Point | $P = 60x + 20y$ |
|---|---|
| $(0,0)$ | $P = 60(0) + 20(0) = 0$ |
| $(0,1500)$ | $P = 60(0) + 20(1,500) = 30,000$ |
| $(400,700)$ | $P = 60(400) + 20(700) = 38,000$ |
| $(575,0)$ | $P = 60(575) + 20(0) = 34,500$ |

The maximum profit is \$380 when 400 gallons are produced using the old process and 700 gallons using the new process.

(C) The mathematical model for this problem is:

Maximize $P = 60x + 20y$

Subject to: $20x + 5y \le 7,200$

$40x + 20y \le 30,000$

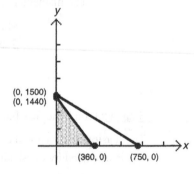

The feasible region $S$ for this problem is indicated by the shading in the graph at the right. The corner points are $(0, 0)$, $(0, 1440)$, and $(360, 0)$.

The value of $P$ at each corner point is:

| Corner Point | $P = 60x + 20y$ |
|---|---|
| $(0,0)$ | $P = 60(0) + 20(0) = 0$ |
| $(0,1440)$ | $P = 60(0) + 20(1,440) = 28,800$ |
| $(360,0)$ | $P = 60(360) + 20(0) = 21,600$ |

The maximum profit is \$288 when 1,440 gallons are produced by the new process exclusively.

Copyright © 2011 Pearson Education, Inc. Publishing as Prentice Hall.

**41.**  Construct the mathematical model

a.  Decision variables:

Let $x$ = Number of bags of Brand $A$

$y$ = Number of bags of Brand $B$

b.  Relevant material in table form:

|  | Brand A | Brand B |  |
|---|---|---|---|
| Amounts |  |  |  |
| Phosphoric acid | 4 | 4 | 1000 |
| Chlorine | 2 | 1 | 400 |
| Nitrogen | 8 lbs. | 3 lbs. |  |

c.  Objective function:

Maximize the amount of nitrogen $N = 8x + 3y$

d.  Problem constraints:

$4x + 4y \geq 1,000$  [Phosphoric acid constraint]

$2x + y \leq 400$  [Chlorine constraint]

e.  Non-negative constraints

$x \geq 0, y \geq 0$

(A) The mathematical model for this problem is:

Maximize $N = 8x + 3y$

Subject to: $4x + 4y \geq 1000$

$2x + y \leq 400$

$x \geq 0, y \geq 0$

The feasible region $S$ is the solution set of the system of inequalities, and is indicated by the shading in the graph at the right. The corner points are $(0, 250)$, $(0, 400)$, and $(150, 100)$.

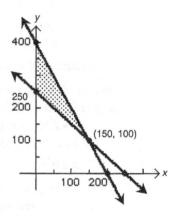

Since $S$ is bounded, $N$ has a maximum value by <u>4</u>(a).

The value of $N$ at each corner point is given in the table below:

| Corner Point | $N = 8x + 3y$ |
|---|---|
| $(0, 250)$ | $N = 8(0) + 3(250) = 750$ |
| $(150, 100)$ | $N = 8(150) + 3(100) = 1500$ |
| $(0, 400)$ | $N = 8(0) + 3(400) = 1200$ |

Thus, the maximum occurs when $x = 150$ and $y = 100$. That is, the grower should use 150 bags of Brand A and 100 bags of Brand B. The maximum number of pounds of nitrogen is 1500.

(B)  The mathematical model for this problem is:

Minimize $N = 8x + 3y$

Subject to: $4x + 4y \geq 1000$

$2x + y \leq 400$

$x \geq 0, y \geq 0$

The feasible region $S$ and the corner points are the same as in part (A). Thus, the minimum occurs when $x = 0$ and $y = 250$. That is, the grower should use 0 bags of Brand A and 250 bags of Brand B. The minimum number of pounds of nitrogen is 750.

**43.**  Construct the mathematical model

Copyright © 2011 Pearson Education, Inc. Publishing as Prentice Hall.

a.  Decision variables:

Let $x$ = Number of cubic yards of mix $A$

$y$ = Number of cubic yards of mix $B$

b.  Relevant material in table form:

|  | Amount per Cubic Yard (in pounds) | | Minimum monthly requirement |
|---|---|---|---|
|  | Mix A | Mix B |  |
| Phosphoric acid | 20 | 10 | 460 |
| Nitrogen | 30 | 30 | 960 |
| Potash | 5 | 10 | 220 |
| Cost/cubic yd. | $30 | $35 |  |

c.  Objective function:

Minimize the cost $C = 30x + 35y$

d.  Problem constraints:

$20x + 10y \geq 460$   [Phosphoric acid constraint]

$30x + 30y \geq 960$   [Nitrogen constraint]

$5x + 10y \geq 220$   [Potash constraint]

e.  Non-negative constraints

$x \geq 0, y \geq 0$

The mathematical model for this problem is:

Minimize $C = 30x + 35y$

Subject to: $20x + 10y \geq 460$

$30x + 30y \geq 960$

$5x + 10y \geq 220$

$x \geq 0, y \geq 0$

The feasible region $S$ is the solution set of the given inequalities and is indicated by the shading in the graph at the right.

The corner points are (0, 46), (14, 18), (20, 12), and (44, 0).

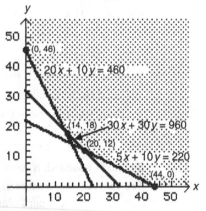

Since $S$ is unbounded and $a = 30 > 0$, $b = 35 > 0$, $C$ has a minimum value by 4(b).

The value of $C$ at each corner point is:

| Corner Point | $C = 30x + 35y$ |
|---|---|
| (0, 46) | $C = 30(0) + 35(46) = 1610$ |
| (14, 18) | $C = 30(14) + 35(18) = 1050$ |
| (20, 12) | $C = 30(20) + 35(12) = 1020$ |
| (44, 0) | $C = 30(44) + 35(0) = 1320$ |

Thus, the minimum occurs when the amount of mix A used is 20 cubic yards and the amount of mix B used is 12 cubic yards. The minimum cost is $C = \$1020$.

Copyright © 2011 Pearson Education, Inc. Publishing as Prentice Hall.

**45.** Construct the mathematical model
  a.   Decision variables:
     Let $x$ = Number of mice used
       $y$ = Number of rats used
  c.   Objective function:
     Maximize the number of mice and rats used $P = x + y$
  d.   Problem constraints:
       $10x + 20y \le 800$   [Box $A$ constraint]
       $20x + 10y \le 640$   [Box $B$ constraint]
  e.   Non-negative constraints
       $x \ge 0, y \ge 0$

The mathematical model for this problem is:

Maximize $P = x + y$
Subject to: $10x + 20y \le 800$
   $20x + 10y \le 640$
   $x \ge 0, y \ge 0$

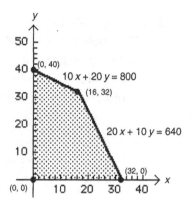

The feasible region $S$ is the solution set of the given inequalities, and is indicated by the shading in the graph at the right.

The corner points are (0, 0), (0, 40), (16, 32), and (32, 0).

Since $S$ is bounded, $P$ has a maximum value by 4(a).

The value of $P$ at each corner point is:

| Corner Point | $P = x + y$ |
|---|---|
| (0, 0) | $P = 0 + 0 = 0$ |
| (0, 40) | $P = 0 + 40 = 40$ |
| (16, 32) | $P = 16 + 32 = 48$ |
| (32, 0) | $P = 32 + 0 = 32$ |

Thus, the maximum occurs when the number of mice used is 16 and the number of rats used is 32. The maximum number of mice and rats that can be used is 48.

## CHAPTER 5 REVIEW

**1.**   $x > 2y - 3$ or $x - 2y > -3$
    Graph the line $x - 2y = -3$ as a dashed line. Substituting $x = 0$,
    $y = 0$ in the inequality produces a true statement, so (0, 0) is in the solution set.

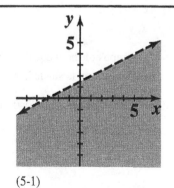

(5-1)

Copyright © 2011 Pearson Education, Inc.  Publishing as Prentice Hall.

2.   $3y - 5x \leq 30$

Graph the line $3y - 5x = 30$ as a solid line. Substituting $x = 0$, $y = 0$ into the inequality produces a true statement, so $(0, 0)$ is in the solution set.

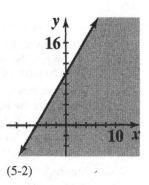

(5-2)

3.   $5x + 9y \leq 90$, $x \geq 0$, $y \geq 0$

The graph of $5x + 9y \leq 90$ is the half-plane below the line $5x + 9y = 90$, including the line. With $x \geq 0$, $y \geq 0$, the graph of the system is the shaded region. The solution region is bounded. The coordinates of the corner points are: $(0, 0)$, $(18, 0)$, $(0, 10)$.

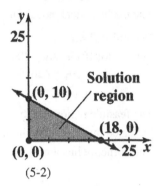

(5-2)

4.   $15x + 16y \geq 1{,}200$, $x \geq 0$, $y \geq 0$

The graph of $15x + 16y \geq 1200$ is the half-plane above the line $15x + 16y = 1{,}200$, including the line. With $x \geq 0$, $y \geq 0$, the graph of the system is the shaded region. The solution region is unbounded. The coordinates of the corner points are $(80, 0)$, $(0, 75)$.

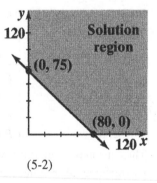

(5-2)

5.   $2x + y \leq 8$
     $3x + 9y \leq 27$
     $x, y \geq 0$

The graphs of the inequalities are shown at the right. The solution region is shaded; it is *bounded*.

The corner points are: $(0, 0)$, $(0, 3)$, $(3, 2)$, $(4, 0)$

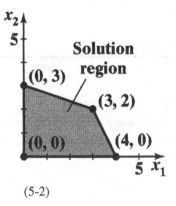

(5-2)

Copyright © 2011 Pearson Education, Inc. Publishing as Prentice Hall.

**6.** $3x + y \geq 9$

$2x + 4y \geq 16$

$x, y \geq 0$

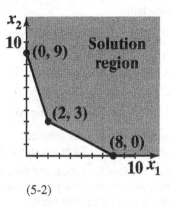

The graphs of the inequalities are shown at the right.  The solution region is shaded; it is *unbounded*.

The corner points are:  (0, 9), (2, 3), (8, 0)

(5-2)

**7.** The boundary line passes through (6, 0) and (0, –4).

slope: $m = \dfrac{0 - (-4)}{6 - 0} = \dfrac{2}{3}$

$y$ intercept: $b = -4$

Boundary line equation: $y = \dfrac{2}{3}x - 4$

$3y = 2x - 12$

$2x - 3y = 12$

Since (0, 0) is in the shaded region and the boundary line is solid, the graph is the graph of $2x - 3y \leq 12$.
(5-1)

**8.** The boundary line passes through (2, 0) and (0, 8).

slope: $m = \dfrac{0 - 8}{2 - 0} = -4$

$y$ intercept: $b = 8$

Boundary line equation: $y = -4x + 8$

$4x + y = 8$

Since (0, 0) is not in the shaded region and the boundary line is solid, the graph is the graph of $4x + y \geq 8$.     (5-1)

**9.** Step (1): Graph the feasible region and find the corner points. The feasible region $S$ is the solution set of the given inequalities. This region is indicated by the shading in the graph at the right.

The corner points are: (0, 0), (0, 4), (4, 2), and (5, 0).

Since $S$ is bounded it follows from 4(a) that $P$ has a maximum value.

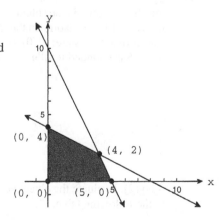

Copyright © 2011 Pearson Education, Inc.  Publishing as Prentice Hall.

Step (2): Evaluate the objective function at each corner point. The value of $P$ at each corner point is given in the following table.

| Corner Point | $P = 2x + 6y$ |
|---|---|
| $(0,0)$ | $P = 2(0) + 6(0) = 0$ |
| $(0,4)$ | $P = 2(0) + 6(4) = 24$ |
| $(4,2)$ | $P = 2(4) + 6(2) = 20$ |
| $(5,0)$ | $P = 2(5) + 6(0) = 10$ |

Step (3): Determine the optimal solution.
The maximum value of $P$ is 24 at $x = 0, y = 4$.      (5-3)

10.  Step (1): Graph the feasible region and find the corner points. The feasible region $S$ is the solution set of the given inequalities. This region is indicated by the shading in the graph at the right.

The corner points are: $(0, 20)$, $(9, 2)$, and $(15, 0)$.

Since $S$ is unbounded and the coefficients of $C$ are positive it follows from 4(b) that $C$ has a minimum value.

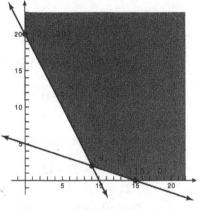

Step (2): Evaluate the objective function at each corner point. The value of $C$ at each corner point is given in the following table.

| Corner Point | $C = 5x + 2y$ |
|---|---|
| $(0, 20)$ | $C = 5(0) + 2(20) = 40$ |
| $(9, 2)$ | $C = 5(9) + 2(2) = 49$ |
| $(15, 0)$ | $C = 5(15) + 2(0) = 75$ |

Step (3): Determine the optimal solution.
The minimum value of $C$ is 40 at $x = 0, y = 20$.      (5-3)

11.  Step (1): Graph the feasible region and find the corner points. The feasible region $S$ is the solution set of the given inequalities. This region is indicated by the shading in the graph at the right.
The corner points are: $(0, 0)$, $(0, 6)$,  $(2, 5)$, $(3, 4)$ and $(5, 0)$.
Since $S$ is bounded it follows from 4(a) that $P$ has a maximum value.

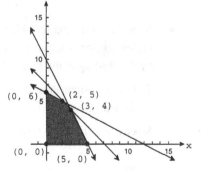

Step (2): Evaluate the objective function at each corner point. The value of $P$ at each corner point is given in the following table.

Copyright © 2011 Pearson Education, Inc. Publishing as Prentice Hall.

| Corner Point | $P = 3x + 4y$ |
|---|---|
| $(0,0)$ | $P = 3(0) + 4(0) = 0$ |
| $(0,6)$ | $P = 3(0) + 4(6) = 24$ |
| $(2,5)$ | $P = 3(2) + 4(5) = 26$ |
| $(3,4)$ | $P = 3(3) + 4(4) = 25$ |
| $(5,0)$ | $P = 3(5) + 4(0) = 15$ |

Step (3): Determine the optimal solution.
The maximum value of $P$ is 26 at $x = 2$, $y = 5$.      (5-3)

12.  Step (1): Graph the feasible region and find the corner points. The feasible region $S$ is the solution set of the given inequalities. This region is indicated by the shading in the graph at the right.

The corner points are: (3, 9), (5, 5), and (10, 0).

Since $S$ is unbounded and the coefficients of $C$ are positive, it follows from 4(b) that $C$ has a minimum value.

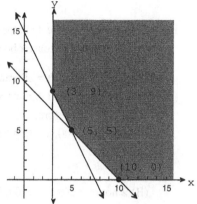

Step (2): Evaluate the objective function at each corner point. The value of $C$ at each corner point is given in the following table.

| Corner Point | $C = 8x + 3y$ |
|---|---|
| $(3,9)$ | $C = 8(3) + 3(9) = 51$ |
| $(5,5)$ | $C = 8(5) + 3(5) = 55$ |
| $(10,0)$ | $C = 8(10) + 3(0) = 80$ |

Step (3): Determine the optimal solution.
The minimum value of $C$ is 51 at $x = 3$, $y = 9$.      (5-3)

13.  Step (1): Graph the feasible region and find the corner points. The feasible region $S$ is the solution set of the given inequalities. This region is indicated by the shading in the graph at the right.

The corner points are: (0, 0), $\left(0, \frac{26}{3}\right)$, (8,6), (10, 2), and (10, 0).

Since $S$ is bounded it follows from 4(a) that $P$ has a maximum value.

Step (2): Evaluate the objective function at each corner point.

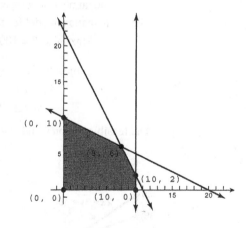

The value of $P$ at each corner point is given in the following table.

Copyright © 2011 Pearson Education, Inc.  Publishing as Prentice Hall.

| Corner Point | $P = 3x + 2y$ |
|---|---|
| $(0,0)$ | $P = 3(0) + 2(0) = 0$ |
| $\left(0, \frac{26}{3}\right)$ | $P = 3(0) + 2\left(\frac{26}{3}\right) = \frac{52}{3} = 17\frac{1}{3}$ |
| $(8,6)$ | $P = 3(8) + 2(6) = 36$ |
| $(10,2)$ | $P = 3(10) + 2(2) = 34$ |
| $(10,0)$ | $P = 3(10) + 2(0) = 30$ |

Step (3): Determine the optimal solution.  The maximum value of $P$ is 36 at $x = 8$, $y = 6$.     (5-3)

14.    Let $x$ = number of calculator boards

      $y$ = number of toaster boards

(A) 5 hours = 5(60) = 300 minutes

The oven is available for 300

minutes. Therefore,

    $4x + 3y \le 300$, $x \ge 0$, $y \ge 0$.

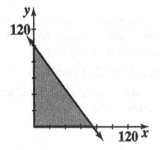

(B) 2 hours = 2(60) = 120 minutes

The wave machine is available for 120 minutes.

Therefore,

    $2x + y \le 120$, $x \ge 0$, $y \ge 0$

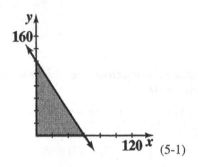

(5-1)

15.    (A)    Let $x$ = the number of regular sails

and $y$ = the number of competition sails.

The mathematical model for this problem is:

    Maximize $P = 100x + 200y$

    Subject to: $2x + 3y \le 150$

               $4x + 10y \le 380$

                  $x, y \ge 0$

The feasible region is indicated by the shading in the graph below.

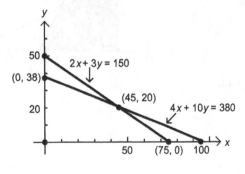

The corner points are (0, 0), (0, 38), (45, 20), (75, 0).

The value $P$ at each corner point is:

| Corner point | $P = 100x + 200y$ |
|---|---|
| $(0,0)$ | $P = 100(0) + 200(0) = 0$ |
| $(0,38)$ | $P = 100(0) + 200(38) = 7,600$ |
| $(45,20)$ | $P = 100(45) + 200(20) = 8,500$ |
| $(75,0)$ | $P = 100(75) + 200(0) = 7,500$ |

Copyright © 2011 Pearson Education, Inc.  Publishing as Prentice Hall.

Optimal solution: max $P$ = $8,500 when 45 regular and 20 competition sails are produced.

(B)    The mathematical model for this problem is:

Maximize $P = 100x + 260y$

Subject to: $2x + 3y \leq 150$

$4x + 10y \leq 380$

$x, y \geq 0$

The feasible region and the corner points are the same as in part (A). The value of $P$ at each corner point is:

| Corner point | $P = 100x + 260y$ |
|---|---|
| $(0,0)$ | $P = 100(0) + 260(0) = 0$ |
| $(0,38)$ | $P = 100(0) + 260(38) = 9,880$ |
| $(45,20)$ | $P = 100(45) + 260(20) = 9,700$ |
| $(75,0)$ | $P = 100(75) + 260(0) = 7,500$ |

The maximum profit increases to $9,880 when 38 competition and 0 regular sails are produced.

(C)    The mathematical model for this problem is:

Maximize $P = 100x + 140y$

Subject to: $2x + 3y \leq 150$

$4x + 10y \leq 380$

$x, y \geq 0$

The feasible region and the corner points are the same as in parts (A) and (B). The value of $P$ at each corner point is:

| Corner point | $P = 100x + 140y$ |
|---|---|
| $(0,0)$ | $P = 100(0) + 140(0) = 0$ |
| $(0,38)$ | $P = 100(0) + 140(38) = 5,320$ |
| $(45,20)$ | $P = 100(45) + 140(20) = 7,300$ |
| $(75,0)$ | $P = 100(75) + 140(0) = 7,500$ |

The maximum profit decreases to $7,500 when 0 competition and 75 regular sails are produced. (5-3)

16.    Let $x$ = number of grams of mix $A$

$y$ = number of grams of mix $B$

The constraints are:

vitamins: $2x + 5y \geq 850$

$2x + 4y \geq 800$

$4x + 5y \geq 1,150$

$x, y \geq 0$

The feasible region is indicated by the shading in the graph at the  right. The corner points are:

$(0, 230)$, $(100, 150)$, $(300, 50)$, $(425, 0)$.

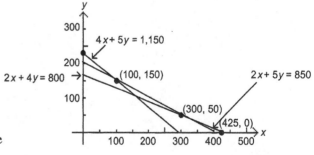

Copyright © 2011 Pearson Education, Inc.  Publishing as Prentice Hall.

(A)    The mathematical model for this problem is:

minimize $C = 0.04x + 0.09y$ subject to the constraints given above.

The value of $C$ at each corner point is:

| Corner point | $C = 0.04x + 0.09y$ |
|---|---|
| $(0, 230)$ | $C = 0.04(0) + 0.09(230) = 20.70$ |
| $(100, 150)$ | $C = 0.04(100) + 0.09(150) = 17.50$ |
| $(300, 50)$ | $C = 0.04(300) + 0.09(50) = 16.50$ |
| $(425, 0)$ | $C = 0.04(425) + 0.09(0) = 17.00$ |

The minimum cost is \$16.50 when 300 grams of mix $A$ and 50 grams of mix $B$ are used.

(B)    The mathematical model for this problem is:

minimize $C = 0.04x + 0.06y$ subject to the constraints given above.

The value of $C$ at each corner point is:

| Corner point | $C = 0.04x + 0.06y$ |
|---|---|
| $(0, 230)$ | $C = 0.04(0) + 0.06(230) = 13.80$ |
| $(100, 150)$ | $C = 0.04(100) + 0.06(150) = 13.00$ |
| $(300, 50)$ | $C = 0.04(300) + 0.06(50) = 15.00$ |
| $(425, 0)$ | $C = 0.04(425) + 0.06(0) = 17.00$ |

The minimum cost decreases to \$13.00 when 100 grams of mix $A$ and 150 grams of mix $B$ are used.

(C)    The mathematical model for this problem is:

minimize $C = 0.04x + 0.12y$ subject to the constraints given above.

The value of $C$ at each corner point is:

| Corner point | $C = 0.04x + 0.12y$ |
|---|---|
| $(0, 230)$ | $C = 0.04(0) + 0.12(230) = 27.60$ |
| $(100, 150)$ | $C = 0.04(100) + 0.12(150) = 22.00$ |
| $(300, 50)$ | $C = 0.04(300) + 0.12(50) = 18.00$ |
| $(425, 0)$ | $C = 0.04(425) + 0.12(0) = 17.00$ |

The minimum cost increases to \$17.00 when 425 grams of mix $A$ and 0 grams of mix $B$ are used. (5-3)

Copyright © 2011 Pearson Education, Inc. Publishing as Prentice Hall.

## 6   LINEAR PROGRAMMING: THE SIMPLEX METHOD

---

### EXERCISE 6-1

*Things to remember:*

1. STANDARD MAXIMIZATION PROBLEM IN STANDARD FORM

   A linear programming problem is said to be a STANDARD MAXIMIZATION PROBLEM IN STANDARD FORM if its mathematical model is:

   Maximize $P = c_1 x_1 + c_2 x_2 + \cdots + c_n x_n$

   Subject to problem constraints of the form:

   $a_1 x_1 + a_2 x_2 + \cdots + a_n x_n \leq b, \ b \geq 0$

   with nonnegative constraints:

   $x_1, x_2, \ldots, x_n \geq 0.$

   [Note: The coefficients of the objective function can be any real numbers.]

2. SLACK VARIABLES

   Given a linear programming problem. SLACK VARIABLES are nonnegative quantities that are introduced to convert problem constraint inequalities into equations.

3. BASIC VARIABLES AND NONBASIC VARIABLES; BASIC SOLUTIONS AND BASIC FEASIBLE SOLUTIONS

   Given a system of linear equations associated with a linear programming problem (such a system will always have more variables than equations).

   The variables are divided into two (mutually exclusive) groups, called BASIC VARIABLES and NONBASIC VARIABLES, as follows: Basic variables are selected arbitrarily with the one restriction that there are as many basic variables as there are equations. The remaining variables are called nonbasic variables.

   A solution found by setting the nonbasic variables equal to zero and solving for the basic variables is called a BASIC SOLUTION. If a basic solution has no negative values, it is a BASIC FEASIBLE SOLUTION.

4. FUNDAMENTAL THEOREM OF LINEAR PROGRAMMING

   If the optimal value of the objective function in a linear programming problem exists, then that value must occur at one (or more) of the basic feasible solutions.

---

1. (A) Since there are 2 problem constraints, 2 slack variables are introduced.

   (B) Since there are two equations (from the two problem constraints) and three decision variables, there are two basic variables and three nonbasic variables.

Copyright © 2011 Pearson Education, Inc. Publishing as Prentice Hall.

(C)   There will be two linear equations and two variables.

**3.**  (A)   There are 5 constraint equations; the number of equations is the same as the number of slack variables.

(B)   There are 4 decision variables since there are 9 variables altogether, and 5 of them are slack variables.

(C)   There are 5 basic variables and 4 nonbasic variables; the number of basic variables equals the number of equations.

(D)   Five linear equations with 5 variables.

**5.**

|     | Nonbasic | Basic | Feasible? |
|-----|----------|-------|-----------|
| (A) | $x_1, x_2$ | $s_1, s_2$ | Yes, all values are nonnegative. |
| (B) | $x_1, s_1$ | $x_2, s_2$ | Yes, all values are nonnegative. |
| (C) | $x_1, s_2$ | $x_2, s_1$ | No, $s_1 = -12 < 0.$ |
| (D) | $x_2, s_1$ | $x_1, s_2$ | No, $s_2 = -12 < 0.$ |
| (E) | $x_2, s_2$ | $x_1, s_1$ | Yes, all values are nonnegative. |
| (F) | $s_1, s_2$ | $x_1, x_2$ | Yes, all values are nonnegative. |

Evaluate $z$ at each basic feasible solution; the maximum value of
$z = 2x_1 + 3x_2$ is 24 at $x_1 = 0$, $x_2 = 8$; at $x_1 = 6$, $x_2 = 4$; at every point of the line segment joining (0, 8) and (6, 4).

**7.**

|     | $x_1$ | $x_2$ | $s_1$ | $s_2$ | Feasible? |
|-----|-------|-------|-------|-------|-----------|
| (A) | 0 | 0 | 50 | 40 | Yes, all values are nonnegative. |
| (B) | 0 | 50 | 0 | -60 | No, $s_2 = -60 < 0.$ |
| (C) | 0 | 20 | 30 | 0 | Yes, all values are nonnegative. |
| (D) | 25 | 0 | 0 | 15 | Yes, all values are nonnegative. |
| (E) | 40 | 0 | -30 | 0 | No, $s_1 = -30 < 0.$ |
| (F) | 20 | 10 | 0 | 0 | Yes, all values are nonnegative. |

Copyright © 2011 Pearson Education, Inc.  Publishing as Prentice Hall.

9.

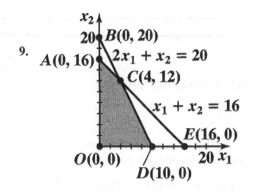

Introduce slack variables $s_1$ and $s_2$ to obtain the system of equations:

$$x_1 + x_2 + s_1 \qquad = 16$$
$$2x_1 + x_2 \qquad + s_2 = 20$$

| $x_1$ | $x_2$ | $s_1$ | $s_2$ | Intersection Point | Feasible? |
|---|---|---|---|---|---|
| 0 | 0 | 16 | 20 | $O$ | Yes |
| 0 | 16 | 0 | 4 | $A$ | Yes |
| 0 | 20 | –4 | 0 | $B$ | No, $s_1 = -4 < 0$ |
| 16 | 0 | 0 | –12 | $E$ | No, $s_2 = -12 < 0$ |
| 10 | 0 | 6 | 0 | $D$ | Yes |
| 4 | 12 | 0 | 0 | $C$ | Yes |

11.

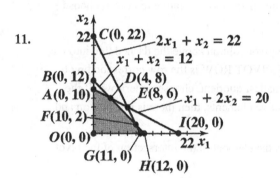

Introduce slack variables $s_1$, $s_2$, and $s_3$ to obtain the system of equations:

$$2x_1 + x_2 + s_1 \qquad\qquad = 22$$
$$x_1 + x_2 \qquad + s_2 \qquad = 12$$
$$x_1 + 2x_2 \qquad\qquad + s_3 = 20$$

| $x_1$ | $x_2$ | $s_1$ | $s_2$ | $s_3$ | Intersection Point | Feasible? |
|---|---|---|---|---|---|---|
| 0 | 0 | 22 | 12 | 20 | $O$ | Yes |
| 0 | 22 | 0 | –10 | –24 | $C$ | No |
| 0 | 12 | 10 | 0 | –4 | $B$ | No |
| 0 | 10 | 12 | 2 | 0 | $A$ | Yes |
| 11 | 0 | 0 | 1 | 9 | $G$ | Yes |
| 12 | 0 | –2 | 0 | 8 | $H$ | No |
| 20 | 0 | –18 | –8 | 0 | $I$ | No |
| 10 | 2 | 0 | 0 | 6 | $F$ | Yes |
| 8 | 6 | 0 | –2 | 0 | $E$ | No |
| 4 | 8 | 6 | 0 | 0 | $D$ | Yes |

Copyright © 2011 Pearson Education, Inc. Publishing as Prentice Hall.

---

---

*Things to remember:*

1. PROCEDURE: SELECTING BASIC AND NONBASIC VARIABLES FOR THE SIMPLEX PROCESS
   Given a simplex tableau,

   *Step 1.* NUMBERS OF VARIABLES: Determine the number of basic and the number of nonbasic variables. These numbers do not change during the simplex process.

   *Step 2.* SELECTING BASIC VARIABLES: A variable can be selected as a basic variable only if it corresponds to a column in the tableau that has exactly one nonzero element (usually 1) and the nonzero element in the column is not in the same row as the nonzero element in the column of another basic variable. This procedure always selects $P$ as a basic variable, since the $P$ column never changes during the simplex process.

   *Step 3.* SELECTING NONBASIC VARIABLES: After the basic variables are selected in Step 2, the remaining variables are selected as the nonbasic variables. The tableau columns under the nonbasic variables will usually contain more than one nonzero element.

2. PROCEDURE: SELECTING THE PIVOT ELEMENT

   *Step 1.* Locate the most negative indicator in the bottom row of the tableau to the left of the $P$ column (the negative number with the largest absolute value). The column containing this element is the PIVOT COLUMN. If there is a tie for the most negative indicator, choose either column.

   *Step 2.* Divide each POSITIVE element in the pivot column above the dashed line into the corresponding element in the last column. The PIVOT ROW is the row corresponding to the smallest quotient. If there is a tie for the smallest quotient, choose either row. If the pivot column above the dashed line has no positive elements, then there is no solution and we stop.

   *Step 3.* The PIVOT (or PIVOT ELEMENT) is the element in the intersection of the pivot column and pivot row.

   [Note: The pivot element is always positive and never appears in the bottom row.]

   [Remember: The entering variable is at the top of the pivot column and the exiting variable is at the left of the pivot row.]

3. PROCEDURE: PERFORMING THE PIVOT OPERATION

   A PIVOT OPERATION or PIVOTING consists of performing row operations as follows:

   *Step 1.* Multiply the pivot row by the reciprocal of the pivot element to transform the pivot element into a 1. (If the pivot element is already a 1, omit this step.)

   *Step 2.* Add multiples of the pivot row to other rows in the tableau to transform all other nonzero elements in the pivot column into 0's.

Copyright © 2011 Pearson Education, Inc. Publishing as Prentice Hall.

[Note: **In a pivot operation, you can never interchange two rows.**]

4.   SIMPLEX ALGORITHM FOR STANDARD MAXIMIZATION PROBLEMS
Problem constraints are of the ≤ form with nonnegative constants on the right hand side.
The coefficients of the objective function can be any real numbers.

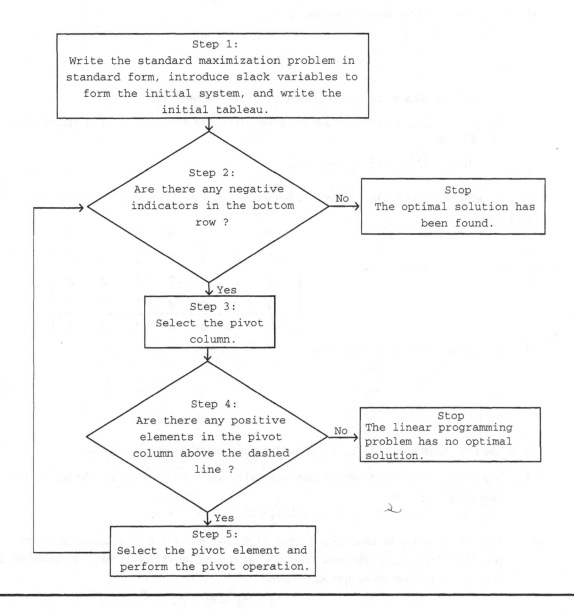

Copyright © 2011 Pearson Education, Inc. Publishing as Prentice Hall.

1. Given the simplex tableau:

$$\begin{array}{ccccc} x_1 & x_2 & s_1 & s_2 & P \\ \left[\begin{array}{ccccc|c} 2 & 1 & 0 & 3 & 0 & 12 \\ 3 & 0 & 1 & -2 & 0 & 15 \\ \hline -4 & 0 & 0 & 4 & 1 & 50 \end{array}\right] \end{array}$$

which corresponds to the system of equations:

(I) $\begin{cases} 2x_1 + x_2 + 3s_2 = 12 \\ 3x_1 + s_1 - 2s_2 = 15 \\ -4x_1 + 4s_2 + P = 50 \end{cases}$

(A) The basic variables are $x_2$, $s_1$, and $P$, and the nonbasic variables are $x_1$ and $s_2$.

(B) The corresponding basic feasible solution is found by setting the nonbasic variables equal to 0 in system (I). This yields:

$$x_1 = 0, \ x_2 = 12, \ s_1 = 15, \ s_2 = 0, P = 50$$

(C) An additional pivot is required, since the last row of the tableau has a negative indicator, the $-4$ in the first column.

3. Given the simplex tableau:

$$\begin{array}{ccccccc} x_1 & x_2 & x_3 & s_1 & s_2 & s_3 & P \\ \left[\begin{array}{ccccccc|c} -2 & 0 & 1 & 3 & 1 & 0 & 0 & 5 \\ 0 & 1 & 0 & -2 & 0 & 0 & 0 & 15 \\ -1 & 0 & 0 & 4 & 1 & 1 & 0 & 12 \\ \hline -4 & 0 & 0 & 2 & 4 & 0 & 1 & 45 \end{array}\right] \end{array}$$

which corresponds to the system of equations:

(I) $\begin{cases} -2x_1 + x_3 + 3s_1 + s_2 = 5 \\ x_2 - 2s_1 = 15 \\ -x_1 + 4s_1 + s_2 + s_3 = 12 \\ -4x_1 + 2s_1 + 4s_2 + P = 45 \end{cases}$

(A) The basic variables are $x_2$, $x_3$, $s_3$, $P$, and the nonbasic variables are $x_1$, $s_1$, $s_2$.

(B) The corresponding basic feasible solution is found by setting the nonbasic variables equal to 0 in system (I). This yields:

$$x_1 = 0, \ x_2 = 15, \ x_3 = 5, \ s_1 = 0, \ s_2 = 0, \ s_3 = 12, \ P = 45$$

(C) Since the last row of the tableau has a negative indicator, the $-4$ in the first column, an additional pivot should be required. However, since there are no positive elements in the pivot column (the first column), the problem has *no optimal solution*.

5. Given the simplex tableau:

$$\begin{array}{ccccc} x_1 & x_2 & s_1 & s_2 & P \\ \left[\begin{array}{ccccc|c} 1 & 4 & 1 & 0 & 0 & 4 \\ 3 & 5 & 0 & 1 & 0 & 24 \\ \hline -8 & -5 & 0 & 0 & 1 & 0 \end{array}\right] \end{array}$$

The most negative indicator is $-8$ in the first column. Thus, the first column is the pivot column. Now, $\dfrac{4}{1}$

$= 4$ and $\dfrac{24}{3} = 8$. Thus, the first row is the pivot row and the pivot element is the element in the first row, first column. These are indicated in the following tableau.

Copyright © 2011 Pearson Education, Inc. Publishing as Prentice Hall.

Enter

$$
\begin{array}{c}
\text{Exit } s_1 \\
s_2 \\
\\
P
\end{array}
\left[
\begin{array}{ccccc|c}
x_1 & x_2 & s_1 & s_2 & P & \\
① & 4 & 1 & 0 & 0 & 4 \\
3 & 5 & 0 & 1 & 0 & 24 \\
\hline
-8 & -5 & 0 & 0 & 1 & 0
\end{array}
\right]
\begin{array}{l}
\frac{4}{1} = 4 \text{ (minimum)} \\
\frac{24}{3} = 8
\end{array}
$$

$$
\left[
\begin{array}{ccccc|c}
① & 4 & 1 & 0 & 0 & 4 \\
3 & 5 & 0 & 1 & 0 & 24 \\
\hline
-8 & -5 & 0 & 0 & 1 & 0
\end{array}
\right]
\sim
\left[
\begin{array}{ccccc|c}
1 & 4 & 1 & 0 & 0 & 4 \\
0 & -7 & -3 & 1 & 0 & 12 \\
\hline
0 & 27 & 8 & 0 & 1 & 32
\end{array}
\right]
$$

$$(-3)R_1 + R_2 \to R_2$$
$$8R_1 + R_3 \to R_3$$

7.  Given the simplex tableau:

$$
\left[
\begin{array}{cccccc|c}
x_1 & x_2 & s_1 & s_2 & s_3 & P & \\
2 & 1 & 1 & 0 & 0 & 0 & 4 \\
3 & 0 & 1 & 1 & 0 & 0 & 8 \\
0 & 0 & 2 & 0 & 1 & 0 & 2 \\
\hline
-4 & 0 & -3 & 0 & 0 & 1 & 5
\end{array}
\right]
$$

The most negative indicator is –4. Thus, the first column is the pivot column. Now, $\frac{4}{2} = 2$, $\frac{8}{3} = 2\frac{2}{3}$.

Thus, the first row is the pivot row, and the pivot element is the element in the first row, first column. These are indicated in the tableau.

Enter

$$
\begin{array}{c}
\text{Exit } x_2 \\
s_2 \\
s_3 \\
P
\end{array}
\left[
\begin{array}{cccccc|c}
x_1 & x_2 & s_1 & s_2 & s_3 & P & \\
② & 1 & 1 & 0 & 0 & 0 & 4 \\
3 & 0 & 1 & 1 & 0 & 0 & 8 \\
0 & 0 & 2 & 0 & 1 & 0 & 2 \\
\hline
-4 & 0 & -3 & 0 & 0 & 1 & 5
\end{array}
\right]
\begin{array}{l}
\frac{4}{2} = 2 \text{ (minimum)} \\
\frac{8}{3} = 2\frac{2}{3}
\end{array}
$$

$$
\left[
\begin{array}{cccccc|c}
② & 1 & 1 & 0 & 0 & 0 & 4 \\
3 & 0 & 1 & 1 & 0 & 0 & 8 \\
0 & 0 & 2 & 0 & 1 & 0 & 2 \\
\hline
-4 & 0 & -3 & 0 & 0 & 1 & 5
\end{array}
\right]
\sim
\left[
\begin{array}{cccccc|c}
x_1 & x_2 & s_1 & s_2 & s_3 & P & \\
① & \frac{1}{2} & \frac{1}{2} & 0 & 0 & 0 & 2 \\
3 & 0 & 1 & 1 & 0 & 0 & 8 \\
0 & 0 & 2 & 0 & 1 & 0 & 2 \\
\hline
-4 & 0 & -3 & 0 & 0 & 1 & 5
\end{array}
\right]
$$

$$\frac{1}{2}R_1 \to R_1$$

$$(-3)R_1 + R_2 \to R_2, \quad 4R_1 + R_4 \to R_4$$

Copyright © 2011 Pearson Education, Inc. Publishing as Prentice Hall.

$$\sim \begin{bmatrix} 1 & \frac{1}{2} & \frac{1}{2} & 0 & 0 & 0 & 2 \\ 0 & -\frac{3}{2} & -\frac{1}{2} & 1 & 0 & 0 & 2 \\ 0 & 0 & 2 & 0 & 1 & 0 & 2 \\ \hdashline 0 & 2 & -1 & 0 & 0 & 1 & 13 \end{bmatrix}$$

9. (A) Introduce slack variables $s_1$ and $s_2$ to obtain:

Maximize $P = 15x_1 + 10x_2$

Subject to: $2x_1 + x_2 + s_1 \qquad = 10$
$\qquad\qquad x_1 + 3x_2 \qquad + s_2 = 10$
$\qquad\qquad x_1,\ x_2,\ s_1,\ s_2 \geq 0$

This system can be written in initial form:

$$2x_1 + x_2 + s_1 \qquad\qquad = 10$$
$$x_1 + 3x_2 \qquad + s_2 \qquad = 10$$
$$-15x_1 - 10x_2 \qquad\qquad + P = 0$$
$$x_1,\ x_2,\ s_1,\ s_2 \geq 0$$

(B) The simplex tableau for this problem is:

Enter

Exit $\quad s_1 \begin{bmatrix} \textcircled{2} & 1 & 1 & 0 & 0 & 10 \\ 1 & 3 & 0 & 1 & 0 & 10 \\ \hdashline -15 & -10 & 0 & 0 & 1 & 0 \end{bmatrix}$

with columns labeled $x_1 \; x_2 \; s_1 \; s_2 \; P$ and rows $s_1$, $s_2$, $P$.

$\dfrac{10}{2} = 5$ (minimum)

$\dfrac{10}{1} = 10$

Column 1 is the pivot column ($-15$ is the most negative indicator). Row 1 is the pivot row (5 is the smallest positive quotient). Thus, the pivot element is the circled 2.

(C) We use the simplex method as outlined above. The pivot elements are circled.

$\begin{array}{c} \\ s_1 \\ s_2 \\ P \end{array} \begin{bmatrix} \textcircled{2} & 1 & 1 & 0 & 0 & 10 \\ 1 & 3 & 0 & 1 & 0 & 10 \\ \hdashline -15 & -10 & 0 & 0 & 1 & 0 \end{bmatrix} \sim \begin{bmatrix} \textcircled{1} & \frac{1}{2} & \frac{1}{2} & 0 & 0 & 5 \\ 1 & 3 & 0 & 1 & 0 & 10 \\ \hdashline -15 & -10 & 0 & 0 & 1 & 0 \end{bmatrix} \sim$

with columns $x_1 \; x_2 \; s_1 \; s_2 \; P$.

$\frac{1}{2}R_1 \to R_1$

$(-1)R_1 + R_2 \to R_2$
$15R_1 + R_3 \to R_3$

Copyright © 2011 Pearson Education, Inc. Publishing as Prentice Hall.

$$\sim \begin{bmatrix} 1 & \frac{1}{2} & \frac{1}{2} & 0 & 0 & | & 5 \\ 0 & \boxed{\frac{5}{2}} & -\frac{1}{2} & 1 & 0 & | & 5 \\ \text{-----} & & & & & & \\ 0 & -\frac{5}{2} & \frac{15}{2} & 0 & 1 & | & 75 \end{bmatrix} \quad \sim \begin{bmatrix} 1 & \frac{1}{2} & \frac{1}{2} & 0 & 0 & | & 5 \\ 0 & \boxed{1} & -\frac{1}{5} & \frac{2}{5} & 0 & | & 2 \\ \text{-----} & & & & & & \\ 0 & -\frac{5}{2} & \frac{15}{2} & 0 & 1 & | & 75 \end{bmatrix}$$

$$\frac{2}{5}R_2 \rightarrow R_2 \qquad\qquad \left(-\frac{1}{2}\right)R_2 + R_1 \rightarrow R_1$$

$$\left(\frac{5}{2}\right)R_2 + R_3 \rightarrow R_3$$

$$\sim \begin{array}{c} \\ x_1 \\ x_2 \\ \\ P \end{array} \begin{array}{c} \begin{array}{ccccc} x_1 & x_2 & s_1 & s_2 & P \end{array} \\ \begin{bmatrix} 1 & 0 & \frac{3}{5} & -\frac{1}{5} & 0 & | & 4 \\ 0 & 1 & -\frac{1}{5} & \frac{2}{5} & 0 & | & 2 \\ \text{-----} & & & & & & \\ 0 & 0 & 7 & 1 & 1 & | & 80 \end{bmatrix} \end{array}$$

All elements in the last row are nonnegative. Thus, max $P = 80$ at $x_1 = 4$, $x_2 = 2$, $s_1 = 0$, $s_2 = 0$.

**11.** **(A)** Introduce slack variables $s_1$ and $s_2$ to obtain:

Maximize  $P = 30x_1 + x_2$
Subject to:  $2x_1 + x_2 + s_1 = 10$
$\qquad\qquad x_1 + 3x_2 + s_2 = 10$
$\qquad\qquad x_1,\ x_2,\ s_1,\ s_2 \geq 0$

This system can be written in the initial form:

$$2x_1 + x_2 + s_1 \qquad\qquad = 10$$
$$x_1 + 3x_2 \qquad + s_2 \qquad = 10$$
$$-30x_1 - x_2 \qquad\qquad + P = 0$$

**(B)** The simplex tableau for this problem is

$$\begin{array}{c} \\ \text{Exit } s_1 \\ s_2 \\ \\ P \end{array} \begin{array}{c} \qquad\quad \text{Enter} \\ \begin{array}{ccccc} x_1 & x_2 & s_1 & s_2 & P \end{array} \\ \begin{bmatrix} \boxed{2} & 1 & 1 & 0 & 0 & | & 10 \\ 1 & 3 & 0 & 1 & 0 & | & 10 \\ \text{-----} & & & & & & \\ -30 & -1 & 0 & 0 & 1 & | & 0 \end{bmatrix} \\ \quad\uparrow \\ \text{pivot} \\ \text{column} \end{array} \qquad \begin{array}{l} \dfrac{10}{2} = 5 \text{ (minimum)} \\[2mm] \dfrac{10}{1} = 10 \end{array}$$

$$\begin{array}{c} \\ s_1 \\ s_2 \\ \\ P \end{array} \begin{array}{c} \begin{array}{ccccc} x_1 & x_2 & s_1 & s_2 & P \end{array} \\ \begin{bmatrix} \boxed{2} & 1 & 1 & 0 & 0 & | & 10 \\ 1 & 3 & 0 & 1 & 0 & | & 10 \\ \text{-----} & & & & & & \\ -30 & -1 & 0 & 0 & 1 & | & 0 \end{bmatrix} \end{array} \quad \sim \begin{bmatrix} \boxed{1} & \frac{1}{2} & \frac{1}{2} & 0 & 0 & | & 5 \\ 1 & 3 & 0 & 1 & 0 & | & 10 \\ \text{-----} & & & & & & \\ -30 & -1 & 0 & 0 & 1 & | & 0 \end{bmatrix}$$

$$\frac{1}{2}R_1 \rightarrow R_1 \qquad\qquad (-1)R_1 + R_2 \rightarrow R_2$$

$$30R_1 + R_3 \rightarrow R_3$$

Copyright © 2011 Pearson Education, Inc. Publishing as Prentice Hall.

$$
\begin{array}{c}
\begin{array}{ccccc} x_1 & x_2 & s_1 & s_2 & P \end{array} \\
\sim
\begin{array}{c} x_1 \\ s_2 \\ P \end{array}
\left[
\begin{array}{ccccc|c}
1 & \frac{1}{2} & \frac{1}{2} & 0 & 0 & 5 \\
0 & \frac{5}{2} & -\frac{1}{2} & 1 & 0 & 5 \\
\hline
0 & 14 & 15 & 0 & 1 & 150
\end{array}
\right]
\end{array}
$$

All the elements in the last row are nonnegative. Thus, max $P = 150$ at $x_1 = 5$, $x_2 = 0$, $s_1 = 0$, $s_2 = 5$.

13.   The simplex tableau for this problem is:

$$
\begin{array}{ccccccc}
& & & \text{Enter} & & & \\
& x_1 & x_2 & s_1 & s_2 & s_3 & P
\end{array}
$$

$$
\begin{array}{c}
s_1 \\ s_2 \\ s_3 \\ P
\end{array}
\left[
\begin{array}{cccccc|c}
2 & 1 & 1 & 0 & 0 & 0 & 10 \\
1 & 1 & 0 & 1 & 0 & 0 & 7 \\
1 & ② & 0 & 0 & 1 & 0 & 12 \\
\hline
-30 & -40 & 0 & 0 & 0 & 1 & 0
\end{array}
\right]
\quad
\begin{array}{l}
10 \\
7 \\
\frac{12}{2} = 6 \text{ (minimum)}
\end{array}
$$

[Note: The pivot elements have been circled.]

pivot row → $s_3$ Exit

pivot column    $\frac{1}{2}R_3 \to R_3$

$$
\sim
\left[
\begin{array}{cccccc|c}
2 & 1 & 1 & 0 & 0 & 0 & 10 \\
1 & 1 & 0 & 1 & 0 & 0 & 7 \\
\frac{1}{2} & ① & 0 & 0 & \frac{1}{2} & 0 & 6 \\
\hline
-30 & -40 & 0 & 0 & 0 & 1 & 0
\end{array}
\right]
$$

$$(-1)R_3 + R_1 \to R_1, \quad (-1)R_3 + R_2 \to R_2, \quad \text{and } 40R_3 + R_4 \to R_4$$

$$
\begin{array}{c}
\text{pivot} \to \\ \text{row} \quad \sim
\end{array}
\left[
\begin{array}{cccccc|c}
\frac{3}{2} & 0 & 1 & 0 & -\frac{1}{2} & 0 & 4 \\
① & 0 & 0 & 1 & -\frac{1}{2} & 0 & 1 \\
\frac{1}{2} & 1 & 0 & 0 & \frac{1}{2} & 0 & 6 \\
\hline
-10 & 0 & 0 & 0 & 20 & 1 & 240
\end{array}
\right]
\quad
\begin{array}{l}
\frac{4}{3/2} = \frac{8}{3} \\
\frac{1}{1/2} = 2 \text{ (minimum)} \\
\frac{6}{1/2} = 12
\end{array}
$$

pivot column    $2R_2 \to R_2$

$$
\sim
\left[
\begin{array}{cccccc|c}
\frac{3}{2} & 0 & 1 & 0 & -\frac{1}{2} & 0 & 4 \\
① & 0 & 0 & 2 & -1 & 0 & 2 \\
\frac{1}{2} & 1 & 0 & 0 & \frac{1}{2} & 0 & 6 \\
\hline
-10 & 0 & 0 & 0 & 20 & 1 & 240
\end{array}
\right]
$$

$$\left(-\frac{3}{2}\right)R_2 + R_1 \to R_1, \quad \left(-\frac{1}{2}\right)R_2 + R_3 \to R_3, \quad \text{and } 10R_2 + R_4 \to R_4$$

Copyright © 2011 Pearson Education, Inc.  Publishing as Prentice Hall.

$$\begin{array}{c} \\ s_1 \\ \sim \ x_1 \\ \sim \ x_2 \\ \\ \end{array}
\begin{array}{cccccc}
x_1 & x_2 & s_1 & s_2 & s_3 & P \\
\end{array}
\left[\begin{array}{cccccc|c}
0 & 0 & 1 & -3 & 1 & 0 & 1 \\
1 & 0 & 0 & 2 & -1 & 0 & 2 \\
0 & 1 & 0 & -1 & 1 & 0 & 5 \\
\hline
0 & 0 & 0 & 20 & 10 & 1 & 260 \\
\end{array}\right]$$

Optimal solution: max $P = 260$ at $x_1 = 2$, $x_2 = 5$, $s_1 = 1$, $s_2 = 0$, $s_3 = 0$.

**15.** The simplex tableau for this problem is:

Enter

Exit

pivot row → $s_1$

$$\begin{array}{c}
 \\
 \\
\to s_1 \\
s_2 \\
s_3 \\
P \\
\end{array}
\begin{array}{cccccc}
x_1 & x_2 & s_1 & s_2 & s_3 & P \\
\end{array}
\left[\begin{array}{cccccc|c}
-2 & ① & 1 & 0 & 0 & 0 & 2 \\
-1 & 1 & 0 & 1 & 0 & 0 & 5 \\
0 & 1 & 0 & 0 & 1 & 0 & 6 \\
\hline
-2 & -3 & 0 & 0 & 0 & 1 & 0 \\
\end{array}\right]
\begin{array}{l}
\frac{2}{1} = 2 \ (\text{minimum}) \\
\frac{5}{1} = 5 \\
\frac{6}{1} = 6 \\
\end{array}$$

↑ pivot column

$(-1)R_1 + R_2 \to R_2$, $(-1)R_1 + R_3 \to R_3$, and $3R_1 + R_4 \to R_4$

pivot row →

$$\sim \left[\begin{array}{cccccc|c}
-2 & 1 & 1 & 0 & 0 & 0 & 2 \\
1 & 0 & -1 & 1 & 0 & 0 & 3 \\
② & 0 & -1 & 0 & 1 & 0 & 4 \\
\hline
-8 & 0 & 3 & 0 & 0 & 1 & 6 \\
\end{array}\right]
\begin{array}{l}
\frac{3}{1} = 3 \\
\frac{4}{2} = 2 \ (\text{minimum}) \\
\end{array}$$

↑ pivot column

$\frac{1}{2}R_3 \to R_3$

$$\sim \left[\begin{array}{cccccc|c}
-2 & 1 & 1 & 0 & 0 & 0 & 2 \\
1 & 0 & -1 & 1 & 0 & 0 & 3 \\
① & 0 & -\frac{1}{2} & 0 & \frac{1}{2} & 0 & 2 \\
\hline
-8 & 0 & 3 & 0 & 0 & 1 & 6 \\
\end{array}\right]$$

$2R_3 + R_1 \to R_1$, $(-1)R_3 + R_2 \to R_2$, and $8R_3 + R_4 \to R_4$

$$\begin{array}{cccccc}
x_1 & x_2 & s_1 & s_2 & s_3 & P \\
\end{array}$$

$$\sim \left[\begin{array}{cccccc|c}
0 & 1 & 0 & 0 & 1 & 0 & 6 \\
0 & 0 & -\frac{1}{2} & 1 & -\frac{1}{2} & 0 & 1 \\
1 & 0 & -\frac{1}{2} & 0 & \frac{1}{2} & 0 & 2 \\
\hline
0 & 0 & -1 & 0 & 4 & 1 & 22 \\
\end{array}\right]$$

↑ pivot column

Since there are no positive elements in the pivot column (above the dashed line), we conclude that there is no optimal solution.

Copyright © 2011 Pearson Education, Inc.  Publishing as Prentice Hall.

**17.** The simplex tableau for this problem is:

$$
\begin{array}{c}
\text{pivot} \to \\
\text{row}
\end{array}
\begin{array}{c}
s_1 \\ s_2 \\ s_3 \\ \hline P
\end{array}
\begin{bmatrix}
\begin{array}{ccccccc}
x_1 & x_2 & s_1 & s_2 & s_3 & P & \\
-1 & \textcircled{1} & 1 & 0 & 0 & 0 & 2 \\
-1 & 3 & 0 & 1 & 0 & 0 & 12 \\
1 & -4 & 0 & 0 & 1 & 0 & 4 \\
\hline
1 & -2 & 0 & 0 & 0 & 1 & 0
\end{array}
\end{bmatrix}
\begin{array}{l}
\frac{2}{1} = 2 \text{ (minimum)} \\[4pt]
\frac{12}{3} = 4 \\
\end{array}
$$

pivot column

$(-3)R_1 + R_2 \to R_2, \; 4R_1 + R_3 \to R_3,$ and $2R_1 + R_4 \to R_4$

$$
\begin{array}{c}
\text{pivot} \to \\
\text{row}
\end{array}
\sim
\begin{bmatrix}
-1 & 1 & 1 & 0 & 0 & 0 & 2 \\
\textcircled{2} & 0 & -3 & 1 & 0 & 0 & 6 \\
-3 & 0 & 4 & 0 & 1 & 0 & 12 \\
\hline
-1 & 0 & 2 & 0 & 0 & 1 & 4
\end{bmatrix}
\begin{array}{l}
\frac{6}{2} = 3 \leftarrow \text{pivot row} \\[6pt]
[\underline{\text{Note}}\text{: We only use the} \\
\textit{positive} \text{ elements above} \\
\text{the dashed line in the} \\
\text{pivot column.}]
\end{array}
$$

pivot column    $\frac{1}{2}R_2 \to R_2$

$$
\sim
\begin{bmatrix}
-1 & 1 & 1 & 0 & 0 & 0 & 2 \\
\textcircled{1} & 0 & -\frac{3}{2} & \frac{1}{2} & 0 & 0 & 3 \\
-3 & 0 & 4 & 0 & 1 & 0 & 12 \\
\hline
-1 & 0 & 2 & 0 & 0 & 1 & 4
\end{bmatrix}
$$

$R_2 + R_1 \to R_1, \; 3R_2 + R_3 \to R_3,$
and $R_2 + R_4 \to R_4$

$$
\sim
\begin{array}{c}
x_2 \\ x_1 \\ s_3 \\ \hline {}
\end{array}
\begin{bmatrix}
\begin{array}{ccccccc}
x_1 & x_2 & s_1 & s_2 & s_3 & P & \\
0 & 1 & -\frac{1}{2} & \frac{1}{2} & 0 & 0 & 5 \\
1 & 0 & -\frac{3}{2} & \frac{1}{2} & 0 & 0 & 3 \\
0 & 0 & -\frac{1}{2} & \frac{3}{2} & 1 & 0 & 21 \\
\hline
0 & 0 & \frac{1}{2} & \frac{1}{2} & 0 & 1 & 7
\end{array}
\end{bmatrix}
$$

Optimal solution: max $P = 7$ at $x_1 = 3$, $x_2 = 5$, $s_1 = 0$, $s_2 = 0$, $s_3 = 21$.

**19.** The simplex tableau for this problem is:

$$
\begin{array}{c}
\text{pivot} \to \\
\text{row}
\end{array}
\begin{array}{c}
s_1 \\ s_2 \\ \hline P
\end{array}
\begin{bmatrix}
\begin{array}{cccccc}
x_1 & x_2 & x_3 & s_1 & s_2 & P \\
\textcircled{1} & 1 & -1 & 1 & 0 & 0 & 10 \\
2 & 4 & 3 & 0 & 1 & 0 & 30 \\
\hline
-5 & -2 & 1 & 0 & 0 & 1 & 0
\end{array}
\end{bmatrix}
\begin{array}{l}
\frac{10}{1} = 10 \text{ (minimum)} \\[4pt]
\frac{30}{2} = 15 \\
\end{array}
$$

pivot column    $(-2)R_1 + R_2 \to R_2, \; 5R_1 + R_3 \to R_3$

$$
\sim
\begin{bmatrix}
1 & 1 & -1 & 1 & 0 & 0 & 10 \\
0 & 2 & \textcircled{5} & -2 & 1 & 0 & 10 \\
0 & 3 & -4 & 5 & 0 & 1 & 50
\end{bmatrix}
\qquad
\sim
\begin{bmatrix}
1 & 1 & -1 & 1 & 0 & 0 & 10 \\
0 & \frac{2}{5} & \textcircled{1} & -\frac{2}{5} & \frac{1}{5} & 0 & 2 \\
0 & 3 & -4 & 5 & 0 & 1 & 50
\end{bmatrix}
$$

$\frac{1}{5}R_2 \to R_2$

$R_2 + R_1 \to R_1, \; 4R_2 + R_3 \to R_3$

Copyright © 2011 Pearson Education, Inc. Publishing as Prentice Hall.

$$\sim \begin{array}{c} \\ x_1 \\ x_3 \\ \\ P \end{array} \left[ \begin{array}{ccccccc|c} x_1 & x_2 & x_3 & s_1 & s_2 & P & \\ 1 & \frac{7}{5} & 0 & \frac{3}{5} & \frac{1}{5} & 0 & 12 \\ 0 & \frac{2}{5} & 1 & -\frac{2}{5} & \frac{1}{5} & 0 & 2 \\ \hline 0 & \frac{23}{5} & 0 & \frac{17}{5} & \frac{4}{5} & 1 & 58 \end{array} \right]$$

Optimal solution: max $P = 58$
$x_1 = 12$, $x_2 = 0$, $x_3 = 2$, $s_1 = s_2 = 0$.

**21.** The simplex tableau for this problem is:

$$\begin{array}{c} \\ s_1 \\ \text{pivot} \rightarrow s_2 \\ \text{row} \\ P \end{array} \left[ \begin{array}{cccccc|c} x_1 & x_2 & x_3 & s_1 & s_2 & P & \\ 1 & 0 & 1 & 1 & 0 & 0 & 4 \\ 0 & 1 & \textcircled{1} & 0 & 1 & 0 & 3 \\ \hline -2 & -3 & -4 & 0 & 0 & 1 & 0 \end{array} \right] \begin{array}{l} \frac{4}{1} = 4 \\ \frac{3}{1} = 3 \text{ (minimum)} \end{array}$$

$$\underset{\text{column}}{\text{pivot}} \quad (-1)R_2 + R_1 \rightarrow R_1, \quad 4R_2 + R_3 \rightarrow R_3$$

$$\sim \left[ \begin{array}{cccccc|c} \textcircled{1} & -1 & 0 & 1 & -1 & 0 & 1 \\ 0 & 1 & 1 & 0 & 1 & 0 & 3 \\ \hline -2 & 1 & 0 & 0 & 4 & 1 & 12 \end{array} \right] \sim \left[ \begin{array}{cccccc|c} 1 & -1 & 0 & 1 & -1 & 0 & 1 \\ 0 & \textcircled{1} & 1 & 0 & 1 & 0 & 3 \\ \hline 0 & -1 & 0 & 2 & 2 & 1 & 14 \end{array} \right]$$

$$2R_1 + R_3 \rightarrow R_3 \qquad\qquad R_2 + R_1 \rightarrow R_1 \text{ and } R_2 + R_3 \rightarrow R_3$$

$$\sim \left[ \begin{array}{cccccc|c} x_1 & x_2 & x_3 & s_1 & s_2 & P & \\ 1 & 0 & 1 & 1 & 0 & 0 & 4 \\ 0 & 1 & 1 & 0 & 1 & 0 & 3 \\ \hline 0 & 0 & 1 & 2 & 3 & 1 & 17 \end{array} \right]$$

Optimal solution: max $P = 17$ at
$x_1 = 4$, $x_2 = 3$, $x_3 = 0$, $s_1 = 0$,
$s_2 = 0$.

**23.** The simplex tableau for this problem is:

$$\begin{array}{c} \\ s_1 \\ \text{pivot} \rightarrow s_2 \\ \text{row} \quad s_3 \\ \\ \end{array} \left[ \begin{array}{ccccccc|c} x_1 & x_2 & x_3 & s_1 & s_2 & s_3 & P & \\ 3 & 2 & 5 & 1 & 0 & 0 & 0 & 23 \\ \textcircled{2} & 1 & 1 & 0 & 1 & 0 & 0 & 8 \\ 1 & 1 & 2 & 0 & 0 & 1 & 0 & 7 \\ \hline -4 & -3 & -2 & 0 & 0 & 0 & 1 & 0 \end{array} \right] \begin{array}{l} \frac{23}{3} = 7\frac{2}{3} \\ \frac{8}{2} = 4 \text{ (minimum)} \\ \frac{7}{1} = 7 \end{array}$$

$$\underset{\text{column}}{\text{pivot}} \quad \frac{1}{2}R_2 \rightarrow R_2$$

$$\sim \begin{bmatrix} 3 & 2 & 5 & 1 & 0 & 0 & 0 & | & 23 \\ \textcircled{1} & \frac{1}{2} & \frac{1}{2} & 0 & \frac{1}{2} & 0 & 0 & | & 4 \\ 1 & 1 & 2 & 0 & 0 & 1 & 0 & | & 7 \\ \hline -4 & -3 & -2 & 0 & 0 & 0 & 1 & | & 0 \end{bmatrix} \sim \begin{bmatrix} 0 & \frac{1}{2} & \frac{7}{2} & 1 & -\frac{3}{2} & 0 & 0 & | & 11 \\ 1 & \frac{1}{2} & \frac{1}{2} & 0 & \frac{1}{2} & 0 & 0 & | & 4 \\ 0 & \textcircled{$\frac{1}{2}$} & \frac{3}{2} & 0 & -\frac{1}{2} & 1 & 0 & | & 3 \\ \hline 0 & -1 & 0 & 0 & 2 & 0 & 1 & | & 16 \end{bmatrix}$$

$(-3)R_2 + R_1 \rightarrow R_1, \ (-1) \ R_2 + R_3 \rightarrow R_3, \qquad 2R_3 \rightarrow R_3$

$4R_2 + R_4 \rightarrow R_4$

$$\sim \begin{bmatrix} 0 & \frac{1}{2} & \frac{7}{2} & 1 & -\frac{3}{2} & 0 & 0 & | & 11 \\ 1 & \frac{1}{2} & \frac{1}{2} & 0 & \frac{1}{2} & 0 & 0 & | & 4 \\ 0 & \textcircled{1} & 3 & 0 & -1 & 2 & 0 & | & 6 \\ \hline 0 & -1 & 0 & 0 & 2 & 0 & 1 & | & 16 \end{bmatrix} \sim$$

$\left(-\dfrac{1}{2}\right) R_3 + R_1 \rightarrow R_1, \ \left(-\dfrac{1}{2}\right) R_3 + R_2 \rightarrow R_2, \ \text{ and}$

$R_3 + R_4 \rightarrow R_4$

|       | $x_1$ | $x_2$ | $x_3$ | $s_1$ | $s_2$ | $s_3$ | $P$ |    |
|-------|-------|-------|-------|-------|-------|-------|-----|----|
| $s_1$ | 0 | 0 | 2 | 1 | -1 | -1 | 0 | 8 |
| $x_1$ | 1 | 0 | -1 | 0 | 1 | -1 | 0 | 1 |
| $x_2$ | 0 | 1 | 3 | 0 | -1 | 2 | 0 | 6 |
| $P$   | 0 | 0 | 3 | 0 | 1 | 2 | 1 | 22 |

Optimal solution: max $P = 22$ at $x_1 = 1$, $x_2 = 6$, $x_3 = 0$, $s_1 = 8$, $s_2 = 0$, $s_3 = 0$.

25.  Multiply the first problem constraint by $\dfrac{10}{6}$, the second by 100, and the third by 10 to clear the fractions.

Then, the simplex tableau for this problem is:

|       | $x_1$ | $x_2$ | $s_1$ | $s_2$ | $s_3$ | $P$ |       |
|-------|-------|-------|-------|-------|-------|-----|-------|
| $s_1$ | 1 | ② | 1 | 0 | 0 | 0 | 1,600 |
| $s_2$ | 3 | 4 | 0 | 1 | 0 | 0 | 3,600 |
| $s_3$ | 3 | 2 | 0 | 0 | 1 | 0 | 2,700 |
| $P$   | -20 | -30 | 0 | 0 | 0 | 1 | 0 |

$\dfrac{1,600}{2} = 800$

$\dfrac{3,600}{4} = 900$

$\dfrac{2,700}{2} = 1,350$

$\dfrac{1}{2} R_1 \rightarrow R_1$

Copyright © 2011 Pearson Education, Inc.  Publishing as Prentice Hall.

$$\sim \begin{bmatrix} \frac{1}{2} & ① & \frac{1}{2} & 0 & 0 & 0 & 800 \\ 3 & 4 & 0 & 1 & 0 & 0 & 3{,}600 \\ 3 & 2 & 0 & 0 & 1 & 0 & 2{,}700 \\ \hdashline -20 & -30 & 0 & 0 & 0 & 1 & 0 \end{bmatrix}$$

$(-4)R_1 + R_2 \rightarrow R_2,\ (-2)R_1 + R_3 \rightarrow R_3,$ and $30R_1 + R_4 \rightarrow R_4$

$$\sim \begin{bmatrix} \frac{1}{2} & 1 & \frac{1}{2} & 0 & 0 & 0 & 800 \\ ① & 0 & -2 & 1 & 0 & 0 & 400 \\ 2 & 0 & -1 & 0 & 1 & 0 & 1{,}100 \\ \hdashline -5 & 0 & 15 & 0 & 0 & 1 & 24{,}000 \end{bmatrix} \quad \begin{array}{l} \dfrac{800}{1/2} = 1{,}600 \\[2mm] \dfrac{400}{1} = 400 \\[2mm] \dfrac{1{,}100}{2} = 550 \end{array}$$

$\left(-\dfrac{1}{2}\right)R_2 + R_1 \rightarrow R_1,\ (-2)R_2 + R_3 \rightarrow R_3,$ and $5R_2 + R_4 \rightarrow R_4$

|       | $x_1$ | $x_2$ | $s_1$ | $s_2$ | $s_3$ | $P$ |        |
|-------|-------|-------|-------|-------|-------|-----|--------|
| $x_2$ | 0 | 1 | $\frac{3}{2}$ | $-\frac{1}{2}$ | 0 | 0 | 600 |
| $x_1$ | 1 | 0 | -2 | 1 | 0 | 0 | 400 |
| $s_3$ | 0 | 0 | 3 | -2 | 1 | 0 | 300 |
| $P$ | 0 | 0 | 5 | 5 | 0 | 1 | 26,000 |

Optimal solution: max $P = 26{,}000$ at $x_1 = 400$, $x_2 = 600$, $s_1 = 0$, $s_2 = 0$, $s_3 = 300$.

**27.** The simplex tableau for this problem is:

|       | $x_1$ | $x_2$ | $x_3$ | $s_1$ | $s_2$ | $s_3$ | $P$ |        |
|-------|-------|-------|-------|-------|-------|-------|-----|--------|
| $s_1$ | 2 | 2 | ⑧ | 1 | 0 | 0 | 0 | 600 |
| $s_2$ | 1 | 3 | 2 | 0 | 1 | 0 | 0 | 600 |
| $s_3$ | 3 | 2 | 1 | 0 | 0 | 1 | 0 | 400 |
| $P$ | -1 | -2 | -3 | 0 | 0 | 0 | 1 | 0 |

$\dfrac{600}{8} = 75$

$\dfrac{600}{2} = 300$

$\dfrac{400}{1} = 400$

$\dfrac{1}{8}R_1 \rightarrow R_1$

$$\sim \begin{bmatrix} \frac{1}{4} & \frac{1}{4} & ① & \frac{1}{8} & 0 & 0 & 0 & 75 \\ 1 & 3 & 2 & 0 & 1 & 0 & 0 & 600 \\ 3 & 2 & 1 & 0 & 0 & 1 & 0 & 400 \\ \hdashline -1 & -2 & -3 & 0 & 0 & 0 & 1 & 0 \end{bmatrix}$$

$(-2)R_1 + R_2 \rightarrow R_2,\ (-1)R_1 + R_3 \rightarrow R_3,$ and $3R_1 + R_4 \rightarrow R_4$

Copyright © 2011 Pearson Education, Inc. Publishing as Prentice Hall.

$$\sim \begin{bmatrix} \frac{1}{4} & \frac{1}{4} & 1 & \frac{1}{8} & 0 & 0 & 0 & 75 \\ \frac{1}{2} & \frac{5}{2} & 0 & -\frac{1}{4} & 1 & 0 & 0 & 450 \\ \frac{11}{4} & \frac{7}{4} & 0 & -\frac{1}{8} & 0 & 1 & 0 & 325 \\ \hdashline -\frac{1}{4} & -\frac{5}{4} & 0 & \frac{3}{8} & 0 & 0 & 1 & 225 \end{bmatrix} \begin{matrix} \frac{75}{1/4} = 300 \\ \frac{450}{5/2} = 180 \\ \frac{325}{7/4} = 185.71 \\ \\ \end{matrix}$$

$$\frac{2}{5} R_2 \to R_2$$

$$\sim \begin{bmatrix} \frac{1}{4} & \frac{1}{4} & 1 & \frac{1}{8} & 0 & 0 & 0 & 75 \\ \frac{1}{5} & 1 & 0 & -\frac{1}{10} & \frac{2}{5} & 0 & 0 & 180 \\ \frac{11}{4} & \frac{7}{4} & 0 & -\frac{1}{8} & 0 & 1 & 0 & 325 \\ \hdashline -\frac{1}{4} & -\frac{5}{4} & 0 & \frac{3}{8} & 0 & 0 & 1 & 225 \end{bmatrix}$$

$$\left(-\frac{1}{4}\right) R_2 + R_1 \to R_1, \quad \left(-\frac{7}{4}\right) R_2 + R_3 \to R_3, \quad \text{and} \quad \frac{5}{4} R_2 + R_4 \to R_4$$

$$\sim \begin{array}{c} x_3 \\ x_2 \\ s_3 \\ P \end{array} \begin{bmatrix} x_1 & x_2 & x_3 & s_1 & s_2 & s_3 & P & \\ \frac{1}{5} & 0 & 1 & \frac{3}{20} & -\frac{1}{10} & 0 & 0 & 30 \\ \frac{1}{5} & 1 & 0 & -\frac{1}{10} & \frac{2}{5} & 0 & 0 & 180 \\ \frac{12}{5} & 0 & 0 & \frac{1}{20} & -\frac{7}{10} & 1 & 0 & 10 \\ \hdashline 0 & 0 & 0 & \frac{1}{4} & \frac{1}{2} & 0 & 1 & 450 \end{bmatrix}$$

Optimal solution: max $P = 450$ at $x_1 = 0$, $x_2 = 180$, $x_3 = 30$,
$s_1 = 0$, $s_2 = 0$, $s_3 = 10$.

29.  The simplex tableau for this problem is:

$$\begin{array}{c} s_1 \\ s_2 \\ s_3 \\ s_4 \\ P \end{array} \begin{bmatrix} x_1 & x_2 & s_1 & s_2 & s_3 & s_4 & P & \\ 1 & 2 & 1 & 0 & 0 & 0 & 0 & 40 \\ 1 & 3 & 0 & 1 & 0 & 0 & 0 & 48 \\ 1 & 4 & 0 & 0 & 1 & 0 & 0 & 60 \\ 0 & 1 & 0 & 0 & 0 & 1 & 0 & 14 \\ \hdashline -2 & -5 & 0 & 0 & 0 & 0 & 1 & 0 \end{bmatrix} \begin{matrix} \frac{40}{2} = 20 \\ \frac{48}{3} = 16 \\ \frac{60}{4} = 15 \\ \frac{14}{1} = 14 \\ \\ \end{matrix}$$

$$(-2) R_4 + R_1 \to R_1, \quad (-3) R_4 + R_2 \to R_2, \quad (-4) R_4 + R_3 \to R_3,$$
$$\text{and } 5 R_4 + R_5 \to R_5$$

Copyright © 2011 Pearson Education, Inc. Publishing as Prentice Hall.

$$\sim \begin{bmatrix} 1 & 0 & 1 & 0 & 0 & -2 & 0 & | & 12 \\ 1 & 0 & 0 & 1 & 0 & -3 & 0 & | & 6 \\ ① & 0 & 0 & 0 & 1 & -4 & 0 & | & 4 \\ 0 & 1 & 0 & 0 & 0 & 1 & 0 & | & 14 \\ \hline -2 & 0 & 0 & 0 & 0 & 5 & 1 & | & 70 \end{bmatrix} \begin{matrix} \frac{12}{1} = 12 \\ \frac{6}{1} = 6 \\ \frac{4}{1} = 4 \\ \\ \end{matrix}$$

$(-1)R_3 + R_1 \rightarrow R_1$, $(-1)R_3 + R_2 \rightarrow R_2$, and $2R_3 + R_5 \rightarrow R_5$

$$\sim \begin{bmatrix} 0 & 0 & 1 & 0 & -1 & 2 & 0 & | & 8 \\ 0 & 0 & 0 & 1 & -1 & ① & 0 & | & 2 \\ 1 & 0 & 0 & 0 & 1 & -4 & 0 & | & 4 \\ 0 & 1 & 0 & 0 & 0 & 1 & 0 & | & 14 \\ \hline 0 & 0 & 0 & 0 & 2 & -3 & 1 & | & 78 \end{bmatrix} \begin{matrix} \frac{8}{2} = 4 \\ \frac{2}{1} = 2 \\ \\ \frac{14}{1} = 14 \\ \end{matrix}$$

$(-2)R_2 + R_1 \rightarrow R_1$, $4R_2 + R_3 \rightarrow R_3$, $(-1)R_2 + R_4 \rightarrow R_4$,
and $3R_2 + R_5 \rightarrow R_5$

$$\sim \begin{bmatrix} 0 & 0 & 1 & -2 & ① & 0 & 0 & | & 4 \\ 0 & 0 & 0 & 1 & -1 & 1 & 0 & | & 2 \\ 1 & 0 & 0 & 4 & -3 & 0 & 0 & | & 12 \\ 0 & 1 & 0 & -1 & 1 & 0 & 0 & | & 12 \\ \hline 0 & 0 & 0 & 3 & -1 & 0 & 1 & | & 84 \end{bmatrix} \begin{matrix} \frac{4}{1} = 4 \\ \\ \\ \frac{12}{1} = 12 \\ \end{matrix}$$

$R_1 + R_2 \rightarrow R_2$, $3R_1 + R_3 \rightarrow R_3$, $(-1)R_1 + R_4 \rightarrow R_4$, and $R_1 + R_5 \rightarrow R_5$

$$\sim \begin{matrix} & x_1 & x_2 & s_1 & s_2 & s_3 & s_4 & P & \\ s_3 & \begin{bmatrix} 0 & 0 & 1 & -2 & 1 & 0 & 0 & | & 4 \\ s_4 & 0 & 0 & 1 & -1 & 0 & 1 & 0 & | & 6 \\ x_1 & 1 & 0 & 3 & -2 & 0 & 0 & 0 & | & 24 \\ x_2 & 0 & 1 & -1 & 1 & 0 & 0 & 0 & | & 8 \\ \hline P & 0 & 0 & 1 & 1 & 0 & 0 & 1 & | & 88 \end{bmatrix} \end{matrix}$$

Optimal solution: max $P = 88$ at $x_1 = 24$, $x_2 = 8$, $s_1 = 0$, $s_2 = 0$, $s_3 = 4$, $s_4 = 6$.

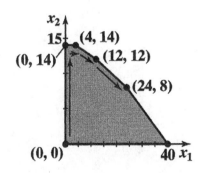

**31.**    Simplex Method:

The simplex tableau for this problem is:

$$
\begin{array}{c}
\phantom{s_1}\\
s_1\\
s_2
\end{array}
\begin{array}{c}
\begin{array}{ccccc}
x_1 & x_2 & s_1 & s_2 & P
\end{array}\\
\left[\begin{array}{ccccc|c}
-2 & \boxed{1} & 1 & 0 & 0 & 4\\
0 & 1 & 0 & 1 & 0 & 10\\
\hline
-2 & -3 & 0 & 0 & 1 & 0
\end{array}\right]
\end{array}
\begin{array}{c}
\frac{4}{1}=4\\[4pt]
\frac{10}{1}=10
\end{array}
$$

$$(-1)R_1 + R_2 \rightarrow R_2,\quad 3R_1 + R_3 \rightarrow R_3$$

$$
\sim\left[\begin{array}{ccccc|c}
-2 & 1 & 1 & 0 & 0 & 4\\
\boxed{2} & 0 & -1 & 1 & 0 & 6\\
\hline
-8 & 0 & 3 & 0 & 1 & 12
\end{array}\right]\ \frac{6}{2}=3
$$

$$\frac{1}{2}R_2 \rightarrow R_2$$

$$
\sim\left[\begin{array}{ccccc|c}
-2 & 1 & 0 & 0 & 0 & 4\\
\boxed{1} & 0 & -\frac{1}{2} & \frac{1}{2} & 0 & 3\\
\hline
-8 & 0 & 3 & 0 & 1 & 12
\end{array}\right]
$$

$$2R_2 + R_1 \rightarrow R_1,\quad 8R_2 + R_3 \rightarrow R_3$$

$$
\sim\left[\begin{array}{ccccc|c}
0 & 1 & 0 & 1 & 0 & 10\\
1 & 0 & -\frac{1}{2} & \frac{1}{2} & 0 & 3\\
\hline
0 & 0 & -1 & 4 & 1 & 36
\end{array}\right]
$$

No positive elements in the pivot column; no optimal solution exists.

Geometric Method:

Step (1): Graph the feasible region and find the corner points. The feasible region $S$ is the solution set of the inequalities. This region is indicated by the shading in the graph at the right.

The corner points are  $(0, 0)$,  $(0, 4)$, and  $(3, 10)$.

Since $S$ is unbounded and the coefficients of the objective function are positive, $P$ does not have a maximum value.

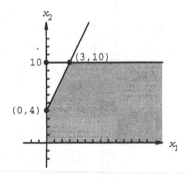

**33.**    The simplex tableau for this problem is:

$$
\begin{array}{c}
\phantom{s_1}\\[2pt]
s_1\\[6pt]
s_2\\[6pt]
s_3\\[2pt]
P
\end{array}
\begin{array}{c}
\begin{array}{cccccc}
x_1 & x_2 & s_1 & s_2 & s_3 & P
\end{array}\\
\left[\begin{array}{cccccc|c}
2 & 1 & 1 & 0 & 0 & 0 & 16\\[4pt]
1 & 0 & 0 & 1 & 0 & 0 & 6\\[4pt]
0 & 1 & 0 & 0 & 1 & 0 & 10\\
\hdashline
-1 & -1 & 0 & 0 & 0 & 1 & 0
\end{array}\right]
\end{array}
$$

Copyright © 2011 Pearson Education, Inc.  Publishing as Prentice Hall.

(A)　Solution using the first column as the pivot column

$$
\begin{array}{cccccc}
x_1 & x_2 & s_1 & s_2 & s_3 & P
\end{array}
$$

$$
\left[\begin{array}{cccccc|c}
2 & 1 & 1 & 0 & 0 & 0 & 16 \\
\textcircled{1} & 0 & 0 & 1 & 0 & 0 & 6 \\
0 & 1 & 0 & 0 & 1 & 0 & 10 \\
\hline
-1 & -1 & 0 & 0 & 0 & 1 & 0
\end{array}\right]
\begin{array}{l}
\frac{16}{2} = 8 \\[4pt]
\frac{6}{1} = 6
\end{array}
$$

$$(-2)R_2 + R_1 \rightarrow R_1,\ R_2 + R_4 \rightarrow R_4$$

$$
\sim
\left[\begin{array}{cccccc|c}
0 & \textcircled{1} & 1 & -2 & 0 & 0 & 4 \\
1 & 0 & 0 & 1 & 0 & 0 & 6 \\
0 & 1 & 0 & 0 & 1 & 0 & 10 \\
\hline
0 & -1 & 0 & 1 & 0 & 1 & 6
\end{array}\right]
\begin{array}{l}
\frac{4}{1} = 4 \\[10pt]
\frac{10}{1} = 10
\end{array}
$$

$$(-1)R_1 + R_3 \rightarrow R_3,\ R_1 + R_4 \rightarrow R_4$$

$$
\sim
\left[\begin{array}{cccccc|c}
0 & 1 & 1 & -2 & 0 & 0 & 4 \\
1 & 0 & 0 & 1 & 0 & 0 & 6 \\
0 & 0 & -1 & \textcircled{2} & 1 & 0 & 6 \\
\hline
0 & 0 & 1 & -1 & 0 & 1 & 10
\end{array}\right]
\begin{array}{l}
\frac{6}{1} = 6 \\[4pt]
\frac{6}{2} = 3
\end{array}
$$

$$\tfrac{1}{2}R_3 \rightarrow R_3$$

$$
\sim
\left[\begin{array}{cccccc|c}
0 & 1 & 1 & -2 & 0 & 0 & 4 \\
1 & 0 & 0 & 1 & 0 & 0 & 6 \\
0 & 0 & -\frac{1}{2} & 1 & \frac{1}{2} & 0 & 3 \\
\hline
0 & 0 & 1 & -1 & 0 & 1 & 10
\end{array}\right]
$$

$$2R_3 + R_1 \rightarrow R_1,\ (-1)R_3 + R_2 \rightarrow R_2,\ R_3 + R_4 \rightarrow R_4$$

$$
\begin{array}{cccccc}
& x_1 & x_2 & s_1 & s_2 & s_3 & P
\end{array}
$$

$$
\sim
\begin{array}{c}
x_2 \\ x_1 \\ s_2 \\ P
\end{array}
\left[\begin{array}{cccccc|c}
0 & 1 & 0 & 0 & 1 & 0 & 10 \\
1 & 0 & \frac{1}{2} & 0 & -\frac{1}{2} & 0 & 3 \\
0 & 0 & -\frac{1}{2} & 1 & \frac{1}{2} & 0 & 3 \\
\hline
0 & 0 & \frac{1}{2} & 0 & \frac{1}{2} & 1 & 13
\end{array}\right]
$$

Optimal solution: max $P = 13$ at $x_1 = 3$, $x_2 = 10$, $s_1 = 0$, $s_2 = 3$, $s_3 = 0$

Copyright © 2011 Pearson Education, Inc. Publishing as Prentice Hall.

(B)    Solution using the second column as the pivot column

$$
\begin{array}{c}
\begin{array}{cccccc}
x_1 & x_2 & s_1 & s_2 & s_3 & P
\end{array}\\
\begin{array}{c}
s_1\\
s_2\\
s_3\\
P
\end{array}
\left[\begin{array}{cccccc|c}
2 & 1 & 1 & 0 & 0 & 0 & 16\\
1 & 0 & 0 & 1 & 0 & 0 & 6\\
0 & ① & 0 & 0 & 1 & 0 & 10\\
\hdashline
-1 & -1 & 0 & 0 & 0 & 1 & 0
\end{array}\right]
\begin{array}{l}
\frac{16}{1}=16\\[6pt]
\\[4pt]
\frac{10}{1}=10
\end{array}
\end{array}
$$

$$(-1)R_3 + R_1 \rightarrow R_1, \quad R_3 + R_4 \rightarrow R_4$$

$$
\sim
\left[\begin{array}{cccccc|c}
② & 0 & 1 & 0 & -1 & 0 & 6\\
1 & 0 & 0 & 1 & 0 & 0 & 6\\
0 & 1 & 0 & 0 & 1 & 0 & 10\\
\hdashline
-1 & 0 & 0 & 0 & 1 & 1 & 10
\end{array}\right]
\begin{array}{l}
\frac{6}{2}=3\\[6pt]
\frac{6}{1}=6
\end{array}
$$

$$\frac{1}{2}R_1 \rightarrow R_1$$

$$
\sim
\left[\begin{array}{cccccc|c}
① & 0 & \frac{1}{2} & 0 & -\frac{1}{2} & 0 & 3\\
1 & 0 & 0 & 1 & 0 & 0 & 6\\
0 & 1 & 0 & 0 & 1 & 0 & 10\\
\hdashline
-1 & 0 & 0 & 0 & 1 & 1 & 10
\end{array}\right]
$$

$$(-1)R_1 + R_2 \rightarrow R_2, \quad R_1 + R_4 \rightarrow R_4$$

$$
\begin{array}{c}
\begin{array}{cccccc}
x_1 & x_2 & s_1 & s_2 & s_3 & P
\end{array}\\
\sim
\begin{array}{c}
x_1\\
s_2\\
x_2\\
P
\end{array}
\left[\begin{array}{cccccc|c}
1 & 0 & \frac{1}{2} & 0 & -\frac{1}{2} & 0 & 3\\
0 & 0 & -\frac{1}{2} & 1 & \frac{1}{2} & 0 & 3\\
0 & ① & 0 & 0 & 1 & 0 & 10\\
\hdashline
0 & 0 & \frac{1}{2} & 0 & \frac{1}{2} & 1 & 13
\end{array}\right]
\end{array}
$$

```
Choosing either solution produces
the same optimal solution.
```

```
Optimal solution: max P = 13 at x₁ = 3, x₂ = 10, s₁ = 0,
s₂ = 3, s₃ = 0
```

35.    The simplex tableau for this problem is:

$$
\begin{array}{c}
\begin{array}{cccccc}
x_1 & x_2 & x_3 & s_1 & s_2 & P
\end{array}\\
\begin{array}{c}
s_1\\
s_2\\
P
\end{array}
\left[\begin{array}{cccccc|c}
1 & 1 & 2 & 1 & 0 & 0 & 20\\
2 & 1 & 4 & 0 & 1 & 0 & 32\\
\hdashline
-3 & -3 & -2 & 0 & 0 & 1 & 0
\end{array}\right]
\end{array}
$$

Copyright © 2011 Pearson Education, Inc. Publishing as Prentice Hall.

(A)    Solution using the first column as the pivot column

$$
\begin{array}{c}
\begin{array}{cccccc} x_1 & x_2 & x_3 & s_1 & s_2 & P \end{array} \\
\left[\begin{array}{cccccc|c}
1 & 1 & 2 & 1 & 0 & 0 & 20 \\
\textcircled{2} & 1 & 4 & 0 & 1 & 0 & 32 \\
\hline
-3 & -3 & -2 & 0 & 0 & 1 & 0
\end{array}\right]
\begin{array}{l} \frac{20}{1} = 20 \\[4pt] \frac{32}{2} = 16 \end{array}
\end{array}
$$

$$\frac{1}{2}R_2 \rightarrow R_2$$

$$
\sim \left[\begin{array}{cccccc|c}
1 & 1 & 2 & 1 & 0 & 0 & 20 \\
1 & \frac{1}{2} & 2 & 0 & \frac{1}{2} & 0 & 16 \\
\hline
-3 & -3 & -2 & 0 & 0 & 1 & 0
\end{array}\right]
$$

$$(-1)R_2 + R_1 \rightarrow R_1, \quad 3R_2 + R_3 \rightarrow R_3$$

$$
\sim \left[\begin{array}{cccccc|c}
0 & \boxed{\tfrac{1}{2}} & 0 & 1 & -\frac{1}{2} & 0 & 4 \\
1 & \frac{1}{2} & 2 & 0 & \frac{1}{2} & 0 & 16 \\
\hline
0 & -\frac{3}{2} & 4 & 0 & \frac{3}{2} & 1 & 48
\end{array}\right]
\begin{array}{l} \frac{4}{1/2} = 8 \\[4pt] \frac{16}{1/2} = 32 \end{array}
$$

$$2R_1 \rightarrow R_1$$

$$
\sim \left[\begin{array}{cccccc|c}
0 & 1 & 0 & 2 & -1 & 0 & 8 \\
1 & \frac{1}{2} & 2 & 0 & \frac{1}{2} & 0 & 16 \\
\hline
0 & -\frac{3}{2} & 4 & 0 & \frac{3}{2} & 1 & 48
\end{array}\right]
$$

$$\left(-\frac{1}{2}\right)R_1 + R_2 \rightarrow R_2, \quad \frac{3}{2}R_1 + R_3 \rightarrow R_3$$

$$
\begin{array}{c}
\begin{array}{cccccc} x_1 & x_2 & x_3 & s_1 & s_2 & P \end{array} \\
\sim \begin{array}{c} x_2 \\ x_1 \\ P \end{array} \left[\begin{array}{cccccc|c}
0 & 1 & 0 & 2 & -1 & 0 & 8 \\
1 & 0 & 2 & -1 & 1 & 0 & 12 \\
\hline
0 & 0 & 4 & 3 & 0 & 1 & 60
\end{array}\right]
\end{array}
$$

Optimal solution: max $P = 60$ at $x_1 = 12$, $x_2 = 8$, $x_3 = 0$, $s_1 = 0$, $s_2 = 0$

(B)    Solution using the second column as the pivot column

$$
\begin{array}{c}
\begin{array}{cccccc} x_1 & x_2 & x_3 & s_1 & s_2 & P \end{array} \\
\begin{array}{c} s_1 \\ s_2 \\ P \end{array} \left[\begin{array}{cccccc|c}
1 & \textcircled{1} & 2 & 1 & 0 & 0 & 20 \\
2 & 1 & 4 & 0 & 1 & 0 & 32 \\
\hline
-3 & -3 & -2 & 0 & 0 & 1 & 0
\end{array}\right]
\begin{array}{l} \frac{20}{1} = 20 \\[4pt] \frac{32}{1} = 32 \end{array}
\end{array}
$$

$$(-1)R_1 + R_2 \rightarrow R_2, \quad 3R_1 + R_3 \rightarrow R_3$$

Copyright © 2011 Pearson Education, Inc. Publishing as Prentice Hall.

$$
\begin{array}{c}
\begin{array}{cccccc}
x_1 & x_2 & x_3 & s_1 & s_2 & P
\end{array}\\
\sim\;
\begin{array}{c}
x_2\\[4pt]
s_2\\[4pt]
P
\end{array}
\left[
\begin{array}{cccccc|c}
1 & 1 & 2 & 1 & 0 & 0 & 20\\
1 & 0 & 2 & -1 & 1 & 0 & 12\\
\hline
0 & 0 & 4 & 3 & 0 & 1 & 60
\end{array}
\right]
\end{array}
$$

Optimal solution: max $P = 60$
at $x_1 = 0$, $x_2 = 20$, $x_3 = 0$,
$s_1 = 0$, $s_2 = 12$

The maximum value of $P$ is 60. Since the optimal solution is obtained at two corner points, $(12, 8, 0)$ and $(0, 20, 0)$, every point on the line segment connecting these points is also an optimal solution.

**37.**  Let $x_1$ = the number of $A$ components

$x_2$ = the number of $B$ components

$x_3$ = the number of $C$ components

The mathematical model for this problem is:
Maximize $P = 7x_1 + 8x_2 + 10x_3$

Subject to $2x_1 + 3x_2 + 2x_3 \le 1000$

$x_1 + x_2 + 2x_3 \le 800$

$x_1,\; x_2,\; x_3 \ge 0$

We introduce slack variables $s_1$, $s_2$ to obtain the equivalent form:

$$
\begin{aligned}
2x_1 + 3x_2 + 2x_3 + s_1 &= 1000\\
x_1 + x_2 + 2x_3 \quad + s_2 &= 800\\
-7x_1 - 8x_2 - 10x_3 \qquad\quad + P &= 0
\end{aligned}
$$

The simplex tableau for this problem is:

$$
\begin{array}{c}
\begin{array}{cccccc}
x_1 & x_2 & x_3 & s_1 & s_2 & P
\end{array}\\
\begin{array}{c}
s_1\\[4pt]
s_2\\[4pt]
P
\end{array}
\left[
\begin{array}{cccccc|c}
2 & 3 & 2 & 1 & 0 & 0 & 1000\\
1 & 1 & \textcircled{2} & 0 & 1 & 0 & 800\\
\hline
-7 & -8 & -10 & 0 & 0 & 1 & 0
\end{array}
\right]
\end{array}
\quad
\begin{array}{l}
\dfrac{1000}{2} = 500\\[10pt]
\dfrac{800}{2} = 400
\end{array}
$$

$\dfrac{1}{2} R_2 \rightarrow R_2$

$$
\sim
\left[
\begin{array}{cccccc|c}
2 & 3 & 2 & 1 & 0 & 0 & 1000\\
\frac{1}{2} & \frac{1}{2} & \textcircled{1} & 0 & \frac{1}{2} & 0 & 400\\
\hline
-7 & -8 & -10 & 0 & 0 & 1 & 0
\end{array}
\right]
$$

$(-2) R_2 + R_1 \rightarrow R_1,\quad 10 R_2 + R_3 \rightarrow R_3$

$$
\sim
\left[
\begin{array}{cccccc|c}
1 & \textcircled{2} & 0 & 1 & -1 & 0 & 200\\
\frac{1}{2} & \frac{1}{2} & 1 & 0 & \frac{1}{2} & 0 & 400\\
\hline
-2 & -3 & 0 & 0 & 5 & 1 & 4000
\end{array}
\right]
\quad
\begin{array}{l}
\dfrac{200}{2} = 100\\[10pt]
\dfrac{400}{1/2} = 800
\end{array}
$$

$\dfrac{1}{2} R_1 \rightarrow R_1$

Copyright © 2011 Pearson Education, Inc. Publishing as Prentice Hall.

$$\sim \begin{bmatrix} \frac{1}{2} & ① & 0 & \frac{1}{2} & -\frac{1}{2} & 0 & | & 100 \\ \frac{1}{2} & \frac{1}{2} & 1 & 0 & \frac{1}{2} & 0 & | & 400 \\ \hdashline -2 & -3 & 0 & 0 & 5 & 1 & | & 4000 \end{bmatrix}$$

$$\left(-\frac{1}{2}\right)R_1 + R_2 \rightarrow R_2, \quad 3R_1 + R_3 \rightarrow R_3$$

$$\sim \begin{bmatrix} \frac{1}{2} & 1 & 0 & \frac{1}{2} & -\frac{1}{2} & 0 & | & 100 \\ \frac{1}{4} & 0 & 1 & -\frac{1}{4} & \frac{3}{4} & 0 & | & 350 \\ \hdashline -\frac{1}{2} & 0 & 0 & \frac{3}{2} & \frac{7}{2} & 1 & | & 4300 \end{bmatrix} \begin{matrix} \frac{100}{1/2} = 200 \\ \frac{350}{1/4} = 1400 \end{matrix}$$

$$2R_1 \rightarrow R_1$$

$$\sim \begin{bmatrix} 1 & 2 & 0 & 1 & -1 & 0 & | & 200 \\ \frac{1}{4} & 0 & 1 & -\frac{1}{4} & \frac{3}{4} & 0 & | & 350 \\ \hdashline -\frac{1}{2} & 0 & 0 & \frac{3}{2} & \frac{7}{2} & 1 & | & 4300 \end{bmatrix}$$

$$\left(-\frac{1}{4}\right)R_1 + R_2 \rightarrow R_2, \quad \frac{1}{2}R_1 + R_3 \rightarrow R_3$$

$$\sim \begin{array}{c} \\ x_1 \\ x_3 \\ P \end{array} \begin{bmatrix} x_1 & x_2 & x_3 & s_1 & s_2 & P & \\ 1 & 2 & 0 & 1 & -1 & 0 & | & 200 \\ 0 & -\frac{1}{2} & 1 & -\frac{1}{2} & 1 & 0 & | & 300 \\ \hdashline 0 & 1 & 0 & 2 & 3 & 1 & | & 4400 \end{bmatrix}$$

Optimal solution: the maximum profit is $4400 when 200 *A* components, 0 *B* components and 300 *C* components are manufactured.

**39.** Let $x_1$ = the amount invested in government bonds,

$x_2$ = the amount invested in mutual funds,

$x_3$ = the amount invested in money market funds.

The mathematical model for this problem is:

Maximize $P = .08x_1 + .13x_2 + .15x_3$

Subject to: $x_1 + x_2 + x_3 \leq 100{,}000$

$\qquad\qquad x_2 + x_3 \leq x_1$

$\qquad\qquad x_1, \; x_2, \; x_3 \geq 0$

We introduce slack variables $s_1$ and $s_2$ to obtain the equivalent form:

$$\begin{aligned} x_1 + \quad x_2 + \quad x_3 + s_1 \qquad\qquad &= 100{,}000 \\ -x_1 + \quad x_2 + \quad x_3 \qquad + s_2 \quad &= 0 \\ -.08x_1 - .13x_2 - .15x_3 \qquad\qquad + P &= 0 \end{aligned}$$

Copyright © 2011 Pearson Education, Inc. Publishing as Prentice Hall.

The simplex tableau for this problem is:

$$
\begin{array}{c}
\begin{array}{cccccc}
x_1 & x_2 & x_3 & s_1 & s_2 & P
\end{array} \\
\begin{array}{c}
s_1 \\
s_2 \\
P
\end{array}
\left[
\begin{array}{cccccc|c}
1 & 1 & 1 & 1 & 0 & 0 & 100{,}000 \\
-1 & 1 & ① & 0 & 1 & 0 & 0 \\
\hline
-.08 & -.13 & -.15 & 0 & 0 & 1 & 0
\end{array}
\right]
\end{array}
\quad
\begin{array}{c}
\dfrac{100{,}000}{1} = 100{,}000 \\[2ex]
\dfrac{0}{1} = 0
\end{array}
$$

$$(-1)R_2 + R_1 \rightarrow R_1 \text{ and } .15R_2 + R_3 \rightarrow R_3$$

$$
\sim
\left[
\begin{array}{cccccc|c}
② & 0 & 0 & 1 & -1 & 0 & 100{,}000 \\
-1 & 1 & 1 & 0 & 1 & 0 & 0 \\
\hline
-.23 & .02 & 0 & 0 & .15 & 1 & 0
\end{array}
\right]
\quad
\sim
\left[
\begin{array}{cccccc|c}
① & 0 & 0 & \frac{1}{2} & -\frac{1}{2} & 0 & 50{,}000 \\
-1 & 1 & 1 & 0 & 1 & 0 & 0 \\
\hline
-.23 & .02 & 0 & 0 & .15 & 1 & 0
\end{array}
\right]
$$

$$\tfrac{1}{2}R_1 \rightarrow R_1 \qquad\qquad R_1 + R_2 \rightarrow R_2 \text{ and } .23R_1 + R_3 \rightarrow R_3$$

$$
\begin{array}{c}
\begin{array}{cccccc}
x_1 & x_2 & x_3 & s_1 & s_2 & P
\end{array} \\
\begin{array}{c}
x_1 \\
\sim\ x_2 \\
P
\end{array}
\left[
\begin{array}{cccccc|c}
1 & 0 & 0 & \frac{1}{2} & -\frac{1}{2} & 0 & 50{,}000 \\
0 & 1 & 1 & \frac{1}{2} & \frac{1}{2} & 0 & 50{,}000 \\
\hline
0 & .02 & 0 & .115 & .035 & 1 & 11{,}500
\end{array}
\right]
\end{array}
$$

Optimal solution: the maximum return is $11,500 when $x_1 = \$50{,}000$ is invested in government bonds, $x_2 = \$0$ is invested in mutual funds, and $x_3 = \$50{,}000$ is invested in money market funds.

**41.** Let $x_1$ = the number of daytime ads,

$x_2$ = the number of prime-time ads,

$x_3$ = the number of late-night ads.

The mathematical model for this problem is:

$$\text{Maximize } P = 14{,}000x_1 + 24{,}000x_2 + 18{,}000x_3$$

$$
\begin{aligned}
\text{Subject to: } 1000x_1 + 2000x_2 + 1500x_3 &\le 20{,}000 \\
x_1 + x_2 + x_3 &\le 15 \\
x_1,\ x_2,\ x_3 &\ge 0
\end{aligned}
$$

We introduce slack variables to obtain the following initial form:

$$
\begin{aligned}
1000x_1 + 2000x_2 + 1500x_3 + s_1 \qquad\qquad &= 20{,}000 \\
x_1 + x_2 + x_3 \qquad + s_2 \qquad &= 15 \\
-14{,}000x_1 - 24{,}000x_2 - 18{,}000x_3 \qquad\qquad + P &= 0
\end{aligned}
$$

The simplex tableau for this problem is:

$$
\begin{array}{c}
\begin{array}{cccccc}
x_1 & x_2 & x_3 & s_1 & s_2 & P
\end{array} \\
\begin{array}{c}
s_1 \\
s_2 \\
P
\end{array}
\left[
\begin{array}{cccccc|c}
1000 & ⟨2000⟩ & 1500 & 1 & 0 & 0 & 20{,}000 \\
1 & 1 & 1 & 0 & 1 & 0 & 15 \\
\hline
-14{,}000 & -24{,}000 & -18{,}000 & 0 & 0 & 1 & 0
\end{array}
\right]
\end{array}
\quad
\begin{array}{c}
\dfrac{20{,}000}{2000} = 10 \\[2ex]
\dfrac{15}{1} = 15
\end{array}
$$

$$\dfrac{1}{2000}R_1 \rightarrow R_1$$

Copyright © 2011 Pearson Education, Inc. Publishing as Prentice Hall.

$$\sim \begin{bmatrix} \frac{1}{2} & \textcircled{1} & \frac{3}{4} & \frac{1}{2000} & 0 & 0 & 10 \\ \hline 1 & 1 & 1 & 0 & 1 & 0 & 15 \\ \hline -14{,}000 & -24{,}000 & -18{,}000 & 0 & 0 & 1 & 0 \end{bmatrix}$$

$(-1)R_1 + R_2 \rightarrow R_2, \quad 24{,}000R_1 + R_3 \rightarrow R_3$

$$\sim \begin{bmatrix} \frac{1}{2} & 1 & \frac{3}{4} & \frac{1}{2000} & 0 & 0 & 10 \\ \textcircled{$\frac{1}{2}$} & 0 & \frac{1}{4} & -\frac{1}{2000} & 1 & 0 & 5 \\ \hline -2000 & 0 & 0 & 12 & 0 & 1 & 240{,}000 \end{bmatrix}$$

$2R_2 \rightarrow R_2$

$$\sim \begin{bmatrix} \frac{1}{2} & 1 & \frac{3}{4} & \frac{1}{2000} & 0 & 0 & 10 \\ \textcircled{1} & 0 & \frac{1}{2} & -\frac{1}{1000} & 2 & 0 & 10 \\ \hline -2000 & 0 & 0 & 12 & 0 & 1 & 240{,}000 \end{bmatrix}$$

$\left(-\frac{1}{2}\right)R_2 + R_1 \rightarrow R_1, \quad 2000R_2 + R_3 \rightarrow R_3$

$$\sim \begin{array}{c} \\ x_2 \\ x_1 \\ P \end{array} \begin{bmatrix} x_1 & x_2 & x_3 & s_1 & s_2 & P & \\ 0 & 1 & \frac{1}{2} & \frac{1}{1000} & -1 & 0 & 5 \\ 1 & 0 & \frac{1}{2} & -\frac{1}{1000} & 2 & 0 & 10 \\ \hline 0 & 0 & 1000 & 10 & 4000 & 1 & 260{,}000 \end{bmatrix}$$

Optimal solution: maximum number of potential customers is 260,000 when $x_1 = 10$ daytime ads, $x_2 = 5$ prime-time ads, and $x_3 = 0$ late-night ads are placed.

**43.** Let $x_1$ = the number of colonial houses,

$x_2$ = the number of split-level houses,

$x_3$ = the number of ranch-style houses.

(A)   The mathematical model for this problem is:

Maximize $P = 20{,}000x_1 + 18{,}000x_2 + 24{,}000x_3$

Subject to:
$$\frac{1}{2}x_1 + \frac{1}{2}x_2 + \phantom{xxx} x_3 \le 30$$
$$60{,}000x_1 + 60{,}000x_2 + 80{,}000x_3 \le 3{,}200{,}000$$
$$4{,}000x_1 + 3{,}000x_2 + 4{,}000x_3 \le 180{,}000$$
$$x_1, \ x_2, \ x_3 \ge 0$$

Copyright © 2011 Pearson Education, Inc. Publishing as Prentice Hall.

We simplify the inequalities and then introduce slack variables to obtain the initial form:

$$\frac{1}{2}x_1 + \frac{1}{2}x_2 + x_3 + s_1 \qquad\qquad = 30$$
$$6x_1 + 6x_2 + 8x_3 \qquad + s_2 \qquad\quad = 320$$
$$4x_1 + 3x_2 + 4x_3 \qquad\qquad + s_3 \quad = 180$$
$$-20{,}000x_1 - 18{,}000x_2 - 24{,}000x_3 \qquad\qquad\quad + P = 0$$

[Note: This simplification will change the interpretation of the slack variables.]
The simplex tableau for this problem is:

|  | $x_1$ | $x_2$ | $x_3$ | $s_1$ | $s_2$ | $s_3$ | $P$ |  |  |
|---|---|---|---|---|---|---|---|---|---|
| $s_1$ | $\frac{1}{2}$ | $\frac{1}{2}$ | ① | 1 | 0 | 0 | 0 | 30 | $\frac{30}{1} = 30$ |
| $s_2$ | 6 | 6 | 8 | 0 | 1 | 0 | 0 | 320 | $\frac{320}{8} = 40$ |
| $s_3$ | 4 | 3 | 4 | 0 | 0 | 1 | 0 | 180 | $\frac{180}{4} = 45$ |
| $P$ | $-20{,}000$ | $-18{,}000$ | $-24{,}000$ | 0 | 0 | 0 | 1 | 0 |  |

$(-8)R_1 + R_2 \rightarrow R_2,\ (-4)R_1 + R_3 \rightarrow R_3,\ 24{,}000R_1 + R_4 \rightarrow R_4$

$$\sim \begin{bmatrix} \frac{1}{2} & \frac{1}{2} & 1 & 1 & 0 & 0 & 0 & 30 \\ 2 & 2 & 0 & -8 & 1 & 0 & 0 & 80 \\ ② & 1 & 0 & -4 & 0 & 1 & 0 & 60 \\ \hline -8000 & -6000 & 0 & 24{,}000 & 0 & 0 & 1 & 720{,}000 \end{bmatrix}$$

$\frac{1}{2}R_3 \rightarrow R_3$

$$\sim \begin{bmatrix} \frac{1}{2} & \frac{1}{2} & 1 & 1 & 0 & 0 & 0 & 30 \\ 2 & 2 & 0 & -8 & 1 & 0 & 0 & 80 \\ ① & \frac{1}{2} & 0 & -2 & 0 & \frac{1}{2} & 0 & 30 \\ \hline -8000 & -6000 & 0 & 24{,}000 & 0 & 0 & 1 & 720{,}000 \end{bmatrix}$$

$\left(-\frac{1}{2}\right)R_3 + R_1 \rightarrow R_1,\ (-2)R_3 + R_2 \rightarrow R_2,\ 8000R_3 + R_4 \rightarrow R_4$

$$\sim \begin{bmatrix} 0 & \frac{1}{4} & 1 & 2 & 0 & -\frac{1}{4} & 0 & 15 \\ 0 & ① & 0 & -4 & 1 & -1 & 0 & 20 \\ 1 & \frac{1}{2} & 0 & -2 & 0 & \frac{1}{2} & 0 & 30 \\ \hline 0 & -2000 & 0 & 8000 & 0 & 4000 & 1 & 960{,}000 \end{bmatrix}$$

$\left(-\frac{1}{4}\right)R_2 + R_1 \rightarrow R_1,\ \left(-\frac{1}{2}\right)R_2 + R_3 \rightarrow R_3,\ 2000R_2 + R_4 \rightarrow R_4$

Copyright © 2011 Pearson Education, Inc. Publishing as Prentice Hall.

$$\begin{array}{c} \\ x_3 \\ x_2 \\ \sim \quad x_1 \\ \\ P \end{array}\begin{array}{c} x_1 \quad x_2 \quad x_3 \quad s_1 \quad s_2 \quad s_3 \quad P \\ \left[\begin{array}{ccccccc|c} 0 & 0 & 1 & 3 & -\frac{1}{4} & 0 & 0 & 10 \\ 0 & 1 & 0 & -4 & 1 & -1 & 0 & 20 \\ 1 & 0 & 0 & 0 & -\frac{1}{2} & 1 & 0 & 20 \\ \hdashline 0 & 0 & 0 & 0 & 2000 & 2000 & 1 & 1{,}000{,}000 \end{array}\right] \end{array}$$

Optimal solution: maximum profit is \$1,000,000 when $x_1 = 20$ colonial houses, $x_2 = 20$ split-level houses, and $x_3 = 10$ ranch-style houses are built.

**45.**   Refer to Problem 43.  The mathematical model for this problem is:

Maximize  $P = 17{,}000x_1 + 18{,}000x_2 + 24{,}000x_3$

Subject to:
$$\frac{1}{2}x_1 + \frac{1}{2}x_2 + \quad x_3 \le 30$$
$$60{,}000x_1 + 60{,}000x_2 + 80{,}000x_3 \le 3{,}200{,}000$$
$$4{,}000x_1 + 3{,}000x_2 + 4{,}000x_3 \le 180{,}000$$

Following the solution in Problem 43, we obtain the simplex tableau:

$$\begin{array}{c} \\ s_1 \\ s_2 \\ s_3 \\ P \end{array}\begin{array}{c} x_1 \qquad x_2 \qquad x_3 \quad s_1 \quad s_2 \quad s_3 \quad P \\ \left[\begin{array}{ccccccc|c} \frac{1}{2} & \frac{1}{2} & ① & 1 & 0 & 0 & 0 & 30 \\ 6 & 6 & 8 & 0 & 1 & 0 & 0 & 320 \\ 4 & 3 & 4 & 0 & 0 & 1 & 0 & 180 \\ \hdashline -17{,}000 & -18{,}000 & -24{,}000 & 0 & 0 & 0 & 1 & 0 \end{array}\right] \end{array} \begin{array}{l} \frac{30}{1} = 30 \\[4pt] \frac{320}{8} = 40 \\[4pt] \frac{180}{4} = 45 \end{array}$$

$$(-8)R_1 + R_2 \to R_2, \quad (-4)R_1 + R_3 \to R_3, \quad 24{,}000R_1 + R_4 \to R_4$$

$$\sim \left[\begin{array}{ccccccc|c} \frac{1}{2} & \frac{1}{2} & 1 & 1 & 0 & 0 & 0 & 30 \\ 2 & ② & 0 & -8 & 1 & 0 & 0 & 80 \\ 2 & 1 & 0 & -4 & 0 & 1 & 0 & 60 \\ \hdashline -5000 & -6000 & 0 & 24{,}000 & 0 & 0 & 1 & 720{,}000 \end{array}\right] \begin{array}{l} \frac{30}{1/2} = 60 \\[4pt] \frac{80}{2} = 40 \\[4pt] \frac{60}{1} = 60 \end{array}$$

$$\frac{1}{2}R_2 \to R_2$$

$$\sim \left[\begin{array}{ccccccc|c} \frac{1}{2} & \frac{1}{2} & 1 & 1 & 0 & 0 & 0 & 30 \\ 1 & ① & 0 & -4 & \frac{1}{2} & 0 & 0 & 40 \\ 2 & 1 & 0 & -4 & 0 & 1 & 0 & 60 \\ \hdashline -5000 & -6000 & 0 & 24{,}000 & 0 & 0 & 1 & 720{,}000 \end{array}\right]$$

$$\left(-\frac{1}{2}\right)R_2 + R_1 \to R_1, \quad (-1)R_2 + R_3 \to R_3, \quad 6{,}000R_2 + R_4 \to R_4$$

Copyright © 2011 Pearson Education, Inc.  Publishing as Prentice Hall.

$$
\begin{array}{c}
\begin{array}{ccccccc}
x_1 & x_2 & x_3 & s_1 & s_2 & s_3 & P
\end{array}\\
\begin{array}{c}
x_3\\
x_2\\
\sim\quad s_3\\
\\
P
\end{array}
\left[
\begin{array}{ccccccc|c}
0 & 0 & 1 & 3 & -\tfrac{1}{4} & 0 & 0 & 10\\
1 & 1 & 0 & -4 & \tfrac{1}{2} & 0 & 0 & 40\\
1 & 0 & 0 & 0 & -\tfrac{1}{2} & 1 & 0 & 20\\
\hline
1000 & 0 & 0 & 0 & 3000 & 0 & 1 & 960{,}000
\end{array}
\right]
\end{array}
$$

```
Optimal solution: maximum profit is $960,000 when x  = 0 colonial
                                                    1
houses, x  = 40 split level houses and x  = 10 ranch houses are
         2                              3
built. In this case, s  = 20 (thousand) labor hours are not used.
                      3
```

**47.** Refer to Problem 43. The mathematical model for this problem is:

Maximize $P = 25{,}000x_1 + 18{,}000x_2 + 24{,}000x_3$

Subject to:
$$\tfrac{1}{2}x_1 + \tfrac{1}{2}x_2 + x_3 \le 30$$
$$60{,}000x_1 + 60{,}000x_2 + 80{,}000x_3 \le 3{,}200{,}000$$
$$4{,}000x_1 + 3{,}000x_2 + 4{,}000x_3 \le 180{,}000$$

Following the solutions in Problems 43 and 45, we obtain the simplex tableau:

$$
\begin{array}{c}
\begin{array}{cccccccc}
x_1 & x_2 & x_3 & s_1 & s_2 & s_3 & P
\end{array}\\
\begin{array}{c}
s_1\\
s_2\\
s_3\\
\\
P
\end{array}
\left[
\begin{array}{ccccccc|c}
\tfrac{1}{2} & \tfrac{1}{2} & 1 & 1 & 0 & 0 & 0 & 30\\
6 & 6 & 8 & 0 & 1 & 0 & 0 & 320\\
④ & 3 & 4 & 0 & 0 & 1 & 0 & 180\\
\hline
-25{,}000 & -18{,}000 & -24{,}000 & 0 & 0 & 0 & 1 & 0
\end{array}
\right]
\begin{array}{l}
\tfrac{30}{1/2}=60\\[4pt]
\tfrac{320}{6}=53.33\\[4pt]
\tfrac{180}{4}=45
\end{array}
\end{array}
$$

$$\tfrac{1}{4}R_3 \to R_3$$

$$
\sim
\left[
\begin{array}{ccccccc|c}
\tfrac{1}{2} & \tfrac{1}{2} & 1 & 1 & 0 & 0 & 0 & 30\\
6 & 6 & 8 & 0 & 1 & 0 & 0 & 320\\
① & \tfrac{3}{4} & 1 & 0 & 0 & \tfrac{1}{4} & 0 & 45\\
\hline
-25{,}000 & -18{,}000 & -24{,}000 & 0 & 0 & 0 & 1 & 0
\end{array}
\right]
$$

$$\left(-\tfrac{1}{2}\right)R_3 + R_1 \to R_1,\quad -6R_3 + R_2 \to R_2,\quad 25{,}000R_3 + R_4 \to R_4$$

$$
\begin{array}{c}
\begin{array}{ccccccc}
x_1 & x_2 & x_3 & s_1 & s_2 & s_3 & P
\end{array}\\
\begin{array}{c}
s_1\\
s_2\\
\sim\quad x_1\\
\\
P
\end{array}
\left[
\begin{array}{ccccccc|c}
0 & \tfrac{1}{8} & \tfrac{1}{2} & 1 & 0 & -\tfrac{1}{8} & 0 & 7.5\\
0 & \tfrac{3}{2} & 2 & 0 & 1 & \tfrac{3}{2} & 0 & 50\\
1 & \tfrac{3}{4} & 1 & 0 & 0 & \tfrac{1}{4} & 0 & 45\\
\hline
0 & 750 & 1000 & 0 & 0 & 6250 & 1 & 1{,}125{,}000
\end{array}
\right]
\end{array}
$$

```
Optimal solution: maximum profit is $1,125,000 when x  = 45
                                                     1
colonial houses, x  = 0 split level houses and x  = 0 ranch houses
                  2                             3
are built. In this case, s  = 7.5 acres of land, and s  =
                          1                            2
50(10,000) = $500,000 of capital are not used.
```

Copyright © 2011 Pearson Education, Inc. Publishing as Prentice Hall.

**49.**   Let $x_1$ = the number of grams of food A,

$x_2$ = the number of grams of food B,

$x_3$ = the number of grams of food C.

The mathematical model for this problem is:

Maximize $P = 3x_1 + 4x_2 + 5x_3$

Subject to:  $x_1 + 3x_2 + 2x_3 \le 30$

$2x_1 + x_2 + 2x_3 \le 24$

$x_1,\ x_2,\ x_3 \ge 0$

We introduce slack variables $s_1$ and $s_2$ to obtain the initial form:

$$x_1 + 3x_2 + 2x_3 + s_1 \qquad\qquad = 30$$
$$2x_1 + x_2 + 2x_3 \qquad + s_2 \qquad = 24$$
$$-3x_1 - 4x_2 - 5x_3 \qquad\qquad + P = 0$$

The simplex tableau for this problem is:

$$
\begin{array}{c} \\ s_1 \\ s_2 \\ \\ P \end{array}
\begin{array}{c} x_1 \quad x_2 \quad x_3 \quad s_1 \quad s_2 \quad P \\
\left[\begin{array}{cccccc|c}
1 & 3 & 2 & 1 & 0 & 0 & 30 \\
2 & 1 & \textcircled{2} & 0 & 1 & 0 & 24 \\
\hline
-3 & -4 & -5 & 0 & 0 & 1 & 0
\end{array}\right]
\end{array}
\begin{array}{l} \\ \dfrac{30}{2} = 15 \\ \dfrac{24}{2} = 12 \\ \\ \end{array}
$$

$$\tfrac{1}{2}R_2 \to R_2$$

$$
\sim \left[\begin{array}{cccccc|c}
1 & 3 & 2 & 1 & 0 & 0 & 30 \\
1 & \tfrac{1}{2} & \textcircled{1} & 0 & \tfrac{1}{2} & 0 & 12 \\
\hline
-3 & -4 & -5 & 0 & 0 & 1 & 0
\end{array}\right]
$$

$$(-2)R_2 + R_1 \to R_1,\ \ 5R_2 + R_3 \to R_3$$

$$
\sim \left[\begin{array}{cccccc|c}
-1 & \textcircled{2} & 0 & 1 & -1 & 0 & 6 \\
1 & \tfrac{1}{2} & 1 & 0 & \tfrac{1}{2} & 0 & 12 \\
\hline
2 & -\tfrac{3}{2} & 0 & 0 & \tfrac{5}{2} & 1 & 60
\end{array}\right]
\begin{array}{l} \dfrac{6}{2} = 3 \\ \dfrac{12}{1/2} = 24 \\ \\ \end{array}
$$

$$\tfrac{1}{2}R_1 \to R_1$$

$$
\sim \left[\begin{array}{cccccc|c}
-\tfrac{1}{2} & \textcircled{1} & 0 & \tfrac{1}{2} & -\tfrac{1}{2} & 0 & 3 \\
1 & \tfrac{1}{2} & 1 & 0 & \tfrac{1}{2} & 0 & 12 \\
\hline
2 & -\tfrac{3}{2} & 0 & 0 & \tfrac{5}{2} & 1 & 60
\end{array}\right]
$$

$$\left(-\tfrac{1}{2}\right)R_1 + R_2 \to R_2,\ \tfrac{3}{2}R_1 + R_3 \to R_3$$

$$
\begin{array}{c} \\ x_2 \\ x_3 \\ \\ P \end{array}
\begin{array}{c} x_1 \quad x_2 \quad x_3 \quad s_1 \quad s_2 \quad P \\
\sim \left[\begin{array}{cccccc|c}
-\tfrac{1}{2} & 1 & 0 & \tfrac{1}{2} & -\tfrac{1}{2} & 0 & 3 \\
\tfrac{5}{4} & 0 & 1 & -\tfrac{1}{4} & \tfrac{3}{4} & 0 & \tfrac{21}{2} \\
\hline
\tfrac{5}{4} & 0 & 0 & \tfrac{3}{4} & \tfrac{7}{4} & 1 & \tfrac{129}{2}
\end{array}\right]
\end{array}
$$

```
Optimal solution: the maximum amount of protein is 64.5 units when
x₁ = 0 grams of food A, x₂ = 3 grams of food B and x₃ = 10.5 grams of
food C are used.
```

Copyright © 2011 Pearson Education, Inc. Publishing as Prentice Hall.

$$\begin{array}{c} \\ s_1 \\ s_2 \\ P \end{array}
\begin{array}{cccccc} x_1 & x_2 & x_3 & s_1 & s_2 & P \end{array}
\left[\begin{array}{cccccc|c}
1 & 3 & 2 & 1 & 0 & 0 & 30 \\
2 & 1 & ② & 0 & 1 & 0 & 24 \\
\hline
-3 & -4 & -5 & 0 & 0 & 1 & 0
\end{array}\right]
\begin{array}{l} \frac{30}{2}=15 \\[4pt] \frac{24}{2}=12 \end{array}$$

$$\frac{1}{2}R_2 \rightarrow R_2$$

$$\sim \left[\begin{array}{cccccc|c}
1 & 3 & 2 & 1 & 0 & 0 & 30 \\
1 & \frac{1}{2} & ① & 0 & \frac{1}{2} & 0 & 12 \\
\hline
-3 & -4 & -5 & 0 & 0 & 1 & 0
\end{array}\right]
\quad \sim \left[\begin{array}{cccccc|c}
-1 & ② & 0 & 1 & -1 & 0 & 6 \\
1 & \frac{1}{2} & 1 & 0 & \frac{1}{2} & 0 & 12 \\
\hline
2 & -\frac{3}{2} & 0 & 0 & \frac{5}{2} & 1 & 60
\end{array}\right]
\begin{array}{l} \frac{6}{2}=3 \\[4pt] \frac{12}{1/2}=24 \end{array}$$

$$(-2)R_2 + R_1 \rightarrow R_1, \; 5R_2 + R_3 \rightarrow R_3 \qquad\qquad \frac{1}{2}R_1 \rightarrow R_1$$

$$\begin{array}{c} \\ \\ \\ \end{array}\hspace{5cm}\begin{array}{cccccc} x_1 & x_2 & x_3 & s_1 & s_2 & P \end{array}$$

$$\sim \left[\begin{array}{cccccc|c}
-\frac{1}{2} & ① & 0 & \frac{1}{2} & -\frac{1}{2} & 0 & 3 \\
1 & \frac{1}{2} & 1 & 0 & \frac{1}{2} & 0 & 12 \\
\hline
2 & -\frac{3}{2} & 0 & 0 & \frac{5}{2} & 1 & 60
\end{array}\right]
\quad \sim \begin{array}{c} x_2 \\ x_3 \\ P \end{array}
\left[\begin{array}{cccccc|c}
-\frac{1}{2} & 1 & 0 & \frac{1}{2} & -\frac{1}{2} & 0 & 3 \\
\frac{5}{4} & 0 & 1 & -\frac{1}{4} & \frac{3}{4} & 0 & \frac{21}{2} \\
\hline
\frac{5}{4} & 0 & 0 & \frac{3}{4} & \frac{7}{4} & 1 & \frac{129}{2}
\end{array}\right]$$

$$\left(-\frac{1}{2}\right)R_1 + R_2 \rightarrow R_2, \; \frac{3}{2}R_1 + R_3 \rightarrow R_3$$

Optimal solution: the maximum amount of protein is 64.5 units when $x_1 = 0$ grams of food $A$, $x_2 = 3$ grams of food $B$ and $x_3 = 10.5$ grams of food $C$ are used.

51.  Let $x_1$ = the number of undergraduate students,

$x_2$ = the number of graduate students,

$x_3$ = the number of faculty members.

The mathematical model for this problem is:

Maximize $P = 18x_1 + 25x_2 + 30x_3$

Subject to:    $x_1 + x_2 + x_3 \leq 20$

$100x_1 + 150x_2 + 200x_3 \leq 3200$

$x_1, \; x_2, \; x_3 \geq 0$

Divide the second inequality by 50 to simplify the arithmetic. Then introduce slack variables $s_1$ and $s_2$ to obtain the initial form.

$$x_1 + x_2 + x_3 + s_1 \qquad\qquad = 20$$
$$2x_1 + 3x_2 + 4x_3 \qquad + s_2 \qquad = 64$$
$$-18x_1 - 25x_2 - 30x_3 \qquad\qquad + P = 0$$

Copyright © 2011 Pearson Education, Inc.  Publishing as Prentice Hall.

The simplex tableau for this problem is:

$$
\begin{array}{c}
\\
s_1 \\
s_2 \\
P
\end{array}
\begin{array}{cccccc}
x_1 & x_2 & x_3 & s_1 & s_2 & P \\
\end{array}
\left[
\begin{array}{cccccc|c}
1 & 1 & 1 & 1 & 0 & 0 & 20 \\
2 & 3 & ④ & 0 & 1 & 0 & 64 \\
\hline
-18 & -25 & -30 & 0 & 0 & 1 & 0
\end{array}
\right]
\begin{array}{l}
\frac{20}{1} = 20 \\
\frac{64}{4} = 16
\end{array}
$$

$$\tfrac{1}{4} R_2 \to R_2$$

$$
\sim
\left[
\begin{array}{cccccc|c}
1 & 1 & 1 & 1 & 0 & 0 & 20 \\
\frac{1}{2} & \frac{3}{4} & ① & 0 & \frac{1}{4} & 0 & 16 \\
\hline
-18 & -25 & -30 & 0 & 0 & 1 & 0
\end{array}
\right]
\sim
\left[
\begin{array}{cccccc|c}
⑴\!/\!2 & \frac{1}{4} & 0 & 1 & -\frac{1}{4} & 0 & 4 \\
\frac{1}{2} & \frac{3}{4} & 1 & 0 & \frac{1}{4} & 0 & 16 \\
\hline
-3 & -\frac{5}{2} & 0 & 0 & \frac{15}{2} & 1 & 480
\end{array}
\right]
\begin{array}{l}
\frac{4}{1/2} = 8 \\
\frac{16}{1/2} = 32
\end{array}
$$

$$(-1)R_2 + R_1 \to R_1, \quad 30R_2 + R_3 \to R_3 \qquad\qquad 2R_1 \to R_1$$

$$
\sim
\left[
\begin{array}{cccccc|c}
① & \frac{1}{2} & 0 & 2 & -\frac{1}{2} & 0 & 8 \\
\frac{1}{2} & \frac{3}{4} & 1 & 0 & \frac{1}{4} & 0 & 16 \\
\hline
-3 & -\frac{5}{2} & 0 & 0 & \frac{15}{2} & 1 & 480
\end{array}
\right]
\sim
\left[
\begin{array}{cccccc|c}
1 & ⑴\!/\!2 & 0 & 2 & -\frac{1}{2} & 0 & 8 \\
0 & \frac{1}{2} & 1 & -1 & \frac{1}{2} & 0 & 12 \\
\hline
0 & -1 & 0 & 6 & 6 & 1 & 504
\end{array}
\right]
\begin{array}{l}
\frac{8}{1/2} = 16 \\
\frac{12}{1/2} = 24
\end{array}
$$

$$\left(-\tfrac{1}{2}\right) R_1 + R_2 \to R_2, \quad 3R_1 + R_3 \to R_3 \qquad\qquad 2R_1 \to R_1$$

$$
\sim
\left[
\begin{array}{cccccc|c}
2 & 1 & 0 & 4 & -1 & 0 & 16 \\
0 & \frac{1}{2} & 1 & -1 & \frac{1}{2} & 0 & 12 \\
\hline
0 & -1 & 0 & 6 & 6 & 1 & 504
\end{array}
\right]
\sim
\begin{array}{c}
\\
x_2 \\
x_3 \\
P
\end{array}
\begin{array}{cccccc}
x_1 & x_2 & x_3 & s_1 & s_2 & P \\
\end{array}
\left[
\begin{array}{cccccc|c}
2 & 1 & 0 & 4 & -1 & 0 & 16 \\
-1 & 0 & 1 & -3 & 1 & 0 & 4 \\
\hline
2 & 0 & 0 & 10 & 5 & 1 & 520
\end{array}
\right]
$$

$$\left(-\tfrac{1}{2}\right)R_1 + R_2 \to R_2, \quad R_1 + R_3 \to R_3$$

```
Optimal solution: the maximum number of interviews is 520 when x₁ = 0
undergraduates, x₂ = 16 graduate students, and x₃ = 4 faculty members
are hired.
```

---

## EXERCISE 6-3

*Things to remember:*

1.  Given a matrix $A$. The transpose of $A$, denoted $A^T$, is the matrix formed by interchanging the rows and corresponding columns of $A$ (first row with first column, second row with second column, and so on.)

2.  FORMATION OF THE DUAL PROBLEM

    Given a minimization problem with $\geq$ problem constraints:

    *Step 1.* Use the coefficients and constants in the problem constraints and the objective function to form a matrix $A$ with the coefficients of the objective function in the last row.

    *Step 2.* Interchange the rows and columns of matrix $A$ to form the matrix $A^T$, the transpose of $A$.

    *Step 3.* Use the rows of $A^T$ to form a maximization problem with $\leq$ problem constraints.

Copyright © 2011 Pearson Education, Inc. Publishing as Prentice Hall.

3.    THE FUNDAMENTAL PRINCIPLE OF DUALITY

A minimization problem has a solution if and only if its dual problem has a solution. If a solution exists, then the optimal value of the minimization problem is the same as the optimal value of the dual problem.

4.    SOLUTION OF A MINIMIZATION PROBLEM

Given a minimization problem with nonnegative coefficients in the objective function:

*Step 1.* Write all problem constraints as $\geq$ inequalities. (This may introduce negative numbers on the right side of the problem constraints.)

*Step 2.* Form the dual problem.

*Step 3.* Write the initial system of the dual problem, using the variables from the minimization problem as the slack variables.

*Step 4.* Use the simplex method to solve the dual problem.

*Step 5.* Read the solution of the minimization problem from the bottom row of the final simplex tableau in Step 4.

[Note: If the dual problem has no solution, then the minimization problem has no solution.]

---

1.    $A = [-5 \ 0 \ 3 \ -1 \ 8]; A^T = \begin{bmatrix} -5 \\ 0 \\ 3 \\ -1 \\ 8 \end{bmatrix}$     3. $A = \begin{bmatrix} 1 \\ -2 \\ 0 \\ 4 \end{bmatrix}; A^T = [1 \ -2 \ 0 \ 4]$

5.    $A = \begin{bmatrix} 2 & 1 & -6 & 0 & -1 \\ 5 & 2 & 0 & 1 & 3 \end{bmatrix}; \ A^T = \begin{bmatrix} 2 & 5 \\ 1 & 2 \\ -6 & 0 \\ 0 & 1 \\ -1 & 3 \end{bmatrix}$

7.    $A = \begin{bmatrix} 1 & 2 & -1 \\ 0 & 2 & -7 \\ 8 & 0 & 1 \\ 4 & -1 & 3 \end{bmatrix}; \ A^T = \begin{bmatrix} 1 & 0 & 8 & 4 \\ 2 & 2 & 0 & -1 \\ -1 & -7 & 1 & 3 \end{bmatrix}$

9.    (A)    Given the minimization problem:

Minimize  $C = 8x_1 + 9x_2$

Subject to:   $x_1 + 3x_2 \geq 4$

$\qquad\qquad 2x_1 + x_2 \geq 5$

$\qquad\qquad x_1, \ x_2 \geq 0$

The matrix $A$ corresponding to this problem is: $A = \begin{bmatrix} 1 & 3 & | & 4 \\ 2 & 1 & | & 5 \\ \hline 8 & 9 & | & 1 \end{bmatrix}$

Copyright © 2011 Pearson Education, Inc. Publishing as Prentice Hall.

The matrix $A^T$ corresponding to the dual problem has the rows of $A$ as its columns. Thus:

$$A^T = \begin{bmatrix} 1 & 2 & 8 \\ 3 & 1 & 9 \\ \hline 4 & 5 & 1 \end{bmatrix}$$

The dual problem is: Maximize $P = 4y_1 + 5y_2$

Subject to:  $y_1 + 2y_2 \le 8$
$3y_1 + y_2 \le 9$
$y_1, y_2 \ge 0$

(B)   Letting $x_1$ and $x_2$ be slack variables, the initial system for the dual problem is:

$y_1 + 2y_2 + x_1 \qquad = 8$
$3y_1 + y_2 \qquad + x_2 \qquad = 9$
$-4y_1 - 5y_2 \qquad\qquad + P = 0$

(C)   The simplex tableau for this problem is:

$$\begin{array}{c} \\ x_1 \\ x_2 \\ P \end{array} \begin{array}{ccccc} y_1 & y_2 & x_1 & x_2 & P \\ \left[\begin{array}{ccccc|c} 1 & 2 & 1 & 0 & 0 & 8 \\ 3 & 1 & 0 & 1 & 0 & 9 \\ \hline -4 & -5 & 0 & 0 & 1 & 0 \end{array}\right] \end{array}$$

11.   From the final simplex table

$$\begin{array}{c} \\ y_2 \\ y_1 \\ P \end{array} \begin{array}{ccccc} y_1 & y_2 & x_1 & x_2 & P \\ \left[\begin{array}{ccccc|c} 0 & 1 & 5 & -2 & 0 & 5 \\ 1 & 0 & -7 & 3 & 0 & 3 \\ \hline 0 & 0 & 1 & 2 & 1 & 121 \end{array}\right] \end{array}$$

(A) the optimal solution of the dual problem is:
    maximum value of $P = 121$ at $y_1 = 3$ and $y_2 = 5$;

(B) the optimal solution of the minimization problem is:
    minimum value of $C = 121$ at $x_1 = 1$, $x_2 = 2$.

13.   (A)   The matrix corresponding to the given problem is: $A = \begin{bmatrix} 4 & 1 & 13 \\ 3 & 1 & 12 \\ \hline 9 & 2 & 1 \end{bmatrix}$

The matrix $A^T$ corresponding to the dual problem has the rows of $A$ as its columns, that is:

$$A^T = \begin{bmatrix} 4 & 3 & 9 \\ 1 & 1 & 2 \\ \hline 13 & 12 & 1 \end{bmatrix}$$

Thus, the dual problem is: Maximize $P = 13y_1 + 12y_2$

Copyright © 2011 Pearson Education, Inc. Publishing as Prentice Hall.

Subject to:  $4y_1 + 3y_2 \leq 9$

$$y_1 + y_2 \leq 2$$

$$y_1, y_2 \geq 0$$

(B)    We introduce slack variables $x_1$ and $x_2$ to obtain the initial system for the dual problem:

$$4y_1 + 3y_2 + x_1 \qquad = 9$$

$$y_1 + y_2 \qquad + x_2 \qquad = 2$$

$$-13y_1 - 12y_2 \qquad \qquad + P = 0$$

The simplex tableau for this problem is:

$$
\begin{array}{c}
\quad\;\; y_1 \quad y_2 \quad x_1 \quad x_2 \quad P \\
\begin{array}{c} x_1 \\ x_2 \\ P \end{array}
\left[
\begin{array}{ccccc|c}
4 & 3 & 1 & 0 & 0 & 9 \\
① & 1 & 0 & 1 & 0 & 2 \\
\hline
-13 & -12 & 0 & 0 & 1 & 0
\end{array}
\right]
\begin{array}{c} \frac{9}{4} = 2.25 \\ \frac{2}{1} = 2 \end{array}
\end{array}
$$

$$
\sim
\begin{array}{c}
\quad\;\; y_1 \quad y_2 \quad x_1 \quad x_2 \quad P \\
\begin{array}{c} x_1 \\ y_1 \\ P \end{array}
\left[
\begin{array}{ccccc|c}
0 & -1 & 1 & -4 & 0 & 1 \\
1 & 1 & 0 & 1 & 0 & 2 \\
\hline
0 & 1 & 0 & 13 & 1 & 26
\end{array}
\right]
\end{array}
$$

$(-4)R_2 + R_1 \rightarrow R_1$ and $13R_2 + R_3 \rightarrow R_3$

Optimal solution: min $C = 26$ at $x_1 = 0$, $x_2 = 13$.

15.  (A)    The matrix corresponding to the given problem is: $A = \begin{bmatrix} 2 & 3 & 15 \\ 1 & 2 & 8 \\ \hline 7 & 12 & 1 \end{bmatrix}$

The matrix $A^T$ corresponding to the dual problem has the rows of $A$ as its columns, that is:

$$A^T = \begin{bmatrix} 2 & 1 & 7 \\ 3 & 2 & 12 \\ \hline 15 & 8 & 1 \end{bmatrix}$$

Thus, the dual problem is: Maximize $P = 15y_1 + 8y_2$

Subject to:  $2y_1 + y_2 \leq 7$

$$3y_1 + 2y_2 \leq 12$$

$$y_1, y_2 \geq 0$$

(B)    We introduce slack variables $x_1$ and $x_2$ to obtain the initial system for the dual problem:

$$2y_1 + y_2 + x_1 \qquad = 7$$

$$3y_1 + 2y_2 \qquad + x_2 \qquad = 12$$

$$-15y_1 - 8y_2 \qquad \qquad + P = 0$$

Copyright © 2011 Pearson Education, Inc. Publishing as Prentice Hall.

The simplex tableau for this problem is:

$$\begin{array}{c} \\ x_1 \\ x_2 \\ P \end{array} \begin{array}{ccccc} y_1 & y_2 & x_1 & x_2 & P \end{array}$$

$$\begin{array}{c} x_1 \\ x_2 \\ P \end{array} \left[\begin{array}{ccccc|c} ② & 1 & 1 & 0 & 0 & 7 \\ 3 & 2 & 0 & 1 & 0 & 12 \\ \hline -15 & -8 & 0 & 0 & 1 & 0 \end{array}\right] \begin{array}{l} \frac{7}{2} = 3.5 \\ \frac{12}{3} = 4 \end{array} \sim \left[\begin{array}{ccccc|c} ① & \frac{1}{2} & \frac{1}{2} & 0 & 0 & \frac{7}{2} \\ 3 & 2 & 0 & 1 & 0 & 12 \\ \hline -15 & -8 & 0 & 0 & 1 & 0 \end{array}\right]$$

$$\frac{1}{2}R_1 \to R_1 \qquad\qquad\qquad (-3)R_1 + R_2 \to R_2 \text{ and } 15R_1 + R_3 \to R_3$$

$$\sim \left[\begin{array}{ccccc|c} 1 & \frac{1}{2} & \frac{1}{2} & 0 & 0 & \frac{7}{2} \\ 0 & ⑫\frac{1}{2} & -\frac{3}{2} & 1 & 0 & \frac{3}{2} \\ \hline 0 & -\frac{1}{2} & \frac{15}{2} & 0 & 1 & \frac{105}{2} \end{array}\right] \begin{array}{l} \frac{7/2}{1/2} = 7 \\ \frac{3/2}{1/2} = 3 \end{array} \sim \left[\begin{array}{ccccc|c} 1 & \frac{1}{2} & \frac{1}{2} & 0 & 0 & \frac{7}{2} \\ 0 & ① & -3 & 2 & 0 & 3 \\ \hline 0 & -\frac{1}{2} & \frac{15}{2} & 0 & 1 & \frac{105}{2} \end{array}\right]$$

$$2R_2 \to R_2 \qquad\qquad\qquad \left(-\frac{1}{2}\right)R_2 + R_1 \to R_1 \text{ and } \frac{1}{2}R_2 + R_3 \to R_3$$

$$\begin{array}{c} \\ y_1 \\ \sim y_2 \\ P \end{array} \begin{array}{ccccc} y_1 & y_2 & x_1 & x_2 & P \end{array}$$

$$\begin{array}{c} y_1 \\ \sim y_2 \\ P \end{array} \left[\begin{array}{ccccc|c} 1 & 0 & 2 & -1 & 0 & 2 \\ 0 & 1 & -3 & 2 & 0 & 3 \\ \hline 0 & 0 & 6 & 1 & 1 & 54 \end{array}\right]$$

Optimal solution: min $C = 54$ at $x_1 = 6$, $x_2 = 1$.

17. (A)    The matrices corresponding to the given problem and to the dual problem are:

$$A = \left[\begin{array}{cc|c} 2 & 1 & 8 \\ -2 & 3 & 4 \\ \hline 11 & 4 & 1 \end{array}\right] \quad \text{and} \quad A^T = \left[\begin{array}{cc|c} 2 & -2 & 11 \\ 1 & 3 & 4 \\ \hline 8 & 4 & 1 \end{array}\right] \text{ respectively.}$$

Thus, the dual problem is: Maximize $P = 8y_1 + 4y_2$

Subject to: $2y_1 - 2y_2 \le 11$

$y_1 + 3y_2 \le 4$

$y_1, y_2 \ge 0$

(B)    We introduce slack variables $x_1$ and $x_2$ to obtain the initial system for the dual problem:

$$2y_1 - 2y_2 + x_1 \qquad\qquad = 11$$
$$y_1 + 3y_2 \qquad + x_2 \qquad = 4$$
$$-8y_1 - 4y_2 \qquad\qquad + P = 0$$

Copyright © 2011 Pearson Education, Inc. Publishing as Prentice Hall.

The simplex tableau for this problem is:

$$
\begin{array}{c}
\phantom{x_1} \\
x_1 \\
\sim x_2 \\
P
\end{array}
\begin{array}{c}
\begin{array}{ccccc}
y_1 & y_2 & x_1 & x_2 & P
\end{array} \\
\left[
\begin{array}{ccccc|c}
2 & -2 & 1 & 0 & 0 & 11 \\
\textcircled{1} & 3 & 0 & 1 & 0 & 4 \\
\hline
-8 & -4 & 0 & 0 & 1 & 0
\end{array}
\right]
\end{array}
\quad
\begin{array}{c}
\frac{11}{2} = 5.5 \\[6pt]
\frac{4}{1} = 4
\end{array}
\quad
\begin{array}{c}
\phantom{x_1} \\
x_1 \\
\sim y_1 \\
P
\end{array}
\begin{array}{c}
\begin{array}{ccccc}
y_1 & y_2 & x_1 & x_2 & P
\end{array} \\
\left[
\begin{array}{ccccc|c}
0 & -8 & 1 & -2 & 0 & 3 \\
1 & 3 & 0 & 1 & 0 & 4 \\
\hline
0 & 20 & 0 & 8 & 1 & 32
\end{array}
\right]
\end{array}
$$

$$(-2)R_2 + R_1 \rightarrow R_1 \text{ and } 8R_2 + R_3 \rightarrow R_3$$

Optimal solution: min $C$ = 32 at $x_1$ = 0, $x_2$ = 8.

19.  (A)    The matrices corresponding to the given problem and the dual problem are:

$$
A = \left[
\begin{array}{cc|c}
-3 & 1 & 6 \\
1 & -2 & 4 \\
\hline
7 & 9 & 1
\end{array}
\right]
\quad \text{and} \quad
A^T = \left[
\begin{array}{cc|c}
-3 & 1 & 7 \\
1 & -2 & 9 \\
\hline
6 & 4 & 1
\end{array}
\right]
$$

respectively.

Thus, the dual problem is:

$$\text{Maximize } P = 6y_1 + 4y_2$$

$$
\text{Subject to: } -3y_1 + y_2 \le 7
$$
$$
y_1 - 2y_2 \le 9
$$
$$
y_1, \; y_2 \ge 0
$$

(B)    We introduce slack variables $x_1$ and $x_2$ to obtain the initial system for the dual problem:

$$
\begin{aligned}
-3y_1 + y_2 + x_1 \qquad &= 7 \\
y_1 - 2y_2 \qquad + x_2 &= 9 \\
-6y_1 - 4y_2 \qquad \qquad + P &= 0
\end{aligned}
$$

The simplex tableau for this problem is:

$$
\begin{array}{c}
\phantom{x_1} \\
x_1 \\
x_2 \\
P
\end{array}
\begin{array}{c}
\begin{array}{ccccc}
y_1 & y_2 & x_1 & x_2 & P
\end{array} \\
\left[
\begin{array}{ccccc|c}
-3 & 1 & 1 & 0 & 0 & 7 \\
\textcircled{1} & -2 & 0 & 1 & 0 & 9 \\
\hline
-6 & -4 & 0 & 0 & 1 & 0
\end{array}
\right]
\end{array}
\sim
\begin{array}{c}
\begin{array}{ccccc}
y_1 & y_2 & x_1 & x_2 & P
\end{array} \\
\left[
\begin{array}{ccccc|c}
0 & -5 & 1 & 3 & 0 & 34 \\
1 & -2 & 0 & 1 & 0 & 9 \\
\hline
0 & -16 & 0 & 6 & 1 & 54
\end{array}
\right]
\end{array}
$$

$3R_2 + R_1 \rightarrow R_1$ and $6R_2 + R_3 \rightarrow R_3$

The negative elements in the second column above the dashed line indicate that the problem does not have a solution.

Copyright © 2011 Pearson Education, Inc.  Publishing as Prentice Hall.

**21.** The matrices corresponding to the given problem and the dual problem are:

$$A = \begin{bmatrix} 2 & 1 & 8 \\ 1 & 2 & 8 \\ \hline 3 & 9 & 1 \end{bmatrix} \quad \text{and} \quad A^T = \begin{bmatrix} 2 & 1 & 3 \\ 1 & 2 & 9 \\ \hline 8 & 8 & 1 \end{bmatrix}$$

respectively. Thus, the dual problem is:

Maximize $P = 8y_1 + 8y_2$

Subject to: $2y_1 + y_2 \le 3$

$y_1 + 2y_2 \le 9$

$y_1,\ y_2 \ge 0$

We introduce slack variables $x_1$ and $x_2$ to obtain the initial system:

$$2y_1 + y_2 + x_1 \qquad\qquad = 3$$
$$y_1 + 2y_2 \qquad + x_2 \qquad = 9$$
$$-8y_1 - 8y_2 \qquad\qquad + P = 0$$

The simplex tableau for this problem is:

$$
\begin{array}{c}
\begin{array}{cccccc}
y_1 & y_2 & x_1 & x_2 & P & \\
\end{array} \\
\begin{array}{c}
x_1 \\ x_2 \\ P
\end{array}
\left[
\begin{array}{ccccc|c}
2 & \textcircled{1} & 1 & 0 & 0 & 3 \\
1 & 2 & 0 & 1 & 0 & 9 \\
\hline
-8 & -8 & 0 & 0 & 1 & 0
\end{array}
\right]
\begin{array}{l}
\frac{3}{1} = 3 \\[4pt]
\frac{9}{2} = 4.5
\end{array}
\end{array}
\sim
\begin{array}{c}
\begin{array}{cccccc}
y_1 & y_2 & x_1 & x_2 & P & \\
\end{array} \\
\begin{array}{c}
y_2 \\ x_2 \\ P
\end{array}
\left[
\begin{array}{ccccc|c}
2 & 1 & 1 & 0 & 0 & 3 \\
-3 & 0 & -2 & 1 & 0 & 3 \\
\hline
8 & 0 & 8 & 0 & 1 & 24
\end{array}
\right]
\end{array}
$$

$(-2)R_1 + R_2 \to R_2$ and $8R_1 + R_3 \to R_3$

Optimal solution: min $C = 24$
at $x_1 = 8$, $x_2 = 0$.

[Note: We could use either column 1 or column 2 as the pivot column. Column 2 involves slightly simpler calculations.]

**23.** The matrices corresponding to the given problem and the dual problem are:

$$A = \begin{bmatrix} 1 & 1 & 4 \\ 1 & -2 & -8 \\ -2 & 1 & -8 \\ \hline 7 & 5 & 1 \end{bmatrix} \quad \text{and} \quad A^T = \begin{bmatrix} 1 & 1 & -2 & 7 \\ 1 & -2 & 1 & 5 \\ \hline 4 & -8 & -8 & 1 \end{bmatrix} \text{ respectively.}$$

Thus, the dual problem is: Maximize $P = 4y_1 - 8y_2 - 8y_3$

Subject to: $y_1 + y_2 - 2y_3 \le 7$

$y_1 - 2y_2 + y_3 \le 5$

$y_1,\ y_2,\ y_3 \ge 0$

We introduce slack variables $x_1$ and $x_2$ to obtain the initial system:

$$y_1 + y_2 - 2y_3 + x_1 \qquad\qquad = 7$$
$$y_1 - 2y_2 + y_3 \qquad + x_2 \qquad = 5$$
$$-4y_1 + 8y_2 + 8y_3 \qquad\qquad + P = 0$$

Copyright © 2011 Pearson Education, Inc. Publishing as Prentice Hall.

The simplex tableau for this problem is:

$$
\begin{array}{c}
\begin{array}{cccccc} y_1 & y_2 & y_3 & x_1 & x_2 & P \end{array} \\
\begin{array}{c} x_1 \\ x_2 \\ P \end{array}
\left[
\begin{array}{cccccc|c}
1 & 1 & -2 & 1 & 0 & 0 & 7 \\
\textcircled{1} & -2 & 1 & 0 & 1 & 0 & 5 \\
\hline
-4 & 8 & 8 & 0 & 0 & 1 & 0
\end{array}
\right]
\end{array}
\quad
\begin{array}{c}
\frac{7}{1} = 7 \\
\frac{5}{1} = 5
\end{array}
\quad \sim \quad
\begin{array}{c}
\begin{array}{cccccc} y_1 & y_2 & y_3 & x_1 & x_2 & P \end{array} \\
\begin{array}{c} x_1 \\ y_1 \\ P \end{array}
\left[
\begin{array}{cccccc|c}
0 & 3 & -3 & 1 & -1 & 0 & 2 \\
1 & -2 & 1 & 0 & 1 & 0 & 5 \\
\hline
0 & 0 & 12 & 0 & 4 & 1 & 20
\end{array}
\right]
\end{array}
$$

$(-1)R_2 + R_1 \rightarrow R_1$ and $4R_2 + R_3 \rightarrow R_3$

Optimal solution: min $C = 20$
at $x_1 = 0$, $x_2 = 4$.

25. The matrices corresponding to the given problem and the dual problem are:

$$
A = \begin{bmatrix}
2 & 1 & | & 16 \\
1 & 1 & | & 12 \\
1 & 2 & | & 14 \\
\hline
10 & 30 & | & 1
\end{bmatrix}
\quad \text{and} \quad
A^T = \begin{bmatrix}
2 & 1 & 1 & | & 10 \\
1 & 1 & 2 & | & 30 \\
\hline
16 & 12 & 14 & | & 1
\end{bmatrix}
\text{ respectively.}
$$

Thus, the dual problem is: Maximize $P = 16y_1 + 12y_2 + 14y_3$

Subject to: $2y_1 + y_2 + y_3 \le 10$

$y_1 + y_2 + 2y_3 \le 30$

$y_1,\ y_2,\ y_3 \ge 0$

We introduce slack variables $x_1$ and $x_2$ to obtain the initial system:

$$
\begin{aligned}
2y_1 + y_2 + y_3 + x_1 \qquad\qquad &= 10 \\
y_1 + y_2 + 2y_3 \qquad + x_2 \quad &= 30 \\
-16y_1 - 12y_2 - 14y_3 \qquad\qquad + P &= 0
\end{aligned}
$$

The simplex tableau for this problem is:

$$
\begin{array}{c}
\begin{array}{cccccc} y_1 & y_2 & y_3 & x_1 & x_2 & P \end{array} \\
\begin{array}{c} x_1 \\ x_2 \\ P \end{array}
\left[
\begin{array}{cccccc|c}
\textcircled{2} & 1 & 1 & 1 & 0 & 0 & 10 \\
1 & 1 & 2 & 0 & 1 & 0 & 30 \\
\hline
-16 & -12 & -14 & 0 & 0 & 1 & 0
\end{array}
\right]
\end{array}
\quad
\begin{array}{c}
\frac{10}{2} = 5 \\
\frac{30}{1} = 30
\end{array}
\quad \sim \quad
\left[
\begin{array}{cccccc|c}
\textcircled{1} & \frac{1}{2} & \frac{1}{2} & \frac{1}{2} & 0 & 0 & 5 \\
1 & 1 & 2 & 0 & 1 & 0 & 30 \\
\hline
-16 & -12 & -14 & 0 & 0 & 1 & 0
\end{array}
\right]
$$

$\frac{1}{2}R_1 \rightarrow R_1$

$(-1)R_1 + R_2 \rightarrow R_2$ and $16R_1 + R_3 \rightarrow R_3$

$$
\sim
\left[
\begin{array}{cccccc|c}
1 & \frac{1}{2} & \textcircled{$\frac{1}{2}$} & \frac{1}{2} & 0 & 0 & 5 \\
0 & \frac{1}{2} & \frac{3}{2} & -\frac{1}{2} & 1 & 0 & 25 \\
\hline
0 & -4 & -6 & 8 & 0 & 1 & 80
\end{array}
\right]
\quad
\begin{array}{c}
\frac{5}{1/2} = 10 \\
\frac{25}{3/2} = 16.66
\end{array}
\quad \sim \quad
\left[
\begin{array}{cccccc|c}
2 & 1 & \textcircled{1} & 1 & 0 & 0 & 10 \\
0 & \frac{1}{2} & \frac{3}{2} & -\frac{1}{2} & 1 & 0 & 25 \\
\hline
0 & -4 & -6 & 8 & 0 & 1 & 80
\end{array}
\right]
$$

$2R_1 \rightarrow R_1$

$\left(-\frac{3}{2}\right)R_1 + R_2 \rightarrow R_2$ and $6R_1 + R_3 \rightarrow R_3$

Copyright © 2011 Pearson Education, Inc. Publishing as Prentice Hall.

$$\begin{array}{c} \\ y_3 \\ \sim \ x_2 \\ P \end{array} \begin{array}{cccccc} y_1 & y_2 & y_3 & x_1 & x_2 & P \\ \left[\begin{array}{cccccc|c} 2 & 1 & 1 & 1 & 0 & 0 & 10 \\ -3 & -1 & 0 & -2 & 1 & 0 & 10 \\ \hline 12 & 2 & 0 & 14 & 0 & 1 & 140 \end{array}\right] \end{array}$$

Optimal solution: min $C = 140$ at $x_1 = 14$, $x_2 = 0$.

$$\sim \left[\begin{array}{cccccc|c} 1 & \frac{1}{2} & \textcircled{\tfrac{1}{2}} & \frac{1}{2} & 0 & 0 & 5 \\ 0 & \frac{1}{2} & \frac{3}{2} & -\frac{1}{2} & 1 & 0 & 25 \\ \hline 0 & -4 & -6 & 8 & 0 & 1 & 80 \end{array}\right] \begin{array}{l} \frac{5}{1/2} = 10 \\[4pt] \frac{25}{3/2} = 16.66 \\[10pt] \end{array} \sim \left[\begin{array}{cccccc|c} 2 & 1 & \textcircled{1} & 1 & 0 & 0 & 10 \\ 0 & \frac{1}{2} & \frac{3}{2} & -\frac{1}{2} & 1 & 0 & 25 \\ \hline 0 & -4 & -6 & 8 & 0 & 1 & 80 \end{array}\right]$$

$$2R_1 \to R_1 \qquad\qquad\qquad\qquad \left(-\frac{3}{2}\right)R_1 + R_2 \to R_2 \text{ and } 6R_1 + R_3 \to R_3$$

$$\begin{array}{c} \\ y_3 \\ \sim \ x_2 \\ P \end{array} \begin{array}{cccccc} y_1 & y_2 & y_3 & x_1 & x_2 & P \\ \left[\begin{array}{cccccc|c} 2 & 1 & 1 & 1 & 0 & 0 & 10 \\ -3 & -1 & 0 & -2 & 1 & 0 & 10 \\ \hline 12 & 2 & 0 & 14 & 0 & 1 & 140 \end{array}\right] \end{array}$$

Optimal solution: min $C = 140$ at $x_1 = 14$, $x_2 = 0$.

**27.** The matrices corresponding to the given problem and the dual problem are:

$$A = \left[\begin{array}{cc|c} 1 & 0 & 4 \\ 1 & 1 & 8 \\ 1 & 2 & 10 \\ \hline 5 & 7 & 1 \end{array}\right] \quad \text{and} \quad A^T = \left[\begin{array}{ccc|c} 1 & 1 & 1 & 5 \\ 0 & 1 & 2 & 7 \\ \hline 4 & 8 & 10 & 1 \end{array}\right] \text{ respectively.}$$

Thus, the dual problem is: Maximize $P = 4y_1 + 8y_2 + 10y_3$

Subject to: $y_1 + y_2 + y_3 \le 5$

$$y_2 + 2y_3 \le 7$$

$$y_1, \ y_2, \ y_3 \ge 0$$

We introduce slack variables $x_1$ and $x_2$ to obtain the initial system:

$$y_1 + y_2 + y_3 + x_1 \qquad\quad = 5$$
$$y_2 + 2y_3 \qquad + x_2 \ = 7$$
$$-4y_1 - 8y_2 - 10y_3 \qquad\quad\ + P = 0$$

The simplex tableau for this problem is:

Copyright © 2011 Pearson Education, Inc. Publishing as Prentice Hall.

$$
\begin{array}{c}
\begin{array}{cccccc}
y_1 & y_2 & y_3 & x_1 & x_2 & P
\end{array}\\
\begin{array}{c}
x_1\\ x_2\\ P
\end{array}
\left[
\begin{array}{cccccc|c}
1 & 1 & 1 & 1 & 0 & 0 & 5\\
0 & 1 & ② & 0 & 1 & 0 & 7\\
\hline
-4 & -8 & -10 & 0 & 0 & 1 & 0
\end{array}
\right]
\begin{array}{l}
\frac{5}{1}=5\\[4pt]
\frac{7}{2}=3.5
\end{array}
\sim
\left[
\begin{array}{cccccc|c}
1 & 1 & 1 & 1 & 0 & 0 & 5\\
0 & \frac{1}{2} & ① & 0 & \frac{1}{2} & 0 & \frac{7}{2}\\
\hline
-4 & -8 & -10 & 0 & 0 & 1 & 0
\end{array}
\right]
\end{array}
$$

$$\tfrac{1}{2}R_2 \to R_2 \qquad\qquad (-1)R_2 + R_1 \to R_1 \ \text{ and } \ 10R_2 + R_3 \to R_3$$

$$
\sim
\left[
\begin{array}{cccccc|c}
① & \frac{1}{2} & 0 & 1 & -\frac{1}{2} & 0 & \frac{3}{2}\\
0 & \frac{1}{2} & 1 & 0 & \frac{1}{2} & 0 & \frac{7}{2}\\
\hline
-4 & -3 & 0 & 0 & 5 & 1 & 35
\end{array}
\right]
\sim
\left[
\begin{array}{cccccc|c}
1 & ①\!\!/②\ & 0 & 1 & -\frac{1}{2} & 0 & \frac{3}{2}\\
0 & \frac{1}{2} & 1 & 0 & \frac{1}{2} & 0 & \frac{7}{2}\\
\hline
0 & -1 & 0 & 4 & 3 & 1 & 41
\end{array}
\right]
\begin{array}{l}
\frac{3/2}{1/2}=3\\[4pt]
\frac{7/2}{1/2}=7
\end{array}
$$

$$4R_1 + R_3 \to R_3 \qquad\qquad\qquad 2R_1 \to R_1$$

$$
\begin{array}{c}
\\
\sim
\left[
\begin{array}{cccccc|c}
2 & ① & 0 & 2 & -1 & 0 & 3\\
0 & \frac{1}{2} & 1 & 0 & \frac{1}{2} & 0 & \frac{7}{2}\\
\hline
0 & -1 & 0 & 4 & 3 & 1 & 41
\end{array}
\right]
\end{array}
\qquad
\begin{array}{c}
\begin{array}{cccccc}
y_1 & y_2 & y_3 & x_1 & x_2 & P
\end{array}\\
\begin{array}{c}
y_2\\ y_3\\ P
\end{array}
\left[
\begin{array}{cccccc|c}
2 & 1 & 0 & 2 & -1 & 0 & 3\\
-1 & 0 & 1 & -1 & 1 & 0 & 2\\
\hline
2 & 0 & 0 & 6 & 2 & 1 & 44
\end{array}
\right]
\end{array}
$$

$$\left(-\tfrac{1}{2}\right)R_1 + R_2 \to R_2 \qquad\qquad \text{Optimal solution: min } C = 44 \text{ at}$$
$$\text{and } R_3 + R_1 \to R_3 \qquad\qquad\qquad x_1 = 6,\ x_2 = 2.$$

**29.** The matrices corresponding to the given problem and the dual problem are:

$$
A = \left[
\begin{array}{ccc|c}
1 & 1 & 2 & 7\\
2 & 1 & 1 & 4\\
\hline
10 & 7 & 12 & 1
\end{array}
\right]
\quad\text{and}\quad
A^T = \left[
\begin{array}{cc|c}
1 & 2 & 10\\
1 & 1 & 7\\
2 & 1 & 12\\
\hline
7 & 4 & 1
\end{array}
\right]
\quad\text{respectively.}
$$

Thus, the dual problem is: Maximize $P = 7y_1 + 4y_2$

$$\text{Subject to: } y_1 + 2y_2 \le 10$$
$$y_1 + y_2 \le 7$$
$$2y_1 + y_2 \le 12$$
$$y_1,\ y_2 \ge 0$$

We introduce slack variables $x_1, x_2,$ and $x_3$ to obtain the initial system:

$$
\begin{aligned}
y_1 + 2y_2 + x_1 \qquad\qquad &= 10\\
y_1 + y_2 \qquad + x_2 \qquad &= 7\\
2y_1 + y_2 \qquad\qquad + x_3 &= 12\\
-7y_1 - 4y_2 \qquad\qquad\qquad + P &= 0
\end{aligned}
$$

The simplex tableau for this problem is:

Copyright © 2011 Pearson Education, Inc.  Publishing as Prentice Hall.

$$
\begin{array}{c}
\begin{array}{cccccc} y_1 & y_2 & x_1 & x_2 & x_3 & P \end{array} \\
\begin{array}{c} x_1 \\ x_2 \\ \sim \ x_3 \\ P \end{array}
\left[
\begin{array}{cccccc|c}
1 & 2 & 1 & 0 & 0 & 0 & 10 \\
1 & 1 & 0 & 1 & 0 & 0 & 7 \\
\boxed{2} & 1 & 0 & 0 & 1 & 0 & 12 \\
\hline
-7 & -4 & 0 & 0 & 0 & 1 & 0
\end{array}
\right]
\end{array}
\begin{array}{c}
\frac{10}{1} = 10 \\[6pt]
\frac{7}{1} = 7 \\[6pt]
\frac{12}{2} = 6
\end{array}
\quad \sim
\left[
\begin{array}{cccccc|c}
1 & 2 & 1 & 0 & 0 & 0 & 10 \\
1 & 1 & 0 & 1 & 0 & 0 & 7 \\
\boxed{1} & \frac{1}{2} & 0 & 0 & \frac{1}{2} & 0 & 6 \\
\hline
-7 & -4 & 0 & 0 & 0 & 1 & 0
\end{array}
\right]
$$

$$\frac{1}{2}R_3 \to R_3 \qquad\qquad\qquad (-1)R_3 + R_1 \to R_1, \ (-1) \ R_3 + R_2 \to R_2,$$
$$\text{and} \quad 7R_3 + R_4 \to R_4$$

$$
\sim
\left[
\begin{array}{cccccc|c}
0 & \frac{3}{2} & 1 & 0 & -\frac{1}{2} & 0 & 4 \\
0 & \boxed{\frac{1}{2}} & 0 & 1 & -\frac{1}{2} & 0 & 1 \\
1 & \frac{1}{2} & 0 & 0 & \frac{1}{2} & 0 & 6 \\
\hline
0 & -\frac{1}{2} & 0 & 0 & \frac{7}{2} & 1 & 42
\end{array}
\right]
\begin{array}{c}
\frac{4}{3/2} = \frac{8}{3} \\[6pt]
\frac{1}{1/2} = 2 \\[6pt]
\frac{6}{1/2} = 12
\end{array}
\quad \sim
\left[
\begin{array}{cccccc|c}
0 & \frac{3}{2} & 1 & 0 & -\frac{1}{2} & 0 & 4 \\
0 & \boxed{1} & 0 & 2 & -1 & 0 & 2 \\
1 & \frac{1}{2} & 0 & 0 & \frac{1}{2} & 0 & 6 \\
\hline
0 & -\frac{1}{2} & 0 & 0 & \frac{7}{2} & 1 & 42
\end{array}
\right]
$$

$$2R_2 \to R_2 \qquad\qquad \left(-\frac{3}{2}\right)R_2 + R_1 \to R_1, \ \left(-\frac{1}{2}\right)R_2 + R_3 \to R_3,$$
$$\text{and} \quad \frac{1}{2}R_2 + R_4 \to R_4$$

$$
\begin{array}{c}
\begin{array}{cccccc} y_1 & y_2 & x_1 & x_2 & x_3 & P \end{array} \\
\begin{array}{c} x_1 \\ \sim \ y_2 \\ y_1 \\ P \end{array}
\left[
\begin{array}{cccccc|c}
0 & 0 & 1 & -3 & 1 & 0 & 1 \\
0 & 1 & 0 & 2 & -1 & 0 & 2 \\
1 & 0 & 0 & -1 & 1 & 0 & 5 \\
\hline
0 & 0 & 0 & 1 & 3 & 1 & 43
\end{array}
\right]
\end{array}
$$

Optimal solution: min $C = 43$ at $x_1 = 0$, $x_2 = 1$, $x_3 = 3$.

$\cdot$

**31.**   The matrices corresponding to the given problem and the dual problem are:

$$
A =
\left[
\begin{array}{ccc|c}
1 & -4 & 1 & 6 \\
-1 & 1 & -2 & 4 \\
\hline
5 & 2 & 2 & 1
\end{array}
\right]
\quad \text{and} \quad
A^T =
\left[
\begin{array}{cc|c}
1 & -1 & 5 \\
-4 & 1 & 2 \\
1 & -2 & 2 \\
\hline
6 & 4 & 1
\end{array}
\right]
$$

Thus, the dual problem is: Maximize $P = 6y_1 + 4y_2$

$$\text{Subject to:} \quad y_1 - y_2 \le 5$$
$$-4y_1 + y_2 \le 2$$
$$y_1 - 2y_2 \le 2$$
$$y_1, \ y_2 \ge 0$$

We introduce slack variables $x_1, x_2$, and $x_3$ to obtain the initial system:

$$
\begin{aligned}
y_1 - y_2 + x_1 \qquad\qquad &= 5 \\
-4y_1 + y_2 \quad + x_2 \qquad &= 2 \\
y_1 - 2y_2 \qquad + x_3 \quad &= 2 \\
-6y_1 - 4y_2 \qquad\qquad + P &= 0
\end{aligned}
$$

The simplex tableau for this problem is:

Copyright © 2011 Pearson Education, Inc. Publishing as Prentice Hall.

$$
\begin{array}{c}
\begin{array}{cccccc}
y_1 & y_2 & x_1 & x_2 & x_3 & P
\end{array} \\
\begin{array}{c}
x_1 \\ x_2 \\ x_3 \\ P
\end{array}
\left[
\begin{array}{cccccc|c}
1 & -1 & 1 & 0 & 0 & 0 & 5 \\
-4 & 1 & 0 & 1 & 0 & 0 & 2 \\
\textcircled{1} & -2 & 0 & 0 & 1 & 0 & 2 \\
\hdashline
-6 & -4 & 0 & 0 & 0 & 1 & 0
\end{array}
\right]
\begin{array}{l}
\frac{5}{1} = 5 \\ \\
\frac{2}{1} = 2
\end{array}
\sim
\left[
\begin{array}{cccccc|c}
0 & \textcircled{1} & 1 & 0 & -1 & 0 & 3 \\
0 & -7 & 0 & 1 & 4 & 0 & 10 \\
1 & -2 & 0 & 0 & 1 & 0 & 2 \\
\hdashline
0 & -16 & 0 & 0 & 6 & 1 & 12
\end{array}
\right]
\end{array}
$$

$(-1)R_3 + R_1 \rightarrow R_1$, $4R_3 + R_2 \rightarrow R_2$,
and $6R_3 + R_4 \rightarrow R_4$

$7R_1 + R_2 \rightarrow R_2$, $2R_1 + R_3 \rightarrow R_3$,
and $16R_1 + R_4 \rightarrow R_4$

$$
\begin{array}{c}
\begin{array}{cccccc}
y_1 & y_2 & x_1 & x_2 & x_3 & P
\end{array} \\
\begin{array}{c}
y_2 \\ x_2 \\ y_1 \\ P
\end{array}
\left[
\begin{array}{cccccc|c}
0 & 1 & 1 & 0 & -1 & 0 & 3 \\
0 & 0 & 7 & 1 & -3 & 0 & 31 \\
1 & 0 & 2 & 0 & -1 & 0 & 8 \\
\hdashline
0 & 0 & 16 & 0 & -10 & 1 & 60
\end{array}
\right]
\end{array}
$$

Since all the entries above the dashed line in the pivot column, the $x_3$ column, are negative, the problem does not have an optimal solution.

33.   The dual problem has 2 variables and 4 problem constraints.

35.   The original problem must have two problem constraints, and any number of variables.

37.   No. The dual problem will not be a standard maximization problem (one of the elements in the last column will be negative.)

39.   Yes. Multiply both sides of the inequality by $-1$.

41.   The matrices corresponding to the given problem and the dual problem are:

$$
A = \left[
\begin{array}{ccc|c}
3 & 2 & 2 & 16 \\
4 & 3 & 1 & 14 \\
5 & 3 & 1 & 12 \\
\hline
16 & 8 & 4 & 1
\end{array}
\right]
\quad \text{and} \quad
A^T = \left[
\begin{array}{ccc|c}
3 & 4 & 5 & 16 \\
2 & 3 & 3 & 8 \\
2 & 1 & 1 & 4 \\
\hline
16 & 14 & 12 & 1
\end{array}
\right]
\quad \text{respectively.}
$$

Thus, the dual problem is:   Maximize $P = 16y_1 + 14y_2 + 12y_3$

Subject to:   $3y_1 + 4y_2 + 5y_3 \leq 16$
$2y_1 + 3y_2 + 3y_3 \leq 8$
$2y_1 + y_2 + y_3 \leq 4$
$y_1, y_2, y_3 \geq 0$

We introduce slack variables $x_1$, $x_2$, and $x_3$ to obtain the initial system:

$$
\begin{aligned}
3y_1 + 4y_2 + 5y_3 + x_1 &= 16 \\
2y_1 + 3y_2 + 3y_3 \phantom{+} + x_2 &= 8 \\
2y_1 + y_2 + y_3 \phantom{+} + x_3 &= 4 \\
-16y_1 - 14y_2 - 12y_3 \phantom{++} + P &= 0
\end{aligned}
$$

Copyright © 2011 Pearson Education, Inc.  Publishing as Prentice Hall.

The simplex tableau for this problem is:

$$
\begin{array}{c}
\phantom{x_1} \\
x_1 \\
x_2 \\
x_3 \\
P
\end{array}
\begin{array}{c}
\begin{array}{ccccccc}
y_1 & y_2 & y_3 & x_1 & x_2 & x_3 & P
\end{array} \\
\left[
\begin{array}{ccccccc|c}
3 & 4 & 5 & 1 & 0 & 0 & 0 & 16 \\
2 & 3 & 3 & 0 & 1 & 0 & 0 & 8 \\
② & 1 & 1 & 0 & 0 & 1 & 0 & 4 \\
\hdashline
-16 & -14 & -12 & 0 & 0 & 0 & 1 & 0
\end{array}
\right]
\end{array}
\begin{array}{l}
\frac{16}{3} = 5.33 \\[4pt]
\frac{8}{2} = 4 \\[4pt]
\frac{4}{2} = 2
\end{array}
$$

$$\frac{1}{2}R_3 \rightarrow R_3$$

$$
\sim
\left[
\begin{array}{ccccccc|c}
3 & 4 & 5 & 1 & 0 & 0 & 0 & 16 \\
2 & 3 & 3 & 0 & 1 & 0 & 0 & 8 \\
① & \frac{1}{2} & \frac{1}{2} & 0 & 0 & \frac{1}{2} & 0 & 2 \\
\hdashline
-16 & -14 & -12 & 0 & 0 & 0 & 1 & 0
\end{array}
\right]
$$

$$(-3)R_3 + R_1 \rightarrow R_1, \quad (-2)R_3 + R_2 \rightarrow R_2, \quad \text{and } 16R_3 + R_4 \rightarrow R_4$$

$$
\sim
\left[
\begin{array}{ccccccc|c}
0 & \frac{5}{2} & \frac{7}{2} & 1 & 0 & -\frac{3}{2} & 0 & 10 \\
0 & ② & 2 & 0 & 1 & -1 & 0 & 4 \\
1 & \frac{1}{2} & \frac{1}{2} & 0 & 0 & \frac{1}{2} & 0 & 2 \\
\hdashline
0 & -6 & -4 & 0 & 0 & 8 & 1 & 32
\end{array}
\right]
\begin{array}{l}
\frac{10}{5/2} = 4 \\[4pt]
\frac{4}{2} = 2 \\[4pt]
\frac{2}{1/2} = 4
\end{array}
\sim
\left[
\begin{array}{ccccccc|c}
0 & \frac{5}{2} & \frac{7}{2} & 1 & 0 & -\frac{3}{2} & 0 & 10 \\
0 & ① & 1 & 0 & \frac{1}{2} & -\frac{1}{2} & 0 & 2 \\
1 & \frac{1}{2} & \frac{1}{2} & 0 & 0 & \frac{1}{2} & 0 & 2 \\
\hdashline
0 & -6 & -4 & 0 & 0 & 8 & 1 & 32
\end{array}
\right]
$$

$$\frac{1}{2}R_2 \rightarrow R_2$$

$$\left(-\frac{5}{2}\right)R_2 + R_1 \rightarrow R_1, \quad \left(-\frac{1}{2}\right)R_2 + R_3 \rightarrow R_3,$$
$$\text{and } 6R_2 + R_4 \rightarrow R_4$$

$$
\sim
\begin{array}{c}
\phantom{y_3} \\
y_3 \\
y_2 \\
y_1 \\
P
\end{array}
\begin{array}{c}
\begin{array}{ccccccc}
y_1 & y_2 & y_3 & x_1 & x_2 & x_3 & P
\end{array} \\
\left[
\begin{array}{ccccccc|c}
0 & 0 & 1 & 1 & -\frac{5}{4} & -\frac{1}{4} & 0 & 5 \\
0 & 1 & 1 & 0 & \frac{1}{2} & -\frac{1}{2} & 0 & 2 \\
1 & 0 & 0 & 0 & -\frac{1}{4} & \frac{3}{4} & 0 & 1 \\
\hdashline
0 & 0 & 2 & 0 & 3 & 5 & 1 & 44
\end{array}
\right]
\end{array}
$$

Optimal solution: min $C = 44$
at $x_1 = 0$, $x_2 = 3$, $x_3 = 5$.

**43.** The first and second inequalities must be rewritten before forming the dual.

Minimize $C = 5x_1 + 4x_2 + 5x_3 + 6x_4$

Subject to:
$$
\begin{aligned}
-x_1 - x_2 \qquad\qquad &\geq -12 \\
-x_3 - x_4 &\geq -25 \\
x_1 \qquad + x_3 \qquad &\geq 20 \\
x_2 \qquad\quad + x_4 &\geq 15 \\
x_1, x_2, x_3, x_4 &\geq 0
\end{aligned}
$$

The matrices corresponding to the given problem and the dual problem are:

Copyright © 2011 Pearson Education, Inc. Publishing as Prentice Hall.

$$A = \begin{bmatrix} -1 & -1 & 0 & 0 & | & -12 \\ 0 & 0 & -1 & -1 & | & -25 \\ 1 & 0 & 1 & 0 & | & 20 \\ 0 & 1 & 0 & 1 & | & 15 \\ \hline 5 & 4 & 5 & 6 & | & 1 \end{bmatrix} \quad \text{and} \quad A^T = \begin{bmatrix} -1 & 0 & 1 & 0 & | & 5 \\ -1 & 0 & 0 & 1 & | & 4 \\ 0 & -1 & 1 & 0 & | & 5 \\ 0 & -1 & 0 & 1 & | & 6 \\ \hline -12 & -25 & 20 & 15 & | & 1 \end{bmatrix}$$

The dual problem is: Maximize $P = -12y_1 - 25y_2 + 20y_3 + 15y_4$

$$\text{Subject to: } \begin{array}{rcl} -y_1 + y_3 & \leq & 5 \\ -y_1 + y_4 & \leq & 4 \\ -y_2 + y_3 & \leq & 5 \\ -y_2 + y_4 & \leq & 6 \\ y_1, y_2, y_3, y_4 & \geq & 0 \end{array}$$

We introduce the slack variables $x_1, x_2, x_3,$ and $x_4$ to obtain the initial system:

$$\begin{array}{rcl} -y_1 + y_3 + x_1 & = & 5 \\ -y_1 + y_4 + x_2 & = & 4 \\ -y_2 + y_3 + x_3 & = & 5 \\ -y_2 + y_4 + x_4 & = & 6 \\ +12y_1 + 25y_2 - 20y_3 - 15y_4 + P & = & 0 \end{array}$$

The simplex tableau for this problem is:

|       | $y_1$ | $y_2$ | $y_3$ | $y_4$ | $x_1$ | $x_2$ | $x_3$ | $x_4$ | $P$ |     |               |
|-------|-------|-------|-------|-------|-------|-------|-------|-------|-----|-----|---------------|
| $x_1$ | -1    | 0     | 1     | 0     | 1     | 0     | 0     | 0     | 0   | 5   | $\frac{5}{1}=5$ |
| $x_2$ | -1    | 0     | 0     | 1     | 0     | 1     | 0     | 0     | 0   | 4   |               |
| $x_3$ | 0     | -1    | ①     | 0     | 0     | 0     | 1     | 0     | 0   | 5   | $\frac{5}{1}=5$ |
| $x_4$ | 0     | -1    | 0     | 1     | 0     | 0     | 0     | 1     | 0   | 6   |               |
| $P$   | 12    | 25    | -20   | -15   | 0     | 0     | 0     | 0     | 1   | 0   |               |

$$(-1)R_3 + R_1 \to R_1 \quad \text{and} \quad 20R_3 + R_5 \to R_5$$

$$\sim \begin{bmatrix} -1 & 1 & 0 & 0 & 1 & 0 & -1 & 0 & 0 & | & 0 \\ -1 & 0 & 0 & ① & 0 & 1 & 0 & 0 & 0 & | & 4 \\ 0 & -1 & 1 & 0 & 0 & 0 & 1 & 0 & 0 & | & 5 \\ 0 & -1 & 0 & 1 & 0 & 0 & 0 & 1 & 0 & | & 6 \\ \hline 12 & 5 & 0 & -15 & 0 & 0 & 20 & 0 & 1 & | & 100 \end{bmatrix} \begin{array}{l} \\ \frac{4}{1}=4 \\ \\ \frac{6}{1}=6 \\ \\ \end{array}$$

$$(-1)R_2 + R_4 \to R_4 \quad \text{and} \quad 15R_2 + R_5 \to R_5$$

Copyright © 2011 Pearson Education, Inc. Publishing as Prentice Hall.

$O_2$: Selecting the second digit

$N_2$: 9 ways

$O_3$: Selecting the third digit

$N_3$: 8 ways

$O_5$: Selecting the fifth digit

$N_5$: 6 ways

Thus, there are

$N_1 \cdot N_2 \cdot N_3 \cdot N_4 \cdot N_5 = 10 \cdot 9 \cdot 8 \cdot 7 \cdot 6 = 30{,}240$

possible combinations.

(B)    Number of five-digit combinations, allowing repetition.

$O_1$: Selecting the first digit

$N_1$: 10 ways

$O_2$: Selecting the second digit

$N_2$: 10 ways

$O_3$: Selecting the third digit

$N_3$: 10 ways

$O_4$: Selecting the fourth digit

$N_4$: 10 ways

$O_5$: Selecting the fifth digit

$N_5$: 10 ways

Thus, there are

$N_1 \cdot N_2 \cdot N_3 \cdot N_4 \cdot N_5 = 10 \cdot 10 \cdot 10 \cdot 10 \cdot 10 = 10^5 = 100{,}000$

possible combinations.

(C)    Number of five digit combinations, if successive digits must be different.

$O_1$: Selecting the first digit

$N_1$: 10 ways

$O_2$: Selecting the second digit

$N_2$: 9 ways

$O_3$: Selecting the third digit

$N_3$: 9 ways

$O_4$: Selecting the fourth digit

$N_4$: 9 ways

$O_5$: Selecting the fifth digit

$N_5$: 9 ways

Thus, there are

$N_1 \cdot N_2 \cdot N_3 \cdot N_4 \cdot N_5 = 10 \cdot 9 \cdot 9 \cdot 9 \cdot 9 = 10 \cdot 9^4 = 65{,}610$ possible combinations.

**41.**    (A)    Letters and/or digits may be repeated.

$O_1$: Selecting the first letter

$N_1$: 26 ways

$O_2$: Selecting the second letter

$N_2$: 26 ways

$O_3$: Selecting the third letter

$N_3$: 26 ways

$O_4$: Selecting the first digit

$N_4$: 10 ways

$O_5$: Selecting the second digit

$N_5$: 10 ways

$O_6$: Selecting the third digit

$N_6$: 10 ways

Thus, there are

$N_1 \cdot N_2 \cdot N_3 \cdot N_4 \cdot N_5 \cdot N_6 = 26 \cdot 26 \cdot 26 \cdot 10 \cdot 10 \cdot 10 = 17{,}576{,}000$

different license plates.

Copyright © 2011 Pearson Education, Inc. Publishing as Prentice Hall.

(B)    No repeated letters and no repeated digits are allowed.

$O_1$: Select the three letters, no letter repeated

$N_1$: $26 \cdot 25 \cdot 24 = 15{,}600$ ways

$O_2$: Select the three numbers, no number repeated

$N_2$: $10 \cdot 9 \cdot 8 = 720$ ways

Thus, there are
$N_1 \cdot N_2 = 15{,}600 \cdot 720 = 11{,}232{,}000$

different license plates with no letter or digit repeated.

**43.**

There are 8 combined choices; the color can be chosen in 2 ways and the size can be chosen in 4 ways. The number of possible combined choices is $2 \cdot 4 = 8$ just as in Example 3.

**45.**    $z = 0$; if $B \subset A$, then there are no elements in $B$ which are not in $A$.

**47.**    $x = 0$ and $z = 0$; $A \varnothing B = A \cap B$ implies $A = B$.

**49.**    Let $T$ = the people who play tennis, and
$G$ = the people who play golf.

Then $n(T) = 32$, $n(G) = 37$, $n(T \cap G) = 8$ and $n(U) = 75$.
Thus, $n(T \cup G) = n(T) + n(G) - n(T \cap G) = 32 + 37 - 8 = 61$

The set of people who play neither tennis nor golf is represented by $T' \cap G'$. Since $U = (T \cup G) \cup (T' \cap G')$ and $(T \cup G) \cap (T' \cap G') = \varnothing$, it follows that

$n(T' \cap G') = n(U) - n(T \cup G) = 75 - 61 = 14$.

There are 14 people who play neither tennis nor golf.

**51.**    Let $F$ = the people who speak French, and
$G$ = the people who speak German.

Then $n(F) = 42$, $n(G) = 55$, $n(F' \cap G') = 17$ and $n(U) = 100$. Since
$U = (F \cup G) \cup (F' \cap G')$ and $(F \cup G) \cap (F' \cap G') = \varnothing$, it follows that
$n(F \cup G) = n(U) - n(F' \cap G') = 100 - 17 = 83$

Copyright © 2011 Pearson Education, Inc. Publishing as Prentice Hall.

Now   $n(F \cup G) = n(F) + n(G) - n(F \cap G)$, so
$$n(F \cap G) = n(F) + n(G) - n(F \cup G) = 42 + 55 - 83 = 14$$
There are 14 people who speak both French and German.

**53.** (A)

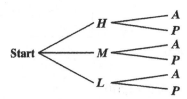

(B)    Operation 1:  Test scores can be classified into three groups, high, middle, or low:

$$N_1 = 3$$

Operation 2:  Interviews can be classified into two groups, aggressive or passive:

$$N_2 = 2$$

The total possible combined classifications is:

$$N_1 \cdot N_2 = 3 \cdot 2 = 6$$

**55.**   $O_1$: Travel from home to airport and back        $O_3$: Fly to second city

$N_1$: 2 ways                                              $N_3$: 2 ways

$O_2$: Fly to first city                                   $O_4$: Fly to third city

$N_2$: 3 ways                                              $N_4$: 1 way

Thus, there are

$$N_1 \cdot N_2 \cdot N_3 \cdot N_4 = 2 \cdot 3 \cdot 2 \cdot 1 = 12$$

different travel plans.

**57.**   Let $U$ = the group of people surveyed

   $H$ = people who own an HDTV, and

   $D$ = people who own a DVD.

Then $n(U) = 1200$, $n(H) = 850$,
$n(D) = 740$ and $n(H \cap D) = 580$.

Now draw a Venn diagram.

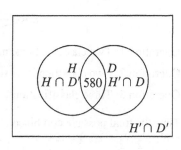

From this diagram, we see that
$$n(H \cap D') = n(H) - n(H \cap D) = 850 - 580 = 270$$
$$n(H' \cap D) = n(D) - n(H \cap D) = 740 - 580 = 160$$
$$n(H \cup D) = n(H \cap D') + n(H \cap D) + n(H' \cap D) = 580 + 270 + 160 = 1010$$
and
$$n(H' \cap D') = n(U) - n(H \cup D) = 1200 - 1010 = 190$$
Thus,
(A)  $n(H \cup D) = 1010$    (B)  $n(H' \cap D') = 190$    (C)  $n(H \cap D') = 270$

Copyright © 2011 Pearson Education, Inc.  Publishing as Prentice Hall.

**59.** Let $U$ = group of people surveyed

$H$ = group of people who receive HBO

$S$ = group of people who receive Showtime.

Then, $n(U) = 8,000$, $n(H) = 2,450$,

$n(S) = 1,940$ and $n(H' \cap S') = 5,180$

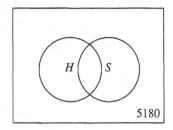

5180

Now, $n(H \cup S) = n(U) - n(H' \cap S') = 8,000 - 5,180 = 2,820$

Since $n(H \cup S) = n(H) + n(S) - n(H \cap S)$, we have

$n(H \cap S) = n(H) + n(S) - n(H \cup S) = 2,450 + 1,940 - 2,820 = 1,570$

Thus, 1,570 subscribers receive both channels.

**61.** From the table:

(A) The number of males aged 20-24 *and* below minimum wage is: 102 (the element in the (2, 2) position in the body of the table); 102,000.

(B) The number of females aged 20 or older *and* at minimum wage is:

$186 + 503 = 689$ (the sum of the elements in the (3, 2) and (3, 3) positions); 689,000.

(C) The number of workers who are *either* aged 16-19 *or* are males at minimum wage is:

$343 + 118 + 367 + 251 + 154 + 237 = 1,470$; 1,470,000.

(D) The number of workers below minimum wage is: $379 + 993 = 1,372$; 1,372,000.

**63.** (A)

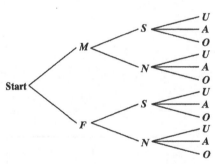

(B) Operation 1: Two classifications, male and female; $N_1 = 2$.

Operation 2: Two classifications, smoker and nonsmoker; $N_2 = 2$.

Operation 3: Three classifications, underweight, average weight, and overweight; $N_3 = 3$.

Thus the total possible combined classifications

$= N_1 \cdot N_2 \cdot N_3 = 2 \cdot 2 \cdot 3 = 12$

**65.** $F$ = number of individuals who contributed to the first campaign

$S$ = number of individuals who contributed to the second campaign.

Then $n(F) = 1,475$, $n(S) = 2,350$ and $n(F \cap S) = 920$.

Now, $n(F \cup S) = n(F) + n(S) - n(F \cap S) = 1,475 + 2,350 - 920 = 2,905$

Thus, 2,905 individuals contributed to either the first campaign or the second campaign.

Copyright © 2011 Pearson Education, Inc. Publishing as Prentice Hall.

EXERCISE 7-4

*Things to remember:*

1.   FACTORIAL

For $n$ a natural number,

$$n! = n(n-1)(n-2) \cdot \ \ldots \ \cdot 3 \cdot 2 \cdot 1$$
$$0! = 1$$
$$n! = n(n-1)!$$

[NOTE: Many calculators have an $\boxed{n!}$ key or its equivalent.]

2.   PERMUTATIONS

A PERMUTATION of a set of distinct objects is an arrangement of the objects in a specific order, without repetition.  The NUMBER OF PERMUTATIONS of $n$ distinct objects without repetition, denoted by $P_{n,n}$, is:

$$P_{n,n} = n(n-1) \cdot \ \ldots \ \cdot 3 \cdot 2 \cdot 1 = n! \quad (n \text{ factors})$$

3.   PERMUTATIONS OF $n$ OBJECTS TAKEN $r$ AT A TIME

A permutation of a set of $n$ distinct objects taken $r$ at a time without repetition is an arrangement of the $r$ objects in a specific order.  The NUMBER OF PERMUTATIONS of $n$ objects taken $r$ at a time, denoted by $P_{n,r}$, is given by:

$$P_{n,r} = n(n-1)(n-2) \cdot \ \ldots \ \cdot (n-r+1)$$
$$(r \text{ factors})$$

or $\quad P_{n,r} = \dfrac{n!}{(n-r)!} \quad 0 \le r \le n$

[Note: $P_{n,n} = \dfrac{n!}{(n-n)!} = \dfrac{n!}{0!} = n! = \dfrac{n!}{0!} = n!$, the number of permutations of $n$ objects taken $n$ at a time.

Remember, by definition, $0! = 1$.]

4.   COMBINATIONS OF $n$ OBJECTS TAKEN $r$ AT A TIME

A combination of a set of $n$ distinct objects taken $r$ at a time without repetition is an $r$-element subset of the set of $n$ objects.  (The arrangement of the elements in the subset does not matter.)  The NUMBER OF COMBINATIONS of $n$ objects taken $r$ at a time, denoted by $C_{n,r}$ or by $\dbinom{n}{r}$ is given by:

$$C_{n,r} = \binom{n}{r} = \dfrac{P_{n,r}}{r!} = \dfrac{n!}{r!(n-r)!} \quad 0 \le r \le n$$

5.   NOTE: In a permutation, the ORDER of the objects counts.  In a combination, order does not count.

---

1.   $7! = 7 \cdot 6 \cdot 5 \cdot 4 \cdot 3 \cdot 2 \cdot 1 = 5{,}040$

3.   $(5+6)! = 11! = 11 \cdot 10 \cdot 9 \cdot 8 \cdot 7 \cdot 6 \cdot 5 \cdot 4 \cdot 3 \cdot 2 \cdot 1 = 39{,}916{,}800$

Copyright © 2011 Pearson Education, Inc.  Publishing as Prentice Hall.

**5.** $5! + 6! = 5 \cdot 4 \cdot 3 \cdot 2 \cdot 1 + 6 \cdot 5 \cdot 4 \cdot 3 \cdot 2 \cdot 1 = 5 \cdot 4 \cdot 3 \cdot 2 \cdot 1(1 + 6) = 120 \cdot 7 = 840$

**7.** $\dfrac{8!}{4!} = \dfrac{(8 \cdot 7 \cdot 6 \cdot 5)4!}{4!} = 1,680$

**9.** $\dfrac{8!}{4!(8-4)!} = \dfrac{8!}{4!4!} = \dfrac{(8 \cdot 7 \cdot 6 \cdot 5)4!}{4!4!} = \dfrac{8 \cdot 7 \cdot 6 \cdot 5}{4 \cdot 3 \cdot 2 \cdot 1} = 70$

**11.** $\dfrac{500!}{498!} = \dfrac{(500 \cdot 499)498!}{498!} = 500 \cdot 499 = 249,500$

**13.** $P_{12,7} = \dfrac{12!}{(12-7)!} = \dfrac{12!}{5!} = \dfrac{(12 \cdot 11 \cdot 10 \cdot 9 \cdot 8 \cdot 7 \cdot 6)5!}{5!} = 3,991,680$

**15.** $C_{12,7} = \dfrac{12!}{7!(12-7)!} = \dfrac{(12 \cdot 11 \cdot 10 \cdot 9 \cdot 8)7!}{7!(5 \cdot 4 \cdot 3 \cdot 2 \cdot 1)} = \dfrac{95,040}{120} = 792$

**17.** $\dfrac{C_{13,4}}{C_{52,4}} = \dfrac{\dfrac{13!}{4!(13-4)!}}{\dfrac{52!}{4!(52-4)!}} = \dfrac{13!}{4! \cdot 9!} \cdot \dfrac{4! \cdot 48!}{52!} = \dfrac{13! \cdot 48!}{9! \cdot 52!} = \dfrac{13 \cdot 12 \cdot 11 \cdot 10}{52 \cdot 51 \cdot 50 \cdot 49} \approx 0.0026$

**19.** $\dfrac{P_{12,6}}{12^6} = \dfrac{\dfrac{12!}{6!}}{12^6} = \dfrac{12 \cdot 11 \cdot 10 \cdot 9 \cdot 8 \cdot 7}{12^6} \approx 0.2228$

**21.** $\dfrac{n!}{(n-2)!} = \dfrac{n(n-1)(n-2)!}{(n-2)!} = n(n-1)$

**23.** $\dfrac{(n+1)!}{2!(n-1)!} = \dfrac{(n+1)(n)(n-1)!}{2 \cdot 1(n-1)!} = \dfrac{n(n+1)}{2}$

**25.** Permutation; order of selection counts.

**27.** Combination; order of selection does not count.

**29.** Neither a permutation nor a combination.

**31.** The number of different finishes (win, place, show) for the ten horses is the number of permutations of 10 objects 3 at a time. This is:

$$P_{10,3} = \dfrac{10!}{(10-3)!} = \dfrac{10!}{7!} = \dfrac{10 \cdot 9 \cdot 8 \cdot 7!}{7!} = 720$$

**33.** (A)    The number of ways that a three-person subcommittee can be selected      from a seven-member committee is the number of combinations (since order *is not* important in selecting a subcommittee) of 7 objects 3 at a time. This is:

$$C_{7,3} = \dfrac{7!}{3!(7-3)!} = \dfrac{7!}{3!4!} = \dfrac{7 \cdot 6 \cdot 5 \cdot 4!}{3 \cdot 2 \cdot 1 \cdot 4!} = 35$$

(B)    The number of ways a president, vice-president, and secretary can be chosen from a committee of 7 people is the number of permutations (since order *is* important in choosing 3 people for the positions) of 7 objects 3 at a time. This is:

Copyright © 2011 Pearson Education, Inc.  Publishing as Prentice Hall.

$$P_{7,3} = \frac{7!}{(7-3)!} = \frac{7!}{4!} = \frac{7 \cdot 6 \cdot 5 \cdot 4!}{4!} = 7 \cdot 6 \cdot 5 = 210$$

**35.** Calculate $x!$, $3^x$, $x^3$ for $x = 1, 2, 3, \ldots$

For $x \geq 7$, $x! > 3^x$; for $x \geq 1$, $3^x > x^3$ (except for $x = 3$). In general, $x!$ grows much faster than $3^x$, and $3^x$ grows much faster than $x^3$.

**37.** This is a "combinations" problem. We want the number of ways to select 5 objects from 13 objects with order not counting. This is:

$$C_{13,5} = \frac{13!}{5!(13-5)!} = \frac{13!}{5!8!} = \frac{13 \cdot 12 \cdot 11 \cdot 10 \cdot 9 \cdot 8!}{5 \cdot 4 \cdot 3 \cdot 2 \cdot 1 \cdot 8!} = 1,287$$

**39.** The five spades can be selected in $C_{13,5}$ ways and the two hearts can be selected in $C_{13,2}$ ways. Applying the Multiplication Principle, we have:

Total number of hands $= C_{13,5} \cdot C_{13,2} = \dfrac{13!}{5!(13-5)!} \cdot \dfrac{13!}{2!(13-2)!} = \dfrac{13!}{5!8!} \cdot \dfrac{13!}{2!11!} = 100,386$

**41.** The three appetizers can be selected in $C_{8,3}$ ways. The four main courses can be selected in $C_{10,4}$ ways. The two desserts can be selected in $C_{7,2}$ ways. Now, applying the Multiplication Principle, the total number of ways in which the above can be selected is given by:

$$C_{8,3} \cdot C_{10,4} \cdot C_{7,2} = \frac{8!}{3!(8-3)!} \cdot \frac{10!}{4!(10-4)!} \cdot \frac{7!}{2!(7-2)!} = 246,960$$

**43.** $C_{n,n-r} = \dfrac{n!}{(n-r)!(n-[n-r])!} = \dfrac{n!}{(n-r)!r!} = \dfrac{n!}{r!(n-r)!} = C_{n,r}$

The sequence of numbers is the same read top to bottom or bottom to top.

**45.** True; $(n+1)! = (n+1)n! > n!$

**47.** False; $P_{n,n-1} = \dfrac{n!}{[n-(n-1)]!} = \dfrac{n!}{1!} = n!$

$$P_{n,n} = \frac{n!}{(n-n)!} = \frac{n!}{0!} = \frac{n!}{1} = n!$$

Note: $P_{n,r} < P_{n,r+1}$ for $0 \leq r < n - 1$

**49.** True; $C_{n,n-r} = \dfrac{n!}{(n-[n-r])!(n-r)!} = \dfrac{n!}{r!(n-r)!} = C_{n,r}$

**51.** (A) A line segment joins two distinct points. Thus, the total number of line segments is given by:

$$C_{8,2} = \frac{8!}{2!(8-2)!} = \frac{8!}{2!6!} = \frac{8 \cdot 7 \cdot 6!}{2 \cdot 1 \cdot 6!} = 28$$

(B) Each triangle requires three distinct points. Thus, there are

Copyright © 2011 Pearson Education, Inc. Publishing as Prentice Hall.

$$C_{8,3} = \frac{8!}{3!(8-3)!} = \frac{8!}{3!5!} = \frac{8 \cdot 7 \cdot 6 \cdot 5!}{3 \cdot 2 \cdot 1 \cdot 5!} = 56 \text{ triangles.}$$

(C)    Each quadrilateral requires four distinct points.  Thus, there are

$$C_{8,4} = \frac{8!}{4!(8-4)!} = \frac{8!}{4!4!} = \frac{8 \cdot 7 \cdot 6 \cdot 5 \cdot 4!}{4 \cdot 3 \cdot 2 \cdot 1 \cdot 4!} = 70 \text{ quadrilaterals.}$$

**53.**  (A)    Two people.

$O_1$: First person selects a chair         $O_2$: Second person selects a chair
$N_1$: 5 ways                                $N_2$: 4 ways

Thus, there are

$$N_1 \cdot N_2 = 5 \cdot 4 = 20$$

ways to seat two people in a row of 5 chairs.
Note that this is $P_{5,2}$.

(B)    Three people. There will be $P_{5,3}$ ways to seat 3 people in a
row of 5 chairs:

$$P_{5,3} = \frac{5!}{(5-3)!} = \frac{5!}{2!} = \frac{5 \cdot 4 \cdot 3 \cdot 2!}{2!} = 60$$

(C)    Four people. The number of ways to seat 4 people in a row of 5  chairs is given by:

$$P_{5,4} = \frac{5!}{(5-4)!} = \frac{5!}{1!} = 5 \cdot 4 \cdot 3 \cdot 2 = 120$$

(D)    Five people. The number of ways to seat 5 people in a row of 5  chairs is given by:

$$P_{5,5} = \frac{5!}{(5-5)!} = \frac{5!}{0!} = 5! = 120$$

**55.**  (A)    The distinct positions are taken into consideration.
The number of starting teams is given by:

$$P_{8,5} = \frac{8!}{(8-5)!} = \frac{8!}{3!} = \frac{8 \cdot 7 \cdot 6 \cdot 5 \cdot 4 \cdot 3!}{3!} = 6,720$$

(B)    The distinct positions are not taken into consideration.
The number of starting teams is given by:

$$C_{8,5} = \frac{8!}{5!(8-5)!} = \frac{8!}{5!3!} = \frac{8 \cdot 7 \cdot 6 \cdot 5!}{5! \cdot 3 \cdot 2 \cdot 1} = 56$$

(C)    Either Mike or Ken, but not both, must start; distinct positions are not taken into consideration.

$O_1$: Select either Mike or Ken

$N_1$: 2 ways

$O_2$: Select 4 players from the remaining 6

$N_2$: $C_{6,4}$

Copyright © 2011 Pearson Education, Inc.  Publishing as Prentice Hall.

Thus, the number of starting teams is given by:

$$N_1 \cdot N_2 = 2 \cdot C_{6,4} = 2 \cdot \frac{6!}{4!(6-4)!} = 2 \cdot \frac{6 \cdot 5 \cdot 4!}{4! \cdot 2 \cdot 1} = 30$$

**57.** For many calculators, $k = 69$, but your calculator may be different. Note that $k!$ may also be calculated as $P_{k,k}$. On a TI-85, the largest integer $k$ for which $k!$ can be calculated using $P_{k,k}$ is 449.

**59.** The largest value will be $C_{24,12} = \dfrac{24!}{12!12!} = 2{,}704{,}156$.

**61.** (A)  Three printers are to be selected for the display. The *order* of selection does not count. Thus, the number of ways to select the 3 printers from 24 is:

$$C_{24,3} = \frac{24!}{3!(24-3)!} = \frac{24 \cdot 23 \cdot 22(21!)}{3 \cdot 2 \cdot 1(21!)} = 2{,}024$$

(B)  Nineteen of the 24 printers are not defective.

Thus, the number of ways to select 3 non-defective printers is:

$$C_{19,3} = \frac{19!}{3!(19-3)!} = \frac{19 \cdot 18 \cdot 17(16!)}{3 \cdot 2 \cdot 1(16!)} = 969$$

**63.** (A)  There are $8 + 12 + 10 = 30$ stores in all. The jewelry store chain will select 10 of these stores to close. Since order does not count here, the total number of ways to select the 10 stores to close is:

$$C_{30,10} = \frac{30!}{10!(30-10)!} = \frac{30 \cdot 29 \cdot 28 \cdot 27 \cdot 26 \cdot 25 \cdot 24 \cdot 23 \cdot 22 \cdot 21(20!)}{10 \cdot 9 \cdot 8 \cdot 7 \cdot 6 \cdot 5 \cdot 4 \cdot 3 \cdot 2 \cdot 1(20!)}$$
$$= 30{,}045{,}015$$

(B)  The number of ways to close 2 stores in Georgia is: $C_{8,2}$

The number of ways to close 5 stores in Florida is: $C_{12,5}$

The number of ways to close 3 stores in Alabama is: $C_{10,3}$

By the multiplication principle, the total number of ways to select the 10 stores for closing is:

$$C_{8,2} \cdot C_{12,5} \cdot C_{10,3} = \frac{8!}{2!(8-2)!} \cdot \frac{12!}{5!(12-5)!} \cdot \frac{10!}{3!(10-3)!}$$
$$= \frac{8 \cdot 7 \cdot 6!}{2 \cdot 1 \cdot 6!} \cdot \frac{12 \cdot 11 \cdot 10 \cdot 9 \cdot 8(7!)}{5 \cdot 4 \cdot 3 \cdot 2 \cdot 1(7!)} \cdot \frac{10 \cdot 9 \cdot 8(7!)}{3 \cdot 2 \cdot 1(7!)}$$
$$= 28 \cdot 792 \cdot 120 = 2{,}661{,}120$$

**65.** (A)  Three females can be selected in $C_{6,3}$ ways. Two males can be selected in $C_{5,2}$ ways. Applying the Multiplication Principle, we have:

Total number of ways $= C_{6,3} \cdot C_{5,2} = \dfrac{6!}{3!(6-3)!} \cdot \dfrac{5!}{2!(5-2)!} = 200$

(B)  Four females and one male can be selected in $C_{6,4} \cdot C_{5,1}$ ways. Thus,

$$C_{6,4} \cdot C_{5,1} = \frac{6!}{4!(6-4)!} \cdot \frac{5!}{1!(5-1)!} = 75$$

(C)  Number of ways in which 5 females can be selected is:

Copyright © 2011 Pearson Education, Inc. Publishing as Prentice Hall.

$$C_{6,5} = \frac{6!}{5!(6-5)!} = 6$$

(D)    Number of ways in which 5 people can be selected is:

$$C_{6+5,5} = C_{11,5} = \frac{11!}{5!(11-5)!} = 462$$

(E)    At least four females includes four females and five females.  Four females and one male can be selected in 75 ways [see part (B)].  Five females can be selected in 6 ways [see part (C)].  Thus, total number of ways $= C_{6,4} \cdot C_{5,1} + C_{6,5} = 75 + 6 = 81$

67.  (A)    Select 3 samples from 8 blood types, no two samples having the same type.  This is a permutation problem.  The number of different examinations is:

$$P_{8,3} = \frac{8!}{(8-3)!} = \frac{8!}{5!} = \frac{8 \cdot 7 \cdot 6 \cdot 5!}{5!} = 336$$

(B)    Select 3 samples from 8 blood types, repetition is allowed.

$O_1$: Select the first sample                    $O_3$: Select the third sample

$N_1$: 8 ways                                         $N_3$: 8 ways

$O_2$: Select the second sample

$N_2$: 8 ways

Thus, the number of different examinations in this case is:

$N_1 \cdot N_2 \cdot N_3 = 8 \cdot 8 \cdot 8 = 8^3 = 512$

69.    This is a permutations problem.  The number of buttons is given by:

$$P_{4,2} = \frac{4!}{(4-2)!} = \frac{4!}{2!} = \frac{4 \cdot 3 \cdot 2!}{2!} = 12$$

## CHAPTER 7 REVIEW

1.    $3^4$ is not less than $4^3$; true.    (7-1)

2.    $2^3$ is less than $3^2$ or $3^4$ is less than $4^3$; true.    (7-1)

3.    $2^3$ is less than $3^2$ and $3^4$ is less than $4^3$; false.    (7-1)

4.    If $2^3$ is less than $3^2$, then $3^4$ is less than $4^3$; false.    (7-1)

5.    If $3^4$ is less than $4^3$, then $2^3$ is less than $3^2$; true.    (7-1)

6.    If $3^4$ is not less than $4^3$, then $2^3$ is not less than $3^2$; false.    (7-1)

7.    T; $\{5, 6, 7\}$ and $\{6, 7, 5\}$ have exactly the same elements.    (7-2)

8.    F; $5 \notin \{55, 555\}$    (7-2)

9.    T; $\{9, 27\}$ is a subset of $\{3, 9, 27, 81\}$.    (7-2)

Copyright © 2011 Pearson Education, Inc.  Publishing as Prentice Hall.

10.  F; $\{1, 2\} \in \{1, \{1, 2\}\}$.    (7-2)

11.  *If* 9 is prime, *then* 10 is odd: conditional; true (9 is not prime so any conclusion is true).    (7-1)

12.  7 is even *or* 8 is odd: disjunction; false (7 is not even and 8 is not odd).    (7-1)

13.  53 is prime *and* 57 is prime: conjunction; false ($57 = 19 \cdot 3$ is not prime).    (7-1)

14.  51 is *not* prime: negation; true ($51 = 17 \cdot 3$ is not prime).    (7-1)

15.  Converse: If the square matrix A does not have an inverse, then the square matrix A has a row of zeros. Contrapositive: If the square matrix A has an inverse, then the square matrix A does not have a row of zeros.    (7-1)

16.  Converse: If the square matrix A has an inverse, then the square matrix A is an identity matrix. Contrapositive: If the square matrix A does not have an inverse, then the square matrix A is not an identity matrix.    (7-1)

17.  $\{1, 2, 3, 4\} \cup \{2, 3, 4, 5\} = \{1, 2, 3, 4, 5\}$    (7-2)

18.  $\{1, 2, 3, 4\} \cap \{2, 3, 4, 5\} = \{2, 3, 4\}$    (7-2)

19.  $\{1, 2, 3, 4\} \cap \{5, 6\} = \varnothing$    (7-2)

20.  (A)  We construct the tree diagram at the right for the experiment:

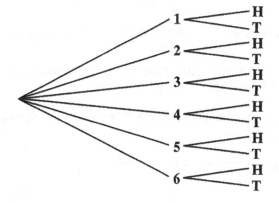

Total combined outcomes = 12.

(B)  Operation 1: Six possible outcomes, 1, 2, 3, 4, 5, or 6; $N_1 = 6$.

Operation 2: Two possible outcomes, heads (H) or tails (T); $N_2 = 2$.

Using the Multiplication Principle, the total combined outcomes = $N_1 \cdot N_2 = 6 \cdot 2 = 12$.    (7-3)

21.  (A)  $n(A) = 30 + 35 = 65$    (B)  $n(B) = 35 + 40 = 75$
(C)  $n(A \cap B) = 35$    (D)  $n(A \cup B) = 65 + 75 - 35 = 105$
or $n(A \cup B) = 30 + 35 + 40 = 105$

(E)  $n(U) = 30 + 35 + 40 + 45 = 150$
(F)  $n(A') = n(U) - n(A) = 150 - 65 = 85$
(G)  $n([A \cap B]') = n(U) - n(A \cap B)$    (H)  $n([A \cup B]') = n(U) - n(A \cup B)$
$= 150 - 35 = 115$    $= 150 - 105 = 45$    (7-3)

22.  $11! = 11 \cdot 10 \cdot 9 \cdot 8 \cdot 7 \cdot 6 \cdot 5 \cdot 4 \cdot 3 \cdot 2 \cdot 1 = 39{,}916{,}800$    (7-4)

23.  $(10 - 6)! = 4! = 24$    (7-4)    24.  $10! - 6! = 3{,}628{,}800 - 720 = 3{,}628{,}080$    (7-4)

Copyright © 2011 Pearson Education, Inc. Publishing as Prentice Hall.

**25.** $\dfrac{15!}{10!} = \dfrac{(15 \cdot 14 \cdot 13 \cdot 12 \cdot 11)10!}{10!} = 360,360$    (7-4)

**26.** $\dfrac{15!}{10!5!} = \dfrac{(15 \cdot 14 \cdot 13 \cdot 12 \cdot 11)10!}{10!(120)} = \dfrac{360,360}{120} = 3,003$    (7-4)

**27.** $C_{8,5} = \dfrac{8!}{5!(8-5)!} = \dfrac{(8 \cdot 7 \cdot 6)5!}{5!3!} = \dfrac{8 \cdot 7 \cdot 6}{6} = 56$    (7-4)

**28.** $P_{8,5} = \dfrac{8!}{(8-5)!} = \dfrac{8!}{3!} = \dfrac{(8 \cdot 7 \cdot 6 \cdot 5 \cdot 4)3!}{3!} = 6,720$    (7-4)

**29.**
| | |
|---|---|
| Operation 1: | First person can choose the seat in 6 different ways; $N_1 = 6$. |
| Operation 2: | Second person can choose the seat in 5 different ways; $N_2 = 5$. |
| Operation 3: | Third person can choose the seat in 4 different ways; $N_3 = 4$. |
| Operation 4: | Fourth person can choose the seat in 3 different ways; $N_4 = 3$. |
| Operation 5: | Fifth person can choose the seat in 2 different ways; $N_5 = 2$. |
| Operation 6: | Sixth person can choose the seat in 1 way; $N_6 = 1$. |

Using the Multiplication Principle, the total number of different arrangements that can be made is $6 \cdot 5 \cdot 4 \cdot 3 \cdot 2 \cdot 1 = 720$.    (7-3)

**30.** This is a permutations problem.  The permutations of 6 objects taken 6 at a time is:

$$P_{6,6} = \dfrac{6!}{(6-6)!} = 6! = 720 \quad (7\text{-}4)$$

**31.**

| $p$ | $q$ | $p \to q$ | $q \to p$ | $(p \to q) \wedge (q \to p)$ |
|---|---|---|---|---|
| T | T | T | T | T |
| T | F | F | T | F |
| F | T | T | F | F |
| F | F | T | T | T |

Contingency    (7-1)

**32.**

| $p$ | $q$ | $q \to p$ | $p \vee (q \to p)$ |
|---|---|---|---|
| T | T | T | T |
| T | F | T | T |
| F | T | F | F |
| F | F | T | T |

Contingency    (7-1)

Copyright © 2011 Pearson Education, Inc.  Publishing as Prentice Hall.

**33.**

| $p$ | $q$ | $\neg p$ | $\neg q$ | $p \vee \neg p$ | $q \wedge \neg q$ | $(p \vee \neg p) \rightarrow (q \wedge \neg q)$ |
|---|---|---|---|---|---|---|
| T | T | F | F | T | F | F |
| T | F | F | T | T | F | F |
| F | T | T | F | T | F | F |
| F | F | T | T | T | F | F |

Contradiction    (7-1)

**34.**

| $p$ | $q$ | $\neg q$ | $p \rightarrow q$ | $\neg q \wedge (p \rightarrow q)$ |
|---|---|---|---|---|
| T | T | F | T | F |
| T | F | T | F | F |
| F | T | F | T | F |
| F | F | T | T | T |

Contingency    (7-1)

**35.**

| $p$ | $q$ | $\neg p$ | $p \rightarrow q$ | $\neg p \rightarrow (p \rightarrow q)$ |
|---|---|---|---|---|
| T | T | F | T | T |
| T | F | F | F | T |
| F | T | T | T | T |
| F | F | T | T | T |

Tautology    (7-1)

**36.**

| $p$ | $Q$ | $\neg q$ | $p \vee \neg q$ | $\neg(p \vee \neg q)$ |
|---|---|---|---|---|
| T | T | F | T | F |
| T | F | T | T | F |
| F | T | F | F | T |
| F | F | T | T | F |

Contingency    (7-1)

**37.** $E \cup K$ is infinite; both $E$ and $K$ are infinite.    (7-2)

**38.** $M \cap K = \{n \in Z \mid 10^3 < n < 10^6\}$ is finite.    (7-2)

**39.** $K' = \{n \in Z \mid n \leq 10^3\}$ is infinite.    (7-2)

**40.** $E \cap M = \{n \in Z \mid n$ is even and $n < 10^6\}$ is infinite.    (7-2)

**41.** $M' = \{n \in Z \mid n \geq 10^6\}$, $K' = \{n \in Z \mid n \leq 10^3\}$; disjoint.    (7-2)

Copyright © 2011 Pearson Education, Inc.  Publishing as Prentice Hall.

**42.** $M = \{n \in Z \mid n < 10^6\}$, $E' = \{n \in Z \mid n \text{ is odd}\}$
$M$ and $E'$ are not disjoint (e.g., $1 \in M \cap E'$)     (7-2)

**43.** $K$ and $K'$ are disjoint by definition.     (7-2)

**44.** Using the multiplication principle, the man has 5 children, $5 \cdot 3 = 15$ grandchildren, and $5 \cdot 3 \cdot 2 = 30$ great-grandchildren, for a total of $5 + 15 + 30 = 50$ descendents.     (7-3)

**45.**

| Operation | Number of ways of completing operation under condition: | | |
| --- | --- | --- | --- |
| | No letter repeated | Letters can be repeated | Adjacent letters not alike |
| $O_1$ | 8 | 8 | 8 |
| $O_2$ | 7 | 8 | 7 |
| $O_3$ | 6 | 8 | 7 |

Total outcomes, without repeating letters $= 8 \cdot 7 \cdot 6 = 336$.
Total outcomes, with repeating letters $= 8 \cdot 8 \cdot 8 = 512$.
Total outcomes, with adjacent letters not alike $= 8 \cdot 7 \cdot 7 = 392$.   (7-3)

**46.** (A)  This is a permutations problem.
$$P_{6,3} = \frac{6!}{(6-3)!} = \frac{6 \cdot 5 \cdot 4 \cdot 3!}{3!} = 120$$

(B)  This is a combinations problem.
$$C_{5,2} = \frac{5!}{2!(5-2)!} = \frac{5 \cdot 4 \cdot 3!}{2 \cdot 1 \cdot 3!} = 10$$     (7-4)

**47.** The largest value of $C_{25,r}$ is $C_{25,12} = C_{25,13} = 5{,}200{,}300$.     (7-4)

**48.** By the multiplication principle, there are
$$N_1 \cdot N_2 \cdot N_3$$
branches in the tree diagram.     (7-3)

**49.** $x^3 - x = 0$
$x(x^2 - 1) = 0$
$\quad\quad x = -1, 0, 1$
$\{x \mid x^3 - x = 0\} = \{-1, 0, 1\}$     (7-2)

**50.** $4! = 24$, $5! = 120$
$\{x \mid x \text{ is a positive integer and } x! < 100\} = \{1, 2, 3, 4\}$.     (7-2, 7-4)

**51.** $\{x \mid x \text{ is a perfect square and } x < 50\} = \{1, 4, 9, 16, 25, 36, 49\}$.     (7-2)

Copyright © 2011 Pearson Education, Inc. Publishing as Prentice Hall.

**52.** (A)   This is a permutations problem.

$$P_{10,3} = \frac{10!}{(10-3)!} = \frac{10!}{7!} = 10 \cdot 9 \cdot 8 = 720$$

(B)   The number of ways in which women are selected for all three positions is given by:

$$P_{6,3} = \frac{6!}{(6-3)!} = \frac{6!}{3!} = 6 \cdot 5 \cdot 4 = 120$$

Thus, $P$(three women are selected) $= \dfrac{P_{6,3}}{P_{10,3}} = \dfrac{120}{720} = \dfrac{1}{6}$

(C)   This is a combinations problem.

$$C_{10,3} = \frac{10!}{3!(10-3)!} = \frac{10 \cdot 9 \cdot 8 \cdot 7!}{3 \cdot 2 \cdot 1 \cdot 7!} = 120 \qquad (7\text{-}3, 7\text{-}4)$$

**53.** Draw a Venn diagram with: $A$ = Chess players, $B$ = Checker players.

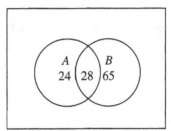

Now, $n(A \cap B) = 28$, $n(A \cap B') = n(A) - n(A \cap B) = 52 - 28 = 24$

$n(B \cap A') = n(B) - n(A \cap B) = 93 - 28 = 65$

Since there are 150 people in all,

$n(A' \cap B') = n(U) - [n(A \cap B') + n(A \cap B) + n(B \cap A')]$

$\qquad\qquad = 150 - (24 + 28 + 65) = 150 - 117 = 33 \qquad (7\text{-}3)$

**54.** $x = 0$; if $A \subset B$, then there are no elements in $A$ which are not in $B$.      (7-3)

**55.** $x = z = w = 0$; if $A \cap B = U$, then $A = B = U$ and there are no elements which are not in $A$, or not in $B$, or not in $A \cup B$.      (7-3)

**56.** True; for $1 < r < n$, $C_{n,r} = \dfrac{n!}{r!(n-r)!} < \dfrac{n!}{(n-r)!} = P_{n,r}$.      (7-4)

**57.** False; $P_{n,n-1} = \dfrac{n!}{[n-(n-1)]!} = \dfrac{n!}{1!} = n!$

Note: $P_{n,r} < n!$ for $0 \le r < n - 1$      (7-4)

**58.** True; for $1 < r < n$, $C_{n,r} = \dfrac{n!}{r!(n-r)!} < n!$      (7-4)

Copyright © 2011 Pearson Education, Inc. Publishing as Prentice Hall.

**59.**

| $p$ | $q$ | $p \wedge q$ |
|-----|-----|--------------|
| T | T | T |
| T | F | F |
| F | T | F |
| F | F | F |

$p \wedge q \Rightarrow p$ since $p$ is true whenever $p \wedge q$ is true.
(7-1)

**60.**

| $p$ | $q$ | $p \rightarrow q$ |
|-----|-----|-------------------|
| T | T | T |
| T | F | F |
| F | T | T |
| F | F | T |

$q \Rightarrow p \rightarrow q$ since $p \rightarrow q$ is true whenever $q$ is true.
(7-1)

**61.**

| $p$ | $q$ | $\neg p$ | $q \wedge \neg q$ | $\neg p \rightarrow (q \wedge \neg q)$ |
|-----|-----|----------|-------------------|----------------------------------------|
| T | T | F | F | T |
| T | F | F | F | T |
| F | T | T | F | F |
| F | F | T | F | F |

$p \equiv \neg p \rightarrow (q \wedge \neg q)$ since $p$ and $\neg p \rightarrow (q \wedge \neg q)$ have the same truth tables. (7-1)

**62.**

| $p$ | $q$ | $p \vee q$ | $\neg p$ | $\neg p \rightarrow q$ |
|-----|-----|------------|----------|------------------------|
| T | T | T | F | T |
| T | F | T | F | T |
| F | T | T | T | T |
| F | F | F | T | F |

$p \vee q \equiv \neg p \rightarrow q$ since $p \vee q$ and $\neg p \rightarrow q$ have the same truth tables. (7-1)

**63.**

| $p$ | $q$ | $p \rightarrow q$ | $p \wedge (p \rightarrow q)$ |
|-----|-----|-------------------|------------------------------|
| T | T | T | T |
| T | F | F | F |
| F | T | T | F |
| F | F | T | F |

$p \wedge (p \rightarrow q) \Rightarrow q$ since $q$ is true whenever $p \wedge (p \rightarrow q)$ is true. (7-1)

**64.**

| $p$ | $q$ | $\neg q$ | $p \wedge \neg q$ | $\neg(p \wedge \neg q)$ | $p \rightarrow q$ |
|-----|-----|----------|-------------------|-------------------------|-------------------|
| T | T | F | F | T | T |
| T | F | T | T | F | F |
| F | T | F | F | T | T |
| F | F | T | F | T | T |

$\neg(p \wedge \neg q) \equiv p \rightarrow q$ since $\neg(p \wedge \neg q)$ and $p \rightarrow q$ have the same truth tables. (7-1)

**65.** Operation 1: Two possible outcomes, boy or girl, $N_1 = 2$.

Operation 2: Two possible outcomes, boy or girl, $N_2 = 2$.

Operation 3: Two possible outcomes, boy or girl, $N_3 = 2$.

Operation 4: Two possible outcomes, boy or girl, $N_4 = 2$.

Operation 5: Two possible outcomes, boy or girl, $N_5 = 2$.

Copyright © 2011 Pearson Education, Inc. Publishing as Prentice Hall.

Using the Multiplication Principle, the total combined outcomes is:
$N_1 \cdot N_2 \cdot N_3 \cdot N_4 \cdot N_5 = 2 \cdot 2 \cdot 2 \cdot 2 \cdot 2 = 32$.

If order pattern is not taken into account, there would be only 6 possible outcomes: families with 0, 1, 2, 3, 4, or 5 boys.    (7-3)

**66.**  $C_{n,r} = \dfrac{n!}{r!(n-r)!}$ and $P_{n,r} = \dfrac{n!}{(n-r)!}$;

$\dfrac{n!}{r!(n-r)!} = \dfrac{n!}{(n-r)!}$ when $r = 0$ or $r = 1$; otherwise $C_{n,r} \neq P_{n,r}$    (7-4)

**67.**  The number of routes starting from $A$ and visiting each of the 5 stores exactly once is the number of permutations of 5 objects taken 5 at a time, i.e.,

$P_{5,5} = \dfrac{5!}{(5-5)!} = 120.$    (7-3)

**68.**  Draw a Venn diagram with:

  $S$ = people who have invested in stocks, and

  $B$ = people who have invested in bonds.

Then $n(U) = 1000$, $n(S) = 340$,
$n(B) = 480$ and $n(S \cap B) = 210$

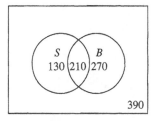

$n(S \cap B') = 340 - 210 = 130$   $n(B \cap S') = 480 - 210 = 270$
$n(S' \cap B') = 1000 - (130 + 210 + 270) = 1000 - 610 = 390$

(A)   $n(S \cup B) = n(S) + n(B) - n(S \cap B) = 340 + 480 - 210 = 610$.

(B)   $n(S' \cap B') = 390$

(C)   $n(B \cap S') = 270$    (7-3)

**69.**  Since the order of selection does not matter, the number of ways to select 6 from 40 is

$C_{40,6} = \dfrac{40!}{6!34!} = \dfrac{40 \cdot 39 \cdot 38 \cdot 37 \cdot 36 \cdot 35 \cdot 34!}{6!34!} = \dfrac{40 \cdot 39 \cdot 38 \cdot 37 \cdot 36 \cdot 35}{6 \cdot 5 \cdot 4 \cdot 3 \cdot 2 \cdot 1} = 10 \cdot 39 \cdot 38 \cdot 37 \cdot 7 = 3{,}838{,}380$    (7-4)

**70.**  (A) The total number of ways that the 67 names can be ordered is
  $67! \approx 3.647 \times 10^{94}$.

(B) It is not possible to print 67! ballots.    (7-4)

Copyright © 2011 Pearson Education, Inc. Publishing as Prentice Hall.

# 8    PROBABILITY

## EXERCISE 8-1

*Things to remember:*

1.    SAMPLE SPACE AND EVENTS

If the formulation of a set $S$ of outcomes (events) of an experiment is such that in each trial of the experiment one and only one of the outcomes in the set $S$ will occur, then $S$ is called a SAMPLE SPACE for the experiment. Each element in $S$ is called a SIMPLE OUTCOME, or SIMPLE EVENT.

An EVENT $E$ is any subset of $S$ (including the empty set $\varnothing$ and the sample space $S$). Event $E$ is a SIMPLE EVENT if it contains only one element and a COMPOUND EVENT if it contains more than one element. Event $E$ OCCURS if the result of performing the experiment is one of the simple events in $E$.

2.    CHOOSING SAMPLE SPACES

There is no one correct sample space for a given experiment. When specifying a sample space for an experiment, include as much detail as necessary to answer all questions of interest regarding the outcomes of the experiment. When in doubt, choose a sample space with more elements rather than fewer.

3.    PROBABILITIES FOR SIMPLE EVENTS

Given a sample space

$S = \{e_1, e_2, ..., e_n\}.$

To each simple event $e_i$ assign a real number denoted by $P(e_i)$, called the PROBABILITY OF THE EVENT $e_i$. These numbers can be assigned in an arbitrary manner provided the following two conditions are satisfied:

(a)    The probability of a simple event is a number between 0 and 1, inclusive. That is,

$$0 \le p(e_i) \le 1$$

(b)    The sum of the probabilities of all simple events in the sample space is 1. That is,

$$P(e_1) + P(e_2) + ... + P(e_n) = 1$$

Any probability assignment that satisfies these two conditions is called an ACCEPTABLE PROBABILITY ASSIGNMENT.

4.    PROBABILITY OF AN EVENT $E$

Given an acceptable probability assignment for the simple events in a sample space $S$, the probability of an arbitrary event $E$, denoted $P(E)$, is defined as follows:

(a)    $P(E) = 0$ if $E$ is the empty set.

(b)    If $E$ is a simple event, then $P(E)$ has already been assigned.

(c)    If $E$ is a compound event, then $P(E)$ is the sum of the probabilities of all the simple events

Copyright © 2011 Pearson Education, Inc.  Publishing as Prentice Hall.

in $E$.

(d)    If $E = S$, then $P(E) = P(S) = 1$ [this follows from 3(b)].

5.    STEPS FOR FINDING THE PROBABILITY OF AN EVENT $E$

(a)    Set up an appropriate sample space $S$ for the experiment.

(b)    Assign acceptable probabilities to the simple events in $S$.

(c)    To obtain the probability of an arbitrary event $E$, add the probabilities of the simple events in $E$.

6.    EMPIRICAL PROBABILITY

If an experiment is conducted $n$ times and event $E$ occurs with FREQUENCY $f(E)$, then the ratio $f(E)/n$ is called the RELATIVE FREQUENCY of the occurrence of event $E$ in $n$ trials. The EMPIRICAL PROBABILITY of $E$, denoted by $P(E)$, is given by the number (if it exists) that the relative frequency $f(E)/n$ approaches as $n$ gets larger and larger. For any particular $n$, the relative frequency $f(E)/n$ is also called the APPROXIMATE EMPIRICAL PROBABILITY of event $E$:

$$P(E) \approx \frac{\text{Frequency of occurrence of } E}{\text{Total number of trials}} = \frac{f(E)}{n}$$

(The larger $n$ is, the better the approximation.)

7.    PROBABILITIES UNDER AN EQUALLY LIKELY ASSUMPTION

If, in a sample space

$$S = \{e_1, e_2, ..., e_n\},$$

each simple event $e_i$ is as likely to occur as any other, then $P(e_i) = \dfrac{1}{n}$, for $i = 1, 2, ..., n$, i.e. assign the same probability, $1/n$, to each simple event.  The probability of an arbitrary event $E$ in this case is:

$$P(E) = \frac{\text{Number of elements in } E}{\text{Number of elements in } S} = \frac{n(E)}{n(S)}$$

---

1.    $P(E) = 1$ means that the occurrence of event $E$ is certain.

*In Problems 3 – 9 the spinner will land on any one of 12 equally likely sectors.*

3.    $n(\text{blue}) = 4$;  $P(\text{blue}) = \dfrac{4}{12} = \dfrac{1}{3}$.

5.    $n(\text{yellow or green}) = 3$;  $P(\text{yellow or green}) = \dfrac{3}{12} = \dfrac{1}{4}$.

7.    $n(\text{orange}) = 0$;  $P(\text{orange}) = \dfrac{0}{12} = 0$.

Copyright © 2011 Pearson Education, Inc.  Publishing as Prentice Hall.

9.  $n$(blue, red, yellow or green) = 12;  $P$(blue, red, yellow or green) $= \dfrac{12}{12} = 1$.

*In problems 11 − 19 there are 52 equally likely outcomes.*

11.  $n$(club) = 13;  $P$(club) $= \dfrac{13}{52} = \dfrac{1}{4}$.

13.  $n$(heart or diamond) = 26;  $P$(heart or diamond) $= \dfrac{26}{52} = \dfrac{1}{2}$.

15.  $P$(jack of clubs) $= \dfrac{1}{52}$.

17.  $n$(king or spade) = $n$(kings) + $n$(spades) − $n$(king of spades) = 4 + 13 − 1 = 16;

$P$(king or spade) $= \dfrac{16}{52} = \dfrac{4}{13}$.

19.  $n$(black diamonds) = 0;   $P$(black diamond) = 0.

21.  Let $B$ = boy and $G$ = girl.  Then

$S = \{(B, B), (B, G), (G, B), (G, G)\}$

where $(B, B)$ means both children are boys, $(B, G)$ means the first child is a boy, the second is a girl, and so on.  The event $E$ corresponding to having two children of opposite sex is $E = \{(B, G), (G, B)\}$.  Since the simple events are equally likely,

$P(E) = \dfrac{n(E)}{n(S)} = \dfrac{2}{4} = \dfrac{1}{2}$.

23.  (A)   Reject; $P(G) = -0.35$  —  no probability can be negative.

(B)   Reject; $P(J) + P(G) + P(P) + P(S) = 0.32 + 0.28 + 0.24 + 0.30$
$= 1.14 \neq 1$

(C)   Acceptable; each probability is between 0 and 1 (inclusive), and the sum of the probabilities is 1.

25.  $P(\{J, P\}) = P(J) + P(P) = 0.26 + 0.30 = 0.56$.

27.  $S = \{(B, B, B), (B, B, G), (B, G, B), (B, G, G), (G, B, B), (G, B, G),$
$\qquad (G, G, B), (G, G, G)\}$
$E = \{(B, B, G)\}$

Since the events are equally likely and $n(S) = 8$, $P(E) = \dfrac{1}{8}$.

29.  The number of three-digit sequences with no digit repeated is $P_{10,3}$.  Since the possible opening combinations are equally likely, the probability of guessing the right combination is:

$\dfrac{1}{P_{10,3}} = \dfrac{1}{10 \cdot 9 \cdot 8} = \dfrac{1}{720} \approx 0.0014$

31.  Let $S$ = the set of five-card hands.  Then $n(S) = C_{52,5}$.

Let $A$ = "five black cards."  Then $n(A) = C_{26,5}$.

Since individual hands are equally likely to occur:

Copyright © 2011 Pearson Education, Inc.  Publishing as Prentice Hall.

$$P(A) = \frac{n(A)}{n(S)} = \frac{C_{26,5}}{C_{52,5}} = \frac{\dfrac{26!}{5!21!}}{\dfrac{52!}{5!47!}} = \frac{26 \cdot 25 \cdot 24 \cdot 23 \cdot 22}{52 \cdot 51 \cdot 50 \cdot 49 \cdot 48} \approx 0.025$$

**33.**  $S$ = set of five-card hands; $n(S) = C_{52,5}$.

$F$ = "five face cards"; $n(F) = C_{12,5}$.

Since individual hands are equally likely to occur:

$$P(F) = \frac{n(F)}{n(S)} = \frac{C_{12,5}}{C_{52,5}} = \frac{\dfrac{12!}{5!7!}}{\dfrac{52!}{5!47!}} = \frac{12 \cdot 11 \cdot 10 \cdot 9 \cdot 8}{52 \cdot 51 \cdot 50 \cdot 49 \cdot 48} \approx 0.000305$$

**35.**  $S$ = {all the days in a year} (assume 365 days; exclude leap year); Equivalently, number the days of the year, beginning with January 1. Then $S = \{1, 2, 3, \ldots, 365\}$.

Assume that each day is as likely as any other day for a person to be born. Then the probability of each simple event is: $\dfrac{1}{365}$.

**37.**  $n(S) = P_{5,5} = 5! = 120$

Let $A$ = all notes inserted into the correct envelopes. Then $n(A) = 1$ and

$$P(A) = \frac{n(A)}{n(S)} = \frac{1}{120} \approx 0.00833$$

**39.**  Using the sample space shown in Figure 2, we have

$n(S) = 36$, $n(A) = 1$, where Event $A$ = "Sum being 2":

$$P(A) = \frac{n(A)}{n(S)} = \frac{1}{36}$$

**41.**  Let $E$ = "Sum being 6." Then $n(E) = 5$. Thus, $P(E) = \dfrac{n(E)}{n(S)} = \dfrac{5}{36}$.

**43.**  Let $E$ = "Sum being less than 5." Then $n(E) = 6$. Thus,

$$P(E) = \frac{n(E)}{n(S)} = \frac{6}{36} = \frac{1}{6}.$$

**45.**  Let $E$ = "Sum not 7 or 11." Then $n(E) = 28$ and $P(E) = \dfrac{n(E)}{n(S)} = \dfrac{28}{36} = \dfrac{7}{9}$.

**47.**  $E$ = "Sum being 1" is not possible.  Thus, $P(E) = 0$.

**49.**  Let $E$ = "Sum is divisible by 3" = "Sum is 3, 6, 9, or 12."  Then

$$n(E) = 12 \text{ and } P(E) = \frac{n(E)}{n(S)} = \frac{12}{36} = \frac{1}{3}.$$

**51.**  Let $E$ = "Sum is 7 or 11." Then $n(E) = 8$. Thus, $P(E) = \dfrac{n(E)}{n(S)} = \dfrac{8}{36} = \dfrac{2}{9}$.

**53.**  Let $E$ = "Sum is divisible by 2 or 3" = "Sum is 2, 3, 4, 6, 8, 9, 10, 12." Then $n(E) = 24$, and

$$P(E) = \frac{n(E)}{n(S)} = \frac{24}{36} = \frac{2}{3}.$$

*For Problems 55—59, the sample space S is given by:*

Copyright © 2011 Pearson Education, Inc.  Publishing as Prentice Hall.

$S = \{(H, H, H), (H, H, T), (H, T, H), (H, T, T)\}$

*The outcomes are equally likely and n(S) = 4.*

**55.** Let $E$ = "1 head." Then $n(E) = 1$ and $P(E) = \dfrac{n(E)}{n(S)} = \dfrac{1}{4}$.

**57.** Let $E$ = "3 heads." Then $n(E) = 1$ and $P(E) = \dfrac{n(E)}{n(S)} = \dfrac{1}{4}$.

**59.** Let $E$ = "More than 1 head." Then $n(E) = 3$ and $P(E) = \dfrac{n(E)}{n(S)} = \dfrac{3}{4}$.

**61.** No. The events "heart," "not a heart" are not equally likely;
$P(H) = \dfrac{1}{4}$, $P(N) = \dfrac{3}{4}$.

**63.** Yes. The events "even," "odd" are equally likely; $P(E) = P(O) = \dfrac{1}{2}$.

**65.** Yes. Since the seven sectors have equal area, the events are equally likely; each event has probability $\dfrac{1}{7}$.

**67.** (A)   Yes. If we flip a fair coin 20 times, a representation of the sample space is $\{0, 1, 2, \ldots , 19, 20\}$ where each element denotes the number of heads. Each of these outcomes is possible; 10 is the outcome with the highest probability.

(B)   Yes. On average we would expect 20 heads in 40 flips of a fair coin. Based on the evidence, it would appear that
$$P(H) = \frac{37}{40} \text{ and } P(T) = \frac{3}{40}.$$

*For Problems 69—75, the sample space S is given by:*
$$S = \begin{Bmatrix} (1,1),(1,2),(1,3) \\ (2,1),(2,2),(2,3) \\ (3,1),(3,2),(3,3) \end{Bmatrix}$$
*The outcomes are equally likely and n(S) = 9.*

**69.** Let $E$ = "Sum is 2." Then $n(E) = 1$ and $P(E) = \dfrac{n(E)}{n(S)} = \dfrac{1}{9}$.

**71.** Let $E$ = "Sum is 4." Then $n(E) = 3$ and $P(E) = \dfrac{n(E)}{n(S)} = \dfrac{3}{9} = \dfrac{1}{3}$.

**73.** Let $E$ = "Sum is 6." Then $n(E) = 1$ and $P(E) = \dfrac{n(E)}{n(S)} = \dfrac{1}{9}$.

**75.** Let $E$ = "Sum is odd" = "Sum is 3 or 5." Then $n(E) = 4$ and
$$P(E) = \frac{n(E)}{n(S)} = \frac{4}{9}.$$

*For Problems 77—83, the sample space is the set of all 5-card hands.*
*The outcomes are equally likely and n(S) = $C_{52,5}$.*

**77.** Let $E$ = "5 cards, jacks through aces." Then $n(E) = C_{16,5}$. Thus

Copyright © 2011 Pearson Education, Inc. Publishing as Prentice Hall.

$$P(E) = \frac{C_{16,5}}{C_{52,5}} = \frac{\frac{16!}{5!11!}}{\frac{52!}{5!47!}} = \frac{16 \cdot 15 \cdot 14 \cdot 13 \cdot 12}{52 \cdot 51 \cdot 50 \cdot 49 \cdot 48} \approx 0.00168.$$

**79.** Let $E$ = "4 aces." Then $n(E) = 48$ (the remaining card can be any one of the 48 cards which are not aces). Thus,

$$P(E) = \frac{48}{C_{52,5}} = \frac{48}{\frac{52!}{5!47!}} = \frac{48 \cdot 5!}{52 \cdot 51 \cdot 50 \cdot 49 \cdot 48} = \frac{5 \cdot 4 \cdot 3 \cdot 2}{52 \cdot 51 \cdot 50 \cdot 49} \approx 0.0000185$$

**81.** Let $E$ = "Straight flush, ace high." Then $n(E) = 4$ (one such hand in each suit). Thus,

$$P(E) = \frac{4}{C_{52,5}} = \frac{4 \cdot 5!}{52 \cdot 51 \cdot 50 \cdot 49 \cdot 48} = \frac{480}{52 \cdot 51 \cdot 50 \cdot 49 \cdot 48} \approx 0.0000015$$

**83.** Let $E$ = "2 aces and 3 queens." The number of ways to get 2 aces is $C_{4,2}$ and the number of ways to get 3 queens is $C_{4,3}$.

Thus, $n(E) = C_{4,2} \cdot C_{4,3} = \frac{4!}{2!2!} \cdot \frac{4!}{3!1!} = \frac{4 \cdot 3}{2} \cdot \frac{4}{1} = 24$
and

$$P(E) = \frac{n(E)}{n(S)} = \frac{24}{C_{52,5}} = \frac{24 \cdot 5!}{52 \cdot 51 \cdot 50 \cdot 59 \cdot 48} \approx 0.000009.$$

**85.** (A)  From the plot, $P(6) = \frac{7}{50} = 0.14$.

  (B)  If the outcomes are equally likely, $P(6) = \frac{1}{6} = 0.167$.

  (C)  The answer here depends on the results of your simulation.

**87.** (A)  Represent the outcomes $H$ and $T$ by the numbers 1 and 2, respectively, and select 500 random integers from the set $\{1, 2\}$.

  (B)  The answer depends on the results of your simulation.

  (C)  If the outcomes are equally likely, then $P(H) = P(T) = \frac{1}{2}$.

**89.** (A)  The sample space $S$ is the set of all possible permutations of the 12 brands taken 4 at a time, and $n(S) = P_{12,4}$. Thus, the probability of selecting 4 brands and identifying them correctly, with no answer repeated, is:

$$P(E) = \frac{1}{P_{12,4}} = \frac{1}{\frac{12!}{(12-4)!}} = \frac{1}{12 \cdot 11 \cdot 10 \cdot 9} \approx 0.000084$$

  (B)  Allowing repetition, $n(S) = 12^4$ and the probability of identifying them correctly is:

$$P(F) = \frac{1}{12^4} \approx 0.000048$$

**91.** (A)  Total number of applicants = $6 + 5 = 11$.

$$n(S) = C_{11,5} = \frac{11!}{5!(11-5)!} = 462$$

The number of ways that three females and two males can be selected is:

Copyright © 2011 Pearson Education, Inc. Publishing as Prentice Hall.

$$C_{6,3} \cdot C_{5,2} = \frac{6!}{3!(6-3)!} \cdot \frac{5!}{2!(5-2)!} = 20 \cdot 10 = 200$$

Thus, $P(A) = \dfrac{C_{6,3} \cdot C_{5,2}}{C_{11,5}} = \dfrac{200}{462} = 0.433$

(B)    $P(\text{4 females and 1 male}) = \dfrac{C_{6,4} \cdot C_{5,1}}{C_{11,5}} = 0.162$

(C)    $P(\text{5 females}) = \dfrac{C_{6,5}}{C_{11,5}} = 0.013$

(D)    $P(\text{at least four females}) = P(\text{4 females and 1 male}) + P(\text{5 females})$

$$= \frac{C_{6,4} \cdot C_{5,1}}{C_{11,5}} + \frac{C_{6,5}}{C_{11,5}}$$

$$= 0.162 + 0.013 \text{ [refer to parts (B) and (C)]} = 0.175$$

**93.** (A)  The sample space $S$ consists of the number of permutations of the 8 blood types chosen 3 at a time. Thus, $n(S) = P_{8,3}$ and the probability of guessing the three types in a sample correctly is:

$$P(E) = \frac{1}{P_{8,3}} = \frac{1}{\dfrac{8!}{(8-3)!}} = \frac{1}{8 \cdot 7 \cdot 6} \approx 0.0030$$

(B)    Allowing repetition, $n(S) = 8^3$ and the probability of guessing the three types in a sample correctly is:

$$P(E) = \frac{1}{8^3} \approx 0.0020$$

**95.** (A)  The total number of ways of selecting a president and a vice-president from the 11 members of the council is:

$$P_{11,2}, \text{ i.e., } n(S) = P_{11,2}.$$

The total number of ways of selecting the president and the vice-president from the 6 democrats is

$P_{6,2}$. Thus, if $E$ is the event "The president and vice-president are both Democrats," then

$$P(E) = \frac{P_{6,2}}{P_{11,2}} = \frac{\dfrac{6!}{(6-2)!}}{\dfrac{11!}{(11-2)!}} = \frac{6 \cdot 5}{11 \cdot 10} = \frac{30}{110} \approx 0.273.$$

(B)    The total number of ways of selecting a committee of 3 from the 11 members of the council is:

$$C_{11,3}, \text{ i.e., } n(S) = C_{11,3} = \frac{11!}{3!(11-3)!} = \frac{11 \cdot 10 \cdot 9 \cdot 8!}{3 \cdot 2 \cdot 1 \cdot 8!} = 165.$$

If we let $F$ be the event "The majority are Republicans," which is the same as having either 2 Republicans and 1 Democrat or all 3 Republicans, then

$$n(F) = C_{5,2} \cdot C_{6,1} + C_{5,3} = \frac{5!}{2!(5-2)!} \cdot \frac{6!}{1!(6-1)!} + \frac{5!}{3!(5-3)!} = 10 \cdot 6 + 10 = 70.$$

Thus,  $P(F) = \dfrac{n(F)}{n(S)} = \dfrac{70}{165} \approx 0.424.$

Copyright © 2011 Pearson Education, Inc.  Publishing as Prentice Hall.

EXERCISE 8-2

*Things to remember:*

1.  UNION AND INTERSECTION OF EVENTS

    If $A$ and $B$ are two events in a sample space $S$, then the UNION of $A$ and $B$, denoted by $A \cup B$, and the INTERSECTION of $A$ and $B$, denoted by $A \cap B$, are defined as follows:

    $A \cup B = \{e \in S \mid e \in A \text{ OR } e \in B\}$       $A \cap B = \{e \in S \mid e \in A \text{ AND } e \in B\}$

    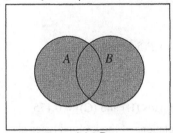
    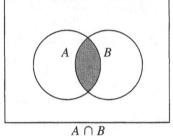

    $A \cup B$               $A \cap B$

    Furthermore, we define:

    The **event $A$ or $B$** to be $A \cup B$.

    The **event $A$ and $B$** to be $A \cap B$.

2.  PROBABILITY OF A UNION OF TWO EVENTS

    For any events $A$ and $B$,

    (a)    $P(A \cup B) = P(A) + P(B) - P(A \cap B)$.

    If $A$ and $B$ are MUTUALLY EXCLUSIVE $(A \cap B = \varnothing)$, then

    (b)    $P(A \cup B) = P(A) + P(B)$.

3.  PROBABILITY OF COMPLEMENTS

    For any event $E$, $E \cup E' = S$ and $E \cap E' = \varnothing$. Thus,

    $$P(E) = 1 - P(E')$$

    $$P(E') = 1 - P(E)$$

4.  PROBABILITY TO ODDS

    If $P(E)$ is the probability of the event $E$, then:

    (a)    Odds for $E = \dfrac{P(E)}{1 - P(E)} = \dfrac{P(E)}{P(E')}$    $[P(E) \neq 1]$

    (b)    Odds against $E = \dfrac{P(E')}{P(E)}$    $[P(E) \neq 0]$

    The ratio $\dfrac{P(E)}{P(E')}$, giving odds for $E$, is usually expressed as an equivalent ratio $\dfrac{a}{b}$ of whole numbers (by multiplying numerator and denominator by the same number), and written "a to b" or "a:b". In this case, the odds against $E$ are written "b to a" or "b:a".

Copyright © 2011 Pearson Education, Inc. Publishing as Prentice Hall.

5.    ODDS TO PROBABILTY
   If the odds for an event $E$ are "$a$  to  $b$", then the probability of $E$ is:

$$P(E) = \frac{a}{a+b}$$

---

**1.**  $P(A \cap B) = \dfrac{39}{100} = 0.39.$

**3.**  $P(A' \cup B) = \dfrac{39 + 21 + 26}{100} = \dfrac{86}{100} = 0.86.$

**5.**  $P(A \cup B) = \dfrac{14 + 39 + 21}{100} = \dfrac{74}{100} = 0.74;$   $P[(A \cup B)'] = 1 - P(A \cup B) = 0.26.$

**7.**  $n(D) = 13, P(D) = \dfrac{n(D)}{n(S)} = \dfrac{13}{52} = \dfrac{1}{4}$

**9.**  $n(F) = 12, n(F') = 40; P(F') = \dfrac{40}{52} = \dfrac{10}{13}$

**11.**  $D \cap F =$ jack, queen, king of diamonds; $n(D \cap F) = 3$

   $P(D \cap F) = \dfrac{3}{52}$

**13.**  $n(D \cup F) = 13 + 12 - 3 = 22$ (number of diamonds + number of face cards) – (jack, queen , king of diamonds)

   $P(D \cup F) = \dfrac{22}{52} = \dfrac{11}{26}$

**15.**  $D \cap F' = 2 - 10$ of diamonds and the ace of diamonds.

   $n(D \cap F') = 10;$  $P(D \cap F') = \dfrac{10}{52} = \dfrac{5}{26}$

**17.**  $n(D \cup F') = n(D) + n(F') - n(D \cap F') = 13 + 40 - 10 = 43$

   $P(D \cup F') = \dfrac{n(D \cup F')}{n(S)} = \dfrac{43}{52}$

**19.**  Let $E =$ number is less than 6
   $F =$ number is greater than 19
   Since $E \cup F = \varnothing$ , we use equation (1):
   $n(E) = 5,$  $n(F) = 6$

   $P(E \cup F) = P(E) + P(F) = \dfrac{5}{25} + \dfrac{6}{25} = \dfrac{11}{25} = 0.44$

**21.**  Let $E =$ number is divisible by 3 = {3, 6, 9, 12, 15, 18, 21, 24}
   $F =$ number is divisible by 4 = {4, 8, 12, 16, 20, 24}
   Since $E \cap F \neq 0$, we use equation (2):

Copyright © 2011 Pearson Education, Inc. Publishing as Prentice Hall.

$n(E) = 8, \; n(F) = 6, \; n(E \cap F) = 2,$

$P(E \cup F) = P(E) + P(F) - P(E \cap F) = \dfrac{8}{25} + \dfrac{6}{25} - \dfrac{2}{25} = \dfrac{12}{25} = 0.48$

23.  Let $E$ = number is even
     $F$ = number is divisible by 5 = {5, 10, 15, 20, 25}
     Since $E \cap F \neq \varnothing$, we use equation (2):
     $n(E) = 12, \; n(F) = 5, \; n(E \cap F) = 2$

     $P(E \cup F) = P(E) + P(F) - P(E \cap F) = \dfrac{12}{25} + \dfrac{5}{25} - \dfrac{2}{25} = \dfrac{15}{25} = \dfrac{3}{5} = 0.6$

25.  $P(\text{loss}) = 1 - P(\text{win}) = 1 - 0.51 = 0.49$

27.  $A$ = sum $\leq 5$, $\; n(A) = 10, \quad P(A) = \dfrac{10}{36} = \dfrac{5}{18}.$

29.  $A$ = number the first die = 6, $n(A) = 6$;

     $B$ = number on the second die = 3; $n(B) = 6$; $n(A \cap B) = 1$;

     $P(A \cup B) = P(A) + P(B) - P(A \cap B) = \dfrac{6}{36} + \dfrac{6}{36} - \dfrac{1}{36} = \dfrac{11}{36}.$

31.  Use $\underline{3}$ to find the odds for Event $E$.

(A)  $P(E) = \dfrac{3}{8}, \; P(E') = 1 - P(E) = \dfrac{5}{8}$

Odds for $E = \dfrac{P(E)}{P(E')}$

$= \dfrac{3/8}{5/8} = \dfrac{3}{5}$ (3 to 5)

Odds against $E = \dfrac{P(E')}{P(E)}$

$= \dfrac{5/8}{3/8} = \dfrac{5}{3}$ (5 to 3)

(B)  $P(E) = \dfrac{1}{4}, \; P(E') = 1 - P(E) = \dfrac{3}{4}$

Odds for $E = \dfrac{P(E)}{P(E')}$

$= \dfrac{1/4}{3/4} = \dfrac{1}{3}$ (1 to 3)

Odds against $E = \dfrac{P(E')}{P(E)}$

$= \dfrac{3/4}{1/4} = \dfrac{3}{1}$ (3 to 1)

(C)  $P(E) = .4, \; P(E') = 1 - P(E) = .6$

Odds for $E = \dfrac{P(E)}{P(E')}$

$= \dfrac{.4}{.6} = \dfrac{2}{3}$ (2 to 3)

Odds against $E = \dfrac{P(E')}{P(E)}$

$= \dfrac{.6}{.4} = \dfrac{3}{2}$ (3 to 2)

(D)  $P(E) = .55, \; P(E') = 1 - P(E) - .45$

Odds for $E = \dfrac{P(E)}{P(E')}$

$= \dfrac{.55}{.45} = \dfrac{11}{9}$ (11 to 9)

Odds against $E = \dfrac{P(E')}{P(E)}$

$= \dfrac{.45}{.55} = \dfrac{9}{11}$
(9 to 11)

33.  Use $\underline{4}$ to find the probability of event $E$.

Copyright © 2011 Pearson Education, Inc. Publishing as Prentice Hall.

(A)    Odds for $E = \dfrac{3}{8}$

   $P(E) = \dfrac{3}{3+8} = \dfrac{3}{11}$

(B)    Odds for $E = \dfrac{11}{7}$

   $P(E) = \dfrac{11}{11+7} = \dfrac{11}{18}$

(C)    Odds for $E = \dfrac{4}{1}$

   $P(E) = \dfrac{4}{4+1} = \dfrac{4}{5} = .8$

(D)    Odds for $E = \dfrac{49}{51}$

   $P(E) = \dfrac{49}{49+51} = \dfrac{49}{100} = .49$

**35.**    False. Odds for $E = \dfrac{P(E)}{P(E')}$ ;

   odds against $E' = \dfrac{P([E']')}{P(E')} = \dfrac{P(E)}{P(E')}$

   So, odds for $E$ = odds against $E'$ independent of $P(E)$.

**37.**    False. Since $P(E \cup F) = P(E) + P(F) - P(E \cap F)$ for *any* events $E$, $F$,

   $P(E) + P(F) = P(E \cup F) + P(E \cap F)$

   whether or not $E$ and $F$ are disjoint.

**39.**    True. $E$ and $F$ are complementary, then $F = E'$ and $E \cap F = \varnothing$ . Therefore, $E$ and $F$ are mutually exclusive.

**41.**    Odds for $E = \dfrac{P(E)}{P(E')} = \dfrac{1/2}{1/2} = 1$.

   The odds in favor of getting a head in a single toss of a coin are 1 to 1.

**43.**    The sample space for this problem is:

   $S = \{$HHH, HHT, THH, HTH, TTH, HTT, THT, TTT$\}$

   Let Event $E$ = "getting at least 1 head."
   Let Event $E'$ = "getting no heads."

   Thus, $\dfrac{P(E)}{P(E')} = \dfrac{7/8}{1/8} = \dfrac{7}{1}$

   The odds in favor of getting at least 1 head are 7 to 1.

**45.**    Let Event $E$ = "getting a number greater than 4."
   Let Event $E'$ = "not getting a number greater than 4."

   Thus, $\dfrac{P(E')}{P(E)} = \dfrac{4/6}{2/6} = \dfrac{2}{1}$

   The odds against getting a number greater than 4 in a single roll of a die are 2 to 1.

**47.**    Let Event $E$ = "getting 3 or an even number" = $\{2, 3, 4, 6\}$.
   Let Event $E'$ = "not getting 3 or an even number" = $\{1, 5\}$.

   Thus, $\dfrac{P(E')}{P(E)} = \dfrac{2/6}{4/6} = \dfrac{1}{2}$

   The odds against getting 3 or an even number are 1 to 2.

Copyright © 2011 Pearson Education, Inc. Publishing as Prentice Hall.

**49.**  Let $E$ = "rolling a sum of five." Then $P(E) = \dfrac{n(E)}{n(S)} = \dfrac{4}{36} = \dfrac{1}{9}$ and $P(E') = \dfrac{8}{9}$.

   (A)   Odds for $E = \dfrac{1/9}{8/9} = \dfrac{1}{8}$  (1 to 8)

   (B)   The house should pay $8 for the game to be fair (see Example 6).

**51.**  (A)   Let $E$ = "sum is less than 4 or greater than 9." Then

   $$P(E) = \frac{10+30+120+80+70}{1000} = \frac{310}{1000} = \frac{31}{100} = .31 \text{ and } P(E') = \frac{69}{100}.$$

   Thus,

   $$\text{Odds for } E = \frac{31/100}{69/100} = \frac{31}{69}$$

   (B)   Let $F$ = "sum is even or divisible by 5." Then

   $$P(F) = \frac{10+50+110+170+120+70+70}{1000} = \frac{600}{1000} = \frac{6}{10} = .6$$

   and $P(F') = \dfrac{4}{10}$. Thus,

   $$\text{Odds for } F = \frac{6/10}{4/10} = \frac{6}{4} = \frac{3}{2}$$

**53.**  Let $A$ = "drawing a face card" (Jack, Queen, King)
   and $B$ = "drawing a club."

   Then $P(A \cup B) = P(A) + P(B) - P(A \cap B) = \dfrac{12}{52} + \dfrac{13}{52} - \dfrac{3}{52} = \dfrac{22}{52} = \dfrac{11}{26}$

   $P[(A \cup B)'] = \dfrac{15}{26}$

   Odds for $A \cup B = \dfrac{11/26}{15/26} = \dfrac{11}{15}$

**55.**  Let $A$ = "drawing a black card"
   and $B$ = "drawing an ace."

   $P(A \cup B) = P(A) + P(B) - P(A \cap B) = \dfrac{26}{52} + \dfrac{4}{52} - \dfrac{2}{52} = \dfrac{28}{52} = \dfrac{7}{13}$

   $P[(A \cup B)'] = \dfrac{6}{13}$

   Odds for $A \cup B = \dfrac{7/13}{6/13} = \dfrac{7}{6}$

**57.**  The sample space $S$ is the set of all 5-card hands and $n(S) = C_{52,5}$

   Let $E$ = "getting at least one diamond."
   Then $E'$ = "no diamonds" and $n(E) = C_{39,5}$.

Copyright © 2011 Pearson Education, Inc.  Publishing as Prentice Hall.

Thus, $P(E') = \dfrac{C_{39,5}}{C_{52,5}}$ , and

$$P(E) = 1 - \frac{C_{39,5}}{C_{52,5}} = 1 - \frac{\dfrac{39!}{5!34!}}{\dfrac{52!}{5!47!}} = 1 - \frac{39 \cdot 38 \cdot 37 \cdot 36 \cdot 35}{52 \cdot 51 \cdot 50 \cdot 49 \cdot 48} \approx 1 - .22 = .78.$$

**59.** The number of numbers less than or equal to 1000 which are divisible by 6 is the largest integer in $\dfrac{1000}{6}$ or 166.

The number of numbers less than or equal to 1000 which are divisible by 8 is the largest integer in $\dfrac{1000}{8}$ or 125.

The number of numbers less than or equal to 1000 which are divisible by both 6 and 8 is the same as the number of numbers which are divisible by 24. This is the largest integer in $\dfrac{1000}{24}$ or 41.

Thus, if $A$ is the event "selecting a number which is divisible by either 6 or 8," then

$n(A) = 166 + 125 - 41 = 250$ and $P(A) = \dfrac{250}{1000} = .25.$

**61.** In general, for three events $A$, $B$, and $C$,
$$P(A \cup B \cup C) = P(A) + P(B) + P(C) - P(A \cap B) - P(A \cap C) - P(B \cap C) + P(A \cap B \cap C)$$
Therefore,

(*)    $P(A \cup B \cup C) = P(A) + P(B) + P(C) - P(A \cap B)$

will hold if $A$ and $C$, and $B$ and $C$ are mutually exclusive. Note that, if either $A \cap C = \varnothing$, or $B \cap C = \varnothing$, then $A \cap B \cap C = \varnothing$.

Equation (*) will also hold if $A$, $B$, and $C$ are mutually exclusive, in which case
$$P(A \cup B \cup C) = P(A) + P(B) + P(C)$$

**63.** From Example 5,
$$P(E) = 1 - \frac{365!}{365^n(365-n)!} = 1 - \frac{1}{365^n} \cdot \frac{365!}{(365-n)!}$$
$$= 1 - \frac{1}{(365)^n} \cdot P_{365,n}$$
$$= 1 - \frac{P_{365,n}}{(365)^n}$$

For calculators with a $P_{n,r}$ key, this form involves fewer calculator steps. Also, 365! produces an overflow error on many calculators, while $P_{365,n}$ does not produce an overflow error for many values of $n$.

**65.** $S$ = set of all lists of $n$ birth months, $n \le 12$. Then
$n(S) = 12 \cdot 12 \cdot \ \dots \ \cdot 12 \ (n \text{ times}) = 12^n.$

Let $E$ = "at least two people have the same birth month."
Then $E'$ = "no two people have the same birth month."

Copyright © 2011 Pearson Education, Inc. Publishing as Prentice Hall.

$n(E') = 12 \cdot 11 \cdot 10 \cdot \ldots \cdot [12 - (n - 1)]$

$$= \frac{12 \cdot 11 \cdot 10 \cdot \ldots \cdot [12 - (n-1)](12 - n)[12 - (n+1)] \cdot \ldots \cdot 3 \cdot 2 \cdot 1}{(12 - n)[12 - (n+1)] \cdot \ldots \cdot 3 \cdot 2 \cdot 1} = \frac{12!}{(12 - n)!}$$

Thus, $P(E') = \dfrac{\dfrac{12!}{(12-n)!}}{12^n} = \dfrac{12!}{12^n(12-n)!}$ and $P(E) = 1 - \dfrac{12!}{12^n(12-n)!}$.

**67.** Odds for $E = \dfrac{P(E)}{P(E')} = \dfrac{P(E)}{1 - P(E)} = \dfrac{a}{b}$. Therefore,

$bP(E) = a[1 - P(E)] = a - aP(E)$.

Thus, $aP(E) + bP(E) = a$

$(a + b)P(E) = a$

$P(E) = \dfrac{a}{a+b}$

**69.** (A) From the plot, 7 and 8 each came up 10 times.

Therefore, $P(7 \text{ or } 8) = \dfrac{10}{50} + \dfrac{10}{50} = \dfrac{20}{50} = 0.4$.

(B) Theoretical probability: $P(7) = \dfrac{1}{6}$, $P(8) = \dfrac{5}{36}$

so $P(7 \text{ or } 8) = \dfrac{1}{6} + \dfrac{5}{36} = \dfrac{6}{36} + \dfrac{5}{36} = \dfrac{11}{36} \approx 0.306$.

(C) The answer depends on the results of your simulation.

**71.** Venn diagram:

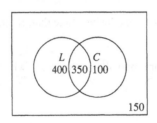

Let $L$ be the event that the student owns a laptop and $C$ be the event that the student owns a car.

The table corresponding to the given data is as follows:

|  | C | C' | Total |
|---|---|---|---|
| L | 350 | 400 | 750 |
| L' | 100 | 150 | 250 |
| Total | 450 | 550 | 1000 |

The corresponding probabilities are:

|  | C | C' | Total |
|---|---|---|---|
| L | .35 | .40 | .75 |
| L' | .10 | .15 | .25 |
| Total | .45 | .55 | 1.00 |

From the above table:

(A) $P(C \text{ or } L) = P(C \cup L) = P(C) + P(L) - P(C \cap L) = .45 + .75 - .35 = .85$

(B) $P(C' \cap L') = .15$

Copyright © 2011 Pearson Education, Inc. Publishing as Prentice Hall.

**73.** (A)  Using the table, we have:

$$P(M_1 \text{ or } A) = P(M_1 \cup A) = P(M_1) + P(A) - P(M_1 \cap A) = .2 + .3 - .05 = .45$$

(B)  $P[(M_2 \cap A') \cup (M_3 \cap A')] = P(M_2 \cap A') + P(M_3 \cap A')$

$$= .2 + .35 \text{ (from the table)}$$
$$= .55$$

**75.**  The sample space $S$ is the set of all possible 10-element samples from the 60 game players, and $n(S) = C_{60,10}$. Let $E$ be the event that a sample contains at least one defective game player. Then $E'$ is the event that a sample contains no game players. Now, $n(E') = C_{51,10}$.

Thus, $P(E') = \dfrac{C_{51,10}}{C_{60,10}} = \dfrac{\dfrac{51!}{10!41!}}{\dfrac{60!}{10!50!}} \approx .17$ and $P(E) \approx 1 - .17 = .83$.

Therefore, the probability that a sample will be returned is .83.

**77.**  The given information is displayed in the Venn diagram:

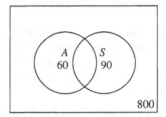

$A$ = suffers from loss of appetite
$S$ = suffers from loss of sleep

Thus, we can conclude that $n(A \cap S) = 1000 - (60 + 90 + 800) = 50$.

$$P(A \cap S) = \dfrac{50}{1000} = .05$$

**79.** (A)  "Unaffiliated or no preference" = $U \cup N$.

$$P(U \cup N) = P(U) + P(N) - P(U \cap N)$$

$$= \dfrac{150}{1000} + \dfrac{85}{1000} - \dfrac{15}{1000} = \dfrac{220}{1000} = \dfrac{11}{50} = .22$$

Therefore, $P[(U \cup N)'] = 1 - \dfrac{11}{50} = \dfrac{39}{50}$ and

Odds for $U \cup N = \dfrac{11/50}{39/50} = \dfrac{11}{39}$

(B)  "Affiliated with a party and prefers candidate $A$" = $(D \cup R) \cap A$.

$$P[(D \cup R) \cap A] = \dfrac{300}{1000} = \dfrac{3}{10} = .3$$

The odds against this event are:
$$\dfrac{1 - 3/10}{3/10} = \dfrac{7/10}{3/10} = \dfrac{7}{3}$$

Copyright © 2011 Pearson Education, Inc. Publishing as Prentice Hall.

<u>EXERCISE 8-3</u>

*Things to remember:*

1.  CONDITIONAL PROBABILITY

    For events $A$ and $B$ in a sample space $S$, the CONDITIONAL PROBABILITY of $A$ given $B$, denoted $P(A \mid B)$, is defined by

    $$P(A \mid B) = \frac{P(A \cap B)}{P(B)}, \quad P(B) \neq 0$$

2.  PRODUCT RULE

    For events $A$ and $B$, $P(A) \neq 0$, $P(B) \neq 0$, in a sample space $S$,

    $$P(A \cap B) = P(A) \cdot P(B \mid A) = P(B) \cdot P(A \mid B).$$

    [Note: We can use either $P(A) \cdot P(B \mid A)$ or $P(B) \cdot P(A \mid B)$ to compute $P(A \cap B)$.]

3.  PROBABILITY TREES

    Given a sequence of probability experiments. To compute the probabilities of combined outcomes:

    *Step 1.* Draw a tree diagram corresponding to all combined outcomes of the sequence of experiments.

    *Step 2.* Assign a probability to each tree branch. (This is the probability of the occurrence of the event on the right end of the branch subject to the occurrence of all events on the path leading to the event on the right end of the branch. The probability of the occurrence of a combined outcome that corresponds to a path through the tree is the product of all branch probabilities on the path.)

    *Step 3.* Use the results in Steps 1 and 2 to answer various questions related to the sequence of experiments as a whole.

4.  INDEPENDENCE

    Let $A$ and $B$ be any events in a sample space $S$. Then $A$ and $B$ are INDEPENDENT if and only if

    $$P(A \cap B) = P(A) \cdot P(B).$$

    Otherwise, $A$ and $B$ are DEPENDENT.

5.  INDEPENDENT SET OF EVENTS

    A set of events is said to be INDEPENDENT if for each finite subset $\{E_1, E_2, ..., E_k\}$

    $$P(E_1 \cap E_2 \cap ... \cap E_k) = P(E_1)P(E_2) \cdot ... \cdot P(E_k)$$

1.  $P(B) = 0.03 + 0.05 + 0.02 = 0.10$

3.  $P(B \cap D) = 0.03$

5.  $P(D \mid B) = \dfrac{P(D \cap B)}{P(B)} = \dfrac{0.03}{0.10} = 0.30$

7.  $P(B \mid D) = \dfrac{P(B \cap D)}{P(D)} = \dfrac{0.03}{0.30} = 0.10$

9.  $P(D \mid C) = \dfrac{P(D \cap C)}{P(C)} = \dfrac{0.07}{0.20} = 0.35$

11. $A \cap C = \varnothing$. Therefore, $P(A \mid C) = \dfrac{P(A \cap C)}{P(C)} = \dfrac{0}{0.20} = 0.$

Copyright © 2011 Pearson Education, Inc. Publishing as Prentice Hall.

13. Events $A$ and $D$ are independent if $P(A \cap D) = P(A) \cdot P(D)$:
$P(A) = 0.70$, $P(D) = 0.30$, $P(A \cap D) = 0.20$; $P(A) \cdot P(D) = 0.21 \neq P(A \cap D)$; $A$ and $D$ are dependent.

15. $P(B) = 0.10$, $P(D) = 0.30$, $P(B \cap D) = 0.03$; $P(B) \cdot P(D) = 0.03 = P(B \cap D)$; $B$ and $D$ are independent.

17. $P(B) = 0.10$, $P(F) = 0.30$, $P(B \cap F) = 0.02$; $P(B) \cdot P(F) = 0.03 \neq P(B \cap F)$; $B$ and $F$ are dependent.

19. $P(A) = 0.70$, $P(B) = 0.10$, $A \cap B = \varnothing$, $P(A \cap B) = 0$; $P(A) \cdot P(B) = 0.07 \neq P(A \cap B)$; $A$ and $B$ are dependent.

21. (A) Let $H_8 = $ "a head on the eighth toss." Since each toss is independent of the other tosses, $P(H_8) = \dfrac{1}{2}$.

   (B) Let $H_i = $ "a head on the $i$th toss." Since the tosses are independent,

   $$P(H_1 \cap H_2 \cap \cdots \cap H_8) = P(H_1)P(H_2) \cdots P(H_8) = \left(\dfrac{1}{2}\right)^8 = \dfrac{1}{2^8} = \dfrac{1}{256}.$$

   Similarly, if $T_i = $ "a tail on the $i$th toss," then

   $$P(T_1 \cap T_2 \cap \cdots \cap T_8) = P(T_1)P(T_2) \cdots P(T_8) = \dfrac{1}{2^8} = \dfrac{1}{256}.$$ Finally, if

   $H = $ "all heads" and $T = $ "all tails," then $H \cap T = \varnothing$ and

   $$P(H \cup T) = P(H) + P(T) = \dfrac{1}{256} + \dfrac{1}{256} = \dfrac{2}{256} = \dfrac{1}{128} \approx .00781.$$

23. Given the table:

| $e_i$ | 1 | 2 | 3 | 4 | 5 |
|---|---|---|---|---|---|
| $P_i$ | .3 | .1 | .2 | .3 | .1 |

   $E = $ "pointer lands on an even number" $= \{2, 4\}$.
   $F = $ "pointer lands on a number less than 4" $= \{1, 2, 3\}$.

   (A) $P(F \mid E) = \dfrac{P(F \cap E)}{P(E)} = \dfrac{P(2)}{P(2) + P(4)} = \dfrac{.1}{.1 + .3} = \dfrac{.1}{.4} = \dfrac{1}{4}$

   (B) $P(E \cap F) = P(2) = .1$
   $P(E) = .4$,
   $P(F) = P(1) + P(2) + P(3) = .3 + .1 + .2 = .6$,
   and
   $P(E)P(F) = (.4)(.6) = .24 \neq P(E \cap F)$.
   Thus, $E$ and $F$ are dependent.

25. From the probability tree,
   (A) $P(M \cap S) = (.3)(.6) = .18$
   (B) $P(R) = P(N \cap R) + P(M \cap R) = (.7)(.2) + (.3)(.4) = .14 + .12 = .26$

27. $E_1 = \{HH, HT\}$ and $P(E_1) = \dfrac{1}{2}$

   $E_2 = \{TH, TT\}$ and $P(E_2) = \dfrac{1}{2}$

   $E_4 = \{HH, TH\}$ and $P(E_4) = \dfrac{1}{2}$

Copyright © 2011 Pearson Education, Inc. Publishing as Prentice Hall.

(A)   Since $E_1 \cap E_4 = \{HH\} \neq \varnothing$, $E_1$ and $E_4$ **are not** mutually exclusive.

Since $P(E_1 \cap E_4) = P(HH) = \dfrac{1}{4} = P(E_1) \cdot P(E_4)$, $E_1$ and $E_4$ are independent.

(B)   Since $E_1 \cap E_2 = \varnothing$, $E_1$ and $E_2$ **are** mutually exclusive.

Since $P(E_1 \cap E_2) = 0$ and $P(E_1) \cdot P(E_2) = \dfrac{1}{4}$, $P(E_1 \cap E_2) \neq P(E_1) \cdot P(E_2)$.

Therefore, $E_1$ and $E_2$ are dependent.

**29.**   Let $E_i = $ "even number on the $i$th throw," $i = 1, 2$,

and $O_i = $ "odd number on the $i$th throw," $i = 1, 2$.

Then $P(E_i) = \dfrac{1}{2}$ and $P(O_i) = \dfrac{1}{2}$, $i = 1, 2$.

The probability tree for this experiment is shown at the right.

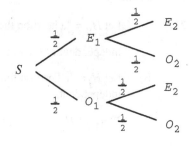

$$P(E_1 \cap E_2) = \left(\dfrac{1}{2}\right)\left(\dfrac{1}{2}\right) = \dfrac{1}{4}$$

$$P(E_1 \cup E_2) = P(E_1) + P(E_2) - P(E_1 \cap E_2) = \dfrac{1}{2} + \dfrac{1}{2} - \dfrac{1}{4} = \dfrac{3}{4}.$$

**31.**   Let $C = $ "first card is a club,"

and $H = $ "second card is a heart."

(A)   Without replacement, the probability tree is as shown at the
right.

Thus,   $P(C \cap H) = \left(\dfrac{1}{4}\right)\left(\dfrac{13}{51}\right) \approx .0637.$

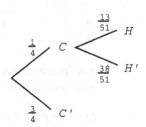

(B)   With replacement, the draws are independent and

$$P(C \cap H) = \left(\dfrac{1}{4}\right)\left(\dfrac{1}{4}\right) = \dfrac{1}{16} = 0.0625.$$

**33.**   $G = $ "the card is black" = {spade or club} and $P(G) = \dfrac{1}{2}$.

$H = $ "the card is divisible by 3" = {3, 6, or 9}.   $P(H) = \dfrac{12}{52} = \dfrac{3}{13}$

$P(H \cap G) = $ {3, 6, or 9 of clubs or spades} $= \dfrac{6}{52} = \dfrac{3}{26}$

(A)   $P(H \mid G) = \dfrac{P(H \cap G)}{P(G)} = \dfrac{3/26}{1/2} = \dfrac{6}{26} = \dfrac{3}{13}$

(B)   $P(H \cap G) = \dfrac{3}{26} = P(H) \cdot P(G)$

Thus, $H$ and $G$ *are* independent.

Copyright © 2011 Pearson Education, Inc. Publishing as Prentice Hall.

**35.** (A)   $S = \{BB, BG, GB, GG\}$

$A = \{BB, GG\}$ and $P(A) = \dfrac{2}{4} = \dfrac{1}{2}$

$B = \{BG, GB, GG\}$ and $P(B) = \dfrac{3}{4}$

$A \cap B = \{GG\}$.

$P(A \cap B) = \dfrac{1}{4}$ and $P(A) \cdot P(B) = \dfrac{1}{2} \cdot \dfrac{3}{4} = \dfrac{3}{8}$

Thus, $P(A \cap B) \neq P(A) \cdot P(B)$ and the events are dependent.

(B)   $S = \{BBB, BBG, BGB, BGG, GBB, GBG, GGB, GGG\}$

$A = \{BBB, GGG\}$

$B = \{BGG, GBG, GGB, GGG\}$

$A \cap B = \{GGG\}$

$P(A) = \dfrac{2}{8} = \dfrac{1}{4}$, $P(B) = \dfrac{4}{8} = \dfrac{1}{2}$, and $P(A \cap B) = \dfrac{1}{8}$

Since $P(A \cap B) = \dfrac{1}{8} = P(A) \cdot P(B)$, $A$ and $B$ are independent.

**37.** (A)   The probability tree with replacement is as follows:

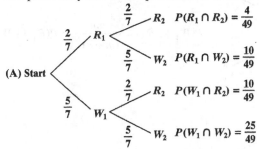

(B)   The probability tree without replacement is as follows:

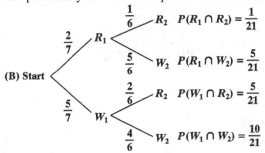

**39.** Let $E$ = At least one ball was red = $\{R_1 \cap R_2, R_1 \cap W_2, W_1 \cap R_2\}$.

(A)   With replacement [see the probability tree in Problem 37(A)]:

$P(E) = P(R_1 \cap R_2) + P(R_1 \cap W_2) + P(W_1 \cap R_2)$

$= \dfrac{4}{49} + \dfrac{10}{49} + \dfrac{10}{49} = \dfrac{24}{49}$

Copyright © 2011 Pearson Education, Inc. Publishing as Prentice Hall.

(B)    Without replacement [see the probability tree in Problem 37(B)]:

$$P(E) = P(R_1 \cap R_2) + P(R_1 \cap W_2) + P(W_1 \cap R_2) = \frac{1}{21} + \frac{5}{21} + \frac{5}{21} = \frac{11}{21}$$

**41.**    False.  Flip a fair coin twice.  Let $A$ = two heads and $B$ = a head on the first toss.  Then

$$P(A \mid B) = \frac{P(A \cap B)}{P(B)} = \frac{1/4}{1/2} = \frac{1}{2} = P(B)$$

But $P(A \cap B) = \frac{1}{4}$ and $P(A) \cdot P(B) = \frac{1}{4} \cdot \frac{1}{2} = \frac{1}{8}$ so $A$ and $B$ are not independent.

**43.**    True.  $P(A \mid B) = \dfrac{P(A \cap B)}{P(B)} = \dfrac{P(A)}{P(B)}$ since $A \subseteq B$ implies $A \cap B = A$

$\dfrac{P(A)}{P(B)} \geq \dfrac{P(A)}{1} = P(A)$ since $P(B) \leq 1$

Therefore, if $A \subseteq B$, then $P(A \mid B) \geq P(A)$.

**45.**    False.  Suppose $0 < P(A) < 1$.  Let $B = A'$.  Then $0 < P(B) < 1$.  Now
$P(A \cap B) = P(A \cap A') = 0$  and  $P(A) \cdot P(B) = P(A) \cdot P(A') \neq 0$

**47.**    True.

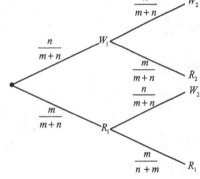

$$P(W_1 \cap R_2) = \frac{mn}{(m+n)^2} = P(R_1 \cap W_2)$$

**49.**    Total number of balls $2 + 3 + 4 = 9$;  $n(S) = C_{9,2} = \dfrac{9!}{2!(9-2)!} = \dfrac{9 \cdot 8 \cdot 7!}{2 \cdot 1 \cdot 7!} = 36$

Let $A$ = Both balls are the same color.

$n(A)$ = (No. of ways 2 red balls are selected)

    + (No. of ways 2 white balls are selected)

    + (No. of ways 2 green balls are selected)

    $= C_{2,2} + C_{3,2} + C_{4,2}$

    $= \dfrac{2!}{2!(2-2)!} + \dfrac{3!}{2!(3-2)!} + \dfrac{4!}{2!(4-2)!}$

    $= 1 + 3 + 6 = 10$

$P(A) = \dfrac{n(A)}{n(S)} = \dfrac{10}{36} = \dfrac{5}{18}$

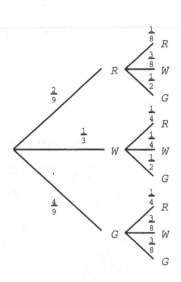

Alternatively, the probability tree for this experiment is shown above,

Copyright © 2011 Pearson Education, Inc.  Publishing as Prentice Hall.

and $P(RR, WW, \text{or } GG) = P(RR) + P(WW) + P(GG)$

$$= \left(\frac{2}{9}\right)\left(\frac{1}{8}\right) + \left(\frac{1}{3}\right)\left(\frac{1}{4}\right) + \left(\frac{4}{9}\right)\left(\frac{3}{8}\right) = \frac{2}{72} + \frac{1}{12} + \frac{12}{72} = \frac{20}{72} = \frac{5}{18}$$

**51.** The probability tree for this experiment is:

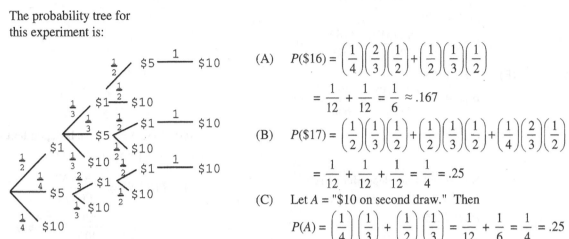

(A)  $P(\$16) = \left(\frac{1}{4}\right)\left(\frac{2}{3}\right)\left(\frac{1}{2}\right) + \left(\frac{1}{2}\right)\left(\frac{1}{3}\right)\left(\frac{1}{2}\right)$

$$= \frac{1}{12} + \frac{1}{12} = \frac{1}{6} \approx .167$$

(B)  $P(\$17) = \left(\frac{1}{2}\right)\left(\frac{1}{3}\right)\left(\frac{1}{2}\right) + \left(\frac{1}{2}\right)\left(\frac{1}{3}\right)\left(\frac{1}{2}\right) + \left(\frac{1}{4}\right)\left(\frac{2}{3}\right)\left(\frac{1}{2}\right)$

$$= \frac{1}{12} + \frac{1}{12} + \frac{1}{12} = \frac{1}{4} = .25$$

(C)  Let $A$ = "\$10 on second draw." Then

$$P(A) = \left(\frac{1}{4}\right)\left(\frac{1}{3}\right) + \left(\frac{1}{2}\right)\left(\frac{1}{3}\right) = \frac{1}{12} + \frac{1}{6} = \frac{1}{4} = .25$$

**53.** Assume that $A$ and $B$ are independent events with $P(A) \neq 0$, $P(B) \neq 0$. Then, by definition
(*)    $P(A \cap B) = P(A) \cdot P(B)$.
Now,

$$P(A \mid B) = \frac{P(A \cap B)}{P(B)} \quad \text{(definition of conditional probability)}$$

$$= \frac{P(A) \cdot P(B)}{P(B)} \quad \text{(by *)}$$

$$= P(A)$$

Also,

$$P(B \mid A) = \frac{P(B \cap A)}{P(A)} = \frac{P(A \cap B)}{P(A)} = \frac{P(A) \cdot P(B)}{P(A)} = P(B)$$

**55.** Assume $P(A) \neq 0$. Then $P(A \mid A) = \dfrac{P(A \cap A)}{P(A)} = \dfrac{P(A)}{P(A)} = 1$.

**57.** If $A$ and $B$ are mutually exclusive, then $A \cap B = \varnothing$ and $P(A \cap B) = P(\varnothing) = 0$. Also, if $P(A) \neq 0$ and $P(B) \neq 0$, then $P(A) \cdot P(B) \neq 0$. Therefore, $P(A \cap B) = 0 \neq P(A) \cdot P(B)$, and events $A$ and $B$ are dependent.

**59.** (A)

|  | Hourly $H$ | Salary $S$ | Salary + bonus $B$ | Total |
|---|---|---|---|---|
| To strike |  |  |  |  |
| Yes ($Y$) | .400 | .180 | .020 | .600 |
| No ($N$) | .150 | .120 | .130 | .400 |
| Totals | .550 | .300 | .150 | 1.000 |

[Note: The probability table above was derived from the table given in the problem by dividing each entry by 1000.]

Referring to the table in part (A):

Copyright © 2011 Pearson Education, Inc.  Publishing as Prentice Hall.

(B)   $P(Y\,|\,H\,) = \dfrac{P(Y\cap H)}{P(H)} = \dfrac{.400}{.55} \approx .727$

(C)   $P(Y\,|\,B\,) = \dfrac{P(Y\cap B)}{P(B)} = \dfrac{.02}{.15} \approx .133$

(D)   $P(S) = .300$

$P(S\,|\,Y\,) = \dfrac{P(S\cap Y)}{P(Y)} = \dfrac{.180}{.60} = .300$

(E)   $P(H) = .550$

$P(H\,|Y) = \dfrac{P(H\cap Y)}{P(Y)} = \dfrac{.400}{.600} \approx .667$

(F)   $P(B\cap N) = .130$

(G)   Yes, $S$ and $Y$ are **independent** since
$P(S\,|\,Y\,) = P(S) = .300$

(H)   No, $H$ and $Y$ are **dependent** since
$P(H\,|\,Y\,) \approx .667$ is not equal to
$P(H) = .550$.

(I)   $P(B\,|\,N\,) = \dfrac{P(B\cap N)}{P(N)}$

$= \dfrac{.130}{.400}$  (from table)

$= .325$

and $P(B) = .150.$  Since
$P(B\,|\,N\,) \neq P(B),\ B$ and $N$ are
**dependent**.

61.   The probability tree for this experiment is:

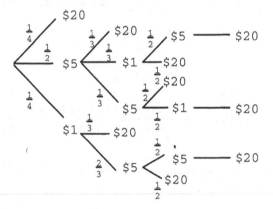

(A)   $P(\$26{,}000) = \left(\dfrac{1}{2}\right)\left(\dfrac{1}{3}\right)\left(\dfrac{1}{2}\right) + \left(\dfrac{1}{4}\right)\left(\dfrac{2}{3}\right)\left(\dfrac{1}{2}\right)$

$= \dfrac{1}{12} + \dfrac{1}{12} = \dfrac{1}{6} \approx 0.167$

(B)   $P(\$31{,}000) = \left(\dfrac{1}{2}\right)\left(\dfrac{1}{3}\right)\left(\dfrac{1}{2}\right) + \left(\dfrac{1}{2}\right)\left(\dfrac{1}{3}\right)\left(\dfrac{1}{2}\right) + \left(\dfrac{1}{4}\right)\left(\dfrac{2}{3}\right)\left(\dfrac{1}{2}\right) = \dfrac{3}{12} = \dfrac{1}{4} = 0.25$

(C)   Let $A$ = "\$20 on third draw." Then

$P(A) = \left(\dfrac{1}{4}\right)\left(\dfrac{2}{3}\right)\left(\dfrac{1}{2}\right) + \left(\dfrac{1}{2}\right)\left(\dfrac{1}{3}\right)\left(\dfrac{1}{2}\right) + \left(\dfrac{1}{2}\right)\left(\dfrac{1}{3}\right)\left(\dfrac{1}{2}\right) = \dfrac{3}{12} = \dfrac{1}{4} = 0.25$

Copyright © 2011 Pearson Education, Inc. Publishing as Prentice Hall.

**63.** (A)

|     | $C$  | $C'$ | Totals |
|-----|------|------|--------|
| $R$ | 0.06 | 0.44 | 0.50 |
| $R'$ | 0.02 | 0.48 | 0.50 |
| Total | 0.08 | 0.92 | 1.00 |

(B)  $P(C) = 0.08$, $P(R) = 0.50$, $P(R \cap C) = 0.06$

Since $0.06 = P(R \cap C) \neq P(R) \cdot P(C) = 0.04$, $R$ and $C$ are **dependent.**

(C)  $P(C \mid R) = \dfrac{P(C \cap R)}{P(R)} = \dfrac{0.06}{0.50} = 0.12$ and $P(C) = 0.08$

Since $P(C \mid R) > P(C)$, cancer is more likely to be developed if the red dye is used. The FDA should ban the use of the red dye.

(D) The new probability table is

|     | $C$  | $C'$ | Totals |
|-----|------|------|--------|
| $R$ | 0.02 | 0.48 | 0.50 |
| $R'$ | 0.06 | 0.44 | 0.50 |
| Total | 0.08 | 0.92 | 1.00 |

Now $P(C \mid R) = \dfrac{P(C \cap R)}{P(R)} = \dfrac{0.02}{0.5} = 0.04$ and $P(C) = 0.08$.   Since $P(C \mid R) < P(C)$ it appears

that the red dye reduces the development of cancer. Therefore, the use of the dye should not be banned.

**65.** (A)

|          | $A$  | $B$  | $C$  | Total |
|----------|------|------|------|-------|
| Female ($F$) | .130 | .286 | .104 | .520 |
| Male ($F'$)  | .120 | .264 | .096 | .480 |
| Total    | .250 | .550 | .200 | 1.000 |

[Note: The probability table above was derived from the table given in the problem by dividing each entry by 1000.]

Referring to the table in part (A):

(B)   $P(A \mid F) = \dfrac{P(A \cap F)}{P(F)} = \dfrac{.130}{.520} \approx .250,$   $P(A \mid F') = \dfrac{P(A \cap F')}{P(F')} = \dfrac{.120}{.480} = .250$

(C)   $P(C \mid F) = \dfrac{P(C \cap F)}{P(F)} = \dfrac{.104}{.520} \approx .200,$   $P(C \mid F') = \dfrac{P(C \cap F')}{P(F')} = \dfrac{.096}{.480} = .200$

(D)   $P(A) = .250$

(E)   $P(B) = .550,$   $P(B \mid F') = \dfrac{P(B \cap F')}{P(F')} = \dfrac{.264}{.480} = .550$

(F)   $P(F \cap C) = .104$

(G)   No, the results in parts (B), (C), (D), and (E) imply that $A$, $B$, and $C$ are independent of $F$ and $F'$.

Copyright © 2011 Pearson Education, Inc. Publishing as Prentice Hall.

---

EXERCISE 8-4

*Things to remember:*

1.   BAYES' FORMULA

Let $U_1$, $U_2$, ..., $U_n$ be $n$ mutually exclusive events whose union is the sample space $S$. Let $E$ be an arbitrary event in $S$ such that $P(E) \neq 0$. Then

$$P(U_1|E) = \frac{P(U_1 \cap E)}{P(E)}$$

$$= \frac{P(U_1 \cap E)}{P(U_1 \cap E) + P(U_2 \cap E) + \cdots + P(U_n \cap E)}$$

$$= \frac{P(E|U_1)P(U_1)}{P(E|U_1)P(U_1) + \cdots + P(E|U_n)P(U_n)}$$

Similar results hold for $U_2$, $U_3$, ..., $U_n$.

2.   BAYES' FORMULA AND PROBABILITY TREES

$$P(E|U_1) = \frac{\text{product of branch probabilities leading to } E \text{ through } U_1}{\text{sum of all branch probabilities leading to } E}$$

Similar results hold for $U_2$, $U_3$, ..., $U_n$.

---

**1.**   $P(M \cap A) = P(M) \cdot P(A \mid M) = (.6)(.8) = .48$

**3.**   $P(A) = P(M \cap A) + P(N \cap A) = P(M)P(A \mid M) + P(N)P(A \mid N) = (.6)(.8) + (.4)(.3) = .60$

**5.**   $P(M \mid A) = \dfrac{P(M \cap A)}{P(M \cap A) + P(N \cap A)} = \dfrac{.48}{.60} = .80$   (see Problems 1 and 3)

**7.**   Referring to the Venn diagram:

$$P(U_1 \mid R) = \frac{P(U_1 \cap R)}{P(R)} = \frac{\dfrac{25}{100}}{\dfrac{60}{100}} = \frac{25}{60} = \frac{5}{12} \approx .417$$

Using Bayes' formula:

Copyright © 2011 Pearson Education, Inc. Publishing as Prentice Hall.

$$P(U_1 \mid R) = \frac{P(U_1 \cap R)}{P(U_1 \cap R) + P(U_2 \cap R)} = \frac{P(U_1)P(R \mid U_1)}{P(U_1)P(R \mid U_1) + P(U_2)P(R \mid U_2)}$$

$$= \frac{\left(\dfrac{40}{100}\right)\left(\dfrac{25}{40}\right)}{\left(\dfrac{40}{100}\right)\left(\dfrac{25}{40}\right) + \left(\dfrac{60}{100}\right)\left(\dfrac{35}{60}\right)} = \frac{.25}{.25 + .35} = \frac{.25}{.60} = \frac{5}{12} \approx .417$$

**9.** $P(U_1 \mid R') = \dfrac{P(U_1 \cap R')}{P(R')} = \dfrac{\dfrac{15}{100}}{1 - P(R)}$  (from the Venn diagram)

$$= \frac{\dfrac{15}{100}}{1 - \dfrac{60}{100}} = \frac{\dfrac{15}{100}}{\dfrac{40}{100}} = \frac{3}{8} = .375$$

Using Bayes' formula:

$$P(U_1 \mid R') = \frac{P(U_1 \cap R')}{P(R')} = \frac{P(U_1)P(R' \mid U_1)}{P(U_1 \cap R') + P(U_2 \cap R')}$$

$$= \frac{P(U_1)P(R' \mid U_1)}{P(U_1)P(R' \mid U_1) + P(U_2)P(R' \mid U_2)} = \frac{\left(\dfrac{40}{100}\right)\left(\dfrac{15}{40}\right)}{\left(\dfrac{40}{100}\right)\left(\dfrac{15}{40}\right) + \left(\dfrac{60}{100}\right)\left(\dfrac{25}{60}\right)}$$

$$= \frac{.15}{.15 + .25} = \frac{15}{40} = \frac{3}{8} = .375$$

**11.** $P(U \mid C) = \dfrac{P(U \cap C)}{P(C)} = \dfrac{P(U \cap C)}{P(U \cap C) + P(V \cap C) + P(W \cap C)}$

$$= \frac{(.1)(.4)}{(.1)(.4) + (.6)(.2) + (.3)(.7)} = \frac{.04}{.37} \approx .1081$$

[Note: Recall $P(A \cap B)$
$= P(A) \cdot P(B \mid A)$.]

**13.** $P(W \mid C) = \dfrac{P(W \cap C)}{P(C)} = \dfrac{P(W \cap C)}{P(U \cap C) + P(V \cap C) + P(W \cap C)}$

$$= \frac{(.3)(.7)}{(.1)(.4) + (.6)(.2) + (.3)(.7)} = \frac{.21}{.37} \approx .568$$

**15.** $P(V \mid C) = \dfrac{P(V \cap C)}{P(C)} = \dfrac{P(V \cap C)}{P(U \cap C) + P(V \cap C) + P(W \cap C)}$

$$= \frac{(.6)(.2)}{(.1)(.4) + (.6)(.2) + (.3)(.7)} = \frac{.12}{.37} \approx .324$$

Copyright © 2011 Pearson Education, Inc. Publishing as Prentice Hall.

**17.**  From the Venn diagram,

$$P(U_1 \mid R) = \frac{5}{5+15+20} = \frac{5}{40} = \frac{1}{8} = .125$$

or

$$= \frac{P(U_1 \cap R)}{P(R)} = \frac{\frac{5}{100}}{\frac{40}{100}} = .125$$

Using Bayes' formula:

$$P(U_1 \mid R) = \frac{P(U_1 \cap R)}{P(U_1 \cap R) + P(U_2 \cap R) + P(U_3 \cap R)} = \frac{\frac{5}{100}}{\frac{5}{100} + \frac{15}{100} + \frac{20}{100}}$$

$$= \frac{.05}{.05 + .15 + .2} = \frac{.05}{.40} = .125$$

**19.**  From the Venn diagram,

$$P(U_3 \mid R) = \frac{20}{5+15+20} = \frac{20}{40} = .5$$

Using Bayes' formula:

$$P(U_3 \mid R) = \frac{P(U_3 \cap R)}{P(U_1 \cap R) + P(U_2 \cap R) + P(U_3 \cap R)} = \frac{\frac{20}{100}}{\frac{5}{100} + \frac{15}{100} + \frac{20}{100}}$$

$$= \frac{.2}{.05 + .15 + .2} = \frac{.2}{.4} = .5$$

**21.**  From the Venn diagram,

$$P(U_2 \mid R) = \frac{15}{5+15+20} = \frac{15}{40} = .375$$

Using Bayes' formula:

$$P(U_2 \mid R) = \frac{P(U_2 \cap R)}{P(U_1 \cap R) + P(U_2 \cap R) + P(U_3 \cap R)} = \frac{\frac{15}{100}}{\frac{5}{100} + \frac{15}{100} + \frac{20}{100}} = \frac{.15}{.05 + .15 + .2} = \frac{.15}{.40} = .375$$

**23.**  From the given tree diagram, we have:

$$P(A) = \frac{1}{4} \qquad\qquad P(A') = \frac{3}{4}$$

$$P(B \mid A) = \frac{1}{5} \qquad\qquad P(B \mid A') = \frac{3}{5}$$

$$P(B' \mid A) = \frac{4}{5} \qquad\qquad P(B' \mid A') = \frac{2}{5}$$

We want to find the following:

$$P(B) = P(B \cap A) + P(B \cap A') = P(A)P(B \mid A) + P(A')P(B \mid A')$$

$$= \left(\frac{1}{4}\right)\left(\frac{1}{5}\right) + \left(\frac{3}{4}\right)\left(\frac{3}{5}\right) = \frac{1}{20} + \frac{9}{20} = \frac{10}{20} = \frac{1}{2}$$

Copyright © 2011 Pearson Education, Inc.  Publishing as Prentice Hall.

$$P(B') = 1 - P(B) = 1 - \frac{1}{2} = \frac{1}{2}$$

$$P(A \mid B) = \frac{P(A \cap B)}{P(B)} = \frac{P(A)P(B \mid A)}{P(B)} = \frac{\left(\frac{1}{4}\right)\left(\frac{1}{5}\right)}{\frac{1}{2}} = \frac{\frac{1}{20}}{\frac{1}{2}} = \frac{1}{10}$$

Thus, $P(A' \mid B) = 1 - P(A \mid B) = 1 - \frac{1}{10} = \frac{9}{10}$.

$$P(A \mid B') = \frac{P(A \cap B')}{P(B')} = \frac{P(A)P(B' \mid A)}{P(B')}$$

$$= \frac{\left(\frac{1}{4}\right)\left(\frac{4}{5}\right)}{\frac{1}{2}} = \frac{\frac{4}{20}}{\frac{1}{2}} = \frac{2}{5}$$

Thus, $P(A' \mid B') = 1 - P(A \mid B') = 1 - \frac{2}{5} = \frac{3}{5}$.

Therefore, the tree diagram for this problem
is as shown at the right.

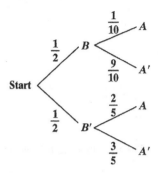

*The following tree diagram is to be used for Problems 25 and 27.*

25.   $P(U_1 \mid W) = \dfrac{P(U_1 \cap W)}{P(W)} = \dfrac{P(U_1 \cap W)}{P(U_1 \cap W) + P(U_2 \cap W)}$

$$= \frac{P(U_1)P(W \mid U_1)}{P(U_1)P(W \mid U_1) + P(U_2)P(W \mid U_2)} = \frac{(.5)(.2)}{(.5)(.2) + (.5)(.6)} = \frac{.1}{.4} = .25$$

27.   $P(U_2 \mid R) = \dfrac{P(U_2 \cap R)}{P(R)} = \dfrac{P(U_2 \cap R)}{P(U_2 \cap R) + P(U_1 \cap R)}$

$$= \frac{P(U_2)P(R \mid U_2)}{P(U_2)P(R \mid U_2) + P(U_1)P(R \mid U_1)} = \frac{(.5)(.4)}{(.5)(.4) + (.5)(.8)} = \frac{.4}{1.2} = \frac{1}{3} \approx .333$$

Copyright © 2011 Pearson Education, Inc. Publishing as Prentice Hall.

**29.** $P(W_1 \mid W_2) = \dfrac{P(W_1 \cap W_2)}{P(W_2)} = \dfrac{P(W_1)P(W_2 \mid W_1)}{P(R_1 \cap W_2) + P(W_1 \cap W_2)}$

$\qquad = \dfrac{P(W_1)P(W_2 \mid W_1)}{P(R_1)P(W_2 \mid R_1) + P(W_1)P(W_2 \mid W_1)} = \dfrac{\left(\dfrac{5}{9}\right)\left(\dfrac{4}{8}\right)}{\left(\dfrac{4}{9}\right)\left(\dfrac{5}{8}\right) + \left(\dfrac{5}{9}\right)\left(\dfrac{4}{8}\right)} = \dfrac{\dfrac{20}{72}}{\dfrac{20}{72} + \dfrac{20}{72}}$

$\qquad\qquad\qquad\qquad\qquad\qquad\qquad\qquad\qquad\qquad\qquad\qquad = \dfrac{20}{40} = \dfrac{1}{2} \text{ or } .5$

**31.** $P(U_{R_1} \mid U_{R_2}) = \dfrac{P(U_{R_1} \cap U_{R_2})}{P(U_{R_2})} = \dfrac{P(U_{R_1})P(U_{R_2} \mid U_{R_1})}{P(U_{W_1} \cap U_{R_2}) + P(U_{R_1} \cap U_{R_2})}$

$\qquad = \dfrac{P(U_{R_1})P(U_{R_2} \mid U_{R_1})}{P(U_{W_1})P(U_{R_2} \mid U_{W_1}) + P(U_{R_1})P(U_{R_2} \mid U_{R_1})}$

$\qquad = \dfrac{\left(\dfrac{7}{10}\right)\left(\dfrac{5}{10}\right)}{\left(\dfrac{3}{10}\right)\left(\dfrac{4}{10}\right) + \left(\dfrac{7}{10}\right)\left(\dfrac{5}{10}\right)} = \dfrac{.35}{.12 + .35} = \dfrac{.35}{.47} = \dfrac{35}{47} \approx .745$

The tree diagram follows:

where $U_{R_1}$ is red from urn one,

$U_{R_2}$ is red from urn two,

$U_{W_1}$ is white from urn one,

and $U_{W_2}$ is white from urn two.

**33.** Suppose $c = e$. Then
$\qquad P(M) = ac + be = ac + bc = c(a + b) = c \quad (a + b = 1)$
and

$\qquad P(M \mid U) = \dfrac{P(M \cap U)}{P(U)} = \dfrac{ac}{a} = c$

Therefore, $M$ and $U$ are independent.
Alternatively, note that
$\qquad P(M) = c, \; P(U) = a \;$ and $\; P(M \cap U) = ac = P(M) \cdot P(U),$
which implies that $M$ and $U$ are independent.

**35.** Draw a tree diagram to verify the probabilities given below.

(A)  With replacement--True.

$\qquad P(B_2 \mid B_1) = \dfrac{m}{m + n}$

Copyright © 2011 Pearson Education, Inc. Publishing as Prentice Hall.

$$P(B_1| B_2) = \frac{P(B_1 \cap B_2)}{P(B_1 \cap B_2) + P(W_1 \cap B_2)}$$

$$= \frac{\dfrac{m^2}{(m+n)^2}}{\dfrac{m^2}{(m+n)^2} + \dfrac{mn}{(m+n)^2}} = \frac{m^2}{m^2 + mn} = \frac{m}{m+n} = P(B_2| B_1)$$

(B)    Without replacement--True.

$$P(B_2| B_1) = \frac{m-1}{m+n-1}$$

$$P(B_1| B_2) = \frac{P(B_1 \cap B_2)}{P(B_1 \cap B_2) + P(W_1 \cap B_2)} = \frac{\dfrac{m(m-1)}{(m+n)(m+n-1)}}{\dfrac{m(m-1)}{(m+n)(m+n-1)} + \dfrac{nm}{(m+n)(m+n-1)}}$$

$$= \frac{m(m-1)}{m(m-1) + mn} = \frac{m-1}{m+n-1} = P(B_2| B_1)$$

**37.**

$$P(H_1| H_2) = \frac{P(H_1 \cap H_2)}{P(H_2)} = \frac{P(H_1 \cap H_2)}{P(H_1 \cap H_2) + P(\overline{H_1} \cap H_2)}$$

$$= \frac{P(H_1)P(H_2 | H_1)}{P(H_1)P(H_2 | H_1) + P(\overline{H_1})P(H_2 | \overline{H_1})} = \frac{\dfrac{13}{52} \cdot \dfrac{12}{51}}{\dfrac{13}{52} \cdot \dfrac{12}{51} + \dfrac{39}{52} \cdot \dfrac{13}{51}}$$

$$= \frac{13(12)}{13(12) + 39(13)} = \frac{12}{51} \approx .235$$

For the 3-card hand in Problems 39-44, let

$E_3$ = all three cards are clubs

$E_2$ = only two of the cards are clubs

$E_1$ = only one of the cards is a club

$E_0$ = none of the cards is a club

Let $E$ = card chosen is a club.

Copyright © 2011 Pearson Education, Inc. Publishing as Prentice Hall.

The probabilities are:

$$P(E_3) = \frac{C_{13,3}}{C_{52,3}} = \frac{286}{22,100} \approx 0.0129$$

$$P(E_2) = \frac{C_{13,2} \cdot C_{39,1}}{C_{52,3}} = \frac{78 \cdot 39}{22,100} \approx 0.1377$$

$$P(E_1) = \frac{13 \cdot C_{39,2}}{C_{52,3}} = \frac{13 \cdot 741}{22,100} \approx 0.4359$$

$$P(E_0) = \frac{C_{39,3}}{C_{52,3}} = \frac{9,139}{22,100} \approx 0.4135$$

$$P(E \mid E_3) = 1, \quad P(E \mid E_2) = \frac{2}{3}, \quad P(E \mid E_1) = \frac{1}{3}, \quad P(E \mid E_0) = 0$$

$$P(E) = P(E_3) \cdot P(E \mid E_3) + P(E_2) \cdot P(E \mid E_2) + P(E_1) \cdot P(E \mid E_1)$$

$$= 0.0129(1) + 0.1377\left(\frac{2}{3}\right) + 0.4359\left(\frac{1}{3}\right) \approx 0.2500$$

**39.** $P(E \mid E_1) = \dfrac{1}{3}$

**41.** $P(E_1 \mid E) = \dfrac{P(E_1 \cap E)}{P(E)} = \dfrac{P(E_1) \cdot P(E \mid E_1)}{P(E)} = \dfrac{0.4359(1/3)}{0.2500} \approx 0.581$

**43.** $P(E_3 \mid E) = \dfrac{P(E_3 \cap E)}{P(E)} = \dfrac{P(E_3) \cdot P(E \mid E_3)}{P(E)} = \dfrac{0.0129(1)}{0.2500} \approx 0.052$

**45.** Consider the following Venn diagram:

$$P(U_1 \mid R) = \frac{P(U_1 \cap R)}{P(U_1 \cap R) + P(U_1' \cap R)}$$

and

$$P(U_1' \mid R) = \frac{P(U_1' \cap R)}{P(U_1 \cap R) + P(U_1' \cap R)}$$

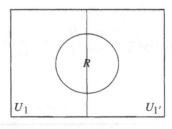

Adding these two equations, we obtain:

$$P(U_1 \mid R) + P(U_1' \mid R) = \frac{P(U_1 \cap R)}{P(U_1 \cap R) + P(U_1' \cap R)} + \frac{P(U_1' \cap R)}{P(U_1 \cap R) + P(U_1' \cap R)}$$

$$= \frac{P(U_1 \cap R) + P(U_1' \cap R)}{P(U_1 \cap R) + P(U_1' \cap R)} = 1$$

Copyright © 2011 Pearson Education, Inc. Publishing as Prentice Hall.

**47.** Consider the following tree diagram:

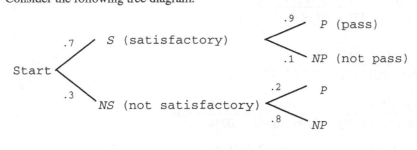

$$P(S \mid P) = \frac{P(S \cap P)}{P(S \cap P) + P(NS \cap P)} = \frac{P(S)P(P \mid S)}{P(S)P(P \mid S) + P(NS)P(P \mid NS)}$$

$$= \frac{(.7)(.9)}{(.7)(.9) + (.3)(.2)} = \frac{.63}{.69} \approx .913$$

$$P(S \mid NP) = \frac{P(S \cap NP)}{P(NP)} = \frac{P(S \cap NP)}{P(S \cap NP) + P(NS \cap NP)} = \frac{(.7)(.1)}{(.7)(.1) + (.3)(.8)} = \frac{.07}{.31} \approx .226$$

**49.** Consider the following tree diagram:

$$P(A \mid D) = \frac{P(A \cap D)}{P(D)} , \text{ where}$$

$$P(D) = P(A \cap D) + P(B \cap D) + P(C \cap D)$$

$$= P(A)P(D \mid A) + P(B)P(D \mid B) + P(C)P(D \mid C)$$

$$= (.2)(.01) + (.40)(.03) + (.40)(.02) = .002 + .012 + .008 = .022$$

Thus, $P(A \mid D) = \dfrac{P(A \cap D)}{P(D)} = \dfrac{P(A)P(D \mid A)}{P(D)} = \dfrac{(.20)(.01)}{.022} = \dfrac{.002}{.022} = \dfrac{2}{22}$ or .091

Similarly,

$$P(B \mid D) = \frac{P(B \cap D)}{P(D)} = \frac{P(B)P(D \mid B)}{P(D)} = \frac{(.40)(.03)}{.022} = \frac{.012}{.022} = \frac{6}{11} \text{ or } .545,$$

and $P(C \mid D) = \dfrac{P(C \cap D)}{P(D)} = \dfrac{P(C)P(D \mid C)}{P(D)} = \dfrac{(.40)(.02)}{.022} = \dfrac{.008}{.022} = \dfrac{4}{11} \text{ or } .364.$

**51.** Consider the following tree diagram:

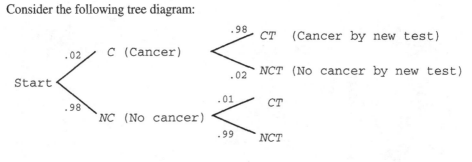

Copyright © 2011 Pearson Education, Inc.  Publishing as Prentice Hall.

$$P(C \mid CT) = \frac{P(C \cap CT)}{P(CT)} = \frac{P(C)P(CT \mid C)}{P(C \cap CT) + P(NC \cap CT)}$$

$$= \frac{P(C)P(CT \mid C)}{P(C)P(CT \mid C) + P(NC)P(CT \mid NC)}$$

$$= \frac{(.02)(.98)}{(.02)(.98) + (.98)(.01)} = \frac{.0196}{.0196 + .0098} = \frac{.0196}{.0294} = .667$$

$$P(C \mid NCT) = \frac{P(C \cap NCT)}{P(NCT)} = \frac{P(C)P(NCT \mid C)}{P(C)P(NCT \mid C) + P(NC)P(NCT \mid NC)}$$

$$= \frac{(.02)(.02)}{(.02)(.02) + (.98)(.99)} \approx .000412$$

**53.**  Consider the following tree diagram.

$$P(L \mid HD) = \frac{P(L \cap HD)}{P(HD)} = \frac{P(L)P(HD \mid L)}{P(L \cap HD) + P(NL \cap HD)}$$

$$= \frac{P(L)P(HD \mid L)}{P(L)P(HD \mid L) + P(NL)P(HD \mid NL)}$$

$$= \frac{(.07)(.4)}{(.07)(.4) + (.93)(.1)} \quad \text{(from the tree diagram)}$$

$$= \frac{.028}{.028 + .093} = \frac{.028}{.121} = \frac{28}{121} \approx .231$$

$$P(L \mid ND) = \frac{P(L \cap ND)}{P(ND)} = \frac{P(L)P(ND \mid L)}{P(L \cap ND) + P(NL \cap ND)} = \frac{P(L)P(ND \mid L)}{P(L)P(ND \mid L) + P(NL)P(ND \mid NL)}$$

$$= \frac{(.07)(.1)}{(.07)(.1) + (.93)(.2)} \quad \text{(from the tree diagram)}$$

$$= \frac{.007}{.007 + .186} = \frac{.007}{.194} = \frac{7}{194} \approx .036$$

Copyright © 2011 Pearson Education, Inc.  Publishing as Prentice Hall.

**55.**  Consider the following tree diagram.

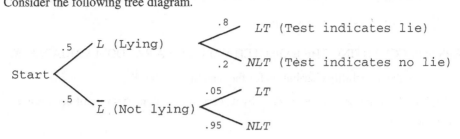

$$P(L \mid LT) = \frac{P(L \cap LT)}{P(LT)} = \frac{P(L \cap LT)}{P(L \cap LT) + P(\bar{L} \cap LT)} = \frac{(.5)(.8)}{(.5)(.8) + (.5)(.05)}$$

$$= \frac{.4}{.425} \approx .941$$

If the test indicates that the subject was lying, then he was lying with a probability of 0.941.

$$P(\bar{L} \mid LT) = \frac{P(\bar{L} \cap LT)}{P(LT)} = \frac{(.5)(.05)}{(.5)(.8) + (.5)(.05)}$$

$$= \frac{.05}{.85} \approx .0588$$

If the test indicates that the subject was lying, there is still a probability of 0.0588 that he was not lying.

---

## EXERCISE 8-5

*Things to remember:*

1.  RANDOM VARIABLE

    A RANDOM VARIABLE is a function that assigns a numerical value to each simple event in a sample space $S$.

2.  PROBABILITY DISTRIBUTION OF A RANDOM VARIABLE $X$

    The PROBABILITY DISTRIBUTION OF A RANDOM VARIABLE $X$, denoted

    $$P(X = x) = p(x),$$

    satisfies

    (a)  $0 \le p(x) \le 1, x \in \{x_1, x_2, ..., x_n\}$,

    (b)  $p(x_1) + p(x_2) + \cdots + p(x_n) = 1$,

    where $\{x_1, x_2, ..., x_n\}$ are the (range) values of $X$.

3.  EXPECTED VALUE OF A RANDOM VARIABLE $X$

    Given the probability distribution for the random variable $X$:

    | $x_i$ | $x_1$ | $x_2$ | $\cdots$ | $x_n$ |
    |-------|-------|-------|----------|-------|
    | $p_i$ | $p_1$ | $p_2$ | $\cdots$ | $p_n$ |

    where $p_i = p(x_i)$.

Copyright © 2011 Pearson Education, Inc. Publishing as Prentice Hall.

The expected value of $X$, denoted by $E(X)$, is given by the formula:

$$E(X) = x_1 p_1 + x_2 p_2 + \cdots + x_m \ p_m$$

4.    STEPS FOR COMPUTING THE EXPECTED VALUE OF A RANDOM VARIABLE $X$.

*Step 1.* Form the probability distribution for the random variable $X$.

*Step 2.* Multiply each image value of $X$, $x_i$, by its corresponding probability of occurrence, $p_i$, then add the results.

---

1.    Expected value of $X$:

$$E(X) = -3(.3) + 0(.5) + 4(.2) = -0.1$$

3.    Assign the number 0 to the event of observing zero heads, the number 1 to the event of observing one head, and the number 2 to the event of observing two heads.  The probability distribution for $X$, then, is:

| $x_i$ | 0 | 1 | 2 |
|-------|---|---|---|
| $p_i$ | $\frac{1}{4}$ | $\frac{1}{2}$ | $\frac{1}{4}$ |

[Note: One head can occur two ways out of a total of four different ways (HT, TH).]

Hence,  $E(X) = 0 \cdot \dfrac{1}{4} + 1 \cdot \dfrac{1}{2} + 2 \cdot \dfrac{1}{4} = 1$.

5.    Assign a payoff of \$1 to the event of observing a head and –\$1 to the event of observing a tail.  Thus, the probability distribution for $X$ is:

| $x_i$ | 1 | –1 |
|-------|---|----|
| $p_i$ | $\frac{1}{2}$ | $\frac{1}{2}$ |

Hence,  $E(X) = 1 \cdot \dfrac{1}{2} + (-1) \cdot \dfrac{1}{2} = 0$.  The game is fair.

7.    The table shows a payoff or probability distribution for the game.

| Net gain | $x_i$ | –3 | –2 | –1 | 0 | 1 | 2 |
|----------|-------|----|----|----|---|---|---|
|          | $p_i$ | $\frac{1}{6}$ | $\frac{1}{6}$ | $\frac{1}{6}$ | $\frac{1}{6}$ | $\frac{1}{6}$ | $\frac{1}{6}$ |

[Note: A payoff valued at –\$3 is assigned to the event of observing a "1" on the die, resulting in a net gain of –\$3, and so on.]

Hence,  $E(X) = -3 \cdot \dfrac{1}{6} - 2 \cdot \dfrac{1}{6} - 1 \cdot \dfrac{1}{6} + 0 \cdot \dfrac{1}{6} + 1 \cdot \dfrac{1}{6} + 2 \cdot \dfrac{1}{6} = -\dfrac{1}{2}$  or  –\$0.50.

The game is not fair.

9.    The probability distribution is:

| Number of heads | Gain, $x_i$ | Probability, $p_i$ |
|-----------------|-------------|--------------------|
| 0 | 2 | $\frac{1}{4}$ |
| 1 | –3 | $\frac{1}{2}$ |
| 2 | 2 | $\frac{1}{4}$ |

The expected value is:

$$E(X) = 2 \cdot \frac{1}{4} + (-3) \cdot \frac{1}{2} + 2 \cdot \frac{1}{4} = 1 - \frac{3}{2} = -\frac{1}{2} \text{ or } -\$0.50.$$

Copyright © 2011 Pearson Education, Inc.  Publishing as Prentice Hall.

11. In 4 rolls of a die, the total number of possible outcomes is
$6 \cdot 6 \cdot 6 \cdot 6 = 6^4$. Thus, $n(S) = 6^4 = 1296$. The total number of outcomes that contain *no* 6's is $5 \cdot 5 \cdot 5 \cdot 5 = 5^4$.
Thus, if $E$ is the event
"At least one 6," then $n(E) = 6^4 - 5^4 = 671$ and

$$P(E) = \frac{n(E)}{n(S)} = \frac{671}{1296} \approx 0.5178.$$

First, we compute the expected value to you.

The payoff table is:

| $x_i$ | −$1 | $1 |
|-------|------|------|
| $P_i$ | 0.5178 | 0.4822 |

The expected value to you is:

$E(X) = (-1)(0.5178) + 1(0.4822) = -0.0356$ or −$0.036

The expected value to her is:

$E(X) = 1(0.5178) + (-1)(0.4822) = 0.0356$ or $0.036

13. Let $x$ = amount you should lose if a 6 turns up.
The payoff table is:

| | 1 | 2 | 3 | 4 | 5 | 6 |
|-------|------|------|------|------|------|------|
| $x_i$ | $5 | $5 | $10 | $10 | $10 | $x |
| $P_i$ | $\frac{1}{6}$ | $\frac{1}{6}$ | $\frac{1}{6}$ | $\frac{1}{6}$ | $\frac{1}{6}$ | $\frac{1}{6}$ |

Now $E(X) = 5\left(\dfrac{1}{6}\right) + 5\left(\dfrac{1}{6}\right) + 10\left(\dfrac{1}{6}\right) + 10\left(\dfrac{1}{6}\right) + 10\left(\dfrac{1}{6}\right) + x\left(\dfrac{1}{6}\right) = \dfrac{40}{6} + \dfrac{x}{6}$

The game is fair if and only if $E(X) = 0$:

solving $\dfrac{40}{6} + \dfrac{x}{6} = 0$

gives $x = -40$

Thus, you should **lose** $40 for the game to be fair.

15. $P(\text{sum} = 7) = \dfrac{6}{36} = \dfrac{1}{6}$

$P(\text{sum} = 11 \text{ or } 12) = P(\text{sum} = 11) + P(\text{sum} = 12) = \dfrac{2}{36} + \dfrac{1}{36} = \dfrac{3}{36} = \dfrac{1}{12}$

$P(\text{sum other than 7, 11, or 12}) = 1 - P(\text{sum} = 7, 11, \text{ or } 12) = 1 - \dfrac{9}{36} = \dfrac{27}{36} = \dfrac{3}{4}$

Let $x_1$ = sum is 7, $x_2$ = sum is 11 or 12, $x_3$ = sum is not 7, 11, or 12, and let $t$ denote the amount you "win"
if $x_3$ occurs. Then the payoff table is:

| $x_i$ | −$10 | $11 | $t$ |
|-------|------|------|------|
| $p_i$ | $\frac{1}{6}$ | $\frac{1}{12}$ | $\frac{3}{4}$ |

Copyright © 2011 Pearson Education, Inc. Publishing as Prentice Hall.

The expected value is:

$$E(X) = -10\left(\frac{1}{6}\right) + 11\left(\frac{1}{12}\right) + t\left(\frac{3}{4}\right) = \frac{-10}{6} + \frac{11}{12} + \frac{3t}{4}$$

The game is fair if $E(X) = 0$, i.e., if

$$\frac{-10}{6} + \frac{11}{12} + \frac{3}{4}t = 0 \text{ or } \frac{3}{4}t = \frac{10}{6} - \frac{11}{12} = \frac{20}{12} - \frac{11}{12} = \frac{9}{12} = \frac{3}{4}$$

Therefore, $t = \$1$.

17. Let $K$ = card is a King. Then $P(K) = \frac{4}{52} = \frac{1}{13}$ and $P(K') = \frac{12}{13}$.

    The payoff table is:

    | | $K$ | $K'$ |
    |---|---|---|
    | $x_i$ | \$10 | −\$1 |
    | $p_i$ | $\frac{1}{13}$ | $\frac{12}{13}$ |

    Now, $E(X) = 10\left(\frac{1}{13}\right) - 1\left(\frac{12}{13}\right) = -\frac{2}{13} = -\$0.154$

19. Let $K$ = hand contains at least one King. Then
    $K'$ = hand contains no Kings. The probabilities are:

    $$P(K') = \frac{C_{48,5}}{C_{52,5}} \approx 0.65884$$

    $P(K) = 1 - P(K') \approx 0.34116$
    The payoff table is:

    | | $K$ | $K'$ |
    |---|---|---|
    | $x_i$ | \$10 | −\$1 |
    | $p_i$ | 0.34116 | 0.65884 |

    Now, $E(X) = 10(0.34116) - 1(0.65884) \approx \$2.75$

21. Course $A_1$: $E(X) = (-200)(.1) + 100(.2) + 400(.4) + 100(.3)$
    $$= -20 + 20 + 160 + 30$$
    $$= \$190$$

    Course $A_2$: $E(X) = (-100)(.1) + 200(.2) + 300(.4) + 200(.3)$
    $$= -10 + 40 + 120 + 60$$
    $$= \$210$$

    $A_2$ will produce the largest expected value, and that value is \$210.

23. The probability of winning \$35 is $\frac{1}{38}$ and the probability of losing \$1 is $\frac{37}{38}$. Thus, the payoff table is:

    | $x_i$ | \$35 | −\$1 |
    |---|---|---|
    | $p_i$ | $\frac{1}{38}$ | $\frac{37}{38}$ |

    The expected value of the game is:

    $$E(X) = 35\left(\frac{1}{38}\right) + (-1)\left(\frac{37}{38}\right) = \frac{35-37}{38} = \frac{-1}{19} \approx -0.0526 \text{ or } -5.26\cent$$

Copyright © 2011 Pearson Education, Inc. Publishing as Prentice Hall.

**25.** Let $p$ = probability of winning. Then $1 - p$ is the probability of losing and the payoff table is:

| | W | L |
|---|---|---|
| $x_i$ | 99,900 | −100 |
| $p_i$ | $p$ | $1-p$ |

Since $E(X) = 100$, we have
$$99,900(p) - 100(1 - p) = 100$$
$$99,900p - 100 + 100p = 100$$
$$100,000p = 200$$
$$p = 0.002$$

The probability of winning is 0.002. Since the expected value is positive, you should play the game. In the long run, you will win $100 per game.

**27.**

| $p_i$ | | $x_i$ |
|---|---|---|
| $\dfrac{1}{5000}$ | chance of winning | $499 |
| $\dfrac{3}{5000}$ | chance of winning | $99 |
| $\dfrac{5}{5000}$ | chance of winning | $19 |
| $\dfrac{20}{5000}$ | chance of winning | $4 |
| $\dfrac{4971}{5000}$ | chance of losing   $1 | [Note: $5000 - (1 + 3 + 5 + 20) = 4971$.] |

The payoff table is:

| $x_i$ | $499 | $99 | $19 | $4 | −$1 |
|---|---|---|---|---|---|
| $P_i$ | 0.0002 | 0.0006 | 0.001 | 0.004 | 0.9942 |

Thus,

$$E(X) = 499(0.0002) + 99(0.0006) + 19(0.001) + 4(0.004) - 1(0.9942) = -0.80$$
or  $E(X) = -\$0.80$ or $-80¢$

**29.**  (A)  Total number of simple events $= n(S) = C_{10,2} = \dfrac{10!}{2!(10-2)!} = \dfrac{10!}{2!8!} = \dfrac{10 \cdot 9}{2} = 45$

$P(\text{zero defective}) = P(0) = \dfrac{C_{7,2}}{45}$   [Note: None defective means 2 selected from 7 nondefective.]

$$= \dfrac{\frac{7!}{2!5!}}{45} = \dfrac{21}{45} = \dfrac{7}{15}$$

$P(\text{one defective}) = P(1) = \dfrac{C_{3,1} \cdot C_{7,1}}{45} = \dfrac{21}{45} = \dfrac{7}{15}$

$P(\text{two defective}) = P(2) = \dfrac{C_{3,2}}{45}$   [Note: Two defectives selected from 3 defectives.]

$$= \dfrac{3}{45} = \dfrac{1}{15}$$

The probability distribution is as follows:

| $x_i$ | 0 | 1 | 2 |
|---|---|---|---|
| $P_i$ | $\frac{7}{15}$ | $\frac{7}{15}$ | $\frac{1}{15}$ |

Copyright © 2011 Pearson Education, Inc. Publishing as Prentice Hall.

(B) $E(X) = 0\left(\dfrac{7}{15}\right) + 1\left(\dfrac{7}{15}\right) + 2\left(\dfrac{1}{15}\right) = \dfrac{9}{15} = \dfrac{3}{5} = 0.6$

**31.** (A) The total number of simple events $= n(S) = C_{1000,5}$.

$$P(0 \text{ winning tickets}) = P(0) = \frac{C_{997,5}}{C_{1000,5}}$$

$$= \frac{997 \cdot 996 \cdot 995 \cdot 994 \cdot 993}{1000 \cdot 999 \cdot 998 \cdot 997 \cdot 996} \approx 0.985$$

$$P(1 \text{ winning ticket}) = P(1) = \frac{C_{3,1} \cdot C_{997,4}}{C_{1000,5}} = \frac{3 \cdot \dfrac{997!}{4!(993)!}}{\dfrac{1000!}{5!(995)!}} \approx 0.0149$$

$$P(2 \text{ winning tickets}) = P(2) = \frac{C_{3,2} \cdot C_{997,3}}{C_{1000,5}} = \frac{3 \cdot \dfrac{997!}{3!(994)!}}{\dfrac{1000!}{5!(995)!}} \approx 0.0000599$$

$$P(3 \text{ winning tickets}) = P(3) = \frac{C_{3,3} \cdot C_{997,2}}{C_{1000,5}} = \frac{1 \cdot \dfrac{997!}{2!(995)!}}{\dfrac{1000!}{5!(995)!}} \approx 0.00000006$$

The payoff table is as follows:

| $x_i$ | −$5 | $195 | $395 | $595 |
|-------|------|-------|-----------|------------|
| $P_i$ | 0.985 | 0.0149 | 0.0000599 | 0.00000006 |

(B) The expected value to you is:

$E(X) = (-5)(0.985) + 195(0.0149) + 395(0.0000599) + 595(0.00000006) \approx -\$2.00$

**33.** (A) From the statistical plot, the number 13 came up 3 times in 200 games. If $1 was bet on 13 in each of the 200 games, then the result is:

$3(35) - 1(197) = 105 - 197 = -\$92$

(B) Based on the simulation, the value per game is: $-\dfrac{92}{200} = -\$0.46$;  you lose 46 cents per game. The (theoretical) expected value of the game is:

$$E(X) = \frac{1}{38}(35) + \frac{37}{38}(-1) = \frac{35}{38} - \frac{37}{38} = -\frac{1}{19} \approx -\$0.0526;$$

You will lose 5 cents per game.

(C) The simulated gain or loss depends on the results of your simulation. From (B), the expected loss is: $-\dfrac{1}{19}(500) \approx -\$26.32$.

Copyright © 2011 Pearson Education, Inc. Publishing as Prentice Hall.

**35.** Let $D_i$ = exactly $i$ diamonds in the 3-card hand, $i = 0, 1, 2, 3$. The probabilities are

$$P(D_0) = \frac{C_{39,3}}{C_{52,3}} \approx 0.4135, \quad P(D_1) = \frac{13 \cdot C_{39,2}}{C_{52,3}} \approx 0.4359$$

$$P(D_2) = \frac{C_{13,2} \cdot 39}{C_{52,3}} \approx 0.01377, \quad P(D_3) = \frac{C_{13,3}}{C_{52,3}} \approx 0.0129$$

Let $x$ be the amount you should lose. The payoff table is:

| $x_i$ | $-x$ | 20 | 40 | 60 |
|-------|------|------|--------|--------|
| $P_i$ | 0.4135 | 0.4359 | 0.1377 | 0.0129 |

The expected value is
$E(X) = -x(0.4135) + 20(0.4359) + 40(0.1377) + 60(0.0129)$
$\quad = -x(0.4135) + 15$

If the game is fair, $-x(0.4135) + 15 = 0$; $x = \dfrac{15}{0.4135} = \$36.27$.

**37.** The payoff table is as follows:

Gain

| $x_i$ | \$4850 | $-\$150$ |
|-------|--------|----------|
| $p_i$ | 0.01 | 0.99 |

[Note: $5000 - 150 = 4850$, the gain with probability of 0.01 if stolen.]

Hence, $E(X) = 4850(0.01) - 150(0.99) = -\$100$

**39.** The payoff table for site A is as follows:

| $x_i$ | 30 million | −3 million |
|-------|------------|------------|
| $p_i$ | 0.2 | 0.8 |

Hence $E(X) = 30(0.2) - 3(0.8)$
$\quad = 6 - 2.4$
$\quad = \$3.6$ million

The payoff table for site B is as follows:

| $x_i$ | 70 million | −4 million |
|-------|------------|------------|
| $p_i$ | 0.1 | 0.9 |

Hence, $E(X) = 70(0.1) + (-4)(0.9)$
$\quad = 7 - 3.6$
$\quad = \$3.4$ million

The company should choose site A with $E(X) = \$3.6$ million.

**41.** Using 4,
$E(X) = 0(0.12) + 1(0.36) + 2(0.38) + 3(0.14) = 1.54$

**43.** Action $A_1$: $E(X) = 10(0.3) + 5(0.2) + 0(0.5) = \$4.00$
Action $A_2$: $E(X) = 15(0.3) + 3(0.1) + 0(0.6) = \$4.80$
Action $A_2$ is the better choice.

## CHAPTER 8 REVIEW

**1.** First, we calculate the number of 5-card combinations that can be dealt from 52 cards:

$$n(S) = C_{52,5} = \frac{52!}{5! \cdot 47!} = 2,598,960$$

We then calculate the number of 5-club combinations that can be obtained from 13 clubs:

$$n(E) = C_{13,5} = \frac{13!}{5! \cdot 8!} = 1287$$

Thus, $P(5 \text{ clubs}) = P(E) = \dfrac{n(E)}{n(S)} = \dfrac{1287}{2,598,960} \approx 0.0005$. $\quad$ (8-1)

Copyright © 2011 Pearson Education, Inc. Publishing as Prentice Hall.

2. $n(S)$ is computed by using the permutation formula:

$$n(S) = P_{15,2} = \frac{15!}{(15-2)!} = 15 \cdot 14 = 210.$$

Thus, the probability that Brittani will be president and Ramon will be treasurer is:

$$\frac{n(E)}{n(S)} = \frac{1}{210} \approx 0.0048. \qquad (8\text{-}1)$$

3. (A) The total number of ways of drawing 3 cards from 10 with order taken into account is given by:

$$P_{10,3} = \frac{10!}{(10-3)!} = \frac{10 \cdot 9 \cdot 8 \cdot 7!}{7!} = 720$$

Thus, the probability of drawing the code word "dig" is:

$$P(\text{"dig"}) = \frac{1}{720} \approx 0.0014$$

(B) The total number of ways of drawing 3 cards from 10 without regard to order is given by:

$$C_{10,3} = \frac{10!}{3!(10-3)!} = \frac{10 \cdot 9 \cdot 8 \cdot 7!}{3!7!} = 120$$

Thus, the probability of drawing the 3 cards "d," "i," and "g" (in some order) is:

$$P(\text{"d," "i," "g"}) = \frac{1}{120} \approx 0.0083. \qquad (8\text{-}1)$$

4. $P(\text{person having side effects}) = \dfrac{f(E)}{n} = \dfrac{50}{1000} = 0.05. \qquad (8\text{-}1)$

5. The payoff table is as follows:

| $x_i$ | $-\$2$ | $-\$1$ | $\$0$ | $\$1$ | $\$2$ |
|---|---|---|---|---|---|
| $p_i$ | $\frac{1}{5}$ | $\frac{1}{5}$ | $\frac{1}{5}$ | $\frac{1}{5}$ | $\frac{1}{5}$ |

Hence, $E(X) = (-2) \cdot \dfrac{1}{5} + (-1) \cdot \dfrac{1}{5} + 0 \cdot \dfrac{1}{5} + 1 \cdot \dfrac{1}{5} + 2 \cdot \dfrac{1}{5} = 0$

The game is fair. (8-5)

6. $P(A) = .3, P(B) = .4, P(A \cap B) = .1$

(A) $P(A') = 1 - P(A) = 1 - .3 = .7$

(B) $P(A \cup B) = P(A) + P(B) - P(A \cap B) = .3 + .4 - .1 = .6. \qquad (8\text{-}2)$

7. Since the spinner cannot land on $R$ and $G$ simultaneously, $R \cap G = \varnothing$. Thus,
$P(R \cup G) = P(R) + P(G) = .3 + .5 = .8$

The odds for an event $E$ are: $\dfrac{P(E)}{P(E')}$

Thus, the odds for landing on either $R$ or $G$ are: $\dfrac{P(R \cup G)}{P[(R \cup G)']} = \dfrac{.8}{.2} = \dfrac{8}{2}$ or the odds are 8 to 2. (8-2)

8. If the odds for an event $E$ are $a$ to $b$, then $P(E) = \dfrac{a}{a+b}$. Thus, the probability of rolling an 8 before rolling

a 7 is: $\dfrac{5}{11} \approx .455. \qquad (8\text{-}2)$

Copyright © 2011 Pearson Education, Inc. Publishing as Prentice Hall.

**9.** $P(T) = .27$    (8-3)

**10.** $P(Z) = .20$    (8-3)

**11.** $P(T \cap Z) = .02$    (8-3)

**12.** $P(R \cap Z) = .03$    (8-3)

**13.** $P(R \mid Z) = \dfrac{P(R \cap Z)}{P(Z)} = \dfrac{.03}{.20} = .15$    (8-3)

**14.** $P(Z \mid R) = \dfrac{P(Z \cap R)}{P(R)} = \dfrac{.03}{.23} \approx .1304$    (8-3)

**15.** $P(T \mid Z) = \dfrac{P(T \cap Z)}{P(Z)} = \dfrac{.02}{.20} = .10$    (8-3)

**16.** No, because $P(T \cap Z) = .02 \neq P(T) \cdot P(Z) = (.27)(.20) = .054.$    (8-3)

**17.** Yes, because $P(S \cap X) = .10 = P(S) \cdot P(X) = (.5)(.2).$    (8-3)

**18.** $P(A) = .4$ from the tree diagram.    (8-3)

**19.** $P(B \mid A) = .2$ from the tree diagram.    (8-3)

**20.** $P(B \mid A') = .3$ from the tree diagram.    (8-3)

**21.** $P(A \cap B) = P(A)P(B \mid A) = (.4)(.2) = .08$    (8-3)

**22.** $P(A' \cap B) = P(A')P(B \mid A) = (.6)(.3) = .18$    (8-3)

**23.** $P(B) = P(A \cap B) + P(A' \cap B)$
$= P(A)P(B \mid A) + P(A')P(B \mid A')$
$= (.4)(.2) + (.6)(.3) = .08 + .18 = .26$    (8-3)

**24.** $P(A \mid B) = \dfrac{P(A \cap B)}{P(B)} = \dfrac{P(A)P(B \mid A)}{P(A \cap B) + P(A' \cap B)} = \dfrac{P(A)P(B \mid A)}{P(A)P(B \mid A) + P(A')P(B \mid A')}$

$= \dfrac{(.4)(.2)}{(.4)(.2) + (.6)(.3)}$    (from the tree diagram)

$= \dfrac{.08}{.26} = \dfrac{8}{26}$ or $.307 \approx .31$    (8-4)

**25.** $P(A \mid B') = \dfrac{P(A \cap B')}{P(B')} = \dfrac{P(A)P(B' \mid A)}{1 - P(B)} = \dfrac{(.4)(.8)}{1 - .26}$ [$P(B) = .26$, see Problem 23.]

$= \dfrac{.32}{.74} = \dfrac{16}{37}$ or $.432$    (8-4)

**26.** Let $E$ = "born in June, July or August."

(A)    Empirical Probability:
$P(E) = \dfrac{f(E)}{n} = \dfrac{10}{32} = \dfrac{5}{16}$

(B)    Theoretical Probability:
$P(E) = \dfrac{n(E)}{n(S)} = \dfrac{3}{12} = \dfrac{1}{4}$

(C)    As the sample size in part (A) increases, the approximate empirical probability of event $E$ approaches the theoretical probability of event $E$.    (8-1)

Copyright © 2011 Pearson Education, Inc. Publishing as Prentice Hall.

27.   No. The total number of 3-card hands is $C_{52,3}$. The number of hands containing 3 red cards is $C_{26,3} =$ 2600; the number of hands containing 2 red cards and one black card is $C_{26,2} \cdot C_{26,1} = 8,450$. These events are not equally likely.   (8-1)

28.   Yes. The number of hands containing either 2 or 3 red cards equals the number of hands containing 2 or 3 black cards.   (8-1)

29.   $S = \{HH, HT, TH, TT\}$.

The probabilities for 2 "heads," 1 "head," and 0 "heads" are, respectively, $\dfrac{1}{4}$, $\dfrac{1}{2}$, and $\dfrac{1}{4}$. Thus, the payoff table is:

| $x_i$ | $5 | –$4 | $2 |
|-------|-----|------|-----|
| $P_i$ | 0.25 | 0.5 | 0.25 |

$E(X) = 0.25(5) + 0.5(-4) + 0.25(2) = -0.25$ or $-\$0.25$
The game is not fair.   (8-5)

30.   $S = \{(1,1), (2,2), (3,3), (1,2), (2,1), (1,3), (3,1), (2,3), (3,2)\}$;   $n(S) = 3 \cdot 3 = 9$

   (A)   $P(A) = \dfrac{n(A)}{n(S)} = \dfrac{3}{9} = \dfrac{1}{3}$   $[A = \{(1,1), (2,2), (3,3)\}]$

   (B)   $P(B) = \dfrac{n(B)}{n(S)} = \dfrac{2}{9}$   $[B = \{(2,3), (3,2)\}]$   (8-3)

31.   (A)   $P(\text{jack or queen}) = P(\text{jack}) + P(\text{queen}) = \dfrac{4}{52} + \dfrac{4}{52} = \dfrac{8}{52} = \dfrac{2}{13}$

   [Note: jack $\cap$ queen $= \varnothing$.]

   The odds for drawing a jack or queen are 2 to 11.

   (B)   $P(\text{jack or spade}) = P(\text{jack}) + P(\text{spade}) - P(\text{jack and spade})$

   $= \dfrac{4}{52} + \dfrac{13}{52} - \dfrac{1}{52} = \dfrac{16}{52} = \dfrac{4}{13}$

   The odds for drawing a jack or a spade are 4 to 9.

   (C)   $P(\text{ace}) = \dfrac{4}{52} = \dfrac{1}{13}$. Thus,

   $P(\text{card other than an ace}) = 1 - P(\text{ace}) = 1 - \dfrac{1}{13} = \dfrac{12}{13}$
   The odds for drawing a card other than a ace are 12 to 1.   (8-2)

32.   (A)   The probability of rolling a 5 is $\dfrac{4}{36} = \dfrac{1}{9}$.
   Thus, the odds for rolling a five are 1 to 8.

   (B)   Let $x$ = amount house should pay (and return the $1 bet).
   Then, for the game to be fair,

   $E(X) = x\left(\dfrac{1}{9}\right) + (-1)\left(\dfrac{8}{9}\right) = \dfrac{x}{9} - \dfrac{8}{9} = 0$

   $x = 8$

   Thus, the house should pay $8.   (8-2)

Copyright © 2011 Pearson Education, Inc. Publishing as Prentice Hall.

**33.**  Event $E_1$ = 2 heads; $f(E_1)$ = 210.

Event $E_2$ = 1 head; $f(E_2)$ = 480.

Event $E_3$ = 0 heads; $f(E_3)$ = 310.

Total number of trials = 1000.

(A)  The empirical probabilities for the events above are as follows:

$$P(E_1) = \frac{210}{1000} = 0.21$$

$$P(E_2) = \frac{480}{1000} = 0.48$$

$$P(E_3) = \frac{310}{1000} = 0.31$$

(B)  Sample space $S$ = {HH, HT, TH, TT}.

$$P(2 \text{ heads}) = \frac{1}{4} = 0.25$$

$$P(1 \text{ head}) = \frac{2}{4} = 0.5$$

$$P(0 \text{ heads}) = \frac{1}{4} = 0.25$$

(C)  Using part (B), the expected frequencies for each outcome are as follows:

$$2 \text{ heads} = 1000 \cdot \frac{1}{4} = 250$$

$$1 \text{ head} = 1000 \cdot \frac{2}{4} = 500$$

$$0 \text{ heads} = 1000 \cdot \frac{1}{4} = 250 \qquad (8\text{-}1, 8\text{-}5)$$

**34.**  The individual tosses of a coin are independent events (the coin has no memory). Therefore, $P(H) = \frac{1}{2}$.

(8-3)

**35.**  (A)  The sample space $S$ is given by:

$$
\begin{aligned}
S = \{ &(1,1), (1,2), (1,3), (1,4), (1,5), (1,6), \\
&(2,1), (2,2), (2,3), (2,4), (2,5), (2,6), \\
&(3,1), (3,2), (3,3), (3,4), (3,5), (3,6), \\
&(4,1), (4,2), (4,3), (4,4), (4,5), (4,6), \\
&(5,1), (5,2), (5,3), (5,4), (5,5), (5,6), \\
&(6,1), (6,2), (6,3), (6,4), (6,5), (6,6) \}
\end{aligned}
$$

Sum 2, Sum 3, Sum 4, Sum 5

[Note: Event (2,3) means 2 on the first die and 3 on the second die.]

The probability distribution corresponding to this sample space  is:

| Sum $x_i$ | 2 | 3 | 3 | 5 | 6 | 7 | 8 | 9 | 10 | 11 | 12 |
|---|---|---|---|---|---|---|---|---|---|---|---|
| Probability $p_i$ | $\frac{1}{36}$ | $\frac{2}{36}$ | $\frac{3}{36}$ | $\frac{4}{36}$ | $\frac{5}{36}$ | $\frac{6}{36}$ | $\frac{5}{36}$ | $\frac{4}{36}$ | $\frac{3}{36}$ | $\frac{2}{36}$ | $\frac{1}{36}$ |

Copyright © 2011 Pearson Education, Inc.  Publishing as Prentice Hall.

(B)  $E(X) = 2\left(\dfrac{1}{36}\right) + 3\left(\dfrac{2}{36}\right) + 4\left(\dfrac{3}{36}\right) + 5\left(\dfrac{4}{36}\right) + 6\left(\dfrac{5}{36}\right) + 7\left(\dfrac{6}{36}\right) + 8\left(\dfrac{5}{36}\right)$

$\qquad\qquad + 9\left(\dfrac{4}{36}\right) + 10\left(\dfrac{3}{36}\right) + 11\left(\dfrac{2}{36}\right) + 12\left(\dfrac{1}{36}\right) = 7 \qquad (8\text{-}5)$

**36.** The event $A$ that corresponds to the sum being divisible by 4 includes sums 4, 8, and 12. This set is:
$A = \{(1, 3), (2, 2), (3, 1), (2, 6), (3, 5), (4, 4), (5, 3), (6, 2), (6, 6)\}$

The event $B$ that corresponds to the sum being divisible by 6 includes sums 6 and 12. This set is:

$B = \{(1, 5), (2, 4), (3, 3), (4, 2), (5, 1), (6, 6)\}$

$\qquad P(A) = \dfrac{n(A)}{n(S)} = \dfrac{9}{36} = \dfrac{1}{4}$

$\qquad P(B) = \dfrac{n(B)}{n(S)} = \dfrac{6}{36} = \dfrac{1}{6}$

$P(A \cap B) = \dfrac{1}{36}$ [Note: $A \cap B = \{(6, 6)\}$]

$P(A \cup B) = \dfrac{14}{36}$ or $\dfrac{7}{18}$  [Note: $A \cup B = \{(1, 3), (2, 2), (3, 1), (2, 6),$
$\qquad\qquad\qquad\qquad\qquad\qquad (3, 5), (4, 4), (5, 3), (6, 2), (6, 6),$
$\qquad\qquad\qquad\qquad\qquad\qquad (1, 5), (2, 4), (3, 3), (4, 2), (5, 1)\}]$  (8-2)

**37.** The function $P$ cannot be a probability function because:

(a)  $P$ cannot be negative. [Note: $P(e_2) = -0.2$.]

(b)  $P$ cannot have a value greater than 1. [Note: $P(e_4) = 2$.]

(c)  The sum of the values of $P$ must equal 1.
[Note: $P(e_1) + P(e_2) + P(e_3) + P(e_4) = 0.1 + (-0.2) + 0.6 + 2 = 2.5 \neq 1$.]  (8-1)

**38.** Since $n(A \cup B) = n(A) + n(B) - n(A \cap B)$, we have
$80 = 50 + 45 - n(A \cap B)$ which implies $n(A \cap B) = 15$
Now, $n(B') = n(U) - n(B) = 100 - 45 = 55$
$\qquad n(A') = n(U) - n(A) = 100 - 50 = 50, \quad n(A \cap B') = 50 - 15 = 35$
$\qquad n(B \cap A') = 45 - 15 = 30, \quad n(A' \cap B') = 55 - 35 = 20$
Thus,

|    | $A$ | $A'$ | Totals |
|----|-----|------|--------|
| $B$ | 15 | 30 | 45 |
| $B'$ | 35 | 20 | 55 |
| Totals | 50 | 50 | 100 |

(8-2)

**39.** (A)  $P(\text{odd number}) = P(1) + P(3) + P(5) = .2 + .3 + .1 = .6$

(B)  Let $E = $ "number less than 4,"
and $F = $ "odd number."

Copyright © 2011 Pearson Education, Inc. Publishing as Prentice Hall.

Now, $E \cap F = \{1, 3\}$,  $F = \{1, 3, 5\}$.

$$P(E \mid F) = \frac{P(E \cap F)}{P(F)} = \frac{.2 + .3}{.6} = \frac{5}{6} \qquad (8\text{-}3)$$

**40.**  Let $E$ = "card is red" and $F$ = "card is an ace."  Then $F \cap E$ = "card is a red ace."

(A)  $P(F \mid E) = \dfrac{P(F \cap E)}{P(E)} = \dfrac{2/52}{26/52} = \dfrac{1}{13}$

(B)  $P(F \cap E) = \dfrac{1}{26}$,  and  $P(E) = \dfrac{1}{2}$,  $P(F) = \dfrac{1}{13}$.  Thus,

$P(F \cap E) = P(E) \cdot P(F)$, and $E$ and $F$ are independent.     (8-3)

**41.**  (A)  The tree diagram with replacement is:    (B)  The tree diagram without replacement is:

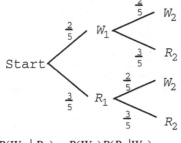

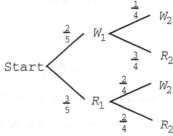

$$P(W_1 \mid R_2) = P(W_1)P(R_2 \mid W_1)$$
$$= \frac{2}{5} \cdot \frac{3}{5} = \frac{6}{25} \approx .24$$

$$P(W_1 \mid R_2) = P(W_1)P(R_2 \mid W_1)$$
$$= \frac{2}{5} \cdot \frac{3}{4} = \frac{6}{20} = \frac{3}{10}$$

(8-3)

**42.**  Part (B) involves dependent events because

$$P(R_2 \mid W_1) = \frac{3}{4}$$

$$P(R_2) = P(W_1 \cap R_2) + P(R_1 \cap R_2) = \frac{6}{20} + \frac{6}{20} = \frac{12}{20} = \frac{3}{5}$$

and  $P(R_2 \mid W_1) \neq P(R_2)$.  The events in part (A) are independent.     (8-3)

**43.**  (A)  Using the tree diagram in Problem 41(A), we have:

$P(\text{zero red balls}) = P(W_1 \cap W_2) = P(W_1)P(W_2) = \dfrac{2}{5} \cdot \dfrac{2}{5} = \dfrac{4}{25} = .16$

$P(\text{one red ball}) = P(W_1 \cap R_2) + P(R_1 \cap W_2) = P(W_1)P(R_2) + P(R_1)P(W_2)$

$$= \frac{2}{5} \cdot \frac{3}{5} + \frac{3}{5} \cdot \frac{2}{5} = \frac{12}{25} = .48$$

$P(\text{two red balls}) = P(R_1 \cap R_2) = P(R_1)P(R_2) = \dfrac{3}{5} \cdot \dfrac{3}{5} = \dfrac{9}{25} = .36$

Thus, the probability distribution is:

Copyright © 2011 Pearson Education, Inc.  Publishing as Prentice Hall.

| Number of red balls $x_i$ | Probability $p_i$ |
|---|---|
| 0 | .16 |
| 1 | .48 |
| 2 | .36 |

The expected number of red balls is:

$$E(X) = 0(.16) + 1(.48) + 2(.36) = .48 + .72 = 1.2$$

(B)  Using the tree diagram in Problem 41(B), we have:

$$P(\text{zero red balls}) = P(W_1 \cap W_2) = P(W_1)P(W_2|\ W_1) = \frac{2}{5} \cdot \frac{1}{4} = \frac{1}{10} = .1$$

$$\begin{aligned}P(\text{one red ball}) &= P(W_1 \cap R_2) + P(R_1 \cap W_2) \\ &= P(W_1)P(R_2|\ W_1) + P(R_1)P(W_2|\ R_1) \\ &= \frac{2}{5} \cdot \frac{3}{4} + \frac{3}{5} \cdot \frac{2}{4} = \frac{12}{20} = \frac{3}{5} = .6\end{aligned}$$

$$P(\text{two red balls}) = P(R_1 \cap R_2) = P(R_1)P(R_2|\ R_1) = \frac{3}{5} \cdot \frac{2}{4} = \frac{6}{20} = .3$$

Thus, the probability distribution is:

| Number of red balls $x_i$ | Probability $p_i$ |
|---|---|
| 0 | .1 |
| 1 | .6 |
| 2 | .3 |

The expected number of red balls is:

$$E(X) = 0(.1) + 1(.6) + 2(.3) = 1.2. \qquad (8\text{-}5)$$

44.  The tree diagram for this problem is as follows:

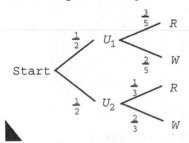

The probability of selecting urn $U_1$ is .5 and that of selecting urn $U_2$ is .5.

(A)  $P(R\,|\,U_1) = \dfrac{3}{5}$ \qquad\qquad (B) $P(R\,|\,U_2) = \dfrac{1}{3}$

(C)  $\begin{aligned}P(R) &= P(R \cap U_1) + P(R \cap U_2) \\ &= P(U_1)P(R\,|\,U_1) + P(U_2)P(R\,|\,U_2) \\ &= \frac{1}{2} \cdot \frac{3}{5} + \frac{1}{2} \cdot \frac{1}{3} = \frac{28}{60} = \frac{7}{15} \approx .4667\end{aligned}$

Copyright © 2011 Pearson Education, Inc.  Publishing as Prentice Hall.

(D)　$P(U_1 \mid R) = \dfrac{P(U_1 \cap R)}{P(R)} = \dfrac{P(U_1)P(R \mid U_1)}{P(U_1)P(R \mid U_1) + P(U_2)P(R \mid U_2)}$

$$= \dfrac{\frac{1}{2} \cdot \frac{3}{5}}{\frac{1}{2} \cdot \frac{3}{5} + \frac{1}{2} \cdot \frac{1}{3}} = \dfrac{\frac{3}{10}}{\frac{7}{15}} = \dfrac{9}{14} \approx .6429$$

(E)　$P(U_2 \mid W) = \dfrac{P(U_2 \cap W)}{P(W)} = \dfrac{P(U_2)P(W \mid U_2)}{P(U_2)P(W \mid U_2) + P(U_1)P(W \mid U_1)}$

$$= \dfrac{\frac{1}{2} \cdot \frac{2}{3}}{\frac{1}{2} \cdot \frac{2}{3} + \frac{1}{2} \cdot \frac{2}{5}} = \dfrac{\frac{2}{3}}{\frac{16}{15}} = \dfrac{5}{8} = .625$$

(F)　$P(U_1 \cap R) = P(U_1)P(R \mid U_1) = \dfrac{1}{2} \cdot \dfrac{3}{5} = \dfrac{3}{10} = .3$

[Note: In parts (A)—(F), we derived the values of the probabilities from the tree diagram.]　　(8-3, 8-4)

**45.**　No, because $P(R \mid U_1) \neq P(R)$. (See Problem 44.)　　(8-3)

**46.**　$n(S) = C_{52,5}$

(A)　Let $A$ be the event "all diamonds." Then $n(A) = C_{13,5}$. Thus,

$$P(A) = \frac{n(A)}{n(S)} = \frac{C_{13,5}}{C_{52,5}}.$$

(B)　Let $B$ be the event "3 diamonds and 2 spades." Then $n(B) = C_{13,3} \cdot C_{13,2}$. Thus,

$$P(B) = \frac{n(B)}{n(S)} = \frac{C_{13,3} \cdot C_{13,2}}{C_{52,5}}.　　(8-1)$$

**47.**　$n(S) = C_{10,4} = \dfrac{10!}{4!(10-4)!} = \dfrac{10 \cdot 9 \cdot 8 \cdot 7 \cdot 6!}{4 \cdot 3 \cdot 2 \cdot 1 \cdot 6!} = 210$

Let $A$ be the event "The married couple is in the group of 4 people." Then

$$n(A) = C_{2,2} \cdot C_{8,2} = 1 \cdot \frac{8!}{2!(8-2)!} = \frac{8 \cdot 7 \cdot 6!}{2 \cdot 1 \cdot 6!} = 28.$$

Thus, $P(A) = \dfrac{n(A)}{n(S)} = \dfrac{28}{210} = \dfrac{2}{15} \approx 0.1333.$　　(8-1)

**48.**　Events $S$ and $H$ are mutually exclusive. Hence, $P(S \cap H) = 0$, while $P(S) \neq 0$ and $P(H) \neq 0$. Therefore,

$$P(S \cap H) \neq P(S) \cdot P(H)$$

which implies that $S$ and $F$ are dependent.　　(8-3)

**49.**　(A)　From the plot, $P(2) = \dfrac{9}{50} = 0.18.$

Copyright © 2011 Pearson Education, Inc. Publishing as Prentice Hall.

(B)    The event $A$ = "the minimum of the two numbers is 2" contains the
simple events (2, 2), (2, 3), (3, 2), (2, 4), (4, 2), (2, 5),

(5, 2), (2, 6), (6, 2). Thus $n(A) = 9$ and $P(A) = \dfrac{9}{36} = \dfrac{1}{4} = 0.25$.

(C)    The empirical probability depends on the results of your simulation.
For the theoretical probability, let $A$ = "minimum of the two numbers is 4".

Then $A = \{(4, 4), (4, 5), (5, 4), (4, 6), (6, 4)\}$,

$n(A) = 5$ and $P(A) = \dfrac{5}{36} \approx 0.139$.      (8-1)

**50.**    The empirical probability depends on the results of your simulation.

Since there are 2 black jacks in a standard 52-card deck, the theoretical probability of drawing a black jack

is: $\dfrac{2}{52} = \dfrac{1}{26} \approx 0.038$.      (8-3)

**51.**    False. If $P(E) = 1$, then $P(E') = 0$ and the odds for $E = \dfrac{P(E)}{P(E')} = \dfrac{1}{0}$; $\dfrac{1}{0}$ is undefined.      (8-2)

**52.**    True. In general, $P(E \cup F) = P(E) + P(F) - P(E \cap F)$.
If $E = F'$, then $E \cap F = F' \cap F = \varnothing$ and $P(E \cap F) = 0$.      (8-2)

**53.**    False. Let $E$ and $F$ be complementary events with $0 < P(E) < 1$, $0 < P(F) < 1$. Then $E \cap F = E \cap E' = \varnothing$
and $P(E \cap F) = 0$ while $P(E) \cdot P(F) = P(E)[1 - P(E)] \neq 0$.      (8-3)

**54.**    False. Counterexample: Roll a fair die; $S = \{1, 2, 3, 4, 5, 6\}$.
Let $E$ = the number that turns up is $\geq 2$;
   $F$ = the number that turns up is $\leq 4$.
Then $E \cup F = \{1, 2, 3, 4, 5, 6\}$ and $P(E \cup F) = 1$ but $F \neq E'$.      (8-2)

**55.**    True. This is the definition of independent events.      (8-3)

**56.**    False. If $E$ and $F$ are mutually exclusive, then $E \cap F = \varnothing$ and the example in Problem 53 is a
counterexample here.      (8-2)

**57.**    Let $E_2$ be the event "2 heads."

(A)    From the table, $f(E_2) = 350$. Thus, the approximate empirical probability of obtaining 2 heads is:

$P(E_2) \approx \dfrac{f(E_2)}{n} = \dfrac{350}{1000} = 0.350$

(B)    $S = \{HHH, HHT, HTH, HTT, THH, THT, TTH, TTT\}$
The theoretical probability of obtaining 2 heads is:

$P(E_2) = \dfrac{n(E_2)}{n(S)} = \dfrac{3}{8} = 0.375$

(C)    The expected frequency of obtaining 2 heads in 1000 tosses of 3 fair coins is:
$f(E_2) = 1000(0.375) = 375$.      (8-1)

Copyright © 2011 Pearson Education, Inc. Publishing as Prentice Hall.

**58.** On one roll of the dice, the probability of getting a double six is $\dfrac{1}{36}$ and the probability of not getting a double six is $\dfrac{35}{36}$.

On two rolls of the dice there are $(36)^2$ possible outcomes. There are 71 ways to get at least one double six, namely a double six on the first roll and any one of the 35 other outcomes on the second roll, or a double six on the second roll and any one of the 35 other outcomes on the first roll, or a double six on both rolls.

Thus, the probability of at least one double six on two rolls is $\dfrac{71}{(36)^2}$ and the probability of no double sixes is:

$$1 - \frac{71}{(36)^2} = \frac{(36)^2 - 2\cdot36 + 1}{(36)^2} = \frac{(36-1)^2}{(36)^2} = \left(\frac{35}{36}\right)^2$$

Let $E$ be the event "At least one double six." Then $E'$ is the event "No double sixes." Continuing with the reasoning above, we conclude that, in 24 rolls of the die,

$$P(E') = \left(\frac{35}{36}\right)^{24} \approx 0.5086$$

Therefore, $P(E) = 1 - 0.5086 = 0.4914$.

The payoff table is:

| $x_i$ | 1 | −1 |
|---|---|---|
| $P_i$ | 0.4914 | 0.5086 |

and $E(X) = 1(0.4914) + (-1)(0.5086) = 0.4914 - 0.5086 = -0.0172$

Thus, your expectation is −\$0.0172. Your friend's expectation is \$0.0172. The game is not fair.    (8-5)

**59.** The total number of ways that 3 people can be selected from a group of 10 is:

$$C_{10,3} = \frac{10!}{3!(10-3)!} = \frac{10\cdot9\cdot8\cdot7!}{3\cdot2\cdot1\cdot7!} = 120$$

The number of ways of selecting *no* women is:

$$C_{7,3} = \frac{7!}{3!(7-3)!} = \frac{7\cdot6\cdot5\cdot4!}{3\cdot2\cdot1\cdot4!} = 35$$

Thus, the number of samples of 3 people that contain at least one woman is $120 - 35 = 85$.

Therefore, if event $A$ is "At least one woman is selected," then

$$P(A) = \frac{n(A)}{n(S)} = \frac{85}{120} = \frac{17}{24} \approx 0.708.$$    (8-1)

**60.** $P(\text{second heart}|\text{first heart}) = P(H_2|H_1) = \dfrac{12}{51} \approx .235$

[Note: One can see that $P(H_2|H_1) = \dfrac{12}{51}$ directly.]    (8-3)

**61.** $P(\text{first heart}|\text{second heart}) = P(H_1|H_2) = \dfrac{P(H_1 \cap H_2)}{P(H_2)} = \dfrac{P(H_1)P(H_2 \mid H_1)}{P(H_2)}$

Copyright © 2011 Pearson Education, Inc. Publishing as Prentice Hall.

$$= \frac{P(H_1)P(H_2 \mid H_1)}{P(H_1 \cap H_2) + P(H'1 \cap H_2)} = \frac{P(H_1)P(H_2 \mid H_1)}{P(H_1)P(H_2 \mid H_1) + P(H_1')P(H_2 \mid H_1')}$$

$$= \frac{\dfrac{13}{52} \cdot \dfrac{12}{51}}{\dfrac{13}{52} \cdot \dfrac{12}{51} + \dfrac{39}{52} \cdot \dfrac{13}{51}} = \frac{12}{51} \approx .235 \qquad (8\text{-}3)$$

**62.** Since each die has 6 faces, there are $6 \cdot 6 = 36$ possible pairs for the two up faces.

A sum of 2 corresponds to having $(1, 1)$ as the up faces. This sum can be obtained in $3 \cdot 3 = 9$ ways (3 faces on the first die, 3 faces on the second). Thus,

$$P(2) = \frac{9}{36} = \frac{1}{4}.$$

A sum of 3 corresponds to the two pairs $(2, 1)$ and $(1, 2)$. The number of such pairs is $2 \cdot 3 + 3 \cdot 2 = 12$. Thus,

$$P(3) = \frac{12}{36} = \frac{1}{3}.$$

A sum of 4 corresponds to the pairs $(3, 1)$, $(2, 2)$, $(1, 3)$.
There are $1 \cdot 3 + 2 \cdot 2 + 3 \cdot 1 = 10$ such pairs. Thus,

$$P(4) = \frac{10}{36}.$$

A sum of 5 corresponds to the pairs $(2, 3)$ and $(3, 2)$.
There are $2 \cdot 1 + 1 \cdot 2 = 4$ such pairs. Thus,

$$P(5) = \frac{4}{36} = \frac{1}{9}.$$

A sum of 6 corresponds to the pair $(3, 3)$ and there is one such pair. Thus,

$$P(6) = \frac{1}{36}.$$

(A) The probability distribution for $X$ is:

| $x_i$ | 2 | 3 | 4 | 5 | 6 |
|-------|---|---|---|---|---|
| $P_i$ | $\frac{9}{36}$ | $\frac{12}{36}$ | $\frac{10}{36}$ | $\frac{4}{36}$ | $\frac{1}{36}$ |

(B) The expected value is:

$$E(X) = 2\left(\frac{9}{36}\right) + 3\left(\frac{12}{36}\right) + 4\left(\frac{10}{36}\right) + 5\left(\frac{4}{36}\right) + 6\left(\frac{1}{36}\right) = \frac{120}{36} = \frac{10}{3} \qquad (8\text{-}5)$$

**63.** The payoff table is:

| $x_i$ | $-\$1.50$ | $-\$0.50$ | $\$0.50$ | $\$1.50$ | $\$2.50$ |
|-------|-----------|-----------|----------|----------|----------|
| $P_i$ | $\frac{9}{36}$ | $\frac{12}{36}$ | $\frac{10}{36}$ | $\frac{4}{36}$ | $\frac{1}{36}$ |

and $E(X) = \dfrac{9}{36}(-1.50) + \dfrac{12}{36}(-0.50) + \dfrac{10}{36}(0.50) + \dfrac{4}{36}(1.50) + \dfrac{1}{36}(2.50)$

$$= -0.375 - 0.167 + 0.139 + 0.167 + 0.069$$

$$= -0.167 \text{ or } -\$0.167 \approx \$0.17$$

The game is not fair. The game would be fair if you paid $\$3.50 - \$0.17 = \$3.33$ to play. $\qquad (8\text{-}5)$

Copyright © 2011 Pearson Education, Inc. Publishing as Prentice Hall.

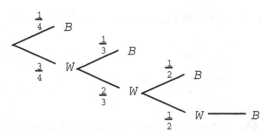

**64.** The tree diagram for this experiment is:

(A)     $P$(black on the fourth draw)

$$= \frac{3}{4} \cdot \frac{2}{3} \cdot \frac{1}{2} = \frac{1}{4}$$

The odds for black on the fourth draw are 1 to 3.

(B)     Let $x$ = amount house should pay (and return the $1 bet).
Then, for the game to be fair:

$$E(X) = x\left(\frac{1}{4}\right) + (-1)\left(\frac{3}{4}\right) = \frac{x}{4} - \frac{3}{4} = 0; \quad x = 3$$

Thus, the house should pay $3.     (8-2, 8-4)

**65.**  $n(S) = 10 \cdot 10 \cdot 10 \cdot 10 \cdot 10 = 10^5$

Let event $A$ = "at least two people identify the same book." Then $A'$ = "each person identifies a different book," and

$$n(A') = 10 \cdot 9 \cdot 8 \cdot 7 \cdot 6 = \frac{10!}{5!}$$

Thus, $P(A') = \dfrac{\dfrac{10!}{5!}}{10^5} = \dfrac{10!}{5!10^5}$ and $P(A) = 1 - \dfrac{10!}{5!10^5} \approx 1 - .3 = .7.$     (8-2)

**66.**  $P(A \mid B) = \dfrac{P(A \cap B)}{P(B)}$ , $P(B \mid A) = \dfrac{P(A \cap B)}{P(A)}$ .

Now, $P(A \mid B) = P(B \mid A)$ if and only if $\dfrac{P(A \cap B)}{P(B)} = \dfrac{P(A \cap B)}{P(A)}$

which implies $P(A) = P(B)$ or $P(A \cap B) = 0.$     (8-3)

**67.**

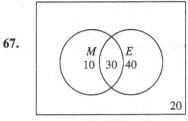

Event $M$ = Reads the morning paper.

Event $E$ = Watches evening news.

(A)     $P$(reads the paper or watches the news) = $P(M$ or $E) = P(M \cup E)$

$$= P(M) + P(E) - P(M \cap E)$$

$$= \frac{40}{100} + \frac{70}{100} - \frac{30}{100} = .8$$

(B)     $P$(does neither) = $\dfrac{20}{100} = .20$   (from the Venn diagram)

or                     $= 1 - P(M \cup E)$     {i.e., $P[(M \cup E)']$}

$$= 1 - .8 = .20$$

Copyright © 2011 Pearson Education, Inc. Publishing as Prentice Hall.

(C)   $P(\text{does exactly one}) = \dfrac{10+40}{100} = .50$  (from the Venn diagram)

or   $= P[(M \cap E') \text{ or } (M' \cap E)] = P(M \cap E') + P(M' \cap E) = \dfrac{10}{100} + \dfrac{40}{100} = .50$

(8-2)

**68.**   Let $A$ be the event that a person has seen the advertising and $P$ be the event that the person purchased the product.  Given:

$P(A) = .4$  and  $P(P \mid A) = .85$

We want to find:

$P(A \cap P) = P(A)P(P \mid A) = (.4)(.85) = .34.$     (8-3)

**69.**   (A)   $P(A) = \dfrac{290}{1000} = 0.290$

$P(B) = \dfrac{290}{1000} = 0.290$

$P(A \cap B) = \dfrac{100}{1000} = 0.100$

$P(A \mid B) = \dfrac{100}{290} = 0.345$

$P(B \mid A) = \dfrac{100}{290} = 0.345$

(B)   $A$ and $B$ are **not** independent because
$0.100 = P(A \cap B) \neq P(A) \cdot P(B) = (0.290)(0.290) = 0.084$

(C)   $P(C) = \dfrac{880}{1000} = 0.880$

$P(D) = \dfrac{120}{1000} = 0.120$

$P(C \cap D) = 0;$   $P(C \mid D) = P(D \mid C) = 0$

(D)   $C$ and $D$ are mutually exclusive since $C \cap D = \varnothing$. $C$ and $D$ are dependent since
$0 = P(C \cap D) \neq P(C) \cdot P(D) = (0.120)(0.880) = 0.106.$     (8-3)

**70.**   The payoff table for plan A is:

| $x_i$ | 10 million | −2 million |
|---|---|---|
| $p_1$ | 0.8 | 0.2 |

Hence, $E(X) = 10(0.8) - 2(0.2) = 8 - 0.4 = \$7.6$ million.

The payoff table for plan B is:

| $x_i$ | 12 million | −2 million |
|---|---|---|
| $p_1$ | 0.7 | 0.3 |

Hence, $E(X) = 12(0.7) - 2(0.3) = 8.4 - 0.6 = \$7.8$ million.

Plan B should be chosen.     (8-5)

Copyright © 2011 Pearson Education, Inc.  Publishing as Prentice Hall.

**71.** The payoff table is:

| $x_i$ | 1830 | −170 |
|-------|------|------|
| $p_1$ | 0.8  | 0.92 |

[Note: $2000 - 170 = 1830$ is the "gain" if the bicycle is stolen]

Hence, $E(X) = 1830(0.08) - 170(0.92) = 146.4 - 156.4 = -\$10.$     (8-5)

**72.** $n(S) = C_{12,4} = \dfrac{12!}{4!(12-4)!} = \dfrac{12 \cdot 11 \cdot 10 \cdot 9 \cdot 8!}{4 \cdot 3 \cdot 2 \cdot 1 \cdot 8!} = 495$

The number of samples that contain *no* substandard parts is:

$C_{10,4} = \dfrac{10!}{4!(10-4)!} = \dfrac{10 \cdot 9 \cdot 8 \cdot 7 \cdot 6!}{4 \cdot 3 \cdot 2 \cdot 1 \cdot 6!} = 210$

Thus, the number of samples that have at least one defective part is $495 - 210 = 285$. If $E$ is the event "The shipment is returned," then

$P(E) = \dfrac{n(E)}{n(S)} = \dfrac{285}{495} \approx 0.576.$     (8-2)

**73.** $n(S) = C_{12,3} = \dfrac{12!}{3!(12-3)!} = \dfrac{12 \cdot 11 \cdot 10 \cdot 9!}{3 \cdot 2 \cdot 1 \cdot 9!} = 220$

A sample will either have 0, 1, or 2 tablet computers.

$P(0) = \dfrac{C_{10,3}}{C_{12,3}} = \dfrac{\dfrac{10!}{3!(10-3)!}}{220} = \dfrac{\dfrac{10 \cdot 9 \cdot 8 \cdot 7!}{3 \cdot 2 \cdot 1 \cdot 7!}}{220} = \dfrac{120}{220} = \dfrac{12}{22}$

$P(1) = \dfrac{C_{2,1} \cdot C_{10,1}}{C_{12,3}} = \dfrac{2 \cdot \dfrac{10!}{2!(10-2)!}}{220} = \dfrac{90}{220} = \dfrac{9}{22}$

$P(2) = \dfrac{C_{2,2} \cdot C_{10,1}}{C_{12,3}} = \dfrac{10}{220} = \dfrac{1}{22}$

(A)   The probability distribution of $X$ is:

| $x_i$ | 0 | 1 | 2 |
|-------|---|---|---|
| $p_i$ | $\frac{12}{22}$ | $\frac{9}{22}$ | $\frac{1}{22}$ |

(B)   $E(X) = 0\left(\dfrac{12}{22}\right) + 1\left(\dfrac{9}{22}\right) + 2\left(\dfrac{1}{22}\right)$

$= \dfrac{11}{22} = \dfrac{1}{2}$     (8-5)

**74.** Let Event   $NH$ = individual with normal heart,

Event   $MH$ = individual with minor heart problem,

Event   $SH$ = individual with severe heart problem,

and Event   $P$ = individual passes the cardiogram test.

Then, using the notation given above, we have:

$P(NH) = .82$

$P(MH) = .11$

$P(SH) = .07$

$P(P \mid NH) = .95$

$P(P \mid MH) = .30$

$P(P \mid SH) = .05$

Copyright © 2011 Pearson Education, Inc.  Publishing as Prentice Hall.

We want to find $P(NH|P) = \dfrac{P(NH \cap P)}{P(P)} = \dfrac{P(NH)P(P|NH)}{P(NH \cap P) + P(MH \cap P) + P(SH \cap P)}$

$$= \dfrac{P(NH)P(P|NH)}{P(NH)P(P|NH) + P(MN)P(P|MH) + P(SH)P(P|SH)}$$

$$= \dfrac{(.82)(.95)}{(.82)(.95) + (.11)(.30) + (.07)(.05)} = .955 \qquad (8\text{-}4)$$

**75.**   The tree diagram for this problem is as follows:

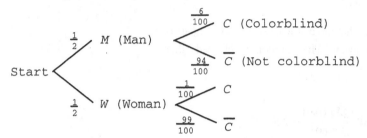

We now compute

$P(M|C) \quad = \dfrac{P(M \cap C)}{P(C)} = \dfrac{P(M \cap C)}{P(M \cap C) + P(W \cap C)} = \dfrac{P(M)P(C|M)}{P(M)P(C|M) + P(W)P(C|W)}$

$$= \dfrac{\dfrac{1}{2} \cdot \dfrac{6}{100}}{\dfrac{1}{2} \cdot \dfrac{6}{100} + \dfrac{1}{2} \cdot \dfrac{1}{100}} = \dfrac{6}{7} \approx .857. \qquad (8\text{-}4)$$

**76.**   According to the empirical probabilities, candidate $A$ should have won the election. Since candidate $B$ won the election one week later, either some of the students changed their minds during the week, or the 30 students in the math class were not representative of the student body.

Copyright © 2011 Pearson Education, Inc.  Publishing as Prentice Hall.

## 9  MARKOV CHAINS

---

### EXERCISE 9-1

*Things to remember:*

1. MARKOV CHAINS

   A MARKOV CHAIN, or PROCESS, is a sequence of experiments, trials, or observations such that the transition probability matrix from one state to the next is constant.

   Given a Markov chain with $n$ states, a $k$th STATE MATRIX is a matrix of the form
   $$S_k = [s_{k1} \ s_{k2} \ \cdots \ s_{kn}]$$

   Each entry $s_{ki}$ is the proportion of the population that are in state $i$ after the $k$th trial, or, equivalently, the probability of a randomly selected element of the population being in state $i$ after the $k$th trial. The sum of all the entries in the $k$th state matrix $S_k$ must be 1.

   A TRANSITION MATRIX is a constant square matrix $P$ of order $n$ such that the entry in the $i$th row and $j$th column indicates the probability of the system moving from the $i$th state to the $j$th state on the next observation or trial. The sum of the entries in each row must be 1.

2. COMPUTING STATE MATRICES FOR A MARKOV CHAIN

   If $S_0$ is the initial state matrix and $P$ is the transition matrix for a Markov chain, then the subsequent state matrices are given by:

   $S_1 = S_0 P$    First-state matrix

   $S_2 = S_1 P$    Second-state matrix

   $S_3 = S_2 P$    Third-state matrix

   $\vdots$

   $S_k = S_{k-1} P$    $k$th-state matrix

3. POWERS OF A TRANSITION MATRIX

   If $P$ is the transition matrix and $S_0$ is an initial state matrix for a Markov chain, then the $k$th state matrix is given by
   $$S_k = S_0 P^k$$

   The entry in the $i$th row and $j$th column of $P^k$ indicates the probability of the system moving from the $i$th state to the $j$th state in $k$ observations or trials. The sum of the entries in each row of $P^k$ is 1.

---

Copyright © 2011 Pearson Education, Inc. Publishing as Prentice Hall.

**1.**  $S_1 = S_0 P = \begin{bmatrix} 1 & 0 \end{bmatrix} \begin{bmatrix} .8 & .2 \\ .4 & .6 \end{bmatrix} \overset{A \ B}{=} [.8 \ .2]$

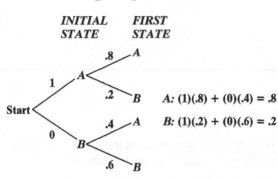

INITIAL        FIRST
STATE          STATE

A: (1)(.8) + (0)(.4) = .8

B: (1)(.2) + (0)(.6) = .2

**3.**  $S_1 = S_0 P = \begin{bmatrix} .5 & .5 \end{bmatrix} \begin{bmatrix} .8 & .2 \\ .4 & .6 \end{bmatrix} \overset{A \ B}{=} [.6 \ .4]$

INITIAL        FIRST
STATE          STATE

A: (.5)(.8) + (.5)(.4) = .6

B: (.5)(.2) + (.5)(.6) = .4

**5.**  $S_2 = S_1 P = [.8 \ .2] \begin{bmatrix} .8 & .2 \\ .4 & .6 \end{bmatrix}$ (from Problem 1)

$\qquad\qquad \overset{A \quad B}{= [.72 \ .28]}$

The probability of being in state $A$ after two trials is .72; the probability of being in state $B$ after two trials is .28.

**7.**  $S_2 = S_1 P = [.6 \ .4] \begin{bmatrix} .8 & .2 \\ .4 & .6 \end{bmatrix}$ (from Problem 3)

$\qquad\qquad \overset{A \quad B}{= [.64 \ .36]}$

The probability of being in state $A$ after two trials is .64; the probability of being in state $B$ after two trials is .36.

Copyright © 2011 Pearson Education, Inc. Publishing as Prentice Hall.

The transition matrix corresponding to the transition diagram for Problems 9 – 16 is:

$$P = \begin{bmatrix} .7 & .3 \\ .9 & .1 \end{bmatrix}$$

**9.** $S_1 = S_0 P = \begin{bmatrix} .2 & .8 \end{bmatrix} \begin{bmatrix} .7 & .3 \\ .9 & .1 \end{bmatrix} = \begin{bmatrix} .86 & .14 \end{bmatrix}.$   **11.**   $S_1 = S_0 P = \begin{bmatrix} .9 & .1 \end{bmatrix} \begin{bmatrix} .7 & .3 \\ .9 & .1 \end{bmatrix} = \begin{bmatrix} .72 & .28 \end{bmatrix}.$

**13.** From Problem 9, $S_1 = \begin{bmatrix} .86 & .14 \end{bmatrix}.$ Thus, $S_2 = \begin{bmatrix} .86 & .14 \end{bmatrix} \begin{bmatrix} .7 & .3 \\ .9 & .1 \end{bmatrix} = \begin{bmatrix} .728 & .272 \end{bmatrix}.$

**15.** From Problem 11, $S_1 = \begin{bmatrix} .72 & .28 \end{bmatrix}.$ Thus, $S_2 = \begin{bmatrix} .72 & .28 \end{bmatrix} \begin{bmatrix} .7 & .3 \\ .9 & .1 \end{bmatrix} = \begin{bmatrix} .756 & .244 \end{bmatrix}.$

**17.** $\begin{bmatrix} .3 & .7 \\ 1 & 0 \end{bmatrix}$    Yes; each entry is nonnegative, the sum of the entries in each row is 1.

**19.** $\begin{bmatrix} .5 & .5 \\ .7 & -.3 \end{bmatrix}$    No; the entry in the (2, 2) position is not nonnegative, the sum of the entries in the second row is not 1.

**21.** No. A transition matrix must be a square matrix.

**23.** $\begin{bmatrix} .5 & .1 & .4 \\ 0 & .5 & .5 \\ .2 & .1 & .7 \end{bmatrix}$    Yes; each entry is nonnegative, the sum of the entries in each row is 1.

**25.**

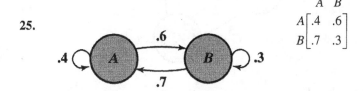

$$\begin{array}{c} \phantom{A}\begin{array}{cc} A & B \end{array} \\ \begin{array}{c} A \\ B \end{array} \begin{bmatrix} .4 & .6 \\ .7 & .3 \end{bmatrix} \end{array}$$

Copyright © 2011 Pearson Education, Inc. Publishing as Prentice Hall.

**27.** No. Choose any $x$, $0 \leq x \leq 1$, then

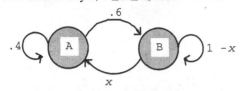

is an acceptable transition diagram, and

$$\begin{array}{c} \\ A \\ B \end{array} \begin{array}{cc} A & B \\ \left[ \begin{array}{cc} .4 & .6 \\ x & 1-x \end{array} \right] \end{array}$$ is the corresponding transition matrix

**29.**

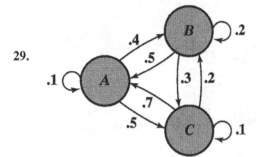

$$\begin{array}{c} \\ A \\ B \\ C \end{array} \begin{array}{ccc} A & B & C \\ \left[ \begin{array}{ccc} .1 & .4 & .5 \\ .5 & .2 & .3 \\ .7 & .2 & .1 \end{array} \right] \end{array}$$

**31.**  $0 + .5 + a = 1$  implies $a = .5$
   $b + 0 + .4 = 1$  implies $b = .6$
   $.2 + c + .1 = 1$  implies $c = .7$

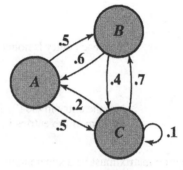

**33.**  $0 + a + .3 = 1$  implies $a = .7$
   $0 + b + 0 = 1$  implies $b = 1$
   $c + .8 + 0 = 1$  implies $c = .2$

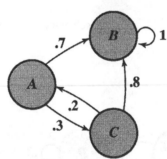

**35.** No. Choose any $x$, $0 \leq x \leq .4$ and let $a = x$, $b = .1$,
   $c = 1 - (x + .4) = .6 - x$

$$\begin{array}{c} \\ A \\ B \\ C \end{array} \begin{array}{ccc} A & B & C \\ \left[ \begin{array}{ccc} .2 & .1 & .7 \\ x & .4 & .6-x \\ .5 & .1 & .4 \end{array} \right] \end{array}$$ is a transition matrix.

Copyright © 2011 Pearson Education, Inc.  Publishing as Prentice Hall.

**37.** The probability of staying in state $A$ is .3.

The probability of staying in state $B$ is .1.

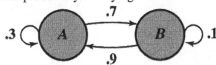

$$\begin{array}{c} \\ A \\ B \end{array} \begin{array}{cc} A & B \\ \begin{bmatrix} .3 & .7 \\ .9 & .1 \end{bmatrix} \end{array}$$

**39.** The probability of staying in state $A$ is .6.

The probability of staying in state $B$ is .3.

Since the probability of staying in state $C$ is 1, the probability of going from state $C$ to state $A$ is 0 and the probability of going from state $C$ to state $B$ is 0.

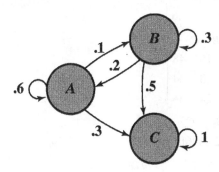

$$\begin{array}{c} \\ A \\ B \\ C \end{array} \begin{array}{ccc} A & B & C \\ \begin{bmatrix} .6 & .1 & .3 \\ .2 & .3 & .5 \\ 0 & 0 & 1 \end{bmatrix} \end{array}$$

**41.** Using $P^2$, the probability of going from state $A$ to state $B$ in two trials is the element in the (1,2) position is .35.

**43.** Using $P^3$, the probability of going from state $C$ to state $A$ in three trials is the number in the (3,1) position is .212.

**45.** $S_2 = S_0 P^2 = [1 \ 0 \ 0] \begin{bmatrix} .43 & .35 & .22 \\ .25 & .37 & .38 \\ .17 & .27 & .56 \end{bmatrix} \begin{array}{ccc} A & B & C \\ \\ \\ \end{array} = [.43 \ .35 \ .22]$

These are the probabilities of going from state $A$ to states $A$, $B$, and $C$, respectively, in two trials.

**47.** $S_3 = S_0 P^3 = [0 \ 0 \ 1] \begin{bmatrix} .35 & .348 & .302 \\ .262 & .336 & .402 \\ .212 & .298 & .49 \end{bmatrix} \begin{array}{ccc} A & B & C \\ \\ \\ \end{array} = [.212 \ .298 \ .49]$

These are the probabilities of going from state $C$ to states $A$, $B$, and $C$, respectively, in three trials.

**49.** $n = 9$

Copyright © 2011 Pearson Education, Inc.  Publishing as Prentice Hall.

**51.**  $P = \begin{bmatrix} .1 & .9 \\ .6 & .4 \end{bmatrix}$,  $P^2 = \begin{bmatrix} .1 & .9 \\ .6 & .4 \end{bmatrix}\begin{bmatrix} .1 & .9 \\ .6 & .4 \end{bmatrix} = \begin{bmatrix} .55 & .45 \\ .3 & .7 \end{bmatrix}$;

$$P^4 = P^2 \cdot P^2 = \begin{bmatrix} .55 & .45 \\ .3 & .7 \end{bmatrix}\begin{bmatrix} .55 & .45 \\ .3 & .7 \end{bmatrix} = \begin{matrix} & A & B \\ A & \\ B & \end{matrix}\begin{bmatrix} .4375 & .5625 \\ .375 & .625 \end{bmatrix}$$

$$S_4 = S_0 P^4 = [.8 \ .2]\begin{bmatrix} .4375 & .5625 \\ .375 & .625 \end{bmatrix} = [.425 \ .575]$$

**53.**  $P = \begin{bmatrix} 0 & .4 & .6 \\ 0 & 0 & 1 \\ 1 & 0 & 0 \end{bmatrix}$,  $P^2 = \begin{bmatrix} 0 & .4 & .6 \\ 0 & 0 & 1 \\ 1 & 0 & 0 \end{bmatrix}\begin{bmatrix} 0 & .4 & .6 \\ 0 & 0 & 1 \\ 1 & 0 & 0 \end{bmatrix} = \begin{bmatrix} .6 & 0 & .4 \\ 1 & 0 & 0 \\ 0 & .4 & .6 \end{bmatrix}$;

$$P^4 = P^2 \cdot P^2 = \begin{bmatrix} .6 & 0 & .4 \\ 1 & 0 & 0 \\ 0 & .4 & .6 \end{bmatrix}\begin{bmatrix} .6 & 0 & .4 \\ 1 & 0 & 0 \\ 0 & .4 & .6 \end{bmatrix} = \begin{matrix} & A & B & C \\ A & \\ B & \\ C & \end{matrix}\begin{bmatrix} .36 & .16 & .48 \\ .6 & 0 & .4 \\ .4 & .24 & .36 \end{bmatrix}$$

$$S_4 = S_0 P^4 = [.2 \ .3 \ .5]\begin{bmatrix} .36 & .16 & .48 \\ .6 & 0 & .4 \\ .4 & .24 & .36 \end{bmatrix} = \begin{matrix} A & B & C \end{matrix} [.452 \ .152 \ .396]$$

**55.**  $S_k = S_0 P^k = [1 \ \ 0]P^k$; The entries in $S_k$ are the entries in the first row of $P^k$.

**57.**  False.  For $P$ to be a transition matrix its entries must be nonnegative numbers and the sum of the entries in each row must be 1.

$$P = \begin{bmatrix} .7 & .3 \\ .9 & .1 \end{bmatrix}$$

is a transition matrix and the sum of the entries in the first column is 1.6.

**59.**  True.  The entries in a transition matrix are nonnegative numbers.  Therefore the product of the entries in each column is nonnegative.

**61.**  False.  See the transition diagram for Problems 9 – 16.  The sum of the probabilities on the arrows going into $A$ is $.7 + .9 = 1.6 \neq 1$.

**63.**  (A)  $P^2 = \begin{bmatrix} .2 & .2 & .3 & .3 \\ 0 & 1 & 0 & 0 \\ .2 & .2 & .1 & .5 \\ 0 & 0 & 0 & 1 \end{bmatrix}\begin{bmatrix} .2 & .2 & .3 & .3 \\ 0 & 1 & 0 & 0 \\ .2 & .2 & .1 & .5 \\ 0 & 0 & 0 & 1 \end{bmatrix} = \begin{bmatrix} .1 & .3 & .09 & .51 \\ 0 & 1 & 0 & 0 \\ .06 & .26 & .07 & .61 \\ 0 & 0 & 0 & 1 \end{bmatrix}$

$P^4 = \begin{bmatrix} .1 & .3 & .09 & .51 \\ 0 & 1 & 0 & 0 \\ .06 & .26 & .07 & .61 \\ 0 & 0 & 0 & 1 \end{bmatrix}\begin{bmatrix} .1 & .3 & .09 & .51 \\ 0 & 1 & 0 & 0 \\ .06 & .26 & .07 & .61 \\ 0 & 0 & 0 & 1 \end{bmatrix}$

Copyright © 2011 Pearson Education, Inc.  Publishing as Prentice Hall.

$$
=\begin{array}{c} \\ A \\ B \\ C \\ D \end{array}\begin{array}{cccc} A & B & C & D \\ \begin{bmatrix} .0154 & .3534 & .0153 & .6159 \\ 0 & 1 & 0 & 0 \\ .0102 & .2962 & .0103 & .6833 \\ 0 & 0 & 0 & 1 \end{bmatrix} \end{array}
$$

(B)    The probability of going from state $A$ to state $D$ in 4 trials is the element in the (1,4) position: .6159.

(C)    The element in the (3,2) position: .2962.

(D)    The element in the (2,1) position: 0.

**65.**    If $P = \begin{bmatrix} a & 1-a \\ 1-b & b \end{bmatrix}$ is a probability matrix, then $0 \le a \le 1, 0 \le b \le 1$

$$
P^2 = \begin{bmatrix} a & 1-a \\ 1-b & b \end{bmatrix}\begin{bmatrix} a & 1-a \\ 1-b & b \end{bmatrix} = \begin{bmatrix} a^2 + (1-a)(1-b) & a(1-a)+(1-a)b \\ (1-b)a + b(1-b) & (1-b)(1-a)+b^2 \end{bmatrix}
$$

Now, $a^2 + (1-a)(1-b) \ge 0$ and $a(1-a) + (1-a)b = (1-a)(a+b) \ge 0$ since $0 \le a \le 1$ and $0 \le b \le 1$.

Also, $a^2 + (1-a)(1-b) + (1-a)(a+b) = a^2 + (1-a)[1-b+a+b]$

$$
= a^2 + (1-a)(1+a) = a^2 + 1 - a^2 = 1
$$

Therefore, the elements in the first row of $P^2$ are nonnegative and their sum is 1. The same arguments apply to the elements in the second row of $P^2$. Thus, $P^2$ is a probability matrix.

**67.**    $P = \begin{bmatrix} .4 & .6 \\ .2 & .8 \end{bmatrix}$

(A)    Let $S_0 = [0 \;\; 10]$. Then

$S_2 = S_0 P^2 = [.24 \;\; .76]$

$S_4 = S_0 P^4 = [.2496 \;\; .7504]$

$S_8 = S_0 P^8 = [.24999936 \;\; .7500006]$

$S_k$ is approaching $[.25 \;\; .75]$

(B)    Let $S_0 = [1 \;\; 0]$. Then

$S_2 = S_0 P^2 = [.28 \;\; .72]$

$S_4 = S_0 P^4 = [.2512 \;\; .7488]$

$S_8 = S_0 P^8 = [.25000192 \;\; .7499980]$

$S_k$ is approaching $[.25 \;\; .75]$

(C)    Let $S_0 = [.5 \;\; .5]$. Then

$S_2 = S_0 P^2 = [.26 \;\; .74]$

$S_4 = S_0 P^4 = [.2504 \;\; .7496]$

$S_8 = S_0 P^8 = [.25000064 \;\; .74999936]$

$S_k$ is approaching $[.25 \;\; .75]$

Copyright © 2011 Pearson Education, Inc. Publishing as Prentice Hall.

(D)    $[.25 \ .75] \begin{bmatrix} .4 & .6 \\ .2 & .8 \end{bmatrix} = [.25 \ .75]$

(E)    The state matrices $S_k$ appear to approach the same matrix [.25 .75], regardless of the values in the initial state matrix $S_0$.

69.    $P^2 = \begin{bmatrix} .28 & .72 \\ .24 & .76 \end{bmatrix}$,    $P^4 = \begin{bmatrix} .2512 & .7488 \\ .2496 & .7504 \end{bmatrix}$

$P^8 = \begin{bmatrix} .25000192 & .7499980 \\ .24999936 & .7500006 \end{bmatrix} \cdots$

The matrices $P^k$ are approaching $Q = \begin{bmatrix} .25 & .75 \\ .25 & .75 \end{bmatrix}$; the rows of $Q$ are the same as the matrix $S = [.25 \ .75]$

in Problem 67.

71.    Let $R$ denote "rain" and $R'$ "not rain".

(A)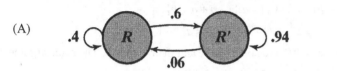

(B)    $\begin{array}{c c} & \begin{array}{c c} R & R' \end{array} \\ \begin{array}{c} R \\ R' \end{array} & \begin{bmatrix} .4 & .6 \\ .06 & .94 \end{bmatrix} \end{array}$

(C)    Rain on Saturday: $P^2 = \begin{array}{c c} & \begin{array}{c c} R & R' \end{array} \\ \begin{array}{c} R \\ R' \end{array} & \begin{bmatrix} .196 & .804 \\ .0804 & .9196 \end{bmatrix} \end{array}$

The probability that it will rain on Saturday is .196.

Rain on Sunday: $P^3 = \begin{array}{c c} & \begin{array}{c c} R & R' \end{array} \\ \begin{array}{c} R \\ R' \end{array} & \begin{bmatrix} .12664 & .87336 \\ .087336 & .912664 \end{bmatrix} \end{array}$

The probability that it will rain on Sunday is .12664.

73.    (A)

(B)    $\begin{array}{c c} & \begin{array}{c c} X & X' \end{array} \\ \begin{array}{c} X \\ X' \end{array} & \begin{bmatrix} .8 & .2 \\ .2 & .8 \end{bmatrix} \end{array}$

(C)    $\begin{array}{c c} X & X' \end{array}$
       $S = [.2 \ .8]$

$S_1 = SP = [.2 \ .8] \begin{bmatrix} .8 & .2 \\ .2 & .8 \end{bmatrix} = [.32 \ .68]$

32% will be using brand $X$ one week later.

Copyright © 2011 Pearson Education, Inc.  Publishing as Prentice Hall.

$$S_2 = SP^2 = [.2 \quad .8] \begin{bmatrix} .68 & .32 \\ .32 & .68 \end{bmatrix} = [.392 \quad .608]$$

39.2% will be using brand $X$ two weeks later.

**75.**  (A)

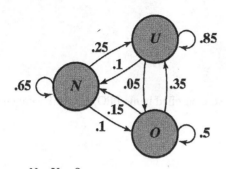

(B)
$$
\begin{array}{c} \\ \\ P = \end{array}
\begin{array}{c} \\ N \\ U \\ O \end{array}
\begin{array}{c} N \quad\ U \quad\ O \\ \begin{bmatrix} .65 & .25 & .10 \\ .10 & .85 & .05 \\ .15 & .35 & .50 \end{bmatrix} \end{array}
$$

(C)  $\quad N \quad U \quad O$
$S = [.50 \ .30 \ .20]$

$\qquad\qquad N \quad\ U \quad\ O$
After one year: $SP = [.385 \ .45 \ .165]$
38.5% will be insured by National Property after one year.

$\qquad\qquad N \quad\ U \quad\ O$
After two years:  $SP^2 = [.32 \ .5365 \ .1435]$
32% will be insured by National Property after two years.

(D)  45% of the homes will be insured by United Family after one year;
53.65% will be insured by United Family after two years.

**77.**  (A)

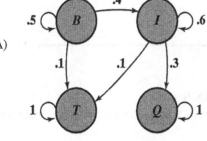

(B)  $P =$
$$
\begin{array}{c} \\ B \\ I \\ T \\ Q \end{array}
\begin{array}{c} B \quad\ I \quad\ T \quad Q \\ \begin{bmatrix} .5 & .4 & .1 & 0 \\ 0 & .6 & .1 & .3 \\ 0 & 0 & 1 & 0 \\ 0 & 0 & 0 & 1 \end{bmatrix} \end{array}
$$

(C)  $\quad B \ I \ T \ Q$
$S = [1 \ 0 \ 0 \ 0]$

$\qquad\qquad B \quad I \ \ T \quad Q$
After one year: $SP^2 = [.25 \ .44 \ .19 \ .12]$
The probability that a beginning agent will be promoted to qualified agent within one year (i.e.,
after 2 reviews) is: .12.

$\qquad\qquad B \qquad I \qquad T \qquad Q$
After two years: $SP^4 = [.0625 \ .2684 \ .3079 \ .3612]$
The probability that a beginning agent will be promoted to qualified agent within two years (i.e.,
after 4 reviews) is: .3612.

Copyright © 2011 Pearson Education, Inc. Publishing as Prentice Hall.

**79. (A)**

$$P = \begin{array}{c} \\ HMO \\ PPO \\ FFS \end{array} \begin{array}{ccc} HMO & PPO & FFS \\ \begin{bmatrix} .80 & .15 & .05 \\ .20 & .70 & .10 \\ .25 & .30 & .45 \end{bmatrix} \end{array}$$

**(B)**

$$\begin{array}{ccc} HMO & PPO & FFS \end{array}$$
$$S = [.20 \ \ .25 \ \ .55]$$
$$\begin{array}{ccc} HMO & PPO & FFS \end{array}$$
$$SP = [.3475 \ \ .37 \ \ .2825]$$

34.75% were enrolled in the HMO; 37% were enrolled in the PPO; 28.25% were enrolled in the FFS.

**(C)**

$$\begin{array}{ccc} HMO & PPO & FFS \end{array}$$
$$SP^2 = [.422625 \ .395875 \ .1815]$$

42.2625% will be enrolled in the HMO; 39.5875% will be enrolled in the PPO; 18.15% will be enrolled in the FFS.

**81. (A)**

$$P = \begin{array}{c} \\ H \\ R \end{array} \begin{array}{cc} H & R \\ \begin{bmatrix} 0.924 & 0.076 \\ 0.108 & 0.892 \end{bmatrix} \end{array}$$

**(B)**

$$\begin{array}{cc} H & R \end{array}$$
$$S = [0.364 \ \ 0.636]; \quad SP = [0.405 \ \ 0.595], \quad 40.5\% \text{ were homeowners in 2000.}$$

**(C)**

$$\begin{array}{cc} H & R \end{array}$$
$$SP^3 = [0.466 \ \ 0.536], \quad 46.6\% \text{ will be homeowners in 2020.}$$

---

## EXERCISE 9-2

*Things to remember:*

1.  STATIONARY MATRIX FOR A MARKOV CHAIN

    The state matrix $S = [s_1 \ s_2 \ \ldots \ s_n]$ is a STATIONARY MATRIX for a Markov chain with transition matrix $P$ if

    $$SP = S$$

    where $s_i \geq 0$, $i = 1, \ldots, n$ and $s_1 + s_2 + \ldots + s_n = 1$.

2.  REGULAR MARKOV CHAINS

    A transition matrix $P$ is REGULAR if some power of $P$ has only positive entries. A Markov chain is a REGULAR MARKOV CHAIN if its transition matrix is regular.

3.  PROPERTIES OF REGULAR MARKOV CHAINS

    Let $P$ be the transition matrix for a regular Markov chain.

    (A)  There is a unique stationary matrix $S$ which can be found by solving the equation

    $$SP = S$$

Copyright © 2011 Pearson Education, Inc. Publishing as Prentice Hall.

(B)    Given any initial state matrix $S_0$, the state matrices $S_k$ approach the stationary matrix $S$.

(C)    The matrices $P^k$ approach a limiting matrix $\overline{P}$ where each row of $\overline{P}$ is equal to the stationary matrix $S$.

---

**1.** $P = \begin{bmatrix} .6 & .4 \\ .4 & .6 \end{bmatrix}$    Yes; $P$ is a transition matrix and all entries are positive.

**3.** $P = \begin{bmatrix} .1 & .9 \\ .5 & .4 \end{bmatrix}$    No; $P$ is not a transition matrix, the sum of the entries in the second row is not 1.

**5.** $P = \begin{bmatrix} .4 & .6 \\ 0 & 1 \end{bmatrix}$    No; $P$ is a transition matrix but the second row of every power of $P$ is [0, 1].

**7.** $P = \begin{bmatrix} 0 & 1 \\ .8 & .2 \end{bmatrix}$    Yes; $P$ is a transition matrix and $P^2 = \begin{bmatrix} .8 & .2 \\ .16 & .84 \end{bmatrix}$ has only positive entries.

**9.** $P = \begin{bmatrix} .6 & .4 \\ .1 & .9 \\ .3 & .7 \end{bmatrix}$    No; $P$ is not a square matrix.

**11.** $P = \begin{bmatrix} 0 & 1 & 0 \\ 0 & 0 & 1 \\ .5 & .5 & 0 \end{bmatrix}$    Yes; $P$ is a transition matrix and

$P^5 = \begin{bmatrix} .25 & .25 & .5 \\ .25 & .5 & .25 \\ .125 & .375 & .5 \end{bmatrix}$ has only positive entries.

**13.** $P = \begin{bmatrix} .1 & .3 & .6 \\ .8 & .1 & .1 \\ 0 & 0 & 1 \end{bmatrix}$    No; $P$ is a transition matrix but the third row of every power of $P$ is [0  0  1].

**15.**    Let $S = [s_1, s_2]$, and solve the system:

$[s_1 \ s_2] \begin{bmatrix} .1 & .9 \\ .6 & .4 \end{bmatrix} = [s_1 \ s_2], s_1 + s_2 = 1$

which is equivalent to

$.1s_1 + .6s_2 = s_1$              $-.9s_1 + .6s_2 = 0$
$.9s_1 + .4s_2 = s_2$     or       $.9s_1 - .6s_2 = 0$
$s_1 + \ s_2 = 1$                  $s_1 + \ s_2 = 1$

The solution is: $s_1 = .4, s_2 = .6$

The stationary matrix $S = [.4 \ .6]$; the limiting matrix $\overline{P} = \begin{bmatrix} .4 & .6 \\ .4 & .6 \end{bmatrix}$.

**17.**    Let $S = [s_1, s_2]$, and solve the system:

$[s_1 \ s_2] \begin{bmatrix} .5 & .5 \\ .3 & .7 \end{bmatrix} = [s_1 \ s_2], s_1 + s_2 = 1$

which is equivalent to

Copyright © 2011 Pearson Education, Inc. Publishing as Prentice Hall.

$$.5s_1 + .3s_2 = s_1 \qquad\qquad -.5s_1 + .3s_2 = 0$$
$$.5s_1 + .7s_2 = s_2 \qquad \text{or} \qquad .5s_1 - .3s_2 = 0$$
$$s_1 + s_2 = 1 \qquad\qquad s_1 + s_2 = 1$$

The solution is: $s_1 = \dfrac{3}{8} = .375$, $s_2 = \dfrac{5}{8} = .625$

The stationary matrix $S = [.375 \;\; .625]$; the limiting matrix

$$\overline{P} = \begin{bmatrix} .375 & .625 \\ .375 & .625 \end{bmatrix}.$$

**19.** Let $S = [s_1 \;\; s_2 \;\; s_3]$, and solve the system:

$$[s_1 \;\; s_2 \;\; s_3] \begin{bmatrix} .5 & .1 & .4 \\ .3 & .7 & 0 \\ 0 & .6 & .4 \end{bmatrix} = [s_1 \;\; s_2 \;\; s_3], \quad s_1 + s_2 + s_3 = 1$$

which is equivalent to

$$.5s_1 + .3s_2 \qquad\quad = s_1 \qquad\qquad -.5s_1 + .3s_2 \qquad\quad = 0$$
$$.1s_1 + .7s_2 + .6s_3 = s_2 \qquad \text{or} \qquad .1s_1 - .3s_2 + .6s_3 = 0$$
$$.4s_1 \qquad\;\; + .4s_3 = s_3 \qquad\qquad .4s_1 \qquad\quad - .6s_3 = 0$$
$$s_1 + s_2 + s_3 = 1 \qquad\qquad s_1 + s_2 + s_3 = 1$$

From the first and third equations, we have $s_2 = \dfrac{5}{3}s_1$, and $s_3 = \dfrac{2}{3}s_1$.

Substituting these values into the fourth equation, we get:

$$s_1 + \frac{5}{3}s_1 + \frac{2}{3}s_1 = 1 \text{ or } \frac{10}{3}s_1 = 1$$

Therefore, $s_1 = .3$, $s_2 = .5$, $s_3 = .2$.

The stationary matrix $S = [.3 \;\; .5 \;\; .2]$;

the limiting matrix $\overline{P} = \begin{bmatrix} .3 & .5 & .2 \\ .3 & .5 & .2 \\ .3 & .5 & .2 \end{bmatrix}.$

**21.** Let $S = [s_1 \;\; s_2 \;\; s_3]$, and solve the system:

$$[s_1 \;\; s_2 \;\; s_3] \begin{bmatrix} .8 & .2 & 0 \\ .5 & .1 & .4 \\ 0 & .6 & .4 \end{bmatrix} = [s_1 \;\; s_2 \;\; s_3], s_1 + s_2 + s_3 = 1$$

which is equivalent to

$$.8s_1 + .5s_2 \qquad\quad = s_1 \qquad\qquad -.2s_1 + .5s_2 \qquad\quad = 0$$
$$.2s_1 + .1s_2 + .6s_3 = s_2 \qquad \text{or} \qquad .2s_1 - .9s_2 + .6s_3 = 0$$
$$.4s_2 + .4s_3 = s_3 \qquad\qquad .4s_2 - .6s_3 = 0$$
$$s_1 + s_2 + s_3 = 1 \qquad\qquad s_1 + s_2 + s_3 = 1$$

Copyright © 2011 Pearson Education, Inc. Publishing as Prentice Hall.

From the first and third equations, we have $s_1 = \frac{5}{2}s_2$, and $s_3 = \frac{2}{3}s_2$. Substituting these values into the fourth equation, we get:

$$\frac{5}{2}s_2 + s_2 + \frac{2}{3}s_2 = 1 \text{ or } \frac{25}{6}s_2 = 1$$

Therefore, $s_2 = \frac{6}{25} = .24$, $s_1 = .6$, $s_3 = .16$.

The stationary matrix $S = [.6\ \ .24\ \ .16]$;   the limiting matrix $\overline{P} = \begin{bmatrix} .6 & .24 & .16 \\ .6 & .24 & .16 \\ .6 & .24 & .16 \end{bmatrix}$.

**23.** False. For the identity matrix $I$, $I^k = I$ for all positive integers $k$; no power of $I$ has all positive entries.

**25.** True. Suppose $P = \begin{bmatrix} a & b \\ c & d \end{bmatrix}$ has two entries equal to $0$. Then either $P$ has a row of zeros, a column of zeros, or $P$ has one of the two forms $P = \begin{bmatrix} a & 0 \\ 0 & d \end{bmatrix}$, or $P = \begin{bmatrix} 0 & b \\ c & 0 \end{bmatrix}$. In each case, $P^k$ has the same form as $P$.

**27.** False. Every state matrix $S = [s_1, s_2]$ is a stationary matrix for the transition matrix $I = \begin{bmatrix} 1 & 0 \\ 0 & 1 \end{bmatrix}$, and $I$ is not regular.

**29.** $P = \begin{bmatrix} .51 & .49 \\ .27 & .73 \end{bmatrix}, P^2 = \begin{bmatrix} .3924 & .6076 \\ .3348 & .6652 \end{bmatrix}, P^4 \approx \begin{bmatrix} .3574 & .6426 \\ .3541 & .6459 \end{bmatrix}$

$P^8 \approx \begin{bmatrix} .3553 & .6447 \\ .3553 & .6447 \end{bmatrix}, P^{16} \approx \begin{bmatrix} .3553 & .6447 \\ .3553 & .6447 \end{bmatrix}$

Therefore, $S \approx [.3553\ \ .6447]$.

**31.** $P = \begin{bmatrix} .5 & .5 & 0 \\ 0 & .5 & .5 \\ .8 & .1 & .1 \end{bmatrix}, P^2 = \begin{bmatrix} .25 & .5 & .25 \\ .4 & .3 & .3 \\ .48 & .46 & .06 \end{bmatrix}, P^4 = \begin{bmatrix} .3825 & .39 & .2275 \\ .364 & .428 & .208 \\ .3328 & .4056 & .2616 \end{bmatrix}$,

$P^8 \approx \begin{bmatrix} .3640 & .4084 & .2277 \\ .3642 & .4095 & .2262 \\ .3620 & .4095 & .2285 \end{bmatrix}, P^{16} \approx \begin{bmatrix} .3636 & .4091 & .2273 \\ .3636 & .4091 & .2273 \\ .3636 & .4091 & .2273 \end{bmatrix}$

Therefore, $S \approx [.3636\ \ .4091\ \ .2273]$.

**33.** (A)

(B) $P = \begin{array}{c} \\ \text{Red} \\ \text{Blue} \end{array} \begin{array}{c} \text{Red Blue} \\ \begin{bmatrix} .4 & .6 \\ .2 & .8 \end{bmatrix} \end{array}$

(C)   Let $S = [s_1\ \ s_2]$ and solve the system:

$$[s_1\ \ s_2]\begin{bmatrix} .4 & .6 \\ .2 & .8 \end{bmatrix} = [s_1\ \ s_2], s_1 + s_2 = 1$$

which is equivalent to

$$.4s_1 + .2s_2 = s_1 \qquad\qquad -.6s_1 + .2s_2 = 0$$

Copyright © 2011 Pearson Education, Inc. Publishing as Prentice Hall.

$.6s_1 + .8s_2 = s_2$   or   $.6s_1 - .2s_2 = 0$
$s_1 + s_2 = 1$      $s_1 + s_2 = 1$

The solution is: $s_1 = .25$,   $s_2 = .75$.

Thus, the stationary matrix $S = [.25\ .75]$. In the long run, the red urn will be selected 25% of the time and the blue urn 75% of the time.

**35.** (A) $s_1 = [.2\ .8]\begin{bmatrix} 0 & 1 \\ 1 & 0 \end{bmatrix} = [.8\ .2];$   $s_2 = s_1P = [.8\ .2]\begin{bmatrix} 0 & 1 \\ 1 & 0 \end{bmatrix} = [.2\ .8];$   $s_3 = s_2P = [.8\ .2]$

and so on.

The state matrices alternate between $[.2\ .8]$ and $[.8\ .2]$; they do not approach a "limiting" matrix.

(B) $s_1 = [.5\ .5]\begin{bmatrix} 0 & 1 \\ 1 & 0 \end{bmatrix} = [.5\ .5];$   $s_2 = s_1P = [.5\ .5]\begin{bmatrix} 0 & 1 \\ 1 & 0 \end{bmatrix} = [.5\ .5]$

and so on.

Thus, $s_1 = s_2 = s_3 = \ldots = [.5\ .5] = s_0$;   $s_0$ is a stationary matrix.

(C) $P = \begin{bmatrix} 0 & 1 \\ 1 & 0 \end{bmatrix}, P^2 = \begin{bmatrix} 1 & 0 \\ 0 & 1 \end{bmatrix}, P^3 = \begin{bmatrix} 0 & 1 \\ 1 & 0 \end{bmatrix}, \ldots$

The powers of $P$ alternate between $P$ and the identity, $I$; they do not approach a limiting matrix.

(D) Parts (B) and (C) are not valid for this matrix. Since $P$ is not regular, this does not contradict Theorem 1.

**37.** (A) $RP = [1\ 0\ 0]\begin{bmatrix} 1 & 0 & 0 \\ .2 & .2 & .6 \\ 0 & 0 & 1 \end{bmatrix} = [1\ 0\ 0]$

Therefore $R$ is a stationary matrix for $P$.

$SP = [0\ 0\ 1]\begin{bmatrix} 1 & 0 & 0 \\ .2 & .2 & .6 \\ 0 & 0 & 1 \end{bmatrix} = [0\ 0\ 1]$

Therefore $S$ is a stationary matrix for $P$.

The powers of $P$ have the form

$\begin{bmatrix} 1 & 0 & 0 \\ a & b & c \\ 0 & 0 & 1 \end{bmatrix}$

Therefore $P$ is not regular. As a result, $P$ may have more than one stationary matrix.

(B) Following the hint, let
$T = a[1\ 0\ 0] + (1-a)[0\ 0\ 1]$,   $0 < a < 1$.
$= [a\ \ 0\ \ 1-a]$

Now, $TP = [a\ \ 0\ \ 1-a]\begin{bmatrix} 1 & 0 & 0 \\ .2 & .2 & .6 \\ 0 & 0 & 1 \end{bmatrix} = [a\ \ 0\ \ 1-a] = T$

Thus, $[a\ \ 0\ \ 1-a]$ is a stationary matrix for $P$ for every $a$ with $0 < a < 1$. Note that if $a = 1$, then $T = R$, and if $a = 0$, $T = S$. If we let $a = .5$, then $T = [.5\ 0\ .5]$ is a stationary matrix.

(C) $P$ has infinitely many stationary matrices.

Copyright © 2011 Pearson Education, Inc. Publishing as Prentice Hall.

**39.** $\overline{P} = \begin{bmatrix} 1 & 0 & 0 \\ .25 & 0 & .75 \\ 0 & 0 & 1 \end{bmatrix}$

Each row of $\overline{P}$ is a stationary matrix for $P$. As we saw in Problem 37, part (B),

$T = [a \quad 0 \quad 1 - a]$

is a stationary matrix for $P$ for each $a$ where $0 \leq a \leq 1$; $a = 1$ gives the first row of $\overline{P}$, $a = .25$ gives the second row; $a = 0$ gives the third row.

**41.** (A) For $P^2, M_2 = .39$; for $P^3, M_3 = .3$; for $P^4, M_4 = .284$; for $P^5, M_5 = .277$.

(B) Each entry of the second column of $P^{k+1}$ is a product of the form

$$ap_{12}^k + bp_{22}^k + cp_{32}^k$$

where $p_{12}^k$, $p_{22}^k$, $p_{32}^k$ are the entries in the second column of $P^k$;

$a, b, c \geq 0$ and $a + b + c = 1$. Thus,

$$M_{k+1} = ap_{12}^k + bp_{22}^k + cp_{32}^k \leq aM_k + bM_k + cM_k = (a + b + c)M_k = M_k$$

Therefore, $M_k \geq M_{k+1}$ for all positive integers $k$.

**43.** The transition matrix is

$$P = \begin{array}{cc} & \begin{array}{cc} H & N \end{array} \\ \begin{array}{c} H \\ N \end{array} & \begin{bmatrix} .89 & .11 \\ .29 & .71 \end{bmatrix} \end{array} \quad \begin{array}{l} H = \text{home trackage} \\ N = \text{national pool} \end{array}$$

Calculating powers of $P$, we have

$$P^2 = \begin{bmatrix} .824 & .176 \\ .464 & .536 \end{bmatrix}, \quad P^4 \approx \begin{bmatrix} .7606 & .2394 \\ .6310 & .3690 \end{bmatrix}, \quad P^8 \approx \begin{bmatrix} .7296 & .2704 \\ .7128 & .2872 \end{bmatrix}, \quad P^{16} \approx \begin{bmatrix} .7251 & .2749 \\ .7248 & .2752 \end{bmatrix}$$

In the long run, 72.5% of the company's box cars will be on its home tracks.

**45.** (A) $S_1 = S_0 P = [.433 \quad .567] \begin{bmatrix} .92 & .08 \\ .2 & .8 \end{bmatrix} = [.51176 \quad .48824] \approx [.512 \quad .488]$

$S_2 = S_0 P^2 = [.433 \quad .567] \begin{bmatrix} .8624 & .1376 \\ .344 & .656 \end{bmatrix} = [.5684672 \quad .4315328] \approx [.568 \quad .432]$

$S_3 = S_0 P^3 = [.433 \quad .567] \begin{bmatrix} .820928 & .179072 \\ .44768 & .55232 \end{bmatrix} = [.609296384 \quad .390703616] \approx [.609 \quad .391]$

(B)

| Year | Data% | Model% |
|------|-------|--------|
| 1970 | 43.3 | 43.3 |
| 1980 | 51.5 | 51.2 |
| 1990 | 57.5 | 56.8 |
| 2000 | 59.8 | 60.9 |

(C) $P^4 \approx \begin{bmatrix} .7911 & .2089 \\ .5223 & .4777 \end{bmatrix}$, $P^8 \approx \begin{bmatrix} .7349 & .2651 \\ .6627 & .3373 \end{bmatrix}$,

$P^{16} \approx \begin{bmatrix} .7158 & .2842 \\ .7106 & .2894 \end{bmatrix}$, $P^{32} \approx \begin{bmatrix} .7143 & .2857 \\ .7143 & .2857 \end{bmatrix}$

In the long run, 71.4% of the female population will be in the labor force.

Copyright © 2011 Pearson Education, Inc. Publishing as Prentice Hall.

47.   The transition matrix for this problem is:

$$
\begin{array}{c}
\quad\; \text{GTT}\;\; \text{NCJ}\;\; \text{Dash} \\
\begin{array}{c} \text{GTT} \\ \text{NCJ} \\ \text{Dash} \end{array}
\begin{bmatrix}
.75 & .05 & .20 \\
.15 & .75 & .1 \\
.05 & .10 & .85
\end{bmatrix}
\end{array}
$$

To find the steady-state matrix, we solve the system

$$
[s_1\; s_2\; s_3\;]
\begin{bmatrix}
.75 & .05 & .20 \\
.15 & .75 & .10 \\
.05 & .10 & .85
\end{bmatrix}
= [s_1\; s_2\; s_3\;], \quad s_1 + s_2 + s_3 = 1
$$

which is equivalent to the system of equations

$$
\begin{aligned}
.75s_1 + .15s_2 + .05s_3 &= s_1 \\
.05s_1 + .75s_2 + .1s_3 &= s_2 \\
.20s_1 + .10s_2 + .85s_3 &= s_3 \\
s_1 + s_2 + s_3 &= 1
\end{aligned}
$$

The solution of this system is $s_1 = .25$, $s_2 = .25$, $s_3 = .5$.

Thus, the expected market share of each company is:   GTT - 25%; NCJ - 25%; and Dash - 50%.

49.   The transition matrix for this problem is:

$$
\begin{array}{c}
\qquad\qquad \text{Poor}\qquad \text{Satisfactory}\qquad \text{Preferred} \\
\begin{array}{c} \text{Poor} \\ \text{Satisfactory} \\ \text{Preferred} \end{array}
\begin{bmatrix}
.60 & .40 & 0 \\
.20 & .60 & .20 \\
0 & .20 & .80
\end{bmatrix}
\end{array}
$$

To find the steady-state matrix, we solve the system

$$
[s_1\; s_2\; s_3]
\begin{bmatrix}
.60 & .40 & 0 \\
.20 & .60 & .20 \\
0 & .20 & .80
\end{bmatrix}
= [s_1\; s_2\; s_3], \quad s_1 + s_2 + s_3 = 1
$$

which is equivalent to the system of equations:

$$
\begin{aligned}
.6s_1 + .2s_2 &= s_1 \\
.4s_1 + .6s_2 + .2s_3 &= s_2 \\
.2s_2 + .8s_3 &= s_3 \\
s_1 + s_2 + s_3 &= 1
\end{aligned}
$$

The solution of this system is $s_1 = .20$, $s_2 = .40$, and $s_3 = .40$.

Thus, the expected percentage in each category is:   poor - 20%;   satisfactory - 40%;   preferred - 40%.

Copyright © 2011 Pearson Education, Inc.  Publishing as Prentice Hall.

**51.** The transition matrix is:

$$P = \begin{bmatrix} .4 & .1 & .3 & .2 \\ .3 & .2 & .2 & .3 \\ .1 & .2 & .2 & .5 \\ .3 & .3 & .1 & .3 \end{bmatrix}$$

$S_0 P = [.3\ .3\ .4\ 0]P = [.25\ .17\ .23\ .35] = S_1$

$S_1 P = [.25\ .17\ .23\ .35]P = [.28\ .21\ .19\ .32] = S_2$

$S_2 P = [.28\ .21\ .19\ .32]P = [.29\ .20\ .20\ .31] = S_3$

$S_3 P = [.29\ .20\ .20\ .31]P = [.29\ .20\ .20\ .31] = S_4$

Thus, $S = [.29\ .20\ .20\ .31]$ is the steady-state matrix. The expected market share for the two Acme soaps is $.20 + .31 = .51$ or 51%.

**53.** To find the stationary solution, we solve the system

$$[s_1\ s_2\ s_3]\begin{bmatrix} .5 & .5 & 0 \\ .25 & .5 & .25 \\ 0 & .5 & .5 \end{bmatrix} = [s_1\ s_2\ s_3], \quad s_1 + s_2 + s_3 = 1,$$

which is equivalent to:

$$
\begin{aligned}
.5s_1 + .25s_2 &= s_1 \\
.5s_1 + .5s_2 + .5s_3 &= s_2 \\
.25s_2 + .5s_3 &= s_3 \\
s_1 + s_2 + s_3 &= 1
\end{aligned}
\qquad \text{or} \qquad
\begin{aligned}
-.5s_1 + .25s_2 &= 0 \\
.5s_1 - .5s_2 + .5s_3 &= 0 \\
.25s_2 - .5s_3 &= 0 \\
s_1 + s_2 + s_3 &= 1
\end{aligned}
$$

The solution of this system is $s_1 = .25, s_2 = .5, s_3 = .25$.

Thus, the stationary matrix is $S = [.25\ .5\ .25]$.

**55.** 

|  | Rapid transit | Auto |
|---|---|---|

(A)  Initial-state matrix $= [.25 \quad .75]$

(B)  Second-state matrix $= [.25\ .75]\begin{bmatrix} .8 & .2 \\ .3 & .7 \end{bmatrix} = [.425\ .575]$

Thus, 42.5% will be using the new system after one month.

Third-state matrix $= [.425\ .575]\begin{bmatrix} .8 & .2 \\ .3 & .7 \end{bmatrix} = [.5125\ .4875]$

Thus, 51.25% will be using the new system after two months.

(C)  To find the stationary solution, we solve the system

$$[s_1\ s_2]\begin{bmatrix} .8 & .2 \\ .3 & .7 \end{bmatrix} = [s_1\ s_2], \quad s_1 + s_2 = 1,$$

which is equivalent to:

Copyright © 2011 Pearson Education, Inc. Publishing as Prentice Hall.

$$.8s_1 + .3s_2 = s_1 \qquad -.2s_1 + .3s_2 = 0$$
$$.2s_1 + .7s_2 = s_2 \quad \text{or} \quad .2s_1 - .3s_2 = 0$$
$$s_1 + s_2 = 1 \qquad s_1 + s_2 = 1$$

The solution of this system of linear equations is $s_1 = .6$ and $s_2 = .4$. Thus, the stationary solution is $S = [.6 \ .4]$, which means that 60% of the commuters will use rapid transit and 40% will travel by automobile after the system has been in service for a long time.

**57.** (A) $S_1 = S_0 P = [.309 \ .691] \begin{bmatrix} .61 & .39 \\ .21 & .79 \end{bmatrix} = [.3336 \ .6664] \approx [.334 \ .666]$

$S_2 = S_0 P^2 = [.309 \ .691] \begin{bmatrix} .454 & .546 \\ .294 & .706 \end{bmatrix} = [.34344 \ .65656] \approx [.343 \ .657]$

$S_3 = S_0 P^3 = [.309 \ .691] \begin{bmatrix} .3916 & .6084 \\ .3276 & .6724 \end{bmatrix} = [.347376 \ .652624] \approx [.347 \ .653]$

(B)

| Year | Data% | Model% |
|------|-------|--------|
| 1970 | 30.9 | 30.9 |
| 1980 | 33.3 | 33.4 |
| 1990 | 34.4 | 34.3 |
| 2000 | 35.6 | 34.7 |

(C) $P^4 \approx \begin{bmatrix} .3666 & .6334 \\ .3410 & .6590 \end{bmatrix}$, $P^8 \approx \begin{bmatrix} .3504 & .6496 \\ .3498 & .6502 \end{bmatrix}$,

$P^{16} \approx \begin{bmatrix} .3500 & .6500 \\ .3500 & .6500 \end{bmatrix}$

In the long run, 35% of the population will live in the South region.

## EXERCISE 9-3

*Things to remember:*

1.   ABSORBING STATES AND TRANSIENT STATES

A state in a Markov chain is an ABSORBING STATE if once the state is entered, it is impossible to leave.

2.   ABSORBING STATES AND TRANSITION MATRICES

A state in a Markov chain is ABSORBING if and only if the row of the transition matrix corresponding to the state has a 1 on the main diagonal and zeros elsewhere.

3.   ABSORBING MARKOV CHAINS

A Markov chain is an ABSORBING CHAIN if

(A)   There is at least one absorbing state.

(B)   It is possible to go from each nonabsorbing state to at least one absorbing state in a finite number of steps.

4.   STANDARD FORMS FOR ABSORBING MARKOV CHAINS

A transition matrix for an absorbing Markov chain is a STANDARD FORM if the rows and columns are labeled so that all the absorbing states precede all the nonabsorbing states. (There may be more than one standard form.) Any standard form can always be partitioned into four submatrices:

Copyright © 2011 Pearson Education, Inc.  Publishing as Prentice Hall.

$$
\begin{array}{c}
\quad A \quad N \\
\begin{array}{c} A \\ N \end{array}
\left[ \begin{array}{c|c} I & 0 \\ \hline R & Q \end{array} \right]
\end{array}
\qquad
\left[ \begin{array}{c} A = \text{all absorbing states} \\ \hline N = \text{all nonabsorbing states} \end{array} \right]
$$

where $I$ is an identity matrix and $0$ is a zero matrix.

5.  LIMITING MATRICES FOR ABSORBING MARKOV CHAINS

If a standard form $P$ for an absorbing Markov chain is partitioned as

$$
P = \left[ \begin{array}{c|c} I & 0 \\ \hline R & Q \end{array} \right]
$$

then $P^k$ approaches a matrix $\overline{P}$ as $k$ increases, where

$$
\overline{P} = \left[ \begin{array}{c|c} I & 0 \\ \hline FR & 0 \end{array} \right]
$$

The matrix $F$ is given by $F = (I - Q)^{-1}$ and is called the FUNDAMENTAL MATRIX for $P$.

The identity matrix used to form the fundamental matrix $F$ must be the same size as the matrix $Q$.

6.  PROPERTIES OF THE LIMITING MATRIX $\overline{P}$

If $P$ is a standard form transition matrix for an absorbing Markov chain, $F$ is the fundamental matrix, and $\overline{P}$ is the limiting matrix, then

(A)  The entry in row $i$ and column $j$ of $\overline{P}$ is the long run probability of going from state $i$ to state $j$. For the nonabsorbing states, these probabilities are also the entries in the matrix $FR$ used to form $\overline{P}$.

(B)  The sum of the entries in each row of the fundamental matrix $F$ is the average number of trials it will take to go from each nonabsorbing state to some absorbing state.

[Note that the rows of both $F$ and $FR$ correspond to the nonabsorbing states in the order given in the standard form $P$.]

---

1.  By 2, states $B$ and $C$ are absorbing states.

3.  By 2, there are no absorbing states.

5.  By 2, states $A$ and $D$ are absorbing states.

7.  $B$ is an absorbing state; the diagram represents an absorbing Markov chain since it is possible to go from states $A$ and $C$ to state $B$ in a finite number of steps.

9.  $C$ is an absorbing state; the diagram does not represent an absorbing Markov chain since it is not possible to go from either states $A$ or $D$ to state $C$.

11.  $P = \begin{bmatrix} 0 & 1 \\ 1 & 0 \end{bmatrix}$  No; $P$ has no absorbing states.

Copyright © 2011 Pearson Education, Inc. Publishing as Prentice Hall.

13.  $P = \begin{array}{c} \\ A \\ B \end{array}\begin{array}{c} A\ \ B \\ \left[\begin{array}{cc} .3 & .7 \\ 0 & 1 \end{array}\right] \end{array}$    Yes; $B$ is an absorbing state and it is possible for state $A$ to go to state $B$ in one step.

15.  $P = \begin{array}{c} \\ A \\ B \\ C \end{array}\begin{array}{c} A\ \ B\ \ C \\ \left[\begin{array}{ccc} 1 & 0 & 0 \\ 0 & 1 & 0 \\ 0 & 0 & 1 \end{array}\right] \end{array}$    Yes; $A$, $B$ and $C$ are each absorbing states, there are no nonabsorbing states.

17.  $P = \begin{array}{c} \\ A \\ B \\ C \end{array}\begin{array}{c} A\ \ B\ \ C \\ \left[\begin{array}{ccc} .9 & .1 & 0 \\ .9 & .1 & 0 \\ 0 & 0 & 1 \end{array}\right] \end{array}$    No; $C$ is an absorbing state but it is impossible to go from either state $A$ or state $B$ to the absorbing state $C$.

19.  $P = \begin{array}{c} \\ A \\ B \\ C \end{array}\begin{array}{c} A\ \ B\ \ C \\ \left[\begin{array}{ccc} .9 & 0 & .1 \\ 0 & 1 & 0 \\ 0 & .2 & .8 \end{array}\right] \end{array}$    Yes; $B$ is an absorbing state and state $C$ can go to state $B$ in one step, state $A$ can go to state $B$ in two steps.

21.  The transition diagram is represented by the matrix:

$\begin{array}{c} \\ A \\ B \\ C \end{array}\begin{array}{c} A\ \ B\ \ C \\ \left[\begin{array}{ccc} .2 & .5 & .3 \\ 0 & 1 & 0 \\ .5 & .1 & .4 \end{array}\right] \end{array}$

A standard form for this matrix is:

$\begin{array}{c} \\ B \\ A \\ C \end{array}\begin{array}{c} B\ \ A\ \ C \\ \left[\begin{array}{ccc} 1 & 0 & 0 \\ .5 & .2 & .3 \\ .1 & .5 & .4 \end{array}\right] \end{array}$

23.  The transition diagram is represented by the matrix

$\begin{array}{c} \\ A \\ B \\ C \\ D \end{array}\begin{array}{c} A\ \ B\ \ C\ \ D \\ \left[\begin{array}{cccc} .3 & .4 & .2 & .1 \\ 0 & 1 & 0 & 0 \\ 0 & .4 & .3 & .3 \\ 0 & 0 & 0 & 1 \end{array}\right] \end{array}$

A standard form for this matrix

$\begin{array}{c} \\ B \\ D \\ A \\ C \end{array}\begin{array}{c} B\ \ D\ \ A\ \ C \\ \left[\begin{array}{cccc} 1 & 0 & 0 & 0 \\ 0 & 1 & 0 & 0 \\ .4 & .1 & .3 & .2 \\ .4 & .3 & 0 & .3 \end{array}\right] \end{array}$

25.  A standard form for

$P = \begin{array}{c} \\ A \\ B \\ C \end{array}\begin{array}{c} A\ \ B\ \ C \\ \left[\begin{array}{ccc} .2 & .3 & .5 \\ 1 & 0 & 0 \\ 0 & 0 & 1 \end{array}\right] \end{array}$

is:

27.  A standard form for

$P = \begin{array}{c} \\ A \\ B \\ C \\ D \end{array}\begin{array}{c} A\ \ B\ \ C\ \ D \\ \left[\begin{array}{cccc} .1 & .2 & .3 & .4 \\ 0 & 1 & 0 & 0 \\ .5 & .2 & .2 & .1 \\ 0 & 0 & 0 & 1 \end{array}\right] \end{array}$

is:

Copyright © 2011 Pearson Education, Inc. Publishing as Prentice Hall.

$$\begin{array}{c} \\ C \\ A \\ B \end{array} \begin{array}{ccc} C & A & B \\ \begin{bmatrix} 1 & 0 & 0 \\ .5 & .2 & .3 \\ 0 & 1 & 0 \end{bmatrix} \end{array} \qquad \begin{array}{c} \\ B \\ D \\ A \\ C \end{array} \begin{array}{cccc} B & D & A & C \\ \begin{bmatrix} 1 & 0 & 0 & 0 \\ 0 & 1 & 0 & 0 \\ .2 & .4 & .1 & .3 \\ .2 & .1 & .5 & .2 \end{bmatrix} \end{array}$$

**29.** For

$$P = \begin{array}{c} \\ A \\ B \\ C \end{array} \begin{array}{ccc} A & B & C \\ \begin{bmatrix} 1 & 0 & 0 \\ 0 & 1 & 0 \\ .1 & .4 & .5 \end{bmatrix} \end{array}$$

we have $R = [.1 \ \ .4]$ and $Q = [.5]$.

The limiting matrix $\overline{P}$ has the form

$$\overline{P} = \left[ \begin{array}{cc|c} 1 & 0 & 0 \\ 0 & 1 & 0 \\ \hline F & R & 0 \end{array} \right]$$

where $F = (I - Q)^{-1} = ([1] - [.5])^{-1} = [.5]^{-1} = [2]$ and $FR = [2][.1 \ \ .4] = [.2 \ \ .8]$.
Thus,

$$\overline{P} = \begin{array}{c} \\ A \\ B \\ C \end{array} \begin{array}{ccc} A & B & C \\ \begin{bmatrix} 1 & 0 & 0 \\ 0 & 1 & 0 \\ .2 & .8 & 0 \end{bmatrix} \end{array}$$

Let $P(i \text{ to } j)$ denote the probability of going from state $i$ to state $j$. Then $P(C \text{ to } A) = .2$, $P(C \text{ to } B) = .8$
Since $F = [2]$, it will take an average of 2 trials to go from $C$ to either $A$ or $B$.

**31.** For

$$P = \begin{array}{c} \\ A \\ B \\ C \end{array} \begin{array}{ccc} A & B & C \\ \begin{bmatrix} 1 & 0 & 0 \\ .2 & .6 & .2 \\ .4 & .2 & .4 \end{bmatrix} \end{array}$$

we have $R = \begin{bmatrix} .2 \\ .4 \end{bmatrix}$ and $Q = \begin{bmatrix} .6 & .2 \\ .2 & .4 \end{bmatrix}$

The limiting matrix $\overline{P}$ has the form

$$\overline{P} = \left[ \begin{array}{c|c} 1 & 0 \ | \ 0 \\ \hline F & R \ | \ 0 \end{array} \right] \text{ where } F = (I - Q)^{-1} = \left( \begin{bmatrix} 1 & 0 \\ 0 & 1 \end{bmatrix} - \begin{bmatrix} .6 & .2 \\ .2 & .4 \end{bmatrix} \right)^{-1} = \begin{bmatrix} .4 & -.2 \\ -.2 & .6 \end{bmatrix}^{-1} = \begin{bmatrix} \frac{2}{5} & -\frac{1}{5} \\ -\frac{1}{5} & \frac{3}{5} \end{bmatrix}^{-1}$$

We use row operations to find the inverse:

$$\begin{bmatrix} \frac{2}{5} & -\frac{1}{5} & 1 & 0 \\ -\frac{1}{5} & \frac{3}{5} & 0 & 1 \end{bmatrix} \sim \begin{bmatrix} 1 & -\frac{1}{2} & \frac{5}{2} & 0 \\ -\frac{1}{5} & \frac{3}{5} & 0 & 1 \end{bmatrix} \sim \begin{bmatrix} 1 & -\frac{1}{2} & \frac{5}{2} & 0 \\ 0 & \frac{1}{2} & \frac{1}{2} & 1 \end{bmatrix} \sim \begin{bmatrix} 1 & -\frac{1}{2} & \frac{5}{2} & 0 \\ 0 & 1 & 1 & 2 \end{bmatrix}$$

$$\left( \frac{5}{2} \right) R_1 \rightarrow R_1 \qquad \left( \frac{1}{5} \right) R_1 + R_2 \rightarrow R_2 \qquad 2R_2 \rightarrow R_2 \qquad \left( \frac{1}{2} \right) R_2 + R_1 \rightarrow R_1$$

$$\sim \begin{bmatrix} 1 & 0 & 3 & 1 \\ 0 & 1 & 1 & 2 \end{bmatrix}$$

Thus, $F = \begin{bmatrix} 3 & 1 \\ 1 & 2 \end{bmatrix}$ and $FR = \begin{bmatrix} 3 & 1 \\ 1 & 2 \end{bmatrix} \begin{bmatrix} .2 \\ .4 \end{bmatrix} = \begin{bmatrix} 1 \\ 1 \end{bmatrix}$

Copyright © 2011 Pearson Education, Inc. Publishing as Prentice Hall.

Now
$$
\begin{array}{c} \\ A \\ \overline{P} = B \\ C \end{array}
\begin{array}{cc} A\ B\ C \\ \begin{bmatrix} 1 & 0 & 0 \\ 1 & 0 & 0 \\ 1 & 0 & 0 \end{bmatrix} \end{array}
$$

$P(B \text{ to } A) = 1$, $P(C \text{ to } A) = 1$

It will take an average of 4 trials to go from $B$ to $A$; it will take an average of 3 trials to go from $C$ to $A$.

**33.** For

$$
\begin{array}{c} \\ A \\ P = \begin{array}{c} B \\ C \\ D \end{array} \end{array}
\begin{array}{cccc} A\ \ B\ \ C\ \ D \\ \begin{bmatrix} 1 & 0 & 0 & 0 \\ 0 & 1 & 0 & 0 \\ .1 & .2 & .6 & .1 \\ .2 & .2 & .3 & .3 \end{bmatrix} \end{array}
$$

we have $R = \begin{bmatrix} .1 & .2 \\ .2 & .2 \end{bmatrix}$ and $Q = \begin{bmatrix} .6 & .1 \\ .3 & .3 \end{bmatrix}$

The limiting matrix $\overline{P}$ has the form

$$
\overline{P} = \left[ \begin{array}{c|c} I & \begin{matrix} 0 & 0 \\ 0 & 0 \end{matrix} \\ \hline FR & 0 \end{array} \right]
$$

where $F = (I - Q)^{-1} = \left( \begin{bmatrix} 1 & 0 \\ 0 & 1 \end{bmatrix} - \begin{bmatrix} .6 & .1 \\ .3 & .3 \end{bmatrix} \right)^{-1} = \begin{bmatrix} .4 & -.1 \\ -.3 & .7 \end{bmatrix}^{-1} = \begin{bmatrix} \frac{2}{5} & -\frac{1}{10} \\ -\frac{3}{10} & \frac{7}{10} \end{bmatrix}^{-1}$

We use row operations to find the inverse:

$$
\begin{bmatrix} \frac{2}{5} & -\frac{1}{10} & 1 & 0 \\ -\frac{3}{10} & \frac{7}{10} & 0 & 1 \end{bmatrix} \sim \begin{bmatrix} 1 & -\frac{1}{4} & \frac{5}{2} & 0 \\ -\frac{3}{10} & \frac{7}{10} & 0 & 1 \end{bmatrix} \sim \begin{bmatrix} 1 & -\frac{1}{4} & \frac{5}{2} & 0 \\ 0 & \frac{5}{8} & \frac{3}{4} & 1 \end{bmatrix} \sim \begin{bmatrix} 1 & -\frac{1}{4} & \frac{5}{2} & 0 \\ 0 & 1 & \frac{6}{5} & \frac{8}{5} \end{bmatrix}
$$

$$
\left( \frac{5}{2} \right) R_1 \to R_1 \qquad \left( \frac{3}{10} \right) R_1 + R_2 \to R_2 \qquad \left( \frac{8}{5} \right) R_2 \to R_2 \qquad \left( \frac{1}{4} \right) R_2 + R_1 \to R_1
$$

$$
\sim \begin{bmatrix} 1 & 0 & \frac{14}{5} & \frac{2}{5} \\ 0 & 1 & \frac{6}{5} & \frac{8}{5} \end{bmatrix}
$$

Thus, $F = \begin{bmatrix} \frac{14}{5} & \frac{2}{5} \\ \frac{6}{5} & \frac{8}{5} \end{bmatrix} = \begin{bmatrix} 2.8 & .4 \\ 1.2 & 1.6 \end{bmatrix}$,

$$
FR = \begin{bmatrix} 2.8 & .4 \\ 1.2 & 1.6 \end{bmatrix} \begin{bmatrix} .1 & .2 \\ .2 & .2 \end{bmatrix} = \begin{bmatrix} .36 & .64 \\ .44 & .56 \end{bmatrix}
$$

and

$$
\begin{array}{c} \\ A \\ \overline{P} = \begin{array}{c} B \\ C \\ D \end{array} \end{array}
\begin{array}{cccc} A\ \ \ B\ \ \ C\ \ D \\ \begin{bmatrix} 1 & 0 & 0 & 0 \\ 0 & 1 & 0 & 0 \\ .36 & .64 & 0 & 0 \\ .44 & .56 & 0 & 0 \end{bmatrix} \end{array}
$$

$P(C \text{ to } A) = .36$, $P(C \text{ to } B) = .64$,
$P(D \text{ to } A) = .44$, $P(D \text{ to } B) = .56$

It will take an average of 3.2 trials to go from $C$ to either $A$ or $B$; it will take an average of 2.8 trials to go from $D$ to either $A$ or $B$.

Copyright © 2011 Pearson Education, Inc.  Publishing as Prentice Hall.

**35.** (A) $S_0 \overline{P} = [0\ 0\ 1]\begin{bmatrix} 1 & 0 & 0 \\ 0 & 1 & 0 \\ .2 & .8 & 0 \end{bmatrix} = [.2\ .8\ 0]$

(B) $S_0 \overline{P} = [.2\ .5\ .3]\begin{bmatrix} 1 & 0 & 0 \\ 0 & 1 & 0 \\ .2 & .8 & 0 \end{bmatrix} = [.26\ .74\ 0]$

**37.** (A) $S_0 \overline{P} = [0\ 0\ 1]\begin{bmatrix} 1 & 0 & 0 \\ 1 & 0 & 0 \\ 1 & 0 & 0 \end{bmatrix} = [1\ 0\ 0]$

(B) $S_0 \overline{P} = [.2\ .5\ .3]\begin{bmatrix} 1 & 0 & 0 \\ 1 & 0 & 0 \\ 1 & 0 & 0 \end{bmatrix} = [1\ 0\ 0]$

**39.** (A) $S_0 \overline{P} = [0\ 0\ 0\ 1]\begin{bmatrix} 1 & 0 & 0 & 0 \\ 0 & 1 & 0 & 0 \\ .36 & .64 & 0 & 0 \\ .44 & .56 & 0 & 0 \end{bmatrix} = [.44\ .56\ 0\ 0]$

(B) $S_0 \overline{P} = [0\ 0\ 1\ 0]\begin{bmatrix} 1 & 0 & 0 & 0 \\ 0 & 1 & 0 & 0 \\ .36 & .64 & 0 & 0 \\ .44 & .56 & 0 & 0 \end{bmatrix} = [.36\ .64\ 0\ 0]$

(C) $S_0 \overline{P} = [0\ 0\ .4\ .6]\begin{bmatrix} 1 & 0 & 0 & 0 \\ 0 & 1 & 0 & 0 \\ .36 & .64 & 0 & 0 \\ .44 & .56 & 0 & 0 \end{bmatrix} = [.408\ .592\ 0\ 0]$

(D) $S_0 \overline{P} = [.1\ .2\ .3\ .4]\begin{bmatrix} 1 & 0 & 0 & 0 \\ 0 & 1 & 0 & 0 \\ .36 & .64 & 0 & 0 \\ .44 & .56 & 0 & 0 \end{bmatrix} = [.384\ .616\ 0\ 0]$

**41.** False. For the transition matrix:

$$\begin{array}{c} \\ A \\ B \\ C \end{array} \begin{array}{c} A\ \ B\ \ C \\ \begin{bmatrix} 1 & 0 & 0 \\ 0 & 0 & 1 \\ 0 & 1 & 0 \end{bmatrix} \end{array}$$

$A$ is an absorbing state; $B$ and $C$ are nonabsorbing states and it is impossible to go from either state $B$ or state $C$ to state $A$.

Copyright © 2011 Pearson Education, Inc.  Publishing as Prentice Hall.

**43.**   False.  The transition matrix in Problem 41 is a counterexample.

**45.**   True.  If every state is absorbing, then there are no nonabsorbing states so 3(<u>B</u>) is vacuously true.

**47.**   False.  A Markov chain whose transition matrix is $I$, the identity matrix, is absorbing but not regular.

**49.**   By Theorem 2, $P$ has a limiting matrix:

$$P^4 \approx \begin{bmatrix} 1 & 0 & 0 & 0 \\ 0 & 1 & 0 & 0 \\ .6364 & .362 & 0 & 0 \\ .7364 & .262 & 0 & 0 \end{bmatrix}, \quad P^8 \approx \begin{bmatrix} 1 & 0 & 0 & 0 \\ 0 & 1 & 0 & 0 \\ .6375 & .3625 & 0 & 0 \\ .7375 & .2625 & 0 & 0 \end{bmatrix}$$

$$P^{16} \approx \begin{bmatrix} 1 & 0 & 0 & 0 \\ 0 & 1 & 0 & 0 \\ .6375 & .3625 & 0 & 0 \\ .7375 & .2625 & 0 & 0 \end{bmatrix}; \quad \overline{P} = \begin{bmatrix} 1 & 0 & 0 & 0 \\ 0 & 1 & 0 & 0 \\ .6375 & .3625 & 0 & 0 \\ .7375 & .2625 & 0 & 0 \end{bmatrix}$$

**51.**   By Theorem 2, $P$ has a limiting matrix:

$$P^4 \approx \begin{bmatrix} 1 & 0 & 0 & 0 & 0 \\ 0 & 1 & 0 & 0 & 0 \\ .0724 & .8368 & .0625 & .011 & .0173 \\ .174 & .7792 & 0 & .0279 & .0189 \\ .4312 & .5472 & 0 & .0126 & .009 \end{bmatrix},$$

$$P^{16} \approx \begin{bmatrix} 1 & 0 & 0 & 0 & 0 \\ 0 & 1 & 0 & 0 & 0 \\ .0875 & .9125 & 0 & 0 & 0 \\ .1875 & .8125 & 0 & 0 & 0 \\ .4375 & .5625 & 0 & 0 & 0 \end{bmatrix}, \quad P^{32} \approx \begin{bmatrix} 1 & 0 & 0 & 0 & 0 \\ 0 & 1 & 0 & 0 & 0 \\ .0875 & .9125 & 0 & 0 & 0 \\ .1875 & .8125 & 0 & 0 & 0 \\ .4375 & .5625 & 0 & 0 & 0 \end{bmatrix};$$

$$\overline{P} = \begin{bmatrix} 1 & 0 & 0 & 0 & 0 \\ 0 & 1 & 0 & 0 & 0 \\ .0875 & .9125 & 0 & 0 & 0 \\ .1875 & .8125 & 0 & 0 & 0 \\ .4375 & .5625 & 0 & 0 & 0 \end{bmatrix}$$

**53.**   <u>*Step 1*.</u>       Transition diagram:

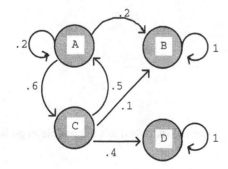

Copyright © 2011 Pearson Education, Inc.  Publishing as Prentice Hall.

Standard form:

$$M = \begin{array}{c} \\ B \\ D \\ A \\ C \end{array}\begin{array}{c} \begin{array}{cccc} B & D & A & C \end{array} \\ \left[\begin{array}{cccc} 1 & 0 & 0 & 0 \\ 0 & 1 & 0 & 0 \\ .2 & 0 & .2 & .6 \\ .1 & .4 & .5 & 0 \end{array}\right] \end{array}$$

*Step 2.*   Limiting matrix:

For $M$, we have $R = \begin{bmatrix} .2 & 0 \\ .1 & .4 \end{bmatrix}$ and $Q = \begin{bmatrix} .2 & .6 \\ .5 & 0 \end{bmatrix}$

The limiting matrix $\overline{M}$ has the form:

$$\overline{M} = \left[\begin{array}{c|c} I & 0 \\ \hline FR & 0 \end{array}\right]$$

where $F = (I - Q)^{-1} = \left( \begin{bmatrix} 1 & 0 \\ 0 & 1 \end{bmatrix} - \begin{bmatrix} .2 & .6 \\ .5 & 0 \end{bmatrix} \right)^{-1} = \begin{bmatrix} .8 & -.6 \\ -.5 & 1 \end{bmatrix}^{-1} = \begin{bmatrix} \frac{4}{5} & -\frac{3}{5} \\ -\frac{1}{2} & 1 \end{bmatrix}^{-1}$

We use row operations to find the inverse:

$$\left[\begin{array}{cc|cc} \frac{4}{5} & -\frac{3}{5} & 1 & 0 \\ -\frac{1}{2} & 1 & 0 & 1 \end{array}\right] \sim \left[\begin{array}{cc|cc} 1 & -\frac{3}{4} & \frac{5}{4} & 0 \\ -\frac{1}{2} & 1 & 0 & 1 \end{array}\right] \sim \left[\begin{array}{cc|cc} 1 & -\frac{3}{4} & \frac{5}{4} & 0 \\ 0 & \frac{5}{8} & \frac{5}{8} & 1 \end{array}\right]$$

$\left(\dfrac{5}{4}\right)R_1 \to R_1 \qquad \left(\dfrac{1}{2}\right)R_1 + R_2 \to R_2 \qquad \left(\dfrac{8}{5}\right)R_2 \to R_2$

$$\sim \left[\begin{array}{cc|cc} 1 & -\frac{3}{4} & \frac{5}{4} & 0 \\ 0 & 1 & 1 & \frac{8}{5} \end{array}\right] \sim \left[\begin{array}{cc|cc} 1 & 0 & 2 & \frac{6}{5} \\ 0 & 1 & 1 & \frac{8}{5} \end{array}\right]$$

$\left(\dfrac{3}{4}\right)R_2 + R_1 \to R_1$

Thus, $F = \begin{bmatrix} 2 & 1.2 \\ 1 & 1.6 \end{bmatrix}$ and $FR = \begin{bmatrix} 2 & 1.2 \\ 1 & 1.6 \end{bmatrix}\begin{bmatrix} .2 & 0 \\ .1 & .4 \end{bmatrix} = \begin{bmatrix} .52 & .48 \\ .36 & .64 \end{bmatrix}$

Therefore, $\overline{M} = \begin{array}{c} \\ B \\ D \\ A \\ C \end{array}\begin{array}{c} \begin{array}{cccc} B & D & A & C \end{array} \\ \left[\begin{array}{cccc} 1 & 0 & 0 & 0 \\ 0 & 1 & 0 & 0 \\ .52 & .48 & 0 & 0 \\ .36 & .64 & 0 & 0 \end{array}\right] \end{array}$

Copyright © 2011 Pearson Education, Inc. Publishing as Prentice Hall.

*Step 3.*        Transition diagram for $\overline{M}$ :

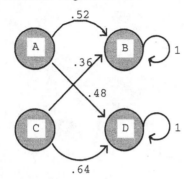

Limiting matrix for $P$:

$$\overline{P} = \begin{array}{c} \\ A \\ B \\ C \\ D \end{array} \begin{array}{cccc} A & B & C & D \\ \begin{bmatrix} 0 & .52 & 0 & .48 \\ 0 & 1 & 0 & 0 \\ 0 & .36 & 0 & .64 \\ 0 & 0 & 0 & 1 \end{bmatrix} \end{array}$$

**55.**  $P^4 \approx \begin{bmatrix} .1276 & .426 & .0768 & .3696 \\ 0 & 1 & 0 & 0 \\ .064 & .29 & .102 & .544 \\ 0 & 0 & 0 & 1 \end{bmatrix}$,   $P^8 \approx \begin{bmatrix} .0212 & .5026 & .0176 & .4585 \\ 0 & 1 & 0 & 0 \\ .0147 & .3468 & .0153 & .6231 \\ 0 & 0 & 0 & 0 \end{bmatrix}$

$P^{32} \approx \begin{bmatrix} 0 & .52 & 0 & .48 \\ 0 & 1 & 0 & 0 \\ 0 & .36 & 0 & .64 \\ 0 & 0 & 0 & 1 \end{bmatrix}$

**57.**   Let $S = [x \quad 1-x \quad 0]$, $0 \le x \le 1$.  Then

$$SP = [x \quad 1-x \quad 0] \begin{bmatrix} 1 & 0 & 0 \\ 0 & 1 & 0 \\ .1 & .5 & .4 \end{bmatrix} = [x \quad 1-x \quad 0]$$

Thus, $S$ is a stationary matrix for $P$.

A stationary matrix for an absorbing Markov chain with two absorbing states and one nonabsorbing state will have one of the forms

   $[x \quad 1-x \quad 0]$, $[x \quad 0 \quad 1-x]$, $[0 \quad x \quad 1-x]$

**59.**  (A)    For $P^2$, $w_2 = .370$;   for $P^4$, $w_4 = .297$;   for $P^8$, $w_8 = .227$;

   for $P^{16}$, $w_{16} = .132$;   for $P^{32}$, $w_{32} = .045$

  (B)    For large $k$, the entries of $Q^k$ are close to 0.

Copyright © 2011 Pearson Education, Inc.  Publishing as Prentice Hall.

**61.** A transition matrix for this problem is:

$$P = \begin{array}{c} \\ F \\ B \\ G \\ A \end{array} \begin{array}{cccc} F & G & A & B \\ \begin{bmatrix} 1 & 0 & 0 & 0 \\ .1 & .8 & .1 & 0 \\ .1 & .4 & .4 & .1 \\ 0 & 0 & 0 & 1 \end{bmatrix} \end{array}$$

A standard form for this matrix is:

$$M = \begin{array}{c} \\ F \\ B \\ G \\ A \end{array} \begin{array}{cccc} F & B & G & A \\ \begin{bmatrix} 1 & 0 & 0 & 0 \\ 0 & 1 & 0 & 0 \\ .1 & 0 & .8 & .1 \\ .1 & .1 & .4 & .4 \end{bmatrix} \end{array}$$

For this matrix, we have:

$$R = \begin{bmatrix} .1 & 0 \\ .1 & .1 \end{bmatrix} \text{ and } Q = \begin{bmatrix} .8 & .1 \\ .4 & .4 \end{bmatrix}$$

The limiting matrix for $M$ has the form:

$$\overline{M} = \left[ \begin{array}{c|c} I & 0 \\ \hline FR & 0 \end{array} \right]$$

where $F = (I - Q)^{-1} = \left( \begin{bmatrix} 1 & 0 \\ 0 & 1 \end{bmatrix} - \begin{bmatrix} .8 & .1 \\ .4 & .4 \end{bmatrix} \right)^{-1} = \begin{bmatrix} .2 & -.1 \\ -.4 & .6 \end{bmatrix}^{-1} = \begin{bmatrix} \frac{1}{5} & -\frac{1}{10} \\ -\frac{2}{5} & \frac{3}{5} \end{bmatrix}^{-1}$

We use row operations to find the inverse:

$$\left[ \begin{array}{cc|cc} \frac{1}{5} & -\frac{1}{10} & 1 & 0 \\ -\frac{2}{5} & \frac{3}{5} & 0 & 1 \end{array} \right] \sim \left[ \begin{array}{cc|cc} 1 & -\frac{1}{2} & 5 & 0 \\ -\frac{2}{5} & \frac{3}{5} & 0 & 1 \end{array} \right] \sim \left[ \begin{array}{cc|cc} 1 & -\frac{1}{2} & 5 & 0 \\ 0 & \frac{2}{5} & 2 & 1 \end{array} \right] \sim \left[ \begin{array}{cc|cc} 1 & -\frac{1}{2} & 5 & 0 \\ 0 & 1 & 5 & \frac{5}{2} \end{array} \right]$$

$$5R_1 \to R_1 \qquad \left(\frac{2}{5}\right)R_1 + R_2 \to R_2 \qquad \left(\frac{5}{2}\right)R_2 \to R_2 \qquad \left(\frac{1}{2}\right)R_2 + R_1 \to R_1$$

$$\sim \left[ \begin{array}{cc|cc} 1 & 0 & \frac{15}{2} & \frac{5}{4} \\ 0 & 1 & 5 & \frac{5}{2} \end{array} \right]$$

Thus, $F = \begin{bmatrix} 7.5 & 1.25 \\ 5 & 2.5 \end{bmatrix}$ and $FR = \begin{bmatrix} .875 & .125 \\ .75 & .25 \end{bmatrix}$

Therefore,

$$\overline{M} = \begin{array}{c} \\ F \\ B \\ G \\ A \end{array} \begin{array}{cccc} F & B & G & A \\ \begin{bmatrix} 1 & 0 & 0 & 0 \\ 0 & 1 & 0 & 0 \\ .875 & .125 & 0 & 0 \\ .75 & .25 & 0 & 0 \end{bmatrix} \end{array}$$

(A)    In the long run, 75% of the accounts in arrears will pay in full.

(B)    In the long run, 12.5% of the accounts in good standing will become bad debts.

(C)    The average number of months that an account in arrears will either be paid in full or classified as a bad debt is:   $5 + 2.5 = 7.5$ months

Copyright © 2011 Pearson Education, Inc. Publishing as Prentice Hall.

**63.** A transition matrix in standard form for this problem is:

$$P = \begin{array}{c} \\ A \\ B \\ C \\ N \end{array} \begin{array}{cccc} A & B & C & N \\ \begin{bmatrix} 1 & 0 & 0 & 0 \\ 0 & 1 & 0 & 0 \\ 0 & 0 & 1 & 0 \\ .6 & .3 & .11 & .8 \end{bmatrix} \end{array}$$

For this matrix, we have $R = [.6 \ .3 \ .11]$ and $Q = [.8]$.

The limiting matrix for $P$ has the form:

$$\overline{P} = \left[ \begin{array}{c|c} I & 0 \\ \hline FR & 0 \end{array} \right]$$

where $F = (I - Q)^{-1} = ([1] - [.8])^{-1} = [.2]^{-1} = 5$

Now, $FR = [5][.6 \ .3 \ .11] = [.3 \ .15 \ .55]$

and

$$\overline{P} = \begin{array}{c} \\ A \\ B \\ C \\ N \end{array} \begin{array}{cccc} A & B & C & N \\ \begin{bmatrix} 1 & 0 & 0 & 0 \\ 0 & 1 & 0 & 0 \\ 0 & 0 & 1 & 0 \\ .3 & .15 & .55 & 0 \end{bmatrix} \end{array}$$

(A)    In the long run, the market share of each company is:

Company $A$: 30%;  Company $B$: 15%; and Company $C$: 55%.

(B)    On the average, it will take 5 years for a department to decide to use a calculator from one of these companies in their courses.

**65.** Let $I$ denote ICU, $C$ denote CCW, $D$ denote "died", and $R$ denote "released". A transition matrix in standard form for this problem is:

$$P = \begin{array}{c} \\ D \\ R \\ I \\ C \end{array} \begin{array}{cccc} D & R & I & C \\ \begin{bmatrix} 1 & 0 & 0 & 0 \\ 0 & 1 & 0 & 0 \\ .02 & 0 & .46 & .52 \\ .01 & .22 & .04 & .73 \end{bmatrix} \end{array}$$

For this matrix, we have

$$R = \begin{bmatrix} .02 & 0 \\ .01 & .22 \end{bmatrix} \text{ and } Q = \begin{bmatrix} .46 & .52 \\ .04 & .73 \end{bmatrix}$$

The limiting matrix for $P$ has the form:

$$\overline{P} = \left[ \begin{array}{c|c} I & 0 \\ \hline FR & 0 \end{array} \right]$$

where $F = (I - Q)^{-1} = \left( \begin{bmatrix} 1 & 0 \\ 0 & 1 \end{bmatrix} - \begin{bmatrix} .46 & .52 \\ .04 & .73 \end{bmatrix} \right)^{-1} = \begin{bmatrix} .54 & -.52 \\ -.04 & .27 \end{bmatrix}^{-1} = \begin{bmatrix} 2.16 & 4.16 \\ .32 & 4.32 \end{bmatrix}$

Copyright © 2011 Pearson Education, Inc. Publishing as Prentice Hall.

Now, $FR = \begin{bmatrix} 2.16 & 4.16 \\ .32 & 4.32 \end{bmatrix} \begin{bmatrix} .02 & 0 \\ .01 & .22 \end{bmatrix} = \begin{bmatrix} .0848 & .9152 \\ .0496 & .9504 \end{bmatrix}$

and

$$\overline{P} = \begin{array}{c} \\ D \\ R \\ I \\ C \end{array} \begin{array}{cccc} D & R & I & C \\ \begin{bmatrix} 1 & 0 & 0 & 0 \\ 0 & 1 & 0 & 0 \\ .0848 & .9152 & 0 & 0 \\ .0496 & .9504 & 0 & 0 \end{bmatrix} \end{array}$$

(A)   In the long run, 91.52% of the patients are released from the hospital.

(B)   In the long run, 4.96% of the patients in the CCW die without being released from the hospital.

(C)   The average number of days a patient in the ICU will stay in the hospital is:

$$2.16 + 4.16 = 6.32 \text{ days}$$

**67.**   A transition matrix in standard form for this problem is:

$$P = \begin{array}{c} \\ L \\ R \\ F \\ B \end{array} \begin{array}{cccc} L & R & F & B \\ \begin{bmatrix} 1 & 0 & 0 & 0 \\ 0 & 1 & 0 & 0 \\ \frac{1}{4} & \frac{1}{4} & 0 & \frac{1}{2} \\ \frac{2}{5} & \frac{1}{5} & \frac{2}{5} & 0 \end{bmatrix} \end{array}$$

For this matrix we have:

$$R = \begin{bmatrix} \frac{1}{4} & \frac{1}{4} \\ \frac{2}{5} & \frac{1}{5} \end{bmatrix} \text{ and } Q = \begin{bmatrix} 0 & \frac{1}{2} \\ \frac{2}{5} & 0 \end{bmatrix}$$

The limiting matrix for $P$ has the form:

$$\overline{P} = \left[ \begin{array}{c|c} I & 0 \\ \hline FR & 0 \end{array} \right]$$

where $F = (I - Q)^{-1} = \left( \begin{bmatrix} 1 & 0 \\ 0 & 1 \end{bmatrix} - \begin{bmatrix} 0 & \frac{1}{2} \\ \frac{2}{5} & 0 \end{bmatrix} \right)^{-1} = \begin{bmatrix} 1 & -\frac{1}{2} \\ -\frac{2}{5} & 1 \end{bmatrix}^{-1}$

We use row operations to find the inverse:

$$\left[ \begin{array}{cc|cc} 1 & -\frac{1}{2} & 1 & 0 \\ -\frac{2}{5} & 1 & 0 & 1 \end{array} \right] \sim \left[ \begin{array}{cc|cc} 1 & -\frac{1}{2} & 1 & 0 \\ 0 & \frac{4}{5} & \frac{2}{5} & 1 \end{array} \right] \sim \left[ \begin{array}{cc|cc} 1 & -\frac{1}{2} & 1 & 0 \\ 0 & 1 & \frac{1}{2} & \frac{5}{4} \end{array} \right] \sim \left[ \begin{array}{cc|cc} 1 & 0 & \frac{5}{4} & \frac{5}{8} \\ 0 & 1 & \frac{1}{2} & \frac{5}{4} \end{array} \right]$$

$$\left(\frac{2}{5}\right) R_1 + R_2 \rightarrow R_2 \quad \left(\frac{5}{4}\right) R_2 \rightarrow R_2 \quad \left(\frac{1}{2}\right) R_2 + R_1 \rightarrow R_1$$

Thus, $F = \begin{bmatrix} \frac{5}{4} & \frac{5}{8} \\ \frac{1}{2} & \frac{5}{4} \end{bmatrix}$ and $FR = \begin{bmatrix} \frac{5}{4} & \frac{5}{8} \\ \frac{1}{2} & \frac{5}{4} \end{bmatrix} \begin{bmatrix} \frac{1}{4} & \frac{1}{4} \\ \frac{2}{5} & \frac{1}{5} \end{bmatrix} = \begin{bmatrix} \frac{9}{16} & \frac{7}{16} \\ \frac{5}{8} & \frac{3}{8} \end{bmatrix}$.

Copyright © 2011 Pearson Education, Inc.  Publishing as Prentice Hall.

Now,

$$\overline{P} = \begin{array}{c} \\ L \\ R \\ F \\ B \end{array} \begin{array}{cccc} L & R & F & B \\ \left[\begin{array}{cccc} 1 & 0 & 0 & 0 \\ 0 & 1 & 0 & 0 \\ \frac{9}{16} & \frac{7}{16} & 0 & 0 \\ \frac{5}{8} & \frac{3}{8} & 0 & 0 \end{array}\right] \end{array}$$

(A)   The long run probability that a rat placed in room $B$ will end up in room $R$ is $\dfrac{3}{8} = .375$.

(B)   The average number of exits that a rat placed in room $B$ will choose until it finds food is:

$$\frac{1}{2} + \frac{5}{4} = \frac{7}{4} = 1.75$$

This yields the system of equations:

$$
\begin{aligned}
.7s_1 + .5s_2 &= s_1 \\
.3s_1 + .5s_2 &= s_2 \\
s_1 + s_2 &= 1
\end{aligned}
\qquad \text{or} \qquad
\begin{aligned}
-.3s_1 + .5s_2 &= 0 \\
.3s_1 - .5s_2 &= 0 \\
s_1 + s_2 &= 1
\end{aligned}
$$

The solution is $s_1 = .625$, $s_2 = .375$. Thus, $S = [.625 \ .375]$.

## CHAPTER 9 REVIEW

1.   $S_1 = S_0 P = [.3 \ .7] \begin{array}{c} \\ \end{array} \overset{\begin{array}{cc} A & B \end{array}}{\begin{bmatrix} .6 & .4 \\ .2 & .8 \end{bmatrix}} = [.32 \ .68]$

   $S_2 = S_1 P = [.32 \ .68] \overset{\begin{array}{cc} A & B \end{array}}{\begin{bmatrix} .6 & .4 \\ .2 & .8 \end{bmatrix}} = [.328 \ .672]$

   The probability of being in state $A$ after one trial is .32; after two trials .328. The probability of being in state $B$ after one trial is .68; after two trials .672.   (9-1)

2.   $A$ is an absorbing state; the chain is absorbing since it is possible to go from state $B$ to state $A$.

3.   There are no absorbing states since there are no 1's on the main diagonal. $P$ is regular since

   $P^2 = \begin{bmatrix} .7 & .3 \\ .21 & .79 \end{bmatrix}$ has only positive entries.   (9-2, 9-3)

4.   $P = \begin{bmatrix} 0 & 1 \\ 1 & 0 \end{bmatrix}$ has no absorbing states. Since $P^k$, $k = 1, 2, 3, \dots$ , alternates between $\begin{bmatrix} 0 & 1 \\ 1 & 0 \end{bmatrix}$ and $\begin{bmatrix} 1 & 0 \\ 0 & 1 \end{bmatrix}$, $P$ is not regular.   (9-2, 9-3)

5.   States $B$ and $C$ are absorbing. The chain is absorbing; it is possible to go from the nonabsorbing state $A$ to the absorbing state $C$ in one step.   (9-2, 9-3)

6.   States $A$ and $B$ are absorbing. The chain is neither absorbing (it is not possible to go from States $C$ and $D$ to either state $A$ or state $B$) nor regular (all powers of $P$ will have the same form).   (9-2, 9-3)

Copyright © 2011 Pearson Education, Inc. Publishing as Prentice Hall.

7.  $P = \begin{array}{c} \\ A \\ B \\ C \end{array} \begin{array}{ccc} A & B & C \\ \left[\begin{array}{ccc} 0 & 1 & 0 \\ .1 & 0 & .9 \\ 0 & 1 & 0 \end{array}\right] \end{array}$

There are no absorbing states.

$P^k$, $k = 1, 2, 3, \ldots$ , alternates between $\begin{bmatrix} 0 & 1 & 0 \\ .1 & 0 & .9 \\ 0 & 1 & 0 \end{bmatrix}$ and $\begin{bmatrix} .1 & 0 & .9 \\ 0 & 1 & 0 \\ .1 & 0 & .9 \end{bmatrix}$. Thus, $P$ is not regular.

(9-1, 9-2, 9-3)

8.  $P = \begin{array}{c} \\ A \\ B \\ C \end{array} \begin{array}{ccc} A & B & C \\ \left[\begin{array}{ccc} 0 & 1 & 0 \\ .1 & .2 & .7 \\ 0 & 0 & 1 \end{array}\right] \end{array}$

$C$ is an absorbing state. The chain is absorbing since it is possible to go from state $A$ to state $C$ (via $B$) and from state $B$ to state $C$.    (9-1, 9-2, 9-3)

9.  $P = \begin{array}{c} \\ A \\ B \\ C \end{array} \begin{array}{ccc} A & B & C \\ \left[\begin{array}{ccc} 0 & 0 & 1 \\ .1 & .2 & .7 \\ 0 & 1 & 0 \end{array}\right] \end{array}$

There are no absorbing states since there are no 1's on the main diagonal.

$P$ is regular since $P^3 = \begin{bmatrix} .1 & .2 & .7 \\ .074 & .388 & .538 \\ .02 & .74 & .24 \end{bmatrix}$ has only positive entries.    (9-1, 9-2, 9-3)

10.  $P = \begin{array}{c} \\ A \\ B \\ C \\ D \end{array} \begin{array}{cccc} A & B & C & D \\ \left[\begin{array}{cccc} .3 & .2 & 0 & .5 \\ 0 & 1 & 0 & 0 \\ 0 & 0 & .2 & .8 \\ 0 & 0 & .3 & .7 \end{array}\right] \end{array}$

$B$ is an absorbing state. The chain is not absorbing since it is not possible to go from state $C$ to $B$, nor is it possible to go from state $D$ to state $B$.    (9-1, 9-2, 9-3)

11.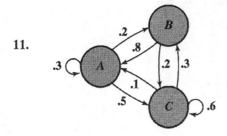

$P = \begin{array}{c} \\ A \\ B \\ C \end{array} \begin{array}{ccc} A & B & C \\ \left[\begin{array}{ccc} .3 & .2 & .5 \\ .8 & 0 & .2 \\ .1 & .3 & .6 \end{array}\right] \end{array}$

(9-1)

Copyright © 2011 Pearson Education, Inc. Publishing as Prentice Hall.

**12.** $P = \begin{array}{c} \\ A \\ B \end{array}\begin{array}{c} A \quad B \\ \begin{bmatrix} .4 & .6 \\ .9 & .1 \end{bmatrix} \end{array}$

(A) $P^2 = \begin{array}{c} \\ A \\ B \end{array}\begin{array}{c} A \quad B \\ \begin{bmatrix} .7 & .3 \\ .45 & .55 \end{bmatrix} \end{array}$

The probability of going from state $A$ to state $B$ in two trials is .3.

(B) $P^3 = \begin{array}{c} \\ A \\ B \end{array}\begin{array}{c} A \quad B \\ \begin{bmatrix} .55 & .45 \\ .675 & .325 \end{bmatrix} \end{array}$

The probability of going from state $B$ to state $A$ in three trials is .675.

(9-1)

**13.** Let $S = [s_1 \quad s_2]$ and solve the system:

$$[s_1 \quad s_2]\begin{bmatrix} .4 & .6 \\ .2 & .8 \end{bmatrix} = [s_1 \quad s_2], \quad s_1 + s_2 = 1$$

which is equivalent to

$$\begin{array}{ccc} .4s_1 + .2s_2 = s_1 & & -.6s_1 + .2s_2 = 0 \\ .6s_1 + .8s_2 = s_2 & \text{or} & .6s_1 - .2s_2 = 0 \\ s_1 + s_2 = 1 & & s_1 + s_2 = 1 \end{array}$$

The solution is: $s_1 = .25, \quad s_2 = .75$.

The stationary matrix $S = \begin{array}{c} A \quad B \\ [.25 \quad .75] \end{array}$

The limiting matrix $\overline{P} = \begin{array}{c} \\ A \\ B \end{array}\begin{array}{c} A \quad B \\ \begin{bmatrix} .25 & .75 \\ .25 & .75 \end{bmatrix} \end{array}$

**14.** Let $S = [s_1 \quad s_2 \quad s_3]$ and solve the system:

$$[s_1 \quad s_2 \quad s_3]\begin{bmatrix} .4 & .6 & 0 \\ .5 & .3 & .2 \\ 0 & .8 & .2 \end{bmatrix} = [s_1 \quad s_2 \quad s_3], \quad s_1 + s_2 + s_3 = 1$$

which is equivalent to

$$\begin{array}{ccc} .4s_1 + .5s_2 = s_1 & & -.6s_1 + .5s_2 = 0 \\ .6s_1 + .3s_2 + .8s_3 = s_2 & \text{or} & .6s_1 - .7s_2 + .8s_3 = 0 \\ .2s_2 + .2s_3 = s_3 & & .2s_2 - .8s_3 = 0 \\ s_1 + s_2 + s_3 = 1 & & s_1 + s_2 + s_3 = 1 \end{array}$$

From the first and third equations, we have $s_1 = \dfrac{5}{6}s_2$ and $s_3 = \dfrac{1}{4}s_2$.

Substituting these values into the fourth equation, we get

$$\frac{5}{6}s_2 + s_2 + \frac{1}{4}s_2 = 1 \text{ or } \frac{25}{12}s_2 = 1 \text{ and } s_2 = .48$$

Therefore, $s_1 = .4, s_2 = .48, s_3 = .12$.

Copyright © 2011 Pearson Education, Inc. Publishing as Prentice Hall.

The stationary matrix $S = [.4\ .48\ .12]$.

The limiting matrix $\overline{P} = \begin{matrix} & A & B & C \\ A & \\ B & \\ C & \end{matrix} \begin{bmatrix} .4 & .48 & .12 \\ .4 & .48 & .12 \\ .4 & .48 & .12 \end{bmatrix}$     (9-2)

**15.** For $P = \begin{matrix} & A & B & C \\ A & \\ B & \\ C & \end{matrix} \begin{bmatrix} 1 & 0 & 0 \\ 0 & 1 & 0 \\ .3 & .1 & .6 \end{bmatrix}$

we have $R = [.3\ .1]$ and $Q = [.6]$. The limiting matrix $\overline{P}$ has the form

$$\overline{P} = \left[\begin{array}{cc|c} 1 & 0 & 0 \\ 0 & 1 & 0 \\ \hline F & R & 0 \end{array}\right]$$

where $F = (I - Q)^{-1} = ([1] - [.6])^{-1} = [.4]^{-1} = \left[\dfrac{5}{2}\right]$

and $FR = \left[\dfrac{5}{2}\right][.3\ .1] = [.75\ .25]$.

Thus, $\overline{P} = \begin{matrix} & A & B & C \\ A & \\ B & \\ C & \end{matrix} \begin{bmatrix} 1 & 0 & 0 \\ 0 & 1 & 0 \\ .75 & .25 & 0 \end{bmatrix}$

$P(C \text{ to } A) = .75$, $P(C \text{ to } B) = .25$. Since $F = \left[\dfrac{5}{2}\right]$, it will take an average of 2.5 trials to go from $C$ to either $A$ or $B$.     (9-3)

**16.** For $P = \begin{matrix} & A & B & C & D \\ A & \\ B & \\ C & \\ D & \end{matrix} \begin{bmatrix} 1 & 0 & 0 & 0 \\ 0 & 1 & 0 & 0 \\ .1 & .5 & .2 & .2 \\ .1 & .1 & .4 & .4 \end{bmatrix}$

we have $R = \begin{bmatrix} .1 & .5 \\ .1 & .1 \end{bmatrix}$ and $Q = \begin{bmatrix} .2 & .2 \\ .4 & .4 \end{bmatrix}$.

The limiting matrix $\overline{P}$ has the form

$$\overline{P} = \left[\begin{array}{c|c} I & 0 \\ \hline FR & 0 \end{array}\right]$$

where $F = (I - Q)^{-1} = \left(\begin{bmatrix} 1 & 0 \\ 0 & 1 \end{bmatrix} - \begin{bmatrix} .2 & .2 \\ .4 & .4 \end{bmatrix}\right)^{-1} = \begin{bmatrix} .8 & -.2 \\ -.4 & .6 \end{bmatrix}^{-1} = \begin{bmatrix} \frac{4}{5} & -\frac{1}{5} \\ -\frac{2}{5} & \frac{3}{5} \end{bmatrix}^{-1}$

Copyright © 2011 Pearson Education, Inc. Publishing as Prentice Hall.

$$\begin{bmatrix} \frac{4}{5} & -\frac{1}{5} & 1 & 0 \\ -\frac{2}{5} & \frac{3}{5} & 0 & 1 \end{bmatrix} \sim \begin{bmatrix} 1 & -\frac{1}{4} & \frac{5}{4} & 0 \\ -\frac{2}{5} & \frac{3}{5} & 0 & 1 \end{bmatrix} \sim \begin{bmatrix} 1 & -\frac{1}{4} & \frac{5}{4} & 0 \\ 0 & \frac{1}{2} & \frac{1}{2} & 1 \end{bmatrix} \sim \begin{bmatrix} 1 & -\frac{1}{4} & \frac{5}{4} & 0 \\ 0 & 1 & 1 & 2 \end{bmatrix}$$

$$\left(\frac{5}{4}\right) R_1 \to R_1 \quad \left(\frac{2}{5}\right) R_1 + R_2 \to R_2 \quad 2R_2 \to R_2 \quad \left(\frac{1}{4}\right) R_2 + R_1 \to R_1$$

$$\sim \begin{bmatrix} 1 & 0 & \frac{3}{2} & \frac{1}{2} \\ 0 & 1 & 1 & 2 \end{bmatrix}$$

Thus, $F = \begin{bmatrix} \frac{3}{2} & \frac{1}{2} \\ 1 & 2 \end{bmatrix} = \begin{bmatrix} 1.5 & .5 \\ 1 & 2 \end{bmatrix}$, $FR = \begin{bmatrix} 1.5 & .5 \\ 1 & 2 \end{bmatrix} \begin{bmatrix} .1 & .5 \\ .1 & .1 \end{bmatrix} = \begin{bmatrix} .2 & .8 \\ .3 & .7 \end{bmatrix}$.

and $\overline{P} = \begin{array}{c} \\ A \\ B \\ C \\ D \end{array} \begin{array}{c} A \quad B \quad C \quad D \\ \begin{bmatrix} 1 & 0 & 0 & 0 \\ 0 & 1 & 0 & 0 \\ .2 & .8 & 0 & 0 \\ .3 & .7 & 0 & 0 \end{bmatrix} \end{array}$

$P(C \text{ to } A) = .2$, $P(C \text{ to } B) = .8$, $P(D \text{ to } A) = .3$, $P(D \text{ to } B) = .7$.

It takes an average of 2 trials to go from $C$ to either $A$ or $B$; it takes an average of three trials to go from $D$ to $A$ or $B$.   (9-3)

**17.** $P = \begin{array}{c} A \\ B \end{array} \begin{array}{c} A \quad B \\ \begin{bmatrix} .4 & .6 \\ .2 & .8 \end{bmatrix} \end{array}$, $P^4 \approx \begin{bmatrix} .2512 & .7488 \\ .2496 & .7504 \end{bmatrix}$, $P^8 \approx \begin{bmatrix} .2500 & .7499 \\ .2499 & .7500 \end{bmatrix}$; $\overline{P} = \begin{array}{c} A \\ B \end{array} \begin{array}{c} A \quad B \\ \begin{bmatrix} .25 & .75 \\ .25 & .75 \end{bmatrix} \end{array}$

(9-3)

**18.** $P = \begin{array}{c} A \\ B \\ C \end{array} \begin{array}{c} A \quad B \quad C \\ \begin{bmatrix} .4 & .6 & 0 \\ .5 & .3 & .2 \\ 0 & .8 & .2 \end{bmatrix} \end{array}$, $P^4 \approx \begin{bmatrix} .4066 & .4722 & .1212 \\ .3935 & .4895 & .117 \\ .404 & .468 & .128 \end{bmatrix}$,

$P^8 \approx \begin{bmatrix} .4001 & .4799 & .1200 \\ .3999 & .4802 & .1199 \\ .4001 & .4798 & .1201 \end{bmatrix}$; $\overline{P} = \begin{array}{c} A \\ B \\ C \end{array} \begin{array}{c} A \quad B \quad C \\ \begin{bmatrix} .4 & .48 & .12 \\ .4 & .48 & .12 \\ .4 & .48 & .12 \end{bmatrix} \end{array}$   (9-3)

**19.** $P = \begin{array}{c} A \\ B \\ C \end{array} \begin{array}{c} A \quad B \quad C \\ \begin{bmatrix} 1 & 0 & 0 \\ 0 & 1 & 0 \\ .3 & .1 & .6 \end{bmatrix} \end{array}$, $P^4 \approx \begin{bmatrix} 1 & 0 & 0 \\ 0 & 1 & 0 \\ .6528 & .2176 & .1296 \end{bmatrix}$, $P^8 \approx \begin{bmatrix} 1 & 0 & 0 \\ 0 & 1 & 0 \\ .7374 & .2458 & .01680 \end{bmatrix}$,

$P^{16} \approx \begin{bmatrix} 1 & 0 & 0 \\ 0 & 1 & 0 \\ .7498 & .2499 & 0 \end{bmatrix}$; $\overline{P} = \begin{array}{c} A \\ B \\ C \end{array} \begin{array}{c} A \quad B \quad C \\ \begin{bmatrix} 1 & 0 & 0 \\ 0 & 1 & 0 \\ .75 & .25 & 0 \end{bmatrix} \end{array}$   (9-3)

Copyright © 2011 Pearson Education, Inc. Publishing as Prentice Hall.

**20.** $P = \begin{array}{c} \\ A \\ B \\ C \\ D \end{array}\begin{array}{c} A \ \ B \ \ C \ \ D \\ \begin{bmatrix} 1 & 0 & 0 & 0 \\ 0 & 1 & 0 & 0 \\ .1 & .5 & .2 & .2 \\ .1 & .1 & .4 & .4 \end{bmatrix} \end{array}$, $P^4 \approx \begin{bmatrix} 1 & 0 & 0 & 0 \\ 0 & 1 & 0 & 0 \\ .1784 & .7352 & .0432 & .0432 \\ .2568 & .5704 & .0864 & .0864 \end{bmatrix}$,

$P^8 \approx \begin{bmatrix} 1 & 0 & 0 & 0 \\ 0 & 1 & 0 & 0 \\ .1972 & .7916 & .0056 & .0056 \\ .2944 & .6832 & .0112 & .0112 \end{bmatrix}$, $P^{16} \approx \begin{bmatrix} 1 & 0 & 0 & 0 \\ 0 & 1 & 0 & 0 \\ .2000 & .7999 & 0 & 0 \\ .2999 & .6997 & 0 & 0 \end{bmatrix}$,

$\overline{P} = \begin{array}{c} \\ A \\ B \\ C \\ D \end{array}\begin{array}{c} A \ \ B \ \ C \ \ D \\ \begin{bmatrix} 1 & 0 & 0 & 0 \\ 0 & 1 & 0 & 0 \\ .2 & .8 & 0 & 0 \\ .3 & .7 & 0 & 0 \end{bmatrix} \end{array}$    (9-3)

**21.** A standard form for the given matrix is:

$P = \begin{array}{c} \\ B \\ D \\ A \\ C \end{array}\begin{array}{c} B \ \ D \ \ A \ \ C \\ \begin{bmatrix} 1 & 0 & 0 & 0 \\ 0 & 1 & 0 & 0 \\ .1 & .1 & .6 & .2 \\ .2 & .2 & .3 & .3 \end{bmatrix} \end{array}$    (9-3)

**22.** We will determine the limiting matrix of:

$P = \begin{array}{c} \\ A \\ B \\ C \end{array}\begin{array}{c} A \ \ B \ \ C \\ \begin{bmatrix} 0 & 1 & 0 \\ 0 & 0 & 1 \\ .2 & .6 & .2 \end{bmatrix} \end{array}$

by solving

$$[s_1 \ \ s_2 \ \ s_3]\begin{bmatrix} 0 & 1 & 0 \\ 0 & 0 & 1 \\ .2 & .6 & .2 \end{bmatrix} = [s_1 \ \ s_2 \ \ s_3], s_1 + s_2 + s_3 = 1.$$

The corresponding system of equations is:

$$\begin{array}{ll} .2s_3 = s_1 & \\ s_1 \qquad + .6s_3 = s_2 & \text{or} \\ s_2 + .2s_3 = s_3 & \\ s_1 + s_2 + s_3 = 1 & \end{array} \qquad \begin{array}{l} s_1 \qquad\quad - .2s_3 = 0 \\ s_1 - s_2 + .6s_3 = 0 \\ s_2 \quad - .8s_3 = 0 \\ s_1 + s_2 + s_3 = 0 \end{array}$$

Copyright © 2011 Pearson Education, Inc. Publishing as Prentice Hall.

From the first and third equations, we have $s_1 = .2s_3$ and $s_2 = .8s_3$. Substituting these values into the fourth equation gives

$$.2s_3 + .8s_3 + s_3 = 1 \text{ and } s_3 = .5$$

It now follows that $s_1 = .1$ and $s_2 = .4$. Thus, $s = [.1 \ .4 \ .5]$ and

$$
\overline{P} = \begin{matrix} & A & B & C \\ A \\ B \\ C \end{matrix} \begin{bmatrix} .1 & .4 & .5 \\ .1 & .4 & .5 \\ .1 & .4 & .5 \end{bmatrix}
$$

(A) $\quad [0 \ 0 \ 1] \begin{bmatrix} .1 & .4 & .5 \\ .1 & .4 & .5 \\ .1 & .4 & .5 \end{bmatrix} \begin{matrix} A & B & C \\ = [.1 \ .4 \ .5] \end{matrix}$

(B) $\quad [.5 \ .3 \ .2] \begin{bmatrix} .1 & .4 & .5 \\ .1 & .4 & .5 \\ .1 & .4 & .5 \end{bmatrix} \begin{matrix} A & B & C \\ = [.1 \ .4 \ .5] \end{matrix} \qquad (9\text{-}3)$

23. The transition matrix:

$$
P = \begin{matrix} & A & B & C \\ A \\ B \\ C \end{matrix} \begin{bmatrix} 1 & 0 & 0 \\ 0 & 1 & 0 \\ .2 & .6 & .2 \end{bmatrix}
$$

is the standard form for an absorbing Markov chain with two absorbing and one nonabsorbing states. For this matrix, we have:

$R = [.2 \ .6]$ and $Q = [.2]$.

The limiting matrix has the form

$$\overline{P} = \left[ \begin{array}{c|c} I & 0 \\ \hline FR & 0 \end{array} \right]$$

where $F = (I - Q)^{-1} = ([1] - [.2])^{-1} = [.8]^{-1} = [1.25]$

Thus, $FR = [1.25][.2 \ .6] = [.25 \ .75]$ and

$$
\overline{P} = \begin{matrix} & A & B & C \\ A \\ B \\ C \end{matrix} \begin{bmatrix} 1 & 0 & 0 \\ 0 & 1 & 0 \\ .25 & .75 & 0 \end{bmatrix}
$$

(A) $\quad [0 \ 0 \ 1] \begin{bmatrix} 1 & 0 & 0 \\ 0 & 1 & 0 \\ .25 & .75 & 0 \end{bmatrix} \begin{matrix} A & B & C \\ = [.25 \ .75 \ 0] \end{matrix}$

(B) $\quad [.5 \ .3 \ .2] \begin{bmatrix} 1 & 0 & 0 \\ 0 & 1 & 0 \\ .25 & .75 & 0 \end{bmatrix} \begin{matrix} A & B & C \\ = [.55 \ .45 \ 0] \end{matrix} \qquad (9\text{-}3)$

Copyright © 2011 Pearson Education, Inc. Publishing as Prentice Hall.

**24.** No. If $P$ is a transition matrix with 2 entries equal to 0, then $P$ has one of the forms: $P_1 = \begin{pmatrix} 1 & 0 \\ 0 & 1 \end{pmatrix}$,

$$P_2 = \begin{pmatrix} 0 & 1 \\ 1 & 0 \end{pmatrix}, \quad P_3 = \begin{pmatrix} 1 & 0 \\ 1 & 0 \end{pmatrix}, \quad P_4 = \begin{pmatrix} 0 & 1 \\ 0 & 1 \end{pmatrix}$$

$P_1^{\,k} = P_1$, $P_2^{\,k} = P_2$, $P_3^{\,k} = P_3$ for all $k$, and $P_2^{\,k} = \begin{pmatrix} 1 & 0 \\ 0 & 1 \end{pmatrix}$ if $k$ is even and $P_2^{\,k} = P_2$ if $k$ is odd. No

power of $P_i$, $i = 1, 2, 3, 4$, has all positive entries.    (9-2)

**25.** Yes; $P = \begin{bmatrix} 0 & .5 & .5 \\ .5 & 0 & .5 \\ .5 & .5 & 0 \end{bmatrix}$ is regular since $P^2 = \begin{bmatrix} .5 & .25 & .25 \\ .25 & .5 & .25 \\ .25 & .25 & .5 \end{bmatrix}$

$P = \begin{bmatrix} 0 & 0 & 1 \\ 0 & 0 & 1 \\ .2 & .3 & .5 \end{bmatrix}$ is regular since $P^2 = \begin{bmatrix} .2 & .3 & .5 \\ .2 & .3 & .5 \\ .1 & .15 & .75 \end{bmatrix}$

**26.** (A)

(B) $P = \begin{array}{c} \\ R \\ B \\ G \end{array} \begin{array}{c} \begin{array}{ccc} R & B & G \end{array} \\ \begin{bmatrix} .5 & .25 & .25 \\ .2 & .6 & .2 \\ .6 & .3 & .1 \end{bmatrix} \end{array}$

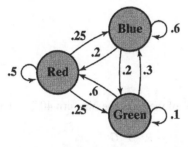

(C) The chain is regular since it has only positive entries.

(D) Let $S = [s_1 \ s_2 \ s_3]$ and solve the system:

$$[s_1 \ s_2 \ s_3] \begin{bmatrix} .5 & .25 & .25 \\ .2 & .6 & .2 \\ .6 & .3 & .1 \end{bmatrix} = [s_1 \ s_2 \ s_3], \quad s_1 + s_2 + s_3 = 1$$

which is equivalent to:

$$\begin{array}{ll} s_1 + s_2 + s_3 = 1 & \qquad s_1 + s_2 + s_3 = 1 \\ .5s_1 + .2s_2 + .6s_3 = s_1 & \qquad -.5s_1 + .2s_2 + .6s_3 = 0 \\ .25s_1 + .6s_2 + .3s_3 = s_2 \qquad \text{or} & \qquad .25s_1 - .4s_2 + .3s_3 = 0 \\ .25s_1 + .2s_2 + .1s_3 = s_3 & \qquad .25s_1 + .2s_2 - .9s_3 = 0 \end{array}$$

We use row operations to solve this system; but first multiply the
second, third and fourth equations by 10 to simplify the
calculations.

Copyright © 2011 Pearson Education, Inc. Publishing as Prentice Hall.

$$\begin{bmatrix} 1 & 1 & 1 & | & 1 \\ -5 & 2 & 6 & | & 0 \\ \frac{5}{2} & -4 & 3 & | & 0 \\ \frac{5}{2} & 2 & -9 & | & 0 \end{bmatrix} \sim \begin{bmatrix} 1 & 1 & 1 & | & 1 \\ 0 & 7 & 11 & | & 5 \\ 0 & -\frac{13}{2} & \frac{1}{2} & | & -\frac{5}{2} \\ 0 & \frac{1}{2} & -\frac{23}{2} & | & -\frac{5}{2} \end{bmatrix} \sim \begin{bmatrix} 1 & 1 & 1 & | & 1 \\ 0 & 1 & 23 & | & 5 \\ 0 & -\frac{13}{2} & \frac{1}{2} & | & -\frac{5}{2} \\ 0 & 7 & 11 & | & 5 \end{bmatrix}$$

$5R_1 + R_2 \rightarrow R_2$                          $-2R_4 \rightarrow R_4$                          $(-1)R_2 + R_1 \rightarrow R_1$

$\left(-\dfrac{5}{2}\right)R_1 + R_3 \rightarrow R_3$                          $R_2 \leftrightarrow R_4$                          $\left(\dfrac{13}{2}\right)R_2 + R_3 \rightarrow R_3$

$\left(-\dfrac{5}{2}\right)R_1 + R_4 \rightarrow R_4$                                                  $(-7)R_2 + R_4 \rightarrow R_4$

$$\sim \begin{bmatrix} 1 & 0 & -22 & | & -4 \\ 0 & 1 & 23 & | & 5 \\ 0 & 0 & 150 & | & 30 \\ 0 & 0 & -150 & | & -30 \end{bmatrix} \sim \begin{bmatrix} 1 & 0 & -22 & | & -4 \\ 0 & 1 & 23 & | & 5 \\ 0 & 0 & 1 & | & \frac{1}{5} \\ 0 & 0 & -150 & | & -30 \end{bmatrix} \sim \begin{bmatrix} 1 & 0 & 0 & | & \frac{2}{5} \\ 0 & 1 & 0 & | & \frac{2}{5} \\ 0 & 0 & 1 & | & \frac{1}{5} \\ 0 & 0 & 0 & | & 0 \end{bmatrix}$$

$\dfrac{1}{150}R_3 \rightarrow R_3$                          $22R_3 + R_1 \rightarrow R_1$

$(-23)R_3 + R_2 \rightarrow R_2$

$150R_3 + R_4 \rightarrow R_4$

The solution is $s_1 = 0.4$, $s_2 = 0.4$, $s_3 = 0.2$ and

$$\overline{P} = \begin{array}{c} \\ R \\ B \\ G \end{array} \begin{array}{c} \begin{array}{ccc} R & B & G \end{array} \\ \begin{bmatrix} .4 & .4 & .2 \\ .4 & .4 & .2 \\ .4 & .4 & .2 \end{bmatrix} \end{array}$$

In the long run, the red urn will be selected 40% of the time, the blue urn 40% of the time, and the green urn 20% of the time.           (9-2)

**27.** (A)                                                      (B) $P = \begin{array}{c} \\ R \\ B \\ G \end{array} \begin{array}{c} \begin{array}{ccc} R & B & G \end{array} \\ \begin{bmatrix} 1 & 0 & 0 \\ .2 & .6 & .2 \\ .6 & .3 & .1 \end{bmatrix} \end{array}$

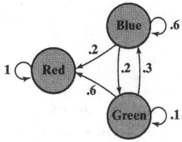

(C)   State $R$ is an absorbing state. The chain is absorbing since it is possible to go from states $B$ and $G$ to state $R$ in a finite number (namely 1) of steps.

Copyright © 2011 Pearson Education, Inc. Publishing as Prentice Hall.

(D)   For $P = \begin{bmatrix} 1 & 0 & 0 \\ .2 & .6 & .2 \\ .6 & .3 & .1 \end{bmatrix}$ we have $R = \begin{bmatrix} .2 \\ .6 \end{bmatrix}$ and $Q = \begin{bmatrix} .6 & .2 \\ .3 & .1 \end{bmatrix}$.

The limiting matrix $\overline{P}$ has the form:

$$\overline{P} = \left[ \begin{array}{c|c} I & 0 \\ \hline FR & 0 \end{array} \right]$$

where $F = (I - Q)^{-1} = \left( \begin{bmatrix} 1 & 0 \\ 0 & 1 \end{bmatrix} - \begin{bmatrix} .6 & .2 \\ .3 & .1 \end{bmatrix} \right)^{-1} = \begin{bmatrix} .4 & -.2 \\ -.3 & .9 \end{bmatrix}^{-1} = \begin{bmatrix} \frac{2}{5} & -\frac{1}{5} \\ -\frac{3}{10} & \frac{9}{10} \end{bmatrix}^{-1}$

We use row operations to find the inverse:

$\left[ \begin{array}{cc|cc} \frac{2}{5} & -\frac{1}{5} & 1 & 0 \\ -\frac{3}{10} & \frac{9}{10} & 0 & 1 \end{array} \right] \sim \left[ \begin{array}{cc|cc} 1 & -\frac{1}{2} & \frac{5}{2} & 0 \\ -\frac{3}{10} & \frac{9}{10} & 0 & 1 \end{array} \right] \sim \left[ \begin{array}{cc|cc} 1 & -\frac{1}{2} & \frac{5}{2} & 0 \\ 0 & \frac{3}{4} & \frac{3}{4} & 1 \end{array} \right]$

$\left( \frac{5}{2} \right) R_1 \rightarrow R_1 \qquad \left( \frac{3}{10} \right) R_1 + R_2 \rightarrow R_2 \qquad \frac{4}{3} R_2 \rightarrow R_2$

$\sim \left[ \begin{array}{cc|cc} 1 & -\frac{1}{2} & \frac{5}{2} & 0 \\ 0 & 1 & 1 & \frac{4}{3} \end{array} \right] \sim \left[ \begin{array}{cc|cc} 1 & 0 & 3 & \frac{2}{3} \\ 0 & 1 & 1 & \frac{4}{3} \end{array} \right]$

$\left( \frac{1}{2} \right) R_2 + R_1 \rightarrow R_1$

Thus, $F = \begin{bmatrix} 3 & \frac{2}{3} \\ 1 & \frac{4}{3} \end{bmatrix}$ and $FR = \begin{bmatrix} 3 & \frac{2}{3} \\ 1 & \frac{4}{3} \end{bmatrix} \begin{bmatrix} \frac{1}{5} \\ \frac{1}{5} \end{bmatrix} = \begin{bmatrix} 1 \\ 1 \end{bmatrix}$.

Now,  $\overline{P} = \begin{array}{c} \\ R \\ B \\ G \end{array} \begin{array}{c} \begin{array}{ccc} R & B & G \end{array} \\ \begin{bmatrix} 1 & 0 & 0 \\ 1 & 0 & 0 \\ 1 & 0 & 0 \end{bmatrix} \end{array}$

Once the red urn is selected, the blue and green urns will never be selected again. It will take an average of 3.67 trials to reach the red urn from the blue urn and an average of 2.33 trials to reach the red urn from the green urn.     (9-3)

**28.**  $[x \;\; y \;\; z \;\; 0] \begin{bmatrix} 1 & 0 & 0 & 0 \\ 0 & 1 & 0 & 0 \\ 0 & 0 & 1 & 0 \\ .1 & .3 & .4 & .2 \end{bmatrix} = [x \;\; y \;\; z \;\; 0]$

Thus, $[x \;\; y \;\; z \;\; 0]$ is a stationary matrix for $P$. If $P$ is a transition matrix for an absorbing chain with three absorbing states and one nonabsorbing state, then $P$ will have exactly three 1's on the main diagonal (and zeros elsewhere in the row containing the 1's) and one row with at least one nonzero entry off the main diagonal. One of the following matrices will be a stationary matrix for $P$:

$[x \;\; y \;\; z \;\; 0], \quad [x \;\; y \;\; 0 \;\; z], \quad [x \;\; 0 \;\; y \;\; z], \quad [0 \;\; x \;\; y \;\; z]$

Copyright © 2011 Pearson Education, Inc.  Publishing as Prentice Hall.

where $x + y + z = 1$. The position of the zero corresponds to the one row of $P$ which has a nonzero entry off the main diagonal.   (9-2, 9-3)

**29.** No such chain exists; if the chain has an absorbing state, then a corresponding transition matrix $P$ will have a row containing a 1 on the main diagonal and zeros elsewhere. All powers of $P$ will have the same row, and so the chain cannot be regular.   (9-2, 9-3)

**30.** No such chain exists. The reasoning in Problem 29 applies here as well.   (9-2, 9-3)

**31.** No such chain exists. By Theorem 1, Section 9-2 a regular Markov chain has a unique stationary matrix.   (9-2)

**32.** $S = [1 \ \ 0 \ \ 0]$ and $S' = [0 \ \ 1 \ \ 0]$ are both stationary matrices for $\quad P = \begin{matrix} & A & B & C \\ A \\ B \\ C \end{matrix}\begin{bmatrix} 1 & 0 & 0 \\ 0 & 1 & 0 \\ .6 & .3 & .1 \end{bmatrix}$   (9-3)

**33.** $P = \begin{matrix} & A & B \\ A \\ B \end{matrix}\begin{bmatrix} 0 & 1 \\ 1 & 0 \end{bmatrix}$ has no limiting matrix; $\quad P^{2k} = \begin{bmatrix} 1 & 0 \\ 0 & 1 \end{bmatrix}$ and $P^{2k+1} = \begin{bmatrix} 0 & 1 \\ 1 & 0 \end{bmatrix}$ for all positive integers $k$.

(9-2, 9-3)

**34.** No such chain exists. By Theorem 1, Section 9-2, a regular Markov chain has a unique limiting matrix.   (9-2)

**35.** No such chain exists. By Theorem 2, Section 9-3, an absorbing Markov chain has a limiting matrix.   (9-3)

**36.** $P = \begin{matrix} & A & B & C & D \\ A \\ B \\ C \\ D \end{matrix}\begin{bmatrix} .2 & .3 & .1 & .4 \\ 0 & 0 & 1 & 0 \\ 0 & .8 & 0 & .2 \\ 0 & 0 & 1 & 0 \end{bmatrix}$   No limiting matrix

For example

$$P^{200} = P^{202} = P^{204} = \dots = \begin{bmatrix} 0 & .2 & .75 & .05 \\ 0 & .8 & 0 & .2 \\ 0 & 0 & 1 & 0 \\ 0 & .8 & 0 & .2 \end{bmatrix}$$

$$P^{201} = P^{203} = P^{205} = \dots = \begin{bmatrix} 0 & .6 & .25 & .15 \\ 0 & 0 & 1 & 0 \\ 0 & .8 & 0 & .2 \\ 0 & 0 & 1 & 0 \end{bmatrix}$$   (9-2, 9-3)

Copyright © 2011 Pearson Education, Inc. Publishing as Prentice Hall.

**37.**  $P = \begin{array}{c} \\ A \\ B \\ C \\ D \end{array} \begin{array}{cccc} A & B & C & D \\ \left[\begin{array}{cccc} 1 & 0 & .3 & .6 \\ .2 & .4 & .1 & .3 \\ .3 & .5 & 0 & .2 \\ .9 & .1 & 0 & 0 \end{array}\right] \end{array}$   Limiting matrix   $\begin{array}{c} \\ A \\ B \\ C \\ D \end{array} \begin{array}{cccc} A & B & C & D \\ \left[\begin{array}{cccc} .392 & .163 & .134 & .311 \\ .392 & .163 & .134 & .311 \\ .392 & .163 & .134 & .311 \\ .392 & .163 & .134 & .311 \end{array}\right] \end{array}$   (9-2)

**38.**  (A)   (B)  $P = \begin{array}{c} \\ x \\ x' \end{array} \begin{array}{cc} x & x' \\ \left[\begin{array}{cc} .7 & .3 \\ .5 & .5 \end{array}\right] \end{array}$   (C)  $S = [.2\ \ .8]$

(D)   $S_1 = SP = [.2\ \ .8]\begin{bmatrix} .7 & .3 \\ .5 & .5 \end{bmatrix} = [.54\ \ .46]$

54% of the consumers will use brand $x$ on the next purchase.

(E)   To find the stationary matrix $S = [s_1\ \ s_2]$, we need to solve:

$$[s_1\ \ s_2]\begin{bmatrix} .7 & .3 \\ .5 & .5 \end{bmatrix} = [s_1\ \ s_2],\ s_1 + s_2 = 1$$

This yields the system of equations:

$$\begin{array}{lll} .7s_1 + .5s_2 = s_1 & & -.3s_1 + .5s_2 = 0 \\ .3s_1 + .5s_2 = s_2 & \text{or} & .3s_1 - .5s_2 = 0 \\ s_1 + s_2 = 1 & & s_1 + s_2 = 1 \end{array}$$

The solution is $s_1 = .625$, $s_2 = .375$. Thus, $S = [.625\ \ .375]$.

(F)   Brand $X$ will have 62.5% of the market in the long run.    (9-2)

**39.**  A transition matrix in standard form for this problem is:

$$P = \begin{array}{c} \\ A \\ B \\ C \\ M \end{array} \begin{array}{cccc} A & B & C & M \\ \left[\begin{array}{cccc} 1 & 0 & 0 & 0 \\ 0 & 1 & 0 & 0 \\ 0 & 0 & 1 & 0 \\ .06 & .08 & .11 & .75 \end{array}\right] \end{array}$$

For this matrix, $R = [.06\ \ .08\ \ .11]$ and $Q = [.75]$.

The limiting matrix for $P$ has the form:  $\overline{P} = \left[\begin{array}{c|c} I & 0 \\ \hline FR & 0 \end{array}\right]$

where  $F = (I - Q)^{-1} = ([1] - [.75])^{-1} = [.25]^{-1} = [4]$.

Copyright © 2011 Pearson Education, Inc. Publishing as Prentice Hall.

Thus, $FR = [4][.06\ .08\ .11] = [.24\ .32\ .44]$

and $\overline{P} = \begin{array}{c} \\ A \\ B \\ C \\ M \end{array} \begin{array}{cccc} A & B & C & M \\ \begin{bmatrix} 1 & 0 & 0 & 0 \\ 0 & 1 & 0 & 0 \\ 0 & 0 & 1 & 0 \\ .24 & .32 & .44 & 0 \end{bmatrix} \end{array}$

(A)   In the long run, brand $A$ will have 24% of the market, brand $B$ will have 32% and brand $C$ will have 44%.

(B)   A company will wait an average of 4 years before converting to one of the new milling machines.
(9-3)

**40.** (A)   $P = \begin{bmatrix} .95 & .05 \\ .40 & .60 \end{bmatrix}, S_0 = [.14\ .86]$

$S_1 = S_0 P = [.14\ .86]\begin{bmatrix} .95 & .05 \\ .40 & .60 \end{bmatrix} = [.48\ .52]$

$S_2 = S_1 P = [.48\ .52]\begin{bmatrix} .95 & .05 \\ .40 & .60 \end{bmatrix} = [.66\ .34]$

(B)

| Year | Data % | Model % |
|------|--------|---------|
| 1995 | 14 | 14 |
| 2000 | 49 | 48 |
| 2005 | 68 | 66 |

(C)   $S_n = S_{n-1}P \approx [.89\ .11]$   for large $n$; 89% will be online in the long run.   (9-2)

**41.**   A transition matrix in standard form for this problem is:

$P = \begin{array}{c} \\ F \\ L \\ T \\ A \end{array} \begin{array}{cccc} F & L & T & A \\ \begin{bmatrix} 1 & 0 & 0 & 0 \\ 0 & 1 & 0 & 0 \\ 0 & .05 & .8 & .15 \\ .17 & .03 & 0 & .8 \end{bmatrix} \end{array}$

where $F$ = "Fellow", $A$ = "Associate", $T$ = "Trainee", $L$ = leaves

For this matrix, $R = \begin{bmatrix} 0 & .05 \\ .17 & .03 \end{bmatrix}$ and $Q = \begin{bmatrix} .8 & .15 \\ 0 & .8 \end{bmatrix}$.

The limiting matrix for $P$ has the form

$\overline{P} = \left[\begin{array}{c|c} I & 0 \\ \hline FR & 0 \end{array}\right]$

where $F = (I - Q)^{-1} = \left(\begin{bmatrix} 1 & 0 \\ 0 & 1 \end{bmatrix} - \begin{bmatrix} .8 & .15 \\ 0 & .8 \end{bmatrix}\right)^{-1} = \begin{bmatrix} .2 & -.15 \\ 0 & .2 \end{bmatrix}^{-1}$

We use row operations to calculate the inverse:

$\left[\begin{array}{cc|cc} .2 & -.15 & 1 & 0 \\ 0 & .2 & 0 & 1 \end{array}\right] \sim \left[\begin{array}{cc|cc} 1 & -.75 & 5 & 0 \\ 0 & 1 & 0 & 5 \end{array}\right] \sim \left[\begin{array}{cc|cc} 1 & 0 & 5 & 3.75 \\ 0 & 1 & 0 & 5 \end{array}\right]$

Copyright © 2011 Pearson Education, Inc.  Publishing as Prentice Hall.

$$5R_1 \rightarrow R_1 \qquad (.75)R_2 + R_1 \rightarrow R_1$$
$$5R_2 \rightarrow R_2$$

Thus, $F = \begin{bmatrix} 5 & 3.75 \\ 0 & 5 \end{bmatrix}$ and $FR = \begin{bmatrix} 5 & 3.75 \\ 0 & 5 \end{bmatrix}\begin{bmatrix} 0 & .05 \\ .17 & .03 \end{bmatrix} = \begin{bmatrix} .6375 & .3625 \\ .85 & .15 \end{bmatrix}$

The limiting matrix is:

$$\overline{P} = \begin{array}{c} \\ F \\ L \\ T \\ A \end{array}\begin{array}{c} \begin{array}{cccc} F & L & T & A \end{array} \\ \begin{bmatrix} 1 & 0 & 0 & 0 \\ 0 & 1 & 0 & 0 \\ .6375 & .3625 & 0 & 0 \\ .85 & .15 & 0 & 0 \end{bmatrix} \end{array}$$

(A)  In the long run, 63.75% of the trainees will become Fellows.

(B)  In the long run, 15% of the Associates will leave the company.

(C)  A trainee remains in the program an average of 5 + 3.75 = 8.75 yrs.    (9-3)

**42.**  We shall find the limiting matrix for:

$$P = \begin{array}{c} \\ R \\ P \\ W \end{array}\begin{array}{c} \begin{array}{ccc} R & P & W \end{array} \\ \begin{bmatrix} 1 & 0 & 0 \\ .5 & .5 & 0 \\ 0 & 1 & 0 \end{bmatrix} \end{array}$$ We have $R = \begin{bmatrix} .5 \\ 0 \end{bmatrix}$ and $Q = \begin{bmatrix} .5 & 0 \\ 1 & 0 \end{bmatrix}$.

The limiting matrix for $P$ will have the form:

$$\overline{P} = \left[\begin{array}{c|c} I & 0 \\ \hline FR & 0 \end{array}\right]$$

where $F = (I - Q)^{-1} = \left(\begin{bmatrix} 1 & 0 \\ 0 & 1 \end{bmatrix} - \begin{bmatrix} .5 & 0 \\ 1 & 0 \end{bmatrix}\right)^{-1} = \begin{bmatrix} .5 & 0 \\ -1 & 1 \end{bmatrix}^{-1} = \begin{bmatrix} \frac{1}{2} & 0 \\ -1 & 1 \end{bmatrix}^{-1}$

We use row operations to find the inverse:

$$\left[\begin{array}{cc|cc} \frac{1}{2} & 0 & 1 & 0 \\ -1 & 1 & 0 & 1 \end{array}\right] \sim \left[\begin{array}{cc|cc} 1 & 0 & 2 & 0 \\ -1 & 1 & 0 & 1 \end{array}\right] \sim \left[\begin{array}{cc|cc} 1 & 0 & 2 & 0 \\ 0 & 1 & 2 & 1 \end{array}\right]$$

$$2R_1 \rightarrow R_1 \qquad R_1 + R_2 \rightarrow R_2$$

Thus, $F = \begin{bmatrix} 2 & 0 \\ 2 & 1 \end{bmatrix}$ and $FR = \begin{bmatrix} 2 & 0 \\ 2 & 1 \end{bmatrix}\begin{bmatrix} .5 \\ 0 \end{bmatrix} = \begin{bmatrix} 1 \\ 1 \end{bmatrix}$

The limiting matrix $\overline{P}$ is:

$$\overline{P} = \begin{array}{c} \\ R \\ P \\ W \end{array}\begin{array}{c} \begin{array}{ccc} R & P & W \end{array} \\ \begin{bmatrix} 1 & 0 & 0 \\ 1 & 0 & 0 \\ 1 & 0 & 0 \end{bmatrix} \end{array}$$

From this matrix, we conclude that eventually all of the flowers will be red.    (9-3)

Copyright © 2011 Pearson Education, Inc. Publishing as Prentice Hall.

**43.** (A) $P = \begin{bmatrix} .74 & .26 \\ .03 & .97 \end{bmatrix}$, $S_0 = [.301 \;\; .699]$

$S_1 = S_0 P = [.301 \;\; .699] \begin{bmatrix} .74 & .26 \\ .03 & .97 \end{bmatrix} = [.244 \;\; .756]$

$S_2 = S_1 P = [.244 \;\; .756] \begin{bmatrix} .74 & .26 \\ .03 & .97 \end{bmatrix} = [.203 \;\; .797]$

(B)

| Year | Data % | Model % |
|------|--------|---------|
| 1985 | 30.1 | 30.1 |
| 1995 | 24.7 | 24.4 |
| 2005 | 20.9 | 20.3 |

(C) $S_n = S_{n-1} P \approx [.103 \;\; .897]$ for large $n$; 10.3% of the adult

U.S. population will be smokers in the long run.     (9-2)

Copyright © 2011 Pearson Education, Inc.  Publishing as Prentice Hall.

## 10  GAMES AND DECISIONS

EXERCISE 10-1

*Things to remember:*

<u>1</u>.    FUNDAMENTAL PRINCIPLE OF GAME THEORY

a. A matrix game is played repeatedly.

b. Player *R* tries to maximize winnings.

c. Player *C* tries to minimize losses.

<u>2</u>.    A SADDLE VALUE is a payoff value that is simultaneously a row minimum and a column maximum of a payoff matrix.

<u>3</u>.    LOCATING SADDLE VALUES:

(a)    Circle the minimum value in each row (it may occur in more than one place).

(b)    Place squares around the maximum value in each column (it may occur in more than one place).

(c)    Any entry with both a circle and a square around it is a saddle value.

<u>4</u>.    EQUALITY OF SADDLE VALUES

If a payoff matrix has saddle values $x$ and $y$, then $x = y$.

<u>5</u>.    STRICTLY DETERMINED MATRIX GAMES

A matrix game is STRICTLY DETERMINED if it has a saddle value.  In a strictly determined game, the OPTIMAL STRATEGIES are:

*R* should choose any row containing a saddle value

*C* should choose any column containing a saddle value

A saddle value is called the VALUE of a strictly determined game.  The game is FAIR if its value is zero.  [Note: In a strictly determined game (assuming both players play their optimal strategy), knowledge of an opponent's move provides no advantage, since the payoff will always be a saddle value.]

<u>6</u>.    A matrix game is NONSTRICTLY DETERMINED if it does not have a saddle value.

**1.**    The game is strictly determined; the element 1 in the 2,2 position is the saddle value.

**3.**    The game is not strictly determined.

**5.**    The game is strictly determined; the element −1 in the 1,2 and 3,2 positions is the saddle value.

**7.**    The game is not strictly determined.

**9.**    The game is fair; the element 0 in the 2,1 position is the saddle value.

**11.**    The game is not fair; the element −1 in the 2,3 position is the saddle value.

**13.**    The game is fair; the element 0 in the 2,2 position is the saddle value.

Copyright © 2011 Pearson Education, Inc.  Publishing as Prentice Hall.

**15.** $\begin{bmatrix} 4 & ⊝3 \\ \boxed{5} & \boxed{④} \end{bmatrix}$

(A) 4 in the 2,2 position.
(B) $R$ plays row 2, $C$ plays column 2.
(C) The value of the game is 4.

**17.** $\begin{bmatrix} ⊝1 & \boxed{0} \\ \boxed{3} & ⊝1 \end{bmatrix}$

The game is not strictly determined.

**19.** $\begin{bmatrix} ⊝1 & ⊝1 \\ ⊝1 & ⊝1 \end{bmatrix}$

(A) –1 in all four positions.
(B) $R$ plays either row 1 or row 2, $C$ plays either
   column 1 or column 2.
(C) The value of the game is –1.

**21.** $\begin{bmatrix} ②2 & ②2 \\ ②2 & \boxed{5} \end{bmatrix}$

The game is strictly determined.

(A)  Both 2's in column 1.

(B)  $R$ plays either row 1 or row 2 and $C$ plays column 1.

(C)  The value of the game is 2.

**23.** $\begin{bmatrix} \boxed{2} & -1 & ⊝5 \\ 1 & ⊝⓪ & 3 \\ -3 & ⊝7 & \boxed{8} \end{bmatrix}$

The game is strictly determined.

(A)  0 (in 2,2 position).

(B)  $R$ plays row 2 and $C$ plays column 2.

(C)  The value of the game is 0.

**25.** $\begin{bmatrix} 3 & ⊝2 \\ ①1 & \boxed{5} \\ ⊝4 & 0 \\ \boxed{5} & ⊝3 \end{bmatrix}$

The game is not strictly determined.

**27.**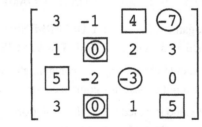

The game is strictly determined.

(A)  Both 0's in column 2 (2,2 and 4,2 positions)

(B)  $R$ plays either row 2 or row 4 and $C$ plays column 2.

(C)  The value of the game is 0.

Copyright © 2011 Pearson Education, Inc. Publishing as Prentice Hall.

**29.**

$$\begin{bmatrix} 0 & 4 & -8 & -3 \\ 2 & 5 & 3 & 2 \\ 1 & -3 & -2 & -9 \\ 2 & 4 & 7 & 2 \end{bmatrix}$$

(A)

Saddle value: 2

(B)    Optimal strategy for $R$: play Row 2 or Row 4.
Optimal strategy for $C$: play Column 1 or Column 4.

(C)    The value of the game: 2

**31.**    Since the value of the game is positive, Player $R$ has the advantage.

**33.**    False.  According to Theorem 1, a payoff matrix has exactly one saddle value

**35.**    False. Consider the payoff matrix in Problem 23; $-7$ is the smallest entry, it is not a saddle value.

**37.**    True. If $A$ is a payoff matrix and the $i$-th row of $A$ is all 0's, and the $j$-th column of $A$ is all 0's, then the element $a_{ij}$ is a saddle value.

**39.**    $\begin{bmatrix} -3 & m \\ \boxed{\textcircled{0}} & 1 \end{bmatrix}$    No; 0 is a saddle value irrespective of the value of $m$.

**41.**

Station $C$

$$\begin{array}{c} \quad\quad \$1.35 \quad \$1.40 \\ \text{Station } R \begin{array}{c} \$1.40 \\ \$1.45 \end{array} \begin{bmatrix} \boxed{\textcircled{50\%}} & 70\% \\ \textcircled{40\%} & 50\% \end{bmatrix} \end{array}$$

The saddle value is 50%, as shown in the upper left-hand corner (1,1 position).

Optimum strategies: $R$ plays row 1 ($1.05) and $C$ plays column 1 ($1.00).

**43.**

Store $C$

| | T.C. | I.V. | S.L.T. |
|---|---|---|---|
| Store $R$ T.C. | 50% | $20\% + \frac{50\%}{2}$ | $20\% + \frac{30\%}{2}$ |
| I.V. | $30\% + \frac{50\%}{2}$ | 50% | $30\% + \frac{20\%}{2}$ |
| S.L.T. | $50\% + \frac{30\%}{2}$ | $50\% + \frac{20\%}{2}$ | 50% |

$$= \begin{bmatrix} 50\% & 45\% & \textcircled{35\%} \\ 55\% & 50\% & \textcircled{40\%} \\ \boxed{65\%} & \boxed{60\%} & \boxed{\textcircled{50\%}} \end{bmatrix}$$

The optimal strategy for both stores is to locate in South Lake Tahoe and split the basin business equally.

Copyright © 2011 Pearson Education, Inc.  Publishing as Prentice Hall.

## EXERCISE 10-2

*Things to remember:*

1. Given the game matrix $M = \begin{bmatrix} a & b \\ c & d \end{bmatrix}$.

   $R$'s strategy is a probability row matrix

   $$P = [p_1 \ p_2], \quad p_1 \geq 0, \quad p_2 \geq 0, \quad p_1 + p_2 = 1.$$

   $C$'s strategy is a probability column matrix

   $$Q = \begin{bmatrix} q_1 \\ q_2 \end{bmatrix}, \quad q_1 \geq 0, \quad q_2 \geq 0, \quad q_1 + q_2 = 1.$$

2. EXPECTED VALUE OF A MATRIX GAME FOR $R$

   For the game matrix $M = \begin{bmatrix} a & b \\ c & d \end{bmatrix}$ and strategies $P = [p_1 \ p_2]$,

   $Q = \begin{bmatrix} q_1 \\ q_2 \end{bmatrix}$ for $R$ and $C$, respectively, the EXPECTED VALUE of the

   game for $R$ is given by

   $$E(P, Q) = PMQ = ap_1q_1 + bp_1q_2 + cp_2q_1 + dp_2q_2.$$

3. FUNDAMENTAL THEOREM OF GAME THEORY: For every $m \times n$ game matrix $M$, there exists strategies $P^*$ and $Q^*$ (not necessarily unique) for $R$ and $C$, respectively, and a unique number $v$ such that:

   $P^*MQ \geq v$ for every strategy $Q$ of $C$;

   $PMQ^* \leq v$ for every strategy $P$ of $R$.

   The number $v$ is called the VALUE of the game. The game is FAIR if $v = 0$. The triplet ($v$, $P^*$, $Q^*$) is called a SOLUTION OF THE GAME. The expected value of the game is

   $$E(P^*, Q^*) = P^*MQ^* = v.$$

4. SOLUTION TO A $2 \times 2$ NONSTRICTLY DETERMINED MATRIX GAME

   For the nonstrictly determined game matrix $M = \begin{bmatrix} a & b \\ c & d \end{bmatrix}$, the optimal strategies $P^*$ and

   $Q^*$ and the value $v$ are given by

Copyright © 2011 Pearson Education, Inc. Publishing as Prentice Hall.

$$P^* = \begin{bmatrix} \dfrac{d-c}{D} & \dfrac{a-b}{D} \end{bmatrix}, \quad Q^* = \begin{bmatrix} \dfrac{d-b}{D} \\ \dfrac{a-c}{D} \end{bmatrix}, \quad \text{and} \quad v = \dfrac{ad-bc}{D},$$

where $D = (a + d) - (b + c)$. [Note: Under the assumption that $M$ is not strictly determined, it can be shown that $D = (a + d) - (b + c)$ will never be 0.]

5.    RECESSIVE ROWS AND COLUMNS

A row of a payoff matrix is a RECESSIVE ROW if there exists another row, called a DOMINANT ROW, such that each element of the dominant row is greater than or equal to ($\geq$) the corresponding element of the recessive row.

A column of a payoff matrix is a RECESSIVE COLUMN if there exists another column, called a DOMINANT COLUMN, such that each element of the dominant column is less than or equal to ($\leq$) the corresponding element of the recessive column.

---

1.    Row 2 dominates row 1, $3 \geq 2$, $5 \geq -1$; row 1 is recessive. No recessive column.

3.    Column $1 \leq$ column 2; column $1 \leq$ column 3. Column 1, dominates columns 2 and 3; coluimns 2 and 3 are recessive columns. No recessive rows.

5.    No recessive rows; no recessive columns.

7.    Row 3 dominates row 2, $3 \geq 0$, $-2 \geq -5$, $1 \geq 1$; row 2 is recessive. No recessive columns.

9.    There are no recessive rows or columns.

11.    $\begin{bmatrix} \boxed{-1} & \boxed{2} \\ \boxed{2} & \boxed{-4} \end{bmatrix}$    `The game is not strictly determined.`

Using 4: The optimal strategy for $R$ is
$$P^* = \begin{bmatrix} \dfrac{d-c}{D} & \dfrac{a-b}{D} \end{bmatrix} = \begin{bmatrix} \dfrac{-4-2}{-9} & \dfrac{-1-2}{-9} \end{bmatrix} = \begin{bmatrix} \dfrac{2}{3} & \dfrac{1}{3} \end{bmatrix}$$
[Note: $D = (a + d) - (b + c) = (-1 - 4) - (2 + 2) = -9$.]

The optimal strategy for $C$ is
$$Q^* = \begin{bmatrix} \dfrac{d-b}{D} \\ \dfrac{a-c}{D} \end{bmatrix} = \begin{bmatrix} \dfrac{-4-2}{-9} \\ \dfrac{-1-2}{-9} \end{bmatrix} = \begin{bmatrix} \dfrac{2}{3} \\ \dfrac{1}{3} \end{bmatrix}$$

The value of the game, $v$, is: $\dfrac{ad-bc}{D} = \dfrac{(-1)(-4) - (2)(2)}{-9} = 0.$

13.    $\begin{bmatrix} \boxed{-1} & \boxed{2} \\ \boxed{1} & \boxed{0} \end{bmatrix}$    `The game is not strictly determined.`

Using 4:
$$P^* = \begin{bmatrix} \dfrac{0-1}{-4} & \dfrac{-1-2}{-4} \end{bmatrix} = \begin{bmatrix} \dfrac{1}{4} & \dfrac{3}{4} \end{bmatrix} \quad [\text{Note: } D = (-1 + 0) - (2 + 1) = -4.]$$

$$Q^* = \begin{bmatrix} \dfrac{0-2}{-4} \\ \dfrac{-1-1}{-4} \end{bmatrix} = \begin{bmatrix} \dfrac{1}{2} \\ \dfrac{1}{2} \end{bmatrix}$$

The value of the game, $v$, is: $\dfrac{(-1)(0) - (2)(1)}{-4} = \dfrac{1}{2}.$

Copyright © 2011 Pearson Education, Inc. Publishing as Prentice Hall.

**15.** $\begin{bmatrix} \boxed{4} & \boxed{-6} \\ \boxed{-2} & \boxed{3} \end{bmatrix}$  The game is not strictly determined.

Using $\underline{4}$:

$P^* = \begin{bmatrix} \frac{3+2}{15} & \frac{4+6}{15} \end{bmatrix} = \begin{bmatrix} \frac{1}{3} & \frac{2}{3} \end{bmatrix}$    $Q^* = \begin{bmatrix} \frac{3+6}{15} \\ \frac{4+2}{15} \end{bmatrix} = \begin{bmatrix} \frac{3}{5} \\ \frac{2}{5} \end{bmatrix}$

The value of the game, $v$, is: $\dfrac{(4)(3) - (-6)(-2)}{15} = 0.$

**17.** $\begin{bmatrix} \boxed{5} & \boxed{-1} \\ 4 & \boxed{1} \end{bmatrix}$  The game is strictly determined.

$P^* = \begin{bmatrix} 0 & 1 \end{bmatrix}$    $Q^* = \begin{bmatrix} 0 \\ 1 \end{bmatrix}$

The value of the game, $v$ = saddle value = 1.

**19.** Using $\underline{5}$, we can eliminate the recessive columns $\begin{bmatrix} 0 \\ 1 \end{bmatrix}$ and $\begin{bmatrix} 2 \\ -1 \end{bmatrix}$.

Thus, we obtain the $2 \times 2$ matrix:

$\begin{bmatrix} \boxed{2} & \boxed{-1} \\ \boxed{-2} & \boxed{1} \end{bmatrix}$ The game is not strictly determined.

Now, using $\underline{4}$:

$P^* = \begin{bmatrix} \frac{1+2}{6} & \frac{2+1}{6} \end{bmatrix} = \begin{bmatrix} \frac{3}{6} & \frac{3}{6} \end{bmatrix} = \begin{bmatrix} \frac{1}{2} & \frac{1}{2} \end{bmatrix}$ [Note: $D = (2+1) - (-1-2) = 6$.]

$Q^* = \begin{bmatrix} \frac{1+1}{6} \\ \frac{2+2}{6} \end{bmatrix} = \begin{bmatrix} \frac{2}{6} \\ \frac{4}{6} \end{bmatrix} = \begin{bmatrix} \frac{1}{3} \\ \frac{2}{3} \end{bmatrix}$ or $\begin{bmatrix} 0 \\ \frac{1}{3} \\ \frac{2}{3} \\ 0 \end{bmatrix}$,  $v = \dfrac{(2)(1) - (-1)(-2)}{6} = 0$

**21.** Using $\underline{5}$, we eliminate the recessive row $\begin{bmatrix} 1 & -4 & -1 \end{bmatrix}$ and the recessive

column $\begin{bmatrix} -1 \\ 0 \\ 2 \end{bmatrix}$. Thus, we obtain the $2 \times 2$ matrix: $\begin{bmatrix} \boxed{2} & \boxed{-3} \\ \boxed{-1} & \boxed{2} \end{bmatrix}$.

The game is nonstrictly determined. Now, using $\underline{4}$:

$P^* = \begin{bmatrix} \frac{2+1}{8} & \frac{2+3}{8} \end{bmatrix} = \begin{bmatrix} \frac{3}{8} & \frac{5}{8} \end{bmatrix}$ or $\begin{bmatrix} 0 & \frac{3}{8} & \frac{5}{8} \end{bmatrix}$ .[Note: $D = (2+2) - (-3-1) = 8$.]

$Q^* = \begin{bmatrix} \frac{2+3}{8} \\ \frac{2+1}{8} \end{bmatrix} = \begin{bmatrix} \frac{5}{8} \\ \frac{3}{8} \end{bmatrix}$ or $\begin{bmatrix} \frac{5}{8} \\ \frac{3}{8} \\ 0 \end{bmatrix}$,  $v = \dfrac{(2)(2) - (-3)(-1)}{8} = \dfrac{1}{8}$

Copyright © 2011 Pearson Education, Inc. Publishing as Prentice Hall.

**23.**

$$\begin{bmatrix} 2 & \boxed{①} & 2 \\ \boxed{3} & 0 & \boxed{-5} \\ 1 & \boxed{-2} & \boxed{7} \end{bmatrix}$$  The game is strictly determined.

$$P* = [1 \quad 0 \quad 0] \quad Q* = \begin{bmatrix} 0 \\ 1 \\ 0 \end{bmatrix} \quad v = \text{saddle value} = 1$$

**25.** True. If $a_{ij}$ is a saddle value of the payoff matrix, then $R$ will always choose row $i$ and $C$ will always choose column $j$. Therefore, the $i$-th entry in the corresponding strategy $P$ (for $R$) is 1 and all other entries are 0. Similarly, the $j$-th in $Q$ (for C) is 1 and all other entries are 0.

**27.** False. Consider the payoff matrix

$$\begin{bmatrix} 0 & 0 & 0 \\ -1 & 1 & 0 \\ 3 & -1 & 1 \end{bmatrix}$$

The first row is not recessive.

**29.** True. Player $C$ puts a square around each entry in column 1 and a square around at least one entry in column 2. Since player $R$ circles at least two entries, at least one entry will have both a circle and a square.

**31.** (A)    $P = \begin{bmatrix} \frac{1}{4} & \frac{1}{4} & \frac{1}{4} & \frac{1}{4} \end{bmatrix}$, $Q = \begin{bmatrix} \frac{1}{4} \\ \frac{1}{4} \\ \frac{1}{4} \\ \frac{1}{4} \\ \frac{1}{4} \end{bmatrix}$

$$PMQ = \begin{bmatrix} \frac{1}{4} & \frac{1}{4} & \frac{1}{4} & \frac{1}{4} \end{bmatrix} \begin{bmatrix} 0 & 2 & 1 & 0 \\ 4 & 3 & 5 & 4 \\ 0 & 2 & 6 & 1 \\ 0 & 1 & 0 & 3 \end{bmatrix} \begin{bmatrix} \frac{1}{4} \\ \frac{1}{4} \\ \frac{1}{4} \\ \frac{1}{4} \end{bmatrix} = [1 \quad 2 \quad 3 \quad 2] \begin{bmatrix} \frac{1}{4} \\ \frac{1}{4} \\ \frac{1}{4} \\ \frac{1}{4} \end{bmatrix} = 2 \quad \text{or } \$2$$

Since you pay C \$3 to play, the expected value is \$2 – \$3 = –\$1.

(B)    The largest of the row minimums is 3 in row 2. Thus, you should play row 2, i.e.,
$P = [0 \quad 1 \quad 0 \quad 0]$. Now

$$PMQ = [0 \quad 1 \quad 0 \quad 0] \begin{bmatrix} 0 & 2 & 1 & 0 \\ 4 & 3 & 5 & 4 \\ 0 & 2 & 6 & 1 \\ 0 & 1 & 0 & 3 \end{bmatrix} \begin{bmatrix} \frac{1}{4} \\ \frac{1}{4} \\ \frac{1}{4} \\ \frac{1}{4} \end{bmatrix} = [4 \quad 3 \quad 5 \quad 4] \begin{bmatrix} \frac{1}{4} \\ \frac{1}{4} \\ \frac{1}{4} \\ \frac{1}{4} \end{bmatrix} = 4 \quad \text{or } \$4$$

Since you pay C \$3 to play, the expected value is \$4 – \$3 = \$1.

(C)    We eliminate the recessive rows [0 2 1 0] and [0 1 0 3], and the recessive columns

Copyright © 2011 Pearson Education, Inc. Publishing as Prentice Hall.

$$\begin{bmatrix} 1 \\ 5 \\ 6 \\ 0 \end{bmatrix} \text{ and } \begin{bmatrix} 0 \\ 4 \\ 1 \\ 3 \end{bmatrix}. \text{ The resulting matrix is: } \begin{bmatrix} \boxed{4} & ③ \\ ⓪ & 2 \end{bmatrix}$$

and the game is strictly determined with a value of 3. You should play row 2 (in the original matrix) and $C$ should play column 2 (in the original matrix).

Thus,

$$P = [0 \quad 1 \quad 0 \quad 0] \text{ and } Q = \begin{bmatrix} 0 \\ 1 \\ 0 \\ 0 \end{bmatrix}$$

$$PMQ = [0 \quad 1 \quad 0 \quad 0] \begin{bmatrix} 0 & 2 & 1 & 0 \\ 4 & 3 & 5 & 4 \\ 0 & 2 & 6 & 1 \\ 0 & 1 & 0 & 3 \end{bmatrix} \begin{bmatrix} 0 \\ 1 \\ 0 \\ 0 \end{bmatrix} = [4 \quad 3 \quad 5 \quad 4] \begin{bmatrix} 0 \\ 1 \\ 0 \\ 0 \end{bmatrix} = 3 \quad \text{or } \$3$$

Since you pay C \$3 to play, the expected value \$3 – \$3 = \$0.

**33.** $PMQ = [p_1 \quad p_2] \begin{bmatrix} a & b \\ c & d \end{bmatrix} \begin{bmatrix} q_1 \\ q_2 \end{bmatrix} = [p_1 \quad p_2] \begin{bmatrix} [a\,b] \cdot \begin{bmatrix} q_1 \\ q_2 \end{bmatrix} \\ [c\,d] \cdot \begin{bmatrix} q_1 \\ q_2 \end{bmatrix} \end{bmatrix}$

$= [p_1 \quad p_2] \begin{bmatrix} aq_1 + bq_2 \\ cq_1 + dq_2 \end{bmatrix} = [p_1(aq_1 + bq_2) + p_2(cq_1 + dq_2)]$

$= [ap_1q_1 + bp_1q_2 + cp_2q_1 + dp_2q_2] = E(P, Q)$

**35.** $P*MQ = \begin{bmatrix} \frac{d-c}{D} & \frac{a-b}{D} \end{bmatrix} \begin{bmatrix} a & b \\ c & d \end{bmatrix} \begin{bmatrix} q_1 \\ q_2 \end{bmatrix}$

In Problem 29, substitute $p_1 = \dfrac{d-c}{D}$ and $p_2 = \dfrac{a-b}{D}$. Thus,

$P*MQ = \dfrac{a(d-c)}{D}q_1 + \dfrac{b(d-c)}{D}q_2 + \dfrac{c(a-b)}{D}q_1 + \dfrac{d(a-b)}{D}q_2$

$= \dfrac{1}{D}[adq_1 - caq_1 + bdq_2 - bcq_2 + caq_1 - cbq_1 + daq_2 - dbq_2]$

$= \dfrac{1}{D}[(ad - cb)q_1 + (ad - bc)q_2] = \dfrac{ad-cb}{D}[q_1 + q_2]$   [Note: $q_1 + q_2 = 1$.]

$= \dfrac{ad-cb}{D}$

Thus,

(A)   $P*MQ = \dfrac{ad-cb}{D}$

Copyright © 2011 Pearson Education, Inc. Publishing as Prentice Hall.

Similarly, $PMQ* = [p_1 \quad p_2]\begin{bmatrix} a & b \\ c & d \end{bmatrix}\begin{bmatrix} \frac{d-b}{D} \\ \frac{a-c}{D} \end{bmatrix}$

$$= ap_1\left(\frac{d-b}{D}\right) + bp_1\left(\frac{a-c}{D}\right) + cp_2\left(\frac{d-b}{D}\right) + dp_2\left(\frac{a-c}{D}\right)$$

$$= \frac{1}{D}(adp_1 - abp_1 + bap_1 - bcp_1 + cdp_2 - cbp_2 + dap_2 - dcp_2)$$

$$= \frac{1}{D}(adp_1 - bcp_1 - cbp_2 + dap_2) = \frac{1}{D}[(ad - bc)p_1 + (da - cb)p_2]$$

$$= \frac{ad - bc}{D}(p_1 + p_2) = \frac{ad - bc}{D} \quad [\text{Note: } p_1 + p_2 = 1.]$$

Thus,

(B)    $PMQ* = \dfrac{ad - bc}{D}$

From (A) and (B), we get $v = \dfrac{ad - bc}{D}$, which satisfies $P*MQ \geq v$ and $PMQ* \leq v$.

**37.** Solve the linear system

$$\frac{d-c}{D} = 0.9$$

$$\frac{a-b}{D} = 0.1$$

$$\frac{d-b}{D} = 0.3$$

$$\frac{a-c}{D} = 0.7$$

where $D = (a + d) - (b + c)$. In standard form, this system is:

$0.9a - 0.9b + 0.1c - 0.1d = 0$
$-0.9a + 0.9b - 0.1c + 0.1d = 0$
$0.3a + 0.7b - 0.3c - 0.7d = 0$
$-0.3a - 0.7b + 0.3c + 0.7d = 0$

which reduces to

$0.9a - 0.9b + 0.1c - 0.1d = 0$
$0.3a + 0.7b - 0.3c - 0.7d = 0$

Choose two of the unknowns arbitrarily and solve for the other two unknowns. For example, setting $d = 0$ and $c = 10$ gives

$0.9a - 0.9b = -1$
$0.3a + 0.7b = 3$

whose solution is $a = \dfrac{20}{9}$, $b = \dfrac{10}{3}$

$$M = \begin{bmatrix} \frac{20}{9} & \frac{10}{3} \\ 10 & 0 \end{bmatrix}$$

Copyright © 2011 Pearson Education, Inc. Publishing as Prentice Hall.

**39.** (A) Replace $p_2$ by $1 - p_1$ and $q_2$ by $1 - q_1$ in $E(P,Q) = ap_1q_1 + bp_1q_2 + cp_2q_1 + dp_2q_2$ and simplify. The result is $E(P,Q) = [Dp_1 - (d - c)]q_1 + (b - d)p_1 + d$ where $D = (a + d) - (b + c)$.

(B) If $Dp_1 - (d - c) = 0$, then $E(P,Q) = (b - d)p_1 + d$ regardless of the value of $q_1$. Therefore,

$$v = \frac{ad - bc}{D} \text{ regardless of the value of } q_1.$$

**41.** Given:

$$
\begin{array}{c c}
 & \begin{array}{cccc} TV & R & P & M \end{array} \\
\begin{array}{c} TV \\ R \\ P \\ M \end{array} &
\left[\begin{array}{cccc}
0 & -1 & -1 & 0 \\
1 & 2 & -1 & -1 \\
0 & -1 & 0 & 1 \\
-1 & -1 & -1 & 0
\end{array}\right]
\end{array}
$$

Eliminate the recessive rows $[0 \;\; -1 \;\; -1 \;\; 0]$ and $[-1 \;\; -1 \;\; -1 \;\; 0]$ and

the recessive columns $\begin{bmatrix} 0 \\ 1 \\ 0 \\ -1 \end{bmatrix}$ and $\begin{bmatrix} 0 \\ -1 \\ 1 \\ 0 \end{bmatrix}$ to obtain the $2 \times 2$ matrix $\begin{bmatrix} \boxed{2} & \boxed{-1} \\ \boxed{-1} & \boxed{0} \end{bmatrix}$.

The game is not strictly determined.

(A) The optimal strategies for $R$ and $C$ are

$$P^* = \begin{bmatrix} \frac{0+1}{4} & \frac{2+1}{4} \end{bmatrix} = \begin{bmatrix} \frac{1}{4} & \frac{3}{4} \end{bmatrix} \quad \text{[Note: } D = 2 + 0 - (-1 - 1) = 4.\text{]}$$

and $Q^* = \begin{bmatrix} \frac{0+1}{4} \\ \frac{2+1}{4} \end{bmatrix} = \begin{bmatrix} \frac{1}{4} \\ \frac{3}{4} \end{bmatrix}$, respectively.

In terms of the original problem:

$$P^* = \begin{bmatrix} 0 & \frac{1}{4} & \frac{3}{4} & 0 \end{bmatrix} \text{ and } Q^* = \begin{bmatrix} 0 \\ \frac{1}{4} \\ \frac{3}{4} \\ 0 \end{bmatrix}$$

The value of the game is: $v = \dfrac{2(0) - (-1)(-1)}{4} = -\dfrac{1}{4}$

(B)    $P = [1 \;\; 0 \;\; 0 \;\; 0]$ and $Q = \begin{bmatrix} 0 \\ \frac{1}{4} \\ \frac{3}{4} \\ 0 \end{bmatrix}$

$$PMQ = [1 \;\; 0 \;\; 0 \;\; 0]\begin{bmatrix}
0 & -1 & -1 & 0 \\
1 & 2 & -1 & -1 \\
0 & -1 & 0 & 1 \\
-1 & -1 & -1 & 0
\end{bmatrix}\begin{bmatrix} 0 \\ \frac{1}{4} \\ \frac{3}{4} \\ 0 \end{bmatrix} = [0 \;\;-1\;\;-1\;\;0]\begin{bmatrix} 0 \\ \frac{1}{4} \\ \frac{3}{4} \\ 0 \end{bmatrix} = -1$$

The expected value is $-1$.

Copyright © 2011 Pearson Education, Inc. Publishing as Prentice Hall.

(C)    $P = \begin{bmatrix} 0 & \frac{1}{4} & \frac{3}{4} & 0 \end{bmatrix}$  and  $Q^* = \begin{bmatrix} 0 \\ 1 \\ 0 \\ 0 \end{bmatrix}$

$$PMQ = \begin{bmatrix} 0 & \frac{1}{4} & \frac{3}{4} & 0 \end{bmatrix} \begin{bmatrix} 0 & -1 & -1 & 0 \\ 1 & 2 & -1 & -1 \\ 0 & -1 & 0 & 1 \\ -1 & -1 & -1 & 0 \end{bmatrix} \begin{bmatrix} 0 \\ 1 \\ 0 \\ 0 \end{bmatrix} = \begin{bmatrix} 0 & \frac{1}{4} & \frac{3}{4} & 0 \end{bmatrix} \begin{bmatrix} -1 \\ 2 \\ -1 \\ -1 \end{bmatrix} = -\frac{1}{4}$$

The expected value is $-\dfrac{1}{4}$.

(D)    $P = \begin{bmatrix} 0 & 0 & 1 & 0 \end{bmatrix}$  and  $Q = \begin{bmatrix} 0 \\ 0 \\ 1 \\ 0 \end{bmatrix}$

$$PMQ = \begin{bmatrix} 0 & 0 & 1 & 0 \end{bmatrix} \begin{bmatrix} 0 & -1 & -1 & 0 \\ 1 & 2 & -1 & -1 \\ 0 & -1 & 0 & 1 \\ -1 & -1 & -1 & 0 \end{bmatrix} \begin{bmatrix} 0 \\ 0 \\ 1 \\ 0 \end{bmatrix} = \begin{bmatrix} 0 & -1 & 0 & 1 \end{bmatrix} \begin{bmatrix} 0 \\ 0 \\ 1 \\ 0 \end{bmatrix} = 0$$

The expected value is 0.

**43.**

                      Player $C$(Fate)
                    Republican    Democrat
          Solar
Player $R$ energy    $1000         $4000
  (You)
           oil       $5000         $3000                The game is not strictly determined.

Thus, using <u>4</u>:

$P^* = \begin{bmatrix} \dfrac{3000-5000}{-5000} & \dfrac{1000-4000}{-5000} \end{bmatrix} = \begin{bmatrix} \dfrac{2}{5} & \dfrac{3}{5} \end{bmatrix}$

[<u>Note:</u> $D = (1000 + 3000) - (4000 + 5000) = -5000$.]

$Q^* = \begin{bmatrix} \dfrac{3000-4000}{-5000} \\ \dfrac{1000-5000}{-5000} \end{bmatrix} = \begin{bmatrix} \dfrac{1}{5} \\ \dfrac{4}{5} \end{bmatrix}$

$v = \dfrac{(1000)(3000) - (4000)(5000)}{-5000} = \dfrac{-17,000}{-5} = \$3400$

This means that you should invest $\dfrac{2}{5}(10,000) = \$4000$ in solar energy

stocks and $\dfrac{3}{5}(10,000) = \$6000$ in oil stocks.  You would then have an

expected gain of $3400 no matter how the election turns out.

Copyright © 2011 Pearson Education, Inc.  Publishing as Prentice Hall.

---

EXERCISE 10-3

*Things to remember:*

2 × 2 MATRIX GAMES AND LINEAR PROGRAMMING   —   GEOMETRIC APPROACH

Given the nonstrictly determined matrix game with payoff matrix

$$M = \begin{bmatrix} a & b \\ c & d \end{bmatrix}.$$

Find $P^* = [p_1 \ p_2]$, $Q^* = \begin{bmatrix} q_1 \\ q_2 \end{bmatrix}$, and $v$, proceed as follows:

STEP 1.    If $M$ is not a positive matrix, convert it into a positive matrix $M_1$ by adding a suitable positive constant $k$ to each element.  Let

$$M_1 = \begin{bmatrix} e & f \\ g & h \end{bmatrix}, \quad \begin{matrix} e = a + k, f = b + k, \\ g = c + k, h = d + k. \end{matrix}$$

This matrix game $M_1$ has the same optimal strategies $P^*$ and $Q^*$ as $M$, and if $v_1$ is the value of the game $M_1$, then $v = v_1 - k$ is the value of the original game $M$.

STEP 2.    Set up the two corresponding linear programming problems:

(A)    Minimize $y = x_1 + x_2$

Subject to: $ex_1 + gx_2 \geq 1$

$fx_1 + hx_2 \geq 1$

$x_1, x_2 \geq 0$

(B)    Maximize $y = z_1 + z_2$

Subject to: $ez_1 + fz_2 \leq 1$

$gz_1 + hz_2 \leq 1$

$z_1, z_2 \geq 0$

STEP 3.    Solve each linear programming problem geometrically.  [Note: Since (B) is the dual of (A), both problems have the same optimal value, i.e., min $y$ in (A) equals max $y$ in (B).]

STEP 4.    Use the solutions in Step 3:

$$v_1 = \frac{1}{y} = \frac{1}{x_1 + x_2} \text{ or } v_1 = \frac{1}{y} = \frac{1}{z_1 + z_2}$$

$$P^* = [p_1 \ p_2] = [v_1 x_1 \ v_1 x_2]$$

$$Q^* = \begin{bmatrix} q_1 \\ q_2 \end{bmatrix} = \begin{bmatrix} v_1 z_1 \\ v_1 z_2 \end{bmatrix}$$

$$v = v_1 - k$$

STEP 5.    The solution found in Step 4 can be checked by showing that

$P^*MQ^* = v$.  (See Theorem 3 of Section 10.2.)

---

Copyright © 2011 Pearson Education, Inc.  Publishing as Prentice Hall.

1. $\begin{bmatrix} -3 & -5 \\ 2 & -6 \end{bmatrix}$. The smallest entry is $-6$; $k = 7$.

3. $\begin{bmatrix} 0 & 1 \\ 2 & 0 \end{bmatrix}$. The smallest entry is $0$; $k = 1$.

5. $\begin{bmatrix} 1 & 2 \\ 4 & 7 \end{bmatrix}$. The entries are all positive already; $k = 0$.

7. <u>Step 1</u>. Convert $M = \begin{bmatrix} 2 & -3 \\ -1 & 2 \end{bmatrix}$ into a positive payoff matrix by adding 4 to each payoff: $M_1 = \begin{bmatrix} 6 & 1 \\ 3 & 6 \end{bmatrix}$

<u>Step 2</u>. Set up the two corresponding linear programming problems:

(A)   Minimize $y = x_1 + x_2$

     Subject to $6x_1 + 3x_2 \geq 1$

               $x_1 + 6x_2 \geq 1$

               $x_1 \geq 0, x_2 \geq 0$

(B)   Maximize $y = z_1 + z_2$

     Subject to $6z_1 + z_2 \leq 1$

               $3z_1 + 6z_2 \leq 1$

               $z_1 \geq 0, z_2 \geq 0$

<u>Step 3</u>. Solve the linear programming problems geometrically:

(A)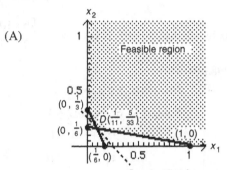

(B)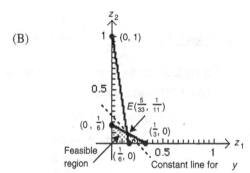

Theorems 1 and 2 in Section 5.1 imply that each problem has a solution that must occur at a corner point.

(A)

| Corner Points | Minimize $y = x_1 + x_2$ |
|---|---|
| $\left(0, \dfrac{1}{3}\right)$ | $\dfrac{1}{3}$ |
| $\left(\dfrac{1}{11}, \dfrac{5}{33}\right)$ | $\dfrac{8}{33}$ |
| $(1, 0)$ | $1$ |

Min $y = \dfrac{8}{33}$ occurs at

$x_1 = \dfrac{1}{11}$, $x_2 = \dfrac{5}{33}$

(B)

| Corner Points | Maximize $y = z_1 + z_2$ |
|---|---|
| $(0, 0)$ | $0$ |
| $\left(0, \dfrac{1}{6}\right)$ | $\dfrac{1}{6}$ |
| $\left(\dfrac{5}{33}, \dfrac{1}{11}\right)$ | $\dfrac{8}{33}$ |
| $\left(\dfrac{1}{6}, 0\right)$ | $\dfrac{1}{6}$ |

Max $y = \dfrac{8}{33}$ occurs at

$z_1 = \dfrac{5}{33}$, $z_2 = \dfrac{1}{11}$

Copyright © 2011 Pearson Education, Inc. Publishing as Prentice Hall.

<u>Step 4</u>. Use the solutions in Step 3.

(A)  $v_1 = \dfrac{1}{x_1 + x_2} = \dfrac{1}{\dfrac{1}{11} + \dfrac{5}{33}} = \dfrac{33}{8}$

$p_1 = v_1 x_1 = \dfrac{33}{8} \cdot \dfrac{1}{11} = \dfrac{3}{8}$

$p_2 = v_1 x_2 = \dfrac{33}{8} \cdot \dfrac{5}{33} = \dfrac{5}{8}$

(B)  $v_1 = \dfrac{1}{z_1 + z_2} = \dfrac{1}{\dfrac{5}{33} + \dfrac{1}{11}} = \dfrac{33}{8}$

$q_1 = v_1 z_1 = \dfrac{33}{8} \cdot \dfrac{5}{33} = \dfrac{5}{8}$

$q_2 = v_1 z_2 = \dfrac{33}{8} \cdot \dfrac{1}{11} = \dfrac{3}{8}$

The optimal strategies for both games $M_1$ and $M_2$ are:

$$p^* = [p_1, p_2] = \left[\dfrac{3}{8}, \dfrac{5}{8}\right], \quad Q^* = \begin{bmatrix} q_1 \\ q_2 \end{bmatrix} = \begin{bmatrix} \dfrac{5}{8} \\ \dfrac{3}{8} \end{bmatrix}$$

The value of the original game is: $v = v_1 - k = \dfrac{33}{8} - 4 = \dfrac{1}{8}$.

<u>Step 5</u>. Check:

$$P^*MQ^* = \left[\dfrac{3}{8}, \dfrac{5}{8}\right] \begin{bmatrix} 2 & -3 \\ -1 & 2 \end{bmatrix} \begin{bmatrix} \dfrac{5}{8} \\ \dfrac{3}{8} \end{bmatrix} = \left[\dfrac{1}{8}, \dfrac{1}{8}\right] \begin{bmatrix} \dfrac{5}{8} \\ \dfrac{3}{8} \end{bmatrix} = \dfrac{1}{8}$$

9.  <u>Step 1</u>. Convert $\begin{bmatrix} -1 & 3 \\ 2 & -6 \end{bmatrix}$ into a positive payoff matrix by adding 7 to each payoff:  $M_1 = \begin{bmatrix} 6 & 10 \\ 9 & 1 \end{bmatrix}$

<u>Step 2</u>. Set up the two corresponding linear programming problems:

(A)  Minimize  $y = x_1 + x_2$

Subject to  $6x_1 + 9x_2 \geq 1$

$10x_1 + x_2 \geq 1$

$x_1 \geq 0, x_2 \geq 0$

(B)  Maximize  $y = z_1 + z_2$

Subject to  $6z_1 + 10z_2 \leq 1$

$9z_1 + z_2 \leq 1$

$z_1 \geq 0, z_2 \geq 0$

<u>Step 3</u>. Solve the linear programming problems geometrically:
The graphs of the inequalities are shown in the following figures.

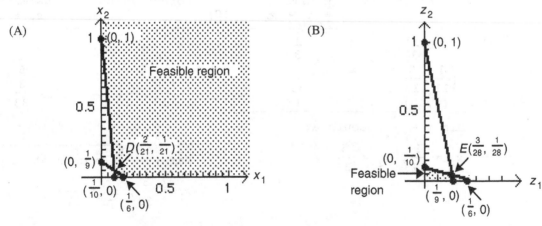

Theorems 1 and 2 in Section 5.1 imply that each problem has a solution that must occur at a corner point.

Copyright © 2011 Pearson Education, Inc.  Publishing as Prentice Hall.

| (A) | Corner Points | Minimize $y = x_1 + x_2$ |
|---|---|---|
| | $(0, 1)$ | $1$ |
| | $\left(\dfrac{2}{21}, \dfrac{1}{21}\right)$ | $\dfrac{1}{7}$ |
| | $\left(\dfrac{1}{6}, 0\right)$ | $\dfrac{1}{6}$ |

Min $y = \dfrac{1}{7}$ occurs at

$$x_1 = \frac{2}{21}, \quad x_2 = \frac{1}{21}$$

(B)

| Corner Points | Maximize $y = z_1 + z_2$ |
|---|---|
| $(0, 0)$ | $0$ |
| $\left(0, \dfrac{1}{10}\right)$ | $\dfrac{1}{10}$ |
| $\left(\dfrac{3}{28}, \dfrac{1}{28}\right)$ | $\dfrac{1}{7}$ |
| $\left(\dfrac{1}{9}, 0\right)$ | $\dfrac{1}{9}$ |

Max $y = \dfrac{1}{7}$ occurs at $z_1 = \dfrac{3}{28}$,

$$z_2 = \frac{1}{28}$$

Step 4. Use the solutions in Step 3:

(A)  $$v_1 = \frac{1}{x_1 + x_2} = \frac{1}{\dfrac{2}{21} + \dfrac{1}{21}} = 7$$

$$p_1 = v_1 x_1 = 7 \cdot \frac{2}{21} = \frac{2}{3}$$

$$p_2 = v_1 x_2 = 7 \cdot \frac{1}{21} = \frac{1}{3}$$

(B)  $$v_1 = \frac{1}{z_1 + z_2} = \frac{1}{\dfrac{3}{28} + \dfrac{1}{28}} = 7$$

$$q_1 = v_1 z_1 = 7 \cdot \frac{3}{28} = \frac{3}{4}$$

$$q_2 = v_1 z_2 = 7 \cdot \frac{1}{28} = \frac{1}{4}$$

The optimal strategies for both games $M_1$ and $M_2$ are:

$$p^* = [p_1, p_2] = \left[\frac{2}{3}, \frac{1}{3}\right], \quad Q^* = \begin{bmatrix} q_1 \\ q_2 \end{bmatrix} = \begin{bmatrix} \dfrac{3}{4} \\ \dfrac{1}{4} \end{bmatrix}$$

The value of the original game is: $v = v_1 - k = 7 - 7 = 0$.

Step 5. Check:

$$P^* M Q^* = \left[\frac{2}{3}, \frac{1}{3}\right] \begin{bmatrix} -1 & 3 \\ 2 & -6 \end{bmatrix} \begin{bmatrix} \dfrac{3}{4} \\ \dfrac{1}{4} \end{bmatrix} = [0, 0] \begin{bmatrix} \dfrac{3}{4} \\ \dfrac{1}{4} \end{bmatrix} = 0$$

11.  Step 1. Convert $\begin{bmatrix} -2 & -1 \\ 5 & 6 \end{bmatrix}$ into a positive payoff matrix by adding 3 to each payoff: $M_1 = \begin{bmatrix} 1 & 2 \\ 8 & 9 \end{bmatrix}$

Step 2. Set up the two corresponding linear programming problems:

(A)  Minimize $y = x_1 + x_2$

Subject to: $x_1 + 8x_2 \geq 1$

$2x_1 + 9x_2 \geq 1$

$x_1, x_2 \geq 0$

(B) Maximize $y = z_1 + z_2$

Subject to: $z_1 + 2z_2 \leq 1$

$8z_1 + 9z_2 \leq 1$

$z_1, z_2 \geq 0$

Copyright © 2011 Pearson Education, Inc. Publishing as Prentice Hall.

<u>Step 3</u>. Solve the linear programming problems geometrically:
The graphs of the inequalities are shown in the following figures:

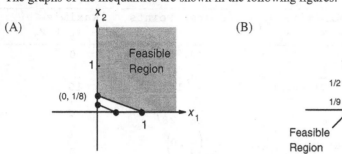

(A)

(B)

Theorems 1 and 2 in Section 5.1 imply that each problem has a solution that must occur at a corner point.

(A)

| Corner Points | Minimize |
|---|---|
| | $y = x_1 + x_2$ |
| $\left(0, \dfrac{1}{8}\right)$ | $\dfrac{1}{8}$ |
| $(1, 0)$ | $1$ |

Min $y = \dfrac{1}{8}$ occurs at $x_1 = 0$,

$x_2 = \dfrac{1}{8}$

(B)

| Corner Points | Maximize |
|---|---|
| | $y = z_1 + z_2$ |
| $(0, 0)$ | $0$ |
| $\left(0, \dfrac{1}{9}\right)$ | $\dfrac{1}{9}$ |
| $\left(\dfrac{1}{8}, 0\right)$ | $\dfrac{1}{8}$ |

Max $y = \dfrac{1}{8}$ occurs at $z_1 = \dfrac{1}{8}$,

$z_2 = 0$

<u>Step 4</u>. Use the solutions in Step 3:

(A)   $v_1 = \dfrac{1}{x_1 + x_2} = \dfrac{1}{0 + \dfrac{1}{8}} = 8$

$p_1 = v_1 x_1 = 8 \cdot 0 = 0$

$p_2 = v_1 x_2 = 8 \cdot \dfrac{1}{8} = 1$

(B)   $v_1 = \dfrac{1}{z_1 + z_2} = \dfrac{1}{\dfrac{1}{8} + 0} = 8$

$q_1 = v_1 z_1 = 8 \cdot \dfrac{1}{8} = 1$

$q_2 = v_1 z_2 = 8 \cdot 0 = 0$

The optimal strategies for both games $M_1$ and $M_2$ are:

$$p^* = [p_1, p_2] = [0, 1], \quad Q^* = \begin{bmatrix} q_1 \\ q_2 \end{bmatrix} = \begin{bmatrix} 1 \\ 0 \end{bmatrix}$$

The value of the original game is: $v = v_1 - k = 8 - 3 = 5$.

<u>Step 5</u>. Check:

$$P^*MQ^* = [0, 1] \begin{bmatrix} -2 & -1 \\ 5 & 6 \end{bmatrix} \begin{bmatrix} 1 \\ 0 \end{bmatrix} = [5, 6] \begin{bmatrix} 1 \\ 0 \end{bmatrix} = 5$$

**13.**   $M = \begin{bmatrix} -2 & -1 \\ 5 & 6 \end{bmatrix}$

Note that the matrix game is strictly determined:

$\begin{bmatrix} \text{ⓧ}{-2} & -1 \\ \boxed{5} & \boxed{6} \end{bmatrix}$; 5 is a saddle value.

Thus, the method of Section 10-1 applies.

**15.**   If $p_i > 0$ and $q_j > 0$, then $v = P^*MQ^* \geq p_i a_{ij} q_j > 0$.

Copyright © 2011 Pearson Education, Inc. Publishing as Prentice Hall.

**17.** True: $P* \begin{bmatrix} 0 & 0 \\ 0 & 0 \end{bmatrix} Q* = 0$ for all strategies $P*$ and $Q*$.

**19.** False: See Problems 7.

**21.** False: For a counterexample, see Problem 7 and 11.

**23.** In the game matrix $\begin{bmatrix} 3 & -6 \\ 4 & -6 \\ -2 & 3 \\ -3 & 0 \end{bmatrix}$ row 1 and row 4 are recessive rows. Eliminating these rows, we get

$M = \begin{bmatrix} 4 & -6 \\ -2 & 3 \end{bmatrix}$.

Step 1. Convert $M$ into a positive payoff matrix by adding 7 to each payoff: $M = \begin{bmatrix} 11 & 1 \\ 5 & 10 \end{bmatrix}$

Step 2. Set up the two corresponding linear programming problems:

(A)    Minimize $y = x_1 + x_2$

      Subject to $11x_1 + 5x_2 \geq 1$

               $x_1 + 10x_2 \geq 1$

               $x_1 \geq 0, x_2 \geq 0$

(B)    Maximize $y = z_1 + z_2$

      Subject to $11z_1 + z_2 \leq 1$

               $5z_1 + 10z_2 \leq 1$

               $z_1 \geq 0, z_2 \geq 0$

Step 3. Solve the linear programming problems geometrically:

The graphs corresponding to the inequalities above are shown in the following figures.

(A) 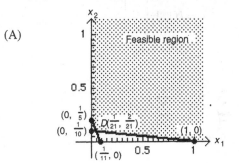     (B)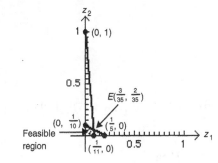

Theorems 1 and 2 in Section 5.1 imply that each problem has a solution that must occur at a corner point.

(A)

| Corner Points | Minimize $y = x_1 + x_2$ |
|---|---|
| $\left(0, \frac{1}{5}\right)$ | $\frac{1}{5}$ |
| $\left(\frac{1}{21}, \frac{2}{21}\right)$ | $\frac{1}{7}$ |
| $(1, 0)$ | $1$ |

Min $y = \frac{1}{7}$ occurs at

$x_1 = \frac{1}{21}$, $x_2 = \frac{2}{21}$

(B)

| Corner Points | Maximize $y = z_1 + z_2$ |
|---|---|
| $(0, 0)$ | $0$ |
| $\left(0, \frac{1}{10}\right)$ | $\frac{1}{10}$ |
| $\left(\frac{3}{35}, \frac{2}{35}\right)$ | $\frac{1}{5}$ |
| $\left(\frac{1}{11}, 0\right)$ | $\frac{1}{11}$ |

Max $y = \frac{1}{5}$ occurs at

$z_1 = \frac{3}{35}$, $z_2 = \frac{2}{35}$

Copyright © 2011 Pearson Education, Inc. Publishing as Prentice Hall.

Step 4. Use the solutions in Step 3:

(A)   $v_1 = \dfrac{1}{x_1 + x_2} = \dfrac{1}{\dfrac{1}{21} + \dfrac{2}{21}} = 7$

$p_1 = v_1 x_1 = 7 \cdot \dfrac{1}{21} = \dfrac{1}{3}$

$p_2 = v_1 x_2 = 7 \cdot \dfrac{2}{21} = \dfrac{2}{3}$

(B)   $v_1 = \dfrac{1}{z_1 + z_2} = \dfrac{1}{\dfrac{3}{35} + \dfrac{2}{35}} = 7$

$q_1 = v_1 z_1 = 7 \cdot \dfrac{3}{35} = \dfrac{3}{5}$

$q_2 = v_1 z_2 = 7 \cdot \dfrac{2}{35} = \dfrac{2}{5}$

The optimal strategies are:

$$p^* = [0, p_1, p_2, 0] = \begin{bmatrix} 0 & \frac{1}{3} & \frac{2}{3} & 0 \end{bmatrix} \text{ (Recall that rows 1 and 4 were recessive)}, \quad Q^* = \begin{bmatrix} \frac{3}{5} \\ \frac{2}{5} \end{bmatrix}$$

The value of the original game is: $v = v_1 - k = 7 - 7 = 0$.

Step 5. Check: $P^* M Q^* = \begin{bmatrix} 0 & \frac{1}{3} & \frac{2}{3} & 0 \end{bmatrix} \begin{bmatrix} 3 & -6 \\ 4 & -6 \\ -2 & 3 \\ -3 & 0 \end{bmatrix} \begin{bmatrix} \frac{3}{5} \\ \frac{2}{5} \end{bmatrix} = [0 \ \ 0] \begin{bmatrix} \frac{3}{5} \\ \frac{2}{5} \end{bmatrix} = 0$

25.   (A)   Let $P = (p_1, p_2)$ and $Q = \begin{bmatrix} q_1 \\ q_2 \end{bmatrix}$, where $p_1 + p_2 = 1$, $q_1 + q_2 = 1$.

Then

$$P(kJ)Q = k(p_1, p_2) \begin{bmatrix} 1 & 1 \\ 1 & 1 \end{bmatrix} \begin{bmatrix} q_1 \\ q_2 \end{bmatrix} = k(p_1 + p_2, p_1 + p_2) \begin{bmatrix} q_1 \\ q_2 \end{bmatrix}$$

$$= k(1, 1) \begin{bmatrix} q_1 \\ q_2 \end{bmatrix}$$

$$= k(q_1 + q_2) = k$$

(B)   Let $P = (p_1, p_2, \dots, p_m)$ and $Q = \begin{bmatrix} q_1 \\ q_2 \\ \vdots \\ q_n \end{bmatrix}$ where $\displaystyle\sum_{i=1}^{m} p_i = 1$, $\displaystyle\sum_{j=1}^{n} q_j = 1$

Then

$$P(kJ)Q = k(p_1, p_2, \dots, p_m) \begin{bmatrix} 1 & 1 & \cdots & 1 \\ 1 & 1 & \cdots & 1 \\ \vdots & \vdots & & \vdots \\ 1 & 1 & \cdots & 1 \end{bmatrix} \begin{bmatrix} q_1 \\ q_2 \\ \vdots \\ q_n \end{bmatrix}$$

$$(m \times n)$$

$$= k\left(\sum_{i=1}^{m} p_i, \sum_{i=1}^{m} p_i, \cdots, \sum_{i=1}^{m} p_i\right) \begin{bmatrix} q_1 \\ q_2 \\ \vdots \\ q_n \end{bmatrix} = k(1, 1, \dots, 1) \begin{bmatrix} q_1 \\ q_2 \\ \vdots \\ q_n \end{bmatrix} = k\sum_{j=1}^{n} q_j = k$$

Copyright © 2011 Pearson Education, Inc. Publishing as Prentice Hall.

**27.** Eliminate the recessive rows $[0 \;-1 \;-1 \;\; 0]$ and $[-1 \;-1 \;-1 \;\; 0]$, and the recessive columns $\begin{bmatrix} 0 \\ 1 \\ 0 \\ -1 \end{bmatrix}$ and

$\begin{bmatrix} 0 \\ -1 \\ 1 \\ 0 \end{bmatrix}$ to obtain the $2 \times 2$ matrix: $\quad M = \begin{bmatrix} 2 & -1 \\ -1 & 0 \end{bmatrix}$

Step 1. Add 2 to each element to get the positive matrix:
$$M_1 = \begin{bmatrix} 4 & 1 \\ 1 & 2 \end{bmatrix}$$

Step 2. Set up the two corresponding linear programming problems:

(A)  Minimize $y = x_1 + x_2$

Subject to $4x_1 + x_2 \geq 1$

$x_1 + 2x_2 \geq 1$

$x_1, x_2 \geq 0$

(B)  Maximize $y = z_1 + z_2$

Subject to $4z_1 + z_2 \leq 1$

$z_1 + 2z_2 \leq 1$

$z_1, z_2 \geq 0$

Step 3. Solve the linear programming problems geometrically:

(A)

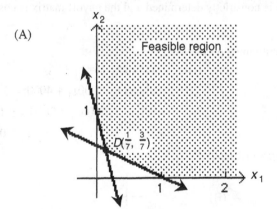

(B)

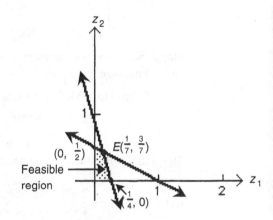

Theorems 1 and 2 in Section 5.1 imply that each problem has a solution that must occur at a corner point.

(A)

| Corner Points | Minimize $y = x_1 + x_2$ |
|---|---|
| $(0, 1)$ | 1 |
| $\left(\dfrac{1}{7}, \dfrac{3}{7}\right)$ | $\dfrac{4}{7}$ |
| $(1, 0)$ | 1 |

Min $y = \dfrac{4}{7}$ occurs at $x_1 = \dfrac{1}{7}$,

$x_2 = \dfrac{3}{7}$

(B)

| Corner Points | Maximize $y = z_1 + z_2$ |
|---|---|
| $(0, 0)$ | 0 |
| $\left(0, \dfrac{1}{2}\right)$ | $\dfrac{1}{2}$ |
| $\left(\dfrac{1}{7}, \dfrac{3}{7}\right)$ | $\dfrac{4}{7}$ |
| $\left(\dfrac{1}{4}, 0\right)$ | $\dfrac{1}{4}$ |

Max $y = \dfrac{4}{7}$ occurs at $z_1 = \dfrac{1}{7}$,

$z_2 = \dfrac{3}{7}$

Copyright © 2011 Pearson Education, Inc. Publishing as Prentice Hall.

<u>Step 4</u>. Use the solutions in Step 3:

(A) $v_1 = \dfrac{1}{x_1 + x_2} = \dfrac{1}{\frac{1}{7} + \frac{3}{7}} = \dfrac{7}{4}$ 　　　　　(B) $v_1 = \dfrac{1}{z_1 + z_2} = \dfrac{1}{\frac{1}{7} + \frac{3}{7}} = \dfrac{7}{4}$

$p_1 = v_1 x_1 = \dfrac{7}{4} \cdot \dfrac{1}{7} = \dfrac{1}{4}$ 　　　　　　$q_1 = v_1 z_1 = \dfrac{7}{4} \cdot \dfrac{1}{7} = \dfrac{1}{4}$

$p_2 = v_1 x_2 = \dfrac{7}{4} \cdot \dfrac{3}{7} = \dfrac{3}{4}$ 　　　　　　$q_2 = v_1 z_2 = \dfrac{7}{4} \cdot \dfrac{3}{7} = \dfrac{3}{4}$

The optimal strategies are:

$$p^* = \begin{bmatrix} 0 & \frac{1}{4} & \frac{3}{4} & 0 \end{bmatrix}, \quad Q^* = \begin{bmatrix} 0 \\ \frac{1}{4} \\ \frac{3}{4} \\ 0 \end{bmatrix} \quad \text{(Recall the recessive rows and columns in the original matrix.)}$$

The value of the original game is: $v = v_1 - k = \dfrac{7}{4} - 2 = -\dfrac{1}{4}$.

<u>Step 5</u>. Check: $P^* M Q^* = \begin{bmatrix} 0 & \frac{1}{4} & \frac{3}{4} & 0 \end{bmatrix} \begin{bmatrix} 0 & -1 & -1 & 0 \\ 1 & 2 & -1 & -1 \\ 0 & -1 & 0 & 1 \\ -1 & -1 & -1 & 0 \end{bmatrix} \begin{bmatrix} 0 \\ \frac{1}{4} \\ \frac{3}{4} \\ 0 \end{bmatrix} = \begin{bmatrix} \frac{1}{4} & -\frac{1}{4} & -\frac{1}{4} & \frac{1}{2} \end{bmatrix} \begin{bmatrix} 0 \\ \frac{1}{4} \\ \frac{3}{4} \\ 0 \end{bmatrix} = -\dfrac{1}{4}$

**29.**

|  | Player $C$ (fate) | |
|---|---|---|
|  | Republican | Democrat |
| Solar | $1000 | $4000 |
| Oil | $5000 | $3000 |

Player $R$ (You)

This game is nonstrictly determined and the payoff matrix is positive.

<u>Step 2</u>. Set up the two corresponding linear programming problems:

(A)　Minimize $y = x_1 + x_2$ 　　　　　(B)　Maximize $y = z_1 + z_2$

　　　Subject to $1000 x_1 + 5000 x_2 \geq 1$ 　　　　　Subject to $1000 z_1 + 4000 z_2 \leq 1$

　　　　　　　$4000 x_1 + 3000 x_2 \geq 1$ 　　　　　　　　$5000 z_1 + 3000 z_2 \leq 1$

　　　　　　　　　$x_1, x_2 \geq 0$ 　　　　　　　　　　　　$z_1, z_2 \geq 0$

<u>Step 3</u>. Solve the linear programming problems geometrically:

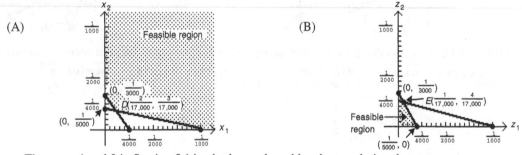

(A)　　　　　　　　　　　　　　　　(B)

Theorems 1 and 2 in Section 5.1 imply that each problem has a solution that must occur at a corner point.

Copyright © 2011 Pearson Education, Inc. Publishing as Prentice Hall.

(A)

| Corner Points | Minimize $y = x_1 + x_2$ |
|---|---|
| $\left(0, \dfrac{1}{3000}\right)$ | $\dfrac{1}{3000}$ |
| $\left(\dfrac{2}{17,000}, \dfrac{3}{17,000}\right)$ | $\dfrac{5}{17,000}$ |
| $\left(\dfrac{1}{1000}, 0\right)$ | $\dfrac{1}{1000}$ |

Min $y = \dfrac{5}{17,000}$ occurs at

$$x_1 = \dfrac{2}{17,000}, \quad x_2 = \dfrac{3}{17,000}$$

(B)

| Corner Points | Maximize $y = z_1 + z_2$ |
|---|---|
| $(0, 0)$ | $0$ |
| $\left(0, \dfrac{1}{4000}\right)$ | $\dfrac{1}{4000}$ |
| $\left(\dfrac{1}{17,000}, \dfrac{4}{17,000}\right)$ | $\dfrac{5}{17,000}$ |
| $\left(\dfrac{1}{5000}, 0\right)$ | $\dfrac{1}{5000}$ |

Max $y = \dfrac{5}{17,000}$ occurs at

$$z_1 = \dfrac{1}{17,000}, \quad z_2 = \dfrac{4}{17,000}$$

Step 4. Use the solutions in Step 3:

(A)   $$v = \frac{1}{x_1 + x_2} = \frac{1}{\dfrac{2}{17,000} + \dfrac{3}{17,000}} = \frac{17,000}{5} = 3400$$

$$p_1 = v_1 x_1 = \frac{17,000}{5} \cdot \frac{2}{17,000} = \frac{2}{5}$$

$$p_2 = v_1 x_2 = \frac{17,000}{5} \cdot \frac{3}{17,000} = \frac{3}{5}$$

(B)   $$v = \frac{1}{z_1 + z_2} = \frac{1}{\dfrac{1}{17,000} + \dfrac{4}{17,000}} = \frac{17,000}{5} = 3400$$

$$q_1 = v_1 z_1 = \frac{17,000}{5} \cdot \frac{1}{17,000} = \frac{1}{5}$$

$$q_2 = v_1 z_2 = \frac{17,000}{5} \cdot \frac{4}{17,000} = \frac{4}{5}$$

The optimal strategies are:

$$P^* = \begin{bmatrix} \dfrac{2}{5} & \dfrac{3}{5} \end{bmatrix}, Q^* = \begin{bmatrix} \dfrac{1}{5} \\ \dfrac{4}{5} \end{bmatrix}$$

The value of the original game is: $v = 3400$ or $\$3400$.

Step 5. Check:

$$P^*MQ^* = \begin{bmatrix} \dfrac{2}{5} & \dfrac{3}{5} \end{bmatrix} \begin{bmatrix} 1000 & 4000 \\ 5000 & 3000 \end{bmatrix} \begin{bmatrix} \dfrac{1}{5} \\ \dfrac{4}{5} \end{bmatrix} = \begin{bmatrix} 3400 & 3400 \end{bmatrix} \begin{bmatrix} \dfrac{1}{5} \\ \dfrac{4}{5} \end{bmatrix} = 3400$$

Copyright © 2011 Pearson Education, Inc. Publishing as Prentice Hall.

---

EXERCISE 10-4

*Things to remember:*

2 × 3 MATRIX GAMES AND LINEAR PROGRAMMING — SIMPLEX METHOD AND THE DUAL PROBLEM.

Given the nonstrictly determined matrix game $M$ free of recessive rows and columns

$$M = \begin{bmatrix} r_1 & r_2 & r_3 \\ s_1 & s_2 & s_3 \end{bmatrix}.$$

To find $P^* = [p_1 \quad p_2]$, $Q^* = \begin{bmatrix} q_1 \\ q_2 \\ q_3 \end{bmatrix}$, and $v$, proceed as follows:

STEP 1. If $M$ is not positive, convert it into a positive matrix $M_1$ by adding a suitable positive constant $k$ to each element. Let

$$M_1 = \begin{bmatrix} a_1 & a_2 & a_3 \\ b_1 & b_2 & b_3 \end{bmatrix}$$

[Note: If $v_1$ is the value of the game $M_1$, then $v = v_1 - k$ is the value of the original game $M$.]

STEP 2. Set up the two linear programming problems:

    (A)    Minimize $y = x_1 + x_2$

            Subject to: $a_1 x_1 + b_1 x_2 \geq 1$
$$a_2 x_1 + b_2 x_2 \geq 1$$
$$a_3 x_1 + b_3 x_2 \geq 1$$
$$x_1, x_2 \geq 0$$

    (B)    Maximize $y = z_1 + z_2 + z_3$

            Subject to: $a_1 z_1 + a_2 z_2 + a_3 z_3 \leq 1$
$$b_1 z_1 + b_2 z_2 + b_3 z_3 \leq 1$$
$$z_1, z_2, z_3 \geq 0$$

[Note: (A) is the dual of (B).]

STEP 3. Solve the maximization problem (B) using the simplex method. Since (A) is the dual of (B), this will also produce the solution of (A).

STEP 4. Using the solutions in Step 3:

$$v_1 = \frac{1}{y} = \frac{1}{x_1 + x_2} \quad \text{or} \quad v_1 = \frac{1}{y} = \frac{1}{z_1 + z_2 + z_3}$$

$$P^* = [p_1 \quad p_2] = [v_1 x_1 \quad v_1 x_2], \qquad Q^* = \begin{bmatrix} q_1 \\ q_2 \\ q_3 \end{bmatrix} = \begin{bmatrix} v_1 z_1 \\ v_1 z_2 \\ v_1 z_3 \end{bmatrix}$$

$$v = v_1 - k$$

STEP 5. The solution found in Step 4 can be checked by showing that $P^* M Q^* = v$.

---

Copyright © 2011 Pearson Education, Inc. Publishing as Prentice Hall.

**1.** Convert $\begin{bmatrix} 1 & 4 & 0 \\ 0 & -1 & 2 \end{bmatrix}$ into a positive payoff matrix by adding 2 to each payoff: $M_1 = \begin{bmatrix} 3 & 6 & 2 \\ 2 & 1 & 4 \end{bmatrix}$

The linear programming problems corresponding to $M_1$ are as follows:

(A)   Minimize $y = x_1 + x_2$

   Subject to  $3x_1 + 2x_2 \geq 1$
   $6x_1 + x_2 \geq 1$
   $2x_1 + 4x_2 \geq 1$
   $x_1 \geq 0, x_2 \geq 0$

(B)   Maximize $y = z_1 + z_2 + z_3$

   Subject to  $3z_1 + 6z_2 + 2z_3 \leq 1$
   $2z_1 + z_2 + 4z_3 \leq 1$
   $z_1 \geq 0, z_2 \geq 0, z_3 \geq 0$

Solve (B) by using the simplex method.  We introduce slack variables $x_1$ and $x_2$ to obtain:

$$3z_1 + 6z_2 + 2z_3 + x_1 \qquad\qquad = 1$$
$$2z_1 + z_2 + 4z_3 \qquad + x_2 \qquad = 1$$
$$-z_1 - z_2 - z_3 \qquad\qquad + y = 0$$
$$z_1, z_2, z_3 \geq 0, \quad x_1, x_2 \geq 0$$

The simplex tableau for this system is:

$$
\begin{array}{cccccc}
z_1 & z_2 & z_3 & x_1 & x_2 & y
\end{array}
$$
$$
\begin{bmatrix}
③ & 6 & 2 & 1 & 0 & 0 & | & 1 \\
2 & 1 & 4 & 0 & 1 & 0 & | & 1 \\
\hline
-1 & -1 & -1 & 0 & 0 & 1 & | & 0
\end{bmatrix}
\begin{array}{c} \frac{1}{3} \\ \frac{1}{2} \\ \\ \end{array}
$$

$$\frac{1}{3}R_1 \to R_1$$

Choose the first column as the pivot column.  Then, following the method outlined in Chapter 6 , we get the pivot element in the 1,1 position, indicated by a circle.

$$
\begin{bmatrix}
① & 2 & \frac{2}{3} & \frac{1}{3} & 0 & 0 & | & \frac{1}{3} \\
2 & 1 & 4 & 0 & 1 & 0 & | & 1 \\
\hline
-1 & -1 & -1 & 0 & 0 & 1 & | & 0
\end{bmatrix}
\sim
\begin{bmatrix}
1 & 2 & \frac{2}{3} & \frac{1}{3} & 0 & 0 & | & \frac{1}{3} \\
0 & -3 & ⑧\!\!\frac{8}{3} & -\frac{2}{3} & 1 & 0 & | & \frac{1}{3} \\
\hline
0 & 1 & -\frac{1}{3} & \frac{1}{3} & 0 & 1 & | & \frac{1}{3}
\end{bmatrix}
\begin{array}{c} \frac{1}{2} \\ \frac{1}{8} \\ \\ \end{array}
$$

$$R_2 + (-2)R_1 \to R_2 \text{ and } R_3 + R_1 \to R_3 \qquad\qquad \frac{3}{8}R_2 \to R_2$$

$$
\sim
\begin{bmatrix}
1 & 2 & \frac{2}{3} & \frac{1}{3} & 0 & 0 & | & \frac{1}{3} \\
0 & -\frac{9}{8} & ① & -\frac{1}{4} & \frac{3}{8} & 0 & | & \frac{1}{8} \\
\hline
0 & 1 & -\frac{1}{3} & \frac{1}{3} & 0 & 1 & | & \frac{1}{3}
\end{bmatrix}
$$

$$
\begin{array}{cccccc}
z_1 & z_2 & z_3 & x_1 & x_2 & y
\end{array}
$$
$$
\sim
\begin{bmatrix}
1 & \frac{11}{4} & 0 & \frac{1}{2} & -\frac{1}{4} & 0 & | & \frac{1}{4} \\
0 & -\frac{9}{8} & 1 & -\frac{1}{4} & \frac{3}{8} & 0 & | & \frac{1}{8} \\
\hline
0 & \frac{5}{8} & 0 & \frac{1}{4} & \frac{1}{8} & 1 & | & \frac{9}{24}
\end{bmatrix}
$$

$$R_1 + \left(-\frac{2}{3}\right)R_2 \to R_1 \text{ and } R_3 + \frac{1}{3}R_2 \to R_3$$

Copyright © 2011 Pearson Education, Inc.  Publishing as Prentice Hall.

The maximum, $y = z_1 + z_2 + z_3 = \dfrac{9}{24}$, occurs at $z_1 = \dfrac{1}{4}$, $z_2 = 0$, $z_3 = \dfrac{1}{8}$.

Thus, $v_1 = \dfrac{1}{z_1 + z_2 + z_3} = \dfrac{1}{y} = \dfrac{1}{\frac{9}{24}} = \dfrac{24}{9} = \dfrac{8}{3}$, and $q_1 = v_1 z_1 = \dfrac{8}{3} \cdot \dfrac{1}{4} = \dfrac{2}{3}$,

$q_2 = v_1 z_2 = \dfrac{8}{3} \cdot 0 = 0$, $q_3 = v_1 z_3 = \dfrac{8}{3} \cdot \dfrac{1}{8} = \dfrac{1}{3}$.

The solution to the minimization problem (A) can be read from the bottom row of the final simplex tableau for the dual problem above. Thus, from the row

$$\begin{array}{cc} x_1 & x_2 \end{array}$$

$$\begin{bmatrix} 0 & \frac{5}{8} & 0 & \frac{1}{4} & \frac{1}{8} & 1 & \frac{9}{24} \end{bmatrix}$$

we conclude that the solution to (A) is:

$$\text{Min } y = x_1 + x_2 = \frac{9}{24} = \frac{3}{8} \text{ at } x_1 = \frac{1}{4},\ x_2 = \frac{1}{8}.$$

Also,

$$p_1 = v_1 x_1 = \frac{8}{3}\left(\frac{1}{4}\right) = \frac{2}{3}, \qquad p_2 = v_1 x_2 = \frac{8}{3}\left(\frac{1}{8}\right) = \frac{1}{3}$$

Finally,

$$P^* = \begin{bmatrix} \frac{2}{3} & \frac{1}{3} \end{bmatrix},\quad Q^* = \begin{bmatrix} \frac{2}{3} \\ 0 \\ \frac{1}{3} \end{bmatrix},\quad \text{and}\quad v = \frac{8}{3} - 2 = \frac{2}{3}.$$

**3.** Convert $\begin{bmatrix} 0 & 1 & -2 \\ -1 & 0 & 3 \\ 2 & -3 & 0 \end{bmatrix}$ into a positive payoff matrix by adding 4 to each payoff: $M_1 = \begin{bmatrix} 4 & 5 & 2 \\ 3 & 4 & 7 \\ 6 & 1 & 4 \end{bmatrix}$

The linear programming problems corresponding to $M_1$ are as follows:

(A)  Minimize $y = x_1 + x_2 + x_3$

Subject to $4x_1 + 3x_2 + 6x_3 \geq 1$

$5x_1 + 4x_2 + x_3 \geq 1$

$2x_1 + 7x_2 + 4x_3 \geq 1$

$x_1,\ x_2,\ x_3 \geq 0$

(B)  Maximize $y = z_1 + z_2 + z_3$

Subject to $4z_1 + 5z_2 + 2z_3 \leq 1$

$3z_1 + 4z_2 + 7z_3 \leq 1$

$6z_1 + z_2 + 4z_3 \leq 1$

$z_1,\ z_2,\ z_3 \geq 0$

We first solve (B) by using the simplex method. Introduce slack variables $x_1$, $x_2$, and $x_3$ to obtain:

$$4z_1 + 5z_2 + 2z_3 + x_1 \qquad\qquad = 1$$
$$3z_1 + 4z_2 + 7z_3 \qquad + x_2 \qquad = 1$$
$$6z_1 + z_2 + 4z_3 \qquad\qquad + x_3 = 1$$
$$-z_1 - z_2 - z_3 \qquad\qquad\qquad + y = 0$$
$$z_1,\ z_2,\ z_3 \geq 0,\ x_1,\ x_2,\ x_3 \geq 0$$

Copyright © 2011 Pearson Education, Inc.  Publishing as Prentice Hall.

The simplex tableau for this system is as follows:

$$\begin{array}{ccccccc} z_1 & z_2 & z_3 & x_1 & x_2 & x_3 & y \end{array}$$

$$\left[\begin{array}{ccccccc|c} 4 & 5 & 2 & 1 & 0 & 0 & 0 & 1 \\ 3 & 4 & \boxed{7} & 0 & 1 & 0 & 0 & 1 \\ 6 & 1 & 4 & 0 & 0 & 1 & 0 & 1 \\ \hdashline -1 & -1 & -1 & 0 & 0 & 0 & 1 & 0 \end{array}\right] \begin{array}{c} \frac{1}{2} \\ \frac{1}{7} \\ \frac{1}{4} \\ \end{array} \qquad \sim \left[\begin{array}{ccccccc|c} 4 & 5 & 2 & 1 & 0 & 0 & 0 & 1 \\ \frac{3}{7} & \frac{4}{7} & \boxed{1} & 0 & \frac{1}{7} & 0 & 0 & \frac{1}{7} \\ 6 & 1 & 4 & 0 & 0 & 1 & 0 & 1 \\ \hdashline -1 & -1 & -1 & 0 & 0 & 0 & 1 & 0 \end{array}\right]$$

$$\frac{1}{7}R_2 \rightarrow R_2 \qquad\qquad\qquad R_1 - 2R_2 \rightarrow R_1,\; R_3 - 4R_2 \rightarrow R_3 \text{ and } R_4 + R_2 \rightarrow R_4$$

$$\begin{array}{ccccccc} z_1 & z_2 & z_3 & x_1 & x_2 & x_3 & y \end{array}$$

$$\sim \left[\begin{array}{ccccccc|c} \frac{22}{7} & \frac{27}{7} & 0 & 1 & -\frac{2}{7} & 0 & 0 & \frac{5}{7} \\ \frac{3}{7} & \frac{4}{7} & 1 & 0 & \frac{1}{7} & 0 & 0 & \frac{1}{7} \\ \boxed{\frac{30}{7}} & -\frac{9}{7} & 0 & 0 & -\frac{4}{7} & 1 & 0 & \frac{3}{7} \\ \hdashline -\frac{4}{7} & -\frac{3}{7} & 0 & 0 & \frac{1}{7} & 0 & 1 & \frac{1}{7} \end{array}\right] \begin{array}{l} \frac{5}{7} \div \frac{22}{7} = \frac{5}{22} \\ \frac{1}{7} \div \frac{3}{7} = \frac{1}{3} \\ \frac{3}{7} \div \frac{30}{7} = \frac{1}{10} \end{array}$$

$$\frac{7}{30}R_3 \rightarrow R_3$$

$$\sim \left[\begin{array}{ccccccc|c} \frac{22}{7} & \frac{27}{7} & 0 & 1 & -\frac{2}{7} & 0 & 0 & \frac{5}{7} \\ \frac{3}{7} & \frac{4}{7} & 1 & 0 & \frac{1}{7} & 0 & 0 & \frac{1}{7} \\ \boxed{1} & -\frac{3}{10} & 0 & 0 & -\frac{2}{15} & \frac{7}{30} & 0 & \frac{1}{10} \\ \hdashline -\frac{4}{7} & -\frac{3}{7} & 0 & 0 & \frac{1}{7} & 0 & 1 & \frac{1}{7} \end{array}\right]$$

$$R_1 + \left(-\frac{22}{7}\right)R_3 \rightarrow R_1,\; R_2 + \left(-\frac{3}{7}\right)R_3 \rightarrow R_2,\; \text{and } R_4 + \frac{4}{7}R_3 \rightarrow R_4$$

$$\sim \left[\begin{array}{ccccccc|c} 0 & \boxed{\frac{24}{5}} & 0 & 1 & \frac{2}{15} & -\frac{11}{15} & 0 & \frac{2}{5} \\ 0 & \frac{7}{10} & 1 & 0 & \frac{1}{5} & -\frac{1}{10} & 0 & \frac{1}{10} \\ 1 & -\frac{3}{10} & 0 & 0 & -\frac{2}{15} & \frac{7}{30} & 0 & \frac{1}{10} \\ \hdashline 0 & -\frac{3}{5} & 0 & 0 & \frac{1}{15} & \frac{2}{15} & 1 & \frac{1}{5} \end{array}\right] \begin{array}{l} \frac{2}{5} \div \frac{24}{5} = \frac{1}{12} \\ \frac{1}{10} \div \frac{7}{10} = \frac{1}{7} \end{array}$$

$$\frac{5}{24}R_1 \rightarrow R_1$$

$$\begin{array}{ccccccc} & & & & & & z_1 \;\; z_2 \;\; z_3 \;\; x_1 \;\; x_2 \;\; x_3 \;\; y \end{array}$$

$$\sim \left[\begin{array}{ccccccc|c} 0 & \boxed{1} & 0 & \frac{5}{24} & \frac{1}{36} & -\frac{11}{72} & 0 & \frac{1}{12} \\ 0 & \frac{7}{10} & 1 & 0 & \frac{1}{5} & -\frac{1}{10} & 0 & \frac{1}{10} \\ 1 & -\frac{3}{10} & 0 & 0 & -\frac{2}{15} & \frac{7}{30} & 0 & \frac{1}{10} \\ \hdashline 0 & -\frac{3}{5} & 0 & 0 & \frac{1}{15} & \frac{2}{15} & 1 & \frac{1}{5} \end{array}\right] \sim \left[\begin{array}{ccccccc|c} 0 & 1 & 0 & \frac{5}{24} & \frac{1}{36} & -\frac{11}{72} & 0 & \frac{1}{12} \\ 0 & 0 & 1 & -\frac{7}{48} & \frac{13}{72} & \frac{1}{144} & 0 & \frac{1}{24} \\ 1 & 0 & 0 & \frac{1}{16} & -\frac{1}{8} & \frac{3}{16} & 0 & \frac{1}{8} \\ \hdashline 0 & 0 & 0 & \frac{1}{8} & \frac{1}{12} & \frac{1}{24} & 1 & \frac{1}{4} \end{array}\right]$$

$$R_2 - \frac{7}{10}R_1 \rightarrow R_2,\; R_3 + \frac{3}{10}R_1 \rightarrow R_3,$$

$$\text{and } R_4 + \frac{3}{5}R_1 \rightarrow R_4$$

We obtain: $z_1 = \dfrac{1}{8}$, $z_2 = \dfrac{1}{12}$, and $z_3 = \dfrac{1}{24}$. Thus, $v_1 = \dfrac{1}{\frac{1}{8} + \frac{1}{12} + \frac{1}{24}} = 4$, and $q_1 = v_1 z_1 = 4 \cdot \dfrac{1}{8} = \dfrac{1}{2}$, $q_2$

$$= v_1 z_2 = 4 \cdot \dfrac{1}{12} = \dfrac{1}{3},\; q_3 = v_1 z_3 = 4 \cdot \dfrac{1}{24} = \dfrac{1}{6}.$$

Copyright © 2011 Pearson Education, Inc. Publishing as Prentice Hall.

The solution to the minimization problem (A) can be read from the bottom row of the final simplex tableau for the dual problem above.  Thus, from the row

$$\begin{matrix} & x_1 & x_2 & x_3 & & & & \\ \end{matrix}$$
$$\begin{bmatrix} 0 & 0 & 0 & \frac{1}{8} & \frac{1}{12} & \frac{1}{24} & 1 & | & \frac{1}{4} \end{bmatrix}$$

we conclude that the solution to (A) is:

$$\text{Min } y = x_1 + x_2 + x_3 = \frac{1}{4} \text{ at } x_1 = \frac{1}{8}, x_2 = \frac{1}{12}, x_3 = \frac{1}{24}.$$

Also,

$$p_1 = v_1 x_1 = 4\left(\frac{1}{8}\right) = \frac{1}{2}$$

$$p_2 = v_1 x_2 = 4\left(\frac{1}{12}\right) = \frac{1}{3}$$

$$p_3 = v_1 x_3 = 4\left(\frac{1}{24}\right) = \frac{1}{6}$$

Finally,

$$P^* = \begin{bmatrix} \frac{1}{2} & \frac{1}{3} & \frac{1}{6} \end{bmatrix}, \quad Q^* = \begin{bmatrix} \frac{1}{2} \\ \frac{1}{3} \\ \frac{1}{6} \end{bmatrix}, \text{ and } v = 4 - 4 = 0.$$

**5.** $M = \begin{bmatrix} 4 & 2 & 2 & 1 \\ 0 & 1 & -1 & 3 \\ -2 & -1 & -3 & 2 \end{bmatrix}$

Eliminate the recessive row $[-2 \ \ -1 \ \ -3 \ \ 2]$ and the recessive columns

$\begin{bmatrix} 4 \\ 0 \\ -2 \end{bmatrix}$ and $\begin{bmatrix} 2 \\ 1 \\ -1 \end{bmatrix}$ to obtain the $2 \times 2$ matrix $\begin{bmatrix} 2 & 1 \\ -1 & 3 \end{bmatrix}$.

Convert this matrix into a positive payoff matrix by adding 2 to each payoff

$$M_1 = \begin{bmatrix} 4 & 3 \\ 1 & 5 \end{bmatrix}$$

The linear programming problems corresponding to $M_1$ are:

(A)    Minimize  $y = x_1 + x_2$
   Subject to:  $4x_1 + x_2 \geq 1$
                  $3x_1 + 5x_2 \geq 1$
                   $x_1, \ x_2 \geq 0$

(B) Maximize  $y = z_1 + z_2$
   Subject to:  $4z_1 + 3z_2 \leq 1$
                    $z_1 + 5z_2 \leq 1$
                     $z_1, \ z_2 \geq 0$

Solve (B) by using the simplex method. We introduce slack variables $x_1, x_2$ to obtain:

$$4z_1 + 3z_2 + x_1 \quad\ \ = 1$$
$$z_1 + 5z_2 \quad\ + x_2 = 1$$
$$z_1, \ z_2, \ x_1, \ x_2 \geq 0$$

Copyright © 2011 Pearson Education, Inc.  Publishing as Prentice Hall.

The simplex tableau for this system is:

$$\begin{array}{ccccc} z_1 & z_2 & x_1 & x_2 & y \\ \end{array}$$

$$\left[\begin{array}{ccccc|c} ④ & 3 & 1 & 0 & 0 & 1 \\ 1 & 5 & 0 & 1 & 0 & 1 \\ \hline -1 & -1 & 0 & 0 & 1 & 0 \end{array}\right] \begin{array}{c} \frac{1}{4} \\ 1 \\ \\ \end{array} \quad \left(\frac{1}{4}\right)R_1 \to R_1$$

We choose the first column as the pivot column. Following the methods of Chapter 6, we get the pivot element in the 1,1 position (circled).

$$\left[\begin{array}{ccccc|c} 1 & \frac{3}{4} & \frac{1}{4} & 0 & 0 & \frac{1}{4} \\ 1 & 5 & 0 & 1 & 0 & 1 \\ \hline -1 & -1 & 0 & 0 & 1 & 0 \end{array}\right] \sim \left[\begin{array}{ccccc|c} 1 & \frac{3}{4} & \frac{1}{4} & 0 & 0 & \frac{1}{4} \\ 0 & ⑰\!\!\frac{17}{4} & -\frac{1}{4} & 1 & 0 & \frac{3}{4} \\ \hline 0 & -\frac{1}{4} & \frac{1}{4} & 0 & 1 & \frac{1}{4} \end{array}\right]$$

$$\begin{array}{c} (-1)R_1 + R_2 \to R_2 \\ R_1 + R_3 \to R_3 \end{array} \qquad \left(\frac{4}{17}\right)R_2 \to R_2$$

$$\sim \left[\begin{array}{ccccc|c} 1 & \frac{3}{4} & \frac{1}{4} & 0 & 0 & \frac{1}{4} \\ 0 & 1 & -\frac{1}{17} & \frac{4}{17} & 0 & \frac{3}{17} \\ \hline 0 & -\frac{1}{4} & 0 & 0 & 1 & \frac{1}{4} \end{array}\right]$$

$$\left(-\frac{3}{4}\right)R_2 + R_1 \to R_1, \quad \left(\frac{1}{4}\right)R_2 + R_3 \to R_3$$

$$\begin{array}{ccccc} z_1 & z_2 & x_1 & x_2 & y \\ \end{array}$$

$$\left[\begin{array}{ccccc|c} 1 & 0 & \frac{5}{17} & -\frac{3}{17} & 0 & \frac{2}{17} \\ 0 & 1 & -\frac{1}{17} & \frac{4}{17} & 0 & \frac{3}{17} \\ \hline 0 & 0 & \frac{4}{17} & \frac{1}{17} & 1 & \frac{5}{17} \end{array}\right]$$

The maximum $y = z_1 + z_2 = \dfrac{5}{17}$ occurs at $z_1 = \dfrac{2}{17}$, $z_2 = \dfrac{3}{17}$.

Thus, $v_1 = \dfrac{1}{z_1 + z_2} = \dfrac{1}{y} = \dfrac{17}{5}$ and $q_1 = v_1 z_1 = \dfrac{2}{5}$, $q_2 = v_1 z_2 = \dfrac{3}{5}$.

The solution to the minimization problem (A) can be read from the bottom row of the final simplex tableau for the dual problem (B). Thus, from the row

$$\begin{array}{ccccc} z_1 & z_2 & x_1 & x_2 & y \\ \end{array}$$

$$\left[\begin{array}{ccccc|c} 0 & 0 & \frac{4}{17} & \frac{1}{17} & 1 & \frac{5}{17} \end{array}\right]$$

we find the solution to (A):

Minimum $y = x_1 + x_2 = \dfrac{5}{17}$ at $x_1 = \dfrac{4}{17}$, $x_2 = \dfrac{1}{17}$

Also, $p_1 = v_1 x_1 = \dfrac{4}{5}$, $p_2 = v_1 x_2 = \dfrac{1}{5}$

Finally,

$$P^* = \left[\begin{array}{ccc} \frac{4}{5} & \frac{1}{5} & 0 \end{array}\right] \qquad Q^* = \left[\begin{array}{c} 0 \\ 0 \\ \frac{2}{5} \\ \frac{3}{5} \end{array}\right] \quad \text{and} \quad v = v_1 - 2 = \dfrac{17}{5} - 2 = \dfrac{7}{5}.$$

Copyright © 2011 Pearson Education, Inc. Publishing as Prentice Hall.

7.  $M = \begin{bmatrix} -2 & -1 & 3 & -1 \\ 1 & 2 & 4 & 0 \\ -1 & 1 & -1 & 1 \\ 0 & 1 & -1 & 2 \end{bmatrix}$

Eliminate the recessive rows $[-2 \ -1 \ 3 \ -1]$, $[-1 \ 1 \ -1 \ 1]$ and recessive column $\begin{bmatrix} -1 \\ 2 \\ 1 \\ 1 \end{bmatrix}$ to obtain the

$2 \times 3$ matrix $\begin{bmatrix} 1 & 4 & 0 \\ 0 & -1 & 2 \end{bmatrix}$. This is the matrix of Problem 1. Thus,

$$P^* = \begin{bmatrix} 0 & \frac{2}{3} & 0 & \frac{1}{3} \end{bmatrix}, \ Q^* = \begin{bmatrix} \frac{2}{3} \\ 0 \\ 0 \\ 0 \\ \frac{1}{3} \end{bmatrix} \text{ and } v = \frac{2}{3}.$$

9.  (A)   The matrix for this game is as follows:

$$\begin{array}{c} \\ \text{Paper} \\ \text{Stone} \\ \text{Scissors} \end{array} \begin{array}{c} \text{Paper Stone Scissors} \\ \begin{bmatrix} 0 & 1 & -1 \\ -1 & 0 & 1 \\ 1 & -1 & 0 \end{bmatrix} \end{array}.$$

(B)   Convert this game matrix into a positive payoff matrix by adding 2 to each payoff:

$$M_1 = \begin{bmatrix} 2 & 3 & 1 \\ 1 & 2 & 3 \\ 3 & 1 & 2 \end{bmatrix}$$

The linear programming problems corresponding to $M_1$ are as follows:

(1)   Minimize $y = x_1 + x_2 + x_3$

   Subject to $2x_1 + x_2 + 3x_3 \geq 1$

   $3x_1 + 2x_2 + x_3 \geq 1$

   $x_1 + 3x_2 + 2x_3 \geq 1$

   $x_1, \ x_2, \ x_3 \geq 0$

(2)   Maximize $y = z_1 + z_2 + z_3$

   Subject to $2z_1 + 3z_2 + z_3 \leq 1$

   $z_1 + 2z_2 + 3z_3 \leq 1$

   $3z_1 + z_2 + 2z_3 \leq 1$

   $z_1, \ z_2, \ z_3 \geq 0$

Solve (2) by using the simplex method. Introduce slack variables $x_1, \ x_2, \ x_3$ to obtain:

$$2z_1 + 3z_2 + z_3 + x_1 \qquad \qquad = 1$$
$$z_1 + 2z_2 + 3z_3 \qquad + x_2 \qquad = 1$$
$$3z_1 + z_2 + 2z_3 \qquad \qquad + x_3 \quad = 1$$
$$-z_1 - z_2 - z_3 \qquad \qquad \qquad + y = 0$$
$$z_1, \ z_2, \ z_3 \geq 0, \ x_1, \ x_2, \ x_3 \geq 0$$

Copyright © 2011 Pearson Education, Inc.  Publishing as Prentice Hall.

The simplex tableau for this system is given below:

$$
\begin{array}{ccccccc}
z_1 & z_2 & z_3 & x_1 & x_2 & x_3 & y
\end{array}
$$

$$
\left[\begin{array}{ccccccc|c}
2 & 3 & 1 & 1 & 0 & 0 & 0 & 1 \\
1 & 2 & ③ & 0 & 1 & 0 & 0 & 1 \\
3 & 1 & 2 & 0 & 0 & 1 & 0 & 1 \\
\hdashline
-1 & -1 & -1 & 0 & 0 & 0 & 1 & 0
\end{array}\right]
\begin{array}{c}
1 \\ \frac{1}{3} \\ \frac{1}{2} \\ \\
\end{array}
\quad \sim \quad
\left[\begin{array}{ccccccc|c}
2 & 3 & 1 & 1 & 0 & 0 & 0 & 1 \\
\frac{1}{3} & \frac{2}{3} & ① & 0 & \frac{1}{3} & 0 & 0 & \frac{1}{3} \\
3 & 1 & 2 & 0 & 0 & 1 & 0 & 1 \\
\hdashline
-1 & -1 & -1 & 0 & 0 & 0 & 1 & 0
\end{array}\right]
$$

$$\frac{1}{3}R_2 \rightarrow R_2$$

$$R_1 + (-1)R_2 \rightarrow R_1, \quad R_3 + (-2)R_2 \rightarrow R_3,$$
$$\text{and } R_4 + R_2 \rightarrow R_4$$

$$
\sim
\left[\begin{array}{ccccccc|c}
\frac{5}{3} & \frac{7}{3} & 0 & 1 & -\frac{1}{3} & 0 & 0 & \frac{2}{3} \\
\frac{1}{3} & \frac{2}{3} & 1 & 0 & \frac{1}{3} & 0 & 0 & \frac{1}{3} \\
\left(\frac{7}{3}\right) & -\frac{1}{3} & 0 & 0 & -\frac{2}{3} & 1 & 0 & \frac{1}{3} \\
\hdashline
-\frac{2}{3} & -\frac{1}{3} & 0 & 0 & \frac{1}{3} & 0 & 1 & \frac{1}{3}
\end{array}\right]
\begin{array}{l}
\frac{2}{3} \div \frac{5}{3} = \frac{2}{5} \\
\frac{1}{3} \div \frac{1}{3} = 1 \\
\frac{1}{3} \div \frac{7}{3} = \frac{1}{7} \\
\\
\end{array}
$$

$$\frac{3}{7}R_3 \rightarrow R_3$$

$$
\begin{array}{ccccccc}
z_1 & z_2 & z_3 & x_1 & x_2 & x_3 & y
\end{array}
$$

$$
\sim
\left[\begin{array}{ccccccc|c}
\frac{5}{3} & \frac{7}{3} & 0 & 1 & -\frac{1}{3} & 0 & 0 & \frac{2}{3} \\
\frac{1}{3} & \frac{2}{3} & 1 & 0 & \frac{1}{3} & 0 & 0 & \frac{1}{3} \\
① & -\frac{1}{7} & 0 & 0 & -\frac{2}{7} & \frac{3}{7} & 0 & \frac{1}{7} \\
\hdashline
-\frac{2}{3} & -\frac{1}{3} & 0 & 0 & \frac{1}{3} & 0 & 1 & \frac{1}{3}
\end{array}\right]
$$

$$R_1 + \left(-\frac{5}{3}\right)R_3 \rightarrow R_1, \quad R_2 + \left(-\frac{1}{3}\right)R_3 \rightarrow R_2, \quad \text{and } R_4 + \frac{2}{3}R_3 \rightarrow R_4$$

$$
\sim
\left[\begin{array}{ccccccc|c}
0 & \left(\frac{54}{21}\right) & 0 & 1 & \frac{3}{21} & -\frac{5}{7} & 0 & \frac{3}{7} \\
0 & \frac{5}{7} & 1 & 0 & \frac{3}{7} & -\frac{1}{7} & 0 & \frac{2}{7} \\
1 & -\frac{1}{7} & 0 & 0 & -\frac{2}{7} & \frac{3}{7} & 0 & \frac{1}{7} \\
\hdashline
0 & -\frac{3}{7} & 0 & 0 & \frac{1}{7} & \frac{2}{7} & 1 & \frac{3}{7}
\end{array}\right]
\begin{array}{l}
\frac{3}{7} \div \frac{54}{21} = \frac{1}{6} \\
\frac{2}{7} \div \frac{5}{7} = \frac{2}{5} \\
\\
\\
\end{array}
$$

$$\frac{21}{54}R_1 \rightarrow R_1$$

$$
\sim
\left[\begin{array}{ccccccc|c}
0 & ① & 0 & \frac{21}{54} & \frac{1}{18} & -\frac{5}{18} & 0 & \frac{1}{6} \\
0 & \frac{5}{7} & 1 & 0 & \frac{3}{7} & -\frac{1}{7} & 0 & \frac{2}{7} \\
1 & -\frac{1}{7} & 0 & 0 & -\frac{2}{7} & \frac{3}{7} & 0 & \frac{1}{7} \\
\hdashline
0 & -\frac{3}{7} & 0 & 0 & \frac{1}{7} & \frac{2}{7} & 1 & \frac{3}{7}
\end{array}\right]
\quad \sim \quad
\begin{array}{ccccccc}
z_1 & z_2 & z_3 & x_1 & x_2 & x_3 & y
\end{array}
$$

$$
\left[\begin{array}{ccccccc|c}
0 & 1 & 0 & \frac{21}{54} & \frac{1}{18} & -\frac{5}{18} & 0 & \frac{1}{6} \\
0 & 0 & 1 & -\frac{5}{18} & \frac{7}{18} & \frac{7}{126} & 0 & \frac{1}{6} \\
1 & 0 & 0 & \frac{1}{18} & -\frac{5}{18} & \frac{49}{126} & 0 & \frac{1}{6} \\
\hdashline
0 & 0 & 0 & \frac{1}{6} & \frac{1}{6} & \frac{1}{6} & 1 & \frac{1}{2}
\end{array}\right]
$$

$$R_2 + \left(-\frac{5}{7}\right)R_1 \rightarrow R_2, \quad R_3 + \frac{1}{7}R_1 \rightarrow R_3,$$

$$\text{and } R_4 + \frac{3}{7}R_1 \rightarrow R_4$$

Copyright © 2011 Pearson Education, Inc.  Publishing as Prentice Hall.

Thus, $z_1 = \dfrac{1}{6}$, $z_2 = \dfrac{1}{6}$, $z_3 = \dfrac{1}{6}$, $v_1 = \dfrac{1}{z_1 + z_2 + z_3} = \dfrac{1}{\dfrac{1}{6} + \dfrac{1}{6} + \dfrac{1}{6}} = 2$, and

$$q_1 = v_1 z_1 = 2 \cdot \dfrac{1}{6} = \dfrac{1}{3}, \quad q_2 = v_1 z_2 = 2 \cdot \dfrac{1}{6} = \dfrac{1}{3}, \quad q_3 = v_1 z_3 = 2 \cdot \dfrac{1}{6} = \dfrac{1}{3}.$$

The solution to the minimization problem (A) can be read from the bottom row of the final simplex tableau for the dual problem shown on the previous page. Thus, from the row

$$\begin{array}{ccc} x_1 & x_2 & x_3 \end{array}$$

$$\begin{bmatrix} 0 & 0 & 0 & \frac{1}{6} & \frac{1}{6} & \frac{1}{6} & 1 \;\big|\; \frac{1}{2} \end{bmatrix}$$

we conclude that the solution to (A) is:

$$\text{Min } y = x_1 + x_2 + x_3 = \dfrac{1}{2} \quad \text{at } x_1 = \dfrac{1}{6}, \quad x_2 = \dfrac{1}{6}, \quad x_3 = \dfrac{1}{6}.$$

Also,

$$p_1 = v_1 x_1 = 2\left(\dfrac{1}{6}\right) = \dfrac{1}{3}, \quad p_2 = v_1 x_2 = 2\left(\dfrac{1}{6}\right) = \dfrac{1}{3}, \quad p_3 = v_1 x_3 = 2\left(\dfrac{1}{6}\right) = \dfrac{1}{3}$$

Finally,

$$P^* = \begin{bmatrix} \frac{1}{3} & \frac{1}{3} & \frac{1}{3} \end{bmatrix}, \quad Q^* = \begin{bmatrix} \frac{1}{3} \\ \frac{1}{3} \\ \frac{1}{3} \\ \frac{1}{3} \end{bmatrix}, \quad \text{and} \quad v = 2 - 2 = 0.$$

11.   (A)                 Economy

$$\begin{array}{c} \\ \text{Deluxe} \\ \text{Standard} \\ \text{Economy} \end{array} \begin{array}{cc} \text{Up} & \text{Down} \\ \begin{bmatrix} 2 & -1 \\ 1 & 1 \\ 0 & 4 \end{bmatrix} \end{array}$$

(B)   Since the payoff matrix is not positive, we add 2 to each element to obtain the matrix:

$$M_1 = \begin{bmatrix} 4 & 1 \\ 3 & 3 \\ 2 & 6 \end{bmatrix}$$

The linear programming problems corresponding to $M_1$ are:

(A)   Minimize $y = x_1 + x_2 + x_3$

     Subject to $4x_1 + 3x_2 + 2x_3 \geq 1$

             $x_1 + 3x_2 + 6x_3 \geq 1$

              $x_1, \; x_2, \; x_3 \geq 0$

(B)   Maximize $y = z_1 + z_2$

     Subject to $4z_1 + z_2 \leq 1$

             $3z_1 + 3z_2 \leq 1$

             $2z_1 + 6z_2 \leq 1$

                $z_1, \; z_2 \geq 0$

Copyright © 2011 Pearson Education, Inc. Publishing as Prentice Hall.

Solve (B) by using the simplex method. Introduce slack variables $x_1$, $x_2$, $x_3$ to obtain:

$$4z_1 + z_2 + x_1 \qquad\qquad = 1$$
$$3z_1 + 3z_2 \quad\; + x_2 \qquad\; = 1$$
$$2z_1 + 6z_2 \qquad\quad + x_3 \;\; = 1$$
$$-z_1 - z_2 \qquad\qquad\quad + y = 0$$

The simplex tableau for this system is:

$$
\begin{array}{cccccc}
z_1 & z_2 & x_1 & x_2 & x_3 & y
\end{array}
$$

$$
\left[\begin{array}{cccccc|c}
④ & 1 & 1 & 0 & 0 & 0 & 1 \\
3 & 3 & 0 & 1 & 0 & 0 & 1 \\
2 & 6 & 0 & 0 & 1 & 0 & 1 \\
\hline
-1 & -1 & 0 & 0 & 0 & 1 & 0
\end{array}\right]
\begin{array}{c} \frac{1}{4} \\ \frac{1}{3} \\ \frac{1}{2} \\ \end{array}
\sim
\left[\begin{array}{cccccc|c}
① & \frac{1}{4} & \frac{1}{4} & 0 & 0 & 0 & \frac{1}{4} \\
3 & 3 & 0 & 1 & 0 & 0 & 1 \\
2 & 6 & 0 & 0 & 1 & 0 & 1 \\
\hline
-1 & -1 & 0 & 0 & 0 & 1 & 0
\end{array}\right]
$$

$$\frac{1}{4}R_1 \rightarrow R_1 \qquad\qquad R_2 + (-3)R_1 \rightarrow R_2,\; R_3 + (-2)R_1 \rightarrow R_3$$
$$R_4 + R_1 \rightarrow R_4$$

$$
\sim
\left[\begin{array}{cccccc|c}
1 & \frac{1}{4} & \frac{1}{4} & 0 & 0 & 0 & \frac{1}{4} \\
0 & \frac{9}{4} & -\frac{3}{4} & 1 & 0 & 0 & \frac{1}{4} \\
0 & ⑪_{\!2} & -\frac{1}{2} & 0 & 1 & 0 & \frac{1}{2} \\
\hline
0 & -\frac{3}{4} & \frac{1}{4} & 0 & 0 & 1 & \frac{1}{4}
\end{array}\right]
\begin{array}{c} 1 \\ \frac{1}{9} \\ \frac{1}{11} \\ \end{array}
\sim
\left[\begin{array}{cccccc|c}
1 & \frac{1}{4} & \frac{1}{4} & 0 & 0 & 0 & \frac{1}{4} \\
0 & \frac{9}{4} & -\frac{3}{4} & 1 & 0 & 0 & \frac{1}{4} \\
0 & ① & -\frac{1}{11} & 0 & \frac{2}{11} & 0 & \frac{1}{11} \\
\hline
0 & -\frac{3}{4} & \frac{1}{4} & 0 & 0 & 1 & \frac{1}{4}
\end{array}\right]
$$

$$\frac{2}{11}R_3 \rightarrow R_3 \qquad\qquad R_1 + \left(-\frac{1}{4}\right)R_3 \rightarrow R_1,\; R_2 + \left(-\frac{9}{4}\right)R_3 \rightarrow R_2$$
$$R_4 + \frac{3}{4}R_3 \rightarrow R_4$$

$$
\begin{array}{cccccc}
z_1 & z_2 & x_1 & x_2 & x_3 & y
\end{array}
$$

$$
\sim
\left[\begin{array}{cccccc|c}
1 & 0 & \frac{12}{44} & 0 & -\frac{2}{44} & 0 & \frac{10}{44} \\
0 & 0 & -\frac{24}{44} & 1 & -\frac{18}{44} & 0 & \frac{2}{44} \\
0 & 1 & -\frac{1}{11} & 0 & \frac{2}{11} & 0 & \frac{1}{11} \\
\hline
0 & 0 & \frac{2}{11} & 0 & \frac{3}{22} & 1 & \frac{14}{44}
\end{array}\right]
$$

Thus, max $y = z_1 + z_2 = \dfrac{14}{44} = \dfrac{7}{22}$ occurs at $z_1 = \dfrac{10}{44} = \dfrac{5}{22}$, $z_2 = \dfrac{1}{11}$.

Now

$$v_1 = \frac{1}{y} = \frac{1}{\frac{7}{22}} = \frac{22}{7} \text{ and } q_1 = v_1 z_1 = \frac{22}{7}\left(\frac{5}{22}\right) = \frac{5}{7},\; q_2 = v_1 z_2 = \frac{22}{7}\left(\frac{2}{22}\right) = \frac{2}{7}.$$

The solution to the minimization problem (A) can be read from the
bottom row of the final simplex tableau for the dual problem above.
Thus, from the row

$$
\begin{array}{ccc}
x_1 & x_2 & x_3
\end{array}
$$
$$
\left[\begin{array}{cccc}
0 & 0 & \frac{2}{11} & 0 \quad \frac{3}{22} \quad 1 \;\Big|\; \frac{7}{22}
\end{array}\right]
$$

Copyright © 2011 Pearson Education, Inc. Publishing as Prentice Hall.

we conclude that the solution to (A) is:

$$\text{Min } y = x_1 + x_2 + x_3 = \frac{7}{22} \text{ at } x_1 = \frac{2}{11}, \ x_2 = 0, \ x_3 = \frac{3}{22}.$$

Also,

$$p_1 = v_1 x_1 = \frac{22}{7}\left(\frac{2}{11}\right) = \frac{4}{7}, \quad p_2 = v_1 x_2 = \frac{22}{7}(0) = 0, \quad p_3 = v_1 x_3 = \frac{22}{7}\left(\frac{3}{22}\right) = \frac{3}{7}$$

Finally, the optimal strategies are

$$P^* = \begin{bmatrix} \frac{4}{7} & 0 & \frac{3}{7} \end{bmatrix}, \quad Q^* = \begin{bmatrix} \frac{5}{7} \\ \frac{2}{7} \end{bmatrix};$$

and the value of the game is: $v = v_1 - 2 = \dfrac{22}{7} - 2 = \dfrac{8}{7}$ million dollars.

(C)    Irrespective of what the economy does, the company's budget should be allocated as follows:

$\dfrac{4}{7}$ of budget on Deluxe, 0 on Standard, and $\dfrac{3}{7}$ on Economy.

(D)    If the company orders only deluxe VCR's and fate plays the strategy "Down", then
    $P = [1, 0, 0]$ and $Q = [0, 1]$
The expected value is

$$PMQ = [1 \ 0 \ 0]\begin{bmatrix} 2 & -1 \\ 1 & 1 \\ 0 & 4 \end{bmatrix}\begin{bmatrix} 0 \\ 1 \end{bmatrix} = [2 \ -1]\begin{bmatrix} 0 \\ 1 \end{bmatrix} = -1.$$

From part (B), the optimal strategy for the company is
$P^* = \begin{bmatrix} \frac{4}{7} & 0 & \frac{3}{7} \end{bmatrix}$. If the company uses this strategy and fate plays

the strategy "Down", then the expected value of the game is

$$P^*MQ = \begin{bmatrix} \frac{4}{7} & 0 & \frac{3}{7} \end{bmatrix}\begin{bmatrix} 2 & -1 \\ 1 & 1 \\ 0 & 4 \end{bmatrix}\begin{bmatrix} 0 \\ 1 \end{bmatrix} = \begin{bmatrix} \frac{8}{7} & \frac{8}{7} \end{bmatrix}\begin{bmatrix} 0 \\ 1 \end{bmatrix} = \frac{8}{7}.$$

## CHAPTER 10 REVIEW

1.  $\begin{bmatrix} \boxed{5} & \textcircled{\small{-3}} \\ \textcircled{\small{-4}} & \boxed{2} \end{bmatrix}$    The game is not strictly determined.    (10-1)

2.  $\begin{bmatrix} \boxed{\textcircled{\small{-1}}} & 4 \\ \textcircled{\small{-2}} & \boxed{1} \end{bmatrix}$    The game is strictly determined; $-1$ in the 1,1 position is the saddle value.    (10-1)

3.  $\begin{bmatrix} \boxed{5} & \textcircled{\small{-2}} \\ 3 & \boxed{\textcircled{0}} \end{bmatrix}$    The value $v = 0$; the game is fair.    (10-1)

Copyright © 2011 Pearson Education, Inc. Publishing as Prentice Hall.

4.  $\begin{bmatrix} \boxed{7} & \boxed{3} \\ -9 & -1 \end{bmatrix}$    The value $v = 3$; the game is not fair.    (10-1)

5.  $\begin{bmatrix} \boxed{9} & \boxed{-3} \\ \boxed{-5} & \boxed{2} \end{bmatrix}$    The game is not strictly determined.

$v = \dfrac{ad - bc}{D} = \dfrac{9(2) - (-3)(-5)}{9 + 2 - (-3 - 5)} = \dfrac{3}{19}$ ; the game is not fair.    (10-2)

6.  $\begin{bmatrix} \boxed{6} & \boxed{-10} \\ \boxed{-3} & \boxed{5} \end{bmatrix}$    The game is not strictly determined.

$v = \dfrac{ad - bc}{D} = \dfrac{6(5) - (-10)(-3)}{6 + 5 - (-10 - 3)} = 0$; the game is fair.    (10-2)

7.  $\begin{bmatrix} \boxed{-8} & \boxed{7} \\ \boxed{-5} & 4 \end{bmatrix}$    The value $v = -5$; the game is not fair.    (10-1)

8.  $\begin{bmatrix} \boxed{-3} & \boxed{4} \\ \boxed{5} & \boxed{-2} \end{bmatrix}$    The game is not strictly determined.

$v = \dfrac{ad - bc}{D} = \dfrac{(-3)(-2) - (4)(5)}{(-3 - 2) - (4 + 5)} = \dfrac{-14}{-14} = 1$ ; the game is not fair.    (10-2)

9.  $\begin{bmatrix} \boxed{-2} & 2 \\ \boxed{-3} & \boxed{7} \end{bmatrix}$    (A) −2 in the 1,1 position.
(B) $R$ plays row 1, $C$ plays column 1.
(C) The value of the game is −2.    (10-1)

10. $\begin{bmatrix} \boxed{-2} & \boxed{4} \\ \boxed{3} & \boxed{-6} \end{bmatrix}$    The game is not strictly determined.    (10-1)

11. $\begin{bmatrix} -3 & -1 & \boxed{5} & \boxed{-8} \\ 1 & \boxed{0} & \boxed{0} & \boxed{2} \\ 3 & \boxed{0} & 1 & \boxed{0} \\ \boxed{6} & -2 & \boxed{-4} & \boxed{2} \end{bmatrix}$

(A)   The two zeros in column 2
(2,2 and 3,2 positions).

(B)   $R$ plays either row 2 or row 3
and $C$ plays column 2.

(C)   The value of the game = 0.

The game is strictly determined.

(10-1)

Copyright © 2011 Pearson Education, Inc. Publishing as Prentice Hall.

**12.**
$$\begin{bmatrix} 1 & \boxed{-2} & \boxed{3} \\ \boxed{-1} & \boxed{2} & 0 \\ \boxed{3} & 0 & \boxed{-4} \end{bmatrix}$$

`The game is not strictly determined.` (10-1)

**13.** The game matrix $\begin{bmatrix} -2 & 3 & 5 \\ -1 & -3 & 0 \\ 0 & -1 & 1 \end{bmatrix}$ has a recessive row 2 and a recessive column 3. Thus, the reduced game

matrix is: $M_1 = \begin{bmatrix} -2 & 3 \\ 0 & -1 \end{bmatrix}$. (10-2)

**14.** $M = \begin{bmatrix} -2 & 1 \\ 0 & -1 \end{bmatrix}$

Optimal strategy for $R = P^* = \begin{bmatrix} \dfrac{d-c}{D} & \dfrac{a-b}{D} \end{bmatrix} = \begin{bmatrix} \dfrac{-1-0}{-4} & \dfrac{-2-1}{-4} \end{bmatrix} = \begin{bmatrix} \dfrac{1}{4} & \dfrac{3}{4} \end{bmatrix}$.

[Note: $D = (a+d) - (b+c) = (-1-2) - (-1+0) = -4$]

Optimal strategy for $C = Q^* = \begin{bmatrix} \dfrac{d-b}{D} \\ \dfrac{a-c}{D} \end{bmatrix} = \begin{bmatrix} \dfrac{-1-1}{-4} \\ \dfrac{-2-0}{-4} \end{bmatrix} = \begin{bmatrix} \dfrac{1}{2} \\ \dfrac{1}{2} \end{bmatrix}$

Value of the game $= v = \dfrac{ad - bc}{D} = \dfrac{(-2)(-1) - (1)(0)}{-4} = -\dfrac{1}{2}$. (10-2)

**15.** Obtain the positive payoff matrix by adding 3 to each payoff:

$M_1 = \begin{bmatrix} 1 & 4 \\ 3 & 2 \end{bmatrix}$

The linear programming problems corresponding to $M_1$ are as follows:

(A) Minimize $y = x_1 + x_2$  
    Subject to: $x_1 + 3x_2 \geq 1$  
               $4x_1 + 2x_2 \geq 1$  
               $x_1,\ x_2 \geq 0$

(B) Maximize $y = z_1 + z_2$  
    Subject to: $z_1 + 4z_2 \leq 1$  
               $3z_1 + 2z_2 \leq 1$  
               $z_1,\ z_2 \geq 0$

(10-3)

Copyright © 2011 Pearson Education, Inc. Publishing as Prentice Hall.

**16.**  The graphs corresponding to the linear programming problems given in the solution to Problem 17 are shown below.

(A)
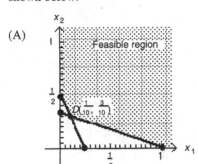

(B)
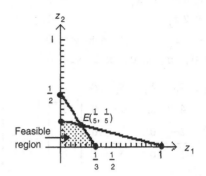

(A)

| Corner Points | Minimize $y = x_1 + x_2$ |
|---|---|
| $\left(0, \dfrac{1}{2}\right)$ | $\dfrac{1}{2}$ |
| $\left(\dfrac{1}{10}, \dfrac{3}{10}\right)$ | $\dfrac{2}{5}$ |
| $(1, 0)$ | $1$ |

Min $y = \dfrac{2}{5}$ occurs $\left(\dfrac{1}{10}, \dfrac{3}{10}\right)$

(B)

| Corner Points | Maximize $y = z_1 + z_2$ |
|---|---|
| $(0, 0)$ | $0$ |
| $\left(0, \dfrac{1}{4}\right)$ | $\dfrac{1}{4}$ |
| $\left(\dfrac{1}{5}, \dfrac{1}{5}\right)$ | $\dfrac{2}{5}$ |
| $\left(0, \dfrac{1}{3}\right)$ | $\dfrac{1}{3}$ |

Max $y = \dfrac{2}{5}$ occurs at $\left(\dfrac{1}{5}, \dfrac{1}{5}\right)$

Now $v = \dfrac{1}{x_1 + x_2} = \dfrac{1}{\dfrac{1}{10} + \dfrac{3}{10}} = \dfrac{10}{4} = \dfrac{5}{2} = \dfrac{1}{z_1 + z_2}$

$$p_1 = v_1 x_1 = \dfrac{5}{2} \cdot \dfrac{1}{10} = \dfrac{1}{4}$$

$$p_2 = v_1 x_2 = \dfrac{5}{2} \cdot \dfrac{3}{10} = \dfrac{3}{4}$$

$$q_1 = v_1 z_1 = \dfrac{5}{2} \cdot \dfrac{1}{5} = \dfrac{1}{2}$$

$$q_2 = v_1 z_2 = \dfrac{5}{2} \cdot \dfrac{1}{5} = \dfrac{1}{2}$$

The optimal strategies are:

$$P^* = \begin{bmatrix} \dfrac{1}{4} & \dfrac{3}{4} \end{bmatrix}, \quad Q^* = \begin{bmatrix} \dfrac{1}{2} \\ \dfrac{1}{2} \end{bmatrix}$$

and the value of the game is: $v = v_1 - k = \dfrac{5}{2} - 3 = -\dfrac{1}{2}$.

Check: $P^*MQ^* = \begin{bmatrix} \dfrac{1}{4} & \dfrac{3}{4} \end{bmatrix}\begin{bmatrix} -2 & 1 \\ 0 & -1 \end{bmatrix}\begin{bmatrix} \dfrac{1}{2} \\ \dfrac{1}{2} \end{bmatrix} = \begin{bmatrix} -\dfrac{1}{2} & -\dfrac{1}{2} \end{bmatrix}\begin{bmatrix} \dfrac{1}{2} \\ \dfrac{1}{2} \end{bmatrix} = -\dfrac{1}{4} - \dfrac{1}{4} = -\dfrac{1}{2}$     (10-3)

Copyright © 2011 Pearson Education, Inc. Publishing as Prentice Hall.

**17.**   First we solve Problem 15(B).  Introduce slack variables $x_1$ and $x_2$ to obtain:

$$z_1 + 4z_2 + x_1 \qquad\qquad = 1$$
$$3z_1 + 2z_2 \qquad + x_2 \qquad = 1$$
$$-z_1 - z_2 \qquad\qquad + y = 0$$
$$z_1 \geq 0, \ z_2 \geq 0, \ x_1, \ x_2 \geq 0$$

$$
\begin{array}{ccccc}
z_1 & z_2 & x_1 & x_2 & y
\end{array}
$$

$$
\left[
\begin{array}{ccccc|c}
1 & 4 & 1 & 0 & 0 & 1 \\
\boxed{3} & 2 & 0 & 1 & 0 & 1 \\
\hline
-1 & -1 & 0 & 0 & 1 & 0
\end{array}
\right]
\begin{array}{c} \frac{1}{1} \\ \frac{1}{3} \\ {} \end{array}
\sim
\left[
\begin{array}{ccccc|c}
1 & 4 & 1 & 0 & 0 & 1 \\
\boxed{1} & \frac{2}{3} & 0 & \frac{1}{3} & 0 & \frac{1}{3} \\
\hline
-1 & -1 & 0 & 0 & 1 & 0
\end{array}
\right]
$$

$$\frac{1}{3}R_2 \to R_2 \qquad\qquad\qquad R_1 + (-1)R_2 \to R_1 \text{ and } R_3 + R_2 \to R_3$$

$$
\sim
\left[
\begin{array}{ccccc|c}
0 & \boxed{\frac{10}{3}} & 1 & -\frac{1}{3} & 0 & \frac{2}{3} \\
1 & \frac{2}{3} & 0 & \frac{1}{3} & 0 & \frac{1}{3} \\
\hline
0 & -\frac{1}{3} & 0 & \frac{1}{3} & 1 & \frac{1}{3}
\end{array}
\right]
\begin{array}{c} \frac{2}{3} \div \frac{10}{3} = \frac{2}{10} \\ \frac{1}{3} \div \frac{2}{3} = \frac{1}{2} \\ {} \end{array}
$$

$$\frac{3}{10}R_1 \to R_1$$

$$
\begin{array}{ccccc}
& & & & z_1 \quad z_2 \quad x_1 \quad x_2 \quad y
\end{array}
$$

$$
\sim
\left[
\begin{array}{ccccc|c}
0 & \boxed{1} & \frac{3}{10} & -\frac{1}{10} & 0 & \frac{1}{5} \\
1 & \frac{2}{3} & 0 & \frac{1}{3} & 0 & \frac{1}{3} \\
\hline
0 & -\frac{1}{3} & 0 & \frac{1}{3} & 1 & \frac{1}{3}
\end{array}
\right]
\sim
\left[
\begin{array}{ccccc|c}
0 & 1 & \frac{3}{10} & -\frac{1}{10} & 0 & \frac{1}{5} \\
1 & 0 & -\frac{1}{5} & \frac{2}{5} & 0 & \frac{1}{5} \\
\hline
0 & 0 & \frac{1}{10} & \frac{3}{10} & 1 & \frac{2}{5}
\end{array}
\right]
$$

$$R_2 + \left(-\frac{2}{3}\right)R_1 \to R_2 \text{ and } R_3 + \frac{1}{3}R_1 \to R_3$$

Thus, $z_1 = \dfrac{1}{5}$, $z_2 = \dfrac{1}{5}$, $y_{\max} = \dfrac{2}{5}$, $v_1 = \dfrac{1}{z_1 + z_2} = \dfrac{1}{\frac{1}{5} + \frac{1}{5}} = \dfrac{5}{2}$, and

$$q_1 = v_1 z_1 = \frac{5}{2} \cdot \frac{1}{5} = \frac{1}{2}, \quad q_2 = v_1 z_2 = \frac{5}{2} \cdot \frac{1}{5} = \frac{1}{2}.$$

The solution to the minimization problem (A) can be read from the bottom row of the final simplex tableau for the dual problem on the previous page.  Thus, from the row

$$
\begin{array}{ccccc}
 & & x_1 & x_2 & 
\end{array}
$$
$$
\left[
\begin{array}{ccccc|c}
0 & 0 & \frac{1}{10} & \frac{3}{10} & 1 & \frac{2}{5}
\end{array}
\right]
$$

we conclude that the solution is:

$$\text{Min } y = x_1 + x_2 = \frac{2}{5} \text{ at } x_1 = \frac{1}{10}, \ x_2 = \frac{3}{10}$$

Also,

$$p_1 = v_1 x_1 = \frac{5}{2}\left(\frac{1}{10}\right) = \frac{1}{4}$$

$$p_2 = v_1 x_2 = \frac{5}{2}\left(\frac{3}{10}\right) = \frac{3}{4}$$

Finally,

$$P^* = \left[\begin{array}{cc} \frac{1}{4} & \frac{3}{4} \end{array}\right], \qquad Q^* = \left[\begin{array}{c} \frac{1}{2} \\ \frac{1}{2} \end{array}\right], \qquad \text{and} \qquad v = \frac{5}{2} - 3 = -\frac{1}{2}. \qquad\qquad (10\text{-}4)$$

Copyright © 2011 Pearson Education, Inc.  Publishing as Prentice Hall.

**18.** False: For a counterexample, see Problem 6.    (10-2)

**19.** True: The saddle value is the value of the game.    (10-1)

**20.** False: $M = \begin{bmatrix} 1 & 2 & 3 \\ 2 & 4 & 6 \\ 3 & 6 & 9 \end{bmatrix}$  has two recessive rows.    (10-2)

**21.** True: Suppose that the payoff matrix $M$ has all negative entries. Since the entries of $P^*$ and $Q^*$ are all nonnegative with at least one entry in each case positive, $v = P^*MQ^* < 0$.    (10-3)

**22.**

$$\begin{bmatrix} \textcircled{-1} & \boxed{2} & \boxed{8} \\ \boxed{0} & \boxed{2} & \textcircled{-4} \\ \boxed{\textcircled{0}} & 1 & 3 \end{bmatrix}$$

This game is strictly determined.

Also, $P^* = [0\ \ 0\ \ 1]$, $Q^* = \begin{bmatrix} 1 \\ 0 \\ 0 \end{bmatrix}$, and the value of the game $v = 0$.    (10-1)

**23.** $M = \begin{bmatrix} -1 & 5 & -3 & 7 \\ -4 & -3 & 2 & -2 \\ 3 & 0 & 2 & 1 \end{bmatrix}$

First, eliminate the recessive row $[-4\ \ -3\ \ 2\ \ -2]$, and the recessive columns

$\begin{bmatrix} -1 \\ -4 \\ 3 \end{bmatrix}$ and $\begin{bmatrix} 7 \\ -2 \\ 1 \end{bmatrix}$ to obtain the $2 \times 2$ matrix $\begin{bmatrix} \textcircled{5} & \boxed{-3} \\ \boxed{0} & \textcircled{2} \end{bmatrix}$

The game is not strictly determined so the solution method of Section 10-2 can be applied. Let

$D = (a + d) - (b + c) = (5 + 2) - (-3 + 0) = 10$. The optimal strategies are:

$P^* = \left[ \dfrac{d-c}{D},\ \ \dfrac{a-b}{D} \right] = \left[ \dfrac{2}{10},\ \ \dfrac{8}{10} \right] = \left[ \dfrac{1}{5},\ \ \dfrac{4}{5} \right]$,    $Q^* = \begin{bmatrix} \dfrac{d-b}{D} \\ \dfrac{a-c}{D} \end{bmatrix} = \begin{bmatrix} \dfrac{5}{10} \\ \dfrac{5}{10} \end{bmatrix} = \begin{bmatrix} \dfrac{1}{2} \\ \dfrac{1}{2} \end{bmatrix}$

and $v = \dfrac{ad - bc}{D} = \dfrac{10}{10} = 1$.

Thus, the optimal strategies for the original matrix and the value $v$ are:

$P^* = \left[ \dfrac{1}{5}\ \ 0\ \ \dfrac{4}{5} \right]$,    $Q^* = \begin{bmatrix} 0 \\ \frac{1}{2} \\ \frac{1}{2} \\ 0 \end{bmatrix}$, $v = 1$.    (10-2)

Copyright © 2011 Pearson Education, Inc. Publishing as Prentice Hall.

24. $\begin{bmatrix} \boxed{0} & \boxed{3} & \boxed{-1} \\ -1 & \boxed{-2} & \boxed{1} \end{bmatrix}$ This game is nonstrictly determined.

We obtain a positive payoff matrix by adding 3 to each payoff:

$$M_1 = \begin{bmatrix} 3 & 6 & 2 \\ 2 & 1 & 4 \end{bmatrix}$$

The linear programming problems corresponding to $M_1$ are as follows:

(A)  Minimize $y = x_1 + x_2$
Subject to: $3x_1 + 2x_2 \geq 1$
$6x_1 + x_2 \geq 1$
$2x_1 + 4x_2 \geq 1$
$x_1, x_2 \geq 0$

(B)  Maximize $y = z_1 + z_2 + z_3$
Subject to: $3z_1 + 6z_2 + 2z_3 \leq 1$
$2z_1 + z_2 + 4z_3 \leq 1$
$z_1, z_2, z_3 \geq 0$

Solve part (B) by using the simplex method. Introduce slack variables $x_1$ and $x_2$ to obtain:

$$\begin{array}{rcl}
3z_1 + 6z_2 + 2z_3 + x_1 & = & 1 \\
2z_1 + z_2 + 4z_3 \quad + x_2 & = & 1 \\
-z_1 - z_2 - z_3 \quad\quad\quad + y & = & 0 \\
z_1, z_2, z_3 \geq 0, \; x_1, x_2 \geq 0
\end{array}$$

$$
\begin{array}{c}
\begin{array}{cccccc}
z_1 & z_2 & z_3 & x_1 & x_2 & y
\end{array} \\
\left[\begin{array}{cccccc|c}
\boxed{3} & 6 & 2 & 1 & 0 & 0 & 1 \\
2 & 1 & 4 & 0 & 1 & 0 & 1 \\
\hdashline
-1 & -1 & -1 & 0 & 0 & 1 & 0
\end{array}\right]
\begin{array}{c} \frac{1}{3} \\ \frac{1}{2} \\ \\ \end{array}
\end{array}
\sim
\left[\begin{array}{cccccc|c}
\boxed{1} & 2 & \frac{2}{3} & \frac{1}{3} & 0 & 0 & \frac{1}{3} \\
2 & 1 & 4 & 0 & 1 & 0 & 1 \\
\hdashline
-1 & -1 & -1 & 0 & 0 & 1 & 0
\end{array}\right]
$$

$$\frac{1}{3}R_1 \to R_1 \qquad\qquad R_2 + (-2)R_1 \to R_2 \text{ and } R_3 + R_1 \to R_3$$

$$\sim
\left[\begin{array}{cccccc|c}
1 & 2 & \frac{2}{3} & \frac{1}{3} & 0 & 0 & \frac{1}{3} \\
0 & -3 & \boxed{\frac{8}{3}} & -\frac{2}{3} & 1 & 0 & \frac{1}{3} \\
\hdashline
0 & 1 & -\frac{1}{3} & \frac{1}{3} & 0 & 1 & \frac{1}{3}
\end{array}\right]
\begin{array}{l} \frac{1}{3} \div \frac{2}{3} = \frac{1}{2} \\ \frac{1}{3} \div \frac{8}{3} = \frac{1}{8} \\ \\ \end{array}
$$

$$\frac{3}{8}R_2 \to R_2$$

Copyright © 2011 Pearson Education, Inc. Publishing as Prentice Hall.

$$\sim \begin{bmatrix} 1 & 2 & \frac{2}{3} & \frac{1}{3} & 0 & 0 & | & \frac{1}{3} \\ 0 & -\frac{9}{8} & \boxed{1} & -\frac{1}{4} & \frac{3}{8} & 0 & | & \frac{1}{8} \\ \hdashline 0 & 1 & -\frac{1}{3} & \frac{1}{3} & 0 & 1 & | & \frac{1}{3} \end{bmatrix} \sim \begin{matrix} z_1 & z_2 & z_3 & x_1 & x_2 & y \\ \begin{bmatrix} 1 & \frac{11}{4} & 0 & \frac{1}{2} & -\frac{1}{4} & 0 & | & \frac{1}{4} \\ 0 & -\frac{9}{8} & 1 & -\frac{1}{4} & \frac{3}{8} & 0 & | & \frac{1}{8} \\ \hdashline 0 & \frac{5}{8} & 0 & \frac{1}{4} & \frac{1}{8} & 1 & | & \frac{3}{8} \end{bmatrix} \end{matrix}$$

$$R_1 + \left(-\frac{2}{3}\right) R_2 \rightarrow R_1 \text{ and } R_3 + \frac{1}{3} R_2 \rightarrow R_3$$

Thus, $z_1 = \frac{1}{4}$, $z_2 = 0$, $z_3 = \frac{1}{8}$, $y_{max} = \frac{3}{8}$, $v_1 = \dfrac{1}{\frac{1}{4} + 0 + \frac{1}{8}} = \frac{8}{3}$,

$q_1 = v_1 z_1 = \frac{8}{3} \cdot \frac{1}{4} = \frac{2}{3}$, $q_2 = v_1 z_2 = \frac{8}{3} \cdot 0 = 0$, $q_3 = v_1 z_3 = \frac{8}{3} \cdot \frac{1}{8} = \frac{1}{3}$.

Therefore, $Q^* = \begin{bmatrix} \frac{2}{3} \\ 0 \\ \frac{1}{3} \end{bmatrix}$ and $v = v_1 - 3 = \frac{8}{3} - 3 = -\frac{1}{3}$.

The solution to the minimization problem (A) can be read from the
bottom row of the final simplex tableau for the dual problem above.
Thus, from the row

$$\begin{matrix} & & & x_1 & x_2 & \\ \begin{bmatrix} 0 & \frac{5}{8} & 0 & \frac{1}{4} & \frac{1}{8} & 1 & | & \frac{3}{8} \end{bmatrix} \end{matrix}$$

we conclude that the solution is:

$$\text{Min } y = x_1 + x_2 = \frac{3}{8} \text{ at } x_1 = \frac{1}{4}, \ x_2 = \frac{1}{8}$$

Now,

$$p_1 = v_1 x_1 = \frac{8}{3}\left(\frac{1}{4}\right) = \frac{2}{3}, \quad p_2 = v_1 x_2 = \frac{8}{3}\left(\frac{1}{8}\right) = \frac{1}{3}$$

and

$$P^* = \begin{bmatrix} \frac{2}{3} & \frac{1}{3} \end{bmatrix}. \qquad (10\text{-}4)$$

**25.** $M = \begin{bmatrix} 2 & 6 & -4 & -7 \\ 4 & 7 & -3 & -5 \\ 3 & 3 & 9 & 8 \end{bmatrix}$

First eliminate the recessive row [2  6  −4  −7], and the recessive columns $\begin{bmatrix} 6 \\ 7 \\ 3 \end{bmatrix}$ and $\begin{bmatrix} -4 \\ -3 \\ 9 \end{bmatrix}$ to obtain the

2×2 matrix $\begin{bmatrix} \boxed{4} & \boxed{-5} \\ \boxed{3} & \boxed{8} \end{bmatrix}$.

The game is not strictly determined so the solution method of Section 10-2 can be applied. Let
$D = (a + d) − (b + c) = (4 + 8) − (−5 + 3) = 14.$

The optimal strategies are

$$P^* = \begin{bmatrix} \frac{d-c}{D}, & \frac{a-b}{D} \end{bmatrix} = \begin{bmatrix} \frac{5}{14}, & \frac{9}{14} \end{bmatrix}, \quad Q^* = \begin{bmatrix} \frac{d-b}{D} \\ \frac{a-c}{D} \end{bmatrix} = \begin{bmatrix} \frac{13}{14} \\ \frac{1}{14} \end{bmatrix} \text{ and } v = \frac{ad-bc}{D} = \frac{47}{14}$$

Copyright © 2011 Pearson Education, Inc. Publishing as Prentice Hall.

Thus, the optimal strategies for the original matrix and the value $v$ are:

$$P^* = \begin{bmatrix} 0 & \frac{5}{14} & \frac{9}{14} \end{bmatrix}, \quad Q^* = \begin{bmatrix} \frac{13}{14} \\ 0 \\ 0 \\ \frac{1}{14} \end{bmatrix}, \quad v = \frac{47}{14} \qquad (10\text{-}2)$$

**26.**
$$\begin{bmatrix} -1 & 1 & \boxed{-2} \\ \boxed{0} & \boxed{-2} & \boxed{2} \\ \boxed{-3} & \boxed{2} & -1 \end{bmatrix}$$

This is a nonstrictly determined game

We obtain the positive payoff matrix by adding 4 to each payoff:

$$M_1 = \begin{bmatrix} 3 & 5 & 2 \\ 4 & 2 & 6 \\ 1 & 6 & 3 \end{bmatrix}$$

The linear programming problems corresponding to $M_1$ are as follows:

(A)   Minimize $y = x_1 + x_2 + x_3$

Subject to: $3x_1 + 4x_2 + x_3 \geq 1$

$5x_1 + 2x_2 + 6x_3 \geq 1$

$2x_1 + 6x_2 + 3x_3 \geq 1$

$x_1, \ x_2, \ x_3 \geq 0$

(B)   Maximize $y = z_1 + z_2 + z_3$

Subject to: $3z_1 + 5z_2 + 2z_3 \leq 1$

$4z_1 + 2z_2 + 6z_3 \leq 1$

$z_1 + 6z_2 + 3z_3 \leq 1$

$z_1, \ z_2, \ z_3 \geq 0$

We use the simplex method to solve part (B).  Introduce slack variables $x_1$, $x_2$, and $x_3$ to obtain:

$$\begin{aligned}
3z_1 + 5z_2 + 2z_3 + x_1 \qquad\qquad\quad &= 1 \\
4z_1 + 2z_2 + 6z_3 \qquad + x_2 \qquad\quad &= 1 \\
z_1 + 6z_2 + 3z_3 \qquad\qquad + x_3 \quad &= 1 \\
-z_1 - z_2 - z_3 \qquad\qquad\qquad + y &= 0 \\
z_1, \ z_2, \ z_3 \geq 0, \ x_1, \ x_2, \ x_3 &\geq 0
\end{aligned}$$

$$\begin{array}{ccccccc}
z_1 & z_2 & z_3 & x_1 & x_2 & x_3 & y
\end{array}$$
$$\left[\begin{array}{ccccccc|c}
3 & 5 & 2 & 1 & 0 & 0 & 0 & 1 \\
④ & 2 & 6 & 0 & 1 & 0 & 0 & 1 \\
1 & 6 & 3 & 0 & 0 & 1 & 0 & 1 \\
\hline
-1 & -1 & -1 & 0 & 0 & 0 & 1 & 0
\end{array}\right]
\begin{array}{l} \frac{1}{3} \\ \frac{1}{4} \\ 1 \\ \\ \end{array}$$

$$\sim \left[\begin{array}{ccccccc|c}
3 & 5 & 2 & 1 & 0 & 0 & 0 & 1 \\
① & \frac{1}{2} & \frac{3}{2} & 0 & \frac{1}{4} & 0 & 0 & \frac{1}{4} \\
1 & 6 & 3 & 0 & 0 & 1 & 0 & 1 \\
\hline
-1 & -1 & -1 & 0 & 0 & 0 & 1 & 0
\end{array}\right]$$

$$\frac{1}{4}R_2 \to R_2$$

$$R_1 + (-3)R_2 \to R_1, \ R_3 + (-1)R_2 \to R_3,$$
$$\text{and } R_4 + R_2 \to R_4$$

$$\sim \left[\begin{array}{ccccccc|c}
0 & ⑦⁄₂ & -\frac{5}{2} & 1 & -\frac{3}{4} & 0 & 0 & \frac{1}{4} \\
1 & \frac{1}{2} & \frac{3}{2} & 0 & \frac{1}{4} & 0 & 0 & \frac{1}{4} \\
0 & \frac{11}{2} & \frac{3}{2} & 0 & -\frac{1}{4} & 1 & 0 & \frac{3}{4} \\
\hline
0 & -\frac{1}{2} & \frac{1}{2} & 0 & \frac{1}{4} & 0 & 1 & \frac{1}{4}
\end{array}\right]
\begin{array}{l} \frac{1}{4} \div \frac{7}{2} = \frac{1}{14} \\ \frac{1}{4} \div \frac{1}{2} = \frac{1}{2} \\ \frac{3}{4} \div \frac{11}{2} = \frac{3}{22} \\ \\ \end{array}$$

$$\frac{2}{7}R_1 \to R_1$$

Copyright © 2011 Pearson Education, Inc.  Publishing as Prentice Hall.

$$
\begin{bmatrix}
z_1 & z_2 & z_3 & x_1 & x_2 & x_3 & y & \\
0 & \boxed{1} & -\frac{5}{7} & \frac{2}{7} & -\frac{3}{14} & 0 & 0 & \frac{1}{14} \\
1 & \frac{1}{2} & \frac{3}{2} & 0 & \frac{1}{4} & 0 & 0 & \frac{1}{4} \\
0 & \frac{11}{2} & \frac{3}{2} & 0 & -\frac{1}{4} & 1 & 0 & \frac{3}{4} \\
\hdashline
0 & -\frac{1}{2} & \frac{1}{2} & 0 & \frac{1}{4} & 0 & 1 & \frac{1}{4}
\end{bmatrix}
\sim
\begin{bmatrix}
z_1 & z_2 & z_3 & x_1 & x_2 & x_3 & y & \\
0 & 1 & -\frac{5}{7} & \frac{2}{7} & -\frac{3}{14} & 0 & 0 & \frac{1}{14} \\
1 & 0 & \frac{13}{7} & -\frac{1}{7} & \frac{5}{14} & 0 & 0 & \frac{3}{14} \\
0 & 0 & \frac{38}{7} & -\frac{11}{7} & \frac{13}{14} & 1 & 0 & \frac{5}{14} \\
\hdashline
0 & 0 & \frac{1}{7} & \frac{1}{7} & \frac{1}{7} & 0 & 1 & \frac{2}{7}
\end{bmatrix}
$$

$$R_2 + \left(-\frac{1}{2}\right)R_1 \to R_2, \quad R_3 + \left(-\frac{11}{2}\right)R_1 \to R_3,$$

$$\text{and } R_4 + \frac{1}{2}R_1 \to R_4$$

Thus, $z_1 = \frac{3}{14}$, $z_2 = \frac{1}{14}$, $z_3 = 0$, and $y_{max} = \frac{2}{7}$.

$$v_1 = \frac{1}{z_1 + z_2 + z_3} = \frac{1}{\frac{3}{14} + \frac{1}{14} + 0} = \frac{14}{4} = \frac{7}{2}.$$

Therefore,

$$q_1 = v_1 z_1 = \frac{7}{2} \cdot \frac{3}{14} = \frac{3}{4}, \quad q_2 = v_1 z_2 = \frac{7}{2} \cdot \frac{1}{14} = \frac{1}{4}, \quad q_3 = v_1 z_3 = \frac{7}{2} \cdot 0 = 0,$$

$$Q^* = \begin{bmatrix} \frac{3}{4} \\ \frac{1}{4} \\ 0 \end{bmatrix}, \quad \text{and } v = v_1 - 4 = \frac{7}{2} - 4 = -\frac{1}{2}.$$

The solution to the minimization problem (A) can be read from the bottom row of the final simplex tableau for the dual problem above. Thus, from the row

$$
\begin{array}{ccc}
x_1 & x_2 & x_3 \\
\end{array}
$$
$$
\begin{bmatrix} 0 & 0 & \frac{1}{7} & \frac{1}{7} & \frac{1}{7} & 0 & 1 & \frac{2}{7} \end{bmatrix}
$$

we conclude that the solution is:

$$\text{Min } y = x_1 + x_2 + x_3 = \frac{2}{7} \text{ at } x_1 = \frac{1}{7}, \quad x_2 = \frac{1}{7}, \quad x_3 = 0$$

Now,

$$p_1 = v_1 x_1 = \frac{7}{2}\left(\frac{1}{7}\right) = \frac{1}{2}, \quad p_2 = v_1 x_2 = \frac{7}{2}\left(\frac{1}{7}\right) = \frac{1}{2}, \quad p_3 = v_1 x_3 = \frac{7}{2}(0) = 0$$

and $P^* = \begin{bmatrix} \frac{1}{2} & \frac{1}{2} & 0 \end{bmatrix}$.     (10-4)

27.  Yes. Let $M = \begin{bmatrix} a & b \\ c & d \end{bmatrix}$ be the payoff matrix for a strictly determined $2 \times 2$ matrix game, and suppose that $a$, the element in the (1, 1) position is a saddle value. Then $a \le b$ since $a$ is the minimum in its row and $a \ge c$ since $a$ is the maximum in its column.

If $c$, the element in the (2, 1) position is minimum in the second row, then $c \le d$ and the column $\begin{bmatrix} b \\ d \end{bmatrix}$ is recessive; $\begin{bmatrix} b \\ d \end{bmatrix} \ge \begin{bmatrix} a \\ c \end{bmatrix}$.

If $d$, the element in the (2, 2) position, is the minimum in the second row, then $b \ge a \ge c \ge d$ and $[c \ d]$ is a recessive row; $[a \ b] \ge [c \ d]$.     (10-1, 10-2)

Copyright © 2011 Pearson Education, Inc. Publishing as Prentice Hall.

**28.** No. Assume that the matrix $\begin{bmatrix} 1 & 3 & 2 \\ 3 & 1 & 2 \\ 2 & 2 & 2 \end{bmatrix}$ is the payoff matrix for a $3 \times 3$ matrix game. The minimum

values in each row and the maximum values in each column are as indicated:

$$\begin{bmatrix} ①  & \boxed{3} & \boxed{2} \\ \boxed{3} & ① & \boxed{2} \\ ② & ② & \boxed{②} \end{bmatrix}$$

The 2 in the (3, 3) position is a saddle value so that the game is strictly determined. There are no recessive rows or columns. (10-1, 10-2)

**29.** (A) The payoff matrix $M$ is:

$$\begin{array}{c} \qquad\qquad \text{Cathy} \\ \qquad\quad 1 \qquad 2 \\ \text{Ron} \begin{array}{c} 1 \\ 2 \end{array}\begin{bmatrix} -2 & 3 \\ 1 & -2 \end{bmatrix} \end{array}$$

(B) Solve using the method in Section 10-2.

 The game is not strictly determined.

Let $D = (a + d) - (b + c) = [-2 + (-2)] - (3 + 1) = -8$.

The optimal strategies are:

$$P^* = [p_1 \ p_2] = \left[ \tfrac{d-c}{D} \quad \tfrac{a-b}{D} \right] = \left[ \tfrac{-2-1}{-8} \quad \tfrac{-2-3}{-8} \right] = \left[ \tfrac{3}{8} \quad \tfrac{5}{8} \right]$$

$$Q^* = \begin{bmatrix} q_1 \\ q_2 \end{bmatrix} = \begin{bmatrix} \tfrac{d-b}{D} \\ \tfrac{a-c}{D} \end{bmatrix} = \begin{bmatrix} \tfrac{-2-3}{-8} \\ \tfrac{-2-1}{-8} \end{bmatrix} = \begin{bmatrix} \tfrac{5}{8} \\ \tfrac{3}{8} \end{bmatrix}$$

The expected value $v$ of the game is given by:

$$v = \frac{ad - bc}{D} = \frac{4 - 3}{-8} = -\frac{1}{8} \qquad (10\text{-}2)$$

**30.** The payoff matrix is $M = \begin{bmatrix} -2 & 3 \\ 1 & -2 \end{bmatrix}$.

Step 1. Convert $M$ into a positive payoff matrix by adding 3 to each entry: $M_1 = \begin{bmatrix} 1 & 6 \\ 4 & 1 \end{bmatrix}$

Step 2. Set up the two corresponding linear programming problems:

(A) Minimize $y = x_1 + x_2$        (B) Maximize $y = z_1 + z_2$

     Subject to: $x_1 + 4x_2 \geq 1$          Subject to: $z_1 + 6z_2 \leq 1$

               $6x_1 + x_2 \geq 1$                  $4z_1 + z_2 \leq 1$

                 $x_1, \ x_2 \geq 0$                      $z_1, \ z_2 \geq 0$

Copyright © 2011 Pearson Education, Inc. Publishing as Prentice Hall.

Step 3. Solve the linear programming problems geometrically:

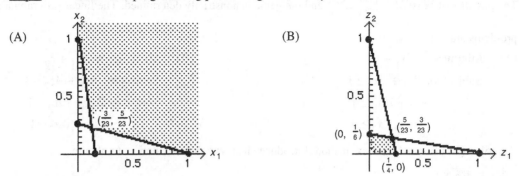

Theorems 1 and 2 in Section 5.1 imply that each problem has a solution that must occur at a corner point.

(A)

| Corner Points | Minimize $y = x_1 + x_2$ |
|---|---|
| $(0,\ 1)$ | $1$ |
| $\left(\dfrac{3}{23},\ \dfrac{5}{23}\right)$ | $\dfrac{8}{23}$ |
| $(1,\ 0)$ | $1$ |

Min $y = \dfrac{8}{23}$ occurs at

$$x_1 = \dfrac{3}{23},\quad x_2 = \dfrac{5}{23}$$

(B)

| Corner Points | Maximize $y = z_1 + z_2$ |
|---|---|
| $(0,\ 0)$ | $0$ |
| $\left(0,\dfrac{1}{6}\right)$ | $\dfrac{1}{6}$ |
| $\left(\dfrac{5}{23},\ \dfrac{3}{23}\right)$ | $\dfrac{8}{23}$ |
| $\left(\dfrac{1}{4},\ 0\right)$ | $\dfrac{1}{4}$ |

Max $y = \dfrac{8}{23}$ occurs at

$$z_1 = \dfrac{5}{23},\quad z_2 = \dfrac{3}{23}$$

Step 4. Use the solutions in Step 3:

(A)
$$v_1 = \frac{1}{x_1 + x_2} = \frac{1}{\dfrac{3}{23} + \dfrac{5}{23}} = \frac{23}{8}$$

$$p_1 = v_1 x_1 = \frac{23}{8} \cdot \frac{3}{23} = \frac{3}{8}$$

$$p_2 = v_1 x_2 = \frac{23}{8} \cdot \frac{5}{23} = \frac{5}{8}$$

(B)
$$v_1 = \frac{1}{z_1 + z_2} = \frac{1}{\dfrac{5}{23} + \dfrac{3}{23}} = \frac{23}{8}$$

$$q_1 = v_1 z_1 = \frac{23}{8} \cdot \frac{5}{23} = \frac{5}{8}$$

$$q_2 = v_1 z_2 = \frac{23}{8} \cdot \frac{3}{23} = \frac{3}{8}$$

The optimal strategies are:

$$P^* = \begin{bmatrix} \dfrac{3}{8} & \dfrac{5}{8} \end{bmatrix},\ Q^* = \begin{bmatrix} \dfrac{5}{8} \\ \dfrac{3}{8} \end{bmatrix}$$

The value of the original game is: $v = v_1 - k = \dfrac{23}{8} - 3 = -\dfrac{1}{8}$.

Step 5. Check:

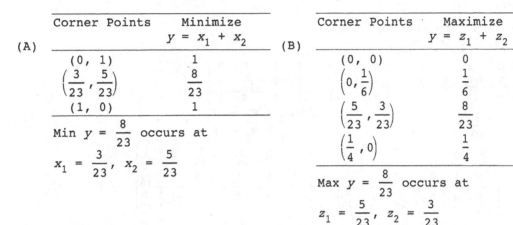

The expected value for Ron is: $-\$\dfrac{1}{8}$;  the expected value for Cathy is: $\$\dfrac{1}{8}$.

(10-3)

Copyright © 2011 Pearson Education, Inc. Publishing as Prentice Hall.

**31.** The payoff matrix is: $M = \begin{bmatrix} 8 & \boxed{4} \\ \boxed{2} & \boxed{10} \end{bmatrix}$ and the game is nonstrictly determined. The linear programming problems are:

(A)  Minimize $y = x_1 + x_2$

Subject to:  $8x_1 + 2x_2 \geq 1$

$4x_1 + 10x_2 \geq 1$

$x_1, \ x_2 \geq 0$

(B)  Maximize $y = z_1 + z_2$

Subject to:  $8z_1 + 4z_2 \leq 1$

$2z_1 + 10z_2 \leq 1$

$z_1, \ z_2 \geq 0$

We solve (B) using the simplex method. Introduce slack variables $x_1$ and $x_2$ to obtain:

$$8z_1 + 4z_2 + x_1 \qquad\quad = 1$$
$$2z_1 + 10z_2 \qquad + x_2 \quad = 1$$
$$-z_1 - z_2 \qquad\qquad + y = 0$$

The simplex tableau for this system is:

$$
\begin{array}{ccccc}
z_1 & z_2 & x_1 & x_2 & y
\end{array}
$$

$$
\left[\begin{array}{ccccc|c}
8 & 4 & 1 & 0 & 0 & 1 \\
2 & \boxed{10} & 0 & 1 & 0 & 1 \\
\hline
-1 & -1 & 0 & 0 & 1 & 0
\end{array}\right]
\sim
\left[\begin{array}{ccccc|c}
8 & 4 & 1 & 0 & 0 & 1 \\
\frac{1}{5} & 1 & 0 & \frac{1}{10} & 0 & \frac{1}{10} \\
\hline
-1 & -1 & 0 & 0 & 1 & 0
\end{array}\right]
$$

$$\frac{1}{10} R_2 \rightarrow R_2 \qquad\qquad\qquad (-4)R_2 + R_1 \rightarrow R_1$$
$$R_2 + R_3 \rightarrow R_3$$

$$
\sim
\left[\begin{array}{ccccc|c}
\frac{36}{5} & 0 & 1 & -\frac{2}{5} & 0 & \frac{3}{5} \\
\boxed{\frac{1}{5}} & 1 & 0 & 1 & 0 & \frac{1}{10} \\
\hline
-\frac{4}{5} & 0 & 0 & \frac{1}{10} & 1 & \frac{1}{10}
\end{array}\right]
\sim
\left[\begin{array}{ccccc|c}
1 & 0 & \frac{5}{36} & -\frac{1}{18} & 0 & \frac{1}{12} \\
\frac{1}{5} & 1 & 0 & 1 & 0 & \frac{1}{10} \\
\hline
-\frac{4}{5} & 0 & 0 & \frac{1}{10} & 1 & \frac{1}{10}
\end{array}\right]
$$

$$\frac{5}{36} R_1 \rightarrow R_1 \qquad\qquad\qquad -\frac{1}{5} R_1 + R_2 \rightarrow R_2$$
$$\frac{4}{5} R_1 + R_3 \rightarrow R_3$$

$$
\begin{array}{ccccc}
z_1 & z_2 & x_1 & x_2 & y
\end{array}
$$

$$
\sim
\left[\begin{array}{ccccc|c}
1 & 0 & \frac{5}{36} & -\frac{1}{18} & 0 & \frac{1}{12} \\
0 & 1 & -\frac{1}{36} & \frac{91}{90} & 0 & \frac{1}{12} \\
\hline
0 & 0 & \frac{1}{9} & \frac{1}{18} & 1 & \frac{1}{6}
\end{array}\right]
$$

The maximum $y = z_1 + z_2 = \frac{1}{6}$ occurs at $z_1 = \frac{1}{12}$, $z_2 = \frac{1}{12}$.

Thus $v = \dfrac{1}{\frac{1}{12} + \frac{1}{12}} = 6$ and

$$q_1 = vz_1 = 6 \cdot \frac{1}{12} = \frac{1}{2}, \quad q_2 = vz_2 = 6 \cdot \frac{1}{12} = \frac{1}{2}$$

The solution to the minimization problem (A) can be read from the bottom row of the final tableau for the dual problem (B):

The minimum $y = x_1 + x_2 = \frac{1}{6}$ occurs at $x_1 = \frac{1}{9}$, $x_2 = \frac{1}{18}$ and

$$p_1 = vx_1 = 6 \cdot \frac{1}{9} = \frac{2}{3}, \quad p_2 = vx_2 = 6 \cdot \frac{1}{18} = \frac{1}{3}$$

The optimal strategies are:

Farmer: $P^* = \begin{bmatrix} \frac{2}{3} & \frac{1}{3} \end{bmatrix}$;  Weather: $Q^* = \begin{bmatrix} \frac{1}{2} \\ \frac{1}{2} \end{bmatrix}$

$(10\text{-}4)$

Copyright © 2011 Pearson Education, Inc. Publishing as Prentice Hall.

**32.** The payoff matrix is: $M = \begin{bmatrix} 8 & 4 \\ 2 & 10 \end{bmatrix}$.

Using the formulas from Section 10-2, the expected value of the game to the farmer is:

$$v = \frac{ad - bc}{D}, \quad \text{where } D = (a + d) - (b + c)$$

$$= \frac{72}{12} = 6 \quad \text{or} \quad \$60,000$$

If the weather plays "dry" for many years and the farmer always plants soybeans, then their strategies are:

farmer $P = [0 \ \ 1]$, weather $\begin{bmatrix} 0 \\ 1 \end{bmatrix}$

and the expected value of the game to the farmer is:

$$E(P,Q) = [0 \ \ 1] \begin{bmatrix} 84 \\ 210 \end{bmatrix} \begin{bmatrix} 0 \\ 1 \end{bmatrix} = [2 \ \ 10] \begin{bmatrix} 0 \\ 1 \end{bmatrix} = 10$$

or $100,000.    (10-2)

**33.** The payoff matrix is: $M = \begin{bmatrix} \boxed{1} & \boxed{-6} \\ \boxed{-5} & 4 \end{bmatrix}$ and the game is not strictly determined.

<u>Step 1</u>. Convert $M$ into a positive payoff matrix by adding 7 to each element: $M_1 = \begin{bmatrix} 8 & 1 \\ 2 & 11 \end{bmatrix}$

<u>Step 2</u>. Set up the two corresponding linear programming problems:

(A)    Minimize $y = x_1 + x_2$

Subject to: $8x_1 + 2x_2 \geq 1$

$x_1 + 11x_2 \geq 1$

$x_1, x_2 \geq 0$

(B)    Maximize $y = z_1 + z_2$

Subject to: $8z_1 + z_2 \leq 1$

$2z_1 + 11z_2 \leq 1$

$z_1, z_2 \geq 0$

<u>Step 3</u>. Solve each linear programming problem geometrically:

(A)

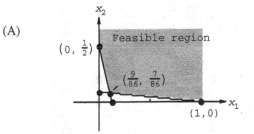

(B)

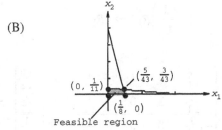

Theorems 1 and 2 in Section 5.1 imply that each problem has a solution that must occur at a corner point.

(A)

| Corner Points | Minimize $y = x_1 + x_2$ |
|---|---|
| $\left(0, \dfrac{1}{2}\right)$ | $\dfrac{1}{2}$ |
| $\left(\dfrac{9}{86}, \dfrac{7}{86}\right)$ | $\dfrac{8}{43}$ |
| $(1, 0)$ | $1$ |

Min $y = \dfrac{8}{43}$ occurs at

$x_1 = \dfrac{9}{86}, \ x_2 = \dfrac{7}{86}$

(B)

| Corner Points | Maximize $y = z_1 + z_2$ |
|---|---|
| $(0, 0)$ | $0$ |
| $\left(0, \dfrac{1}{11}\right)$ | $\dfrac{1}{11}$ |
| $\left(\dfrac{5}{43}, \dfrac{3}{43}\right)$ | $\dfrac{8}{43}$ |
| $\left(\dfrac{1}{8}, 0\right)$ | $\dfrac{1}{8}$ |

Max $y = \dfrac{8}{43}$ occurs at

$z_1 = \dfrac{5}{43}, \ z_2 = \dfrac{3}{43}$

Copyright © 2011 Pearson Education, Inc. Publishing as Prentice Hall.

Step 4. Use the solutions in Step 3:

(A)  $v_1 = \dfrac{1}{x_1 + x_2} = \dfrac{1}{\dfrac{9}{86} + \dfrac{7}{86}} = \dfrac{43}{8}$

$p_1 = v_1 x_1 = \dfrac{43}{8} \cdot \dfrac{9}{86} = \dfrac{9}{16}$

$p_2 = v_1 x_2 = \dfrac{43}{8} \cdot \dfrac{7}{86} = \dfrac{7}{16}$

(B)  $v_1 = \dfrac{1}{z_1 + z_2} = \dfrac{1}{\dfrac{11}{93} + \dfrac{5}{93}} = \dfrac{43}{8}$

$q_1 = v_1 z_1 = \dfrac{43}{8} \cdot \dfrac{5}{43} = \dfrac{5}{8}$

$q_2 = v_1 z_2 = \dfrac{43}{8} \cdot \dfrac{3}{43} = \dfrac{3}{8}$

The optimal strategies for both games $M$ and $M_1$ are:

Store $R$: $P^* = \begin{bmatrix} \dfrac{9}{16} & \dfrac{7}{16} \end{bmatrix}$, store $C$: $Q^* = \begin{bmatrix} \dfrac{5}{8} \\ \dfrac{3}{8} \end{bmatrix}$  (10-3)

**34.** From Problem 33, the linear programming problems are:

(A)  Minimize $y = x_1 + x_2$

Subject to: $8x_1 + 3x_2 \geq 1$

$x_1 + 11x_2 \geq 1$

$x_1, \ x_2 \geq 0$

(B)  Maximize $y = z_1 + z_2$

Subject to: $8z_1 + \ z_2 \leq 1$

$2z_1 + 11z_2 \leq 1$

$z_1, \ z_2 \geq 0$

We solve (B) using the simplex method. Introduce slack variables $x_1$, $x_2$ to obtain:

$$
\begin{array}{ccccc}
z_1 & z_2 & x_1 & x_2 & y
\end{array}
$$

$$
\begin{bmatrix}
8 & 1 & 1 & 0 & 0 & \bigm| & 1 \\
2 & \boxed{11} & 0 & 1 & 0 & \bigm| & 1 \\
-1 & -1 & 0 & 0 & 1 & \bigm| & 0
\end{bmatrix}
\sim
\begin{bmatrix}
8 & 1 & 1 & 0 & 0 & \bigm| & 1 \\
\dfrac{2}{11} & 1 & 0 & \dfrac{1}{11} & 0 & \bigm| & \dfrac{1}{11} \\
-1 & -1 & 0 & 0 & 1 & \bigm| & 0
\end{bmatrix}
$$

$\left(-\dfrac{1}{11}\right) R_1 \ \to \ R_1$

$(-1)R_2 + R_1 \ \to \ R_1$

$R_2 + R_3 \ \to \ R_3$

$$
\sim
\begin{bmatrix}
\dfrac{86}{11} & 0 & 1 & -\dfrac{1}{11} & 0 & \bigm| & \dfrac{10}{11} \\
\dfrac{2}{11} & 1 & 0 & \dfrac{1}{11} & 0 & \bigm| & \dfrac{1}{11} \\
-\dfrac{9}{11} & 0 & 0 & \dfrac{1}{11} & 1 & \bigm| & \dfrac{1}{11}
\end{bmatrix}
\sim
\begin{bmatrix}
1 & 0 & \dfrac{11}{86} & -\dfrac{1}{86} & 0 & \bigm| & \dfrac{5}{43} \\
\dfrac{2}{11} & 1 & 0 & \dfrac{1}{11} & 0 & \bigm| & \dfrac{1}{11} \\
-\dfrac{9}{11} & 0 & 0 & \dfrac{1}{11} & 1 & \bigm| & \dfrac{1}{11}
\end{bmatrix}
$$

$\dfrac{11}{86} R_1 \ \to \ R_1$

$\left(-\dfrac{2}{11}\right) R_1 + R_2 \ \to \ R_2$

$\dfrac{9}{11} R_1 + R_3 \ \to \ R_3$

$$
\begin{array}{ccccc}
z_1 & z_2 & x_1 & x_2 & y
\end{array}
$$

$$
\sim
\begin{bmatrix}
1 & 0 & \dfrac{11}{86} & -\dfrac{1}{86} & 0 & \bigm| & \dfrac{5}{43} \\
0 & 1 & -\dfrac{1}{43} & \dfrac{4}{43} & 0 & \bigm| & \dfrac{3}{43} \\
0 & 0 & \dfrac{9}{86} & \dfrac{7}{86} & 1 & \bigm| & \dfrac{8}{43}
\end{bmatrix}
$$

Copyright © 2011 Pearson Education, Inc.  Publishing as Prentice Hall.

The maximum $y = z_1 + z_2 = \dfrac{8}{43}$ occurs at $z_1 = \dfrac{5}{43}$, $z_2 = \dfrac{3}{43}$.

The expected value $v_1 = \dfrac{1}{z_1 + z_2} = \dfrac{1}{\dfrac{5}{43} + \dfrac{3}{43}} = \dfrac{43}{8}$.

The expected value $v$ of the original game for store $R$ is:

$$v = v_1 - k = \frac{43}{8} - 7 = -\frac{13}{8} = -1.625$$

The minimum $y = x_1 + x_2 = \dfrac{8}{43}$ occurs at $x_1 = \dfrac{9}{86}$, $x_2 = \dfrac{7}{86}$.

Thus, the optimal strategy for store $R$ is $P^* = [p_1,\ p_2]$ where

$$p_1 = v_1 x_1 = \frac{43}{8} \cdot \frac{9}{86} = \frac{9}{16}, \quad p_2 = v_1 x_2 = \frac{43}{8} \cdot \frac{7}{86} = \frac{7}{16}$$

If store $R$ plays its optimal strategy and store $C$ always places a newspaper ad, $Q = \begin{bmatrix} 1 \\ 0 \end{bmatrix}$, the expected value of the game for store $R$ is

$$E(P^*, Q) = \begin{bmatrix} \frac{9}{16} & \frac{7}{16} \end{bmatrix} \begin{bmatrix} 1 & -6 \\ -5 & 4 \end{bmatrix} \begin{bmatrix} 1 \\ 0 \end{bmatrix} = \begin{bmatrix} -\frac{13}{8} & -\frac{13}{8} \end{bmatrix} \begin{bmatrix} 1 \\ 0 \end{bmatrix} = -\frac{13}{8} = -1.625$$

Thus, the expected value for store $C$ is 1.625.                    (10-2, 10-4)

Copyright © 2011 Pearson Education, Inc. Publishing as Prentice Hall.

## 11  DATA DESCRIPTION AND PROBABILITY DISTRIBUTIONS

### EXERCISE 11-1

*Things to remember:*

1.  The following techniques are used to graph qualitative data:

    (a)   Vertical bar graphs.

    (b)   Horizontal bar graphs.

    (c)   Broken-line graphs.

    (d)   Pie graphs.

2.  The following techniques are used to depict quantitative data:

    (a)   Frequency distribution: a table showing class intervals and their corresponding frequencies.

    (b)   Histogram: graphic representation of a frequency distribution.

    (c)   Frequency polygon: a broken-line graph of a frequency distribution.

    (d)   Cumulative frequency table and cumulative frequency polygon (or ogive): the cumulative frequency is plotted over the upper boundary of the corresponding class.

---

1.  (A) and (B)

| Class interval | Tally | Frequency | Relative frequency |
|---|---|---|---|
| 0.5 – 2.5 | \| | 1 | .1 |
| 2.5 – 4.5 | \| | 1 | .1 |
| 4.5 – 6.5 | \|\|\|\| | 4 | .4 |
| 6.5 – 8.5 | \|\|\| | 3 | .3 |
| 8.5 – 10.5 | \| | 1 | .1 |

   (C)   The frequency tables and histograms are the **same** in part (A), but the data set in (B) is more spread out than the data set in (A).

3.  (A)   Let $X$ min $= 1.5$, $X$ max $= 25.5$, change $X$ scl from 1 to 2, and multiply $Y$ max and $Y$ scl by 2; change $X$ scl from 1 to 4 and multiply $Y$ max and $Y$ scl by 4.

    (B)   The shape becomes more symmetrical and more rectangular.

Copyright © 2011 Pearson Education, Inc. Publishing as Prentice Hall.

**5.**

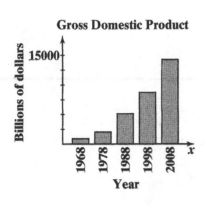

**7.** Greatest percentage increase in production: China – 44.7%

Greatest percentage decrease in production: South Africa – 36.8%

**9.**

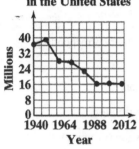

**11.**

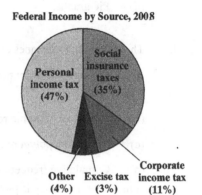

**13.** **(A)**

| CLASS INTERVAL | TALLY | FREQUENCY | RELATIVE FREQUENCY |
|---|---|---|---|
| 30.5-34.5 | \| | 1 | .05 |
| 34.5-38.5 | \|\|\| | 3 | .15 |
| 38.5-42.5 | ⅊ \|\| | 7 | .35 |
| 42.5-46.5 | ⅊ \| | 6 | .30 |
| 46.5-50.5 | \|\| | 2 | .10 |
| 50.5-54.5 | \| | 1 | .05 |

**(B)**

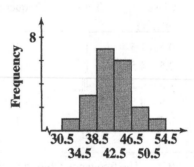

**(C)** Let $A$ = salary above $42,500;
Let $B$ = salary below $38,500.
From the data set itself

$$P(A) = \frac{n(A)}{n(S)} = \frac{9}{20} = 0.45;$$

$$P(B) = \frac{n(B)}{n(S)} = \frac{4}{20} = 0.20$$

**(D)**

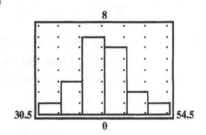

Copyright © 2011 Pearson Education, Inc.  Publishing as Prentice Hall.

**15.** (A) The frequency and relative frequency table for the given data is shown below.

| Class interval | Frequency | Relative frequency |
|---|---|---|
| - 0.5 – 4.5 | 5 | .05 |
| 4.5 – 9.5 | 54 | .54 |
| 9.5 – 14.5 | 25 | .25 |
| 14.5 – 19.5 | 13 | .13 |
| 19.5 – 24.5 | 0 | .00 |
| 24.5 – 29.5 | 1 | .01 |
| 29.5 – 34.5 | 2 | .02 |
| | 100 | 1.00 |

(B) The histogram below is a graphic representation of the tabulated data in part (A).

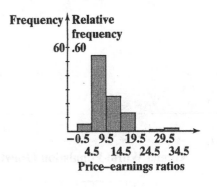

(C) A frequency polygon (broken-line graph) for the tabulated data in part (A) is shown at the left.

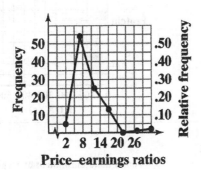

(D) An example of a cumulative and relative cumulative frequency table is shown below. From this table, we note that the probability of a price-earnings ratio drawn at random from the sample lying between 4.5 and 14.5 is 0.84 - 0.05 = 0.79.

| Class interval | Frequency | Cumulative Frequency | Relative cumulative frequency |
|---|---|---|---|
| -0.5 – 4.5 | 5 | 5 | .05 |
| 4.5 – 9.5 | 54 | 59 | .59 |
| 9.5 – 14.5 | 25 | 84 | .84 |
| 14.5 – 19.5 | 13 | 97 | .97 |
| 19.5 – 24.5 | 0 | 97 | .97 |
| 24.5 – 29.5 | 1 | 98 | .98 |
| 29.5 – 34.5 | 2 | 100 | 1.00 |

Copyright © 2011 Pearson Education, Inc.  Publishing as Prentice Hall.

(E)   A cumulative frequency polygon of the
      tabulated data in part (D) is shown at the right.

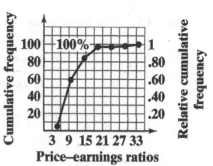

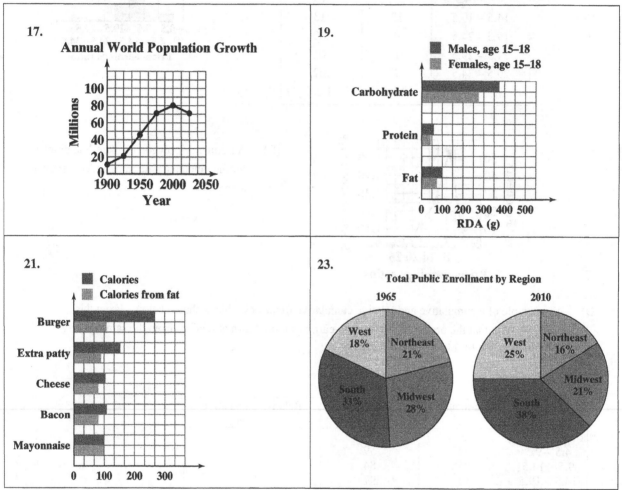

17.

**Annual World Population Growth**

19.

21.

23.

**Total Public Enrollment by Region**

25.   The median age in 1900 was approximately 23; in 2000 the median age was approximately 35.
      The median age decreased in the 1950's and 1960's, but increased in the other decades.

Copyright © 2011 Pearson Education, Inc.  Publishing as Prentice Hall.

**27.** (A)   The frequency and relative frequency table for the given data is shown below.

(B)   A histogram of the tabulated data in part (A) is shown below.

| Class interval | Frequency | Relative Frequency |
|---|---|---|
| 1.95 – 2.15 | 21 | .21 |
| 2.15 – 2.35 | 19 | .19 |
| 2.35 – 2.55 | 17 | .17 |
| 2.55 – 2.75 | 14 | .14 |
| 2.75 – 2.95 | 9 | .09 |
| 2.95 – 3.15 | 6 | .06 |
| 3.15 – 3.35 | 5 | .05 |
| 3.35 – 3.55 | 4 | .04 |
| 3.55 – 3.75 | 3 | .03 |
| 3.75 – 3.95 | 2 | .02 |
|  | 100 | 1.00 |

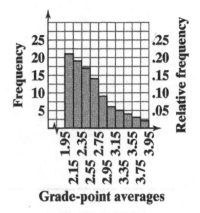

(C)   A frequency polygon of the tabulated data in part (A) is shown at the right.

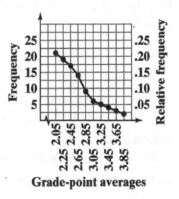

(D)   A cumulative and relative cumulative frequency table for the given data is shown below. The probability of a GPA drawn at random from the sample being over 2.95 is 1 - 0.8 = 0.2.

| Class interval | Frequency | Cumulative frequency | Relative cumulative frequency |
|---|---|---|---|
| 1.95-2.15 | 21 | 21 | .21 |
| 2.15-2.35 | 19 | 40 | .40 |
| 2.35-2.55 | 17 | 57 | .57 |
| 2.55-2.75 | 14 | 71 | .71 |
| 2.75-2.95 | 9 | 80 | .80 |
| 2.95-3.15 | 6 | 86 | .86 |
| 3.15-3.35 | 5 | 91 | .91 |
| 3.35-3.55 | 4 | 95 | .95 |
| 3.55-3.75 | 3 | 98 | .98 |
| 3.75-3.95 | 2 | 100 | 1.00 |

Copyright © 2011 Pearson Education, Inc. Publishing as Prentice Hall.

(E)  A cumulative frequency polygon for the tabulated data in part (D) is shown at the right.

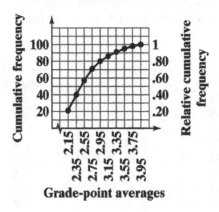

EXERCISE 11-2

*Things to remember:*

1. MEAN (Ungrouped Data)

   If $x_1, x_2, ..., x_n$ is a set of $n$ measurements, then the MEAN of the set of measurements is given by:

   $$[\text{mean}] = \frac{\sum\limits_{i=1}^{n} x_i}{n} = \frac{x_1 + x_2 + \cdots + x_n}{n} \qquad (1)$$

   $\bar{x}$ = [mean] if the data set is a sample
   $\mu$ = [mean] if the data set is the population

2. MEAN (Grouped Data)

   A data set of $n$ measurements is grouped into $k$ classes in a frequency table. If $x_i$ is the midpoint of the $i$th class interval and $f_i$ is the $i$th class frequency, then the MEAN for the grouped data is given by:

   $$[\text{mean}] = \frac{\sum\limits_{i=1}^{k} x_i f_i}{n} = \frac{x_1 f_1 + x_2 f_2 + \cdots + x_k f_k}{n} \qquad (2)$$

   $$n = \sum\limits_{i=1}^{k} f_i = (\text{Total number of measurements})$$

   $\bar{x}$ = [mean] if data set is a sample
   $\mu$ = [mean] if data set is the population

3. MEDIAN

   The MEDIAN of a set of $n$ measurements, arranged in ascending or descending order, is a number such that half the $n$ measurements fall below the median and half fall above.
   To find the median, we proceed as follows:

   (a) If the number of measurements $n$ is odd, the median is the middle number when the measurements are arranged in ascending or descending order.

   (b) If the number of measurements $n$ is even, the median is the mean of the two middle measurements when the measurements are arranged in ascending or descending order.

Copyright © 2011 Pearson Education, Inc. Publishing as Prentice Hall.

<u>4.</u>  MEDIAN (Grouped Data)

The MEDIAN FOR GROUPED DATA with no classes of frequency 0 is the number such that the histogram has the same area to the left of the median as to the right of the median.

<u>5.</u>  MODE

The MODE is the most frequently occurring measurement in a data set. There may be a unique mode, several modes, or, if no measurement occurs more than once, essentially no mode.

---

**1.**  Arrange the given numbers in increasing order:
1, 2, 2, 3, [3, 3,] 3, 4, 4, 5
Using <u>1</u>,

$$\text{Mean} = \bar{x} = \frac{1+2+2+3+3+3+3+4+4+5}{10} = \frac{30}{10} = 3$$

$$\text{Median} = \frac{3+3}{2} = 3$$

$$\text{Mode} = 3$$

**3.**  The mean and median are not suitable for these data. The modal preference for flavor of ice cream is chocolate.

**5.**  We construct a table indicating the class intervals, the class mid-points $x_i$, the frequencies $f_i$, and the products $x_i f_i$.

| Class interval | Midpoint $x_i$ | Frequency $f_i$ | Product $x_i f_i$ |
|---|---|---|---|
| 0.5-2.5 | 1.5 | 2 | 3.0 |
| 2.5-4.5 | 3.5 | 5 | 17.5 |
| 4.5-6.5 | 5.5 | 7 | 38.5 |
| 6.5-8.5 | 7.5 | 1 | 7.5 |
| | | $n = \sum_{i=1}^{4} f_i = 15$ | $\sum_{i=1}^{4} x_i f_i = 66.5$ |

$$\text{Thus, } \bar{x} = \frac{\sum_{i=1}^{n} x_i f_i}{n} = \frac{66.5}{15} \approx 4.433 \approx 4.4$$

**7.**  The median; the measurement 81.2 is an extreme value which distorts the mean.

**9.**  (A)  We would expect the mean and the median to be close to 3.5.

(B)  The answer depends on the results of your simulation.

Copyright © 2011 Pearson Education, Inc. Publishing as Prentice Hall.

**11.** (A)   Let $w, x, y, z$ be the four numbers, with $w \le x \le y \le z$.

Since the mode is 175, $w = x = 175$.
Since the median is 250,

$$\frac{175 + y}{2} = 250$$

$$175 + y = 500$$

$$y = 325$$

Since the mean is 300,

$$\frac{175 + 175 + 325 + z}{4} = 300$$

$$675 + z = 1200$$

$$z = 525$$

The set $\{175, 175, 325, 525\}$ has mean 300, median 250, mode 175.

(B)   Let the four numbers be $u, v, w, x$, where $u, v = m_3$. Choose $w$ such that the mean of $w$ and $m_3$ is $m_2$. Then choose $x$ such that

$$\frac{u + v + w + x}{4} = m_1$$

**13.**   Mean $= \dfrac{5.3 + 12.9 + 10.1 + 8.4 + 18.7 + 16.2 + 35.5 + 10.1}{8}$

$= \dfrac{117.2}{8} \approx 14.7$

Now arrange the data in order (we use increasing):  5.3, 8.4, 10.1, 10.1, 12.9, 16.2, 18.7, 35.5

Median $= \dfrac{10.1 + 12.9}{2} = 11.5;$    Mode $= 10.1$

**15.**

| Class interval | Midpoint $x_i$ | Frequency $f_i$ | Product $x_i f_i$ |
|---|---|---|---|
| 799.5- 899.5 | 849.5 | 3 | 2,548.5 |
| 899.5- 999.5 | 949.5 | 10 | 9,495 |
| 999.5-1099.5 | 1049.5 | 24 | 25,188 |
| 1099.5-1199.5 | 1149.5 | 12 | 13,794 |
| 1199.5-1299.5 | 1249.5 | 1 | 1,249.5 |
| | | $n = \sum\limits_{i=1}^{5} f_i = 50$ | $\sum\limits_{i=1}^{5} x_i f_i = 52{,}275$ |

Mean $= \bar{x} = \dfrac{\sum\limits_{i=1}^{n} x_i f_i}{n} = \dfrac{52{,}275}{50} = 1045.5$ hours.

To find the median, we draw the histogram of the data.
Total area $= 300 + 1000 + 2400 + 1200 + 100$
$= 5000$

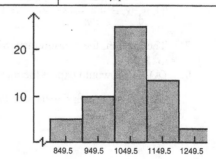

Copyright © 2011 Pearson Education, Inc. Publishing as Prentice Hall.

Let $M$ be the median. Then
$$100(3) + 100(10) + (M - 999.5)24 = 2500$$
$$(M - 999.5)24 = 1200$$
$$M - 999.5 = 50$$
$$M = 1049.5 \text{ hours}$$

17.  Mean: $\displaystyle\sum_{i=1}^{7} x_i = \$9,175.$

$$\overline{x} = \frac{9175}{7} = 1,310.71 \approx \$1,311$$

Median: Arrange the numbers in order:  1,087, 1,215, 1,252, 1,335, 1,394, 1,446, 1,446
Median = \$1,335
Mode: \$1,446

19.

| Class interval | Midpoint $x_i$ | Frequency $f_i$ | Product $x_i f_i$ |
|---|---|---|---|
| 41.5-43.5 | 42.5 | 3 | 127.5 |
| 43.5-45.5 | 44.5 | 7 | 311.5 |
| 45.5-47.5 | 46.5 | 13 | 604.5 |
| 47.5-49.5 | 48.5 | 17 | 824.5 |
| 49.5-51.5 | 50.5 | 19 | 959.5 |
| 51.5-53.5 | 52.5 | 17 | 892.5 |
| 53.5-55.5 | 54.5 | 15 | 817.5 |
| 55.5-57.5 | 56.5 | 7 | 395.5 |
| 57.5-59.5 | 58.5 | 2 | 117.0 |
|  |  | $n = \displaystyle\sum_{i=1}^{9} f_i = 100$ | $\displaystyle\sum_{i=1}^{9} x_i f_i = 5050$ |

$$\text{Mean} = \overline{x} = \frac{\displaystyle\sum_{i=1}^{9} x_i f_i}{100} = \frac{5050}{100} = 50.5\text{g}$$

The histogram for the data is:

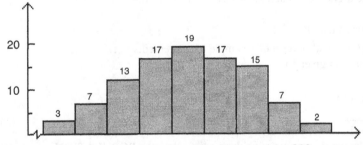

Total area = 6 + 14 + 26 + 34 + 38 + 34 + 30 + 14 + 4 = 200

Let $M$ be the median. Then
$$6 + 14 + 26 + 34 + (M - 49.5)19 = 100$$
$$M - 49.5 = \frac{100 - 80}{19} \approx 1.05$$
$$M = 50.55\text{g}$$

Copyright © 2011 Pearson Education, Inc.  Publishing as Prentice Hall.

**21.** Mean: $\displaystyle\sum_{i=1}^{10} x_i = 22{,}689$

$\bar{x} = \dfrac{22{,}689}{10} = 2{,}268.9$ thousand $= 2{,}268{,}900$

Median: The numbers are arranged in decreasing order.

Median $= \dfrac{1{,}104 + 1{,}101}{2} = 1{,}102.5$ thousand $= 1{,}102{,}500$

Mode: There is no mode.

**23.** The histogram for the data in Table 1 is:

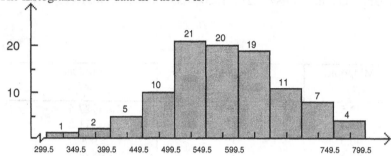

Total area $= 50 + 100 + 250 + 500 + 1050 + 1000 + 950 + 550 + 350 + 200$

$\qquad\qquad = 5000$

Let $M$ be the median. Then

$50 + 100 + 250 + 500 + 1050 + (M - 549.5)20 = 2500$

$M - 549.5 = \dfrac{2500 - 1950}{20} = \dfrac{550}{20} = 27.5$

$M = 577$

---

## EXERCISE 11-3

*Things to remember:*

1.  The RANGE for a set of ungrouped data is the difference between the largest and smallest values in the data set. For a frequency distribution, the range is the difference between the upper boundary of the highest class and the lower boundary of the lowest class.

2.  VARIANCE (Ungrouped Data)*

    The SAMPLE VARIANCE $s^2$ of a set of $n$ sample measurements $x_1$, $x_2, \ldots, x_n$ with mean $\bar{x}$ is given by

    $$S^2 = \frac{\displaystyle\sum_{i=1}^{n}(x_i - \bar{x})^2}{n-1} \qquad\qquad (3)$$

    If $x_1, x_2, \ldots, x_n$ is the whole population with mean $\mu$, then the POPULATION VARIANCE $\sigma^2$ is given by

    $$\sigma^2 = \frac{\displaystyle\sum_{i=1}^{n}(x_i - \mu)^2}{n}$$

    *In this section, our interest is restricted to sample variance.

Copyright © 2011 Pearson Education, Inc. Publishing as Prentice Hall.

3. STANDARD DEVIATION (Ungrouped Data)*

The SAMPLE STANDARD DEVIATION $s$ of a set of $n$ sample measurements $x_1, x_2, \ldots, x_n$ with mean $\bar{x}$ is given by

$$s = \sqrt{\frac{\sum\limits_{i=1}^{n}(x_i - \bar{x})^2}{n-1}} \tag{4}$$

If $x_1, x_2, \ldots, x_n$ is the whole population with mean $\mu$, then the POPULATION STANDARD DEVIATION $\sigma$ is given by

$$\sigma = \sqrt{\frac{\sum\limits_{i=1}^{n}(x_i - \mu)^2}{n}}$$

*In this section, our interest is restricted to sample standard deviation.

4. STANDARD DEVIATION (Grouped Data)*

Suppose a data set of $n$ sample measurements is grouped into $k$ classes in a frequency table where $x_i$ is the midpoint and $f_i$ the frequency of the $i$th class interval. Then the SAMPLE STANDARD DEVIATION $s$ for the grouped data is given by

$$s = \sqrt{\frac{\sum\limits_{i=1}^{k}(x_i - \bar{x})^2 f_i}{n-1}} \tag{5}$$

where $n = \sum\limits_{i=1}^{k} f_i$ = Total number of measurements

If $x_1, x_2, \ldots, x_n$ is the whole population with mean $\mu$, then the POPULATION STANDARD DEVIATION $\sigma$ is given by

$$\sigma = \sqrt{\frac{\sum\limits_{i=1}^{n}(x_i - \mu)^2 f_i}{n}}$$

*In this section, our interest is restricted to sample standard deviation.

---

1. $\bar{x} = 3$, $n = 10$ (see Problem 1, Exercise 11-2)

$$s = \sqrt{\frac{\sum\limits_{i=1}^{n}(x_i - \bar{x})^2}{n-1}}$$

$$= \sqrt{\frac{(1-3)^2 + (2-3)^2 + (2-3)^2 + (3-3)^2 + (3-3)^2 + (3-3)^2 + (3-3)^2 + (4-3)^2 + (4-3)^2 + (5-3)^2}{10-1}}$$

$$= \sqrt{\frac{4+1+1+0+0+0+0+1+1+4}{9}} = \sqrt{\frac{12}{9}} \approx 1.15$$

Copyright © 2011 Pearson Education, Inc.  Publishing as Prentice Hall.

**3.** (A) $\bar{x} = \dfrac{\sum\limits_{i=1}^{10} x_i}{10} = \dfrac{33}{10} = 3.3$

$$s = \sqrt{\dfrac{\sum\limits_{i=1}^{n}(x_i - 3.3)^2}{10-1}}$$

$$= \sqrt{\dfrac{\begin{array}{l}(4-3.3)^2 + (2-3.3)^2 + (3-3.3)^2 + (5-3.3)^2 + (3-3.3)^2 \\ + (1-3.3)^2 + (6-3.3)^2 + (4-3.3)^2 + (2-3.3)^2 + (3-3.3)^2\end{array}}{9}}$$

$$= \sqrt{\dfrac{20.10}{9}} = 1.49$$

The measurements that are in the interval (1.81, 4.79) are within one standard deviation of the mean; 70% of the measurements, {4, 2, 3, 3, 4, 2, 3}, are in this interval.

The measurements that are in the interval (0.32, 6.28) are within 2 standard deviations of the mean; 100% of the measurements are in this interval. It follows immediately that 100% of the measurements are within 3 standard deviations of the mean.

(B) Yes. If the histogram is bell-shaped, then approximately 68% of the data are within one standard deviation of the mean, approximately 95% are within 2 standard deviations of the mean, and almost all the data is within 3 standard deviations of the mean. The data set given here satisfies these criteria.

(C)

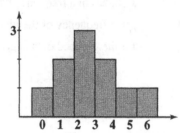

**5.**

| Interval | Midpoint $x_i$ | Frequency $f_i$ | $x_i f_i$ | $(x_i - \bar{x})^2$ | $(x_i - \bar{x})^2 f_i$ |
|---|---|---|---|---|---|
| 0.5– 3.5 | 2 | 2 | 4 | 19.36 | 38.72 |
| 3.5– 6.5 | 5 | 5 | 25 | 1.96 | 9.80 |
| 6.5– 9.5 | 8 | 7 | 56 | 2.56 | 17.92 |
| 9.5–12.5 | 11 | 1 | 11 | 21.16 | 21.16 |
| | | $n = 15$ | $\sum\limits_{i=1}^{4} x_i f_i = 96$ | | $\sum\limits_{i=1}^{4} (x_i - \bar{x})^2 f_i = 87.60$ |

$\bar{x} = \dfrac{96}{15} = 6.4$ $\qquad\qquad s = \sqrt{\dfrac{87.60}{15-1}} \approx 2.5$

**7.** True: Let $S$ be a finite set of numbers and let $T$ be a subset of $S$. The smallest element in $T$ is greater than or equal to the smallest element in $S$; The largest element in $T$ is less than or equal to the largest element in $S$.

**9.** False: If the sample variance is less than 1, then the sample standard deviation is greater than the sample variance.

Copyright © 2011 Pearson Education, Inc. Publishing as Prentice Hall.

**11.** True: Let $\{x_1, x_2\}$ be the sample (assume $x_1 \le x_2$). Then $\bar{x} = \dfrac{x_1 + x_2}{2}$.

$$\text{variance} = \frac{(x_1 - \bar{x})^2 + (x_2 - \bar{x})^2}{2} = \frac{\left(\dfrac{x_1}{2} - \dfrac{x_2}{2}\right)^2 + \left(\dfrac{x_2}{2} - \dfrac{x_1}{2}\right)^2}{2}$$

$$= \frac{x_1^2 - 2x_1x_2 + x_2^2 + x_2^2 - 2x_1x_2 + x_1^2}{8}$$

$$= \frac{2x_1^2 - 4x_1x_2 + 2x_2^2}{8} = \frac{(x_1 - x_2)^2}{4}$$

$$\text{standard deviation} = \sqrt{\frac{(x_2 - x_1)^2}{4}} = \frac{1}{2}(x_2 - x_1) = \frac{1}{2} \text{ the distance from } x_1 \text{ to } x_2.$$

**13.** (A) The first data set. The sums are not equally likely; a sum of 7 is much more likely than a sum of 2 or 12. In the second data set, the numbers 2 through 12 are equally likely and so the numbers will have greater dispersion.

    (B) The answer depends on the results of your simulation.

**15.**

| $x_i$ | $(x_i - \bar{x})^2$ |
|---|---|
| 2.35 | 4.00 |
| 1.42 | 8.58 |
| 8.05 | 13.69 |
| 6.71 | 5.57 |
| 3.11 | 1.54 |
| 2.56 | 3.20 |
| 0.72 | 13.18 |
| 4.17 | .03 |
| 5.33 | .96 |
| 7.74 | 11.49 |
| 3.88 | .22 |
| 6.21 | 3.46 |

$$\sum_{i=1}^{12} x_i = 52.25 \qquad \sum_{i=1}^{12} (x_i - \bar{x})^2 = 65.92$$

$$\bar{x} = \frac{\displaystyle\sum_{i=1}^{12} x_i}{12} = \frac{52.25}{12} \approx 4.35. \text{ Thus, the mean } \bar{x} \approx \$4.35.$$

$$s = \sqrt{\frac{\displaystyle\sum_{i=1}^{12}(x_i - \bar{x})^2}{12-1}} = \sqrt{\frac{65.92}{11}} \approx \sqrt{5.99} \approx 2.45. \text{ Thus, the standard deviation } s \approx \$2.45.$$

Copyright © 2011 Pearson Education, Inc. Publishing as Prentice Hall.

**17.**

| Interval | Midpt. $x_i$ | Freq. $f_i$ | $x_if_i$ | $(x_i - \bar{x})^2$ | $(x_i - \bar{x})^2 f_i$ |
|---|---|---|---|---|---|
| 6.95– 7.45 | 7.20 | 2 | 14.40 | 2.25 | 4.50 |
| 7.45– 7.95 | 7.70 | 10 | 77.00 | 1.00 | 10.00 |
| 7.95– 8.45 | 8.20 | 23 | 188.60 | .25 | 5.75 |
| 8.45– 8.95 | 8.70 | 30 | 261.00 | .00 | .00 |
| 8.95– 9.45 | 9.20 | 21 | 193.20 | .25 | 5.25 |
| 9.45– 9.95 | 9.70 | 13 | 126.10 | 1.00 | 13.00 |
| 9.95–10.45 | 10.20 | 1 | 10.20 | 2.25 | 2.25 |
| | | $n = 100$ | $\sum_{i=1}^{7} x_i f_i = 870.50$ | | $\sum_{i=1}^{7}(x_i - \bar{x})^2 f_i = 40.75$ |

$$\bar{x} = \frac{\sum_{i=1}^{7} x_i f_i}{100} = \frac{870.5}{100} = 8.705$$

Thus, the mean $\bar{x} \approx 8.7$ hrs.

$$s = \sqrt{\frac{\sum_{i=1}^{7}(x_i - \bar{x})^2 f_i}{n-1}} = \sqrt{\frac{40.75}{99}} \approx \sqrt{.4116} \approx .6$$

Thus, the standard deviation $s \approx .6$ hrs.

**19.**

| $x_i$ | $(x_i - \bar{x})^2$ |
|---|---|
| 4.9 | .04 |
| 5.1 | .00 |
| 3.9 | 1.44 |
| 4.2 | .81 |
| 6.4 | 1.69 |
| 3.4 | 2.89 |
| 5.8 | .49 |
| 6.1 | 1.00 |
| 5.0 | .01 |
| 5.6 | .25 |
| 5.8 | .49 |
| 4.6 | .25 |
| $\sum_{i=1}^{12} x_i = 60.8$ | $\sum_{i=1}^{12}(x_i - \bar{x})^2 = 9.36$ |

$$\bar{x} = \frac{\sum_{i=1}^{12} x_i}{12} = \frac{60.8}{12} \approx 5.1$$

Thus, the mean $\bar{x} \approx 5.1$ minutes.

$$s = \sqrt{\frac{\sum_{i=1}^{12}(x_i - \bar{x})^2}{12-1}} = \sqrt{\frac{9.36}{11}}$$

$$\approx \sqrt{0.8509} \approx .92$$

Thus, the standard deviation $s \approx .9$ minutes.

**21.**

| $x_i$ | $(x_i - \bar{x})^2$ |
|---|---|
| 9 | 4 |
| 11 | 0 |
| 11 | 0 |
| 15 | 16 |
| 10 | 1 |
| 12 | 1 |
| 12 | 1 |
| 13 | 4 |
| 8 | 9 |
| 7 | 16 |
| 13 | 4 |
| 12 | 1 |
| $\sum_{i=1}^{12} x_i = 133$ | $\sum_{i=1}^{12}(x_i - \bar{x})^2 = 57$ |

$$\bar{x} = \frac{\sum_{i=1}^{12} x_i}{12} = \frac{133}{12} \approx 11.08$$

Thus, the mean $\bar{x} = 11.1$.

$$s = \sqrt{\frac{\sum_{i=1}^{12}(x_i - \bar{x})^2}{12-1}} = \sqrt{\frac{57}{11}}$$

$$\approx \sqrt{5.18} \approx 2.28$$

Thus, the standard deviation $s \approx 2.3$.

Copyright © 2011 Pearson Education, Inc. Publishing as Prentice Hall.

**EXERCISE 11-4**

*Things to remember:*

1.  BERNOULLI TRIALS

    A sequence of experiments is called a SEQUENCE OF BERNOULLI TRIALS, or a BINOMIAL EXPERIMENT if:

    (a)    Only two outcomes are possible on each trial.

    (b)    The probability of success $p$ for each trial is a constant (the probability of failure is $q = 1 - p$).

    (c)    All trials are independent.

2.  PROBABILITY OF $x$ SUCCESSES IN $n$ BERNOULLI TRIALS

    The probability of exactly $x$ successes in $n$ independent repeated Bernoulli trials, with the probability of success of each trial $p$ (and of failure $q$), is

    $$P(x \text{ successes}) = C_{n,x} p^x q^{n-x}$$

    where $C_{n,x}$ is the number of combinations of $n$ objects taken $x$ at a time.

3.  BINOMIAL FORMULA

    For $n$ a natural number,

    $$(a + b)^n = C_{n,0}a^n + C_{n,1}a^{n-1}b + C_{n,2}a^{n-2}b^2 + \cdots + C_{n,n}b^n$$

4.  BINOMIAL DISTRIBUTION

    $P(X_n = x) = P(x \text{ successes in } n \text{ trials})$

    $$= C_{n,x} p^x q^{n-x} \quad x \in \{0, 1, 2, \dots, n\}$$

    where $p$ is the probability of success and $q$ is the probability of failure on each trial. Informally, $P(X_n = x)$ is written $P(x)$.

5.  MEAN AND STANDARD DEVIATION (RANDOM VARIABLE IN A BINOMIAL DISTRIBUTION)

    Mean:    $\mu = np$

    Standard deviation:    $\sigma = \sqrt{npq}$

---

1.  $n = 5, x = 1, p = \dfrac{1}{2}; q = \dfrac{1}{2}$ and $C_{5,1}\left(\dfrac{1}{2}\right)^1 \left(\dfrac{1}{2}\right)^4 = 5\left(\dfrac{1}{2}\right)\left(\dfrac{1}{16}\right) = \dfrac{5}{32} \approx 0.156$

3.  $n = 6, x = 3, p = 0.4, q = 0.6$ and
    $C_{6,3}(0.4)^3(0.6)^3 = 20(0.064)(0.216) \approx 0.276$

5.  $n = 4, x = 3, p = \dfrac{2}{3}; q = \dfrac{1}{3}$ and $C_{4,3}\left(\dfrac{2}{3}\right)^3 \left(\dfrac{1}{3}\right)^1 = 4\left(\dfrac{8}{27}\right)\left(\dfrac{1}{3}\right) = \dfrac{32}{81} \approx 0.395$

7.  $P(\text{H,T,T,T}) = \dfrac{1}{2} \cdot \dfrac{1}{2} \cdot \dfrac{1}{2} \cdot \dfrac{1}{2} = \dfrac{1}{16}$

9.  Let $p$ = probability of getting a tail = $\dfrac{1}{2}$

    $q$ = probability of getting a head = $\dfrac{1}{2}$

Copyright © 2011 Pearson Education, Inc. Publishing as Prentice Hall.

$n = 4$

$P(\text{at least 3 tails}) = P(x \geq 3) = P(x = 3) + P(x = 4)$

$$= C_{4,3}\left(\frac{1}{2}\right)^3\left(\frac{1}{2}\right) + C_{4,4}\left(\frac{1}{2}\right)^4\left(\frac{1}{2}\right)^0$$

$$= 4 \cdot \frac{1}{16} + \frac{1}{16} = \frac{5}{16}$$

**11.** $P(\text{T,T,T,T}) = \left(\frac{1}{2}\right)^4 = \frac{1}{16}$

**13.** $P(x) = C_{3,x}\left(\frac{1}{4}\right)^x\left(\frac{3}{4}\right)^{3-x}$

| $x$ | 0 | 1 | 2 | 3 |
|---|---|---|---|---|
| $P(x)$ | 0.42 | 0.42 | 0.14 | 0.02 |

The histogram for this distribution is shown at the right.

Mean: $\mu = np = 3\left(\frac{1}{4}\right) = 0.75$

Standard Dev: $\sigma = \sqrt{npq} = \sqrt{3\left(\frac{1}{4}\right)\left(\frac{3}{4}\right)} = \frac{3}{4} = 0.75$

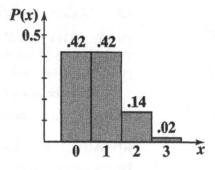

**15.** $P(x) = C_{4,x}\left(\frac{1}{3}\right)^x\left(\frac{2}{3}\right)^{4-x}$

| $x$ | 0 | 1 | 2 | 3 | 4 |
|---|---|---|---|---|---|
| $P(x)$ | 0.20 | 0.40 | 0.30 | 0.10 | 0.01 |

The histogram for this distribution is shown at the right.

$\mu = np = 4\left(\frac{1}{3}\right) = 1.333$

$\sigma = \sqrt{npq} = \sqrt{4\left(\frac{1}{3}\right)\left(\frac{2}{3}\right)} = \frac{\sqrt{8}}{3} \approx 0.943$

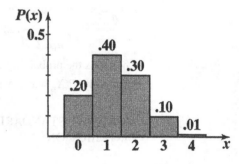

**17.** $\mu = 0, \; \sigma = 0$

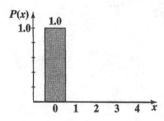

**19.** $P(6,5,6) = \frac{1}{6} \cdot \frac{1}{6} \cdot \frac{1}{6} = \frac{1}{216} \approx 0.005$

**21.** Let $p$ = probability of rolling a 6 in one trial = $\frac{1}{6}$.

$q$ = probability of not getting a 6 = $\frac{5}{6}$.

$n = 3$

Copyright © 2011 Pearson Education, Inc.  Publishing as Prentice Hall.

$$P(\text{at least two 6's}) = P(x \geq 2) = P(x = 2) + P(x = 3)$$

$$= C_{3,2}\left(\frac{1}{6}\right)^2\left(\frac{5}{6}\right) + C_{3,3}\left(\frac{1}{6}\right)^3\left(\frac{5}{6}\right)^0$$

$$= 3 \cdot \frac{1}{36} \cdot \frac{5}{6} + \frac{1}{216} = \frac{16}{216} \approx 0.074$$

**23.** $P(\text{no 6's}) = C_{3,0}\left(\frac{1}{6}\right)^0\left(\frac{5}{6}\right)^3 = \frac{125}{216} \approx 0.579$

**25.** $p = .35, q = 1 - .35 = .65, n = 4$

(A) The probability of getting exactly two hits is given by:
$$P(x = 2) = C_{4,2}(.35)^2(.65)^2 \approx .311$$

(B) The probability of getting at least two hits is given by:
$$P(x \geq 2) = P(2) + P(3) + P(4)$$
$$= C_{4,2}(.35)^2(.65)^2 + C_{4,3}(.35)^3(.65) + C_{4,4}(.35)^4$$
$$= .3105 + .1115 + .0150 \approx .437$$

**27.** $p = \frac{1}{5}, q = \frac{4}{5}$

Let $R$ denote the event "all answers are wrong" and let $S$ denote the event "at least half of the answers are correct." Then

$$P(R) = C_{10,0}\left(\frac{1}{5}\right)^0\left(\frac{4}{5}\right)^{10} = \left(\frac{4}{5}\right)^{10} = 0.107$$

$$P(S) = \sum_{k=5}^{10} C_{10,k}\left(\frac{1}{5}\right)^k\left(\frac{4}{5}\right)^{10-k}$$

$$= \frac{C_{10,5}(4)^5 + C_{10,6}(4)^4 + C_{10,7}(4)^3 + C_{10,8}(4)^2 + C_{10,9}(4) + C_{10,10}}{5^{10}}$$

$$= \frac{252(1024) + 210(256) + 120(64) + 45(16) + 10(4) + 1}{5^{10}} \approx 0.033$$

It is more likely that all guesses are wrong.

**29.** $P(x) = C_{6,x}(.4)^x(.6)^{6-x}$

| $x$ | $P(x)$ |
|-----|--------|
| 0 | .05 |
| 1 | .19 |
| 2 | .31 |
| 3 | .28 |
| 4 | .14 |
| 5 | .04 |
| 6 | .004 |

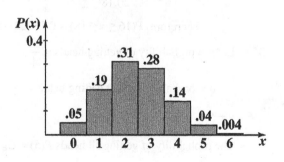

The histogram for this distribution is shown at the right.

$\mu = np = 6 \times .4 = 2.4$

Copyright © 2011 Pearson Education, Inc. Publishing as Prentice Hall.

$$\sigma = \sqrt{npq} = \sqrt{6 \times .4 \times .6} = 1.2$$

**31.** $P(x) = C_{8,x}(.3)^x(.7)^{8-x}$

| $x$ | $P(x)$ |
|---|---|
| 0 | .06 |
| 1 | .20 |
| 2 | .30 |
| 3 | .25 |
| 4 | .14 |
| 5 | .05 |
| 6 | .01 |
| 7 | .0012 |
| 8 | .0001 |

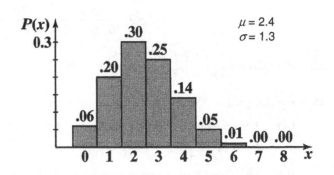

The histogram for this distribution is shown above.

$\mu = np = 8 \times .3 = 2.4$

$\sigma = \sqrt{npq} = \sqrt{8 \times .3 \times .7} \approx 1.3$

**33.** Given $p = 0.85$, $q = 0.15$, $n = 20$

Using a graphing utility, calculate the binomial probability distribution table.

(A)  Mean: $\mu = np = 20(0.85) = 17$

Standard deviation: $\sigma = \sqrt{npq} = \sqrt{20(0.85)(0.15)} \approx 1.597$

(B)  To be within one standard deviation of the mean, we must have
$17 - 1.597 < x < 17 + 1.597$  or  $15.403 < x < 18.597$
Thus, $x = 16$, 17 or 18.

$P(x = 16) = C_{20,16}(0.85)^{16}(0.15)^4 = 4845(0.07425)(0.00051) \approx 0.18212$

$P(x = 17) = C_{20,17}(0.85)^{17}(0.15)^3 = 1140(0.06311)(0.00338) \approx 0.24283$

$P(x = 18) = C_{20,18}(0.85)^{18}(0.15)^2 = 190(0.05365)(0.0225) \approx 0.22934$

Therefore, $P(16 \le x \le 18) \approx 0.18212 + 0.24283 + 0.22934 = 0.65429 \approx 0.654$

**35.** Let $p = $ probability of getting heads $= \dfrac{3}{4}$,

$q = $ probability of not getting heads $= \dfrac{1}{4}$.

$n = 5$, $x = 5$

The probability of getting all heads $P(5) = C_{5,5}\left(\dfrac{3}{4}\right)^5\left(\dfrac{1}{4}\right)^0 = .2373$.

The probability of getting all tails is the same as the probability of getting no heads. Thus,

$P(0) = C_{5,0}\left(\dfrac{3}{4}\right)^0\left(\dfrac{1}{4}\right)^5 = .00098$

Therefore,

$P(\text{all heads or all tails}) = P(5) + P(0) = .2373 + .00098 = .23828 \approx .238$

Copyright © 2011 Pearson Education, Inc. Publishing as Prentice Hall.

**37.** The theoretical probability distribution is obtained by using

$$P(x) = C_{3,x}(.5)^x(.5)^{3-x} = C_{3,x}(.5)^3$$

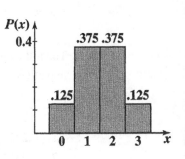

| Frequency of heads in 100 tosses of 3 coins | | | |
|---|---|---|---|
| Number of heads $x$ | $P(x)$ | Theoretical frequency $100P(x)$ | Actual frequency |
| 0 | .125 | 12.5 | (List your |
| 1 | .375 | 37.5 | experimental results |
| 2 | .375 | 37.5 | here.) |
| 3 | .125 | 12.5 | |

The histogram for the theoretical distribution is shown at the right.

**39.** To be symmetric about $x = \dfrac{n}{2}$, we must have $\mu = np = \dfrac{n}{2}$ which implies $p = \dfrac{1}{2}$. Alternatively, for the distribution to be symmetric we must have

$$C_{n,k}p^k q^{n-k} = C_{n,n-k}p^{n-k}q^k$$

for $k = 0, 1, 2, \ldots, n$. This implies $p^k q^{n-k} = p^{n-k}q^k$ which implies $p = q = \dfrac{1}{2}$

**41.** (A) $p = 0.5, q = 0.5, n = 10$
Mean: $\mu = np = 0.5(10) = 5$
Standard deviation: $\sigma = \sqrt{npq} = \sqrt{10(0.5)(0.5)} \approx 1.581$

(B) The answer depends on the results of your simulation.

**43.** (A) Let $p$ = probability of completing the program = .7,
and $q$ = probability of not completing the program = .3.

$n = 7, x = 5$
$$P(5) = C_{7,5}(.7)^5(.3)^2 = 21(.1681)(.09) = .318$$

(B) $P(x \geq 5) = P(5) + P(6) + P(7)$
$$= .318 + C_{7,6}(.7)^6(.3) + C_{7,7}(.7)^7(.3)^0$$
$$= .318 + 7(.1176)(.3) + 1(.0824)(1)$$
$$= .3180 + .2471 + .0824 \approx .647$$

**45.** Let $p$ = probability that an item is defective = .06,
$q$ = probability that an item is not defective = .94.

$n = 10$
$$P(x > 2) = 1 - P[x \leq 2] = 1 - [P(2) + P(1) + P(0)]$$
$$= 1 - [C_{10,2}(.06)^2(.94)^8 + C_{10,1}(.06)^1(.94)^9 + C_{10,0}(.06)^0(.94)^{10}]$$
$$= 1 - [.0988 + .3438 + .5386] = 1 - .9812 \approx .0188$$

A day's output will be inspected with a probability of .0188.

Copyright © 2011 Pearson Education, Inc.  Publishing as Prentice Hall.

**47.** (A)   $p = .05, q = .95, n = 6$

The following function defines the distribution:

$$P(x) = C_{6,x}(.05)^x(.95)^{6-x}$$

(B)   The following table is obtained by using the distribution function in part (A).

| $x$ | $P(x)$ |
|---|---|
| 0 | .735 |
| 1 | .232 |
| 2 | .031 |
| 3 | .002 |
| 4 | .000 |
| 5 | .000 |
| 6 | .000 |

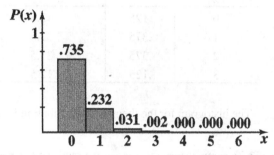

(C)   The histogram for the distribution in part (B) is shown above.

(D)   $\mu = np = 6 \times .05 = .30$

$$\sigma = \sqrt{npq} = \sqrt{6 \times (.05) \times (.95)} = .53$$

**49.** Let $p$ = probability of detecting TB = .8,

and $q$ = probability of not detecting TB = .2.

$n = 4$

The probability that at least one of the specialists will detect TB is:

$P(x \geq 1) = 1 - P(x < 1) = 1 - P(0)$

$\quad = 1 - C_{4,0}(.8)^0(.2)^4$

$\quad = 1 - .0016 = .9984 \approx .998$

**51.** Let $p$ = probability of having a child with brown eyes = .75,

and $q$ = probability of not having a child with brown eyes (i.e., with

blue eyes) = .25.

$n = 5$

(A)   $x = 0$ (all blue-eyed children, i.e., no brown-eyed children)

$P(0) = C_{5,0}(.75)^0(.25)^5 = .00098 \approx .001$

(B)   $x = 3$

$P(3) = C_{5,3}(.75)^3(.25)^2 \approx .264$

(C)   $x \geq 3$

$P(x \geq 3) = P(3) + P(4) + P(5)$

$\quad \approx .264 + C_{5,4}(.75)^4(.25)^1 + C_{5,5}(.75)^5(.25)^0$

$\quad \approx .2640 + .3955 + .2373 = .8968 \approx .897$

Copyright © 2011 Pearson Education, Inc. Publishing as Prentice Hall.

**53.** (A)  $p = .6, q = .4, n = 6$

The following function defines the distribution:

$$P(x) = C_{6,x}(.6)^x(.4)^{6-x}$$

(B)  The following table is obtained by using the distribution function in part (A).

| x | P(x) |
|---|------|
| 0 | .004 |
| 1 | .037 |
| 2 | .138 |
| 3 | .276 |
| 4 | .311 |
| 5 | .187 |
| 6 | .047 |

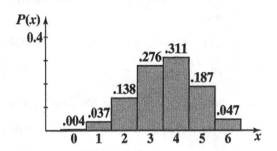

(C)  The histogram for the distribution in part (B) is shown at the right.

(D)  $\mu = np = 6(.6) = 3.6$

$$\sigma = \sqrt{npq} = \sqrt{6 \times .4 \times .6} = 1.2$$

**55.**  Let $p$ = probability of getting the right answer to a question = $\dfrac{1}{5}$,

and $q$ = probability of not getting the right answer to a question = $\dfrac{4}{5}$.

$n = 10, x \geq 7$

$P(x \geq 7) = P(7) + P(8) + P(9) + P(10)$

$$= C_{10,7}\left(\frac{1}{5}\right)^7\left(\frac{4}{5}\right)^3 + C_{10,8}\left(\frac{1}{5}\right)^8\left(\frac{4}{5}\right)^2 + C_{10,9}\left(\frac{1}{5}\right)^9\left(\frac{4}{5}\right) + C_{10,10}\left(\frac{1}{5}\right)^{10} \approx .000864$$

**57.** (A)  $p$ = probability of answer being correct by guessing = $\dfrac{1}{5}$ = .2,

$q = .8, n = 5$

The following function defines the distribution:

$$P(x) = C_{5,x}(.2)^x(.8)^{5-x}$$

(B)  The following table is obtained by using the distribution function in part (A):

| x | P(x) |
|---|------|
| 0 | .328 |
| 1 | .410 |
| 2 | .205 |
| 3 | .051 |
| 4 | .006 |
| 5 | .000 |

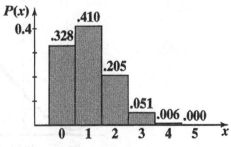

Copyright © 2011 Pearson Education, Inc.  Publishing as Prentice Hall.

(C)   The histogram for part (B) is shown at the right.

(D)   $\mu = np = 5 \times .2 = 1.0$

   $\sigma = \sqrt{npq} = \sqrt{5 \times .2 \times .8} \approx .894$

**59.**   Let $p$ = probability of a divorce within 20 years = .60,
   and $q$ = probability of no divorce within 20 years = .40.

   $n = 6$

(A)   $P(x = 0) = C_{6,0}(.60)^0(.40)^6 \approx .0041$

(B)   $P(x = 6) = C_{6,6}(.60)^6(.40)^0 \approx .0467$

(C)   $P(x = 2) = C_{6,2}(.60)^2(.40)^4 \approx .138$

(D)   $P(x \geq 2) = 1 - P(x < 2) = 1 - [P(0) + P(1)]$

   $= 1 - [.0041 + C_{6,1}(.60)^1(.40)^5]$

   $\approx 1 - [.0041 + .0369] = .959$

## EXERCISE 11-5

*Things to remember:*

1.   NORMAL CURVE PROPERTIES

   (a)   Normal curves are bell-shaped and symmetrical with respect to a vertical line.

   (b)   The mean is the real number at the point where the axis of symmetry intersects the
      horizontal axis.

   (c)   The shape of a normal curve is completely determined by its mean and a positive real
      number called the standard deviation. A small standard deviation indicates a tight
      clustering about the mean and a tall, narrow curve; a large standard deviation indicates a
      large deviation from the mean and a broad, flat curve.

   (d)   Irrespective of the shape, the area between the curve and the $x$ axis is always 1.

   (e)   Irrespective of the shape, 68.3% of the area will lie within an interval of 1 standard
      deviation on either side of the mean, 95.4% within 2 standard deviations on either side,
      and 99.7% within 3 standard deviations on either side (see the figure).

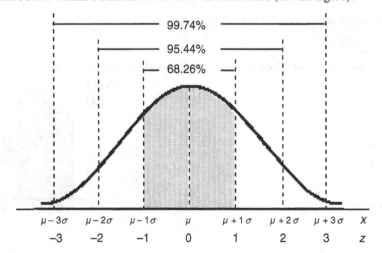

Copyright © 2011 Pearson Education, Inc.  Publishing as Prentice Hall.

2.     If $\mu$ and $\sigma$ are the mean and standard deviation of a normal curve and $x$ is a measurement then the number of standard deviations that $x$ is from the mean is given by:

$$z = \frac{x - \mu}{\sigma}$$

3.     PROPERTIES OF A NORMAL PROBABILITY DISTRIBUTION

      (a)     $P(a \le x \le b) =$ the area under the normal curve from $a$ to $b$.

      (b)     $P(-\infty < x < \infty) = 1 =$ the total area under the normal curve.

      (c)     $P(x = c) = 0$.

4.     RULE OF THUMB TEST

      Use a normal distribution to approximate a binomial distribution only if the interval $[\mu - 3\sigma, \mu + 3\sigma]$ lies entirely in the interval from 0 to $n$.

---

**1.**   0.4772                 **3.**   0.3925                **5.**   0.4970

*In Problems 7 - 11, use* $z = \dfrac{x - \mu}{\sigma}$ *, where* $\mu = 100$, $\sigma = 10$.

**7.**   $x = 115$;   $z = \dfrac{115 - 100}{10} = \dfrac{15}{10} = 1.5$; 115 is 1.5 standard deviations from the mean

**9.**   $x = 90$;   $z = \dfrac{90 - 100}{10} = \dfrac{-10}{10} = -1$; 90 is 1 standard deviation from the mean

**11.**   $x = 124.3$;   $z = \dfrac{124.3 - 100}{10} = \dfrac{24.3}{10} = 2.43$; 124.3 is 2.43 deviations from the mean

**13.**   From Problem 7, $z = 1.5$. From Table I, Appendix C, the area under the normal curve from the mean $\mu = 100$ to 115 is: 0.4332.

**15.**   From Problem 9, 90 is 1 standard deviation from the mean $\mu = 100$. Thus, the area under the normal curve from 90 to 100 is: 0.3413.

**17.**   From Problem 11, $z = 2.43$. Thus, the area under the normal curve from $\mu = 100$ to 124.3 is: 0.4925.

**19.**   $\mu = 60$, $\sigma = 12$

$z$ (for $x = 48$) $= \dfrac{48 - 60}{12} = \dfrac{-12}{12} = -1$

$z$ (for $x = 60$) $= \dfrac{60 - 60}{12} = 0$

Total area $= A = A_1 = 0.3143$.

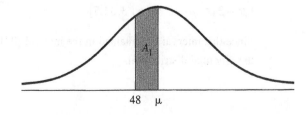

Copyright © 2011 Pearson Education, Inc. Publishing as Prentice Hall.

**21.**   $\mu = 60$, $\sigma = 12$

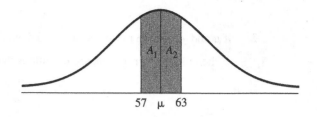

$z$ (for $x = 57$) $= \dfrac{57 - 60}{12} = \dfrac{-3}{12} = -0.25$

$z$ (for $x = 63$) $= \dfrac{63 - 60}{12} = \dfrac{3}{12} = 0.25$

Area $A_1$ = Area $A_2$ = 0.0987

Total area $= A = A_1 + A_2 = 0.1974$.

**23.**   $\mu = 60$, $\sigma = 12$

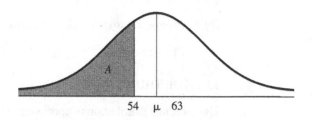

$z$ (for $x = 54$) $= \dfrac{54 - 60}{12} = \dfrac{-6}{12} = -0.5$

Required area $= 0.5 - ($area for $z = 0.5)$

$= 0.5 - 0.1915$

$= 0.3085$.

**25.**   $\mu = 60$, $\sigma = 12$

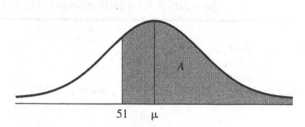

$z$ (for $x = 51$) $= \dfrac{51 - 60}{12} = \dfrac{-9}{12} = -0.75$

Required area $= 0.5 + ($area for $z = 0.75)$

$= 0.5 + 0.2734$

$= 0.7734$

**27.**   False.  If the random variable associated with a normal distribution assumes negative values, then the mean could be less than or equal to  0.

**29.**   True.   See Normal Curve Properties 1 (c).

**31.**   True.   See Normal Curve Properties 1 (d).

**33.**   With $n = 15$, $p = .7$, and $q = .3$, the mean and standard deviation of the binomial distribution are:
$\mu = np = 10.5$

$\sigma = \sqrt{npq} = \sqrt{(15)(.7)(.3)} \approx 1.8$

$[\mu - 3\sigma,\ \mu + 3\sigma] = [5.1,\ 15.9]$

Since this interval is not contained in the interval $[0, 15]$, the normal distribution should *not* be used to approximate the binomial distribution.

**35.**   With $n = 15$, $p = .4$, and $q = .6$, the mean and standard deviation of the binomial distribution are:

$\mu = np = 15(.4) = 6$

$\sigma = \sqrt{npq} = \sqrt{15(.4)(.6)} \approx 1.9$

$[\mu - 3\sigma,\ \mu + 3\sigma] = [.3,\ 11.7]$

Since this interval is contained in the interval $[0, 15]$, the normal distribution *is* a suitable approximation for the binomial distribution.

Copyright © 2011 Pearson Education, Inc.  Publishing as Prentice Hall.

**37.** With $n = 100$, $p = .05$, and $q = .95$, the mean and standard deviation of the binomial distribution are:

$\mu = np = 100(.05) = 5$

$\sigma = \sqrt{npq} = \sqrt{100(.05)(.95)} \approx 2.2$

$[\mu - 3\sigma, \mu + 3\sigma] = [-1.6, 11.6]$

Since this interval is not contained in the interval $[0, 100]$, the normal distribution is *not* a suitable approximation for the binomial distribution.

**39.** With $n = 500$, $p = .05$, and $q = .95$, the mean and standard deviation of the binomial distribution are:

$\mu = np = 500(.05) = 25$

$\sigma = \sqrt{npq} = \sqrt{500(.05)(.95)} \approx 4.9$

$[\mu - 3\sigma, \mu + 3\sigma] = [10.3, 39.7]$

Since this interval is contained in the interval $[0, 500]$, the normal distribution *is* a suitable approximation for the binomial distribution.

**41.** We have $p = 0.1$, $q = 0.9$. If $n$ is the number of trials, then the mean $\mu = np = 0.1n$ and the standard deviation $\sigma = \sqrt{npq} = \sqrt{0.09n} = 0.3\sqrt{n}$.

By the Rule of Thumb test, we must have

$0.1n - 3(0.3\sqrt{n}) \geq 0$   and   $0.1n + 3(0.3\sqrt{n}) \leq n$

The second inequality is automatically satisfied. The first inequality is:

$0.1n - 0.9\sqrt{n} \geq 0$

$\sqrt{n} \geq 9$

$n \geq 81$

*In Problems 43 – 49, $\mu = 500(.4) = 200$, and $\sigma = \sqrt{npq} = \sqrt{500(.4)(.6)} \approx 10.95$. The intervals are adjusted as in Examples 17 and 18.*

**43.** $z$ (for $x = 184.5$) $= \dfrac{184.5 - 200}{10.95} \approx -1.42$

$z$ (for $x = 220.5$) $= \dfrac{220.5 - 200}{10.95} \approx 1.87$

Thus, the probability that the number of successes will be between 185 and 220

$=$ area $A_1$ + area $A_2$

$=$ (area corresponding to $z = -1.42$) + (area corresponding to $z = 1.87$)

$= .4222 + .4693 = .8915 \approx .89$

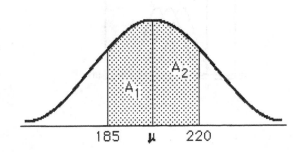

Copyright © 2011 Pearson Education, Inc. Publishing as Prentice Hall.

**45.** $z$ (for $x = 209.5$) $= \dfrac{209.5 - 200}{10.95} \approx .87$

$z$ (for $x = 220.5$) $= \dfrac{220.5 - 200}{10.95} \approx 1.87$

Thus, the probability that the number of successes will
be between 210 and 220
= area $A$
= (area corresponding to $z = 1.87$) –
(area corresponding to $z = .87$)
= .4693 – .3078 = .1615 ≈ .16

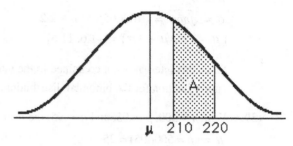

**47.** $z$ (for $x = 224.5$) $= \dfrac{224.5 - 200}{10.95} \approx 2.24$

The probability that the number of
successes will be 225 or more
= area $A$
= .5 – (area corresponding to $z = 2.24$)
= .5 – .4875 = .0125 ≈ .01

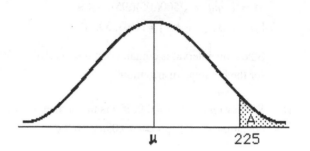

**49.** $z$ (for $x = 175.5$) $= \dfrac{175.5 - 200}{10.95} \approx -2.24$

The probability that the number of
successes will be 175 or less
= area $A$
= .5 – (area corresponding to $z = 2.24$)
= .5 – .4875 = .0125 ≈ .01

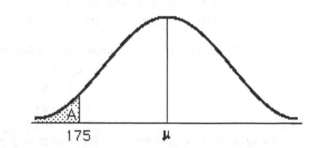

**51.**

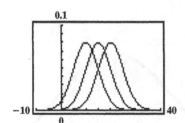

**53.**

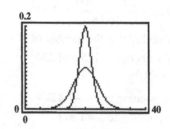

Copyright © 2011 Pearson Education, Inc. Publishing as Prentice Hall.

**55.** 120 scores are selected from a normal distribution with mean $\mu = 75$, and standard deviation $\square = 8$.

(A)    The area under the normal curve between 75 and 83:

$$z = \frac{x - \mu}{\sigma} = \frac{83 - 75}{8} = 1; \text{ Area} = 0.3413 \text{ (Table 1)}$$

By symmetry, the area under the normal curve between 67 and 75 is also 0.3413.    Thus, the area under the normal curve between 67 and 83 is 0.6826.  Now, $120(0.6826) = 81.912$;  approximately 82 scores are expected to  be between 67 and 83.

(B)    The answer depends on the results of your simulation.

**57.**   $\mu = 200,000$, $\sigma = 20,000$, $x \geq 240,000$

$$z \text{ (for } x = 240,000) = \frac{240,000 - 200,000}{20,000} = 2.0$$

Fraction of the salespeople who would be expected to make annual sales of \$240,000 or more

$$= \text{Area } A_1$$
$$= .5 \text{ (area between } \mu \text{ and } 240,000$$
$$= .5 - .4772$$
$$= .0228$$

Thus, the percentage of salespeople expected to make annual sales of  \$240,000 or more is 2.28%.

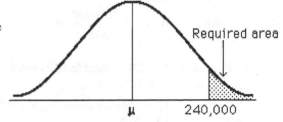

**59.**   $x = 105$, $x = 95$, $\mu = 100$, $\sigma = 2$

$$z \text{ (for } x = 105) = \frac{105 - 100}{2} = 2.5$$

$$z \text{ (for } x = 95) = \frac{95 - 100}{2} = -2.5$$

Fraction of parts to be rejected

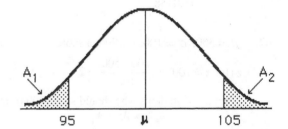

$$= \text{Area } A_1 + A_2$$
$$= 1 - 2(\text{area corresponding to } z = 2.5)$$
$$= 1 - 2(.4938)$$
$$= .0124$$

Thus, the percentage of parts to be rejected is 1.24%.

**61.**   With $n = 40$, $p = .6$, and $q = .4$, the mean and standard deviation of the binomial distribution are:

$$\mu = np = 40(.6) = 24$$

$$\sigma = \sqrt{npq} = \sqrt{40(.6)(.4)} \approx 3.10$$

$$z \text{ (for } x = 15.5) = \frac{15.5 - 24}{3.1} = -2.74$$

The probability that 15 or fewer households use the product

$$= \text{area } A$$

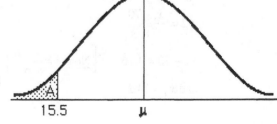

$$= .5 - (\text{area corresponding to } z = 2.74)$$
$$= .5 - .4969$$
$$= .0031$$

Either a rare event has occurred, e.g., the sample was not random, or the company's claim is false.

Copyright © 2011 Pearson Education, Inc.  Publishing as Prentice Hall.

**63.**    $\mu = 240$, $\sigma = 20$

8 days = 192 hours = $x$

$z$ (for $x = 192$) $= \dfrac{192 - 240}{20} = -2.4$

Fraction of people having this incision who would heal in 192
hours or less = Area $A_1$

$\qquad = .5 - (\text{area corresponding to } z = 2.4)$

$\qquad = .5 - .4918 = .0082$

Thus, the percentage of people who would heal in 8 days or less is .82%.

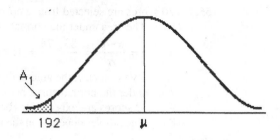

**65.**    $p = .25$, $q = .75$, $n = 1000$

$\mu = np = (1000)(.25) = 250$

$\sigma = \sqrt{npq} \qquad = \sqrt{1000 \times .25 \times .75}$

$\qquad\qquad\quad \approx 13.693 \approx 13.69$

$x = 220.5$ or less

$z$ (for $x = 220.5$) $= \dfrac{220.5 - 250}{13.69} = -2.15$

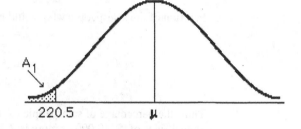

Probability that 220 or fewer families will have two girls

$\qquad = \text{Area } A_1$

$\qquad = 0.5 - (\text{area corresponding to } z = 2.15)$

$\qquad = 0.5 - 0.4842$

$\qquad = .0158$

**67.**    $\mu = 500$, $\sigma = 100$, $x = 700$ or more

$z$ (for $x = 700$) $= \dfrac{700 - 500}{100} = 2$

Fraction of students who should score 700 or more

$\qquad = \text{Area } A_1$

$\qquad = 0.5 - (\text{area corresponding to } z = 2)$

$\qquad = 0.5 - 0.4722$

$\qquad = 0.0228$

Thus, 2.28% should score 700 or more.

**69.**    $\mu = 70$, $\sigma = 8$

We compute $x_1, x_2, x_3$, and $x_4$ corresponding to $z_1, z_2, z_3$, and $z_4$, respectively. The area between $\mu$ and
$x_3$ is .2. Hence, from the table, $z_3 = .52$ (approximately). Thus, we have:

$.52 = \dfrac{x_3 - 70}{8}$

$x_3 - 70 = 4.16 \qquad \left[ \underline{\text{Note:}} \ z = \dfrac{x - \mu}{\sigma} \right] \approx 4.2$

and $x_3 = 74.2$.

Copyright © 2011 Pearson Education, Inc. Publishing as Prentice Hall.

Also, $x_2 = 70 - 4.2 = 65.8$

The area between $\mu$ and $x_4$ is .4. Hence, from the table,
$z_4 = 1.28$ (approximately).

Therefore:

$$1.28 = \frac{x_4 - 70}{8}$$

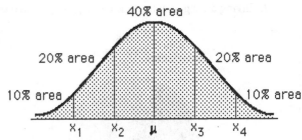

$x_4 - 70 = 10.24 \approx 10.2$ and $x_4 = 70 + 10.2 = 80.2$.

Also, $x_1 = 70 - 10.2 = 59.8$.

Thus, we have $x_1 = 59.8$, $x_2 = 65.8$, $x_3 = 74.2$, $x_4 = 80.2$.

So,
$A$'s = 80.2 or greater, $B$'s = 74.2 to 80.2, $C$'s = 65.8 to 74.2,
$D$'s = 59.8 to 65.8, and $F$'s = 59.8 or lower.

## CHAPTER 11 REVIEW

1.

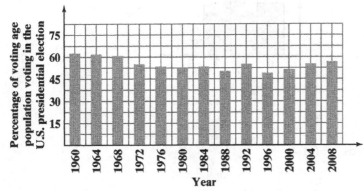

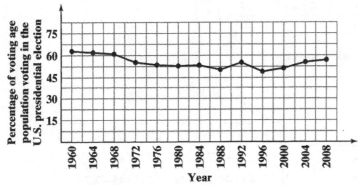

(11-1)

Copyright © 2011 Pearson Education, Inc. Publishing as Prentice Hall.

2. Living arrangements of the elderly, 65 year and over.

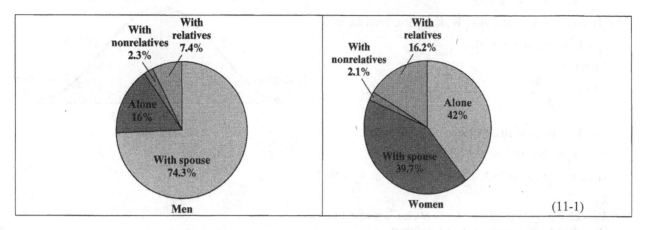

**Men**     **Women**     (11-1)

3. (A) $P(x) = C_{3,x}(.4)^x(.6)^{3-x}$

| $x$ | $P(x)$ |
|-----|--------|
| 0 | .216 |
| 1 | .432 |
| 2 | .288 |
| 3 | .064 |

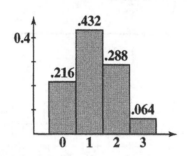

The histogram for this distribution
is shown at the right.

(B) $\mu = np = 3(.4) = 1.2$

$\sigma = \sqrt{npq} = \sqrt{3 \times .4 \times .6} \approx .85$     (11-4)

4. (A) Mean $= \bar{x} = \dfrac{1+1+2+2+2+3+3+4+4+5}{10} = \dfrac{27}{10} = 2.7$

(B) Median $= \dfrac{2+3}{2}$ (2 and 3 are middle scores)

$= 2.5$

(C) Mode = 2

(D) Standard deviation $= s = \sqrt{\dfrac{\sum\limits_{i=1}^{n}(x_i - \bar{x})^2}{n-1}} = \sqrt{\dfrac{\begin{array}{c}(1-2.7)^2 + (1-2.7)^2 + (2-2.7)^2 + (2-2.7)^2 \\ + (2-2.7)^2 + (3-2.7)^2 + (3-2.7)^2 \\ + (4-2.7)^2 + (4-2.7)^2 + (5-2.7)^2\end{array}}{10-1}}$

$\approx 1.34$     (11-2, 11-3)

Copyright © 2011 Pearson Education, Inc. Publishing as Prentice Hall.

**5.** (A)  $\mu = 100$, $\sigma = 10$

$$z \text{ (for } x = 118) = \frac{118 - 100}{10} = 1.8$$

$x = 118$ is 1.8 standard deviations from the mean.

(B)  From the table, the required area
$A_1 = 0.4641$.

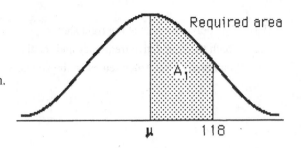

(11-5)

**6.** (A)  The frequency and relative frequency table for the given data is shown below.

| Interval | Frequency | Relative frequency |
|----------|-----------|--------------------|
| 9.5 – 11.5 | 1 | .04 |
| 11.5 – 13.5 | 5 | .20 |
| 13.5 – 15.5 | 12 | .48 |
| 15.5 – 17.5 | 6 | .24 |
| 17.5 – 19.5 | 1 | .04 |
|  | 25 | 1.00 |

(B)  The histogram below shows both frequency and relative frequency scales on the $y$ axis.

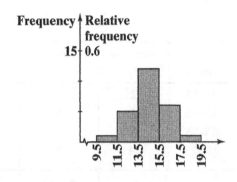

(C)  The polygon graph to the left also shows both frequency and relative frequency scales on the $y$ axis.

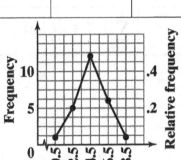

(D)  A cumulative and relative cumulative table for the data is given below.

| Interval | Frequency | Cumulative frequency | Relative cumulative frequency |
|----------|-----------|----------------------|-------------------------------|
| 9.5 – 11.5 | 1 | 1 | .04 |
| 11.5 – 13.5 | 5 | 6 | .24 |
| 13.5 – 15.5 | 12 | 18 | .72 |
| 15.5 – 17.5 | 6 | 24 | .96 |
| 17.5 – 19.5 | 1 | 25 | 1.00 |

Copyright © 2011 Pearson Education, Inc.  Publishing as Prentice Hall.

(E) The polygon graph at the right shows both the cumulative frequency and relative cumulative frequency scales on the $y$-axis.

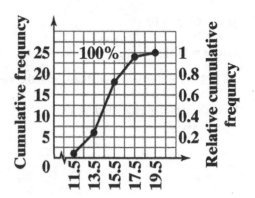

(11-1)

**7.**

| Interval | Midpoint $x_i$ | Frequency $f_i$ | $x_i f_i$ | $(x_i - \bar{x})^2$ | $(x_i - \bar{x})^2 f_i$ |
|---|---|---|---|---|---|
| 0.5–3.5 | 2 | 1 | 2 | 25 | 25 |
| 3.5–6.5 | 5 | 5 | 25 | 4 | 20 |
| 6.5–9.5 | 8 | 7 | 56 | 1 | 7 |
| 9.5–12.5 | 11 | 2 | 22 | 16 | 32 |
| | | $n = 15$ | $\displaystyle\sum_{i=1}^{4} x_i f_i = 105$ | | $\displaystyle\sum_{i=1}^{4} (x_i - \bar{x})^2 f_i = 84$ |

(A) $\bar{x} = \dfrac{105}{15} = 7$ 

(B) $s = \sqrt{\dfrac{84}{14}} \approx 2.45$

(C)  The histogram for the data is

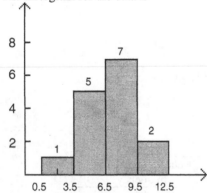

Total area $= 3(1) + 3(5) + 3(7) + 3(2) = 45$.

Let $M$ be the median. Then

$$3(1) + 3(5) + (M - 6.5)7 = 22.5$$
$$(M - 6.5)7 = 4.5$$
$$M - 6.5 = 0.643$$
$$M = 7.143 \qquad \text{(11-2, 11-3)}$$

Copyright © 2011 Pearson Education, Inc.  Publishing as Prentice Hall.

8.   (A)   $P(x) = C_{6,x}(.5)^x(.5)^{6-x} = C_{6,x}(.5)^6$

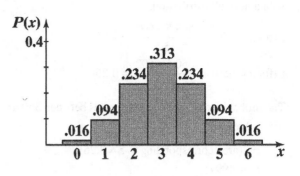

| x | P(x) |
|---|------|
| 0 | .016 |
| 1 | .094 |
| 2 | .234 |
| 3 | .313 |
| 4 | .234 |
| 5 | .094 |
| 6 | .016 |

The histogram for this distribution is shown at the right.

(B)   Mean $= \mu = np = 6 \times .5 = 3$

Standard deviation $= \sigma = \sqrt{npq} = \sqrt{6 \times .5 \times .5} \approx 1.225$     (11-4)

9.   $p = .6, q = .4, n = 1000$

$\mu = np = 1000 \times .6 = 600$

$s = \sqrt{npq} = \sqrt{1000 \times .6 \times .4} = \sqrt{240} \approx 15.49$     (11-4)

10.   (A)   True: $\bar{x} = \dfrac{x_1 + x_2 + \ldots + x_n}{n}$;

$\dfrac{(x_1 + 5) + (x_2 + 5) + \ldots + (x_n + 5)}{n} = \dfrac{x_1 + x_2 + \ldots + x_n + 5n}{n} = \dfrac{x_1 + x_2 + \ldots + x_n}{n} + 5 = \bar{x} + 5.$

(B)   False: Suppose that the data set $x_1, x_2, \ldots, x_n$ has mean $\bar{x}$ and standard deviation 5. Then the data set $x_1 + 5, x_2 + 5, \ldots, x_n + 5$ has mean $\bar{x} + 5$ [part (A)], but the standard deviation $s'$ will remain the same, $s' = s$:

$$s' = \sqrt{\frac{\sum_{i=1}^{n}(\bar{x}_i + 5 - [\bar{x} + 5])^2}{n}} = \sqrt{\frac{\sum_{i=1}^{n}(\bar{x}_i - \bar{x})^2}{n}} = s \quad (11\text{-}2, 11\text{-}3)$$

11.   (A)   False: If $X$ is a binomial random variable with mean $\mu$, then $P(X \geq \mu) = 0.5$ only if $p = q = 0.5$; any binomial distribution with $p \neq 0.5$ will serve as a counter-example.

(B)   True: The normal distribution is symmetric about the mean.

(C)   True: The area is 1 in each case.     (11-4, 11-5)

12.   The mean $\mu$ and the standard deviation $\sigma$ for the binomial distribution are:

$\mu = np = 1000 \times .6 = 600$

$\sigma = \sqrt{npq} = \sqrt{1000 \times .6 \times .4} \approx 15.49 \approx 15.5$

Now, we approximate the binomial distribution

Copyright © 2011 Pearson Education, Inc.  Publishing as Prentice Hall.

with a normal distribution.

$$z \text{ (for } x = 550) = \frac{549.5 - 600}{15.5} = -3.26$$

$$z \text{ (for } x = 650) = \frac{650.5 - 600}{15.5} = 3.26$$

The probability of obtaining successes between 550 and 650 = Area $A$

    = 2(area corresponding to $z = 3.26$)

    = 2(.4994)

    = .9988 ≈ .999    (11-5)

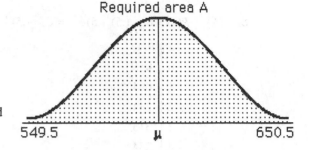

Required area A

13. (A) $\mu = 50$, $\sigma = 6$

$$z \text{ (for } x = 41) = \frac{41 - 50}{6} = -1.5$$

$$z \text{ (for } x = 62) = \frac{62 - 50}{6} = 2.0$$

Required area = $A_1 + A_2$

    = (area corresponding to $z = -1.5$) + (area corresponding to $z = 2$)

    = .4332 + .4772 = .9104

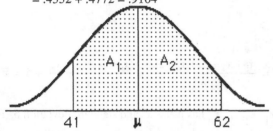

(B) $z \text{ (for } x = 59) = \dfrac{59 - 50}{6} = 1.5$

    Required area = .5 − (area corresponding to $z = 1.5$)

        = .5 − .4332 = .0668

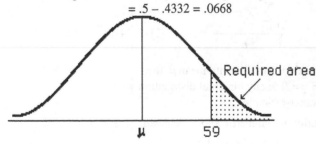

Required area

    (11-5)

14. (A) We would expect the first data set to have smaller standard deviation because the values range between 2 and 12, while in the second data set, the values range between 1 and 36.

    (B) The answer depends on the results of your simulation.    (11-3)

Copyright © 2011 Pearson Education, Inc. Publishing as Prentice Hall.

15. (A)

| $x_i$ | Frequency $f_i$ | $x_i f_i$ | $(x_i - x)^2$ $(x = 14.6)$ | $(x_i - \bar{x})^2 f_i$ |
|---|---|---|---|---|
| 11 | 1 | 11 | 12.96 | 12.96 |
| 12 | 2 | 24 | 6.76 | 13.52 |
| 13 | 3 | 39 | 2.56 | 7.68 |
| 14 | 7 | 98 | .36 | 2.52 |
| 15 | 5 | 75 | .16 | .80 |
| 16 | 3 | 48 | 1.96 | 5.88 |
| 17 | 3 | 51 | 5.76 | 17.28 |
| 19 | 1 | 19 | 19.36 | 19.36 |
| $n = 25$ | $\sum\limits_{i=1}^{8} x_i f_i = 365$ | | $\sum\limits_{i=1}^{8} (x_1 - \bar{x})^2 f_i = 80$ | |

Thus, $\bar{x} = \dfrac{365}{25} = 14.6$    $s = \sqrt{\dfrac{80}{24}} \approx 1.83$

(B)

| Interval | Midpt. $x_i$ | Freq. $f_i$ | $x_i f_i$ | $(x_i - \bar{x})^2$ $(\bar{x} = 14.6)$ | $(x_i - \bar{x})^2 f_i$ |
|---|---|---|---|---|---|
| 9.5–11.5 | 10.5 | 1 | 10.5 | 16.81 | 16.81 |
| 11.5–13.5 | 12.5 | 5 | 62.5 | 4.41 | 22.05 |
| 13.5–15.5 | 14.5 | 12 | 174.0 | .01 | .12 |
| 15.5–17.5 | 16.5 | 6 | 99.0 | 3.61 | 21.66 |
| 17.5–19.5 | 18.5 | 1 | 18.5 | 15.21 | 15.21 |
| | $n = 25$ | | $\sum\limits_{i=1}^{5} x_i f_i = 364.5$ | | $\sum\limits_{i=1}^{5} (x_i - \bar{x})^2 f_i = 75.85$ |

$\bar{x} = \dfrac{364.5}{25} \approx 14.6$    $s = \sqrt{\dfrac{75.85}{24}} \approx 1.78$    (11-2, 11-3)

16. Let $E$ = "rolling a six."   Then $P(E) = \dfrac{1}{6}$, $P(E') = \dfrac{5}{6}$.

Thus, $p = \dfrac{1}{6}$, $q = \dfrac{5}{6}$, $n = 5$.

(A)    $P(X = 3) = P(\text{exactly three 6's}) = C_{5,3}\left(\dfrac{1}{6}\right)^3\left(\dfrac{5}{6}\right)^2 = \dfrac{5!}{3!2!}\left(\dfrac{25}{6^5}\right) \approx .0322$

(B)    $P(\text{at least three 6's}) = P(x \geq 3) = P(3) + P(4) + P(5)$

$$= C_{5,3}\left(\dfrac{1}{6}\right)^3\left(\dfrac{5}{6}\right)^2 + C_{5,4}\left(\dfrac{1}{6}\right)^4\left(\dfrac{5}{6}\right) + C_{5,5}\left(\dfrac{1}{6}\right)^6$$

$$= \dfrac{5!}{3!2!}\left(\dfrac{25}{6^5}\right) + \dfrac{5!}{4!1!}\left(\dfrac{5}{6^5}\right) + \dfrac{5!}{5!0!}\left(\dfrac{1}{6^5}\right) \approx .0355 \qquad (11\text{-}4)$$

Copyright © 2011 Pearson Education, Inc.  Publishing as Prentice Hall.

**17.**   The probability of getting a 7 is: $p = \dfrac{6}{36} = \dfrac{1}{6}$. Thus, the probability of not getting a 7 is: $q = \dfrac{5}{6}$

Now, $P(\text{at least one } 7) = P(x \geq 1) = 1 - P(x < 1)$

$$= 1 - P(0)$$

$$= 1 - C_{3,0}\left(\dfrac{1}{6}\right)^{0}\left(\dfrac{5}{6}\right)^{3}$$

$$= 1 - \dfrac{125}{6^{3}} \approx 1 - .5787 = .4213. \qquad (11\text{-}4)$$

**18.**   **(A)**   The set of scores $\{10, 10, 20, 20, 90, 90, 90, 90, 90, 90\}$ has the specified property:

$$\overline{x} = \dfrac{2(10) + 2(20) + 6(90)}{10} = 60$$

Median = mode = 90

The scores $\{5, 10, 10, 15, 85, 85, 85, 85, 85, 85\}$ also have this property:

$$\overline{x} = 55, \text{ Median = mode = } 85.$$

**(B)**   Let $x_1, x_2, x_3, \ldots, x_{10}$ be the 10 exam scores arranged in increasing order. Then

$$\overline{x} = \dfrac{x_1 + x_2 + x_3 + \ldots + x_{10}}{10} \text{ and } M = \dfrac{x_5 + x_6}{2}$$

If $\overline{x} + 50 = M$, then

$$x_1 + x_2 + x_3 + x_4 + x_5 + x_6 + x_7 + x_8 + x_9 + x_{10} + 500 = 5x_5 + 5x_6$$

and

$$x_1 + x_2 + x_3 + x_4 + x_7 + x_8 + x_9 + x_{10} + 500 = 4x_5 + 4x_6$$

Now, $x_6 \leq x_7 \leq x_8 \leq x_9 \leq x_{10}$, so $4x_6 \leq x_7 + x_8 + x_9 + x_{10}$

Therefore, $x_1 + x_2 + x_3 + x_4 + 500 \leq 4x_5 \leq 400$

This implies $x_1 + x_2 + x_3 + x_4 \leq -100$ which is impossible, since $x_1, x_2, x_3, x_4 \geq 0$.

Thus, the median cannot be 50 points greater than the mean.

The mode could be 50 points greater than the mean.

Consider the exam scores:

$0, 1, 2, 3, 4, 5, 6, 7, 8, 100, 100$

$\overline{x} = 23.6$,  mode = 100       (11-2)

**19.**   **(A)**   This is a binomial experiment with $p = 0.39$, $q = 0.61$, and $n = 20$.

$$P(8) = C_{20,8}(0.39)^{8}(0.61)^{12} \approx 0.179$$

**(B)**   $\mu = 7.8$, $\sigma = 2.18$; $\mu - 3\sigma = 1.26$, $\mu + 3\sigma = 14.34$

The interval $[1.26, 14.34]$ lies in the interval $[0, 20]$. Therefore, the normal distribution can be used to approximate the binomial distribution.

Copyright © 2011 Pearson Education, Inc.  Publishing as Prentice Hall.

(C)    The normal distribution is continuous, not discrete. Therefore, the correct analogue of (A) is:

$$P(7.5 \le x \le 8.5) \approx 0.18$$

using the table in Appendix C.        (11-5)

**20.**    Given $p = 0.9$, $q = 0.1$, $n = 12$

(A)    Mean: $\mu = np = 12(0.9) = 10.8$

Standard deviation: $\sigma = \sqrt{npq} = \sqrt{12(0.9)(0.1)} \approx 1.039$.

(B)    Probability of 12 wins:

$$P(12) = C_{12,12}(0.9)^{12}(0.1)^0 = (0.9)^{12} \approx 0.282$$

(C)    The answer depends on the results of your simulation.        (11-4)

**21.**    Arrange the given numbers in increasing order:
4, 5, 5, 5, 8, 12, 13, 15, 16, 17

(A)    Mean $\bar{x} = \dfrac{4+5+5+5+8+12+13+15+16+17}{10} = 10$

(B)    Median $= \dfrac{8+12}{2} = 10$

(C)    Mode $= 5$

(D)    Standard deviation:

$$s = \sqrt{\dfrac{\begin{array}{c}(4-10)^2 + 3(5-10)^2 + (8-10)^2 + (12-10)^2 \\ + (13-10)^2 + (15-10)^2 + (16-10)^2 + (17-10)^2\end{array}}{10-1}}$$

$$= \sqrt{\dfrac{36+75+4+4+9+25+36+49}{9}} = \sqrt{\dfrac{238}{9}} \approx 5.14 \qquad (11\text{-}2,\ 11\text{-}3)$$

**22.**    The mean and median are not suitable for these data.  The modal preference is soft drink.        (11-2)

**23.**    (A)    The frequency and relative frequency table is as follows:

| Interval | Frequency $f_i$ | Relative frequency |
|---|---|---|
| 29.5 – 31.5 | 3 | .086 |
| 31.5 – 33.5 | 7 | .2 |
| 33.5 – 35.5 | 14 | .4 |
| 35.5 – 37.5 | 7 | .2 |
| 37.5 – 39.5 | 4 | .114 |
| $\sum\limits_{i=1}^{5} f_i = 35$ | Sum $= 1.00$ |

(B)

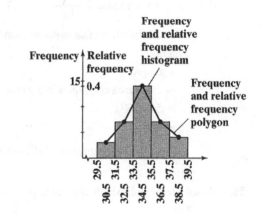

Copyright © 2011 Pearson Education, Inc.  Publishing as Prentice Hall.

(C)

| Interval | Midpt. $x_i$ | Freq. $f_i$ | $x_i f_i$ | $(x_i - \overline{x})^2$ $(\overline{x} = 34.61)$ | $(x_i - \overline{x})^2 f_i$ |
|---|---|---|---|---|---|
| 29.5–31.5 | 30.5 | 3 | 91.5 | 16.89 | 50.67 |
| 31.5–33.5 | 32.5 | 7 | 227.5 | 4.45 | 31.15 |
| 33.5–35.5 | 34.5 | 14 | 483.0 | .01 | .14 |
| 35.5–37.5 | 36.5 | 7 | 255.5 | 3.57 | 24.99 |
| 37.5–39.5 | 38.5 | 4 | 154.0 | 15.13 | 60.53 |
| | | $n = 35$ | $\displaystyle\sum_{i=1}^{5} x_i f_i = 1211.5$ | | $\displaystyle\sum_{i=1}^{5}(x_i - \overline{x})^2 f_i = 167.48$ |

$$\overline{x} = \frac{1211.5}{35} = 34.61, \quad s = \sqrt{\frac{167.48}{34}} = 2.22 \qquad (11\text{-}1, \ 11\text{-}2, \ 11\text{-}3)$$

**24.** $\mu = 100, \ \sigma = 10$

(A) $z \ (\text{for } x = 92) = \dfrac{92 - 100}{10} = -.8$

$z \ (\text{for } x = 108) = \dfrac{108 - 100}{10} = .8$

The probability of an applicant scoring between 92 and 108

    = area $A$

    = 2·area $A_1$

    = 2(area corresponding to $z = .8$)

    = 2(.2881) = .5762

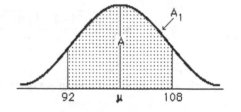

Thus, the percentage of applicants scoring between 92 and 108 is 57.62%.

(B) $z \ (\text{for } x = 115) = \dfrac{115 - 100}{10} = 1.5$

The probability of an applicant scoring 115 or higher

    = area $A$

    = .5 – (area corresponding to $z = 1.5$)

    = .5 – .4332

    = .0668

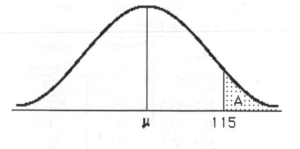

Thus, the percentage of applicants scoring 115 or higher is 6.68%.     (11-5)

**25.** Based on the publisher's claim, the probability that a person selected at random reads the newspaper is: $p = .7$.

Thus, the probability that a randomly selected person does not read the newspaper is: $q = .3$.

Copyright © 2011 Pearson Education, Inc. Publishing as Prentice Hall.

(A)    With $n = 200$, the mean $\mu = 200(.7) = 140$,

and standard deviation $\sigma = \sqrt{200(.7)(.3)} = \sqrt{42} \approx 6.48$.

(B)    $[\mu - 3\sigma, \mu + 3\sigma] = [120.76, 159.24]$ and the interval lies entirely in the interval $[0, 200]$.  Thus, the normal distribution *does* provide an adequate approximation to the binomial distribution.

(C)    $z$ (for $x = 129.5$) $= \dfrac{129.5 - 140}{6.48} = \dfrac{-10.5}{6.48} = -1.62$

$z$ (for $x = 155.5$) $= \dfrac{155.5 - 140}{6.48} = \dfrac{15.5}{6.48} = 2.39$

The probability of finding between 130 and 155 readers in the sample

$=$ area $A_1$ + area $A_2$

$=$ (area corresponding to $z = 1.62$) + (area corresponding to $z = 2.39$)

$= .4474 + .4916$

$= .9390$

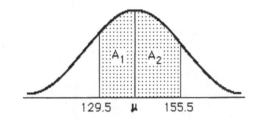

129.5    $\mu$    155.5

(D)    $z$ (for $x = 125.5$) $= \dfrac{125.5 - 140}{6.48} = \dfrac{-14.5}{6.48} = -2.24$

The probability of finding 125 or fewer readers in the sample

$=$ area $A$

$= .5 -$ (area corresponding to $z = 2.24$)

$= .5 - .4875$

$= .0125$

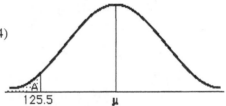

125.5    $\mu$

(E)

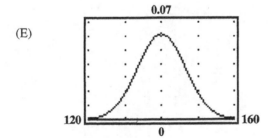

$(11\text{-}4, 11\text{-}5)$

**26.**    $p = .9, q = .1$, and $n = 3$

$P(x \geq 1) = 1 - P(x < 1)$

$= 1 - P(0)$

$= 1 - C_{3,0}\left(\dfrac{9}{10}\right)^0\left(\dfrac{1}{10}\right)^3$

$= 1 - \dfrac{1}{1000} = .999$    $(11\text{-}4)$

Copyright © 2011 Pearson Education, Inc.  Publishing as Prentice Hall.

(a) With $n = 200$, $np = 200(.75) = 150$, and standard deviation $\sqrt{npq} = \sqrt{200(.75)(.25)} = 6.1237$.

(b) $P(X \ge 130) = P(X \ge 129.5)$. Use the normal because $np$ and $nq$ are both $\ge 5$. Use the normal distribution as an appropriate approximation to the binomial distribution.

(c) $z = \dfrac{129.5 - 150}{6.48} = \dfrac{-20.5}{6.48} = -3.16$

(d) $z(129.5) = \dfrac{129.5 - 150}{6.48} = \dfrac{-20.5}{6.48} = -3.16$

The probability of being between 130 and 155 inclusive words:

$z(155.5)$

areas corresponding to $z = 3.16$ that corresponds to $z = 3.16$

$z(155.5)$

(f) $z(134.5) = \dfrac{134.5 - 140}{6.02} = \dfrac{-16.5}{6.02}$

The probability of being at 135 or fewer words is to be found.

The Empirical approximation $P < 1$

Copyright © 2011 Pearson Education, Inc. Publishing as Prentice Hall.

*Things to remember:*

1.    THE SET OF REAL NUMBERS

| SYMBOL | NAME | DESCRIPTION | EXAMPLES |
|---|---|---|---|
| $N$ | Natural numbers | Counting numbers (also called positive integers) | 1, 2, 3, ... |
| $Z$ | Integers | Natural numbers, their negatives, and 0 | ... –2, –1, 0, 1, 2, ... |
| $Q$ | Rational numbers | Any number that can be represented as $\dfrac{a}{b}$, where $a$ and $b$ are integers and $b \neq 0$. Decimal representations are repeating or terminating. | $-4;\ 0;\ 1;\ 25;\ \dfrac{-3}{5};\ \dfrac{2}{3};$ $3.67;\ -0.333\,\overline{3};\ 5.2727\,\overline{27}$ |
| $I$ | Irrational numbers | Any number with a decimal representation that is nonrepeating and non–terminating. | $\sqrt{2}\,;\ \pi;\ \sqrt[3]{7}\,;\ 1.414213...;$ $2.718281828...$ |
| $R$ | Real numbers | Rationals and irrationals | |

2.    BASIC PROPERTIES OF THE SET OF REAL NUMBERS

Let $a$, $b$, and $c$ be arbitrary elements in the set of real numbers $R$.

ADDITION PROPERTIES

    ASSOCIATIVE:    $(a + b) + c = a + (b + c)$

    COMMUTATIVE:    $a + b = b + a$

    IDENTITY:    0 is the additive identity; that is, $0 + a = a + 0$ for all $a$ in $R$, and 0 is the only element in $R$ with this property.

    INVERSE:    For each $a$ in $R$, $-a$ is its unique additive inverse; that is, $a + (-a) = (-a) + a = 0$, and $-a$ is the only element in $R$ relative to $a$ with this property.

MULTIPLICATION PROPERTIES

    ASSOCIATIVE:    $(ab)c = a(bc)$

    COMMUTATIVE:    $ab = ba$

Copyright © 2011 Pearson Education, Inc. Publishing as Prentice Hall.

IDENTITY:    1 is the multiplicative identity; that is, $1a = a1 = a$ for all $a$ in $R$, and 1 is the only element in $R$ with this property.

INVERSE:    For each $a$ in $R$, $a \neq 0$, $\dfrac{1}{a}$ is its unique multiplicative inverse; that is,

$$a\left(\frac{1}{a}\right) = \left(\frac{1}{a}\right)a = 1, \text{ and } \frac{1}{a} \text{ is the only element in } R \text{ relative to } a \text{ with}$$

this property.

## DISTRIBUTIVE PROPERTIES

$$a(b + c) = ab + ac$$

$$(a + b)c = ac + bc$$

3. ## SUBTRACTION AND DIVISION

For all real numbers $a$ and $b$.

SUBTRACTION:    $a - b = a + (-b)$
$7 - (-5) = 7 + [-(-5)] = 7 + 5 = 12$

DIVISION:    $a \div b = a\left(\dfrac{1}{b}\right), b \neq 0$

$$9 \div 4 = 9\left(\frac{1}{4}\right) = \frac{9}{4}$$

NOTE: 0 can never be used as a divisor!

4. ## PROPERTIES OF NEGATIVES

For all real numbers $a$ and $b$.

a. $-(-a) = a$ 

b. $(-a)b = -(ab) = a(-b) = -ab$

c. $(-a)(-b) = ab$ 

d. $(-1)a = -a$

e. $\dfrac{-a}{b} = -\dfrac{a}{b} = \dfrac{a}{-b}, b \neq 0$ 

f. $\dfrac{-a}{-b} = -\dfrac{-a}{b} = -\dfrac{a}{-b} = \dfrac{a}{b}, b \neq 0$

5. ## ZERO PROPERTIES

For all real numbers $a$ and $b$.

a. $a \cdot 0 = 0$

b. $ab = 0$ if and only if $a = 0$ or $b = 0$ (or both)

Copyright © 2011 Pearson Education, Inc. Publishing as Prentice Hall.

<u>6.</u>    FRACTION PROPERTIES

For all real numbers $a$, $b$, $c$, $d$, and $k$ (division by 0 excluded).

a. $\dfrac{a}{b} = \dfrac{c}{d}$ if and only
   if $ad = bc$

b. $\dfrac{ka}{kb} = \dfrac{a}{b}$

c. $\dfrac{a}{b} \cdot \dfrac{c}{d} = \dfrac{ac}{bd}$

d. $\dfrac{a}{b} \div \dfrac{c}{d} = \dfrac{a}{b} \cdot \dfrac{d}{c}$

e. $\dfrac{a}{b} + \dfrac{c}{b} = \dfrac{a+c}{b}$

f. $\dfrac{a}{b} - \dfrac{c}{b} = \dfrac{a-c}{b}$

g. $\dfrac{a}{b} + \dfrac{c}{d} = \dfrac{ad+bc}{bd}$

---

1.  $uv = vu$

3.  $3 + (7 + y) = (3 + 7) + y$

5.  $1(u + v) = u + v$

7.  T; Associative property of multiplication

9.  T; Distributive property

11.  F; $-2(-a)(2x - y) = 2a(2x - y)$

13.  T; Commutative property of addition

15.  T; Property of negatives

17.  T; Multiplicative inverse property

19.  T; Property of negatives

21.  F; $\dfrac{a}{b} + \dfrac{c}{d} = \dfrac{ad+bc}{bd}$

23.  T; Distributive property

25.  T; Zero property

27.  No. For example: $2\left(\dfrac{1}{2}\right) = 1$. In general $a\left(\dfrac{1}{a}\right) = 1$ whenever $a \neq 0$.

29.  (A)  False. For example, $-3$ is an integer but not a natural number.

     (B)  True

     (C)  True. For example, for any natural number $n$, $n = \dfrac{n}{1}$.

31.  $\sqrt{2}$, $\sqrt{3}$, ... ; in general, the square root of any rational number that is not a perfect square; $\pi$, $e$.

33.  (A)  $8 \in N, Z, Q, R$

     (B)  $\sqrt{2} \in R$

     (C)  $-1.414 = -\dfrac{1414}{1000} \in Q, R$

     (D)  $\dfrac{-5}{2} \in Q, R$

Copyright © 2011 Pearson Education, Inc. Publishing as Prentice Hall.

**35.**   (A) F; $a(b-c) = ab - ac$; for example $2(3-1) = 2 \cdot 2 = 4 \neq 2 \cdot 3 - 1 = 5$

(B) F; for example, $(3-7) - 4 = -4 - 4 = -8 \neq 3 - (7-4) = 3 - 3 = 0$.

(C) T; this is the associative property of multiplication.

(D) F; for example, $(12 \div 4) \div 2 = 3 \div 2 = \dfrac{3}{2} \neq 12 \div (4 \div 2) = 12 \div 2 = 6$.

**37.**
$$C = 0.090909\ldots$$
$$100C = 9.090909\ldots$$
$$100C - C = (9.090909\ldots) - (0.090909\ldots)$$
$$99C = 9$$
$$C = \frac{9}{99} = \frac{1}{11}$$

**39.**   (A) $\dfrac{13}{6} \approx 2.166\,666\,667$     (B) $\sqrt{21} \approx 4.582\,575\,695$

(C) $\dfrac{7}{16} = 0.4375$     (D) $\dfrac{29}{111} \approx 0.261\,261\,261$

---

## EXERCISE A–2

*Things to remember:*

1.   NATURAL NUMBER EXPONENT

For $n$ a natural number and $b$ any real number,

$$b^n = b \cdot b \cdot \ldots \cdot b, \ n \text{ factors of } b.$$

For example, $2^3 = 2 \cdot 2 \cdot 2 \ (= 8)$,
$$3^5 = 3 \cdot 3 \cdot 3 \cdot 3 \cdot 3 \ (= 243).$$

In the expression $b^n$, $n$ is called the EXPONENT, and $b$ is called the BASE.

2.   FIRST PROPERTY OF EXPONENTS

For any natural numbers $m$ and $n$, and any real number $b$,

$$b^m \cdot b^n = b^{m+n}.$$

For example, $3^3 \cdot 3^4 = 3^{3+4} = 3^7$.

3.   POLYNOMIALS

a.   A POLYNOMIAL IN ONE VARIABLE $x$ is constructed by adding or subtracting constants and terms of the form $ax^n$, where $a$ is a real number and $n$ is a natural number.

Copyright © 2011 Pearson Education, Inc. Publishing as Prentice Hall.

b.   A POLYNOMIAL IN TWO VARIABLES $x$ AND $y$ is constructed by adding or subtracting constants and terms of the form $ax^m y^n$, where $a$ is a real number and $m$ and $n$ are natural numbers.

c.   Polynomials in more than two variables are defined similarly.

d.   A polynomial with only one term is called a MONOMIAL.
A polynomial with two terms is called a BINOMIAL.
A polynomial with three terms is called a TRINOMIAL.

4.   DEGREE OF A POLYNOMIAL

a.   A term of the form $ax^n$, $a \neq 0$, has degree $n$. A term of the form $ax^m y^n$, $a \neq 0$, has degree $m + n$. A nonzero constant has degree 0.

b.   The DEGREE OF A POLYNOMIAL is the degree of the nonzero term with the highest degree. For example, $3x^4 + \sqrt{2}\, x^3 - 2x + 7$ has degree 4; $2x^3 y^2 - 3x^2 y + 7x^4 - 5y^3 + 6$ has degree 5; the polynomial 4 has degree 0.

c.   The constant 0 is a polynomial but it is not assigned a degree.

5.   Two terms in a polynomial are called LIKE TERMS if they have exactly the same variable factors raised to the same powers. For example, in
$$7x^5 y^2 - 3x^3 y + 2x + 4x^3 y - 1,$$
$-3x^3 y$ and $4x^3 y$ are like terms.

6.   To multiply two polynomials, multiply each term of one by each term of the other, and then combine like terms.

7.   SPECIAL PRODUCTS

a.   $(a - b)(a + b) = a^2 - b^2$

b.   $(a + b)^2 = a^2 + 2ab + b^2$

c.   $(a - b)^2 = a^2 - 2ab + b^2$

8.   ORDER OF OPERATIONS

Multiplication and division precede addition and subtraction, and taking powers precedes multiplication and division.

**1.**   The term of highest degree in $x^3 + 2x^2 - x + 3$ is $x^3$ and the degree of this term is 3.

**3.**   $(2x^2 - x + 2) + (x^3 + 2x^2 - x + 3) = x^3 + 2x^2 + 2x^2 - x - x + 2 + 3 = x^3 + 4x^2 - 2x + 5$

**5.**   $(x^3 + 2x^2 - x + 3) - (2x^2 - x + 2) = x^3 + 2x^2 - x + 3 - 2x^2 + x - 2 = x^3 + 1$

Copyright © 2011 Pearson Education, Inc.  Publishing as Prentice Hall.

**7.** Using a vertical arrangement:

$$x^3 + 2x^2 - x + 3$$
$$\underline{2x^2 - x + 2}$$
$$2x^5 + 4x^4 - 2x^3 + 6x^2$$
$$- x^4 - 2x^3 + x^2 - 3x$$
$$\underline{2x^3 + 4x^2 - 2x + 6}$$
$$2x^5 + 3x^4 - 2x^3 + 11x^2 - 5x + 6$$

**9.** $2(u - 1) - (3u + 2) - 2(2u - 3) = 2u - 2 - 3u - 2 - 4u + 6 = -5u + 2$

**11.** $4a - 2a[5 - 3(a + 2)] = 4a - 2a[5 - 3a - 6] = 4a - 2a[-3a - 1] = 4a + 6a^2 + 2a = 6a^2 + 6a$

**13.** $(a + b)(a - b) = a^2 - b^2$ (Special product $\underline{7}$a)

**15.** $(3x - 5)(2x + 1) = 6x^2 + 3x - 10x - 5 = 6x^2 - 7x - 5$

**17.** $(2x - 3y)(x + 2y) = 2x^2 + 4xy - 3xy - 6y^2 = 2x^2 + xy - 6y^2$

**19.** $(3y + 2)(3y - 2) = (3y)^2 - 2^2 = 9y^2 - 4$ (Special product $\underline{7}$a)

**21.** $-(2x - 3)^2 = -[(2x)^2 - 2(2x)(3) + 3^2] = -[4x^2 - 12x + 9] = -4x^2 + 12x - 9$ (Special product $\underline{7}$c)

**23.** $(4m + 3n)(4m - 3n) = 16m^2 - 9n^2$ (Special product $\underline{7}$a)

**25.** $(3u + 4v)^2 = 9u^2 + 24uv + 16v^2$ (Special product $\underline{7}$b)

**27.** $(a - b)(a^2 + ab + b^2) = a(a^2 + ab + b^2) - b(a^2 + ab + b^2) = a^3 + a^2b + ab^2 - a^2b - ab^2 - b^3$
$$= a^3 - b^3$$

**29.** $[(x - y) + 3z][(x - y) - 3z] = (x - y)^2 - 9z^2 = x^2 - 2xy + y^2 - 9z^2$ (Special product $\underline{7}$a)

**31.** $m - \{m - [m - (m - 1)]\} = m - \{m - [m - m + 1]\} = m - \{m - 1\} = m - m + 1 = 1$

**33.** $(x^2 - 2xy + y^2)(x^2 + 2xy + y^2) = (x - y)^2(x + y)^2 = [(x - y)(x + y)]^2 = [x^2 - y^2]^2 = x^4 - 2x^2y^2 + y^4$

**35.** $(5a - 2b)^2 - (2b + 5a)^2 = 25a^2 - 20ab + 4b^2 - [4b^2 + 20ab + 25a^2] = -40ab$

**37.** $(m - 2)^2 - (m - 2)(m + 2) = m^2 - 4m + 4 - [m^2 - 4] = m^2 - 4m + 4 - m^2 + 4 = -4m + 8$

**39.** $(x - 2y)(2x + y) - (x + 2y)(2x - y) = 2x^2 - 4xy + xy - 2y^2 - [2x^2 + 4xy - xy - 2y^2]$
$$= 2x^2 - 3xy - 2y^2 - 2x^2 - 3xy + 2y^2 = -6xy$$

**41.** $(u + v)^3 = (u + v)(u + v)^2 = (u + v)(u^2 + 2uv + v^2) = u^3 + 3u^2v + 3uv^2 + v^3$

**43.** $(x - 2y)^3 = (x - 2y)(x - 2y)^2 = (x - 2y)(x^2 - 4xy + 4y^2) = x(x^2 - 4xy + 4y^2) - 2y(x^2 - 4xy + 4y^2)$
$$= x^3 - 4x^2y + 4xy^2 - 2x^2y + 8xy^2 - 8y^3$$
$$= x^3 - 6x^2y + 12xy^2 - 8y^3$$

**45.** $[(2x^2 - 4xy + y^2) + (3xy - y^2)] - [(x^2 - 2xy - y^2) + (-x^2 + 3xy - 2y^2)]$
$$= [2x^2 - xy] - [xy - 3y^2] = 2x^2 - 2xy + 3y^2$$

**47.** $[(2x - 1)^2 - x(3x + 1)]^2 = [4x^2 - 4x + 1 - 3x^2 - x]^2 = [x^2 - 5x + 1]^2 = (x^2 - 5x + 1)(x^2 - 5x + 1)$
$$= x^4 - 10x^3 + 27x^2 - 10x + 1$$

Copyright © 2011 Pearson Education, Inc. Publishing as Prentice Hall.

**49.** $2\{(x-3)(x^2-2x+1)-x[3-x(x-2)]\} = 2\{x^3-5x^2+7x-3-x[3-x^2+2x]\}$
$= 2\{x^3-5x^2+7x-3+x^3-2x^2-3x\} = 2\{2x^3-7x^2+4x-3\} = 4x^3-14x^2+8x-6$

**51.** $m+n$

**53.** Given two polynomials, one with degree $m$ and the other of degree $n$, their product will have degree $m+n$ regardless of the relationship between $m$ and $n$.

**55.** Since $(a+b)^2 = a^2+2ab+b^2$, $(a+b)^2 = a^2+b^2$ only when $2ab = 0$; that is, only when $a=0$ or $b=0$.

**57.** Let $x$ = amount invested at 9%.
Then $10,000-x$ = amount invested at 12%.
The total annual income $I$ is:

$$I = 0.09x + 0.12(10,000-x)$$
$$= 1,200 - 0.03x$$

**59.** Let $x$ = number of tickets at $20.
Then $3x$ = number of tickets at $30 and $4,000-x-3x = 4,000-4x$ = number of tickets at $50.
The total receipts $R$ are:

$$R = 20x + 30(3x) + 50(4,000-4x)$$
$$= 20x + 90x + 200,000 - 200x = 200,000 - 90x$$

**61.** Let $x$ = number of kilograms of food $A$.
Then $10-x$ = number of kilograms of food $B$.
The total number of kilograms $F$ of fat in the final food mix is:

$$F = 0.02x + 0.06(10-x) = 0.6 - 0.04x$$

## EXERCISE A-3

*Things to remember:*

1. The discussion is limited to polynomials with integer coefficients.

2. FACTORED FORMS

   A polynomial is in FACTORED FORM if it is written as the product of two or more polynomials. A polynomial with integer coefficients is FACTORED COMPLETELY if each factor cannot be expressed as the product of two or more polynomials with integer coefficients, other than itself and 1.

3. METHODS

   a. Factor out all factors common to all terms, if they are present.

   b. Try grouping terms.

   c. *ac*–Test for polynomials of the form
   $$ax^2+bx+c \quad \text{or} \quad ax^2+bxy+cy^2$$
   If the product $ac$ has two integer factors $p$ and $q$ whose sum is the coefficient $b$ of the middle term, i.e., if integers $p$ and $q$ exist so that
   $$pq = ac \quad \text{and} \quad p+q = b$$
   then the polynomials have first–degree factors with integer coefficients. If no such integers exist then the polynomials will not have first–degree factors with integer coefficients; the polynomials are *not factorable*.

4. SPECIAL FACTORING FORMULAS

   a. $u^2 + 2uv + v^2 = (u+v)^2$ \qquad Perfect square

Copyright © 2011 Pearson Education, Inc. Publishing as Prentice Hall.

b.    $u^2 - 2uv + v^2 = (u - v)^2$        Perfect square

c.    $u^2 - v^2 = (u - v)(u + v)$        Difference of squares

d.    $u^3 - v^3 = (u - v)(u^2 + uv + v^2)$    Difference of cubes

e.    $u^3 + v^3 = (u + v)(u^2 - uv + v^2)$    Sum of cubes

---

1.    $3m^2$ is a common factor: $6m^4 - 9m^3 - 3m^2 = 3m^2(2m^2 - 3m - 1)$

3.    $2uv$ is a common factor: $8u^3v - 6u^2v^2 + 4uv^3 = 2uv(4u^2 - 3uv + 2v^2)$

5.    $(2m - 3)$ is a common factor: $7m(2m - 3) + 5(2m - 3) = (7m + 5)(2m - 3)$

7.    $4ab(2c + d) - (2c + d) = (4ab - 1)(2c + d)$

9.    $2x^2 - x + 4x - 2 = (2x^2 - x) + (4x - 2) = x(2x - 1) + 2(2x - 1) = (2x - 1)(x + 2)$

11.    $3y^2 - 3y + 2y - 2 = (3y^2 - 3y) + (2y - 2) = 3y(y - 1) + 2(y - 1) = (y - 1)(3y + 2)$

13.    $2x^2 + 8x - x - 4 = (2x^2 + 8x) - (x + 4) = 2x(x + 4) - (x + 4) = (x + 4)(2x - 1)$

15.    $wy - wz + xy - xz = (wy - wz) + (xy - xz) = w(y - z) + x(y - z) = (y - z)(w + x)$

   Or   $wy - wz + xy - xz = (wy + xy) - (wz + xz) = y(w + x) - z(w + x) = (w + x)(y - z)$

17.    $am - 3bm + 2na - 6bn = m(a - 3b) + 2n(a - 3b) = (a - 3b)(m + 2n)$

19.    $3y^2 - y - 2$
    $a = 3, b = -1, c = -2$

   Step 1.    Use the $ac$-test to test for factorability
        $ac = (3)(-2) = -6$

        $\underline{pq}$

        $(1)(-6)$

        $(-1)(6)$

        $\overline{(2)(-3)}$

        $(-2)(3)$

        Note that $2 + (-3) = -1 = b$. Thus, $3y^2 - y - 2$ has first-degree factors with integer coefficients.

   Step 2.    Split the middle term using $b = p + q$ and factor by grouping.
        $-1 = -3 + 2$
        $3y^2 - y - 2 = 3y^2 - 3y + 2y - 2 = (3y^2 - 3y) + (2y - 2)$    $= 3y(y - 1) + 2(y - 1)$
                                $= (y - 1)(3y + 2)$

21.    $u^2 - 2uv - 15v^2$
    $a = 1, b = -2, c = -15$

   Step 1.    Use the $ac$–test
        $ac = 1(-15) = -15$

        $\underline{pq}$

        $(1)(-15)$

        $(-1)(15)$

Copyright © 2011 Pearson Education, Inc.  Publishing as Prentice Hall.

$$\boxed{(3)(-5)}$$
$$(-3)(5)$$

Note that $3 + (-5) = -2 = b$. Thus $u^2 - 2uv - 15v^2$ has first- degree factors with integer coefficients.

<u>Step 2.</u>　　Factor by grouping

$-2 = 3 + (-5)$

$u^2 + 3uv - 5uv - 15v^2 = (u^2 + 3uv) - (5uv + 15v^2) = u(u + 3v) - 5v(u + 3v)$

$$= (u + 3v)(u - 5v)$$

**23.**　$m^2 - 6m - 3$

$a = 1, b = -6, c = -3$

<u>Step 1</u>. Use the $ac$–test

$$ac = (1)(-3) = -3 \qquad\qquad \underline{pq}$$

$$(1)(-3)$$
$$(-1)(3)$$

None of the factors add up to $-6 = b$. Thus, this polynomial is *not factorable*.

**25.**　$w^2 x^2 - y^2 = (wx - y)(wx + y)$　(difference of squares)

**27.**　$9m^2 - 6mn + n^2 = (3m - n)^2$　(perfect square)

**29.**　$y^2 + 16$

$a = 1, b = 0, c = 16$

<u>Step 1</u>. Use the $ac$–test

$$ac = (1)(16) \qquad\qquad \underline{pq}$$

$$(1)(16)$$
$$(-1)(-16)$$
$$(2)(8)$$
$$(-2)(-8)$$
$$(4)(4)$$
$$(-4)(-4)$$

None of the factors add up to $0 = b$. Thus this polynomial is *not factorable*.

**31.**　$4z^2 - 28z + 48 = 4(z^2 - 7z + 12) = 4(z - 3)(z - 4)$

**33.**　$2x^4 - 24x^3 + 40x^2 = 2x^2(x^2 - 12x + 20) = 2x^2(x - 2)(x - 10)$

**35.**　$4xy^2 - 12xy + 9x = x(4y^2 - 12y + 9) = x(2y - 3)^2$

**37.**　$6m^2 - mn - 12n^2 = (2m - 3n)(3m + 4n)$

**39.**　$4u^3 v - uv^3 = uv(4u^2 - v^2) = uv[(2u)^2 - v^2] = uv(2u - v)(2u + v)$

**41.**　$2x^3 - 2x^2 + 8x = 2x(x^2 - x + 4)$　　[<u>Note</u>: $x^2 - x + 4$ is *not factorable*.]

**43.**　$8x^3 - 27y^3 = (2x)^3 - (3y)^3 = (2x - 3y)[(2x)^2 + (2x)(3y) + (3y)^2]$

$$= (2x - 3y)[4x^2 + 6xy + 9y^2] \quad \text{(difference of cubes)}$$

**45.**　$x^4 y + 8xy = xy[x^3 + 8] = xy(x + 2)(x^2 - 2x + 4)$

Copyright © 2011 Pearson Education, Inc. Publishing as Prentice Hall.

**47.** $(x + 2)^2 - 9y^2 = [(x + 2) - 3y][(x + 2) + 3y] = (x + 2 - 3y)(x + 2 + 3y)$

**49.** $5u^2 + 4uv - 2v^2$ is *not factorable*.

**51.** $6(x - y)^2 + 23(x - y) - 4 = [6(x - y) - 1][(x - y) + 4] = (6x - 6y - 1)(x - y + 4)$

**53.** $y^4 - 3y^2 - 4 = (y^2)^2 - 3y^2 - 4 = (y^2 - 4)(y^2 + 1) = (y - 2)(y + 2)(y^2 + 1)$

**55.** $15y(x - y)^3 + 12x(x - y)^2 = 3(x - y)^2 [5y(x - y) + 4x] = 3(x - y)^2 [5xy - 5y^2 + 4x]$

**57.** True: $u^n - v^n = (u - v)(u^{n-1} + u^{n-2}v + \ldots + uv^{n-2} + v^{n-1})$

**59.** False; For example, $u^2 + v^2$ cannot be factored.

---

## EXERCISE A-4

*Things to remember:*

**1.**    FUNDAMENTAL PROPERTY OF FRACTIONS

If $a$, $b$, and $k$ are real numbers with $b$, $k \neq 0$, then

$$\frac{ka}{kb} = \frac{a}{b}.$$

A fraction is in LOWEST TERMS if the numerator and denominator have no common factors other than 1 or −1.

**2.**    MULTIPLICATION AND DIVISION

For $a$, $b$, $c$, and $d$ real numbers:

a.    $\dfrac{a}{b} \cdot \dfrac{c}{d} = \dfrac{ac}{bd}, b, d \neq 0$

b.    $\dfrac{a}{b} \div \dfrac{c}{d} = \dfrac{\frac{a}{b}}{\frac{c}{d}} = \dfrac{a}{b} \cdot \dfrac{d}{c}, b, c, d \neq 0$

The same procedures are used to multiply or divide two rational expressions.

**3.**    ADDITION AND SUBTRACTION

For $a$, $b$, and $c$ real numbers:

a.    $\dfrac{a}{b} + \dfrac{c}{b} = \dfrac{a + c}{b}, b \neq 0$

b.    $\dfrac{a}{b} - \dfrac{c}{b} = \dfrac{a - c}{b}, b \neq 0$

The same procedures are used to add or subtract two rational expressions (with the same denominator).

**4.**    THE LEAST COMMON DENOMINATOR (LCD)

Copyright © 2011 Pearson Education, Inc. Publishing as Prentice Hall.

The LCD of two or more rational expressions is found as follows:

a.   Factor each denominator completely, including integer factors.

b.   Identify each different factor from all the denominators.

c.   Form a product using each different factor to the highest power that occurs in any one denominator. This product is the LCD.

The least common denominator is used to add or subtract rational expressions having different denominators.

---

**1.**

$$\frac{d^5}{3a} \div \left( \frac{d^2}{6a^2} \cdot \frac{a}{4d^3} \right) = \frac{d^5}{3a} \div \left( \frac{\cancel{d} \, \cancel{d^2}}{24 \cancel{d^2} \cancel{d^3}} \right) = \frac{d^5}{3a} \div \frac{1}{24ad} = \frac{d^5}{\cancel{3a}} \cdot \frac{\overset{8}{\cancel{24}} \, \cancel{a} d}{1} = 8d^6$$

$$a \quad d$$

**3.**   $\dfrac{x^2}{12} + \dfrac{x}{18} - \dfrac{1}{30} = \dfrac{15x^2}{180} + \dfrac{10x}{180} - \dfrac{6}{180}$

$$= \frac{15x^2 + 10x - 6}{180}$$

We find the LCD of 12, 18, 30:

$12 = 2^2 \cdot 3$, $18 = 2 \cdot 3^2$, $30 = 2 \cdot 3 \cdot 5$

Thus, LCD $= 2^2 \cdot 3^2 \cdot 5 = 180$.

**5.**   $\dfrac{4m-3}{18m^3} + \dfrac{3}{4m} - \dfrac{2m-1}{6m^2}$

$$= \frac{2(4m-3)}{36m^3} + \frac{3(9m^2)}{36m^3} - \frac{6m(2m-1)}{36m^3}$$

$$= \frac{8m - 6 + 27m^2 - 6m(2m-1)}{36m^3}$$

$$= \frac{8m - 6 + 27m^2 - 12m^2 + 6m}{36m^3} = \frac{15m^2 + 14m - 6}{36m^3}$$

Find the LCD of $18m^3$, $4m$, $6m^2$:

$18m^3 = 2 \cdot 3^2 m^3$, $4m = 2^2 m$,

$6m^2 = 2 \cdot 3m^2$

Thus, LCD $= 36m^3$.

**7.**   $\dfrac{x^2-9}{x^2-3x} \div (x^2 - x - 12) = \dfrac{\cancel{(x-3)} \, (x+3)}{x \cancel{(x-3)}} \cdot \dfrac{1}{(x-4)\cancel{(x+3)}} = \dfrac{1}{x(x-4)}$

**9.**   $\dfrac{2}{x} - \dfrac{1}{x-3} = \dfrac{2(x-3)}{x(x-3)} - \dfrac{x}{x(x-3)}$        LCD $= x(x-3)$

$$= \frac{2x - 6 - x}{x(x-3)} = \frac{x-6}{x(x-3)}$$

**11.**   $\dfrac{2}{(x+1)^2} - \dfrac{5}{x^2-x-2} = \dfrac{2}{(x+1)^2} - \dfrac{5}{(x+1)(x-2)}$        LCD $= (x+1)^2(x-2)$

$$= \frac{2(x-2)}{(x+1)^2(x-2)} - \frac{5(x+1)}{(x+1)^2(x-2)} = \frac{2x-4-5x-5}{(x+1)^2(x-2)} = \frac{-3x-9}{(x+1)^2(x-2)}$$

**13.**   $\dfrac{x+1}{x-1} - 1 = \dfrac{x+1}{x-1} - \dfrac{x-1}{x-1} = \dfrac{x+1-(x-1)}{x-1} = \dfrac{2}{x-1}$

Copyright © 2011 Pearson Education, Inc. Publishing as Prentice Hall.

**15.**  $\dfrac{3}{a-1} - \dfrac{2}{1-a} = \dfrac{3}{a-1} - \dfrac{-2}{-(1-a)} = \dfrac{3}{a-1} + \dfrac{2}{a-1} = \dfrac{5}{a-1}$

**17.**  $\dfrac{2x}{x^2-16} - \dfrac{x-4}{x^2+4x} = \dfrac{2x}{(x-4)(x+4)} - \dfrac{x-4}{x(x+4)}$  $\qquad$ LCD $= x(x-4)(x+4)$

$\qquad\qquad = \dfrac{2x(x)-(x-4)(x-4)}{x(x-4)(x+4)} = \dfrac{2x^2-(x^2-8x+16)}{x(x-4)(x+4)} = \dfrac{x^2+8x-16}{x(x-4)(x+4)}$

**19.**  $\dfrac{x^2}{x^2+2x+1} + \dfrac{x-1}{3x+3} - \dfrac{1}{6} = \dfrac{x^2}{(x+1)^2} + \dfrac{x-1}{3(x+1)} - \dfrac{1}{6}$  $\qquad$ LCD $= 6(x+1)^2$

$\qquad\qquad = \dfrac{6x^2}{6(x+1)^2} + \dfrac{2(x+1)(x-1)}{6(x+1)^2} - \dfrac{(x+1)^2}{6(x+1)^2}$

$\qquad\qquad = \dfrac{6x^2+2(x^2-1)-(x^2+2x+1)}{6(x+1)^2} = \dfrac{7x^2-2x-3}{6(x+1)^2}$

**21.**  $\dfrac{1-\dfrac{x}{y}}{2-\dfrac{y}{x}} = \dfrac{\dfrac{y-x}{y}}{\dfrac{2x-y}{x}} = \dfrac{y-x}{y} \cdot \dfrac{x}{2x-y} = \dfrac{x(y-x)}{y(2x-y)}$

**23.**  $\dfrac{c+2}{5c-5} - \dfrac{c-2}{3c-3} + \dfrac{c}{1-c} = \dfrac{c+2}{5(c-1)} - \dfrac{c-2}{3(c-1)} - \dfrac{c}{c-1}$  $\qquad$ LCD $= 15(c-1)$

$\qquad\qquad = \dfrac{3(c+2)}{15(c-1)} - \dfrac{5(c-2)}{15(c-1)} - \dfrac{15c}{15(c-1)} = \dfrac{3c+6-5c+10-15c}{15(c-1)} = \dfrac{-17c+16}{15(c-1)}$

**25.**  $\dfrac{1+\dfrac{3}{x}}{x-\dfrac{9}{x}} = \dfrac{\dfrac{x+3}{x}}{\dfrac{x^2-9}{x}} = \dfrac{x+3}{x} \cdot \dfrac{x}{x^2-9} = \dfrac{\cancel{x+3}}{\cancel{x}} \cdot \dfrac{\cancel{x}}{\cancel{(x+3)}(x-3)} = \dfrac{1}{x-3}$

**27.**  $\dfrac{\dfrac{1}{2(x+h)}-\dfrac{1}{2x}}{h} = \left(\dfrac{1}{2(x+h)}-\dfrac{1}{2x}\right) \div \dfrac{h}{1} = \dfrac{x-x-h}{2x(x+h)} \cdot \dfrac{1}{h} = \dfrac{-\cancel{h}}{2x(x+h)\cancel{h}} = \dfrac{-1}{2x(x+h)}$

**29.**  $\dfrac{\dfrac{x}{y}-2+\dfrac{y}{x}}{\dfrac{x}{y}-\dfrac{y}{x}} = \dfrac{\dfrac{x^2-2xy+y^2}{xy}}{\dfrac{x^2-y^2}{xy}} = \dfrac{(x-y)^2}{\cancel{xy}} \cdot \dfrac{\cancel{xy}}{(x-y)(x+y)} = \dfrac{x-y}{x+y}$

**31.** (A)  $\dfrac{x^2+4x+3}{x+3} = x+4$:  $\quad$ Incorrect

$\qquad$ (B)  $\dfrac{x^2+4x+3}{x+3} = \dfrac{\cancel{(x+3)}(x+1)}{\cancel{x+3}} = x+1 \quad (x \ne -3)$

**33.** (A)  $\dfrac{(x+h)^2-x^2}{h} = 2x+1$:  $\quad$ Incorrect

$\qquad$ (B)  $\dfrac{(x+h)^2-x^2}{h} = \dfrac{x^2+2xh+h^2-x^2}{h} = \dfrac{2xh+h^2}{h} = \dfrac{\cancel{h}(2x+h)}{\cancel{h}} = 2x+h \quad (h \ne 0)$

Copyright © 2011 Pearson Education, Inc.  Publishing as Prentice Hall.

**35.** (A) $\dfrac{x^2-3x}{x^2-2x-3}+x-3=1$:   Incorrect

(B) $\dfrac{x^2-3x}{x^2-2x-3}+x-3=\dfrac{x(x-3)}{(x-3)(x+1)}+x-3=\dfrac{x}{x+1}+x-3=\dfrac{x+(x-3)(x+1)}{x+1}=\dfrac{x^2-x-3}{x+1}$

**37.** (A) $\dfrac{2x^2}{x^2-4}-\dfrac{x}{x-2}=\dfrac{x}{x+2}$:   Correct

$$\dfrac{2x^2}{x^2-4}-\dfrac{x}{x-2}=\dfrac{2x^2}{(x-2)(x+2)}-\dfrac{x}{x-2}=\dfrac{2x^2-x(x+2)}{(x-2)(x+2)}$$

$$=\dfrac{x^2-2x}{(x-2)(x+2)}=\dfrac{x(x-2)}{(x-2)(x+2)}=\dfrac{x}{x+2}$$

**39.** $\dfrac{\dfrac{1}{3(x+h)^2}-\dfrac{1}{3x^2}}{h}=\left[\dfrac{1}{3(x+h)^2}-\dfrac{1}{3x^2}\right]\div\dfrac{h}{1}=\dfrac{x^2-(x+h)^2}{3x^2(x+h)^2}\cdot\dfrac{1}{h}=\dfrac{x^2-(x^2+2xh+h^2)}{3x^2(x+h)^2h}$

$$=\dfrac{-2xh-h^2}{3x^2(x+h)^2h}=\dfrac{-h(2x+h)}{3x^2(x+h)^2h}=-\dfrac{(2x+h)}{3x^2(x+h)^2}=\dfrac{-2x-h}{3x^2(x+h)^2}$$

**41.** $x-\dfrac{2}{1-\dfrac{1}{x}}=x-\dfrac{2}{\dfrac{x-1}{x}}=x-\dfrac{2x}{x-1}$       (LCD $=x-1$)

$$=\dfrac{x(x-1)}{x-1}-\dfrac{2x}{x-1}=\dfrac{x^2-x-2x}{x-1}=\dfrac{x(x-3)}{x-1}$$

---

## EXERCISE A-5

*Things to remember:*

1.  DEFINITION OF $a^n$, where $n$ is an integer and $a$ is a real number:

a.   For $n$ a positive integer,
$a^n = a\cdot a\cdot\cdots\cdot a$, $n$ factors of $a$.

b.   For $n=0$,
$a^0=1$, $a\ne 0$, $0^0$ is not defined.

c.   For $n$ a negative integer,
$a^n=\dfrac{1}{a^{-n}}$, $a\ne 0$.

[Note: If $n$ is negative, then $-n$ is positive.]

Copyright © 2011 Pearson Education, Inc. Publishing as Prentice Hall.

2. PROPERTIES OF EXPONENTS

GIVEN: $n$ and $m$ are integers and $a$ and $b$ are real numbers.

a. $a^m a^n = a^{m+n}$ $\qquad\qquad$ $a^8 a^{-3} = a^{8+(-3)} = a^5$

b. $(a^n)^m = a^{mn}$ $\qquad\qquad$ $(a^{-2})^3 = a^{3(-2)} = a^{-6}$

c. $(ab)^m = a^m b^m$ $\qquad\qquad$ $(ab)^{-2} = a^{-2}b^{-2}$

d. $\left(\dfrac{a}{b}\right)^m = \dfrac{a^m}{b^m}, b \neq 0$ $\qquad\qquad$ $\left(\dfrac{a}{b}\right)^5 = \dfrac{a^5}{b^5}$

e. $\dfrac{a^m}{a^n} = a^{m-n} = \dfrac{1}{a^{n-m}}, a \neq 0$ $\qquad$ $\dfrac{a^{-3}}{a^7} = \dfrac{1}{a^{7-(-3)}} = \dfrac{1}{a^{10}}$

3. SCIENTIFIC NOTATION

Let $r$ be any finite decimal. Then $r$ can be expressed as the product of a number between 1 and 10 and an integer power of 10; that is, $r$ can be written $r = a \times 10^n$, $1 \leq a < 10$, $a$ in decimal form, $n$ an integer A number expressed in this form is said to be in SCIENTIFIC NOTATION.

Examples:

$7 = 7 \times 10^0$ $\qquad\qquad\qquad$ $0.5 = 5 \times 10^{-1}$

$67 = 6.7 \times 10$ $\qquad\qquad\qquad$ $0.45 = 4.5 \times 10^{-1}$

$580 = 5.8 \times 10^2$ $\qquad\qquad\qquad$ $0.0032 = 3.2 \times 10^{-3}$

$43,000 = 4.3 \times 10^4$ $\qquad\qquad\qquad$ $0.000\,045 = 4.5 \times 10^{-5}$

---

**1.** $2x^{-9} = \dfrac{2}{x^9}$ $\qquad\qquad$ **3.** $\dfrac{3}{2w^{-7}} = \dfrac{3w^7}{2}$

**5.** $2x^{-8}x^5 = 2x^{-8+5} = 2x^{-3} = \dfrac{2}{x^3}$ $\qquad$ **7.** $\dfrac{w^{-8}}{w^{-3}} = \dfrac{1}{w^{-3+8}} = \dfrac{1}{w^5}$

**9.** $(2a^{-3})^2 = 2^2(a^{-3})^2 = 4a^{-6} = \dfrac{4}{a^6}$ $\qquad$ **11.** $(a^{-3})^2 = a^{-6} = \dfrac{1}{a^6}$

**13.** $(2x^4)^{-3} = 2^{-3}(x^4)^{-3} = \dfrac{1}{8} \cdot x^{-12} = \dfrac{1}{8x^{12}}$ $\qquad$ **15.** $82,300,000,000 = 8.23 \times 10^{10}$

**17.** $0.783 = 7.83 \times 10^{-1}$ $\qquad\qquad$ **19.** $0.000\,034 = 3.4 \times 10^{-5}$

**21.** $4 \times 10^4 = 40,000$ $\qquad\qquad$ **23.** $7 \times 10^{-3} = 0.007$

**25.** $6.171 \times 10^7 = 61,710,000$ $\qquad\qquad$ **27.** $8.08 \times 10^{-4} = 0.000\,808$

**29.** $(22 + 31)^0 = (53)^0 = 1$

**31.** $\dfrac{10^{-3} \times 10^4}{10^{-11} \times 10^{-2}} = \dfrac{10^{-3+4}}{10^{-11-2}} = \dfrac{10^1}{10^{-13}} = 10^{1+13} = 10^{14}$

**33.** $(5x^2 y^{-3})^{-2} = 5^{-2}x^{-4}y^6 = \dfrac{y^6}{5^2 x^4} = \dfrac{y^6}{25x^4}$

Copyright © 2011 Pearson Education, Inc. Publishing as Prentice Hall.

**35.** $\left(\dfrac{-5}{2x^3}\right)^{-2} = \dfrac{(-5)^{-2}}{(2x^3)^{-2}} = \dfrac{\dfrac{1}{(-5)^2}}{\dfrac{1}{(2x^3)^2}} = \dfrac{\dfrac{1}{25}}{\dfrac{1}{4x^6}} = \dfrac{4x^6}{25}$

**37.** $\dfrac{8x^{-3}y^{-1}}{6x^2 y^{-4}} = \dfrac{4y^{-1+4}}{3x^{2+3}} = \dfrac{4y^3}{3x^5}$

**39.** $\dfrac{7x^5 - x^2}{4x^5} = \dfrac{7x^5}{4x^5} - \dfrac{x^2}{4x^5} = \dfrac{7}{4} - \dfrac{1}{4x^3} = \dfrac{7}{4} - \dfrac{1}{4}x^{-3}$

**41.** $\dfrac{5x^4 - 3x^2 + 8}{2x^2} = \dfrac{5x^4}{2x^2} - \dfrac{3x^2}{2x^2} + \dfrac{8}{2x^2} = \dfrac{5}{2}x^2 - \dfrac{3}{2} + 4x^{-2}$

**43.** $\dfrac{3x^2(x-1)^2 - 2x^3(x-1)}{(x-1)^4} = \dfrac{x^2(x-1)[3(x-1) - 2x]}{(x-1)^4} = \dfrac{x^2(x-3)}{(x-1)^3}$

**45.** $2x^{-2}(x-1) - 2x^{-3}(x-1)^2 = \dfrac{2(x-1)}{x^2} - \dfrac{2(x-1)^2}{x^3} = \dfrac{2x(x-1) - 2(x-1)^2}{x^3}$

$\qquad = \dfrac{2(x-1)[x - (x-1)]}{x^3} = \dfrac{2(x-1)}{x^3}$

**47.** $\dfrac{9,600,000,000}{(1,600,000)(0.00000025)} = \dfrac{9.6 \times 10^9}{(1.6 \times 10^6)(2.5 \times 10^{-7})} = \dfrac{9.6 \times 10^9}{1.6(2.5) \times 10^{6-7}}$

$\qquad = \dfrac{9.6 \times 10^9}{4.0 \times 10^{-1}} = 2.4 \times 10^{9+1} = 2.4 \times 10^{10} = 24,000,000,000$

**49.** $\dfrac{(1,250,000)(0.00038)}{0.0152} = \dfrac{(1.25 \times 10^6)(3.8 \times 10^{-4})}{1.52 \times 10^{-2}} = \dfrac{1.25(3.8) \times 10^{6-4}}{1.52 \times 10^{-2}} = 3.125 \times 10^4 = 31,250$

**51.** On a calculator : $2^{3^2} = 64$. A calculator interprets $2^{3^2}$ as $(2^3)^2 = 8^2 = 64$.

**53.** $a^m a^0 = a^{m+0} = a^m$. Therefore, $a^m a^0 = a^m$ which implies $a^0 = 1$.

**55.** $\dfrac{u+v}{u^{-1} + v^{-1}} = \dfrac{u+v}{\dfrac{1}{u} + \dfrac{1}{v}} = \dfrac{u+v}{\dfrac{v+u}{uv}} = (u+v) \cdot \dfrac{uv}{v+u} = uv$

**57.** $\dfrac{b^{-2} - c^{-2}}{b^{-3} - c^{-3}} = \dfrac{\dfrac{1}{b^2} - \dfrac{1}{c^2}}{\dfrac{1}{b^3} - \dfrac{1}{c^3}} = \dfrac{\dfrac{c^2 - b^2}{b^2 c^2}}{\dfrac{c^3 - b^3}{b^3 c^3}} = \dfrac{(c-b)(c+b)}{b^2 c^2} \cdot \dfrac{b^3 c^3}{(c-b)(c^2 + cb + b^2)} = \dfrac{bc(c+b)}{c^2 + cb + b^2}$

**59.** (A)   Per capita debt: $\dfrac{1.0025 \times 10^{13}}{3.04 \times 10^8} \approx 0.32977 \times 10^5 = 32,977$ or $\$32,977$

Copyright © 2011 Pearson Education, Inc.  Publishing as Prentice Hall.

(B)  Per capita interest:  $\dfrac{4.51\times10^{11}}{3.04\times10^{8}}\approx1.484\times10^{3}=1,484$  or $1,484

(C)  Percentage interest paid on debt:  $\dfrac{4.51\times10^{11}}{1.0025\times10^{13}}\approx4.50\times10^{-2}=0.045$  or 4.5%

**61.**  (A)  $9\text{ ppm}=\dfrac{9}{1,000,000}=\dfrac{9}{10^{6}}=9\times10^{-6}$   (B)  0.000 009   (C)  0.0009%

**63.**  $\dfrac{466}{100,000}\times304,000,000=\dfrac{4.66\times10^{2}}{10^{5}}\times3.04\times10^{8}=14.1664\times10^{5}=1,416,640$

To the nearest thousand, there were 1,417,000 violent crimes committed in 2008.

## EXERCISE A-6

*Things to remember:*

**1.**   *n*th ROOT

Let $b$ be a real number. For any natural number $n$,

$r$ is an *n*th ROOT of $b$ if $r^{n}=b$

If $n$ is odd, then $b$ has exactly one real *n*th root.
If $n$ is even, and $b<0$, then $b$ has NO real *n*th roots.
If $n$ is even, and $b>0$, then $b$ has two real *n*th roots;
  if $r$ is an *n*th root, then $-r$ is also an *n*th root.
0 is an *n*th root of 0 for all $n$

**2.**   NOTATION

Let $b$ be a real number and let $n>1$ be a natural number. If $n$ is odd, then the *n*th root of $b$ is denoted

$b^{1/n}$  or  $\sqrt[n]{b}$

If $n$ is even and $b>0$, then the PRINCIPAL *n*th ROOT OF $b$ is the positive *n*th root; the principal *n*th root is denoted

$b^{1/n}$  or  $\sqrt[n]{b}$

In the $\sqrt[n]{b}$ notation, the symbol $\sqrt{\phantom{x}}$ is called a RADICAL, $n$ is the INDEX of the radical and $b$ is called the RADICAND

**3.**   RATIONAL EXPONENTS

If $m$ and $n$ are natural numbers without common prime factors, $b$ is a real number, and $b$ is nonnegative when $b$ is even, then

$$b^{m/n}=\begin{cases}\left(b^{1/n}\right)^{m}=\left(\sqrt[n]{b}\right)^{m}\\\left(b^{m}\right)^{1/n}=\sqrt[n]{b^{m}}\end{cases}$$

Copyright © 2011 Pearson Education, Inc. Publishing as Prentice Hall.

and $b^{-m/n} = \dfrac{1}{b^{m/n}}$, $b \neq 0$

The two definitions of $b^{m/n}$ are equivalent under the indicated restrictions on $m$, $n$, and $b$.

<u>4.</u>   PROPERTIES OF RADICALS

If $m$ and $n$ are natural numbers greater than or equal to 2 and $x$ and $y$ are positive real numbers, then

a. $\sqrt[n]{x^n} = x$ $\qquad\qquad\qquad\qquad$ $\sqrt[3]{x^3} = x$

b. $\sqrt[n]{xy} = \sqrt[n]{x}\,\sqrt[n]{y}$ $\qquad\qquad\qquad$ $\sqrt[5]{xy} = \sqrt[5]{x}\,\sqrt[5]{y}$

c. $\sqrt[n]{\dfrac{x}{y}} = \dfrac{\sqrt[n]{x}}{\sqrt[n]{y}}$ $\qquad\qquad\qquad$ $\sqrt[4]{\dfrac{x}{y}} = \dfrac{\sqrt[4]{x}}{\sqrt[4]{y}}$

---

**1.** $6x^{3/5} = 6\sqrt[5]{x^3}$

**3.** $(32x^2y^3)^{3/5} = \sqrt[5]{(32x^2y^3)^3}$

**5.** $(x^2 + y^2)^{1/2} = \sqrt{x^2 + y^2}$

[Note: $\sqrt{x^2 + y^2} \neq x + y$.]

**7.** $5\sqrt[4]{x^3} = 5x^{3/4}$

**9.** $\sqrt[5]{(2x^2y)^3} = (2x^2y)^{3/5}$

**11.** $\sqrt[3]{x} + \sqrt[3]{y} = x^{1/3} + y^{1/3}$

**13.** $25^{1/2} = (5^2)^{1/2} = 5$

**15.** $16^{3/2} = (4^2)^{3/2} = 4^3 = 64$

**17.** $-49^{1/2} = -\sqrt{49} = -7$

**19.** $-64^{2/3} = -(\sqrt[3]{64})^2 = -16$

**21.** $\left(\dfrac{4}{25}\right)^{3/2} = \left(\left(\dfrac{2}{5}\right)^2\right)^{3/2} \left(\dfrac{2}{5}\right)^3 = \dfrac{2^3}{5^3} = \dfrac{8}{125}$

**23.** $9^{-3/2} = (3^2)^{-3/2} = 3^{-3} = \dfrac{1}{3^3} = \dfrac{1}{27}$

**25.** $x^{4/5}x^{-2/5} = x^{4/5-2/5} = x^{2/5}$

**27.** $\dfrac{m^{2/3}}{m^{-1/3}} = m^{2/3-(-1/3)} = m^1 = m$

**29.** $(8x^3y^{-6})^{1/3} = (2^3x^3y^{-6})^{1/3} = 2^{3/3}x^{3/3}y^{-6/3} = 2xy^{-2} = \dfrac{2x}{y^2}$

**31.** $\left(\dfrac{4x^{-2}}{y^4}\right)^{-1/2} = \left(\dfrac{2^2x^{-2}}{y^4}\right)^{-1/2} = \dfrac{2^{2(-1/2)}x^{-2(-1/2)}}{y^{4(-1/2)}} = \dfrac{2^{-1}x^1}{y^{-2}} = \dfrac{xy^2}{2}$

**33.** $\dfrac{(8x)^{-1/3}}{12x^{1/4}} = \dfrac{\frac{1}{(8x)^{1/3}}}{12x^{1/4}} = \dfrac{\frac{1}{2x^{1/3}}}{12x^{1/4}} = \dfrac{1}{24x^{1/4+1/3}} = \dfrac{1}{24x^{7/12}}$

**35.** $\sqrt[5]{(2x+3)^5} = [(2x+3)^5]^{1/5} = 2x + 3$

**37.** $\sqrt{6x}\,\sqrt{15x^3}\,\sqrt{30x^7} = \sqrt{6(15)(30)x^{11}} = \sqrt{3(30)^2x^{11}} = 30x^5\sqrt{3x}$

Copyright © 2011 Pearson Education, Inc.  Publishing as Prentice Hall.

**39.** $\dfrac{\sqrt{6x}\sqrt{10}}{\sqrt{15x}} = \sqrt{\dfrac{60x}{15x}} = \sqrt{4} = 2$

**41.** $3x^{3/4}(4x^{1/4} - 2x^8) = 12x^{3/4+1/4} - 6x^{3/4+8} = 12x - 6x^{3/4+32/4} = 12x - 6x^{35/4}$

**43.** $(3u^{1/2} - v^{1/2})(u^{1/2} - 4v^{1/2}) = 3u - 12u^{1/2}v^{1/2} - u^{1/2}v^{1/2} + 4v = 3u - 13u^{1/2}v^{1/2} + 4v$

**45.** $(6m^{1/2} + n^{-1/2})(6m - n^{-1/2}) = 36m^{3/2} + 6mn^{-1/2} - 6m^{1/2}n^{-1/2} - n^{-1} = 36m^{3/2} + \dfrac{6m}{n^{1/2}} - \dfrac{6m^{1/2}}{n^{1/2}} - \dfrac{1}{n}$

**47.** $(3x^{1/2} - y^{1/2})^2 = (3x^{1/2})^2 - 6x^{1/2}y^{1/2} + (y^{1/2})^2 = 9x - 6x^{1/2}y^{1/2} + y$

**49.** $\dfrac{\sqrt[3]{x^2} + 2}{2\sqrt[3]{x}} = \dfrac{x^{2/3} + 2}{2x^{1/3}} = \dfrac{x^{2/3}}{2x^{1/3}} + \dfrac{2}{2x^{1/3}} = \dfrac{1}{2}x^{1/3} + \dfrac{1}{x^{1/3}} = \dfrac{1}{2}x^{1/3} + x^{-1/3}$

**51.** $\dfrac{2\sqrt[4]{x^3} + \sqrt[3]{x}}{3x} = \dfrac{2x^{3/4} + x^{1/3}}{3x} = \dfrac{2x^{3/4}}{3x} + \dfrac{x^{1/3}}{3x} = \dfrac{2}{3}x^{3/4-1} + \dfrac{1}{3}x^{1/3-1} = \dfrac{2}{3}x^{-1/4} + \dfrac{1}{3}x^{-2/3}$

**53.** $\dfrac{2\sqrt[3]{x} - \sqrt{x}}{4\sqrt{x}} = \dfrac{2x^{1/3} - x^{1/2}}{4x^{1/2}} = \dfrac{2x^{1/3}}{4x^{1/2}} - \dfrac{x^{1/2}}{4x^{1/2}} = \dfrac{1}{2}x^{1/3-1/2} - \dfrac{1}{4} = \dfrac{1}{2}x^{-1/6} - \dfrac{1}{4}$

**55.** $\dfrac{12mn^2}{\sqrt{3mn}} = \dfrac{12mn^2}{\sqrt{3mn}} \cdot \dfrac{\sqrt{3mn}}{\sqrt{3mn}} = \dfrac{12mn^2\sqrt{3mn}}{3mn} = 4n\sqrt{3mn}$

**57.** $\dfrac{2(x+3)}{\sqrt{x-2}} = \dfrac{2(x+3)}{\sqrt{x-2}} \cdot \dfrac{\sqrt{x-2}}{\sqrt{x-2}} = \dfrac{2(x+3)\sqrt{x-2}}{x-2}$

**59.** $\dfrac{7(x-y)^2}{\sqrt{x} - \sqrt{y}} = \dfrac{7(x-y)^2}{\sqrt{x} - \sqrt{y}} \cdot \dfrac{\sqrt{x} + \sqrt{y}}{\sqrt{x} + \sqrt{y}} = \dfrac{7(x-y)^2(\sqrt{x} + \sqrt{y})}{x-y} = 7(x-y)(\sqrt{x} + \sqrt{y})$

**61.** $\dfrac{\sqrt{5xy}}{5x^2y^2} = \dfrac{\sqrt{5xy}}{5x^2y^2} \cdot \dfrac{\sqrt{5xy}}{\sqrt{5xy}} = \dfrac{5xy}{5x^2y^2\sqrt{5xy}} = \dfrac{1}{xy\sqrt{5xy}}$

**63.** $\dfrac{\sqrt{x+h} - \sqrt{x}}{h} = \dfrac{\sqrt{x+h} - \sqrt{x}}{h} \cdot \dfrac{\sqrt{x+h} + \sqrt{x}}{\sqrt{x+h} + \sqrt{x}} = \dfrac{x+h-x}{h(\sqrt{x+h} + \sqrt{x})} = \dfrac{h}{h(\sqrt{x+h} + \sqrt{x})} = \dfrac{1}{\sqrt{x+h} + \sqrt{x}}$

**65.** $\dfrac{\sqrt{t} - \sqrt{x}}{t^2 - x^2} = \dfrac{\sqrt{t} - \sqrt{x}}{(t-x)(t+x)} \cdot \dfrac{\sqrt{t} + \sqrt{x}}{\sqrt{t} + \sqrt{x}} = \dfrac{t-x}{(t-x)(t+x)(\sqrt{t} + \sqrt{x})} = \dfrac{1}{(t+x)(\sqrt{t} + \sqrt{x})}$

**67.** $(x+y)^{1/2} \overset{?}{=} x^{1/2} + y^{1/2}$

Let $x = y = 1$. Then

$(1+1)^{1/2} = 2^{1/2} = \sqrt{2} \approx 1.414$

$1^{1/2} + 1^{1/2} = \sqrt{1} + \sqrt{1} = 1 + 1 = 2; \quad \sqrt{2} \ne 2$

Copyright © 2011 Pearson Education, Inc.  Publishing as Prentice Hall.

**71.** $\sqrt{x^2} = x$ for all real numbers $x$: False

$\sqrt{(-2)^2} = \sqrt{4} = 2 \neq -2$

**73.** $\sqrt[3]{x^3} = |x|$ for all real numbers $x$: False

$\sqrt[3]{(-1)^3} = \sqrt[3]{-1} = -1 \neq |-1| = 1$

**75.** False: $(-8)^{1/3} = -2$ since $(-2)^3 = -8$

**77.** True: $r^{1/2} = \sqrt{r}$ and $-r^{1/2} = -\sqrt{r}$ are each square roots of $r$.

**79.** True: $(\sqrt{10})^4 = (10^{1/2})^4 = 10^2 = 100$

$(-\sqrt{10})^4 = (-1)^4(\sqrt{10})^4 = (1)(100) = 100$

**81.** False: $5\sqrt{7} - 6\sqrt{5} \approx -0.1877$; $\sqrt{a}$ is never negative.

**83.** $-\dfrac{1}{2}(x-2)(x+3)^{-3/2} + (x+3)^{-1/2} = \dfrac{-(x-2)}{2(x+3)^{3/2}} + \dfrac{1}{(x+3)^{1/2}} = \dfrac{-x+2+2(x+3)}{2(x+3)^{3/2}} = \dfrac{x+8}{2(x+3)^{3/2}}$

**85.** $\dfrac{(x-1)^{1/2} - x\left(\dfrac{1}{2}\right)(x-1)^{-1/2}}{x-1} = \dfrac{(x-1)^{1/2} - \dfrac{x}{2(x-1)^{1/2}}}{x-1} = \dfrac{\dfrac{2(x-1)^{1/2}(x-1)^{1/2}}{2(x-1)^{1/2}} - \dfrac{x}{2(x-1)^{1/2}}}{x-1}$

$= \dfrac{\dfrac{2(x-1)-x}{2(x-1)^{1/2}}}{x-1} = \dfrac{x-2}{2(x-1)^{3/2}}$

**87.** $\dfrac{(x+2)^{2/3} - x\left(\dfrac{2}{3}\right)(x+2)^{-1/3}}{(x+2)^{4/3}} = \dfrac{(x+2)^{2/3} - \dfrac{2x}{3(x+2)^{1/3}}}{(x+2)^{4/3}} = \dfrac{\dfrac{3(x+2)^{1/3}(x+2)^{2/3} - 2x}{3(x+2)^{1/3}}}{(x+2)^{4/3}}$

$= = \dfrac{3(x+2)-2x}{3(x+2)^{5/3}} = \dfrac{x+6}{3(x+2)^{5/3}}$

**89.** $22^{3/2} = 22^{1.5} \approx 103.2$   or   $22^{3/2} = \sqrt{(22)^3} = \sqrt{10,648} \approx 103.2$

**91.** $827^{-3/8} = \dfrac{1}{827^{3/8}} = \dfrac{1}{827^{0.375}} \approx \dfrac{1}{12.42} \approx 0.0805$

**93.** $37.09^{7/3} \approx 37.09^{2.3333} \approx 4,588$

**95.** (A) $\sqrt{3} + \sqrt{5} \approx 1.732 + 2.236 = 3.968$

(B) $\sqrt{2+\sqrt{3}} + \sqrt{2-\sqrt{3}} \approx 2.449$

(C) $1 + \sqrt{3} \approx 2.732$

(D) $\sqrt[3]{10+6\sqrt{3}} \approx 2.732$

(E) $\sqrt{8+\sqrt{60}} \approx 3.968$

(F) $\sqrt{6} \approx 2.449$

Copyright © 2011 Pearson Education, Inc.  Publishing as Prentice Hall.

(A) and (E) have the same value:

$$\left(\sqrt{3}+\sqrt{5}\right)^2 = 3 + 2\sqrt{3}\sqrt{5} + 5 = 8 + 2\sqrt{15}$$

$$\left[\sqrt{8+\sqrt{60}}\right]^2 = 8 + \sqrt{4\cdot 15} = 8 + 2\sqrt{15}$$

(B) and (F) have the same value.

$$\left(\sqrt{2+\sqrt{3}} + \sqrt{2-\sqrt{3}}\right)^2 = 2 + \sqrt{3} + 2\sqrt{2+\sqrt{3}}\sqrt{2-\sqrt{3}} + 2 - \sqrt{3} = 4 + 2\sqrt{4-3} = 4 + 2 = 6$$

$$\left(\sqrt{6}\right)^2 = 6.$$

(C) and (D) have the same value.

$$\left(1+\sqrt{3}\right)^3 = \left(1+\sqrt{3}\right)^2\left(1+\sqrt{3}\right) = (1 + 2\sqrt{3} + 3)(1 + \sqrt{3}) = (4 + 2\sqrt{3})(1 + \sqrt{3})$$
$$= 4 + 6\sqrt{3} + 6 = 10 + 6\sqrt{3}$$

$$\left(\sqrt[3]{10+6\sqrt{3}}\right)^3 = 10 + 6\sqrt{3}$$

---

## EXERCISE A-7

*Things to remember:*

1. A QUADRATIC EQUATION in one variable is any equation that can be written in the form

    $$ax^2 + bx + c = 0, \ a \neq 0 \quad \text{STANDARD FORM}$$

    where $x$ is a variable and $a$, $b$, and $c$ are constants.

2. Quadratic equations of the form $ax^2 + c = 0$ can be solved by the SQUARE ROOT METHOD. The solutions are:

    $$x = \pm\sqrt{\frac{-c}{a}} \quad \text{provided} \quad \frac{-c}{a} \geq 0;$$

    otherwise, the equation has no real solutions.

3. If the left side of the quadratic equation when written in standard form can be FACTORED,

    $$ax^2 + bx + c = (px + q)(rx + s),$$

    then the solutions are

    $$x = \frac{-q}{p} \quad \text{or} \quad x = \frac{-s}{r}.$$

4. The solutions of the quadratic equation written in standard form are given by the QUADRATIC FORMULA:

    $$x = \frac{-b \pm \sqrt{b^2 - 4ac}}{2a}$$

    The quantity $b^2 - 4ac$ under the radical is called the DISCRIMINANT and the equation:

    (i) Has two real solutions if $b^2 - 4ac > 0$.

    (ii) Has one real solution if $b^2 - 4ac = 0$.

    (iii) Has no real solution if $b^2 - 4ac < 0$.

5. FACTORABILITY THEOREM

Copyright © 2011 Pearson Education, Inc. Publishing as Prentice Hall.

The second-degree polynomial, $ax^2 + bx + c$, with integer coefficients, can be expressed as the product of two first-degree polynomials with integer coefficients if and only if

$\sqrt{b^2 - 4ac}$ is an integer.

<u>6.</u>    FACTOR THEOREM

If $r_1$ and $r_2$ are solutions of $ax^2 + bx + c = 0$, then

$ax^2 + bx + c = a(x - r_1)(x - r_2)$.

---

**1.** $2x^2 - 22 = 0$

$x^2 - 11 = 0$

$x^2 = 11$

$x = \pm\sqrt{11}$

**3.** $(3x - 1)^2 = 25$

$3x - 1 = \pm\sqrt{25} = \pm 5$

$3x = 1 \pm 5 = -4 \text{ or } 6$

$x = -\dfrac{4}{3} \text{ or } 2$

**5.** $2u^2 - 8u - 24 = 0$

$u^2 - 4u - 12 = 0$

$(u - 6)(u + 2) = 0$

$u - 6 = 0 \text{ or } u + 2 = 0$

$u = 6 \text{ or } \quad u = -2$

**7.** $x^2 = 2x$

$x^2 - 2x = 0$

$x(x - 2) = 0$

$x = 0 \text{ or } x - 2 = 0$

$x = 2$

**9.** $x^2 - 6x - 3 = 0$

$x = \dfrac{-b \pm \sqrt{b^2 - 4ac}}{2a}, \quad a = 1, \ b = -6, \ c = -3$

$= \dfrac{-(-6) \pm \sqrt{(-6)^2 - 4(1)(-3)}}{2(1)} = \dfrac{6 \pm \sqrt{48}}{2} = \dfrac{6 \pm 4\sqrt{3}}{2} = 3 \pm 2\sqrt{3}$

**11.** $3u^2 + 12u + 6 = 0$

Since 3 is a factor of each coefficient, divide both sides by 3.

$u^2 + 4u + 2 = 0$

$u = \dfrac{-b \pm \sqrt{b^2 - 4ac}}{2a}, \quad a = 1, \ b = 4, \ c = 2$

$= \dfrac{-4 \pm \sqrt{4^2 - 4(1)(2)}}{2(1)} = \dfrac{-4 \pm \sqrt{8}}{2} = \dfrac{-4 \pm 2\sqrt{2}}{2} = -2 \pm \sqrt{2}$

**13.**    $\dfrac{2x^2}{3} = 5x$

$2x^2 = 15x$

$2x^2 - 15x = 0$

$x(2x - 15) = 0$

$x = 0 \text{ or } 2x - 15 = 0$

$x = \dfrac{15}{2}$

**15.** $4u^2 - 9 = 0$

$4u^2 = 9 \quad \text{(solve by square root method)}$

$u^2 = \dfrac{9}{4}$

$u = \pm\sqrt{\dfrac{9}{4}} = \pm\dfrac{3}{2}$

Copyright © 2011 Pearson Education, Inc. Publishing as Prentice Hall.

**17.**
$$8x^2 + 20x = 12$$
$$8x^2 + 20x - 12 = 0$$
$$2x^2 + 5x - 3 = 0$$
$$(x + 3)(2x - 1) = 0$$
$$x + 3 = 0 \quad \text{or} \quad 2x - 1 = 0$$
$$x = -3 \quad \text{or} \quad 2x = 1$$
$$x = \frac{1}{2}$$

**19.**
$$x^2 = 1 - x$$
$$x^2 + x - 1 = 0$$
$$x = \frac{-b \pm \sqrt{b^2 - 4ac}}{2a}, \quad a = 1, \quad b = 1, \quad c = -1$$
$$= \frac{-1 \pm \sqrt{(1)^2 - 4(1)(-1)}}{2(1)} = \frac{-1 \pm \sqrt{5}}{2}$$

**21.**
$$2x^2 = 6x - 3$$
$$2x^2 - 6x + 3 = 0$$
$$x = \frac{-b \pm \sqrt{b^2 - 4ac}}{2a}, \quad a = 2, \quad b = -6, \quad c = 3$$
$$= \frac{-(-6) \pm \sqrt{(-6)^2 - 4(2)(3)}}{2(2)} = \frac{6 \pm \sqrt{12}}{4} = \frac{6 \pm 2\sqrt{3}}{4} = \frac{3 \pm \sqrt{3}}{2}$$

**23.**
$$y^2 - 4y = -8$$
$$y^2 - 4y + 8 = 0$$
$$y = \frac{-b \pm \sqrt{b^2 - 4ac}}{2a}, \quad a = 1, \quad b = -4, \quad c = 8$$
$$= \frac{-(-4) \pm \sqrt{(-4)^2 - 4(1)(8)}}{2(1)} = \frac{4 \pm \sqrt{-16}}{2}$$

Since $\sqrt{-16}$ is not a real number, there are no real solutions.

**25.**
$$(2x + 3)^2 = 11$$
$$2x + 3 = \pm\sqrt{11}$$
$$2x = -3 \pm \sqrt{11}$$
$$x = -\frac{3}{2} \pm \frac{1}{2}\sqrt{11}$$

**27.**
$$\frac{3}{p} = p$$
$$p^2 = 3$$
$$p = \pm\sqrt{3}$$

**29.**
$$2 - \frac{2}{m^2} = \frac{3}{m}$$
$$2m^2 - 2 = 3m$$
$$2m^2 - 3m - 2 = 0$$
$$(2m + 1)(m - 2) = 0$$
$$m = -\frac{1}{2}, \quad 2$$

Copyright © 2011 Pearson Education, Inc. Publishing as Prentice Hall.

**31.** $x^2 + 40x - 84$

    Step 1.    Test for factorability

$$\sqrt{b^2 - 4ac} = \sqrt{(40)^2 - 4(1)(-84)} = \sqrt{1936} = 44$$

                Since the result is an integer, the polynomial has first-degree factors with integer coefficients.

    Step 2.    Use the factor theorem

$$x^2 + 40x - 84 = 0$$

$$x = \frac{-40 \pm 44}{2} = 2, -42 \text{ (by the quadratic formula)}$$

                Thus, $x^2 + 40x - 84 = (x - 2)(x - [-42]) = (x - 2)(x + 42)$

**33.** $x^2 - 32x + 144$

    Step 1.    Test for factorability

$$\sqrt{b^2 - 4ac} = \sqrt{(-32)^2 - 4(1)(144)} = \sqrt{448} \approx 21.166$$

                Since this is not an integer, the polynomial is not factorable.

**35.** $2x^2 + 15x - 108$

    Step 1.    Test for factorability

$$\sqrt{b^2 - 4ac} = \sqrt{(15)^2 - 4(2)(-108)} = \sqrt{1089} = 33$$

                Thus, the polynomial has first-degree factors with integer coefficients.

    Step 2.    Use the factor theorem

$$2x^2 + 15x - 108 = 0$$

$$x = \frac{-15 \pm 33}{4} = \frac{9}{2}, -12$$

                Thus, $2x^2 + 15x - 108 = 2\left(x - \frac{9}{2}\right)(x - [-12]) = (2x - 9)(x + 12)$

**37.** $4x^2 + 241x - 434$

    Step 1.    Test for factorability

$$\sqrt{b^2 - 4ac} = \sqrt{(241)^2 - 4(4)(-434)} = \sqrt{65025} = 255$$

                Thus, the polynomial has first-degree factors with integer coefficients.

    Step 2.    Use the factor theorem

$$4x^2 + 241x - 434 = 0$$

$$x = \frac{-241 \pm 255}{8} = \frac{14}{8}, -\frac{496}{8} \text{ or } \frac{7}{4}, -62$$

                Thus, $4x^2 + 241x - 434 = 4\left(x - \frac{7}{4}\right)(x + 62) = (4x - 7)(x + 62)$

**39.**

$$A = P(1 + r)^2$$

$$(1 + r)^2 = \frac{A}{P}$$

Copyright © 2011 Pearson Education, Inc. Publishing as Prentice Hall.

$$1 + r = \sqrt{\frac{A}{P}}$$

$$r = \sqrt{\frac{A}{P}} - 1$$

**41.** $x^2 + 4x + c = 0$

The discriminant is: $16 - 4c$

(A)   If $16 - 4c > 0$, i.e., if $c < 4$, then the equation has two distinct real roots.

(B)   If $16 - 4c = 0$, i.e., if $c = 4$, then the equation has one real double root.

(C)   If $16 - 4c < 0$, i.e., if $c > 4$, then there are no real roots.

**43.**   Setting the supply equation equal to the demand equation, we have

$$\frac{x}{450} + \frac{1}{2} = \frac{6,300}{x}$$

$$\frac{1}{450}x^2 + \frac{1}{2}x = 6,300$$

$$x^2 + 225x - 2,835,000 = 0$$

$$x = \frac{-225 \pm \sqrt{(225)^2 - 4(1)(-2,835,000)}}{2} \quad \text{(quadratic formula)}$$

$$= \frac{-225 \pm \sqrt{11,390,625}}{2} = \frac{-225 \pm 3375}{2} = 1,575 \text{ units}$$

Note, we discard the negative root since a negative number of units cannot be produced or sold. Substituting $x = 1,575$ into either equation (we use the demand equation), we get

$$p = \frac{6,300}{1,575} = 4$$

Supply equals demand at $4 per unit.

**45.**   $A = P(1 + r)^2 = P(1 + 2r + r^2) = Pr^2 + 2Pr + P$
Let $A = 625$ and $P = 484$. Then,

$484r^2 + 968r + 484 = 625$

$484r^2 + 968r - 141 = 0$

Using the quadratic formula,

$$r = \frac{-968 \pm \sqrt{(968)^2 - 4(484)(-141)}}{968} = \frac{-968 \pm \sqrt{1,210,000}}{968}$$

$$= \frac{-968 \pm 1100}{968} \approx 0.1364 \text{ or } -2.136$$

Since $r > 0$, we have $r = 0.1364$ or $13.64\%$.

**47.**   $v^2 = 64h$

For $h = 1$, $v^2 = 64(1) = 64$. Therefore, $v = 8$ ft/sec.

For $h = 0.5$, $v^2 = 64(0.5) = 32$.

Therefore, $v = \sqrt{32} = 4\sqrt{2} \approx 5.66$ ft/sec.

Copyright © 2011 Pearson Education, Inc.  Publishing as Prentice Hall.

## APPENDIX B  SPECIAL TOPICS

### EXERCISE B-1

*Things to remember:*

1. **SEQUENCES**

   A SEQUENCE is a function whose domain is a set of successive integers. If the domain of a given sequence is a finite set, then the sequence is called a FINITE SEQUENCE; otherwise, the sequence is an INFINITE SEQUENCE. In general, unless stated to the contrary or the context specifies otherwise, the domain of a sequence will be understood to be the set $N$ of natural numbers.

2. **NOTATION FOR SEQUENCES**

   Rather than function notation $f(n)$, $n$ in the domain of a given sequence $f$, subscript notation $a_n$ is normally used to denote the value in the range corresponding to $n$, and the sequence itself is denoted $\{a_n\}$ rather than $f$ or $f(n)$. The elements in the range, $a_n$, are called the TERMS of the sequence; $a_1$ is the first term, $a_2$ is the second term, and $a_n$ is the $n$th term or general term.

3. **SERIES**

   Given a sequence $\{a_n\}$. The sum of the terms of the sequence,
   $a_1 + a_2 + a_3 + \cdots$ is called a SERIES. If the sequence is finite, the corresponding series is a FINITE SERIES; if the sequence is infinite, then the corresponding series is an INFINITE SERIES.

   Only finite series are considered in this section.

4. **NOTATION FOR SERIES**

   Series are represented using SUMMATION NOTATION.
   If $\{a_k\}$, $k = 1, 2, \ldots, n$ is a finite sequence, then the series

   $$a_1 + a_2 + a_3 + \cdots + a_n$$

   is denoted

   $$\sum_{k=1}^{n} a_k.$$

   The symbol $\sum$ is called the SUMMATION SIGN and $k$ is called the SUMMING INDEX.

5. **ARITHMETIC MEAN**

   If $\{a_k\}$, $k = 1, 2, \ldots, n$, is a finite sequence, then the ARITHMETIC MEAN $\bar{a}$ of the sequence is defined as

   $$\bar{a} = \frac{1}{n} \sum_{k=1}^{n} a_k.$$

Copyright © 2011 Pearson Education, Inc.  Publishing as Prentice Hall.

**1.** $a_n = 2n + 3;$

$a_1 = 2 \cdot 1 + 3 = 5$
$a_2 = 2 \cdot 2 + 3 = 7$
$a_3 = 2 \cdot 3 + 3 = 9$
$a_4 = 2 \cdot 4 + 3 = 11$

**3.** $a_n = \dfrac{n+2}{n+1};$

$a_1 = \dfrac{1+2}{1+1} = \dfrac{3}{2}$

$a_2 = \dfrac{2+2}{2+1} = \dfrac{4}{3}$

$a_3 = \dfrac{3+2}{3+1} = \dfrac{5}{4}$

$a_4 = \dfrac{4+2}{4+1} = \dfrac{6}{5}$

**5.** $a_n = (-3)^{n+1};$

$a_1 = (-3)^{1+1} = (-3)^2 = 9$

$a_2 = (-3)^{2+1} = (-3)^3 = -27$

$a_3 = (-3)^{3+1} = (-3)^4 = 81$

$a_4 = (-3)^{4+1} = (-3)^5 = -243$

**7.** $a_n = 2n + 3; \; a_{10} = 2 \cdot 10 + 3 = 23$

**9.** $a_n = \dfrac{n+2}{n+1}; \; a_{99} = \dfrac{99+2}{99+1} = \dfrac{101}{100}$

**11.** $\displaystyle\sum_{k=1}^{6} k = 1 + 2 + 3 + 4 + 5 + 6 = 21$

**13.** $\displaystyle\sum_{k=4}^{7} (2k - 3) = (2 \cdot 4 - 3) + (2 \cdot 5 - 3) + (2 \cdot 6 - 3) + (2 \cdot 7 - 3) = 5 + 7 + 9 + 11 = 32$

**15.** $\displaystyle\sum_{k=0}^{3} \dfrac{1}{10^k} = \dfrac{1}{10^0} + \dfrac{1}{10^1} + \dfrac{1}{10^2} + \dfrac{1}{10^3} = 1 + \dfrac{1}{10} + \dfrac{1}{100} + \dfrac{1}{1000} = \dfrac{1111}{1000} = 1.111$

**17.** $a_1 = 5, a_2 = 4, a_3 = 2, a_4 = 1, a_5 = 6.$  Here $n = 5$ and the arithmetic mean is given by:

$$\bar{a} = \frac{1}{5} \sum_{k=1}^{5} a_k = \frac{1}{5}(5 + 4 + 2 + 1 + 6) = \frac{18}{5} = 3.6$$

**19.** $a_1 = 96, a_2 = 65, a_3 = 82, a_4 = 74, a_5 = 91, a_6 = 88, a_7 = 87, a_8 = 91, a_9 = 77,$ and $a_{10} = 74.$  Here $n = 10$ and the arithmetic mean is given by:

$$\bar{a} = \frac{1}{10} \sum_{k=1}^{10} a_k = \frac{1}{10}(96 + 65 + 82 + 74 + 91 + 88 + 87 + 91 + 77 + 74) = \frac{825}{10} = 82.5$$

**21.** $a_n = \dfrac{(-1)^{n+1}}{2^n};$

$a_1 = \dfrac{(-1)^2}{2^1} = \dfrac{1}{2}$

$a_2 = \dfrac{(-1)^3}{2^2} = -\dfrac{1}{4}$

$a_3 = \dfrac{(-1)^4}{2^3} = \dfrac{1}{8}$

$a_4 = \dfrac{(-1)^5}{2^4} = -\dfrac{1}{16}$

$a_5 = \dfrac{(-1)^6}{2^5} = \dfrac{1}{32}$

Copyright © 2011 Pearson Education, Inc.  Publishing as Prentice Hall.

**23.**  $a_n = n[1 + (-1)^n]$;    $a_1 = 1[1 + (-1)^1] = 0$

$a_2 = 2[1 + (-1)^2] = 4$

$a_3 = 3[1 + (-1)^3] = 0$

$a_4 = 4[1 + (-1)^4] = 8$

$a_5 = 5[1 + (-1)^5] = 0$

**25.**  $a_n = \left(-\dfrac{3}{2}\right)^{n-1}$;    $a_1 = \left(-\dfrac{3}{2}\right)^0 = 1$

$a_2 = \left(-\dfrac{3}{2}\right)^1 = -\dfrac{3}{2}$

$a_3 = \left(-\dfrac{3}{2}\right)^2 = \dfrac{9}{4}$

$a_4 = \left(-\dfrac{3}{2}\right)^3 = -\dfrac{27}{8}$

$a_5 = \left(-\dfrac{3}{2}\right)^4 = \dfrac{81}{16}$

**27.**  Given $-2, -1, 0, 1, \ldots$ The sequence is the set of successive integers beginning with $-2$. Thus, $a_n = n - 3$, $n = 1, 2, 3, \ldots$ .

**29.**  Given $4, 8, 12, 16, \ldots$ The sequence is the set of positive integer multiples of 4. Thus, $a_n = 4n$, $n = 1, 2, 3, \ldots$ .

**31.**  Given $\dfrac{1}{2}, \dfrac{3}{4}, \dfrac{5}{6}, \dfrac{7}{8}, \ldots$ The sequence is the set of fractions whose numerators are the odd positive integers and whose denominators are the even positive integers. Thus,

$a_n = \dfrac{2n-1}{2n}$, $n = 1, 2, 3, \ldots$ .

**33.**  Given $1, -2, 3, -4, \ldots$ The sequence consists of the positive integers with alternating signs. Thus,

$a_n = (-1)^{n+1}n$, $n = 1, 2, 3, \ldots$ .

**35.**  Given $1, -3, 5, -7, \ldots$ The sequence consists of the odd positive integers with alternating signs. Thus,

$a_n = (-1)^{n+1}(2n - 1)$, $n = 1, 2, 3, \ldots$ .

**37.**  Given $1, \dfrac{2}{5}, \dfrac{4}{25}, \dfrac{8}{125}, \ldots$ The sequence consists of the nonnegative integral powers of $\dfrac{2}{5}$. Thus,

$a_n = \left(\dfrac{2}{5}\right)^{n-1}$, $n = 1, 2, 3, \ldots$ .

**39.**  Given $x, x^2, x^3, x^4, \ldots$ The sequence is the set of positive integral powers of $x$. Thus, $a_n = x^n$, $n = 1, 2, 3, \ldots$ .

Copyright © 2011 Pearson Education, Inc.  Publishing as Prentice Hall.

**41.** Given $x, -x^3, x^5, -x^7, \ldots$ The sequence is the set of positive odd integral powers of $x$ with alternating signs. Thus,

$$a_n = (-1)^{n+1} x^{2n-1}, \quad n = 1, 2, 3, \ldots.$$

**43.** $\displaystyle\sum_{k=1}^{5} (-1)^{k+1}(2k-1)^2 = (-1)^2(2\cdot1-1)^2 + (-1)^3(2\cdot2-1)^2 + (-1)^4(2\cdot3-1)^2 + (-1)^5(2\cdot4-1)^2$

$$+ (-1)^6(2\cdot5-1)^2 = 1 - 9 + 25 - 49 + 81$$

**45.** $\displaystyle\sum_{k=2}^{5} \frac{2^k}{2k+3} = \frac{2^2}{2\cdot2+3} + \frac{2^3}{2\cdot3+3} + \frac{2^4}{2\cdot4+3} + \frac{2^5}{2\cdot5+3} = \frac{4}{7} + \frac{8}{9} + \frac{16}{11} + \frac{32}{13}$

**47.** $\displaystyle\sum_{k=1}^{5} x^{k-1} = x^0 + x^1 + x^2 + x^3 + x^4 = 1 + x + x^2 + x^3 + x^4$

**49.** $\displaystyle\sum_{k=0}^{4} \frac{(-1)^k x^{2k+1}}{2k+1} = \frac{(-1)^0 x}{2\cdot0+1} + \frac{(-1)^1 x^3}{2\cdot1+1} + \frac{(-1)^2 x^5}{2\cdot2+1} + \frac{(-1)^3 x^7}{2\cdot3+1} + \frac{(-1)^4 x^9}{2\cdot4+1} = x - \frac{x^3}{3} + \frac{x^5}{5} - \frac{x^7}{7} + \frac{x^9}{9}$

**51.** (A) $\displaystyle 2 + 3 + 4 + 5 + 6 = \sum_{k=1}^{5} (k+1)$      (B) $\displaystyle 2 + 3 + 4 + 5 + 6 = \sum_{j=0}^{4} (j+2)$

**53.** (A) $\displaystyle 1 - \frac{1}{2} + \frac{1}{3} - \frac{1}{4} = \sum_{k=1}^{4} \frac{(-1)^{k+1}}{k}$      (B) $\displaystyle 1 - \frac{1}{2} + \frac{1}{3} - \frac{1}{4} = \sum_{j=0}^{3} \frac{(-1)^j}{j+1}$

**55.** $\displaystyle 2 + \frac{3}{2} + \frac{4}{3} + \ldots + \frac{n+1}{n} = \sum_{k=1}^{n} \frac{k+1}{k}$

**57.** $\displaystyle \frac{1}{2} - \frac{1}{4} + \frac{1}{8} - \ldots + \frac{(-1)^{n+1}}{2^n} = \sum_{k=1}^{n} \frac{(-1)^{k+1}}{2^k}$

**59.** False: $\displaystyle 1 + \frac{1}{2} + \frac{1}{3} + \frac{1}{4} + \frac{1}{5} + \frac{1}{6} + \ldots + \frac{1}{64}$

$$= 1 + \frac{1}{2} + \left(\frac{1}{3} + \frac{1}{4}\right) + \left(\frac{1}{5} + \frac{1}{6} + \frac{1}{7} + \frac{1}{8}\right) + \left(\frac{1}{9} + \ldots + \frac{1}{16}\right) + \left(\frac{1}{17} + \ldots + \frac{1}{32}\right) + \left(\frac{1}{33} + \ldots + \frac{1}{64}\right)$$

$$> 1 + \frac{1}{2} + \frac{1}{2} + \frac{1}{2} + \frac{1}{2} + \frac{1}{2} + \frac{1}{2} = 4$$

**61.** True: $\displaystyle \frac{1}{2} - \frac{1}{4} + \frac{1}{8} - \frac{1}{16} + \frac{1}{32} - \ldots$

$$= \left(\frac{1}{2} - \frac{1}{4}\right) + \left(\frac{1}{8} - \frac{1}{16}\right) + \left(\frac{1}{32} - \frac{1}{64}\right) + \text{(positive terms)}$$

$$= \frac{1}{4} + \frac{1}{16} + \frac{1}{64} + \ldots > \frac{1}{4}$$

Copyright © 2011 Pearson Education, Inc. Publishing as Prentice Hall.

**63.**  $a_1 = 2$ and $a_n = 3a_{n-1} + 2$

for $n \geq 2$.

$a_1 = 2$

$a_2 = 3 \cdot a_1 + 2 = 3 \cdot 2 + 2 = 8$

$a_3 = 3 \cdot a_2 + 2 = 3 \cdot 8 + 2 = 26$

$a_4 = 3 \cdot a_3 + 2 = 3 \cdot 26 + 2 = 80$

$a_5 = 3 \cdot a_4 + 2 = 3 \cdot 80 + 2 = 242$

**65.**  $a_1 = 1$ and $a_n = 2a_{n-1}$

for $n \geq 2$.

$a_1 = 1$

$a_2 = 2 \cdot a_1 = 2 \cdot 1 = 2$

$a_3 = 2 \cdot a_2 = 2 \cdot 2 = 4$

$a_4 = 2 \cdot a_3 = 2 \cdot 4 = 8$

$a_5 = 2 \cdot a_4 = 2 \cdot 8 = 16$

**67.**  If $a_1 = \dfrac{A}{2}$, $a_n = \dfrac{1}{2}\left(a_{n-1} + \dfrac{A}{a_{n-1}}\right)$, $n \geq 2$, let $A = 2$. Then:

$a_1 = \dfrac{2}{2} = 1$

$a_2 = \dfrac{1}{2}\left(a_1 + \dfrac{A}{a_1}\right) = \dfrac{1}{2}(1 + 2) = \dfrac{3}{2}$

$a_3 = \dfrac{1}{2}\left(a_2 + \dfrac{A}{a_2}\right) = \dfrac{1}{2}\left(\dfrac{3}{2} + \dfrac{2}{3/2}\right) = \dfrac{1}{2}\left(\dfrac{3}{2} + \dfrac{4}{3}\right) = \dfrac{17}{12}$

$a_4 = \dfrac{1}{2}\left(a_3 + \dfrac{A}{a_3}\right) = \dfrac{1}{2}\left(\dfrac{17}{12} + \dfrac{2}{17/12}\right) = \dfrac{1}{2}\left(\dfrac{17}{12} + \dfrac{24}{17}\right) = \dfrac{577}{408} \approx 1.414216$

and $\sqrt{2} \approx 1.414214$

**69.**  $a_1 = 1$, $a_2 = 1$, $a_n = a_{n-1} + a_{n-2}$, $n \geq 3$

$a_3 = a_2 + a_1 = 2$, $a_4 = a_3 + a_2 = 3$, $a_5 = a_4 + a_3 = 5$,

$a_6 = a_5 + a_4 = 8$, $a_7 = a_6 + a_5 = 13$, $a_8 = a_7 + a_6 = 21$

$a_9 = a_8 + a_7 = 34$, $a_{10} = a_9 + a_8 = 55$

---

## EXERCISE B-2

*Things to remember:*

<u>1.</u>    A sequence of numbers $a_1$, $a_2$, $a_3$, ..., $a_n$, ..., is called an ARITHMETIC SEQUENCE if there is constant $d$, called the COMMON DIFFERENCE, such that

$$a_n - a_{n-1} = d,$$

that is,

$$a_n = a_{n-1} + d$$

for all $n > 1$.

<u>2.</u>    A sequence of numbers $a_1$, $a_2$, $a_3$, ..., $a_n$, ..., is called a GEOMETRIC SEQUENCE if there exists a nonzero constant $r$, called the COMMON RATIO, such that

$$\frac{a_n}{a_{n-1}} = r,$$

that is,

$$a_n = ra_{n-1}$$

Copyright © 2011 Pearson Education, Inc. Publishing as Prentice Hall.

for all $n > 1$.

3. $n$TH TERM OF AN ARITHMETIC SEQUENCE

   If $\{a_n\}$ is an arithmetic sequence with common difference $d$, then

   $$a_n = a_1 + (n-1)d$$

   for all $n > 1$.

4. $n$TH TERM OF A GEOMETRIC SEQUENCE

   If $\{a_n\}$ is a geometric sequence with common ratio $r$, then

   $$a_n = a_1 r^{n-1}$$

   for all $n > 1$.

5. SUM FORMULAS FOR FINITE ARITHMETIC SERIES

   The sum $S_n$ of the first $n$ terms of an arithmetic series
   $a_1 + a_2 + a_3 + \ldots + a_n$ with common difference $d$, is given by

   (a)     $S_n = \dfrac{n}{2}[2a_1 + (n-1)d]$     (First Form)

   or by

   (b)     $S_n = \dfrac{n}{2}(a_1 + a_n)$.     (Second Form)

6. SUM FORMULAS FOR FINITE GEOMETRIC SERIES

   The sum $S_n$ of the first $n$ terms of a geometric series
   $a_1 + a_2 + a_3 + a_n$ with common ratio $r$, is given by:

   $$S_n = \frac{a_1(r^n - 1)}{r - 1}, \quad r \neq 1, \quad \text{(First Form)}$$

   or by

   $$S_n = \frac{ra_n - a_1}{r - 1}, \quad r \neq 1. \quad \text{(Second Form)}$$

7. SUM OF AN INFINITE GEOMETRIC SERIES

   If $a_1 + a_2 + a_3 + \ldots + a_n + \ldots,$ is an infinite geometric series with common ratio $r$ having
   the property $-1 < r < 1$, then the sum $S_\infty$ is defined to be:

   $$S_\infty = \frac{a_1}{1 - r}.$$

Copyright © 2011 Pearson Education, Inc. Publishing as Prentice Hall.

1.  (A)  $-11, -16, -21, \ldots$

This is an arithmetic sequence with common difference $d = -5$;
$a_4 = -26$, $a_5 = -31$.

(B)  $2, -4, 8, \ldots$

This is a geometric sequence with common ratio $r = -2$;
$a_4 = -16$, $a_5 = 32$.

(C)  $1, 4, 9, \ldots$

This is neither an arithmetic sequence $(4 - 1 \neq 9 - 4)$ nor a geometric sequence $\left( \dfrac{4}{1} \neq \dfrac{9}{4} \right)$.

(D)  $\dfrac{1}{2}, \dfrac{1}{6}, \dfrac{1}{18}, \ldots$

This is a geometric sequence with common ratio $r = \dfrac{1}{3}$;

$a_4 = \dfrac{1}{54}, a_5 = \dfrac{1}{162}$.

3.  $\displaystyle\sum_{k=1}^{101} (-1)^{k+1} = 1 - 1 + 1 - 1 + \ldots + 1$

This is a geometric series with $a_1 = 1$ and common ratio $r = -1$.

$S_{101} = \dfrac{1[(-1)^{101} - 1]}{-1 - 1} = \dfrac{-2}{-2} = 1$

5.  This series is neither arithmetic nor geometric.

7.  $5 + 4.9 + 4.8 + \ldots + 0.1$ is an arithmetic series with $a_1 = 5$, $a_{50} = 0.1$ and common difference $d = -0.1$:

$S_{50} = \dfrac{50}{2} [5 + 0.1] = 25(5.1) = 127.5$

9.  $a_2 = a_1 + d = 7 + 4 = 11$
$a_3 = a_2 + d = 11 + 4 = 15$  (using $\underline{1}$)

11. $a_{21} = a_1 + (21 - 1)d = 2 + 20 \cdot 4 = 82$  (using $\underline{3}$)

$S_{31} = \dfrac{31}{2} [2a_1 + (31 - 1)d] = \dfrac{31}{2} [2 \cdot 2 + 30 \cdot 4] = \dfrac{31}{2} \cdot 124 = 1922$  [using $\underline{5}$(a)]

13. Using $\underline{5}$(b), $S_{20} = \dfrac{20}{2} (a_1 + a_{20}) = 10(18 + 75) = 930$

15. $a_2 = a_1 r = 3(-2) = -6$
$a_3 = a_1 r^2 = 3(-2)^2 = 3 \cdot 4 = 12$
$a_4 = a_1 r^3 = 3(-2)^3 = 3(-8) = -24$  (using $\underline{4}$)

17. Using $\underline{6}$, $S_7 = \dfrac{-3 \cdot 729 - 1}{-3 - 1} = \dfrac{-2188}{-4} = 547.$

19. Using $\underline{4}$, $a_{10} = 100(1.08)^9 = 199.90.$

Copyright © 2011 Pearson Education, Inc.  Publishing as Prentice Hall.

21.   Using <u>4</u>, $200 = 100\, r^8$. Thus, $r^8 = 2$ and $r = \sqrt[8]{2} \approx \pm 1.09$.

23.   Using <u>6</u>, $S_{10} = \dfrac{500[(0.6)^{10} - 1]}{0.6 - 1} \approx 1242,$

$$S_{\infty} = \frac{500}{1 - 0.6} = 1250.$$

25.   $S_{41} = \displaystyle\sum_{k=1}^{41} 3k + 3$. The sequence of terms is an arithmetic sequence. Therefore,

$$S_{41} = \frac{41}{2}(a_1 + a_{41}) = \frac{41}{2}(6 + 126) = \frac{41}{2}(132) = 41(66) = 2{,}706$$

27.   $S_8 = \displaystyle\sum_{k=1}^{8} (-2)^{k-1}$. The sequence of terms is a geometric sequence with common ratio $r = -2$ and

$$a_1 = (-2)^0 = 1.$$

$$S_8 = \frac{1[(-2)^8 - 1]}{-2 - 1} = \frac{256 - 1}{-3} = -85$$

29.   Let $a_1 = 13$, $d = 2$. Then, using <u>3</u>, we can find $n$:

$$67 = 13 + (n-1)2 \quad \text{or} \quad 2(n-1) = 54$$
$$n - 1 = 27$$
$$n = 28$$

Therefore, using <u>5</u>(b), $S_{28} = \dfrac{28}{2}[13 + 67] = 14\cdot 80 = 1120.$

31.   (A)   $2 + 4 + 8 + \cdots$ . Since $r = \dfrac{4}{2} = \dfrac{8}{4} = \cdots = 2$ and $|2| = 2 > 1$, the sum does not exist.

      (B)   $2, -\dfrac{1}{2}, \dfrac{1}{8}, \ldots$ . In this case, $r = \dfrac{-1/2}{2} = \dfrac{1/8}{-1/2} = \cdots = -\dfrac{1}{4}.$

         Since $|r| < 1$, $S_{\infty} = \dfrac{2}{1 - (-1/4)} = \dfrac{2}{5/4} = \dfrac{8}{5} = 1.6.$

33.   $f(1) = -1$, $f(2) = 1$, $f(3) = 3, \ldots$ . This is an arithmetic progression with $a_1 = -1$, $d = 2$. Thus, using

     <u>5</u>(a),

$$f(1) + f(2) + f(3) + \cdots + f(50) = \frac{50}{2}[2(-1) + 49\cdot 2] = 25\cdot 96 = 2400.$$

35.   $f(1) = \dfrac{1}{2}$, $f(2) = \left(\dfrac{1}{2}\right)^2 = \dfrac{1}{4}$, $f(3) = \left(\dfrac{1}{2}\right)^3 = \dfrac{1}{8}, \ldots$ . This is a geometric progression with $a_1 = \dfrac{1}{2}$ and $r$

     $= \dfrac{1}{2}$. Thus, using <u>6</u>:

Copyright © 2011 Pearson Education, Inc. Publishing as Prentice Hall.

$$f(1) + f(2) + \cdots + f(10) = S_{10} = \frac{\frac{1}{2}\left[\left(\frac{1}{2}\right)^{10} - 1\right]}{\frac{1}{2} - 1} \approx 0.999$$

**37.** Consider the arithmetic progression with $a_1 = 1$, $d = 2$. This progression is the sequence of odd positive integers. Now, using 5(a), the sum of the first $n$ odd positive integers is:

$$S_n = \frac{n}{2}[2 \cdot 1 + (n-1)2] = \frac{n}{2}(2 + 2n - 2) = \frac{n}{2} \cdot 2n = n^2$$

**39.** $S_n = a_1 + a_1 r + \ldots + a_1 r^{n-1}$. If $r = 1$, then $S_n = na_1$.

**41.** No: $\frac{n}{2}(1 + 1.1) = 100$ implies $(2.1)n = 200$ and this equation does not have an integer solution.

**43.** Yes: Solve the equation $6 = \frac{10}{1-r}$ for $r$. This yields $r = -\frac{2}{3}$.

The infinite geometric series: $10 - 10\left(\frac{2}{3}\right) + 10\left(\frac{2}{3}\right)^2 - 10\left(\frac{2}{3}\right)^3 + \ldots$

has sum $S_\infty = 6$.

**45.** Consider the time line:

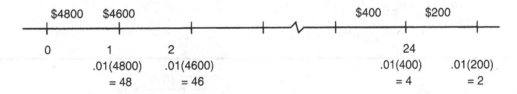

The total cost of the loan is $2 + 4 + 6 + \cdots + 46 + 48$. The terms form an arithmetic progression with $n = 24$, $a_1 = 2$, and $a_{24} = 48$. Thus, using 5(b):

$$S_{24} = \frac{24}{2}(2 + 48) = 24 \cdot 25 = \$600$$

**47.** This is a geometric progression with $a_1 = 3{,}500{,}000$ and $r = 0.7$.
Thus, using 7:

$$S_\infty = \frac{3{,}500{,}000}{1 - 0.7} \approx \$11{,}670{,}000$$

**49.**

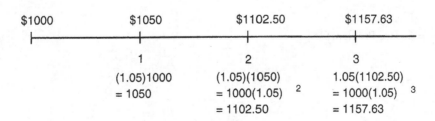

Copyright © 2011 Pearson Education, Inc. Publishing as Prentice Hall.

In general, after $n$ years, the amount $A_n$ in the account is:

$$A_n = 1000(1.05)^n$$

Thus, $A_{10} = 1000(1.05)^{10} \approx \$1628.89$

and $A_{20} = 1000(1.05)^{20} \approx \$2653.30$

## EXERCISE B-3

*Things to remember:*

1. If $n$ is a positive integer, then $n$ FACTORIAL, denoted $n!$, is the product of the integers from 1 to $n$; that is,

$$n! = n \cdot (n-1) \cdot \ldots \cdot 3 \cdot 2 \cdot 1 = n(n-1)!$$

Also, $1! = 1$ and $0! = 1$.

2. If $n$ and $r$ are nonnegative integers and $r \leq n$, then:

$$C_{n,r} = \frac{n!}{r!(n-r)!}$$

3. BINOMIAL THEOREM

For all natural numbers $n$:

$$(a+b)^n = C_{n,0}a^n + C_{n,1}a^{n-1}b + C_{n,2}a^{n-2}b^2$$
$$+ \cdots + C_{n,n-1}ab^{n-1} + C_{n,n}b^n.$$

---

1. $6! = 6 \cdot 5 \cdot 4 \cdot 3 \cdot 2 \cdot 1 = 720$

3. $\dfrac{10!}{9!} = \dfrac{10 \cdot 9!}{9!} = 10$

5. $\dfrac{12!}{9!} = \dfrac{12 \cdot 11 \cdot 10 \cdot 9!}{9!} = 1320$

7. $\dfrac{5!}{2!3!} = \dfrac{5 \cdot 4 \cdot 3!}{2 \cdot 1 \cdot 3!} = 10$

9. $\dfrac{6!}{5!(6-5)!} = \dfrac{6 \cdot 5!}{5!1!} = 6$

11. $\dfrac{20!}{3!17!} = \dfrac{20 \cdot 19 \cdot 18 \cdot 17!}{3!17!}$

$= \dfrac{20 \cdot 19 \cdot 18}{3 \cdot 2 \cdot 1} = 1140$

13. $C_{5,3} = \dfrac{5!}{3!(5-3)!} = \dfrac{5!}{3!2!} = 10$  (see Problem 7)

15. $C_{6,5} = \dfrac{6!}{5!(6-5)!} = 6$  (see Problem 9)

17. $C_{5,0} = \dfrac{5!}{0!(5-0)!} = \dfrac{5!}{1 \cdot 5!} = 1$

19. $C_{18,15} = \dfrac{18!}{15!(18-15)!} = \dfrac{18 \cdot 17 \cdot 16 \cdot 15!}{15!3!} = \dfrac{18 \cdot 17 \cdot 16}{3 \cdot 2 \cdot 1} = 816$

Copyright © 2011 Pearson Education, Inc.  Publishing as Prentice Hall.

**21.** Using 3,

$$(a+b)^4 = C_{4,0}a^4 + C_{4,1}a^3b + C_{4,2}a^2b^2 + C_{4,3}ab^3 + C_{4,4}b^4 = a^4 + 4a^3b + 6a^2b^2 + 4ab^3 + b^4$$

**23.** Using 3,

$$(x-1)^6 = [x + (-1)]^6$$
$$= C_{6,0}x^6 + C_{6,1}x^5(-1) + C_{6,2}x^4(-1)^2 + C_{6,3}x^3(-1)^3 + C_{6,4}x^2(-1)^4 + C_{6,5}x(-1)^5 + C_{6,6}(-1)^6$$
$$= x^6 - 6x^5 + 15x^4 - 20x^3 + 15x^2 - 6x + 1$$

**25.** $(2a-b)^5 = [2a + (-b)]^5$

$$= C_{5,0}(2a)^5 + C_{5,1}(2a)^4(-b) + C_{5,2}(2a)^3(-b)^2 + C_{5,3}(2a)^2(-b)^3 + C_{5,4}(2a)(-b)^4 + C_{5,5}(-b)^5$$
$$= 32a^5 - 80a^4b + 80a^3b^2 - 40a^2b^3 + 10ab^4 - b^5$$

**27.** The fifth term in the expansion of $(x-1)^{18}$ is:

$$C_{18,4}x^{14}(-1)^4 = \frac{18 \cdot 17 \cdot 16 \cdot 15}{4 \cdot 3 \cdot 2 \cdot 1}x^{14} = 3060x^{14}$$

**29.** The seventh term in the expansion of $(p+q)^{15}$ is:

$$C_{15,6}\ p^9q^6 = \frac{15 \cdot 14 \cdot 13 \cdot 12 \cdot 11 \cdot 10}{6 \cdot 5 \cdot 4 \cdot 3 \cdot 2 \cdot 1}p^9q^6 = 5005p^9q^6$$

**31.** The eleventh term in the expansion of $(2x+y)^{12}$ is:

$$C_{12,10}(2x)^2y^{10} = \frac{12 \cdot 11}{2 \cdot 1}4x^2y^{10} = 264x^2y^{10}$$

**33.** $C_{n,0} = \dfrac{n!}{0!(n-0)!} = \dfrac{n!}{1 \cdot n!} = 1, \quad C_{n,n} = \dfrac{n!}{n!(n-n)!} = \dfrac{n!}{n!0!} = 1$

**35.** The next two rows are:

1 5 10 10 5 1   and   1 6 15 20 15 6 1,

respectively. These are the coefficients in the binomial expansions of $(a+b)^5$ and $(a+b)^6$.

**37.** The $n$th row of Pascal's triangle gives the coefficients of $(a+b)^k$.
If we let $a=1$ and $b=-1$, we get

$$0 = (1-1)^n = C_{n,0}1^n + C_{n,1}(1)^{n-1}(-1) + C_{n,2}1^{n-2}(-1)^2 + \ldots + C_{n,n}(-1)^n$$
$$= C_{n,0} - C_{n,1} + C_{n,2} - C_{n,3} + \ldots + (-1)^nC_{n,n}.$$

**39.** $C_{n,r-1} + C_{n,r} = \dfrac{n!}{(r-1)!(n-[r-1])!} + \dfrac{n!}{r!(n-r)!} = \dfrac{n!}{(r-1)!(n-r+1)!} + \dfrac{n!}{r!(n-r)!}$

$$= \frac{r \cdot n! + (n-r+1)n!}{r!(n-r+1)!} = \frac{(n+1)n!}{r!(n-r+1)!} = \frac{(n+1)!}{r!(n+1-r)!} = C_{n+1,r}$$

Copyright © 2011 Pearson Education, Inc.  Publishing as Prentice Hall.